Fundamental Constants

Quantity	Symbol	Approximate Value	Current Best Value[†]
Speed of light in vacuum	c	3.00×10^8 m/s	2.99792458×10^8 m/s
Gravitational constant	G	6.67×10^{-11} N·m²/kg²	$6.67259(85) \times 10^{-11}$ N·m²/kg²
Avogadro's number	N_A	6.02×10^{23} mol⁻¹	$6.0221367(36) \times 10^{23}$ mol⁻¹
Gas constant	R	8.315 J/mol·K = 1.99 cal/mol·K = 0.082 atm·liter/mol·K	8.314510(70) J/mol·K
Boltzmann's constant	k	1.38×10^{-23} J/K	$1.380658(12) \times 10^{-23}$ J/K
Charge on electron	e	1.60×10^{-19} C	$1.60217733(49) \times 10^{-19}$ C
Stefan-Boltzmann constant	σ	5.67×10^{-8} W/m²·K⁴	$5.67051(19) \times 10^{-8}$ W/m²·K⁴
Permittivity of free space	$\epsilon_0 = (1/c^2\mu_0)$	8.85×10^{-12} C²/N·m²	$8.854187817... \times 10^{-12}$ C²/N·m²
Permeability of free space	μ_0	$4\pi \times 10^{-7}$ T·m/A	$1.2566370614... \times 10^{-6}$ T·m/A
Planck's constant	h	6.63×10^{-34} J·s	$6.6260755(40) \times 10^{-34}$ J·s
Electron rest mass	m_e	9.11×10^{-31} kg = 0.000549 u = 0.511 MeV/c^2	$9.1093897(54) \times 10^{-31}$ kg = $5.48579903(13) \times 10^{-4}$ u
Proton rest mass	m_p	1.6726×10^{-27} kg = 1.00728 u = 938.3 MeV/c^2	$1.6726231(10) \times 10^{-27}$ kg = 1.007276470(12) u
Neutron rest mass	m_n	1.6749×10^{-27} kg = 1.008665 u = 939.6 MeV/c^2	$1.6749286(10) \times 10^{-27}$ kg = 1.008664904(14) u
Atomic mass unit (1 u)		1.6605×10^{-27} kg = 931.5 MeV/c^2	$1.6605402(10) \times 10^{-27}$ kg = 931.49432(28) MeV/c^2

[†]Reviewed 1993 by B. N. Taylor, National Institute of Standards and Technology. Numbers in parentheses indicate one standard deviation experimental uncertainties in final digits. Values without parentheses are exact (i.e., defined quantities).

Other Useful Data

Joule equivalent (1 cal)	4.186 J
Absolute zero (0 K)	–273.15°C
Earth: Mass	5.97×10^{24} kg
Radius (mean)	6.38×10^3 km
Moon: Mass	7.35×10^{22} kg
Radius (mean)	1.74×10^3 km
Sun: Mass	1.99×10^{30} kg
Radius (mean)	6.96×10^5 km
Earth-sun distance (mean)	149.6×10^6 km
Earth-moon distance (mean)	384×10^3 km

The Greek Alphabet

Alpha	A	α	Nu	N	ν
Beta	B	β	Xi	Ξ	ξ
Gamma	Γ	γ	Omicron	O	o
Delta	Δ	δ	Pi	Π	π
Epsilon	E	ε	Rho	P	ρ
Zeta	Z	ζ	Sigma	Σ	σ
Eta	H	η	Tau	T	τ
Theta	Θ	θ	Upsilon	Υ	υ
Iota	I	ι	Phi	Φ	ϕ, φ
Kappa	K	κ	Chi	X	χ
Lambda	Λ	λ	Psi	Ψ	ψ
Mu	M	μ	Omega	Ω	ω

Values of Some Numbers

π = 3.1415927	$\sqrt{2}$ = 1.4142136	ln 2 = 0.6931472	$\log_{10}e$ = 0.4342945
e = 2.7182818	$\sqrt{3}$ = 1.7320508	ln 10 = 2.3025851	1 rad = 57.2957795°

Mathematical Signs and Symbols

$\propto$	is proportional to	$\leq$	is less than or equal to
$=$	is equal to	$\geq$	is greater than or equal to
$\approx$	is approximately equal to	Σ	sum of
$\neq$	is not equal to	$\bar{x}$	average value of x
$>$	is greater than	Δx	change in x
$\gg$	is much greater than	$\Delta x \rightarrow 0$	Δx approaches zero
$<$	is less than	$n!$	$n(n-1)(n-2)\ldots(1)$
$\ll$	is much less than		

Unit Conversions (Equivalents)

Length

1 in. = 2.54 cm
1 cm = 0.394 in.
1 ft = 30.5 cm
1 m = 39.37 in. = 3.28 ft
1 mi = 5280 ft = 1.61 km
1 km = 0.621 mi
1 nautical mile (U.S.) = 1.15 mi = 6076 ft = 1.852 km
1 fermi = 1 femtometer (fm) = 10^{-15} m
1 angstrom (Å) = 10^{-10} m
1 light-year (ly) = 9.46×10^{15} m
1 parsec = 3.26 ly = 3.09×10^{16} m

Volume

1 liter (L) = 1000 mL = 1000 cm^3 = 1.0×10^{-3} m^3 =
 1.057 quart (U.S.) = 54.6 in.^{3}
1 gallon (U.S.) = 4 qt (U.S.) = 231 in.^{3} = 3.78 L =
 0.83 gal (Imperial)
1 m^3 = 35.31 ft^3

Speed

1 mi/h = 1.47 ft/s = 1.609 km/h = 0.447 m/s
1 km/h = 0.278 m/s = 0.621 mi/h
1 ft/s = 0.305 m/s = 0.682 mi/h
1 m/s = 3.28 ft/s = 3.60 km/h
1 knot = 1.151 mi/h = 0.5144 m/s

Angle

1 radian (rad) = 57.30° = 57°18′
1° = 0.01745 rad
1 rev/min (rpm) = 0.1047 rad/s

Time

1 day = 8.64×10^4 s
1 year = 3.156×10^7 s

Mass

1 atomic mass unit (u) = 1.6605×10^{-27} kg
1 kg = 0.0685 slug
[1 kg has a weight of 2.20 lb where $g = 9.81$ m/s^2.]

Force

1 lb = 4.45 N
1 N = 10^5 dyne = 0.225 lb

Energy and Work

1 J = 10^7 ergs = 0.738 ft·lb
1 ft·lb = 1.36 J = 1.29×10^{-3} Btu = 3.24×10^{-4} kcal
1 kcal = 4.18×10^3 J = 3.97 Btu
1 eV = 1.602×10^{-19} J
1 kWh = 3.60×10^6 J = 860 kcal

Power

1 W = 1 J/s = 0.738 ft·lb/s = 3.42 Btu/h
1 hp = 550 ft·lb/s = 746 W

Pressure

1 atm = 1.013 bar = 1.013×10^5 N/m^2
  = 14.7 lb/in.^{2} = 760 torr
1 lb/in.^{2} = 6.90×10^3 N/m^2
1 Pa = 1 N/m^2 = 1.45×10^{-4} lb/in.^{2}

SI Derived Units and Their Abbreviations

Quantity	Unit	Abbreviation	In Terms of Base Units[†]
Force	newton	N	kg·m/s^2
Energy and work	joule	J	kg·m^2/s^2
Power	watt	W	kg·m^2/s^3
Pressure	pascal	Pa	kg/(m·s^2)
Frequency	hertz	Hz	s^{-1}
Electric charge	coulomb	C	A·s
Electric potential	volt	V	kg·m^2/(A·s^3)
Electric resistance	ohm	Ω	kg·m^2/(A^2·s^3)
Capacitance	farad	F	A^2·s^4/(kg·m^2)
Magnetic field	tesla	T	kg/(A·s^2)
Magnetic flux	weber	Wb	kg·m^2/(A·s^2)
Inductance	henry	H	kg·m^2/(s^2·A^2)

[†]kg = kilogram (mass), m = meter (length), s = second (time), A = ampere (electric current).

Metric (SI) Multipliers

Prefix	Abbreviation	Value
exa	E	10^{18}
peta	P	10^{15}
tera	T	10^{12}
giga	G	10^9
mega	M	10^6
kilo	k	10^3
hecto	h	10^2
deka	da	10^1
deci	d	10^{-1}
centi	c	10^{-2}
milli	m	10^{-3}
micro	μ	10^{-6}
nano	n	10^{-9}
pico	p	10^{-12}
femto	f	10^{-15}
atto	a	10^{-18}

PHYSICS

PRINCIPLES WITH APPLICATIONS

Fifth Edition

Volume 2

Douglas C. Giancoli

PRENTICE HALL, Upper Saddle River, New Jersey 07458

Editor-in-Chief: Paul F. Corey
Development Editors: Ray Mullaney, David Chelton
Production Editor: Susan Fisher
Executive Editor: Alison Reeves
Editorial Director: Tim Bozik
Director of Marketing: Kelly McDonald
Senior Marketing Manager: Leslie Cavaliere
Marketing Liaison: Ramona Sherman
Assistant Editor: Wendy Rivers
Assistant Vice President of Production and Manufacturing: David W. Riccardi
Executive Managing Editor: Kathleen Schiaparelli
Art Manager: Gus Vibal
Page Layout: Richard Foster, Karen Stephens
Illustrators: Patrice Van Acker, Tamara Newnam Cavallo
Photo Editor: Warren Ramezzana
Creative Director: Paula Maylahn
Art Director: Amy Rosen
Manager, Art and Formatting: John Jordan
Manufacturing Manager: Trudy Pisciotti
Advertising Design Manager: Meghan Dacey
Advertising and Promotions Manager: Elise Schneider
Additional Formatting: Jeff Henn
Production Assistants: Paula Williams, Nancy Gross, Adam Velthaus
Art Assistants: Pat Gutrierrez, Charlie Pelletreau
Photo Research Coordinator: Lorinda Morris-Nantz
Photo Research Administrator: Melinda Reo
Photo Research: Tobi Zausner, Mary Teresa Giancoli, Rhoda Sidney
Editorial Assistants: Marilyn Coco, Jo Marie Jacobs
Cover photo: Malcom Hanes Pressenbild/Adventure Photo & Film

 © 1998, 1995, 1991, 1985, 1980
by Douglas C. Giancoli

Printed in the United States of America

10 9 8 7 6 5 4 3 2

ISBN   0-13-679762-8

Prentice-Hall International (UK) Limited, *London*
Prentice-Hall of Australia Pty. Limited, *Sydney*
Prentice-Hall Canada Inc., *Toronto*
Prentice-Hall Hispanoamericana, S.A., *Mexico City*
Prentice-Hall of India Private Limited, *New Delhi*
Prentice-Hall of Japan, Inc., *Tokyo*
Simon & Schuster Asia Pte. Ltd., *Singapore*
Editora Prentice-Hall do Brasil, Ltda., *Rio de Janeiro*
Prentice-Hall, *Upper Saddle River, New Jersey*

C O N T E N T S

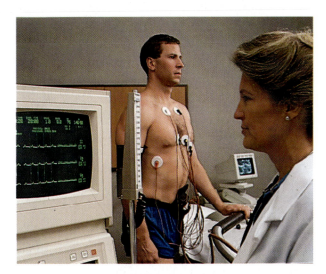

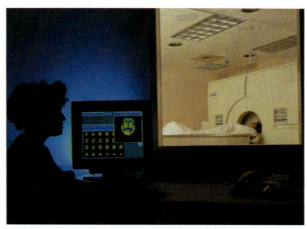

PREFACE

A Fifth Edition?

By the fifth edition of an algebra-based physics text, one might expect that the author has at last gotten it right.

I hope the earlier editions weren't all that wrong. The idea of a new edition is to improve, to bring in material, and perhaps to delete material that makes the book longer but isn't all that useful. All these things have been done:

Physics. Physics itself may not change all that rapidly, but over the span of a few years, there may be some new discoveries to include, such as

- planets revolving around distant stars
- information gathered by the Hubble Space Telescope
- updates in particle physics and cosmology (e.g., age of universe).

Pedagogy. One aspect of physics that is changing fairly rapidly is research on how students learn. As a result, this new edition contains some new elements:

Conceptual Examples, an average of 2 or 3 per chapter, are each a sort of brief Socratic question and answer. It is intended that readers will be stimulated by the questions to think, or reflect, and come up with a response—before reading the Response given. Here are a few:

- Velocity vs. acceleration (Chapter 2)
- What exerts the force on a car? (Chapter 4)
- Apple and the wagon (reference frames and projectile motion, Chapter 3)
- Which object rolls down a plane faster? (Chapter 8)
- Finger on a full straw (Chapter 10)
- Suction (Chapter 10)
- Boiling pasta (Chapter 14)
- Electric shielding/safety from lightning (Chapter 16)
- Which part of the photo is the reflection? (Chapter 23).

Estimating Examples, also a new feature of this edition, are intended to show how to make order-of-magnitude estimates even when the data are scarce, even when you might never have guessed that any result was possible at all.

Problem Solving has not been slighted in the least. There are many new worked-out Examples and here are some highlights:

- Air bags (2-10)
- Bungee jumper (6-14)
- Computer hard drive (8-4)
- Loudspeaker (11-7)
- Photocopier (16-5)
- Age of archeological bone (30-10).

Some of the new Examples have replaced older less useful ones. Many other Examples have been improved by more detailed reasoning, by displaying more mathematical steps, and by improving the ambience to make them more real-world and so more inviting and interesting.

New problems have been added and many of the old ones have undergone change.

<u>Example Titles.</u> Examples of all three types (including Conceptual and Estimating) now have titles (for fun and for easy reference).

<u>Emphasized Equations.</u> The great laws of physics are emphasized not only by setting them off, but also by giving them a marginal note in capitals and in a box. The equations that express the great laws, as well as the major equations that one just can't do without, are emphasized with a tan screen.

New Topics in this fifth edition include (these are only a few)

- Rolling motion (Chapter 8)
- Work in rotational motion ($W = \tau\theta$) (Chapter 8)
- v and a for simple harmonic motion (Chapter 11)
- Highway mirages (Chapter 24)
- Hubble Space Telescope (several places)
- Higgs Boson, Symmetry (explained), Supersymmetry (Chapter 32).

Diagrams. There are many more diagrams (over 200 new ones, for an increase of 20 percent), a lot of them to go with Examples and with Problems. Many of the old diagrams have been improved with more realistic backgrounds and figures, and more detail, and the use of photorealistic art.

Photographs. Many of the chapter opening photos now have vectors or other analysis superimposed on them—to give students a richer feeling for the physics. These are visual images of physics that will be fixed in the students' minds.

Many new and interesting photos have been added in the text to bring home the usefulness of physics, a few of which are: diffusion, images in spherical mirrors, depth of field with a camera lens, Hubble Space Telescope, and DNA X-ray diffraction.

Applications. Relevant applications of physics to everyday life and to biology, medicine, architecture, geology, and other fields has always been a strong feature of this book, and continues to be. Among other things, they answer the students' question, "Why study physics?" New applications have been added (and a few older ones dropped), some of which are

- Elevator and counterweight (Chapter 4)
- Shielding (Chapters 16 and 20)
- Dry cell (Chapter 18)
- Aurora borealis (Chapter 20)
- Induction stove (Chapter 21)
- TV and radio antennas (Chapter 22)
- CD player, laser and disk (Chapter 28)
- Smoke detectors (Chapter 29)

and some already mentioned earlier such as airbags, bungee jumping, photocopiers, highway mirages, and computer hard drives.

Revised Physics. No topic, no paragraph in this book was overlooked in the search to improve the clarity of the presentation. Many changes and clarifications have been made, both small and not so small. Here are just a few of the more important ones:

- New tables of typical lengths, times, masses (Chapter 1) and voltages (Chapter 17)
- Chapter 2: rearranged presentation of displacement, velocity, and reference frames
- New diagrams to aid understanding of velocity and acceleration (Chapter 2)
- Unit conversion moved to Chapter 1
- Relative velocity moved to end of Chapter 3
- Simplified introduction to Newton's second law (Chapter 4)
- Simple machines: pulley (Chapter 4), lever (Chapter 9), hydraulic lift (Chapter 10)
- New Section: "Car rounding a curve" (Chapter 5)
- Period and frequency introduced earlier (Chapter 5)
- Work and energy reworked in general, and potential energy especially dealt with in more detail (Chapter 6)
- Angular momentum simplified a bit, especially vector aspects (Chapter 8)
- More formulas for moment of inertia (Fig. 8-20)
- Rotating reference frames, inertial forces, and Coriolis removed from Chapter 8 to an Appendix.
- Greatly simplified vertical spring derivation (Chapter 11)
- Energy transported by waves simplified, with more difficult parts in optional Sections (Chapters 11 and 12)
- Speed of light measurement moved from Chapter 23 on Optics to Chapter 22 on EM waves
- Magnifying glass reworked (Chapter 23)
- Relativistic momentum reworked and in more detail (Chapter 26)
- New energy state diagrams for complex atoms (Chapter 28)
- New results in elementary particle physics and cosmology (Chapters 32 and 33).

Page Layout: Complete Derivations. Serious attention has been paid to how each page was formatted, especially for page turns. Great effort has been made to keep important derivations and arguments on facing pages. Thus readers don't have to turn back to check. More importantly, throughout the book, readers see before them, on two facing pages, an important slice of phyics. On rare occasions when an argument related to a particular figure requires a page turn, that figure is repeated after the page turn so readers won't have to look back.

Deletions. With all of these additions, something had to go to keep the book from getting too long. Some new Examples simply replaced less interesting old ones. The treatment of quite a few topics was shortened and some were simply dropped. Here are some of the deletions: derived vs. base units; operational definitions (general details); section on "Laws or Definitions" dropped (kept a tiny bit earlier in Chapter 4); vector nature of angular quantities (greatly shortened); Reynolds number, sedimentation, Stokes' equation; flow in tubes (halved); Olber's paradox.

Scope of this Book

This book is written for students. The two motivating factors are to give students a thorough understanding of the basic concepts of physics and, by means of interesting applications, to prepare them to use physics in their own lives and professions. In particular, this book is written for students who are taking a one-year introductory course in physics that uses algebra and trigonometry but not calculus. Many of these students have as their main interest biology, (pre)medicine, architecture, technology, or the earth and environmental sciences. This book contains a wide range of applications to these and other fields, as well as to everyday life. These applications answer that common student query, "Why must I study physics?" The answer, of course, is that physics comes into play in all these fields very importantly, and here they can see how. Physics is all about us. Indeed, it is the goal of this book to help students see the world through eyes that know physics.

Before the applications must come the physics. And this new edition, even more than previous editions, aims to explain physics in a readable and interesting manner that is accessible and clear. It aims to teach students by anticipating their needs and difficulties, but without oversimplifying.

General Approach

This book offers an in-depth presentation of physics, and retains the basic approach of the earlier editions. Rather than using the common, dry, dogmatic approach of treating topics formally and abstractly first, and only later relating the material to the students' own experience, my approach is to recognize that physics is a description of reality and thus to start each topic with concrete observations and experiences that students can directly relate to. Then we move on to the generalizations. Not only does this make the material more interesting and easier to understand, but it is closer to the way physics is actually practiced.

I have sought, where possible, to present the basic concepts of physics in their historical and philosophic context.

As mentioned above, this book includes of a wide range of examples and applications from other fields: biology, medicine, architecture, technology, earth sciences, the environment, and daily life. Some applications serve only as examples of physical principles. Others are treated in depth, with whole Sections devoted to them (among these are the study of medical imaging systems, constructing arches and domes, and the effects of radiation). But applications do not dominate the text—this is, after all, a physics book. They have been carefully chosen and integrated into the text so as not to interfere with the development of the physics but rather illuminate it. You won't find essay sidebars here. The applications are integrated right into the physics. Even when an application gets a separate Section all to itself, it is directly tied to the physics just studied. To make it easy to spot these applications, a new *Physics Applied* marginal note has been added.

Mathematics can be an obstacle to student understanding. To avoid frightening students with an initial chapter on mathematics, I have instead incorporated many important mathematical tools, such as addition of vectors and trigonometry, directly in the text where first needed. In addition, the appendices contain a review of many mathematical topics such as algebra and geometry, as well as dimensional analysis. A few advanced topics are also given an Appendix: Rotating frames of reference, Inertial forces, Coriolis effect; Gauss's law; Galilean and Lorentz transformations.

It is necessary, I feel, to pay careful attention to detail, especially when deriving an important result. I have aimed at including all steps in a derivation, and have tried to make clear which equations are general, and which are not, by explicitly stating the limitations of important equations in brackets next to the equation, such as

$$x = x_0 + v_0 t + \frac{1}{2} a t^2. \qquad \text{[constant acceleration]}$$

Difficult language, too, can hinder understanding: and I have tried to write in a relaxed style, avoiding jargon, and often talking directly to the students. New or unusual terms are carefully defined when first used.

Color is used pedagogically to bring out the physics. Different types of vectors are given different colors (see the chart on page xxiii). There are many new diagrams to illustrate new Examples (and old ones too) and to enrich the text and problems. The fifth edition features new and revised art—including new photorealistic art, more illustrations to accompany the in-text Examples and end-of-chapter problems, and dozens of new photos.

Problem Solving

Strong attention is given to problem solving. Learning how to approach and solve problems is a basic part of a physics course, and is a highly useful skill in itself. Solving problems is also important because the process brings understanding of the physics. Scattered throughout the book are special Sections and special Boxes devoted to how to approach the solving of problems. Many are found in the early chapters, where students first begin wrestling with problem solving; but many are also found later in the book, throughout mechanics, and in electricity, for example, where problem solving is an emphasized issue, as well as in thermodynamics and in optics. These Problem Solving Boxes provide a summary of how to approach problem solving. They do *not* provide a prescription to be followed. Hence they are often placed *after* a few Examples have been done, as a sort of summary of how we have been approaching Problems.

Over 400 Examples are fully worked out in the text. In this new edition, there are three types of Examples: regular worked Examples, Estimating Examples, and Conceptual Examples. The regular Examples are fully worked out in the text, and most are accompanied by analytical drawings. These Examples are designed to help students develop problem-solving skills and range from simple to fairly complicated. Estimating Examples encourage student analysis and understanding by using "back of the envelope" estimations as a problem-solving technique; they increase awareness of the power of analytical thinking. Conceptual Examples, in contrast to numerical problem solving and the application of formulas, challenge students to explore the basic concepts that are fundamental to understanding physics. Many Examples are taken from everyday life and aim at being realistic applications of physics principles.

There are over 3100 end-of-chapter exercises, including more than 700 questions that require verbal answers based on an understanding of the concepts, and about 2400 problems involving mathematical calculation.

Each chapter contains a large group of problems arranged by Section and graded according to difficulty: level I problems are simple, usually plug-in types, designed to give students confidence; level II are normal problems, requiring more thought and often the combination of two different concepts; level III are the most difficult and serve as a challenge to superior students. The arrangement by Section number means only that those problems depend on material up to and including that Section: earlier material may also be relied upon. The ranking of problems by difficulty (I, II, III) is intended only as a guide.

I suggest that instructors assign a significant number of the level I and level II problems, and reserve level III problems to stimulate the best students. Although most level I problems may seem easy, they help to build self-confidence—an important part of learning, especially in physics.

Each chapter also contains a group of "General Problems" which are unranked and not arranged by Section number.

Answers to odd-numbered problems are given at the back of the book. Throughout the text, *Système International* (SI) units are used. Other metric and British units are defined for informational purposes.

Organization

The general outline of this new edition retains a traditional order of topics: mechanics (Chapters 1 to 12), including vibrations, waves, and sound, followed by kinetic theory and thermodynamics (Chapters 13 to 15), electricity and magnetism (Chapters 16 to 22), light (Chapters 23 to 25), and modern physics (Chapters 26 to 33). Nearly all topics customarily taught in introductory physics courses are included here.

The tradition of beginning with mechanics is sensible, I believe, because it was developed first, historically, and because so much else in physics depends on it. Within mechanics, there are various ways to order topics, and this book allows for considerable flexibility. I prefer, for example, to cover statics after dynamics, partly because many students have trouble with the concept of force without motion. Besides, statics is a special case of dynamics—we study statics so that we can prevent structures from becoming dynamic (falling down)—and that sense of being at the limit of dynamics is intuitively helpful. Nonetheless statics (Chapter 9) can be covered earlier, if desired, before dynamics, after a brief introduction to vectors. Another option is light, which I have placed after electricity and magnetism and EM waves. But light could be treated immediately after the chapters on waves (Chapter 11 and 12). Special relativity (Chapter 26), which is located along with the other chapters on modern physics, could instead be treated along with mechanics—say, after Chapter 7.

Not every chapter need be given equal weight. Whereas Chapter 4 or Chapter 21 might require $1\frac{1}{2}$ to 2 weeks of coverage, Chapter 12 or 22 may need only $\frac{1}{2}$ week.

The book contains more material than can be covered in most one-year courses, so instructors have flexibility in choice of topics. Sections marked with a star (asterisk) are considered optional (if not covered in class, they can be a resource for later study). These Sections contain slightly more advanced physics material, often material not usually covered in typical courses, and/or interesting applications. They contain no material needed in later chapters (except perhaps in later optional Sections). This does not imply that all nonstarred sections must be covered: there still remains considerable flexibility in the choice of material. For a brief course, all optional material could be dropped as well as major parts of Chapters 10, 12, 19, 22, 28, 29, 32, and 33, as well as selected parts of Chapters 7, 8, 9, 15, 21, 24, 25, and 31.

Thanks

More than 60 physics professors provided direct feedback on every aspect of the text: organization, content, figures, and suggestions for new Examples and Problems. These reviewers for this fifth edition are listed below. I owe each of them a debt of gratitude:

David B. Aaron (South Dakota State University)
Zaven Altounian (McGill University)
Atam P. Arya (West Virginia University)
David E. Bannon (Chemeketa Community College)
Jacob Becher (Old Dominion University)
Michael S. Berger (Indiana University, Bloomington)
Donald E. Bowen (Stephen F. Austin University)
Neal M. Cason (University of Notre Dame)
H. R. Chandrasekhar (University of Missouri)
Ram D. Chaudhari (SUNY—Oswego)
K. Kelvin Cheng (Texas Tech University)
Lowell O. Christensen (American River College)
Mark W. Plano Clark (Doane College)
Irvine G. Clator (UNC, Wilmington)
Scott Cohen (Portland State University)
Lattie Collins (East Tennessee State University)
Sally Daniels (Oakland University)
Jack E. Denson (Mississippi State University)
Eric Dietz (California State University, Chico)
Paul Draper (University of Texas, Arlington)
Miles J. Dresser (Washington State University)
F. Eugene Dunnam (University of Florida)
Gregory E. Francis (Montana State University)
Philip Gash (California State University, Chico)
J. David Gavenda (University of Texas, Austin)
Grant W. Hart (Brigham Young University)
Melissa Hill (Marquette University)
Mark Hillery (Hunter College)
Hans Hochheimer (Colorado State University)
Alex Holloway (University of Nebraska, Omaha)

James P. Jacobs (University of Montana)
Larry D. Johnson (Northeast Louisiana University)
David Lamp (Texas Tech University)
Paul Lee (University of California, Northridge)
Daniel J. McLaughlin (University of Hartford)
Victor Montemeyer (Middle Tennessee State Univ.)
Dennis Nemeschansky (USC)
Robert Oakley (University of Southern Maine)
Robert Pelcovits (Brown University)
Brian L. Pickering (Laney College)
T.A.K. Pillai (University of Wisconsin, La Crosse)
Michael Ram (University of Buffalo)
David Reid (Eastern Michigan University)
Charles Richardson (University of Arkansas)
Lawrence Rowan (UNC, Chapel Hill)
Roy S. Rubins (University of Texas, Arlington)
Thomas Sayetta (East Carolina University)
Neil Schiller (Ocean County College)
Juergen Schroeer (Illinois State University)
Marc Sher (College of William and Mary)
James P. Sheerin (Eastern Michigan University)
Donald Sparks (Los Angeles Pierce College)
Michael G. Strauss (University of Oklahoma)
Harold E. Taylor (Stockton State University)
Michael Thoennessen (Michigan State University)
Linn D. Van Woerkom (Ohio State University)
S. L. Varghese (University of South Alabama)
Robert A. Walking (University of Southern Maine)
Lowell Wood (University of Houston)
David Wright (Tidewater Community College)

I am grateful also to all those physicist-reviewers of the earlier editions:

Narahari Achar (Memphis State University)
William T. Achor (Western Maryland College)
Arthur Alt (College of Great Falls)
Zaven Altounian (McGill University)
John Anderson (University of Pittsburgh)
Subhash Antani (Edgewood College)
Sirus Aryainejad (Eastern Illinois University)
Charles R. Bacon (Ferris State University)
Arthur Ballato (Brookhaven National Laboratory)
Gene Barnes (California State U., Sacramento)
Isaac Bass
Paul A. Bender (Washington State University)
Joseph Boyle (Miami–Dade Community College)
Peter Brancazio (Brooklyn College, CUNY)
Michael E. Browne (University of Idaho)
Michael Broyles (Collin County Community College)

Anthony Buffa (Cal Poly S.L.O.)
David Bushnell (Northern Illinois University)
Albert C. Claus (Loyola University of Chicago)
Lawrence Coleman (Univ. of California, Davis)
Waren Deshotels (Marquette University)
Frank Drake (Univ. of California, Santa Cruz)
Miles J. Dresser (Washington State University)
Ryan Droste (The College of Charleston)
Frank A. Ferrone (Drexel University)
Len Feuerhelm (Oklahoma Christian University)
Donald Foster (Wichita State University)
Philip Gash (California State University, Chico)
Simon George (California State Univ., Long Beach)
James Gerhart (University of Washington)
Bernard Gerstman (Florida International Univ.)
Charles Glashausser (Rutgers University)

Hershel J. Hausman (Ohio State University)
Laurent Hodges (Iowa State University)
Joseph M. Hoffman (Frostburg State University)
Peter Hoffmann-Pinther (U. of Houston–Downtown)
Fred W. Inman (Mankato State University)
M. Azad Islan (State Univ. of New York—Potsdam)
James P. Jacobs (Seattle University)
Gordon Jones (Mississippi State University)
Rex Joyner (Indiana Institute of Technology)
Sina David Kaviani (El Camino College)
Joseph A. Keane (St. Thomas Aquinas College)
Kirby W. Kemper (Florida State University)
Sanford Kern (Colorado State University)
James E. Kettler (Ohio University–Eastern Campus)
James R. Kirk (Edinboro University)
Alok Kuman (State Univ. of New York—Oswego)
Sung Kyu Kim (Macalester College)
Amer Lahamer (Berea College)
Clement Y. Lam (North Harris College)
Peter Landry (McGill University, Montreal)
Michael Lieber (University of Arkansas)
Bryan H. Long (Columbia State College)
Michael C. LoPresto (Henry Ford Com. College)
James Madsen (University of Wisconsin, River Falls)
Ponn Mahes (Winthrop University)
Robert H. March (University of Wisconsin–Madison)
David Markowitz (University of Connecticut)
E. R. Menzel (Texas Tech University)
Robert Messina
David Mills (College of the Redwoods)
George K. Miner (University of Dayton)
Marina Morrow (Lansing Community College)
Ed Nelson (University of Iowa)
Gregor Novak (Indiana Univ./Purdue Univ.)
Roy J. Peterson (University of Colorado–Boulder)
Frederick M. Phelps (Central Michigan University)

T. A. K. Pillai (University of Wisconsin–La Crosse)
John Polo (Edinboro University of Pennsylvania)
W. Steve Quon (Ventura College)
John Reading (Texas A&M)
William Riley (Ohio State University)
Larry Rowan (University of North Carolina)
R. S. Rubins (University of Texas, Arlington)
D. Lee Rutledge (Oklahoma State University)
Hajime Sakai (Univ. of Massachusetts at Amherst)
Ann Schmiedekamp (Penn State U., Ogontz Campus)
Mark Semon (Bates College)
Eric Sheldon (University of Massachusetts–Lowell)
K. Y. Shen (California State University, Long Beach)
Joseph Shinar (Iowa State University)
Thomas W. Sills (Wilbur Wright College)
Anthony A. Siluidi (Kent State University)
Michael A. Simon (Housatonic Community College)
Upindranath Singh (Embry-Riddle)
Michael I. Sobel (Brooklyn College)
Thor F. Stromberg (New Mexico State University)
James F. Sullivan (University of Cincinnati)
Kenneth Swinney (Bevill State Community College)
John E. Teggins (Auburn Univ. at Montgomery)
Colin Terry (Ventura College)
Jagdish K. Tuli (Brookhaven National Laboratory)
Kwok Yeung Tsang (Georgia Institute of Technology)
Paul Urone (CSU, Sacramento)
Jearl Walker (Cleveland State University)
Jai-Ching Wang (Alabama A&M University)
John C. Wells (Tennessee Technological)
Gareth Williams (San Jose State University)
Thomas A. Weber (Iowa State University)
Wendall S. Williams (Case Western Reserve Univ.)
Jerry Wilson (Metropolitan State College at Denver)
Peter Zimmerman (Louisiana State University)

I owe special thanks to Irv Miller for working out all the problems and for managing the team that also worked out the problems, each checking the others, and finally, producing the answers at the back of the book as well as producing the Solutions Manual.

I am grateful to Paul Draper, Robert Pelcovits, Gregory E. Francis, and James P. Jacobs, who inspired many of the Conceptual Examples, as well as suggestions for Questions and Problems. I wish also to thank Professors Howard Shugart, John Heilbron, Joe Cerny, and Roger Falcone for helpful discussions, and for hospitality at the University of California, Berkeley, Physics Department. Thanks also to Prof. Tito Arecchi at the Istituto Nazionale di Ottica, Florence, Italy, and to the staff of the Institute for the History of Science, Florence, for their kind hospitality.

Finally, I am most grateful to the many people at Prentice Hall with whom I worked on this project, especially Susan Fisher, Marilyn Coco, David Chelton, Tim Bozik, Gary June, Kathleen Schiaparelli, Richard Foster, Patrice Van Acker and Dave Riccardi. And special thanks to Paul Corey for guiding this project at every stage with clarity and that rare gift of "getting things done," and to Ray Mullaney, whose high level of dedication through edition after edition has helped make this a clear and accurate book. The final responsibility for all errors lies with me, of course. I welcome comments and corrections.

Douglas C. Giancoli

Supplements

For the Instructor

Instructor's Solutions Manual by Irvin A. Miller *Print version*
(0-13-627985-6); *Electronic Versions:* Windows (0-13-627993-7);
Macintosh (0-13-628009-9)
Contains detailed, worked solutions to every problem in the text by Irv
Miller of Drexel University.

Answers to Questions
Prepared by Michelle Rallis and Kurt Reibel of The Ohio State University,
Columbus and Gordon Aubrecht of The Ohio State University, Marion,
this supplement contains answers to all end-of-chapter questions.

Transparency Pack (0-13-628041-2)
Includes 400 four-color transparencies—nearly twice the number of images
as the previous edition.

Test Item File by Bo Lou (0-13-628017-X)
Over 2,400 multiple-choice test questions—30% new! Many new conceptual problems have been added for the Fifth Edition.

Prentice Hall Custom Test
Windows (0-13-628025-0); Macintosh (0-13-628033-1)
Based on the powerful testing technology developed by Engineering Soft-
ware Associates, Inc. (ESA); Prentice Hall Custom Test allows instructors to
create and tailor exams to their own needs. With the Online Testing Program,
exams can also be administered online and data can then be automatically
transferred for evaluation. A comprehensive desk reference guide is included along with online assistance.

For the Student

Student Study Guide by Joseph Boyle (0-13-627944-9)
Complements the strong pedagogy in Giancoli's text with overviews, topic
summaries and exercises, key phrases and terms, self-study exams, and questions for review of each chapter.

MCAT Study Guide by Joseph Boone (0-13-627951-1)
A thoroughly revised study resource that references all of the physics topics
on the MCAT to the appropriate sections in the text. Additional review,
review questions, and problems are provided.

Physics on the Internet: A Student's Guide
by Andrew Stull and Carl Adler (0-13-890153-8)
The perfect tool to help students take advantage of the *Physics: Principles
and Applications, Fifth Edition* Web page. This useful resource gives clear
steps to access Prentice Hall's regularly updated physics resources, along with
an overview of general navigation strategies. Available FREE for students
when purchased in a special package with Giancoli's text.

Prentice Hall/*The New York Times***
Themes of theTimes—Physics**
This unique newspaper supplement brings together a collection of the latest physics-related articles from the pages of *The New York Times.*

Media Supplements

Interactive Journey through Physics by Cindy Schwarz, Vassar College, CD-ROM for Windows and Macintosh, ©1997 (0-13-254103-3)
Whether your students are interested in exploring fascinating physics concepts, improving their grades, or reviewing for the MCAT, *Interactive Journey through Physics* will augment the traditional learning experiences of lecture, lab, and text.

Interactive Physics Player Workbook by Cindy Schwarz,
Windows book/disk (0-13-667312-0); Macintosh book/disk (0-13-477670-4)
An easy way to use *Interactive Physics* in your courses, this highly interactive workbook/disk package contains 40 simulation projects of varying degrees of difficulty. Each contains a physics review, simulation details, hints, explanation of results, math help, and a self test.

Physics Explorer Runtime Version by LOGAL,
Windows (0-13-627969-4); Macintosh (0-13-627977-5)
Tailored for use with Giancoli's text, *Physics Explorer Runtime Version* contains simulations of over 100 problems and examples directly from Giancoli's text. Students can conduct experiments, interactively record results on a spreadsheet, and generate graphs using each of ten independent learning models:
Particle Mechanics—One Body, Two Body, Gravity, and Harmonic Motion
Wave Mechanics—Waves, Ripple Tank, Diffraction
Electricity and Magnetism—One Body Electrodynamics, AC/DC Circuits, Electrostatics.

Presentation Manager CD-Rom
This CD-ROM contains all the text art and videos from the Physics You Can See video tape as well as additional lab and demonstration videos and animations from the *Interactive Journey through Physics* CD-ROM.

Physics: Principles and Applications Web Site
http://www.prenhall.com/giancoli
Features include practice tests with on-line feedback/grading keyed to the text.

Physics You Can See *Videos* (0-205-12393-7)
Each two- to five-minute segment demonstrates a classical physics experiment. Includes 11 segments such as "Coin and Feather" (acceleration due to gravity); "Monkey & Gun" (projectile motion); "Swivel Hips" (force pairs); and "Collapse a Can" (atmospheric pressure).

NOTES TO STUDENTS AND INSTRUCTORS ON THE FORMAT

1. Sections marked with a star (*) are considered optional. They can be omitted without interrupting the main flow of topics. No later material depends on them except possibly later starred sections. They may be fun to read though.

2. The customary conventions are used: symbols for quantities (such as m for mass) are italicized, whereas units (such as m for meter) are not italicized. Boldface ($\mathbf{F}$) is used for vectors.

3. Few equations are valid in all situations. Where practical, the limitations of important equations are stated in square brackets next to the equation. The equations that represent the great laws of physics are displayed with a tan background, as are a few other equations that are so useful that they are indispensable.

4. The number of significant figures (see Section 1–4) should not be assumed to be greater than given: if a number is stated as (say) 6, with its units, it is meant to be 6 and not 6.0 or 6.00.

5. At the end of each chapter is a set of questions that students should attempt to answer (to themselves at least). These are followed by problems which are ranked as level I, II, or III, according to estimated difficulty, with level I problems being easiest. These problems are arranged by Section, but problems for a given Section may depend on earlier material as well. There follows a group of General Problems, which are not arranged by Section nor ranked as to difficulty. Questions and problems that relate to optional Sections are starred.

6. Being able to solve problems is a crucial part of learning physics, and provides a powerful means for understanding the concepts and principles of physics. This book contains many aids to problem solving: (a) worked-out Examples and their solutions in the text, which are set off with a vertical blue line in the margin, and should be studied as an integral part of the text; (b) special "Problem-solving boxes" placed throughout the text to suggest ways to approach problem solving for a particular topic—but don't get the idea that every topic has its own "techniques," because the basics remain the same; (c) special problem-solving Sections (marked in blue in the Table of Contents); (d) marginal notes (see below), many of which refer to hints for solving problems, in which case they are so indicated; (e) problems themselves at the end of each chapter (see point 5 above); (f) some of the worked-out Examples are Estimation Examples, which show how rough or approximate results can be obtained even if the given data are sparse (see Section 1–7).

7. Conceptual Examples look like ordinary Examples but are conceptual rather than numerical. Each proposes a question or two, which hopefully starts you to think and come up with a response. Give yourself a little time to come up with your own response before reading the Response given.

8. Marginal notes: brief notes in the margin of almost every page are printed in blue and are of four types: (a) ordinary notes (the majority) that serve as a sort of outline of the text and can help you later locate important concepts and equations; (b) notes that refer to the great laws and principles of physics, and these are in capital letters and in a box for emphasis; (c) notes that refer to a problem-solving hint or technique treated in the text, and these say "Problem Solving"; (d) notes that refer to a physics application in the text or an Example, and these say "Physics Applied."

9. This book is printed in full color. But not simply to make it more attractive. The color is used above all in the figures, to give them greater clarity for our analysis, and to provide easier learning of the physical principles involved. The table on the next page is a summary of how color is used in the figures, and shows which colors are used for the different kinds of vectors, for field lines, and for other symbols and objects. These colors are used consistently throughout the book.

NOTES ON USE OF COLOR

Vectors

A general vector

    resultant vector (sum) is slightly thicker

    components of any vector are dashed

Displacement ($\mathbf{D}$, $\mathbf{r}$)

Velocity ($\mathbf{v}$)

Acceleration ($\mathbf{a}$)

Force ($\mathbf{F}$)

    Force on second or

    third object in same figure

Momentum ($\mathbf{p}$ or $m\mathbf{v}$)

Angular momentum ($\mathbf{L}$)

Angular velocity (ω)

Torque (τ)

Electric field ($\mathbf{E}$)

Magnetic field ($\mathbf{B}$)

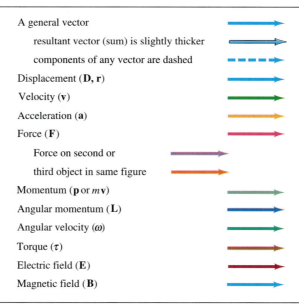

Electricity and magnetism

Electric field lines

Equipotential lines

Magnetic field lines

Electric charge (+)   +   or   • +

Electric charge (–)   −   or   • −

Electric circuit symbols

Wire

Resistor

Capacitor

Inductor

Battery

Optics

Light rays

Object

Real image
(dashed)

Virtual image
(dashed and paler)

Other

Energy level
(atom, etc.)

Measurement lines |←—1.0 m—→|

Path of a moving
object

Direction of motion
or current

xxiii

This comb has been rubbed by a cloth or paper towel to give it a static electric charge. Because the comb is electrically charged, it induces a separation of charge in all those scraps of paper, and thus attracts them.

16 ELECTRIC CHARGE AND ELECTRIC FIELD

The word "electricity" may evoke an image of complex modern technology: computers, lights, motors, electric power. But the electric force would seem to play an even deeper role in our lives: according to atomic theory, the forces that act between atoms and molecules to hold them together to form liquids and solids are electrical forces, and electric forces are also involved in the metabolic processes that occur within our bodies. Many of the forces we have dealt with so far, such as elastic forces, the normal force, and other contact forces (pushes and pulls) are now considered to result from electric forces acting at the atomic level. This does not include gravity, however, which is a separate force.[†]

The earliest studies on electricity date back to the ancients, but it has been only in the past two centuries that electricity was studied in detail. We will discuss the development of ideas about electricity, including practical devices, as well as the relation to magnetism, in the next seven chapters.

[†]As we discussed in Section 5–10, physicists in this century came to recognize four different fundamental forces in nature: (1) gravitational force, (2) electromagnetic force (we will see later that electric and magnetic forces are intimately related), (3) strong nuclear force, and (4) weak nuclear force. The last two forces operate at the level of the nucleus of an atom. The electromagnetic and weak nuclear forces are now thought to have a common origin known as the electroweak force. We will discuss these forces in later chapters.

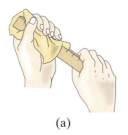
(a)

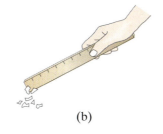
(b)

FIGURE 16–1 Rub a plastic ruler and bring it close to some tiny pieces of paper.

16–1 Static Electricity; Electric Charge and Its Conservation

The word *electricity* comes from the Greek word *elektron*, which means "amber." Amber is petrified tree resin, and the ancients knew that if you rub an amber rod with a piece of cloth, the amber attracts small pieces of leaves or dust. A piece of hard rubber, a glass rod, or a plastic ruler rubbed with a cloth will also display this "amber effect," or **static electricity** as we call it today. You can readily pick up small pieces of paper with a plastic comb or ruler that you've just vigorously rubbed with even a paper towel. See the photo on the previous page and Fig. 16–1. You have probably experienced static electricity when combing your hair or when taking a synthetic blouse or shirt from a clothes dryer. And you may have felt a shock when you touched a metal doorknob after sliding across a car seat or walking across a nylon carpet. In each case, an object becomes "charged" due to a rubbing process and is said to possess a net **electric charge**.

Is all electric charge the same, or is it possible that there is more than one type? In fact, there are two types of electric charge, as the following simple experiments show. A plastic ruler is suspended by a thread and rubbed vigorously with a cloth to charge it. When a second ruler, which has also been charged in the same way, is brought close to the first, it is found that the one ruler *repels* the other. This is shown in Fig. 16–2a. Similarly, if a rubbed glass rod is brought close to a second charged glass rod, again a repulsive force is seen to act, Fig. 16–2b. However, if the charged glass rod is brought close to the charged plastic ruler, it is found that they *attract* each other, Fig. 16–2c. The charge on the glass must therefore be different from that on the plastic. Indeed, it is found experimentally that all charged objects fall into one of two categories. Either they are attracted to the plastic and repelled by the glass, just as glass is; or they are repelled by the plastic and attracted to the glass, just as the plastic ruler is. Thus there seem to be two, and only two, types of electric charge. Each type of charge repels the same type but attracts the opposite type. That is: **unlike charges attract; like charges repel**.

The two types of electric charge were referred to as *positive* and *negative* by the American statesman, philosopher, and scientist Benjamin Franklin (1706–1790). The choice of which name went with which type of charge was of course arbitrary. Franklin's choice sets the charge on the rubbed glass rod to be positive charge, so the charge on a rubbed plastic ruler (or amber) is called negative charge. We still follow this convention today.

Franklin argued that whenever a certain amount of charge is produced on one body in a process, an equal amount of the opposite type of charge is produced on another body. The positive and negative are to be treated *algebraically*, so that during any process, the net change in the

FIGURE 16–2 Unlike charges attract, whereas like charges repel one another.

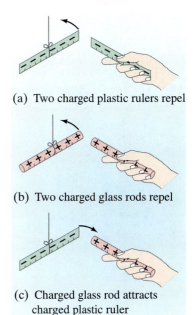

(a) Two charged plastic rulers repel

(b) Two charged glass rods repel

(c) Charged glass rod attracts charged plastic ruler

Likes repel; unlikes attract

amount of charge produced is zero. For example, when a plastic ruler is rubbed with a paper towel, the plastic acquires a negative charge and the towel an equal amount of positive charge. The charges are separated, but the sum of the two is zero. This is an example of a law that is now well established: the **law of conservation of electric charge**, which states that

the net amount of electric charge produced in any process is zero.

If one object or one region of space acquires a positive charge, then an equal amount of negative charge will be found in neighboring areas or objects. No violations have ever been found, and this conservation law is as firmly established as those for energy and momentum.

16–2 Electric Charge in the Atom

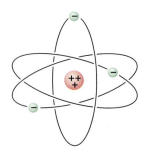

FIGURE 16–3 Simple model of the atom.

Only within the past century has it become clear that electricity starts inside the atom itself. In later chapters we will discuss atomic structure and the ideas that led to our present view of the atom in more detail. But it will help our understanding of electricity if we discuss it briefly now.

Today's view, somewhat simplified, shows the atom as having a heavy, positively charged nucleus surrounded by one or more negatively charged electrons (Fig. 16–3). The nucleus contains protons, which are positively charged, and neutrons, which have no net electric charge. The magnitude of the charge on protons and electrons is exactly the same, but their signs are opposite. Hence, neutral atoms contain equal numbers of protons and electrons. Sometimes, however, an atom may lose one or more of its electrons, or may gain extra electrons. In this case the atom will have a net positive or negative charge, and is called an **ion**.

In solid materials the nuclei tend to remain close to fixed positions, whereas some of the electrons move quite freely. The charging of a solid object by rubbing is explained mainly by the transfer of electrons from one material to the other. When a plastic ruler becomes negatively charged by rubbing with a paper towel, the transfer of electrons from the towel to the plastic leaves the towel with a positive charge equal in magnitude to the negative charge acquired by the plastic. (In liquids and gases, nuclei or ions can move as well as electrons.)

Normally when objects are charged by rubbing, they hold their charge only for a limited time and eventually return to the neutral state. Where does the charge go? In some cases it is neutralized by charged ions in the air (formed, for example, by collisions with charged particles known as cosmic rays that reach the Earth from space). Often more importantly, the charge can "leak off" onto water molecules in the air. This is because water molecules are **polar**—that is, even though they are neutral, their charge is not distributed uniformly, Fig. 16–4. Thus the extra electrons on, say, a charged plastic ruler can "leak off" into the air because they are attracted to the positive end of water molecules. A positively charged object, on the other hand, can be neutralized by transfer of loosely held electrons from water molecules in the air. On dry days, static electricity is much more noticeable since the air contains fewer water molecules to allow leakage. On humid or rainy days, it is difficult to make any object hold its charge for long.

FIGURE 16–4 Diagram of a water molecule. Because it has opposite charges on different ends, it is called a "polar" molecule.

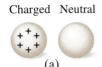 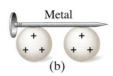 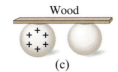

Charged	Neutral	Metal	Wood
(a)		(b)	(c)

FIGURE 16–5 (a) A charged metal sphere and a neutral metal sphere. (b) The two spheres connected by a metal nail, which conducts charge from one to the other. (c) The two spheres connected by an insulator (wood): almost no charge is conducted.

<h2>16–3   Insulators and Conductors</h2>

Suppose we have two metal spheres, one highly charged and the other electrically neutral (Fig. 16–5a). If we now place an iron nail so that it touches both the spheres (Fig. 16–5b), it is found that the previously uncharged sphere quickly becomes charged. If, instead, we connect the two spheres together by a wooden rod or a piece of rubber (Fig. 16–5c), the uncharged ball does not become noticeably charged. Materials like the iron nail are said to be **conductors** of electricity, whereas wood and rubber are **nonconductors** or **insulators**.

Metals are generally good conductors whereas most other materials are insulators (although even insulators conduct electricity very slightly). It is interesting that nearly all natural materials fall into one or the other of these two quite distinct categories. There are a few materials, however (notably silicon, germanium, and carbon), that fall into an intermediate (but distinct) category known as **semiconductors**.

Metals are good conductors

From the atomic point of view, the electrons in an insulating material are bound very tightly to the nuclei. In a good conductor, on the other hand, some of the electrons are bound very loosely and can move about freely within the material (although they cannot *leave* the metal easily) and are often referred to as *free electrons* or *conduction electrons*. When a positively charged object is brought close to or touches a conductor, the free electrons are attracted by this positive charge and move quickly toward it. On the other hand, the free electrons move swiftly away from a negative charge that is brought close. In a semiconductor, there are very few free electrons, and in an insulator, almost none.

<h2>16–4   Induced Charge; the Electroscope</h2>

Suppose a positively charged metal object is brought close to an uncharged metal object. If the two touch, the free electrons in the neutral one are attracted to the positively charged object and some will pass over to it, Fig. 16–6. Since the second object is now missing some of its negative electrons, it will have a net positive charge. This process is called "charging by conduction," or "by contact," and the two objects end up with the same sign of charge.

Now suppose a positively charged object is brought close to a neutral metal rod, but does not touch it. Although the electrons of the metal rod do not leave the rod, they still move within the metal toward the charged object, which leaves a positive charge at the opposite end, Fig. 16–7. A charge is said to have been *induced* at the two ends of the metal rod. Of course no net charge has been created in the rod; charges have merely been *separated*. The net charge on the metal rod is still zero. However, if the metal were broken into two pieces, we could have two charged objects, one charged positively and one charged negatively.

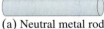

(a) Neutral metal rod

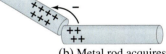

(b) Metal rod acquires charge by contact

FIGURE 16–6 (a) Neutral metal rod acquires a charge (b) when placed in contact with a charged metal object.

FIGURE 16–7 Charging by induction.

(a)   Neutral metal rod

(b)   Metal rod still neutral, but with a separation of charge

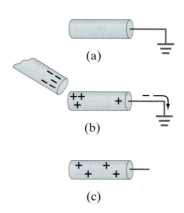

(a)

(b)

(c)

FIGURE 16–8 Inducing a charge on an object connected to ground.

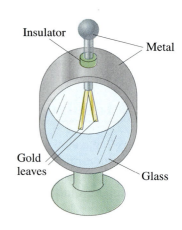

Insulator

Metal

Gold leaves

Glass

FIGURE 16–9 Electroscope.

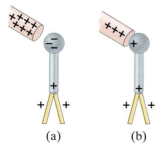

(a)   (b)

FIGURE 16–10 Electroscope charged (a) by induction, (b) by conduction.

FIGURE 16–11 A previously charged electroscope can be used to determine the sign of a given charge.

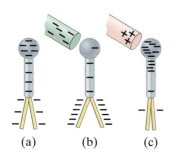

(a)   (b)   (c)

Another way to induce a net charge on a metal object is to connect it with a conducting wire to the ground (or a conducting pipe leading into the ground) as shown in Fig. 16–8a ($\perp$ means "ground"). The object is then said to be "grounded" or "earthed." Now the Earth, since it is so large and can conduct, can easily accept or give up electrons; hence it acts like a reservoir for charge. If a charged object—let's say negative this time—is brought up close to the metal, free electrons in the metal are repelled and many of them move down the wire into the Earth, Fig. 16–8b. This leaves the metal positively charged. If the wire is now cut, the metal will have a positive induced charge on it (Fig. 16–8c). If the wire were cut after the negative object is moved away, the electrons would all have moved back into the metal and it would be neutral.

An **electroscope** is a device that can be used for detecting charge. As shown in Fig. 16–9, inside of a case are two movable leaves, often made of gold. (Sometimes only one leaf is movable.) The leaves are connected by a conductor to a metal ball on the outside of the case, but are insulated from the case itself. If a positively charged object is brought close to the knob, a separation of charge is induced, as electrons are attracted up into the ball, leaving the leaves positively charged, Fig. 16–10a. The two leaves repel each other as shown. If, instead, the knob is charged by conduction, the whole apparatus acquires a net charge as shown in Fig. 16–10b. In either case, the greater the amount of charge, the greater the separation of the leaves.

Note, however, that you cannot tell the sign of the charge in this way, since a negative charge will cause the leaves to separate just as much as an equal-magnitude positive charge—in either case the two leaves repel each other. An electroscope can, however, be used to determine the sign of the charge if it is first charged by conduction, say negatively, as in Fig. 16–11a. Now if a negative object is brought close, as in Fig. 16–11b, more electrons are induced to move down into the leaves and they separate further. On the other hand, if a positive charge is brought close, the electrons are induced to flow upward, leaving the leaves less negative and their separation is reduced, Fig. 16–11c.

The electroscope was much used in the early studies of electricity. The same principle, aided by some electronics, is used in much more sensitive modern **electrometers**.

16–5 Coulomb's Law

We have seen that an electric charge exerts a force on other electric charges. What factors affect the magnitude of this force? To answer this, the French physicist Charles Coulomb (1736–1806) investigated electric forces in the 1780s using a torsion balance (Fig. 16–12) much like that used by Cavendish for his studies of the gravitational force (Section 5–6).

Although precise instruments for the measurement of electric charge were not available in Coulomb's time, he was able to prepare small spheres with different magnitudes of charge in which the *ratio* of the charges was known. He reasoned that if a charged conducting sphere is placed in contact with an identical uncharged sphere, the charge on the first would be shared equally by the two of them because of symmetry. He thus had a way to produce charges equal to $\frac{1}{2}$, $\frac{1}{4}$, and so on, of the original charge. Although he had some difficulty with induced charges, Coulomb was able to argue that the force one tiny charged object exerted on a second tiny charged object is directly proportional to the charge on each of them. That is, if the charge on either one of the objects was doubled, the force was doubled; and if the charge on both of the objects was doubled, the force increased to four times the original value. This was the case when the distance between the two charges remained the same. If the distance between them was allowed to increase, he found that the force decreased with the *square of the distance* between them. That is, if the distance was doubled, the force fell to one-fourth of its original value. Thus, Coulomb concluded, the force one tiny charged object exerts on a second one is proportional to the product of the magnitude of the charge on one, Q_1, times the magnitude of the charge on the other, Q_2, and inversely proportional to the square of the distance r between them (Fig. 16–13). As an equation, we can write **Coulomb's law** as

$$F = k\frac{Q_1 Q_2}{r^2}, \qquad (16\text{–}1)$$

where k is a proportionality constant. The validity of Coulomb's law today rests on precision measurements that are much more sophisticated than Coulomb's original experiment.

Since we are dealing here with a new quantity (electric charge), we could choose its unit so that the proportionality constant k in Eq. 16–1 would be one. Indeed, such a system of units was once common.[†] However, the most widely used unit now is the **coulomb** (C), which is the SI unit. The precise definition of the coulomb today is in terms of electric current and magnetic field, and will be discussed later (Section 20–7). In SI units, k has the value

$$k = 8.988 \times 10^9 \, \text{N·m}^2/\text{C}^2 \approx 9.0 \times 10^9 \, \text{N·m}^2/\text{C}^2.$$

Thus, 1 C is that amount of charge which, if placed on each of two point objects 1.0 m apart, will result in each object exerting a force of $(9.0 \times 10^9 \, \text{N·m}^2/\text{C}^2)(1.0 \, \text{C})(1.0 \, \text{C})/(1.0 \, \text{m})^2 = 9.0 \times 10^9 \, \text{N}$ on the other. This would be an enormous force, equal to the weight of almost a million tons. We don't normally encounter charges as large as a coulomb.

Charges produced by rubbing ordinary objects (such as a comb or plastic ruler) are typically around a microcoulomb ($1 \, \mu\text{C} = 10^{-6} \, \text{C}$) or

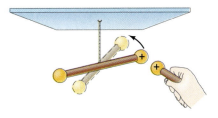

FIGURE 16–12 Schematic diagram of Coulomb's apparatus. It is similar to Cavendish's, which was used for the gravitational force. When a charged sphere is placed close to the one on the suspended bar, the bar rotates slightly. The suspending fiber resists the twisting motion and the angle of twist is proportional to the force applied. By the use of this apparatus, Coulomb investigated how the electric force varies as a function of the magnitude of the charges and of the distance between them.

COULOMB'S LAW

FIGURE 16–13 Coulomb's law, Eq. 16–1, gives the force between two point charges, Q_1 and Q_2, a distance r apart.

Unit for charge: the coulomb

[†]This is a cgs system of units, and the unit of electric charge is called the *electrostatic unit* (esu) or the statcoulomb. One esu is defined as that charge, on each of two point objects 1 cm apart, that gives rise to a force of 1 dyne.

less. The magnitude of the charge on one electron, on the other hand, has been determined to be about 1.602×10^{-19} C, and its sign is negative. This is the smallest known charge,[†] and because of its fundamental nature, it is given the symbol e and is often referred to as the *elementary charge*:

$$e = 1.602 \times 10^{-19} \, \text{C}.$$

Charge on electron (the elementary charge)

Note that e is defined as a positive number, so the charge on the electron is $-e$. (The charge on a proton, on the other hand, is $+e$). Since an object cannot gain or lose a fraction of an electron, the net charge on any object must be an integral multiple of this charge. Electric charge is thus said to be **quantized** (existing only in discrete amounts: $1e, 2e, 3e$, etc.). Because e is so small, however, we normally don't notice this discreteness in macroscopic charges ($1 \, \mu\text{C}$ requires about 10^{13} electrons), which thus seem continuous.

Electric charge is quantized

Equation 16–1 gives the *magnitude* of the electric force that either object exerts on the other, when the magnitudes of the charges Q_1 and Q_2 are given. The *direction* of the electric force *is always along the line joining the two objects*. If the two charges have the same sign, the force on either object is directed away from the other. If the two charges have opposite signs, the force on one is directed toward the other, Fig. 16–14. Notice that the force one charge exerts on the second is equal but opposite to that exerted by the second on the first, in accord with Newton's third law.

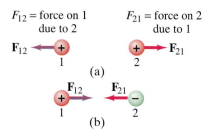

F_{12} = force on 1 due to 2 F_{21} = force on 2 due to 1

(a)

(b)

FIGURE 16–14 Direction of the force depends on whether the charges have (a) the same sign, or (b) opposite sign.

[Note the similarity of Coulomb's law to the law of universal gravitation, Eq. 5–4. Both are inverse square laws ($F \propto 1/r^2$). Both also have a proportionality to a product of a property of each body—mass for gravity, electric charge for electricity. A major difference between the two laws is that gravity is always an attractive force, whereas the electric force can be either attractive or repulsive.]

The constant k in Eq. 16–1 is often written in terms of another constant, ϵ_0, called the **permittivity of free space**. It is related to k by $k = 1/4\pi\epsilon_0$. Coulomb's law can then be written

COULOMB'S LAW (in terms of ϵ_0)

$$F = \frac{1}{4\pi\epsilon_0} \frac{Q_1 Q_2}{r^2}, \tag{16–2}$$

where

$$\epsilon_0 = \frac{1}{4\pi k} = 8.85 \times 10^{-12} \, \text{C}^2/\text{N} \cdot \text{m}^2.$$

Equation 16–2 looks more complicated than Eq. 16–1, but other fundamental equations we haven't seen yet are simpler in terms of ϵ_0 rather than k. It doesn't matter which form we use, of course, since Eqs. 16–1 and 16–2 are equivalent.

It should be recognized that Eqs. 16–1 and 16–2 apply to objects whose size is much smaller than the distance between them. Ideally, it is precise for **point charges** (spatial size negligible compared to other distances). For finite-sized objects, it is not always clear what value to use for r, particularly since the charge may not be distributed uniformly on the objects. If the two objects are spheres and the charge is known to be distributed uniformly on each, then r is the distance between their centers.

Coulomb's law describes the force between two charges when they are at rest. Additional forces come into play when charges are in motion, and these will be discussed in later chapters. In this chapter we discuss only charges at rest, the study of which is called **electrostatics**.

[†]According to the standard model of elementary particle physics, subnuclear particles called quarks have a smaller charge than that on the electron, equal to $\frac{1}{3}e$ or $\frac{2}{3}e$. Quarks have not been detected directly, and theory indicates that free quarks may not be detectable.

When calculating with Coulomb's law, we usually ignore the signs of the charges and determine direction based on whether the force is attractive or repulsive.

EXAMPLE 16–1 **Electric force on electron by proton.** Determine the magnitude of the electric force on the electron of a hydrogen atom exerted by the single proton ($Q_2 = +e$) that is its nucleus. Assume the electron "orbits" the proton at its average distance of $r = 0.53 \times 10^{-10}$ m, Fig. 16–15.

SOLUTION We use Eq. 16–1 with $r = 0.53 \times 10^{-10}$ m, and $Q_1 = Q_2 = 1.6 \times 10^{-19}$ C (ignoring the signs of the charges):

$$F = \frac{(9.0 \times 10^9 \, \text{N·m}^2/\text{C}^2)(1.6 \times 10^{-19} \, \text{C})(1.6 \times 10^{-19} \, \text{C})}{(0.53 \times 10^{-10} \, \text{m})^2}$$

$$= 8.2 \times 10^{-8} \, \text{N}.$$

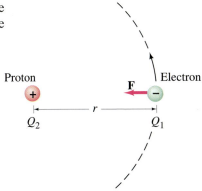

FIGURE 16–15 Example 16–1.

The direction of the force on the electron is toward the proton, since the charges have opposite signs and the force is attractive.

CONCEPTUAL EXAMPLE 16–2 **Which charge exerts the greater force?** Two positive point charges, $Q_1 = 50 \, \mu\text{C}$ and $Q_2 = 1 \, \mu\text{C}$, are separated by a distance l, Fig. 16–16. Which is larger in magnitude, the force that Q_1 exerts on Q_2, or the force that Q_2 exerts on Q_1?

RESPONSE From Coulomb's law, the force on Q_1 exerted by Q_2 is:

$$F_{12} = k\frac{Q_1 Q_2}{l^2}.$$

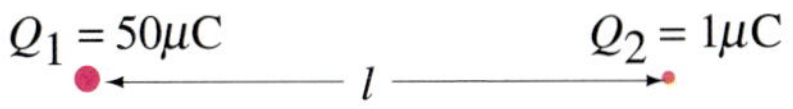

FIGURE 16–16 Example 16–2.

The force on Q_2 exerted by Q_1 is the same except that Q_1 and Q_2 are reversed. The equation is symmetric with respect to the two charges, so $F_{21} = F_{12}$. Newton's third law also tells us that these two forces must have equal magnitude.

It is very important to keep in mind that Eq. 16–1 (or 16–2) gives the force on a charge due to only *one* other charge. If several (or many) charges are present, the *net force on any one of them will be the vector sum of the forces due to each of the others.*

Electric forces add as vectors

16–6 Solving Problems Involving Coulomb's Law and Vectors

The electric force between charged particles at rest (sometimes referred to as the **electrostatic force** or as the **Coulomb force**) is, like all forces, a vector: it has both magnitude and direction. When several forces act on an object (call them $\mathbf{F}_1$, $\mathbf{F}_2$, etc.), the net force $\mathbf{F}_{\text{net}}$ on the object is the vector sum of all the forces acting on it:

$$\mathbf{F}_{\text{net}} = \mathbf{F}_1 + \mathbf{F}_2 + \cdots.$$

(This is sometimes referred to as the principle of superposition for forces). We studied how to add vectors in Chapter 3, and in Chapter 4 we applied the rules for adding vectors to forces. It might be a good idea now to review Sections 3–2, 3–3, 3–4, as well as Section 4–9 on general problem-solving techniques. Here is a brief review of vectors.

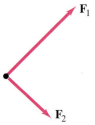

(a) Two forces acting on an object.

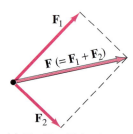

(b) The total, or net, force is $\mathbf{F} = \mathbf{F}_1 + \mathbf{F}_2$ by the tail-to-tip method of adding vectors.

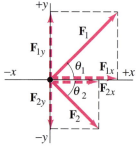

(c) $\mathbf{F} = \mathbf{F}_1 + \mathbf{F}_2$ by the parallelogram method.

(d) $\mathbf{F}_1$ and $\mathbf{F}_2$ resolved into their x and y components.

FIGURE 16–17 Review of vector addition.

Given two vector forces, $\mathbf{F}_1$ and $\mathbf{F}_2$, acting on a body (Fig. 16–17a), they can be added using the tail-to-tip method (Fig. 16–17b) or by the parallelogram method (Fig. 16–17c), as discussed in Section 3–2. These two methods are useful for *understanding* a given problem (for getting a picture in your mind of what is going on), but for *calculating* the direction and magnitude of the resultant sum, it is more precise to use the method of adding components. Figure 16–17d shows the components of our $\mathbf{F}_1$ and $\mathbf{F}_2$ resolved into components along chosen x and y axes (for more details, see Section 3–4). From the definitions of the trigonometric functions (Figs. 3–11 and 3–12), we have

$$F_{1x} = F_1 \cos \theta_1 \qquad F_{2x} = F_2 \cos \theta_2$$
$$F_{1y} = F_1 \sin \theta_1 \qquad F_{2y} = -F_2 \sin \theta_2.$$

We add up the x and y components separately to obtain the components of the resultant force $\mathbf{F}$, which are

$$F_x = F_{1x} + F_{2x} = F_1 \cos \theta_1 + F_2 \cos \theta_2,$$
$$F_y = F_{1y} + F_{2y} = F_1 \sin \theta_1 - F_2 \sin \theta_2.$$

The magnitude of $\mathbf{F}$ is

$$F = \sqrt{F_x^2 + F_y^2}.$$

The direction of $\mathbf{F}$ is specified by the angle θ that $\mathbf{F}$ makes with the x axis, which is given by

$$\tan \theta = \frac{F_y}{F_x}.$$

This review has been necessarily brief; a rereading of the appropriate parts of Chapters 3 and 4 will give more details.

When dealing with several charges, it is often helpful to use subscripts on each of the forces involved. The first subscript refers to the particle *on* which the force acts; the second refers to the particle that exerts the force. For example, if we have three charges, $\mathbf{F}_{31}$ means the force exerted *on* particle 3 *by* particle 1.

As in all problem solving, it is very important to draw a diagram, in particular a free-body diagram for each body (Chapter 4), showing all the forces acting on that body. In applying Coulomb's law, we usually deal with charge magnitudes only (leaving out minus signs) to get the magnitude of each force. Then determine the direction of the force physically (like charges repel, unlike attract), and show the force on the diagram. Finally, add the forces on one object together as vectors.

EXAMPLE 16–3 Three charges in a line. Three charged particles are arranged in a line, as shown in Fig. 16–18a. Calculate the net electrostatic force on particle 3 (the $-4.0 \, \mu\text{C}$ on the right) due to the other two charges.

SOLUTION The net force on particle 3 will be the vector sum of the force $\mathbf{F}_{31}$ exerted by particle 1 and the force $\mathbf{F}_{32}$ exerted by particle 2: $\mathbf{F} = \mathbf{F}_{31} + \mathbf{F}_{32}$. The magnitudes of these two forces are

$$F_{31} = \frac{(9.0 \times 10^9 \, \text{N} \cdot \text{m}^2/\text{C}^2)(4.0 \times 10^{-6} \, \text{C})(8.0 \times 10^{-6} \, \text{C})}{(0.50 \, \text{m})^2} = 1.2 \, \text{N}.$$

$$F_{32} = \frac{(9.0 \times 10^9 \, \text{N} \cdot \text{m}^2/\text{C}^2)(4.0 \times 10^{-6} \, \text{C})(3.0 \times 10^{-6} \, \text{C})}{(0.20 \, \text{m})^2} = 2.7 \, \text{N}.$$

Since we were calculating the magnitudes of the forces, we omitted the signs of the charges; but we must be aware of them to get the direction of each force. Let the line joining the particles be the x axis, and we take it positive to the right. Then, because $\mathbf{F}_{31}$ is repulsive and $\mathbf{F}_{32}$ is attractive, the directions of the forces are as shown in Fig. 16–18b: F_{31} points in the positive x direction and F_{32} points in the negative x direction. The net force on particle 3 is then

$$F = -F_{32} + F_{31} = -2.7\,\text{N} + 1.2\,\text{N} = -1.5\,\text{N}.$$

The magnitude of the net force is 1.5 N, and it points to the left.

Notice in this Example that the charge in the middle (Q_2) in no way blocks the effect of the other charge (Q_1); Q_2 does exert its own force, of course.

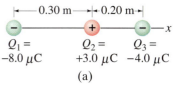

(a)

(b)

FIGURE 16–18 Diagram for Example 16–3.

EXAMPLE 16–4 **Electric force using vector components.** Calculate the net electrostatic force on charge Q_3 shown in Fig. 16–19a due to the charges Q_1 and Q_2.

SOLUTION The forces $\mathbf{F}_{31}$ and $\mathbf{F}_{32}$ have the directions shown in the diagram since Q_1 exerts an attractive force and Q_2 a repulsive force. The magnitudes of $\mathbf{F}_{31}$ and $\mathbf{F}_{32}$ are (ignoring signs since we know the directions)

$$F_{31} = \frac{(9.0 \times 10^9\,\text{N·m}^2/\text{C}^2)(6.5 \times 10^{-5}\,\text{C})(8.6 \times 10^{-5}\,\text{C})}{(0.60\,\text{m})^2} = 140\,\text{N},$$

$$F_{32} = \frac{(9.0 \times 10^9\,\text{N·m}^2/\text{C}^2)(6.5 \times 10^{-5}\,\text{C})(5.0 \times 10^{-5}\,\text{C})}{(0.30\,\text{m})^2} = 330\,\text{N}.$$

We resolve $\mathbf{F}_1$ into its components along the x and y axes, as shown:

$$F_{31x} = F_{31} \cos 30° = 120\,\text{N},$$

$$F_{31y} = -F_{31} \sin 30° = -70\,\text{N}.$$

The force $\mathbf{F}_{32}$ has only a y component. So the net force $\mathbf{F}$ on Q_3 has components

$$F_x = F_{31x} = 120\,\text{N}$$

$$F_y = F_{32} + F_{31y} = 330\,\text{N} - 70\,\text{N} = 260\,\text{N}.$$

Thus the magnitude of the net force is

$$F = \sqrt{F_x^2 + F_y^2} = \sqrt{(120\,\text{N})^2 + (260\,\text{N})^2} = 290\,\text{N};$$

and it acts at an angle θ (see Fig. 16–19b) given by $\tan \theta = F_y/F_x = 260\,\text{N}/120\,\text{N} = 2.2$, so $\theta = 65°$.

FIGURE 16–19 Determining the forces for Example 16–4.

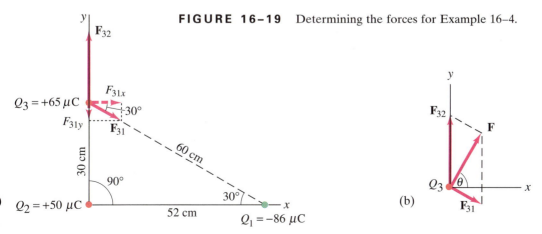

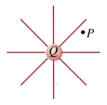

FIGURE 16–20 An electric field surrounds every charge. *P* is an arbitrary point.

FIGURE 16–21 Force exerted by charge $+Q$ on a small test charge, q, placed at points *a*, *b*, and *c*.

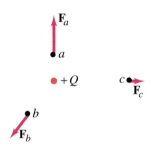

16–7 The Electric Field

Many common forces might be referred to as "contact forces," such as your hands pushing or pulling a cart, or a tennis racket hitting a tennis ball.

In contrast, both the gravitational force and the electrical force act over a distance: there is a force even when the two objects are not touching. The idea of a force *acting at a distance* was a difficult one for early thinkers. Newton himself felt uneasy with this idea when he published his law of universal gravitation. A helpful way to look at the situation uses the idea of the **field**, developed by the British scientist Michael Faraday (1791–1867). In the electrical case, according to Faraday, an *electric field* extends outward from every charge and permeates all of space (Fig. 16–20). When a second charge is placed near the first charge, it feels a force because of the electric field that is there (say, at point *P* in Fig. 16–20). The electric field at the location of the second charge is considered to interact directly with this charge to produce the force. It must be emphasized, however, that a field, as we think of it here, is *not* a kind of matter.

We can investigate the electric field surrounding a charge or group of charges by measuring the force on a small positive **test charge**. By a test charge we mean a charge so small that the force it exerts does not significantly alter the distribution of the charges that create the field being measured. The force on a tiny positive test charge q placed at various locations in the vicinity of a single positive charge Q would be as shown in Fig. 16–21. The force at *b* is less than at *a* because the distance is greater (Coulomb's law); and the force at *c* is smaller still. In each case, the force is directed radially outward from Q. The electric field is defined in terms of the force on such a positive test charge. In particular, the **electric field**, **E**, at any point in space is defined as the force **F** exerted on a tiny positive test charge at that point divided by the magnitude of the test charge q:

Definition of electric field

$$\mathbf{E} = \frac{\mathbf{F}}{q}.$$ **(16–3)**

Ideally, **E** is defined as the limit of $\mathbf{F}/q$ as q is taken smaller and smaller, approaching zero. From this definition (Eq. 16–3), we see that the electric field at any point in space is a vector whose direction is the direction of the force on a positive test charge at that point, and whose magnitude is the *force per unit charge*. Thus **E** is measured in units of newtons per coulomb (N/C).

The reason for defining **E** as $\mathbf{F}/q$ (with $q \rightarrow 0$) is so that **E** does not depend on the magnitude of the test charge q. This means that **E** describes only the effect of the charges creating the electric field at that point.

The electric field at any point in space can be measured, based on the definition, Eq. 16–3. For simple situations involving one or several point charges, we can calculate what **E** will be. For example, the electric field at a distance r from a single point charge Q would have magnitude

$$E = \frac{kqQ/r^2}{q}$$

Electric field due to one point charge

$$= k\frac{Q}{r^2};$$ [single point charge] **(16–4a)**

or, in terms of ϵ_0 (Eq. 16–2)

$$E = \frac{1}{4\pi\epsilon_0}\frac{Q}{r^2}. \qquad \text{[single point charge]} \quad \textbf{(16–4b)}$$

This relation for the electric field due to a single point charge is also (in addition to Eq. 16–1) referred to as Coulomb's law. Notice that E is independent of q—that is, it depends only on the charge Q which produces the field, and not on the value of the test charge q.

EXAMPLE 16–5 **Electrostatic copier.** An electrostatic copier works by selectively arranging positive charges (in a pattern to be copied) on the surface of a nonconducting drum, then gently sprinkling negatively charged dry toner (ink) particles onto the drum. The toner particles temporarily stick to the pattern on the drum and are later transferred to paper and "melted" to produce the copy. Suppose each toner particle has a mass of 9.0×10^{-16} kg and carries an average of 20 extra electrons to provide an electric charge. Assuming that the electric force on a toner particle must exceed twice its weight in order to ensure sufficient attraction, compute the required electric field strength near the surface of the drum. See Fig. 16–22.

SOLUTION The minimum value of electric field satisfies the relation

$$qE = 2\,mg$$

where $q = 20e$. Hence

$$E = \frac{2\,mg}{q} = \frac{2(9.0 \times 10^{-16}\,\text{kg})(9.8\,\text{m/s}^2)}{20(1.6 \times 10^{-19}\,\text{C})}$$

$$= 5.5 \times 10^3\,\text{N/C}.$$

EXAMPLE 16–6 **Electric field of a single point charge.** Calculate the magnitude and direction of the electric field at a point P which is 30 cm to the right of a point charge $Q = -3.0 \times 10^{-6}\,\text{C}$.

SOLUTION The magnitude of the electric field due to a single point charge is given by Eq. 16–4:

$$E = k\frac{Q}{r^2} = \frac{(9.0 \times 10^9\,\text{N}\cdot\text{m}^2/\text{C}^2)(3.0 \times 10^{-6}\,\text{C})}{(0.30\,\text{m})^2} = 3.0 \times 10^5\,\text{N/C}.$$

The direction of the electric field is *toward* the charge Q as shown in Fig. 16–23a since we defined the direction as that of the force on a positive test charge. If Q had been positive, the electric field would have pointed away, as in Fig. 16–23b.

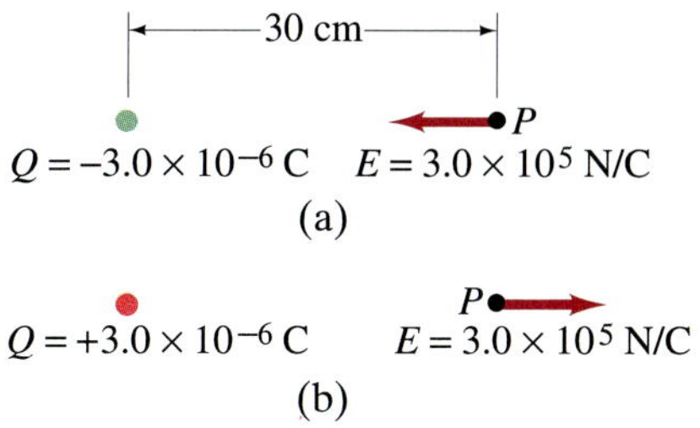

This Example illustrates a general result: The electric field due to a positive charge points away from the charge, whereas **E** due to a negative charge points toward that charge.

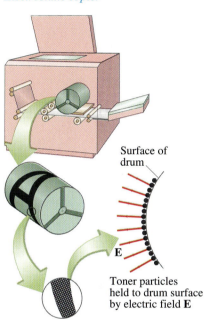

Surface of drum

E

Toner particles held to drum surface by electric field **E**

FIGURE 16–22 Example 16–5.

FIGURE 16–23 Example 16–6. Electric field at point P (a) due to a negative charge Q, and (b) due to a positive charge Q.

|←———— 30 cm ————→|

$Q = -3.0 \times 10^{-6}\,\text{C}$ $E = 3.0 \times 10^5\,\text{N/C}$

(a)

$Q = +3.0 \times 10^{-6}\,\text{C}$ $E = 3.0 \times 10^5\,\text{N/C}$

(b)

If the field is due to more than one charge, the individual fields (call them $\mathbf{E}_1$, $\mathbf{E}_2$, etc.) due to each charge are added vectorially to get the total field at any point:

$$\mathbf{E} = \mathbf{E}_1 + \mathbf{E}_2 + \cdots.$$

The validity of this **superposition principle** for electric fields is fully confirmed by experiment.[†]

EXAMPLE 16–7 **E in between two point charges.** Two point charges are separated by a distance of 10.0 cm. One has a charge of $-25\ \mu C$ and the other $+50\ \mu C$. (*a*) What is the direction and magnitude of the electric field at a point P in between them, that is 2.0 cm from the negative charge (Fig. 16–24a)? (*b*) If an electron is placed at rest at P, what will its acceleration (direction and magnitude) be initially?

SOLUTION (*a*) The field will be a combination of two fields both pointing to the left: the field due to the negative charge Q_1 points toward Q_1, and the field due to the positive charge Q_2 points away from Q_2, again to the left, Fig. 16–24b. Thus, we can add the magnitudes of the two fields together algebraically, ignoring the signs of the charges:

$$E = k\frac{Q_1}{r_1^2} + k\frac{Q_2}{r_2^2} = k\left(\frac{Q_1}{r_1^2} + \frac{Q_2}{r_2^2}\right) = k\frac{Q_1}{r_1^2}\left[1 + \frac{(Q_2/Q_1)}{(r_2^2/r_1^2)}\right]$$

where in the last step we factored out (Q_1/r_1^2). We substitute in $r_1 = 2.0\ \text{cm} = 2.0 \times 10^{-2}\ \text{m}$ and $r_2 = 8.0 \times 10^{-2}\ \text{m}$:

$$E = (9.0 \times 10^9\ \text{N·m}^2/\text{C}^2)\frac{(25 \times 10^{-6}\ \text{C})}{(2.0 \times 10^{-2}\ \text{m})^2}\left[1 + \frac{(50/25)}{(8.0/2.0)^2}\right]$$

$$= 5.6 \times 10^8\ [1 + \tfrac{1}{8}]\ \text{N/C} = 6.3 \times 10^8\ \text{N/C}.$$

Notice how factoring out Q_1/r_1^2 on the first line allowed us to see the relative strengths of the two contributing fields—namely that Q_2's field is only $\tfrac{1}{8}$ of Q_1's (or $\tfrac{1}{9}$ of the total).

(*b*) The electron will feel a force to the *right* since it is negatively charged and the acceleration will therefore be to the right, with a magnitude of

$$a = \frac{F}{m} = \frac{qE}{m} = \frac{(1.60 \times 10^{-19}\ \text{C})(6.3 \times 10^8\ \text{N/C})}{9.1 \times 10^{-31}\ \text{kg}} = 1.1 \times 10^{20}\ \text{m/s}^2.$$

FIGURE 16–24 Example 16–7. In (b), we don't know the relative lengths of $\mathbf{E}_1$ and $\mathbf{E}_2$ until we do the calculation.

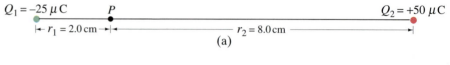

[†]A more general form of Coulomb's law, which allows calculation of the electric field in some useful situations, is Gauss's law, discussed in Appendix D.

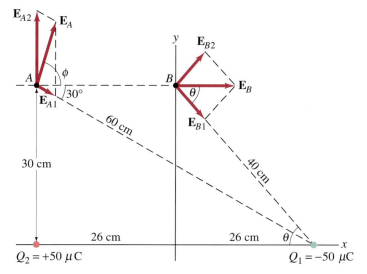

FIGURE 16–25 Calculation of the electric field at points A and B for Example 16–8.

EXAMPLE 16–8 **E above two point charges.** Calculate the total electric field (a) at point A and (b) at point B in Fig. 16–25 due to both charges, Q_1 and Q_2.

SOLUTION (a) The calculation is much like that of Example 16–4, but now we are dealing with electric fields. The electric field at A is the vector sum of the fields $\mathbf{E}_{A1}$ due to Q_1, and $\mathbf{E}_{A2}$ due to Q_2; using Eq. 16–4, $E = kQ/r^2$, they have magnitudes:

➡ **PROBLEM SOLVING**

Ignore signs of charges and determine direction physically, showing directions on diagram

$$E_{A1} = \frac{(9.0 \times 10^9 \text{ N·m}^2/\text{C}^2)(50 \times 10^{-6} \text{ C})}{(0.60 \text{ m})^2} = 1.25 \times 10^6 \text{ N/C},$$

$$E_{A2} = \frac{(9.0 \times 10^9 \text{ N·m}^2/\text{C}^2)(50 \times 10^{-6} \text{ C})}{(0.30 \text{ m})^2} = 5.0 \times 10^6 \text{ N/C}.$$

The directions are as shown, so the total electric field at A, $\mathbf{E}_A$, has components

$$E_{Ax} = E_{A1} \cos 30° = 1.1 \times 10^6 \text{ N/C},$$

$$E_{Ay} = E_{A2} - E_{A1} \sin 30° = 4.4 \times 10^6 \text{ N/C}.$$

Thus the magnitude of $\mathbf{E}_A$ is

$$E_A = \sqrt{(1.1)^2 + (4.4)^2} \times 10^6 \text{ N/C} = 4.5 \times 10^6 \text{ N/C},$$

and its direction is ϕ given by $\tan \phi = E_{Ay}/E_{Ax} = 4.4/1.1 = 4.0$, so $\phi = 76°$.
(b) Because B is equidistant (40 cm by the Pythagorean theorem) from the two equal charges, the magnitudes of E_{B1} and E_{B2} are the same; that is,

➡ **PROBLEM SOLVING**

Use symmetry to save work, when possible

$$E_{B1} = E_{B2} = \frac{kQ}{r^2} = \frac{(9.0 \times 10^9 \text{ N·m}^2/\text{C}^2)(50 \times 10^{-6} \text{ C})}{(0.40 \text{ m})^2}$$

$$= 2.8 \times 10^6 \text{ N/C}.$$

Also, because of the symmetry, the y components are equal and opposite. Hence the total field E_B is horizontal and equals $E_{B1} \cos \theta + E_{B2} \cos \theta = 2E_{B1} \cos \theta$; from the diagram, $\cos \theta = 26 \text{ cm}/40 \text{ cm} = 0.65$. Then

$$E_B = 2E_{B1} \cos \theta = 2(2.8 \times 10^6 \text{ N/C})(0.65) = 3.6 \times 10^6 \text{ N/C},$$

and the direction of $\mathbf{E}_B$ is along the $+x$ direction.

Solving electrostatics problems follows, to a large extent, the general problem-solving procedure discussed in Section 4–9. In particular,

1. Draw a careful diagram—namely, a free-body diagram for each object, showing all the forces acting on that object, or the electric field at a point due to all sources.

2. Apply Coulomb's law to get the magnitude of each force on a charged object, or the electric field at a point. Deal only with magnitudes of charges (leaving out minus signs), and obtain the magnitude of each force or electric field. Then determine the direction of each force or electric field physically (like charges repel each other, unlike charges attract). Show and label each vector force or field on your diagram. Then add vectorially all the forces on an object, or the contributing fields at a point, to get the resultant.

3. Use symmetry (say, in the geometry) whenever possible.

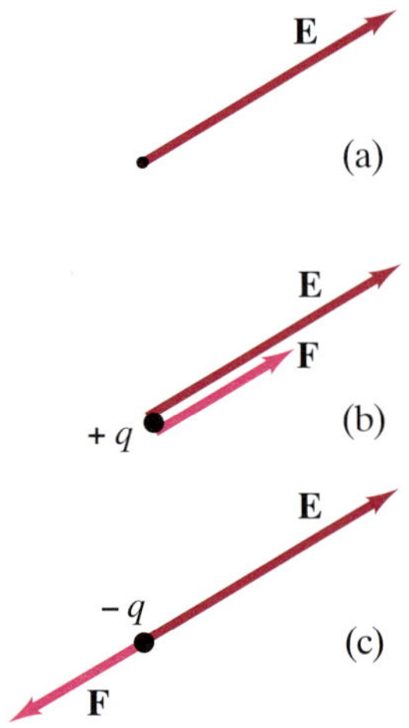

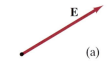

If we are given the electric field **E** at a given point in space, then we can calculate the force **F** on a charge q placed at that point by writing (see Eq. 16–3):

$$\mathbf{F} = q\mathbf{E}.$$

If q is positive, **F** and **E** will point in the same direction. If q is negative, **F** and **E** point in opposite directions. See Fig. 16–26.

16–8 Field Lines

Since the electric field is a vector, it is sometimes referred to as a *vector field*. We could indicate the electric field with arrows at various points in a given situation, such as at a, b, and c in Fig. 16–27. The directions of $\mathbf{E}_a$, $\mathbf{E}_b$, and $\mathbf{E}_c$ are the same as that of the forces shown earlier in Fig. 16–21, but the lengths (magnitudes) are different since we divide by q. However, the relative lengths of $\mathbf{E}_a$, $\mathbf{E}_b$, and $\mathbf{E}_c$ are the same as for the forces since we divide by the same q each time. To indicate the electric field in such a way at *many* points, however, would result in many arrows, which might appear complicated or confusing. To avoid this, we use another technique, that of field lines.

In order to visualize the electric field, we draw a series of lines to indicate the direction of the electric field at various points in space. These **electric field lines** (sometimes called *lines of force*) are drawn so that they indicate the direction of the force due to the given field on a positive test charge. The lines of force due to a single positive charge are shown in Fig. 16–28a and for a single negative charge in Fig. 16–28b. In part (a) the

FIGURE 16–26 (a) Electric field at a given point in space. (b) Force on a positive charge. (c) Force on a negative charge.

FIGURE 16–27 Electric field vector shown at three points, due to a single point charge Q. (Compare to Fig. 16–21.)

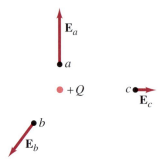

FIGURE 16–28 Electric field lines (a) near a single positive point charge, (b) near a single negative point charge.

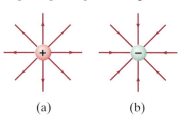

(a)   (b)

lines point radially outward from the charge, and in part (b) they point radially inward toward the charge because that is the direction the force would be on a positive test charge in each case (as in Fig. 16–23). Only a few representative lines are shown. One could just as well draw lines in between those shown since the electric field exists there as well. However, we can always draw the lines so that the *number of lines starting on a positive charge, or ending on a negative charge, is proportional to the magnitude of the charge*. Notice that near the charge, where the force is greatest, the lines are closer together. This is a general property of electric field lines; *the closer the lines are together, the stronger the electric field in that region*. In fact the lines can always be drawn so that the number of lines crossing unit area perpendicular to **E** is proportional to the magnitude of the electric field.

Figure 16–29a shows the electric field lines surrounding two charges of opposite sign. The electric field lines are curved in this case and they are directed from the positive charge to the negative charge. The direction of the field at any point is directed tangentially as shown by the arrow at point P. To satisfy yourself that this is the correct pattern for the electric field lines, you can make a few calculations such as those done in Example 16–8 for just this case (see Fig. 16–25). Figures 16–29b and c show the electric field lines surrounding two equal positive charges (b), and (c) for unequal charges, $+2Q$ and $-Q$; note that twice as many lines leave $+2Q$ as there are lines entering $-Q$ (number of lines is proportional to magnitude of Q). Finally, in (d), we see the field between two oppositely charged parallel plates. Notice that the electric field lines between the two plates start out perpendicular to the surface of the metal plates (we'll see why this is always true in the next Section) and go directly from one plate to the other, as we expect because a positive test charge placed between the plates would feel a strong repulsion from the positive plate and a strong attraction to the negative plate. The field lines between the plates are parallel and equally spaced, except near the edges. Thus, in the central region, the electric field has the same magnitude at all points and we can write

$$E = \text{constant.} \quad \text{[between two closely spaced parallel plates]} \quad \textbf{(16–5)}$$

Although the field fringes near the edges (the lines curve), we can often ignore this, particularly if the separation of the plates is small compared to their size. This should be compared to the field of a single point charge, where the field decreases as the square of the distance, Eq. 16–4.

We summarize the properties of field lines as follows:

1. The field lines indicate the direction of the electric field; the field points in the direction tangent to the field line at any point.

2. The lines are drawn so that the magnitude of the electric field, E, is proportional to the number of lines crossing unit area perpendicular to the lines. The closer the lines, the stronger the field.

3. Electric field lines start on positive charges and end on negative charges; and the number starting or ending is proportional to the magnitude of the charge.

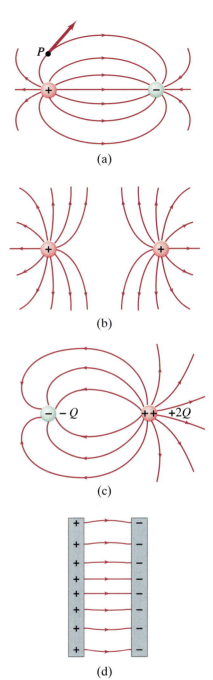

(a)

(b)

(c)

(d)

FIGURE 16–29 Electric field lines for four arrangements of charges.

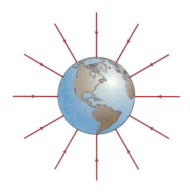

FIGURE 16–30 The Earth's gravitational field.

The field concept can also be applied to the gravitational force. Thus we can say that a **gravitational field** exists for every object that has mass. One object attracts another by means of the gravitational field. The Earth, for example, can be said to possess a gravitational field (Fig. 16–30) which is responsible for the gravitational force on objects. The *gravitational field intensity* is defined as the *force per unit mass*. The magnitude of the Earth's gravitational field intensity at any point is then (GM_E/r^2), where M_E is the mass of the Earth, r is the distance of the point from the Earth's center, and G is the gravitational constant (Chapter 5). At the Earth's surface, r is simply the radius of the Earth and the gravitational field intensity is simply equal to g, the acceleration due to gravity (since $F/m = mg/m = g$). Beyond the Earth, the gravitational field intensity can be calculated at any point as a sum of terms due to Earth, Sun, Moon, and other bodies that contribute significantly.

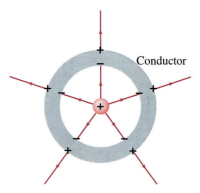

FIGURE 16–31 A charge placed inside a spherical shell. Charges are induced on the conductor surfaces. The electric field exists even beyond the shell but not within the conductor itself.

FIGURE 16–32 If the electric field **E** at the surface of a conductor had a component parallel to the surface, $\mathbf{E}_\parallel$, the latter would accelerate electrons into motion. In the static case (no charges are in motion), $\mathbf{E}_\parallel$ must be zero, and so the electric field must be perpendicular to the conductor's surface: $\mathbf{E} = \mathbf{E}_\perp$.

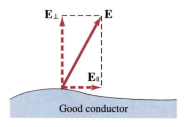

16–9 Electric Fields and Conductors

We now discuss some properties of good conductors. First, *the electric field inside a good conductor is zero in the static situation*—that is, when the charges are at rest. If there were an electric field within a conductor, there would be a force on its free electrons since $\mathbf{F} = q\mathbf{E}$. The electrons would move until they reached positions where the electric field, and therefore the electric force on them, did become zero.

This reasoning has some interesting consequences. For one, *any net charge on a good conductor distributes itself on the surface.* For a negatively charged conductor, you can imagine that the negative charges repel one another and race to the surface to get as far from one another as possible. Another consequence is the following. Suppose that a positive charge Q is surrounded by an isolated uncharged metal conductor whose shape is a spherical shell, Fig. 16–31. Because there can be no field within the metal, the lines leaving the positive charge must end on negative charges on the inner surface of the metal. Thus an equal amount of negative charge, $-Q$, is induced on the inner surface of the spherical shell. Then, since the shell is neutral, a positive charge, $+Q$, of the same magnitude must exist on the outer surface of the shell. Thus, although no field exists in the metal itself, an electric field exists outside of it, as shown in Fig. 16–31, as if the metal were not even there.

A related property of static electric fields and conductors is that *the electric field is always perpendicular to the surface outside of a conductor.* If there were a component of **E** parallel to the surface (Fig. 16–32), electrons at the surface would move along the surface in response to this force until they reached positions where no force was exerted on them—that is, until the electric field was perpendicular to the surface.

These properties pertain only to conductors. Inside a nonconductor, which does not have free electrons, an electric field can exist (Section 17–8). And the electric field outside a nonconductor does not necessarily make an angle of 90° to the surface.

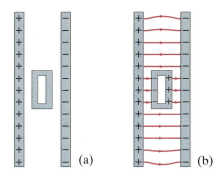

CONCEPTUAL EXAMPLE 16–9 | **Shielding, and safety in a storm.** A hollow metal box is placed between two parallel charged plates as shown in Fig. 16–33a. What's the field like inside the box?

RESPONSE If our metal box were solid, and not hollow, the electrons in the box, even if it were neutral overall, would redistribute themselves along the surface so that the field lines would not penetrate the conducting metal of the box. For a hollow box, the external field is not changed since the electrons in the metal can move just as freely as before to the surface. Hence we conclude that the field inside the hollow metal box is zero. So the field lines are something like those shown in Fig. 16–33b. A conducting box used in this way is an effective device for shielding delicate instruments and electronic circuits from unwanted external electric fields. We also can see that a relatively safe place to be during a lightning storm is inside a car, surrounded by metal.

FIGURE 16–33 Example 16–9.

➥ **PHYSICS APPLIED**

Shielding
Safety in a storm

16–10 Electric Forces in Molecular Biology: DNA Structure and Replication

The study of the structure and functioning of a living cell at the molecular level is known as molecular biology. It is an important area for application of physics. Since the interior of a cell is mainly water, we can imagine it as a vast sea of molecules continually in motion (as in kinetic theory, Chapter 13), colliding with one another with various amounts of kinetic energy. These molecules interact with one another in various ways—chemical reactions (making and breaking of bonds between atoms) and more brief interactions or unions that occur because of *electrostatic attraction* between molecules.

The many processes that occur within the cell are now considered to be the result of *random ("thermal") molecular motion plus the ordering effect of the electrostatic force*. We now use these ideas to analyze some basic cellular processes involving macromolecules (large molecules). The picture we present here has not been seen "in action." Rather, it is a model of what happens based on presently accepted physical theories and a great variety of experimental results.

The genetic information that is passed on from generation to generation in all living objects is contained in the chromosomes, which are made up of genes. Each gene contains the information needed to produce a particular type of protein molecule. The genetic information contained in a gene is built into the principal molecule of a chromosome, the DNA (deoxyribonucleic acid). A DNA molecule consists of a long chain of many small molecules known as nucleotide bases. There are only four types of bases: adenine (A), cytosine (C), guanine (G), and thymine (T).

The DNA in a chromosome generally consists of two long DNA chains wrapped about one another in the shape of a "double helix." As

➥ **PHYSICS APPLIED**

Inside a cell:
Kinetic theory plus
electrostatic force

➥ **PHYSICS APPLIED**

DNA structure

FIGURE 16–34 (a) Section of a DNA double helix. (b) "Close-up" view of the helix, showing how A and T attract each other and how G and C attract each other through electrostatic forces, to hold the double helix together. The red dots are used to indicate the electrostatic attraction (often called a "weak bond" or "hydrogen bond"). Note that there are two weak bonds between A and T, and three between C and G. The distance unit is the angstrom (1 Å = 10^{-10} m).

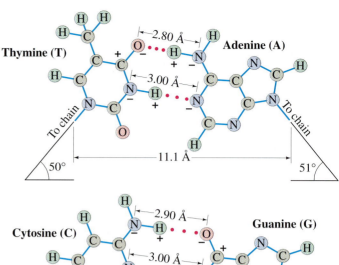

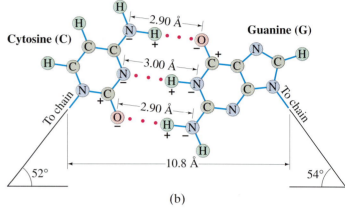

(a)

(b)

shown in Fig. 16–34, the two strands are held together by electrostatic forces—that is, by the attraction of positive charges to negative charges. We see in part (a) that an A (adenine) on one strand is always opposite a T on the other strand; similarly, a G is always opposite a C. This happens because the shapes of the four molecules A, T, C, and G are such that a T fits closely only into an A, and a G into a C; and only in the case of this close proximity of the charged portions is the electrostatic force great enough to hold them together even for a short time (Fig. 16–34b), forming what are often referred to as "weak bonds." The electrostatic force between A and T, and between C and G, exists because these molecules have charged parts due to some electrons in each of these molecules spending more time orbiting one atom than another. For example, the electron normally on the H atom of adenine spends some of its time orbiting the adjacent N atom (more on this in Chapter 29), so the N has a net negative charge and the H a positive charge (upper part of Fig. 16–34b). This H^+ atom of adenine is then attracted to the O^- atom of thymine.

How does the arrangement shown in Fig. 16–34 come about? It occurs when the chromosome replicates (duplicates) itself just before cell division. Indeed, the arrangement of A opposite T and G opposite C ensures that the genetic information is passed on accurately to the next generation. The process of replication is shown in a simplified form in Fig. 16–35. The two strands of the DNA chain separate (with the help of enzymes, which also operate via the electrostatic force), leaving the charged parts of the bases exposed. Without going into the details of how replication starts, let us see how the correct order of bases occurs by focusing our attention on the G molecule indicated by the arrow on the lowest strand in the figure. There are many unattached nucleotide bases of all four kinds bouncing around in the cellular fluid. The only one of the four bases that will experience at-

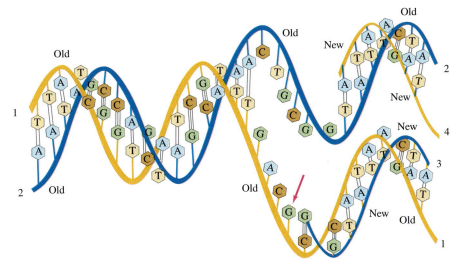

FIGURE 16–35
Replication of DNA.

traction to our G, if it bounces close to it, will be a C. The charges on the other three bases are not arranged so that they can get close to those on the G, and thus there will be no significant attractive force exerted on them—remember that the force decreases rapidly with distance. Because the G attracts an A, T, or G almost not at all, an A, T, or G will be knocked away by collisions with other molecules before enzymes can attach it to the growing (number 3) chain. But the electrostatic force will often hold a C opposite our G long enough so that an enzyme can attach the C to the growing end of the new chain.

Thus we see that electrostatic forces not only hold the two chains together; electrostatic forces also operate to select the bases in the proper order during replication, so the genetic information is passed on accurately to the next generation. Note in Fig. 16–35 that the new number 4 strand has the same order of bases as the old number 1 strand; and the new number 3 strand is the same as the old number 2. So the two new helixes, 1–3 and 2–4, are identical to the original 1–2 helix. The error rate—say a T being incorporated in a new chain opposite a G—is on the order of 1 in 10^4 [and is kept even lower (1 in 10^8 to 10^9) with the aid of special enzymatic "proofreading and repair" mechanisms]. Such an error constitutes a spontaneous mutation and a possible change in some characteristic of the organism. It is important for the survival of the organism that the error rate be low, but it cannot be zero if evolution (which can only occur through mutation) is to take place.

This process of DNA replication is often presented as if it occurred in clockwork fashion—as if each molecule knew its role and went to its assigned place, like bees in a hive. But this is not the case. The forces of attraction between the electric charges of the molecules are rather weak and become significant only when the molecules can come close together and several "weak bonds" can be made. Indeed, if the shapes are not just right, there is almost no electrostatic attraction, which is why there are few mistakes. The point is that there are many molecules in the cell, all jostling about, but only that one type which has the proper shape will be attracted sufficiently so as to remain long enough to become attached to the growing chain. Thus, out of the random motion of the molecules, the electrostatic force acts to bring order out of chaos.

SUMMARY

There are two kinds of **electric charge**, positive and negative. These designations are to be taken algebraically—that is, any charge is plus or minus so many coulombs (C), in SI units.

Electric charge is **conserved**: if a certain amount of one type of charge is produced in a process, an equal amount of the opposite type is also produced; thus the *net* charge produced is zero.

According to the atomic theory, electricity originates in the atom, which consists of a positively charged nucleus surrounded by negatively charged electrons. Each electron has a charge $-e = -1.6 \times 10^{-19}$ C.

Conductors are those materials in which many electrons are relatively free to move, whereas electric **insulators** are those in which very few electrons are free to move.

An object is negatively charged when it has an excess of electrons, and positively charged when it has less than its normal amount of electrons. The charge on any object is thus a whole number times $+e$ or $-e$; that is, charge is **quantized**.

An object can become charged by rubbing (in which electrons are transferred from one material to another), by conduction (which is transfer of charge from one charged object to another by touching), or by induction (the separation of charge within an object because of the close approach of another charged object but without touching).

Electric charges exert a force on each other. If two charges are of opposite types, one positive and one negative, they each exert an attractive force on the other. If the two charges are the same type, each repels the other.

The magnitude of the force one point charge exerts on another is proportional to the product of their charges, and inversely proportional to the square of the distance between them:

$$F = k\frac{Q_1 Q_2}{r^2};$$

this is **Coulomb's law**. In SI units, k is often written as $1/4\pi\epsilon_0$.

We think of an **electric field** as existing in space around any charge or group of charges. The force on another charged object is then said to be due to the electric field present at its location.

The *electric field*, **E**, at any point in space due to one or more charges, is defined as the force per unit charge that would act on a test charge q placed at that point:

$$\mathbf{E} = \frac{\mathbf{F}}{q}.$$

Electric fields are represented by **electric field lines** that start on positive charges and end on negative charges. Their direction indicates the direction the force would be on a tiny positive test charge placed at a point. The lines can be drawn so that the number per unit area is proportional to the magnitude of E.

The static electric field (that is, no charges moving) inside a good conductor is zero, and the electric field lines just outside a charged conductor are perpendicular to its surface.

QUESTIONS

1. If you charge a pocket comb by rubbing it with a silk scarf, how can you determine if the comb is positively or negatively charged?

2. Why does a shirt or blouse taken from a clothes dryer sometimes cling to your body?

3. Explain why fog or rain droplets tend to form around ions or electrons in the air.

4. A positively charged rod is brought close to a neutral piece of paper, which it attracts. Draw a diagram showing the separation of charge and explain why attraction occurs.

5. Why does a plastic ruler that has been rubbed with a cloth have the ability to pick up small pieces of paper? Why is this difficult to do on a humid day?

6. Contrast the *net charge* on a conductor to the "free charges" in the conductor.

7. Figures 16–7 and 16–8 show how a charged rod placed near an uncharged metal object can attract (or repel) electrons. There are a great many electrons in the metal, yet only some of them move as shown. Why not all of them?

8. When an electroscope is charged, the two leaves repel each other and remain at an angle. What balances the electric force of repulsion so that the leaves don't separate further?

9. The form of Coulomb's law is very similar to that for Newton's law of universal gravitation. What are the differences between these two laws? Compare also gravitational mass and electric charge.

10. We are not normally aware of the gravitational or electrical force between two ordinary objects. What is the reason in each case? Give an example where we are aware of each one and why.

11. Is the electric force a conservative force? Why or why not? (See Chapter 6.)

12. What experimental observations mentioned in the text rule out the possibility that the numerator in Coulomb's law contains the sum $(Q_1 + Q_2)$ rather than the product $Q_1 Q_2$?

13. When a charged ruler attracts small pieces of paper, sometimes a piece jumps quickly away after touching the ruler. Explain.

14. Explain why we use *small* test charges when measuring electric fields.

15. When determining an electric field, must we use a *positive* test charge, or would a negative one do as well? Explain.

16. Draw the electric field lines surrounding two negative electric charges a distance l apart.

17. Assume that the two opposite charges in Fig. 16–29a are 12.0 cm apart. Consider the magnitude of the electric field 2.5 cm from the positive charge. On which side of this charge—top, bottom, left, or right—is the electric field the strongest? The weakest?

18. Consider the electric field at the three points indicated by the letters A, B, and C in Fig. 16–36. First draw an arrow at each point indicating the direction of the net force that a positive test charge would experience if placed at that point, then list the letters in order of *decreasing* field strength (strongest first).

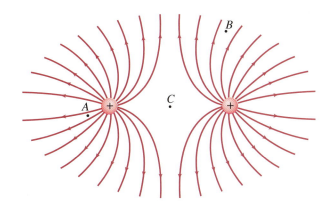

FIGURE 16–36   Question 18.

19. Why can electric field lines never cross?

20. Consider a small positive test charge located on an electric field line at some point, such as point P in Fig. 16–29a. Is the direction of the velocity and/or acceleration of the test charge along this line? Discuss.

21. We wish to determine the electric field at a point near a positively charged metal sphere (a good conductor). We do so by bringing a small test charge, q_0, to this point and measure the force F_0 on it. Will F_0/q_0 be greater than, less than, or equal to, the electric field $\mathbf{E}$ as it was at that point before the test charge was present?

PROBLEMS

SECTIONS 16–5 AND 16–6

1. (I) How many electrons make up a charge of $-30.0 \, \mu C$?

2. (I) Two charged smoke particles exert a force of $4.2 \times 10^{-2} \, N$ on each other. What will be the force if they are moved so they are only one eighth as far apart?

3. (I) Two charged balls are 20.0 cm apart. They are moved, and the force on each of them is found to have been tripled. How far apart are they now?

4. (I) Two charged Ping-Pong balls separated by a distance of 1.50 m exert an electric force of 0.0200 N on each other. What will be the force if the objects are brought closer, to a separation of only 30.0 cm?

5. (I) What is the magnitude of the electric force of attraction between an iron nucleus ($q = +26e$) and its innermost electron if the distance between them is $1.5 \times 10^{-12} \, m$?

6. (I) What is the repulsive electrical force between two protons in a nucleus that are $5.0 \times 10^{-15} \, m$ apart from each other?

7. (I) What is the magnitude of the force a $+15$-μC charge exerts on a $+3.0$-mC charge 40 cm away? ($1 \, \mu C = 10^{-6} \, C$, $1 \, mC = 10^{-3} \, C$.)

8. (II) A person scuffing her feet on a wool rug on a dry day accumulates a net charge of $-60 \, \mu C$. How many excess electrons does this person get, and by how much does her mass increase?

9. (II) Imagine that space invaders could deposit extra electrons in equal amounts on the Earth and on your car, which has a mass of 1050 kg. Note that the rubber tires would provide some insulation. How much charge Q would need to be placed on your car (same amount on the Earth) in order to levitate it (overcome gravity)? [*Hint*: Assume that the Earth's charge is spread uniformly so it acts as if it were located at the Earth's center, and then the separation distance is the radius of the Earth.]

10. (II) What is the total charge of all the electrons in 1.0 kg of H_2O?

11. (II) Particles of charge $+70$, $+48$, and $-80 \, \mu C$ are placed in a line (Fig. 16–37). The center one is 0.35 m from each of the others. Calculate the net force on each charge due to the other two.

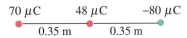

FIGURE 16–37   Problem 11.

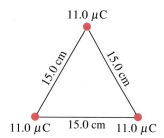

FIGURE 16–38 Problem 12.

12. (II) Three positive particles of charges 11.0 μC are located at the corners of an equilateral triangle of side 15.0 cm (Fig. 16–38). Calculate the magnitude and direction of the net force on each particle.

13. (II) A charge of 6.00 mC is placed at each corner of a square 1.00 m on a side. Determine the magnitude and direction of the force on each charge.

14. (II) Repeat Problem 13 for the case when two of the positive charges, on opposite corners, are replaced by negative charges of the same magnitude (Fig. 16–39).

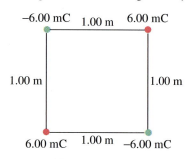

FIGURE 16–39 Problem 14.

15. (II) Compare the electric force holding the electron in orbit around the proton ($r = 0.53 \times 10^{-10}$ m) in the hydrogen nucleus with the gravitational force between the same electron and proton. What is the ratio of these two forces?

16. (II) Suppose that electrical attraction, rather than gravity, were responsible for holding the Moon in orbit around the Earth. If equal and opposite charges Q were placed on the Earth and the Moon, what should be the value of Q to maintain the present orbit? Use these data: mass of Earth = 5.97×10^{24} kg, mass of Moon = 7.35×10^{22} kg, radius of orbit = 3.84×10^8 m. Treat the Earth and Moon as point particles.

17. (II) Two positive point charges are a fixed distance apart. The sum of their charges is Q_T. What charge must each have in order to (a) maximize the electric force between them, and (b) minimize it?

18. (II) In one model of the hydrogen atom, the electron revolves in a circular orbit around the proton with a speed of 1.1×10^6 m/s. Determine the radius of the electron's orbit.

19. (III) A $+5.7\,\mu$C and a $-3.5\,\mu$C charge are placed 25 cm apart. Where can a third charge be placed so that it experiences no net force?

20. (III) Two small nonconducting spheres have a total charge of 80.0 μC. When placed 1.06 m apart, the force each exerts on the other is 12.0 N and is repulsive. What is the charge on each? What if the force were attractive?

SECTIONS 16–7 AND 16–8

21. (I) What is the magnitude of the acceleration experienced by an electron in an electric field of 600 N/C? How does the direction of the acceleration depend on the direction of the field at that point? How does the direction of the acceleration depend on the electron's velocity at that point?

22. (I) What is the magnitude and direction of the electric force on an electron in a uniform electric field of strength 3500 N/C that points due east?

23. (I) A proton is released in a uniform electric field, and it experiences an electric force of 3.2×10^{-14} N toward the south. What are the magnitude and direction of the electric field?

24. (I) A force of 8.4 N is exerted on a -8.8-μC charge in a downward direction. What is the magnitude and direction of the electric field at this point?

25. (I) What is the magnitude and direction of the electric field 30.0 cm directly above a 33.0×10^{-6}-C charge?

26. (II) What is the magnitude and direction of the electric field at a point midway between a -8.0-μC and a $+6.0$-μC charge 4.0 cm apart?

27. (II) An electron is released from rest in a uniform electric field and accelerates to the north at a rate of 125 m/s². What is the magnitude and direction of the electric field?

28. (II) The electric field midway between two equal but opposite point charges is 1750 N/C, and the distance between the charges is 16.0 cm. What is the magnitude of the charge on each?

29. (II) Determine the direction and magnitude of the electric field at the point P shown in Fig. 16–40. The two charges are separated by a distance of $2a$ and the point P is a distance x out on the perpendicular bisector of the line joining them. Express your answers in terms of Q, x, a, and k.

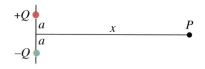

FIGURE 16–40 Problem 29.

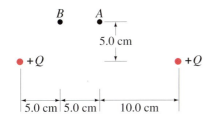

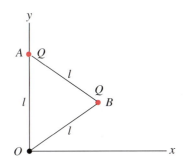

FIGURE 16–41 Problem 30.

30. (II) Use Coulomb's law to determine the magnitude and direction of the electric field at points A and B in Fig. 16–41 due to the two positive charges ($Q = 9.0\,\mu C$) shown. Is your result consistent with Fig. 16–29b?

31. (II) Calculate the electric field at the center of a square 60 cm on a side if one corner is occupied by a $+45.0$-μC charge and the other three are occupied by -31.0-μC charges.

32. (II) Calculate the electric field at one corner of a square 1.00 m on a side if the other three corners are occupied by 2.80×10^{-6}-C charges.

33. (II) (a) Determine the electric field **E** at the origin O in Fig. 16–42 due to the two charges at A and B. (b) Repeat, but let the charge at B be reversed in sign.

FIGURE 16–42
Problem 33.

34. (II) Draw, approximately, the electric field lines about two point charges, $+Q$ and $-3Q$, which are a distance l apart.

35. (II) What is the electric field strength at a point in space where a proton ($m = 1.67 \times 10^{-27}$ kg) experiences an acceleration of 1 million "g's"?

36. (II) A spacecraft makes a trip from the Earth to the Moon, 380,000 km away. At what point in the trip will the gravitational field be zero? The mass of the Moon is about $\frac{1}{81}$ that of the Earth.

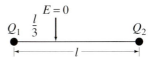

FIGURE 16–43
Problem 37.

37. (III) You are given two unknown point charges, Q_1 and Q_2. At a point on the line joining them, one third of the way from Q_1 to Q_2, the electric field is zero (Fig. 16–43). What can you say about these two charges?

38. (III) An electron (mass $m = 9.11 \times 10^{-31}$ kg) is accelerated in the uniform field **E** ($E = 1.85 \times 10^4$ N/C) between two parallel charged plates. The separation of the plates is 1.20 cm. The electron is accelerated from rest near the negative plate and passes through a tiny hole in the positive plate, Fig. 16–44. (a) With what speed does it leave the hole? (b) Show that the gravitational force can be ignored.

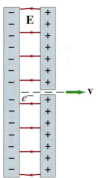

FIGURE 16–44
Problem 38.

39. (III) An electron moving at 1 percent the speed of light to the right enters a uniform electric field region where the field is known to be parallel to its direction of motion. If the electron is to be brought to rest in the space of 5.0 cm, (a) what direction is required for the electric field, and (b) what is the strength of the field?

*SECTION 16–10

*40. (III) The two strands of the helix-shaped DNA molecule are held together by electrostatic forces as shown in Fig. 16–34. Assume that the net average charge indicated on H and N atoms is 0.2e, and on the indicated C and O atoms is 0.4e, that atoms on each molecule are separated by 1.0×10^{-10} m, and that all relevant angles are 120°. Estimate the net force between: (a) a thymine and an adenine; and (b) a cytosine and a guanine. (c) Estimate the total force for a DNA molecule containing 10^5 pairs of such molecules.

GENERAL PROBLEMS

41. How close must two electrons be if the electric force between them is equal to the weight of either at the Earth's surface?

42. A 3.0-g copper penny has a positive charge of 42 μC. What fraction of its electrons has it lost?

43. A proton ($m = 1.67 \times 10^{-27}$ kg) is suspended at rest in a uniform electric field **E**. Take into account gravity and determine **E**.

44. Measurements indicate that there is an electric field surrounding the Earth. Its magnitude is about 150 N/C at the Earth's surface and points inward toward the Earth's center. What is the magnitude of the electric charge on the Earth? Is it positive or negative? [*Hint:* The electric field outside a uniformly charged sphere is the same as if all the charge were concentrated at its center.]

45. A water droplet of radius 0.018 mm remains stationary in the air. If the electric field of the Earth is 150 N/C, how many excess electron charges must the water droplet have?

46. Calculate the magnitude of the electric field at the center of a square with sides 25 cm long if the corners, taken in rotation, have charges of $1.0\,\mu C$, $2.0\,\mu C$, $3.0\,\mu C$, and $4.0\,\mu C$ (all positive).

47. Estimate the net force between the CO group and the HN group shown in Fig. 16–45. The C and O have charges $\pm 0.40e$ and the H and N have charges $\pm 0.20e$ where $e = 1.6 \times 10^{-19}$ C. [*Hint:* Do not include the "internal" forces between C and O, or between H and N.]

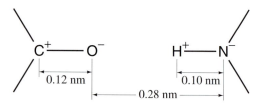

FIGURE 16–45   Problem 47.

48. Two charges, $-Q_0$ and $-3Q_0$, are a distance l apart. These two charges are free to move along the line passing through them both, but do not because there is a third charge nearby. What must be the magnitude of the third charge and its placement in order for the first two to be in equilibrium?

49. A point charge ($m = 1.0$ g) at the end of an insulating string of length 50 cm (Fig. 16–46) is observed to be in equilibrium in a known uniform horizontal electric field, $E = 9200$ N/C, when the pendulum has swung so it is 1.0 cm high. If the field points to the right in Fig. 16–46, determine the magnitude and sign of the point charge.

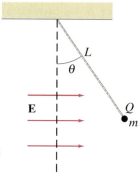

FIGURE 16–46
Problem 49.

50. A positive point charge $Q_1 = 2.5 \times 10^{-5}$ C is fixed at the origin of coordinates, and a negative charge $Q_2 = -5.0 \times 10^{-6}$ C is fixed to the x axis at $x = +2.0$ m. Find the location of the place(s) along the x axis where the electric field due to these two charges is zero.

51. An electron with speed $v_0 = 1.5 \times 10^6$ m/s is traveling parallel to an electric field ($v_0 \parallel E$) of magnitude $E = 7.7 \times 10^3$ N/C. (*a*) How far will it travel before it stops? (*b*) How much time will elapse before it returns to its starting point?

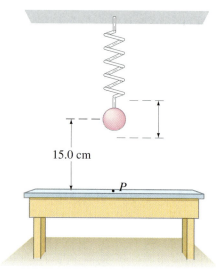

FIGURE 16–47   Problem 52.

52. A small lead ball is encased in insulating plastic and suspended vertically from an ideal spring ($k = 126$ N/m) above a lab table, Fig. 16–47. The total mass of the coated ball is 0.800 kg, and its center lies 15.0 cm above the tabletop when in equilibrium. The ball is pulled down 5.00 cm below equilibrium, an electric charge $Q = -3.00 \times 10^{-6}$ C is deposited on the ball, and the system is released. Using what you know about harmonic oscillation, write an expression for the electric field strength as a function of time that would be measured at the point on the tabletop (P) directly below the ball.

53. A large electroscope is made with "leaves" that are 70-cm-long wires with 24-g balls at the ends. When charged, nearly all the charge resides on the balls. If the wires each make a 30° angle with the vertical (Fig. 16–48), what total charge Q must have been applied to the electroscope?

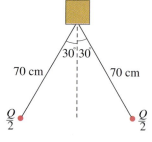

FIGURE 16–48
Problem 53.

54. (II) Dry air will break down and generate a spark if the electric field exceeds about 3×10^6 N/C. How much charge could be packed onto a green pea (diameter 0.75 cm) before the pea spontaneously discharges? [*Hint*: Eq. 16–4 works outside a sphere if r is measured from its center.]

55. Two point charges, $Q_1 = -6.7\ \mu C$ and $Q_2 = 1.3\ \mu C$ are located between two oppositely charged parallel plates, as shown in Fig. 16–49. The two point charges are separated by a distance of $x = 0.34$ m. Assume that the electric field produced by the charged plates is uniform and equal to $E = 73{,}000$ N/C. Calculate the net electrostatic force on Q_1 and give its direction.

56. A point charge of mass 0.210 kg, and net charge $+0.340\ \mu C$, hangs at rest at the end of an insulating string above a single sheet of charge. The horizontal sheet of charge creates a uniform vertical electric field in the vicinity of the point charge. The tension in the string is measured to be 5.67 N. Calculate the magnitude and direction of the electric field due to the sheet of charge (Fig. 16–50).

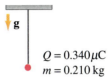

$Q = 0.340\,\mu C$
$m = 0.210$ kg

Uniform sheet of charge

FIGURE 16–50   Problem 56.

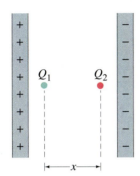

Q_1 Q_2

$\longmapsto x \longmapsto$

FIGURE 16–49   Problem 55.

Lightning: The potential difference (voltage) between clouds and the Earth can become so high that electrons are pulled off atoms of the air by the large electric field. The air becomes a conductor as the ionized atoms and freed electrons flow rapidly, colliding with more atoms, and causing more ionization. The massive flow of charge reduces the potential difference and the "discharge" quickly ceases. The light represents energy released when the ions and electrons recombine to form atoms (Chapter 27).

17 ELECTRIC POTENTIAL AND ELECTRIC ENERGY; CAPACITANCE

We saw in Chapter 6 that the concept of energy was extremely valuable in dealing with mechanical problems. For one thing, energy is a conserved quantity and is thus an important aspect of nature. Furthermore, we saw that many problems could be solved using the energy concept even though a detailed knowledge of the forces involved was not possible, or when a calculation involving Newton's laws would have been too difficult.

The energy point of view can be used in electricity, and it is especially useful. It not only extends the law of conservation of energy, but it gives us another way to view electrical phenomena; and it is a tool in solving problems more easily, in many cases, than by using forces and electric fields.

17–1 Electric Potential and Potential Difference

To apply conservation of energy, we need to define electric potential energy as for other types of potential energy (Chapter 6). That is, we define the change in electric potential energy, $\text{PE}_b - \text{PE}_a$, when a charge q moves from some point b to a second point a, as the negative of the work done by the electric force to move the charge from b to a. For example, consider the electric field between two equally but oppositely charged parallel plates whose separation is small compared to their width and height, so the field will be uniform over most of the region, Fig. 17–1. Now consider a small positive point charge q placed at point b very near the positive plate as shown. If the charge is released, the electric force will do work on the charge and accelerate it toward the negative plate. In the process, the charged particle will have its kinetic energy increased. The potential energy will be decreased by an equal amount, equal to the negative of the work done by the electric force. In accord with the conservation of energy, electric potential energy is transformed into kinetic energy, and the total energy is conserved. Note that the positive charge q has its greatest potential energy at point b, near the positive plate,[†] so $(PE_b - PE_a) > 0$. The reverse is true for a negative charge: its potential energy is greatest near the negative plate.

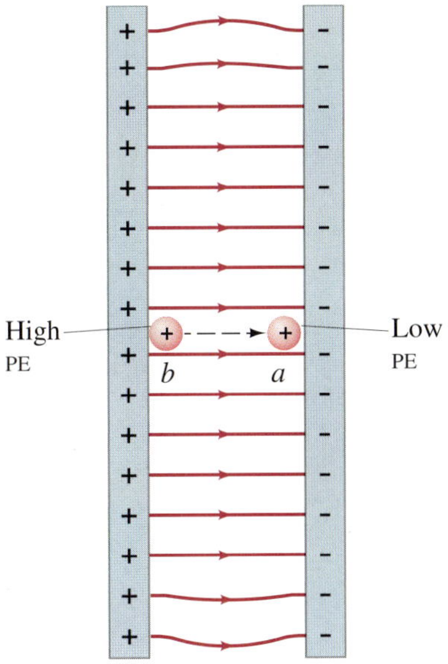

We defined the electric field (Chapter 16) as the force per unit charge. Similarly, it is useful to define the **electric potential** (or simply the **potential** when "electric" is understood) as the *potential energy per unit charge.* Electric potential is given the symbol V. If a point charge q has electric potential energy PE_a at some point a, the electric potential V_a at this point is

$$V_a = \frac{PE_a}{q}.$$

FIGURE 17–1 Work is done by the electric field in moving the positive charge from position b to position a.

As we discussed in Chapter 6 (Section 6–4), only differences in potential energy are physically measurable. Hence only the **difference in potential**, or the **potential difference**, between two points a and b (such as between a and b in Fig. 17–1) is measurable. Since the difference in potential energy, $PE_b - PE_a$, is equal to the work, W_{ba}, done by the electric force to move the charge from point b to point a, we have that the potential difference V_{ba} is

Potential difference

$$V_{ba} = V_b - V_a = \frac{W_{ba}}{q}.$$

The unit of electric potential, and of potential difference, is joules/coulomb and is given a special name, the **volt**, in honor of Alessandro Volta (1745–1827; he is best known for having invented the electric battery, as discussed in Chapter 18). The volt is abbreviated V, so $1\,V = 1\,J/C$. Note from our definition that the positive plate in Fig. 17–1 is at a higher potential than the negative plate. Thus a positively charged object moves naturally from a high potential to a low potential. A negative charge does the reverse. Potential difference, since it is measured in volts, is often referred to as **voltage**.

The volt $(1\,V = 1\,J/C)$

If we wish to speak of the potential, V_a, at some point a, we must be aware that V_a depends on where the potential is chosen to be zero. The zero point for electric potential in a given situation, just as for potential energy, can be chosen arbitrarily since only differences in potential energy can be measured. Often the ground, or a conductor connected directly to the ground, is taken as zero potential, and other potentials are given with respect to ground. (Thus, a point where the voltage is 50 V is one where the difference of potential between it and ground is 50 V.) In other cases, as we shall see, we may choose the potential to be zero at infinity (see Section 17–5).

Voltage = potential difference

V = 0 chosen arbitrarily

[†]At this point it has its greatest ability to do work (on some other object or system).

Since the electric potential is defined as the potential energy per unit charge, then the change in potential energy of a charge q when moved between two points a and b is

Electric potential and potential energy

$$\Delta \text{PE} = \text{PE}_b - \text{PE}_a = qV_{ba}. \tag{17-1}$$

That is, if an object with charge q moves through a potential difference V_{ba}, its potential energy changes by an amount qV_{ba}. For example, if the potential difference between the two plates in Fig. 17–1 is 6 V, then a 1-C charge moved (say by an external force) from a to b will gain $(1\,\text{C})(6\,\text{V}) = 6\,\text{J}$ of electric potential energy. (And it will lose 6 J of electric PE if it moves from b to a.) Similarly, a 2-C charge will gain 12 J, and so on. Thus, electric potential difference is a measure of how much energy an electric charge can acquire in a given situation. And, since energy is the ability to do work, the electric potential difference is also a measure of how much work a given charge can do. The exact amount depends both on the potential difference and on the charge.

Potential likened to height of a cliff

To better understand electric potential, let's make a comparison to the gravitational case when a rock falls from the top of a cliff. The greater the height, h, of a cliff, the more potential energy ($= mgh$) the rock has at the top of the cliff, relative to the bottom, and the more kinetic energy it will have when it reaches the bottom. The actual amount of kinetic energy it will acquire, and the amount of work it can do, depends both on the height of the cliff and the mass m of the rock. A large rock and a small rock can be at the same height h (Fig. 17–2a) and thus have the same "gravitational potential," but the larger rock has the greater potential energy. Similarly, in the electrical case (Fig. 17–2b): the potential energy change, or the work that can be done, depends both on the potential difference (corresponding to the height of the cliff) and on the charge (corresponding to mass), Eq. 17–1.

Practical sources of electrical energy such as batteries and electric generators are meant to maintain a particular potential difference. The actual amount of energy used or transformed depends on how much charge flows. For example, consider an automobile headlight connected to a 12.0-V battery. The amount of energy transformed (into light, and of course thermal energy) is proportional to how much charge flows, which in turn depends on how long the light is on. If over a given period 5.0 C of charge flows through the light, the total energy transformed is $(5.0\,\text{C})(12.0\,\text{V}) = 60\,\text{J}$. If the headlight is left on twice as long, 10.0 C of charge will flow and the energy transformed is $(10.0\,\text{C})(12.0\,\text{V}) = 120\,\text{J}$.

Table 17–1 presents some typical voltages.

TABLE 17–1
Some Typical Voltages

Source	Voltage (approx.)
Thundercloud to ground	10^8 V
High-voltage power line	10^6 V
Power supply for TV tube	10^4 V
Automobile ignition	10^4 V
Household outlet	10^2 V
Automobile battery	12 V
Flashlight battery	1.5 V
Resting potential across nerve membrane	10^{-1} V
Potential changes on skin (EKG and EEG)	10^{-4} V

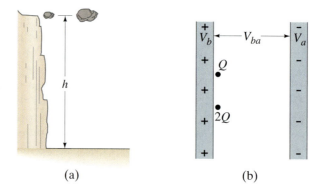

FIGURE 17–2 (a) Two rocks are at the same height. The larger rock has more PE. (b) Two charges have the same electric potential. The $2Q$ charge has more PE.

(a)   (b)

EXAMPLE 17–1 **Electron in TV tube.** Suppose an electron in the picture tube of a television set is accelerated from rest through a potential difference $V_{ba} = +5000\,\text{V}$ (Fig. 17–3). (a) What is the change in potential energy of the electron? (b) What is the speed of the electron ($m = 9.1 \times 10^{-31}\,\text{kg}$) as a result of this acceleration? (c) Repeat for a proton ($m = 1.67 \times 10^{-27}\,\text{kg}$) that accelerates through a potential difference of $V_{ba} = -5000\,\text{V}$.

SOLUTION (a) The charge on an electron is $e = -1.6 \times 10^{-19}\,\text{C}$. Therefore its change in potential energy (Eq. 17–1) is equal to

$$\Delta\text{PE} = qV_{ba} = (-1.6 \times 10^{-19}\,\text{C})(+5000\,\text{V})$$

$$= -8.0 \times 10^{-16}\,\text{J}.$$

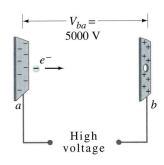

FIGURE 17–3 Electron accelerated in TV picture tube. Example 17–1.

The minus sign in the result indicates that the PE decreases. (The potential difference, V_{ba}, has a positive sign since the final potential is higher than the initial potential; that is, negative electrons are attracted from a negative electrode to a positive one.)

(b) The potential energy lost by the electron becomes kinetic energy. From conservation of energy (see Eq. 6–11), $\Delta\text{KE} + \Delta\text{PE} = 0$, so

$$\Delta\text{KE} = -\Delta\text{PE}$$

$$\tfrac{1}{2}mv^2 - 0 = -qV_{ba},$$

where the initial KE $= 0$ since we assume the electron started from rest. We solve for v and put in the mass of the electron $m = 9.1 \times 10^{-31}\,\text{kg}$:

$$v = \sqrt{-\frac{2qV_{ba}}{m}}$$

$$= \sqrt{-\frac{2(-1.6 \times 10^{-19}\,\text{C})(5000\,\text{V})}{9.1 \times 10^{-31}\,\text{kg}}}$$

$$= 4.2 \times 10^7\,\text{m/s}.$$

(Note: For such a high speed, which is $\tfrac{1}{7}$ the speed of light, we should use the theory of relativity, Chapter 26.)

(c) The proton has the same magnitude of charge as the electron, though of opposite sign. Hence for the same magnitude of V_{ba} we expect the same change in PE, but a lesser speed since the proton's mass is greater. Thus:

$$\Delta\text{PE} = qV_{ba} = (+1.6 \times 10^{-19}\,\text{C})(-5000\,\text{V}) = -8.0 \times 10^{-16}\,\text{J},$$

and

$$v = \sqrt{-\frac{2qV_{ba}}{m}}$$

$$= \sqrt{-\frac{2(1.6 \times 10^{-19}\,\text{C})(-5000\,\text{V})}{(1.67 \times 10^{-27}\,\text{kg})}}$$

$$= 9.8 \times 10^5\,\text{m/s}.$$

Note that the energy doesn't depend on the mass, only on the charge and voltage. The speed *does* depend on *m*.

17–2 Relation Between Electric Potential and Electric Field

The effects of any charge distribution can be described either in terms of electric field or in terms of electric potential. Electric potential is often easier to use since it is a scalar, whereas electric field is a vector. There is an intimate connection between the potential and the field. Let us examine this relation for the case of a uniform electric field, such as that between the parallel plates of Fig. 17–1 whose difference of potential is V_{ba}. We won't worry about signs. The work done by the electric field to move a positive charge q from b to a is

$$W = qV_{ba}.$$

We can also write the work done as the force times distance and recall that the force on q is $F = qE$, where E is the uniform electric field between the plates. Thus

$$W = Fd = qEd$$

where d is the distance (parallel to the field lines) between points a and b. We now set these two expressions for W equal and find $qV_{ba} = qEd$, or

$$V_{ba} = Ed. \qquad \text{[E uniform]} \quad \textbf{(17–2a)}$$

If we solve for E, we find that

$$E = V_{ba}/d. \qquad \text{[E uniform]} \quad \textbf{(17–2b)}$$

From this equation we can see that the units for electric field can be written as volts per meter (V/m) as well as newtons per coulomb (N/C). These are equivalent in general, since $1\,\text{N/C} = 1\,\text{N·m/C·m} = 1\,\text{J/C·m} = 1\,\text{V/m}$.

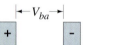

V related to uniform E

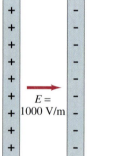

FIGURE 17–4 Example 17–2.

> **EXAMPLE 17–2** **Electric field obtained from voltage.** Two parallel plates are charged to a voltage of 50 V. If the separation between the plates is 0.050 m, calculate the electric field between them (Fig. 17–4).
>
> **SOLUTION** We have from Eq. 17–2,
>
> $$E = \frac{V}{d} = \frac{50\,\text{V}}{0.050\,\text{m}} = 1000\,\text{V/m}.$$

[Optional paragraph: More general relation between E and V]

[In a region where E is not uniform, the connection between E and V takes on a different form than Eq. 17–2. In general, it is possible to show that the electric field in a given direction at any point in space is equal to the *rate at which the electric potential changes over distance in that direction*. Actually, if we take into account direction, this gives the negative of the electric field. For example, the x component of the electric field is given by $E_x = -\Delta V/\Delta x$, where ΔV is the change in potential over the very short distance Δx. Note that this relation resembles Eq. 17–2b except that the distance Δx must be very small—so small that E does not change appreciably over this distance. Similar relations apply for the y and z components of **E**. Another way of stating the relation is this: if we plot V on a graph versus x, the slope of the graph at any point equals the magnitude of the x component of the electric field at that point. And we must insert a minus sign if we want the direction to come out right.]

17–3 Equipotential Lines

The electric potential can be represented diagrammatically by drawing **equipotential lines** or, in three dimensions, **equipotential surfaces**. An equipotential surface is one on which all points are at the same potential. That is, the potential difference between any two points on the surface is zero, and no work is required to move a charge from one point to the other. An *equipotential surface must be perpendicular to the electric field* at any point. If this were not so—that is, if there were a component of **E** parallel to the surface—it would require work to move the charge along the surface against this component of **E**; and this would contradict the idea that it is an *equi*potential surface.

*Equipotentials ⊥ **E***

The fact that the electric field lines and equipotential surfaces are mutually perpendicular helps us locate the equipotentials when the electric field lines are known. In a normal two-dimensional drawing, we show equipotential *lines*, which are the intersections of equipotential surfaces with the plane of the drawing. In Fig. 17–5, a few of the equipotential lines are drawn (dashed green lines) for the electric field (red lines) between two parallel plates at a potential difference of 20 V. The negative plate is arbitrarily chosen to be zero volts and the potential of each equipotential line is indicated. Note that **E** points toward lower values of *V*. The equipotential lines for the case of two equal but oppositely charged particles are shown in Fig. 17–6 as green dashed lines. Equipotential lines and surfaces, unlike field lines, are always continuous and never end, and so continue beyond the borders

FIGURE 17–5 Equipotential lines (the green dashed lines) between two charged parallel plates; note that they are perpendicular to the electric field (solid red lines).

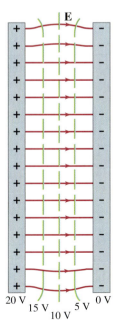

FIGURE 17–6 Equipotential lines (green, dashed) are always perpendicular to the electric field lines (solid red) shown here for two equal but oppositely charged particles.

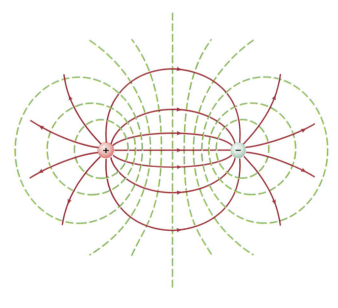

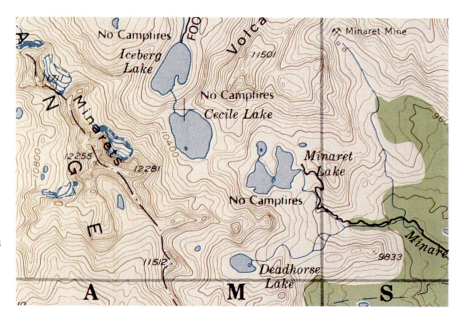

FIGURE 17–7 A topographic map (here, a portion of the Sierra Nevada in California) shows continuous contour lines, each of which is at a fixed height above sea level. Here they are at 80 ft (25 m) intervals. If you walk along one contour line, you neither climb nor descend. If you cross lines, and especially (maximally), if you climb perpendicular to the lines, you will be changing your gravitational potential (rapidly, if the lines are close together).

of Figs. 17–5 and 17–6. A useful analogy is a topographic map: the contour lines are essentially gravitational equipotential lines (Fig. 17–7).

We saw in Section 16–9 that there can be no electric field within a conductor in the static case, for otherwise the free electrons would feel a force and would move. Indeed *a conductor must be entirely at the same potential in the static case*, and the surface of a conductor is then an equipotential surface. (If it weren't, the free electrons at the surface would move, since whenever there is a potential difference between two points, work can be done on charged particles to move them.) This is fully consistent with our result, discussed earlier, that the electric field at the surface of a conductor must be perpendicular to the surface.

Conductors are equipotential surfaces

17–4 The Electron Volt, a Unit of Energy

The joule is a very large unit for dealing with energies of electrons, atoms, or molecules, whether in atomic and nuclear physics or in chemistry or molecular biology (see Example 17–1). For this purpose, the **electron volt** (eV) is used. One electron volt is defined as the energy acquired by a particle carrying a charge equal to that on the electron ($q = e$) as a result of moving through a potential difference of 1 V. Since the charge on an electron has magnitude 1.6×10^{-19} C, and since the change in potential energy equals qV, 1 eV is equal to $(1.6 \times 10^{-19}\,\text{C}) \cdot (1.0\,\text{V}) = 1.6 \times 10^{-19}$ J:

Electron volt

$$1\,\text{eV} = 1.6 \times 10^{-19}\,\text{J}.$$

An electron that accelerates through a potential difference of 1000 V will lose 1000 eV of potential energy and will thus gain 1000 eV or 1 keV (kilo-electron volt) of kinetic energy. On the other hand, if a particle has a charge equal to twice the charge on the electron ($= 2e = 3.2 \times 10^{-19}$ C), when it moves through a potential difference of 1000 V its energy will change by 2000 eV.

Although the electron volt is handy for *stating* the energies of molecules and elementary particles, it is *not* a proper SI unit. For calculations, electron volts should be converted to joules using the conversion factor just given. In Example 17–1, for example, the electron acquired a kinetic energy of 8.0×10^{-16} J. We normally would quote this energy as 5000 eV ($= 8.0 \times 10^{-16}$ J/1.6×10^{-19} J/eV). But in determining the speed in SI units, we have to use the KE in joules (J).

17–5 Electric Potential Due to Point Charges

The electric potential at a distance r from a single point charge Q can be derived from the expression for its electric field (Eq. 16–4) using calculus. The potential in this case is usually taken to be zero at infinity (∞); this is also where the electric field ($E = kQ/r^2$) is zero. The result is

$$V = k\frac{Q}{r}$$
[single point charge] **(17–3)**
$$= \frac{1}{4\pi\epsilon_0}\frac{Q}{r}.$$

Electric potential of point charge ($V = 0$ at $r = \infty$)

We can think of V here as representing the absolute potential, where $V = 0$ at $r = \infty$, or we can think of V as the potential difference between r and infinity. Notice that the potential V decreases with the first power of the distance, whereas the electric field (Eq. 16–4) decreases as the *square* of the distance. The potential near a positive charge is large, and it decreases toward zero at very large distances. For a negative charge, the potential is negative and increases toward zero at large distances (Fig. 17–8).

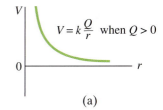

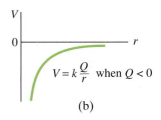

FIGURE 17–8 Potential V as a function of distance r from a single point charge Q when the charge is (a) positive, (b) negative.

EXAMPLE 17–3 **Work to force two + charges close together.** What minimum work is required by an external force to bring a charge $q = 3.00\ \mu$C from a great distance away (take $r = \infty$) to a point 0.500 m from a charge $Q = 20.0\ \mu$C?

SOLUTION The work required is equal to the change in potential energy:

$$W = qV_{ba} = q\left(\frac{kQ}{r_b} - \frac{kQ}{r_a}\right),$$

where $r_b = 0.500$ m and $r_a = \infty$. The second term in parentheses is zero ($1/\infty = 0$) so

$$W = (3.00 \times 10^{-6}\ \text{C})\frac{(9.00 \times 10^9\ \text{N·m}^2/\text{C}^2)(2.00 \times 10^{-5}\ \text{C})}{(0.500\ \text{m})} = 1.08\ \text{J}.$$

[Note that we could not calculate the work done by multiplying force times distance because the force is not constant.]

To determine the electric field surrounding a collection of two or more point charges requires adding up the electric fields due to each charge. Since the electric field is a vector, this can often be a chore. To find the electric potential due to a collection of point charges is far easier, since the electric potential is a scalar, and hence you only need to add numbers together without concern for direction. This is a major advantage in using electric potential. We do have to include the signs of charges, however.

Potentials add as scalars (Fields add as vectors)

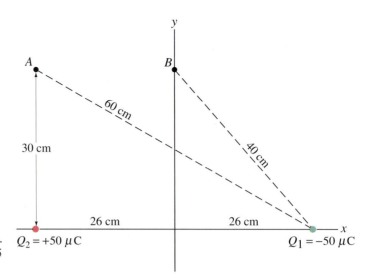

FIGURE 17–9 Example 17–4. (See also Example 16–8, Fig. 16–25 in the previous chapter.)

EXAMPLE 17–4 **Potential above two charges.** Calculate the electric potential at points A and B in Fig. 17–9 due to the two charges shown. (This is the same situation as Example 16–8, Fig. 16–25, where we calculated the electric field at these points.)

SOLUTION The potential at point A is the sum of the potentials due to the + and − charges, and we use Eq. 17–3 for each:

$$V_A = V_{A2} + V_{A1}$$

$$= \frac{(9.0 \times 10^9 \, \text{N·m}^2/\text{C}^2)(5.0 \times 10^{-5} \, \text{C})}{0.30 \, \text{m}}$$

$$+ \frac{(9.0 \times 10^9 \, \text{N·m}^2/\text{C}^2)(-5.0 \times 10^{-5} \, \text{C})}{0.60 \, \text{m}}$$

$$= 1.50 \times 10^6 \, \text{V} - 0.75 \times 10^6 \, \text{V}$$

$$= 7.5 \times 10^5 \, \text{V}.$$

At point B:

$$V_B = V_{B2} + V_{B1}$$

$$= \frac{(9.0 \times 10^9 \, \text{N·m}^2/\text{C}^2)(5.0 \times 10^{-5} \, \text{C})}{0.40 \, \text{m}}$$

$$+ \frac{(9.0 \times 10^9 \, \text{N·m}^2/\text{C}^2)(-5.0 \times 10^{-5} \, \text{C})}{0.40 \, \text{m}}$$

$$= 0 \, \text{V}.$$

It should be clear that the potential will be zero everywhere on the plane equidistant between the two charges. Thus this plane is an equipotential surface with $V = 0$.

A simple summation like these can easily be performed for any number of point charges.

CONCEPTUAL EXAMPLE 17–5 **Potential energies.**   Consider the three pairs of charges in Fig. 17–10. (*a*) Which set has a positive potential energy? (*b*) Which set has the most negative potential energy? (*c*) Which set requires the most work to separate the charges to infinity? Assume the charges all have the same magnitude.

RESPONSE   We can combine Eqs. 17–1 and 17–3, calling the two charges Q_1 and Q_2:

$$\text{PE} = k\frac{Q_1 Q_2}{r}.$$

(*a*) Set (iii) has a positive potential energy because the charges have the same sign. (*b*) Set (i) has the most negative potential energy because the charges are of opposite sign and their separation is less than that for set (ii). That is, r is smaller for (i). (*c*) Set (i) will require the most work for separation to infinity. The more negative the potential energy, the more work required to separate the charges and bring the PE up to zero ($r = \infty$).

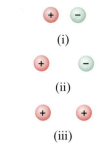

FIGURE 17–10   Example 17–5.

17–6   **Electric Dipoles**

FIGURE 17–11   Electric dipole. Calculation of potential V at point P.

Two equal point charges Q, of opposite sign, separated by a distance l, are called an **electric dipole**. The two charges we saw in Fig. 17–9 constitute an electric dipole. The electric field lines and equipotential surfaces for a dipole were shown in Fig. 17–6. Because electric dipoles occur often in physics, as well as in other fields such as molecular biology, it is useful to examine them more closely.

Let us calculate the electric potential at an arbitrary point P due to a dipole, as shown in Fig. 17–11. Since V is the sum of the potentials due to each of the two charges, we have

$$V = \frac{kQ}{r} + \frac{k(-Q)}{r + \Delta r} = kQ\left(\frac{1}{r} - \frac{1}{r + \Delta r}\right) = kQ\frac{\Delta r}{r(r + \Delta r)},$$

where r is the distance from P to the positive charge and $r + \Delta r$ is the distance to the negative charge. This equation becomes simpler if we consider points P whose distance from the dipole is much larger than the separation of the two charges—that is, for $r \gg l$. From the diagram we can see that in this case, $\Delta r \approx l\cos\theta$; and since $r \gg \Delta r = l\cos\theta$, we can neglect Δr in the denominator as compared to r. Therefore, we obtain

$$V = \frac{kQl\cos\theta}{r^2}. \qquad \text{[dipole; } r \gg l\text{]} \quad \textbf{(17–4a)}$$

Potential far from a dipole

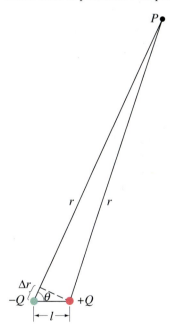

When θ is between $0°$ and $90°$, V is positive. If θ is between $90°$ and $180°$, V is negative (since $\cos\theta$ is then negative). This makes sense since in the first case P is closer to the positive charge and in the second case it is closer to the negative charge. At $\theta = 90°$, the potential is zero ($\cos 90° = 0$), in agreement with the result of Example 17–4 (point B). From Eq. 17–4a, we see that the potential decreases as the *square* of the distance from the dipole, whereas for a single point charge the potential decreases with the first power of the distance (Eq. 17–3). It is not surprising that the potential should fall off faster for a dipole; for when you are far from a dipole, the two equal but opposite charges appear so close together as to tend to neutralize each other.

Dipole moment p = Ql

The product Ql which occurs in Eq. 17–4a is referred to as the **dipole moment**, p, of the dipole. Equation 17–4a can be written in terms of the dipole moment as

$$V = \frac{kp \cos \theta}{r^2}.$$ [dipole; $r \gg l$] **(17–4b)**

A dipole moment has units of coulomb-meters (C·m), although for molecules a smaller unit called a *debye* is sometimes used: 1 debye = 3.33×10^{-30} C·m.

In many molecules, even though they are electrically neutral, the electrons spend more time in the vicinity of one atom than another, which results in a separation of charge. Such molecules have a dipole moment and are called **polar molecules**. We have already seen that water (Fig. 16–4) is a polar molecule, and we have encountered others in our discussion of molecular biology (Section 16–10). Table 17–2 gives the dipole moments for several molecules. The $+$ and $-$ signs indicate on which atoms these charges lie. The last two entries are a part of many organic molecules and play an important role in molecular biology.

TABLE 17–2
Dipole Moments of Selected Molecules

Molecule	Dipole Moment (C·m)
$H_2^{(+)}O^{(-)}$	6.1×10^{-30}
$H^{(+)}Cl^{(-)}$	3.4×10^{-30}
$N^{(-)}H_3^{(+)}$	5.0×10^{-30}
$>N^{(-)}-H^{(+)\ddagger}$	$\approx 3.0 \times 10^{-30}$
$>C^{(+)}{=}O^{(-)\ddagger}$	$\approx 8.0 \times 10^{-30}$

$\ddagger$These groups often appear on larger molecules; hence the value for the dipole moment will vary somewhat, depending on the rest of the molecule.

EXAMPLE 17–6 **The C=O group dipole.** The distance between the carbon ($+$) and oxygen ($-$) atoms in the group C=O is about 1.2×10^{-10} m. Calculate (*a*) the net charge Q on the C (carbon) and O (oxygen) atoms, and (*b*) the potential 9.0×10^{-10} m from the dipole along its axis, with the oxygen being the nearer atom (that is, to the left in Fig. 17–11, so $\theta = 180°$). (*c*) What would the potential be at this point if only the oxygen (O) were charged?

SOLUTION (*a*) The dipole moment $p = Ql$. Therefore $Q = p/l$, and from Table 17–2:

$$Q = \frac{p}{l} = \frac{8.0 \times 10^{-30} \text{ C·m}}{1.2 \times 10^{-10} \text{ m}} = 6.7 \times 10^{-20} \text{ C}.$$

(*b*) Since $\theta = 180°$, we have, using Eq. 17–4:

$$V = \frac{kp \cos \theta}{r^2}$$

$$= \frac{(9.0 \times 10^9 \text{ N·m}^2/\text{C}^2)(8.0 \times 10^{-30} \text{ C·m})(-1.00)}{(9.0 \times 10^{-10} \text{ m})^2}$$

$$= -0.089 \text{ V}.$$

(*c*) If we assume that the oxygen has charge $Q = -6.7 \times 10^{-20}$ C [as in part (*a*) above] and that the carbon is not charged, we use the formula for a single charge, Eq. 17–3:

$$V = \frac{kQ}{r} = \frac{(9.0 \times 10^9 \text{ N·m}^2/\text{C}^2)(-6.7 \times 10^{-20} \text{ C})}{9.0 \times 10^{-10} \text{ m}} = -0.67 \text{ V}.$$

Of course, we expect the potential of a single charge to have greater magnitude than that of a dipole of equal charge at the same distance.

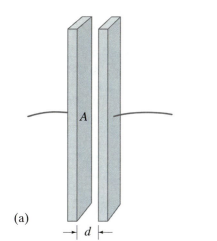
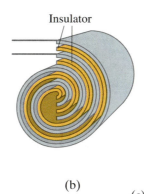

(a)

Insulator

(b)

(c)

$\rightarrow\!|\ d\ |\!\leftarrow$

FIGURE 17–12 Capacitors: Diagrams of (a) parallel plate, (b) cylindrically shaped (rolled up parallel plate). (c) Photo of some real capacitors.

17–7 Capacitance

A **capacitor**, sometimes called a *condenser*, is a device that can store electric charge, and consists of two conducting objects (usually plates or sheets) placed near each other but not touching. Capacitors are widely used in electronic circuits: they store charge for later use, as in a camera flash, and as energy backup in computers if the power fails; capacitors block surges of charge and energy to protect circuits; very tiny capacitors serve as memory for the "ones" and "zeroes" of the binary code in the random access memory (RAM) of computers; and capacitors serve many other applications, some of which we will discuss. A typical capacitor consists of a pair of parallel plates of area A separated by a small distance d (Fig. 17–12a). Often the two plates are rolled into the form of a cylinder with paper or other insulator separating the plates (Fig. 17–12b; Fig. 17–12c is a photo of some actual capacitors used for various applications). In a diagram, a capacitor is represented by the symbol

$$\dashv\!\vdash.\qquad\text{[capacitor symbol]}$$

If a voltage is applied to a capacitor, say by connecting the capacitor to a battery as in Fig. 17–13, it quickly becomes charged. One plate acquires a negative charge, and the other an equal amount of positive charge. For a given capacitor, the amount of charge Q acquired by each plate is proportional to the potential difference V:

$$Q = CV. \qquad \textbf{(17–5)}$$

The constant of proportionality, C, in this relation is called the **capacitance** of the capacitor. The unit of capacitance is coulombs per volt, and this unit is called a **farad** (F). Most capacitors have capacitance in the range 1 pF (picofarad $= 10^{-12}$ F) to 1 μF (microfarad $= 10^{-6}$ F). The relation, Eq. 17–5, was first suggested by Volta in the late eighteenth century.

The capacitance C is a constant for a given capacitor: it does not depend on Q or V. Its value depends only on the structure and dimensions of the capacitor itself. For a parallel-plate capacitor whose plates have area A and are separated by a distance d of air (Fig. 17–12a), the capacitance is given by

$$C = \epsilon_0 \frac{A}{d}. \qquad \text{[parallel-plate capacitor]} \quad \textbf{(17–6)}$$

This relation makes sense intuitively: a larger area A means that for a

PHYSICS APPLIED

Uses of capacitors

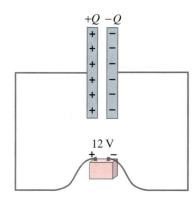

$+Q \ -Q$

12 V

FIGURE 17–13 Parallel-plate capacitor connected to a battery.

Capacitance

Unit is farad $(1\,\text{F} = 1\,\text{C/V})$

given number of charges, there will be less repulsion between them (they're farther apart), so we expect that more charge can be held on each plate. And a greater separation d means the charge on each plate exerts less attractive force on the other plate, so less charge is drawn from the battery, and the capacitance is less.[†] The constant ϵ_0 is the *permittivity of free space* which, as we saw in Chapter 16, has the value 8.85×10^{-12} C²/N·m².

EXAMPLE 17–7 **Capacitor calculations.** (a) Calculate the capacitance of a capacitor whose plates are 20 cm × 3.0 cm and are separated by a 1.0-mm air gap. (b) What is the charge on each plate if the capacitor is connected to a 12-V battery? (c) What is the electric field between the plates?

SOLUTION (a) The area $A = (20 \times 10^{-2}\text{ m})(3.0 \times 10^{-2}\text{ m}) = 6.0 \times 10^{-3}\text{ m}^2$. The capacitance C is then

$$C = \epsilon_0 \frac{A}{d} = (8.85 \times 10^{-12}\text{ C}^2/\text{N·m}^2)\frac{6.0 \times 10^{-3}\text{ m}^2}{1.0 \times 10^{-3}\text{ m}} = 53\text{ pF}.$$

(b) The charge on each plate is

$$Q = CV = (53 \times 10^{-12}\text{ F})(12\text{ V}) = 6.4 \times 10^{-10}\text{ C}.$$

(c) From Eq. 17–2 for a uniform electric field

$$E = \frac{V}{d} = \frac{12\text{ V}}{1.0 \times 10^{-3}\text{ m}} = 1.2 \times 10^4\text{ V/m}.$$

17–8 Dielectrics

TABLE 17–3
Dielectric Constants (20°C)

Material	Dielectric Constant, K
Vacuum	1.0000
Air (1 atm)	1.0006
Paraffin	2.2
Rubber, hard	2.8
Vinyl (plastic)	2.8–4.5
Paper	3–7
Quartz	4.3
Glass	4–7
Porcelain	6–8
Mica	7
Ethyl alcohol	24
Water	80

In most capacitors there is an insulating sheet (such as paper or plastic) called a **dielectric** between the plates. This serves several purposes. First, because higher voltages can be applied without charge passing across the gap, dielectrics break down (charge suddenly starts to flow through them when the voltage is high enough) less readily than air. Furthermore, a dielectric allows the plates to be placed closer together without touching, thus allowing an increased capacitance because d is less in Eq. 17–6. Finally, it is found experimentally that if the dielectric fills the space between the two conductors, the capacitance is increased by a factor K which is known as the **dielectric constant** (Table 17–3). Thus, for a parallel-plate capacitor,

$$C = K\epsilon_0 \frac{A}{d}. \qquad \text{[parallel-plate capacitor]} \quad \textbf{(17–7)}$$

[†]Equation 17–6 is readily derived using the result from Appendix D on Gauss's law, namely that the electric field between two parallel plates is given by Eq. D–4,

$$E = \frac{Q/A}{\epsilon_0}.$$

We combine this with Eq. 17–2, $V = Ed$, to obtain

$$V = \left(\frac{Q}{A\epsilon_0}\right)d.$$

Thus, from Eq. 17–5, the definition of capacitance,

$$C = \frac{Q}{V} = \frac{Q}{(Q/A\epsilon_0)d} = \epsilon_0 \frac{A}{d}$$

which is Eq. 17–6.

This can also be written

$$C = \epsilon \frac{A}{d},$$

where

$$\epsilon = K\epsilon_0$$

is the *permittivity of the material*.

Let us now examine, from the molecular point of view, why the capacitance of a capacitor should increase when a dielectric is inserted between the plates. Consider a capacitor whose plates are separated by an air gap. This capacitor has a charge $+Q$ on one plate and $-Q$ on the other (Fig. 17–14a). The capacitor is isolated (not connected to a battery) so charge cannot flow to or from the plates. The potential difference between the plates, V_0, is given by Eq. 17–5: $Q = C_0 V_0$; the subscripts $(_0)$ refer to the situation when only air is between the plates. Now we insert a dielectric between the plates (Fig. 17–14b). The molecules of the dielectric may be *polar*. That is, although the molecules are neutral, the electrons may not be evenly distributed, so that one part of the molecule is positive and another part negative. Because of the electric field between the plates, the molecules will tend to become oriented as shown. Even if the molecules are not polar, the electric field between the plates will induce some separation of charge in the molecules. Although the electrons do not leave the molecules, they will move slightly within the molecules toward the positive plate. So the situation is still as illustrated in Fig. 17–14b. The net effect in either case is as if there were a net negative charge on the outer edge of the dielectric facing the positive plate, and a net positive charge on the opposite side, as shown in Fig. 17–14c.

Now imagine a positive test charge within the dielectric. The force that it feels is reduced by a factor K, the dielectric constant. This is reflected by the fact that some of the electric field lines actually do not pass through the dielectric, but end (and restart) on the charges induced on the surface of the dielectric (Fig. 17–14c). Because the force on our test charge is reduced by a factor K, the work needed to move it from one plate to the other is reduced by a factor K. (We assume that the dielectric fills all the space between the plates.) The voltage, which is the work done per unit charge, must therefore also have decreased by the factor K. That is, the voltage between the plates is now

$$V = \frac{V_0}{K}.$$

Now the charge Q on the plates has not changed, because they are isolated. So we have

$$Q = CV,$$

where C is the capacitance when the dielectric is present. When we combine this with the relation, $V = V_0/K$, we obtain

$$C = \frac{Q}{V} = \frac{Q}{V_0/K} = \frac{QK}{V_0} = KC_0,$$

since $C_0 = Q/V_0$. Thus we see, from an atomic point of view, why the capacitance is increased by the factor K.

Molecular description of dielectrics

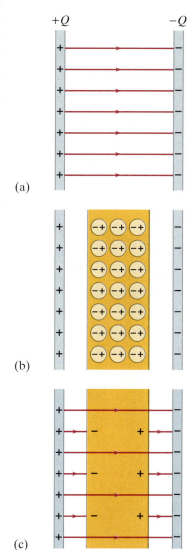

(a)

(b)

(c)

FIGURE 17–14 Molecular view of the effects of a dielectric.

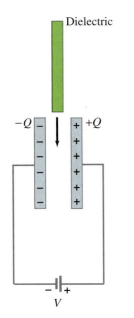

Dielectric

$-Q$ +Q

V

FIGURE 17–15   Conceptual Example 17–8.

FIGURE 17–16   Key on a computer keyboard. Pressing the key reduces the capacitor spacing thus increasing the capacitance which can be detected electronically.

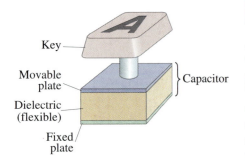

Key

Movable plate

Dielectric (flexible)

Fixed plate

Capacitor

CONCEPTUAL EXAMPLE 17–8   **Inserting a dielectric at constant *V*.**   A capacitor consisting of two plates separated by a distance d is connected to a battery of voltage V and acquires a charge Q. While it is still connected to the battery, a slab of dielectric material is inserted between the plates of the capacitor (Fig. 17–15). Will Q increase, decrease, or stay the same?

RESPONSE   Since the capacitor remains connected to the battery, the voltage stays constant and equal to the V. From the discussion above, we know that the capacitance C must increase when the dielectric material is inserted. From the relation $Q = CV$, if V stays constant, but C increases, Q must increase as well. As the dielectric is inserted, more charge will be pulled from the battery and deposited onto the plates of the capacitor as its capacitance increases.

Many computer keyboards operate by capacitance. As shown in Fig. 17–16, each key is connected to the upper plate of a capacitor. The upper plate moves down when the key is pressed, reducing the spacing between the capacitor plates, and increasing the capacitance (Eq. 17–7: smaller d, larger C). The *change* in capacitance is detected by an electronic circuit. The capacitors are designed so the capacitance change is different for each key. Hence the detected capacitance change is the signature for which key was pressed.

17–9   Storage of Electric Energy

A charged capacitor stores electric energy. The energy stored in a capacitor will be equal to the work done to charge it. The net effect of charging a capacitor is to remove charge from one plate and add it to the other plate. This is what a battery does when it is connected to a capacitor. A capacitor does not become charged instantly. It takes time (Section 19–7). When some charge is on each plate, it requires work to add more charge of the same sign. The more charge already on a plate, the more work is required to add more. The work needed to add a small amount of charge Δq, when a potential difference V is across the plates, is $\Delta W = V \Delta q$. Initially, when the capacitor is uncharged, no work is required to move the first bit of charge over. By the end of the charging process, however, the work needed to add a charge Δq will be equal to $V_f \Delta q$ where V_f is the final voltage ($V_f = Q/C$). If the voltage across the capacitor were constant, the work needed to move charge Q would be $W = QV$. But the voltage across the capacitor is proportional to how much charge it already has accumulated (Eq. 17–5), and so the voltage increases during the charging process from zero to its final value, V_f, at the end. Then the total work done, W, will be equivalent to moving all the charge Q at once across a voltage equal to the average voltage during the whole process. (This is just like calculating the work done to compress a spring, Section 6–4.) The average voltage is $(V_f - 0)/2 = V_f/2$, so

$$W = Q \frac{V_f}{2}.$$

Thus we can say that the energy, U, stored in a capacitor is

$$U = \text{energy} = \tfrac{1}{2}QV,$$

where V is the potential difference between the plates (we have dropped the subscript), and Q is the charge on each plate. Since $Q = CV$, we can also write

$$U = \tfrac{1}{2}QV = \tfrac{1}{2}CV^2 = \tfrac{1}{2}\frac{Q^2}{C}. \tag{17-8}$$

Energy stored in capacitor

EXAMPLE 17–9 **Energy stored in a capacitor.** A camera flash unit stores energy in a $150\,\mu\text{F}$ capacitor at $200\,\text{V}$. How much electric energy can be stored?

SOLUTION From Eq. 17–8, we have

$$U = \text{energy} = \tfrac{1}{2}CV^2 = \tfrac{1}{2}(150 \times 10^{-6}\,\text{F})(200\,\text{V})^2 = 3.0\,\text{J}.$$

Notice how the units work out: $\text{FV}^2 = \left(\dfrac{\text{C}}{\text{V}}\right)(\text{V}^2) = \text{CV} = \text{C}\left(\dfrac{\text{J}}{\text{C}}\right) = \text{J}.$

If this energy could be released in $\frac{1}{1000}$ of a second ($10^{-3}\,\text{s}$) the power output would be equivalent to $3000\,\text{W}$.

➡ **PHYSICS APPLIED**

Camera flash

Energy is not a substance and does not have a definite location. Nonetheless, it is often useful to think of it as being stored in the electric field between the plates. As an example, let us calculate the energy stored in a parallel-plate capacitor in terms of the electric field.

We saw in Eq. 17–2 that the electric field E between two large but close parallel plates is uniform and is related to the potential difference by $V = Ed$, where d is the separation. Also, Eq. 17–6 tells us that $C = \epsilon_0 A/d$. Thus

$$U = \tfrac{1}{2}CV^2 = \tfrac{1}{2}\left(\frac{\epsilon_0 A}{d}\right)(E^2 d^2)$$

$$= \tfrac{1}{2}\epsilon_0 E^2 A d.$$

The quantity Ad is simply the volume between the plates in which the electric field E exists. If we divide both sides by the volume, we obtain an expression for the energy per unit volume or **energy density**:

$$u = \text{energy density} = \frac{\text{energy}}{\text{volume}} = \tfrac{1}{2}\epsilon_0 E^2. \tag{17-9}$$

Energy stored per unit volume in electric field

The energy stored per unit volume is proportional to the square of the electric field in that region. If a dielectric is present, ϵ_0 is replaced by ϵ. We derived Eq. 17–9 for the special case of a capacitor. But it can be shown to be valid for any region of space where there is an electric field.

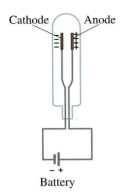

FIGURE 17–17 If the cathode inside the evacuated glass tube is heated to glowing, negatively charged "cathode rays" (electrons) are "boiled off" and flow across to the anode (+) to which they are attracted.

➡ **PHYSICS APPLIED**

CRT

* **17–10** ## Cathode Ray Tube: TV and Computer Monitors, Oscilloscope

An important device that makes use of voltage, and that allows us to "visualize" voltages in the sense of displaying graphically how a voltage changes in time, is the **cathode ray tube (CRT)**. A CRT used in this way is an *oscilloscope*—but an even more common use of a CRT is as the picture tube of television sets and computer monitors.

The operation of a CRT depends first of all on the phenomenon of **thermionic emission**, discovered by Thomas Edison (1847–1931) in the course of experiments on developing the electric light bulb. To understand how thermionic emission occurs, consider two small plates (electrodes) inside an evacuated "bulb" or "tube" as shown in Fig. 17–17, to which is applied a potential difference (by a battery, say). The negative electrode is called the **cathode**[†], the positive one the **anode**. If the negative cathode is heated (usually by an electric current, as in a lightbulb) so that it becomes hot and glowing, it is found that negative charge leaves the cathode and flows to the positive anode. These negative charges are now called electrons, but originally they were called **cathode rays** since they seemed to come from the cathode (see Section 27–1 on the discovery of the electron).

We can understand how electrons might be "boiled off" a hot metal plate if we treat electrons like molecules in a gas. This makes sense if electrons are relatively free to move about inside a metal, which is consistent with metals being good conductors. However, electrons don't readily escape from the metal. If an electron were to escape outside the metal surface, a net positive charge would remain behind, and this would attract the electron back. To escape, an electron needs a certain minimum kinetic energy, just as molecules in a liquid must have a minimum KE to "evaporate" into the gaseous state. We saw in Chapter 13 that the average kinetic energy ($\overline{KE}$) of molecules in a gas is proportional to the absolute temperature T. We can apply this idea, but only very roughly, to free electrons in a metal as if they made up an "electron gas." Of course, some electrons have more KE than average and others less. At room temperature, very few electrons would have sufficient energy to escape. At high temperature, $\overline{KE}$ is larger and many electrons escape—just as molecules evaporate from liquids, which occurs more readily at high temperatures. Thus, significant thermionic emission occurs only at elevated temperatures.

The **cathode-ray tube** (CRT) derives its name from the fact that inside an evacuated glass tube, a beam of cathode rays (electrons) is directed to various parts of a screen to produce a "picture." A simple CRT is diagrammed in Fig. 17–18. Electrons emitted by the heated cathode are accelerated by a high voltage (5,000–50,000 V) applied to the anode. The electrons pass out of this "electron gun" through a small hole in the anode. The inside of the tube face is coated with a fluorescent material that glows

[†]These terms were coined by Michael Faraday and come from the Greek words meaning, respectively, "descent" and "a way up."

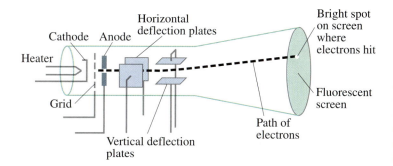

Heater
Cathode Anode
Horizontal
deflection plates
Grid
Vertical deflection
plates
Path of
electrons
Bright spot
on screen
where
electrons hit
Fluorescent
screen

FIGURE 17–18 A cathode-ray tube. Magnetic deflection coils are often used in place of the electric deflection plates. The relative positions of the elements have been exaggerated for clarity.

when struck by electrons. A tiny bright spot is thus visible where the electron beam strikes the screen. Two horizontal and two vertical plates deflect the beam of electrons when a voltage is applied to them. The electrons are deflected toward whichever plate is positive. By varying the voltage on the deflection plates, the bright spot can be placed at any point on the screen. Today it is more usual for CRTs to make use of magnetic deflection coils (Chapter 20) instead of electric plates.

In the picture tube or monitor for a computer or television set, the electron beam is made to sweep over the screen in the manner shown in Fig. 17–19. The beam is swept horizontally by the horizontal deflection plates or coils. When the horizontal deflecting field is maximum in one direction, the beam is at one edge of the screen. As the field decreases to zero, the beam moves to the center; and as the field increases to a maximum in the opposite direction, the beam approaches the opposite edge. When the beam reaches this edge, the voltage or current abruptly changes to return the beam to the opposite side of the screen. Simultaneously, the beam is deflected downward slightly by the vertical deflection plates (or coils), and then another horizontal sweep is made. For television in the United States, 525 lines constitutes a complete sweep over the entire screen. (High-definition TV will provide more than double this number of lines, giving greater picture sharpness. Some European systems already provide significantly more lines than the present U.S. standard.) The complete picture of 525 lines is swept out in $\frac{1}{30}$ s. Actually, a single vertical sweep takes $\frac{1}{60}$ s and involves every other line. The lines in between are then swept out over the next $\frac{1}{60}$ s (called interlacing). We see a picture because the image is retained by the fluorescent screen and by our eyes for about $\frac{1}{20}$ s. The picture we see consists of the varied brightness of the spots on the screen. The brightness at any point is controlled by the grid (a "porous" electrode, such as a wire grid, that allows passage of electrons) which can limit the flow of electrons by means of the voltage applied to it: the more negative this voltage, the more electrons are repelled and the fewer pass through. The voltage on the grid is determined by the video signal (a voltage) sent out by the TV station and received by the TV set. Accompanying this signal are signals that synchronize the grid voltage to the horizontal and vertical sweeps.

PHYSICS APPLIED

TV and computer monitors

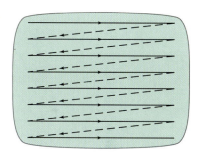

FIGURE 17–19 Electron beam sweeps across a television screen in a succession of horizontal lines.

FIGURE 17–20 An electrocardiogram (ECG) trace displayed on a CRT.

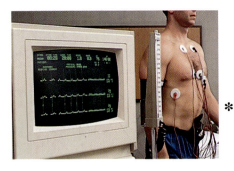

An **oscilloscope** is a device for amplifying, measuring, and visually observing an electrical signal (a "signal" is usually a time-varying voltage), especially rapidly changing signals. The signal is displayed on the screen of a CRT. In normal operation, the electron beam is swept horizontally at a uniform rate in time by the horizontal deflection plates. The signal to be displayed is applied, after amplification, to the vertical deflection plates. The visible "trace" on the screen, which could be an ECG (Fig. 17–20), a voltage in an electronic device being repaired, or a signal from an experiment on nerve conduction, is thus a plot of the signal voltage (vertically) versus time (horizontally).

* 17–11 The Electrocardiogram (ECG or EKG)

Each time the heart beats, changes in electrical potential occur on its surface that can be detected using metal contacts, called "electrodes," which are attached to the skin. The changes in potential are small, on the order of millivolts (mV), and must be amplified. They are displayed either with a chart recorder on paper, or on a cathode-ray-tube oscilloscope screen (Fig. 17–20). The record of the potential changes for a given person's heart is called an electrocardiogram (EKG or ECG). An example is shown in Fig. 17–21. The instrument itself is called an electrocardiograph. We are not so interested now in the electronics, but in the source of these potential changes and their relation to heart activity.

Muscle cells and nerve cells are similar in that both have an electric dipole layer across the cell wall. That is, in the normal situation there is a net positive charge on the exterior surface and a net negative charge on the interior surface, as shown in Fig. 17–22a. The amount of charge depends on the size of the cell, but is approximately 10^{-3} C/m^2 of surface. For a cell whose surface area is 10^{-5} m^2, the total charge on either surface is thus $\approx 10^{-8}$ C. Just before the contraction of heart muscles, changes occur in the cell wall, so that positive ions on the exterior of the cell are able to pass through the wall and neutralize those on the inside, or even make the inside surface slightly positive compared to the exterior, as shown in Fig. 17–22b. This depolarization, as it is called, starts at one end of the cell and progresses toward the opposite end, as indicated by the arrow in part (b), until the whole muscle is depolarized. The muscle then slowly repolarizes to its original state (Fig. 17–22a). The whole process requires less than a second. Figure 17–22c shows rough graphs of the potential V as a function of time at the two points P and P' (on either side of this cell) as the depolarization moves across the cell.

In the heart, the path of depolarization is complicated. Furthermore, after depolarization, the muscles repolarize to the resting state (Fig. 17–22a). Thus the potential difference as a function of time is quite complicated (Fig. 17–21).

FIGURE 17–21 Typical ECG. Two heart beats are shown.

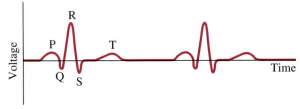

It is standard procedure to divide a typical electrocardiogram into regions corresponding to the various deflections (or "waves" as they are called), as shown in Fig. 17–21. Each of the deflections corresponds to the activity of a particular part of the heart beat (Fig. 10–39). The P wave corresponds to contraction of the atria. The QRS group corresponds to contraction of the ventricles; this group has three main phases because the depolarization follows a complicated path from left to right, and toward the front, then downward to the left and toward the rear. The T wave corresponds to recovery (repolarization) of the heart in preparation for the next cycle.

Electrocardiograms make use of three basic electrodes, one placed on either side of the heart on the hands, and one on the left foot. Sometimes six additional electrodes are placed at other locations. The measurement of so many potential differences provides additional information (some of it redundant), since the heart is a three-dimensional object and depolarization takes place in all three dimensions. A complete electrocardiogram may include as many as 12 graphs.

The ECG is a powerful tool in identifying heart defects. For example, the right side of the heart enlarges if the right ventricle must push against an abnormally large load (as when blood vessels become hardened or clogged). This problem is readily observed on an ECG, since the S wave becomes very large (negatively). *Infarcts*, which are dead regions of the heart muscle that result from heart attacks, are also detected on an ECG because they reflect the depolarization wave.

The interpretation of an ECG depends to a great extent on experience obtained with many patients rather than on theoretical understanding. A good deal of scientific research remains to be done.

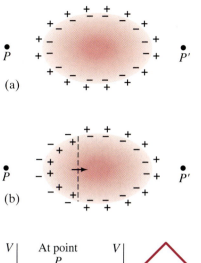

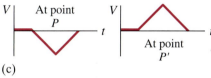

FIGURE 17–22 Heart muscle cell showing: (a) charge dipole layer in resting state; (b) depolarization of cell progressing as muscle begins to contract; and (c) potential V at points P and P' as a function of time.

SUMMARY

The **electric potential** at any point in space is defined as the electric potential energy per unit charge.

The **electric potential difference** between any two points is defined as the work done to move a 1-C electric charge between the two points. Potential difference is measured in volts (1 V = 1 J/C) and is sometimes referred to as **voltage**.

The change in PE of a charge q when it moves through a potential difference V_{ba} is

$$\Delta \text{PE} = qV_{ba}.$$

The potential difference V between two points where a uniform electric field E exists is given by

$$V = Ed,$$

where d is the distance between the two points.

An **equipotential line** or **surface** is all at the same potential, and is perpendicular to the electric field at all points.

The electric potential due to a single point charge Q, relative to zero potential at infinity, is given by

$$V = \frac{kQ}{r}.$$

A **capacitor** is a device used to store charge and consists of two nontouching conductors. The two conductors generally hold equal and opposite charges, Q, and the ratio of this charge to the potential difference V between the conductors is called the **capacitance**, C; so

$$Q = CV.$$

The capacitance of a parallel-plate capacitor is proportional to the area of each plate and inversely proportional to their separation.

The space between the two conductors of a capacitor contains a nonconducting material such as air, paper, or plastic; these materials are referred to as **dielectrics**, and the capacitance is proportional to a property of dielectrics called the *dielectric constant*, K (nearly equal to 1 for air).

A charged capacitor stores an amount of electric energy given by

$$\tfrac{1}{2}QV = \tfrac{1}{2}CV^2 = \tfrac{1}{2}\frac{Q^2}{C}.$$

This energy can be thought of as stored in the electric field between the plates.

The energy stored in any electric field E has a density (energy per unit volume) of $\tfrac{1}{2}\epsilon_0 E^2$.

QUESTIONS

1. If two points are at the same potential, does this mean that no work is done in moving a test charge from one point to the other? Does this imply that no force must be exerted?

2. Can two equipotential lines cross? Explain.

3. Draw a few equipotential lines in Fig. 16–29b.

4. Is there a point along the line joining two equal positive charges where the electric field is zero? Where the electric potential is zero? Explain.

5. An electron is accelerated by a potential difference of, say, 100 V. How much greater would its final speed be if it were accelerated with four times as much voltage?

6. If a negative charge is initially at rest in an electric field, will it move toward a region of higher potential or lower potential? What about a positive charge? How does the potential energy of the charge change in each case?

7. State clearly the difference between: (a) electric potential and electric field, (b) electric potential and electric potential energy.

8. If the potential at a point is zero, must the electric field also be zero? Give an example.

9. What can you say about the electric field in a region of space that has the same potential throughout?

10. How does the Earth's gravitational field change with distance? What about its gravitational potential?

11. Can a particle ever move from a region of low electric potential to one of high potential and yet have its electric potential energy decrease? Explain.

12. When dealing with practical devices, we often take the ground (the earth) to be 0 V. If, instead, we said the ground was -10 V, how would this affect (a) V, and (b) E, at other points?

13. When a battery is connected to a capacitor, why do the two plates acquire charges of the same magnitude? Will this be true if the two conductors are different sizes or shapes?

14. We have seen that the capacitance C depends on the size, shape, and position of the two conductors, as well as on the dielectric constant K. What then did we mean when we said that C is a constant in Eq. 17–5?

15. How does the energy stored in a capacitor change when a dielectric is inserted if (a) the capacitor is isolated so Q doesn't change, (b) the capacitor remains connected to a battery so V doesn't change?

PROBLEMS

SECTIONS 17–1 TO 17–4

1. (I) How much work is needed to move a -8.6-μC charge from ground to a point whose potential is $+75$ V?

2. (I) How much work is needed to move a proton from a point with a potential of $+100$ V to a point where it is -50 V? Express your answer both in joules and electron volts.

3. (I) How much kinetic energy will an electron gain (in joules and eV) if it falls through a potential difference of 21,000 V in a TV picture tube?

4. (I) An electron acquires 3.45×10^{-16} J of kinetic energy when it is accelerated by an electric field in a computer monitor from plate A to plate B. What is the potential difference between the plates, and which plate is at the higher potential?

5. (I) How strong is the electric field between two parallel plates 5.2 mm apart if the potential difference between them is 220 V?

6. (I) An electric field of 640 V/m is desired between two parallel plates 11.0 mm apart. How large a voltage should be applied?

7. (I) What potential difference is needed to give a helium nucleus ($Q = 2e$) 65.0 keV of KE?

8. (II) Two parallel plates, connected to a 100-V power supply, are separated by an air gap. How small can the gap be if the air is not to exceed its breakdown value of $E = 3 \times 10^6$ V/m?

9. (II) The work done by an external force to move a -7.50-μC charge from point a to point b is 25.0×10^{-4} J. If the charge was started from rest and had 4.82×10^{-4} J of kinetic energy when it reached point b, what must be the potential difference between a and b?

10. (II) What is the speed of (a) a 750-eV, and (b) a 3.5 keV, electron?

11. (II) What is the speed of a proton whose kinetic energy is 28.0 MeV?

12. (II) An alpha particle (which is a helium nucleus, $Q = +2e$, $m = 6.64 \times 10^{-27}$ kg) is emitted in a radioactive decay with KE = 5.53 MeV. What is its speed?

SECTION 17–5

13. (I) What is the electric potential 15.0 cm from a 4.00-μC point charge?

14. (I) A charge Q creates an electric potential of $+125$ V at a distance of 15 cm. What is Q?

15. (II) A $+30$-μC charge is placed 32 cm from an identical $+30$-μC charge. How much work would be required to move a $+0.50$-μC test charge from a point midway between them to a point 10 cm closer to either of the charges?

16. (II) (a) What is the electric potential a distance of 2.5×10^{-15} m away from a proton? (b) What is the electric potential energy of a system that consists of two protons 2.5×10^{-15} m apart—as might occur inside a typical nucleus?

17. How much voltage must be used to accelerate a proton (radius 1.2×10^{-15} m) so that it has sufficient energy to just penetrate a silicon nucleus? A silicon nucleus has a charge of $+14e$ and its radius is about 3.6×10^{-15} m. Assume the potential is that for point charges.

18. (II) How much work must be done to bring three electrons from a great distance apart to within 1.0×10^{-10} m from one another?

19. (II) Consider point a which is 70 cm north of a -3.8-μC point charge, and point b which is 80 cm west of the charge (Fig. 17–23). Determine (a) $V_{ba} = V_b - V_a$, and (b) $\mathbf{E}_b - \mathbf{E}_a$ (magnitude and direction).

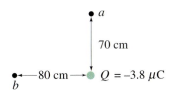

FIGURE 17–23   Problem 19.

20. (II) An electron starts from rest 72.5 cm from a fixed point charge with $Q = -0.125 \, \mu$C. How fast will the electron be moving when it is very far away?

21. (II) Two identical $+7.5$-μC point charges are initially spaced 5.5 cm from each other. If they are released at the same instant from rest, how fast will they be moving when they are very far away from each other? Assume they have identical masses of 1.0 mg.

22. (III) In the Bohr model of the hydrogen atom, an electron orbits a proton (the nucleus) in a circular orbit of radius 0.53×10^{-10} m. (a) What is the electric potential at the electron's orbit due to the proton? (b) What is the kinetic energy of the electron? (c) What is the total energy of the electron in its orbit? (d) What is the *ionization energy*—that is, the energy required to remove the electron from the atom and take it to $r =$ infinity, at rest?

23. (III) Two equal but opposite charges are separated by a distance d, as shown in Fig. 17–24. Determine a formula for $V_{BA} = V_B - V_A$ for points B and A on the line between the charges situated as shown.

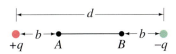

FIGURE 17–24   Problem 23.

*SECTION 17–6

*24. (II) An electron and a proton are 0.53×10^{-10} m apart. (a) What is their dipole moment if they are at rest? (b) What is the average dipole moment if the electron revolves about the proton in a circular orbit?

*25. (II) Calculate the electric potential due to a dipole whose dipole moment is 4.8×10^{-30} C·m at a point 1.1×10^{-9} m away if this point is: (a) along the axis of the dipole nearer the positive charge; (b) 45° above the axis but nearer the positive charge; (c) 45° above the axis but nearer the negative charge.

*26. (II) (a) In Example 17–6, part (b), calculate the electric potential without using the dipole approximation, Eq. 17–4; that is, don't assume $r \gg l$. (b) What is the percent error in this case when the dipole approximation is used?

*27. (III) The dipole moment, considered as a vector, points from the negative to the positive charge. The water molecule, Fig. 17–25, has a dipole moment $\mathbf{p}$ which can be considered as the vector sum of the two dipole moments, $\mathbf{p}_1$ and $\mathbf{p}_2$, as shown. The distance between each H and the O is about 0.96×10^{-10} m. The lines joining the center of the O atom with each H atom make an angle of 104°, as shown, and the net dipole moment has been measured to be $p = 6.1 \times 10^{-30}$ C·m. Determine the charge q on each H atom.

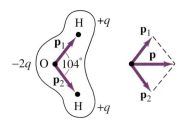

FIGURE 17–25   Problem 27.

*28. (III) Show that if two dipoles with dipole moments p_1 and p_2 are in line with one another (Fig. 17–26), the potential energy of one in the presence of the other (their "interaction energy") is given by

$$\text{PE} = -\frac{2kp_1p_2}{r^3},$$

where r is the distance between the two dipoles. [*Hint:* Assume that r is much greater than the length of either dipole.]

FIGURE 17–26   Problem 28.

* **29.** (III) Show that if an electric dipole is placed in a uniform electric field, then a torque is exerted on it equal to $pE \sin \phi$, where ϕ is the angle between the dipole moment vector and the direction of the electric field as shown in Fig. 17–27. What is the net force on the dipole? How are your answers affected if the field is nonuniform? Note that the dipole moment vector **p** is defined so that its magnitude is Ql and its direction is pointing from the negative end to the positive end as shown.

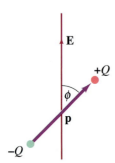

FIGURE 17–27   Problem 29.

SECTIONS 17–7 AND 17–8

30. (I) The two plates of a capacitor hold $+2500 \mu C$ and $-2500 \mu C$ of charge, respectively, when the potential difference is 950 V. What is the capacitance?

31. (I) The potential difference between two parallel wires in air is 120 V. They carry equal and opposite charge of magnitude 95 pC. What is the capacitance of the two wires?

32. (I) A 7500-pF capacitor holds 16.5×10^{-8} C of charge. What is the voltage across the capacitor?

33. (I) How much charge flows from a 12.0-V battery when it is connected to a 9.00-μF capacitor?

34. (I) A 0.20-F capacitor is desired. What area must the plates have if they are to be separated by a 2.2-mm air gap?

35. (I) What is the capacitance of a pair of circular plates with a radius of 5.0 cm separated by 3.2 mm of mica?

36. (II) The charge on a capacitor increases by 15 μC when the voltage across it increases from 97 V to 121 V. What is the capacitance of the capacitor?

37. (II) An electric field of 8.50×10^5 V/m is desired between two parallel plates each of area 35.0 cm^2 and separated by 2.45 mm of air. What charge must be on each plate?

38. (II) If a capacitor has 4.2 μC of charge on it and an electric field of 2.0 kV/mm is desired between the two plates which are separated by 4.0 mm of air, what must each plate's area be?

39. (II) How strong is the electric field between the plates of a 0.80-μF air-gap capacitor if they are 2.0 mm apart and each has a charge of 72 μC?

40. (II) The electric field between the plates of a paper-separated ($K = 3.75$) capacitor is 9.21×10^4 V/m. The plates are 1.95 mm apart and the charge on each plate is 0.775 μC. Determine the capacitance of this capacitor and the area of each plate.

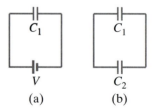

FIGURE 17–28   Problems 41 and 49.

41. (III) A 7.7-μF capacitor is charged by a 125-V battery and then is disconnected from the battery. When this capacitor (C_1) is then connected (Fig. 17–28) to a second (initially uncharged) capacitor, C_2, the voltage on the first drops to 15 V. What is the value of C_2? [*Hint:* Charge is conserved.]

42. (III) A 2.50-μF capacitor is charged to 1000 V and a 6.80-μF capacitor is charged to 650 V. These capacitors are then disconnected from their batteries, and the positive plates are now connected to each other and the negative plates are connected to each other. What will be the potential difference across each and the charge on each? [*Hint:* Charge is conserved.]

SECTION 17–9

43. (I) 550 V is applied to a 7200-pF capacitor. How much energy is stored?

44. (I) A cardiac defibrilator is used to shock a heart that is beating erratically. A capacitor in this device is charged to 6000 V and stores 200 J of energy. What is its capacitance?

45. (II) A homemade capacitor is assembled by placing two 9-in pie pans 10 cm apart and connecting them to the opposite terminals of a 9-V battery. Estimate (*a*) the capacitance, (*b*) the charge on each plate, (*c*) the electric field halfway between the plates, (*d*) the work done by the battery to charge the plates. (*e*) Which of the above values change if a dielectric is inserted?

46. (II) A parallel-plate capacitor has fixed charges $+Q$ and $-Q$. The separation of the plates is then doubled. By what factor does the energy stored in the electric field change?

47. (II) How does the energy stored in a capacitor change if (*a*) the potential difference is doubled, (*b*) the charge on each plate is doubled, and (*c*) the separation of the plates is doubled, as the capacitor remains connected to a battery?

48. (II) A parallel-plate capacitor is isolated with a charge $\pm Q$ on each plate. If the separation of the plates is halved and a dielectric (constant K) is inserted in place of air, by what factor does the energy storage change? To what do you attribute the change in stored potential energy? How does the new value of the electric field between the plates compare with the original value?

49. (III) A 2.70-μF capacitor is charged by a 45.0-V battery. It is disconnected from the battery and then connected to an uncharged 4.00-μF capacitor (Fig. 17–28). Determine the total stored energy (*a*) before the two capacitors are connected, and (*b*) after they are connected. (*c*) What is the change in energy? (*d*) Is energy conserved? Explain.

*SECTION 17–10

*50. (I) Use the ideal gas as a model to estimate the rms speed of a free electron in a metal at 300 K, and at 2500 K (the typical temperature of the cathode in a tube).

*51. (III) In a given CRT, electrons are accelerated horizontally by 15 kV. They then pass through a uniform electric field E for a distance of 2.8 cm which deflects them upward so they reach the top of the screen 22 cm away, 11 cm above the center. Estimate the value of E.

*52. (III) Electrons are accelerated by 14 kV in a CRT. The screen is 30 cm wide and is 34 cm from the 2.6-cm-long deflection plates. Over what range must the horizontally deflecting electric field vary to sweep the beam fully across the screen?

GENERAL PROBLEMS

53. There is an electric field near the Earth's surface whose intensity is about 150 V/m. How much energy is stored per cubic meter in this field?

54. A lightning flash transfers 4.0 C of charge and 4.2 MJ of energy to the Earth. (*a*) Between what potential difference did it travel? (*b*) How much water could this boil, starting from room temperature?

55. Calculate the average translational kinetic energy in eV for (*a*) an oxygen molecule at room temperature, (*b*) a nitrogen molecule at room temperature, (*c*) an iron atom in the Sun's corona where the temperature is about 2 million K, and (*d*) a carbon dioxide molecule in the lower atmosphere of Mars where the temperature is $-50°C$.

56. In a television picture tube, electrons are accelerated by thousands of volts through a vacuum. If a television set were laid on its back, would electrons be able to move upward against the force of gravity? What potential difference, acting over a distance of 3.0 cm, would be needed to balance the downward force of gravity so that an electron would remain stationary? Assume that the electric field is uniform.

57. It takes 8.5 J of energy to move a 3.0-mC charge from one plate of a 9.0-μF capacitor to the other. How much charge is on each plate?

58. An electron starting from rest acquires 5.2 keV of KE in moving from point A to point B. (*a*) How much KE would a proton acquire, starting from rest at B and moving to point A? (*b*) Determine the ratio of their speeds at the end of their respective trajectories.

59. A 2600-pF air-gap capacitor is connected to a 9.0-V battery. If a piece of mica is placed between the plates, how much charge will then flow from the battery?

60. A huge 4.0-F capacitor has enough stored energy to heat 2.5 kg of water from 20°C to 95°C. What is the potential difference across the plates?

61. An uncharged capacitor is connected to a 24.0-V battery until it is fully charged, after which it is disconnected from the battery. A slab of paraffin is then inserted between the plates. What will now be the voltage between the plates?

62. Dry air will break down if the electric field exceeds 3.0×10^6 V/m. What amount of charge can be placed on a capacitor if the area of each plate is 56 cm^2?

63. A 3.4-μC and a -2.0-μC charge are placed 1.5 cm apart. At what points along the line joining them is (*a*) the electric field zero, and (*b*) the potential zero?

64. Three charges are at the corners of an equilateral triangle (side *l*) as shown in Fig. 17–29. Determine the potential at the midpoint of each of the sides.

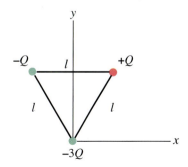

FIGURE 17–29 Problem 64.

65. A capacitor C_1 carries a charge Q_0. It is then connected directly to a second, uncharged, capacitor C_2, as shown in Fig. 17–30. What charge will each carry now? What will be the potential difference across each?

FIGURE 17–30 Problem 65.

66. An electron is accelerated horizontally from rest in a television picture tube by a potential difference of 25,000 V. It then passes between two horizontal plates 6.5 cm long and 1.3 cm apart that have a potential difference of 250 V (Fig. 17–31). At what angle θ will the electron be traveling after it passes between the plates?

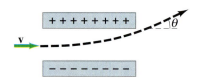

FIGURE 17–31 Problem 66.

67. In a photocell, ultraviolet (UV) light provides enough energy to some electrons in barium metal to eject them from a surface at high speed. See Fig. 17–32. To measure the maximum energy of the electrons, another plate above the barium surface is kept at a negative enough potential that the emitted electrons are slowed down and stopped, and return to the barium surface. If the plate voltage is -3.02 V (compared to the barium) when the fastest electrons are stopped, what was the speed of these electrons when they were emitted?

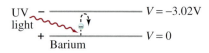

FIGURE 17–32 Problem 67.

68. To get an idea how big a farad is, suppose you wanted to make a 1-F air-filled parallel-plate capacitor for a circuit you were building. (*a*) To make it a reasonable size, suppose you limited the plate area to 1.0 cm^2. What would the gap have to be between the plates? Is this practically achievable? (*b*) Suppose you instead chose the gap between the plates to be 1.0 mm. What would be the area of the plates? Is this a practical solution for your circuit?

69. Near the surface of the Earth there is an electric field of about 150 V/m which points downward. Two identical balls with mass $m = 0.540 \text{ kg}$ are dropped from a height of 2.00 m, but one of the balls is positively charged with $q_1 = 550 \ \mu\text{C}$, and the second is negatively charged with $q_2 = -550 \ \mu\text{C}$. Use conservation of energy to determine the difference in the speed of the two balls when they hit the ground. (Neglect air resistance.)

70. The power supply for a pulsed nitrogen laser has a $0.050 \ \mu\text{F}$ capacitor with a maximum voltage rating of 30 kilovolts. (*a*) Estimate how much energy could be stored in this capacitor. (*b*) If 10 percent of this stored electrical energy is converted to light energy in a pulse that is 10 microseconds long, what is the power of the laser pulse?

71. In lightning storms, the potential difference between the Earth and the bottom of the thunderclouds can be as high as 35,000,000 V. The bottoms of the thunderclouds are typically 1500 m above the Earth, and can have an area of 110 km^2. For the purposes of this problem, model the Earth-cloud system as a huge capacitor and calculate (*a*) the capacitance of the Earth-cloud system, (*b*) the charge stored in the "capacitor," and (*c*) the energy stored in the "capacitor."

The glow of the thin wire filament of a light bulb is caused by the electric current passing through it. Electric energy is transformed to thermal energy (via collisions between moving electrons and atoms of the wire), which causes the wire's temperature to become so high that it glows.

ELECTRIC CURRENTS

In the previous two chapters we have been studying static electricity: electric charges at rest. In this chapter we begin our study of charges in motion as an electric current. Until the year 1800, the technical development of electricity consisted mainly of producing a static charge by friction—for example by machines that could produce rather large potentials such as the one shown in Fig. 18–1. Large sparks could be produced by such machines, but they had little practical value.

In nature itself there were grander displays of electricity such as lightning and "St. Elmo's fire," which is a glow that appeared around the yardarms of ships during storms. That these phenomena were electrical in origin was not recognized until the eighteenth century. For example, it was only in 1752 that Franklin, in his famous kite experiment, showed that lightning was an electric discharge—a giant electric spark.

Finally, in 1800, an event of great practical importance occurred: Alessandro Volta (1745–1827; Fig. 18–2) invented the electric battery, and with it produced the first steady flow of electric charge—that is, a steady electric current. This discovery opened a new era, which transformed our civilization, for today's electrical technology is based on electric current.

FIGURE 18–1 Early electrostatic generator.

FIGURE 18–2 Alessandro Volta. In this portrait, Volta exhibits his battery to Napoleon in 1801.

18–1 The Electric Battery

The events that led to the discovery of the battery are interesting. For not only was this an important discovery, but it also gave rise to a famous scientific debate between Volta and Luigi Galvani (1737–1798), eventually involving many others in the scientific world.

In the 1780s, Galvani, a professor at the University of Bologna (thought to be the world's oldest university still in existence), carried out a long series of experiments on the contraction of a frog's leg muscle through electricity produced by a static-electricity machine. Galvani found that contraction of the muscle could be produced by other means as well: when a brass hook was pressed into the frog's spinal cord and then hung from an iron railing that also touched the frog, the leg muscles again would contract. He found that this phenomenon occurred for other pairs of metals as well.

What was the source of this unusual phenomenon? Galvani believed that the source of the electric charge was in the frog muscle or nerve itself, and that the wire merely transmitted the charge to the proper points; and when he published his work in 1791, he termed it "animal electricity." Many wondered, including Galvani himself, if he had discovered the long-sought "life-force."

Volta, at the University of Pavia 200 km away, was skeptical of Galvani's results. Although he soon confirmed and extended those experiments, Volta still doubted Galvani's idea of animal electricity. Instead he came to believe that the source of the electricity was not in the animal itself, but rather in the *contact between the two metals*. Volta made public his views and soon had many followers, although others still sided with Galvani.

Volta soon realized that a moist conductor, such as a frog muscle or moisture at the contact point of the two dissimilar metals, was necessary in the circuit if it was to be effective. He also saw that the contracting frog muscle was a sensitive instrument for detecting electric "tension" or "electromotive force" (his words for what we now call potential), in fact more sensitive than the best available electroscopes that he and others had developed. Most important, he recognized that a decisive answer to Galvani could be given only if the sensitive frog leg was replaced by an inorganic detector. That is, to cement his view that it was the contact of two dissimilar metals that caused the frog muscle to contract, Volta would have to connect the two dissimilar metals directly to an electroscope and observe a separation of the leaves representing a potential difference. This proved difficult since his most sensitive[†] electroscopes were much less sensitive than the frog muscle. But the eventual success of this experiment vindicated Volta's theory.

Volta's research found that certain combinations of metals produced a greater effect than others, and, using his measurements, he listed them in order of effectiveness. (This "electrochemical series" is still used by chemists today.) And he found that carbon could be used in place of one of the metals.

Volta then conceived what is his greatest contribution to science. Between a disc of zinc and one of silver, he placed a piece of cloth or paper soaked in salt solution or dilute acid and piled a "battery" of such couplings, one on top of another, as shown in Fig. 18–3a. This "pile" or "battery" produced a much increased potential difference. Indeed, when strips

[†]Volta's most sensitive electroscope measured about 40 V per degree (angle of leaf separation). Nonetheless, he was finally able to estimate the potential differences produced by dissimilar metals in contact: for a silver–zinc contact he got about 0.7 V, remarkably close to today's value of 0.78 V.

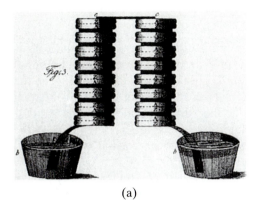

(a)

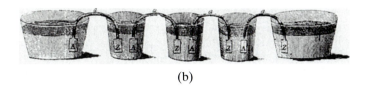

FIGURE 18–3 Two types of voltaic battery: (a) a pile; (b) a "crown of cups." Z stands for zinc and A for silver (*argentum* in Latin). Taken from Volta's original publication.

(b)

of metal connected to the two ends of the pile were brought close, a spark was produced. Volta had designed and built the first electric battery. A second design, known as the "crown of cups," is shown in Fig. 18–3b. Volta made public this great discovery in 1800.

The potential produced by Volta's battery was still weak compared to that produced by the best friction machines of the time, although it could produce considerable charge. (The electrostatic machines were high-potential, low-charge devices.) But it had a great advantage: it was "self-renewing"—it could produce a flow of electric charge continuously for a relatively long period of time. It was not long before even more powerful batteries were constructed.

After Volta's discovery of the electric battery, scientists eventually recognized that a battery produces electricity by transforming chemical energy into electrical energy. Today a great variety of electric cells and batteries are available, from flashlight batteries (sometimes called "dry cells") to the storage battery of a car. The simplest batteries contain two plates or rods made of dissimilar metals (one can be carbon) called **electrodes**. The electrodes are immersed in a solution, such as a dilute acid, called the **electrolyte**. (In a dry cell, the electrolyte is absorbed in a powdery paste.) Such a device is properly called an **electric cell**, and several cells connected together is a **battery** (although today even a single cell is called a battery). The chemical reactions involved in most electric cells are quite complicated. Here we describe how one very simple cell works, emphasizing the physical aspects.

The simple cell shown in Fig. 18–4 uses dilute sulfuric acid as the electrolyte. One of the electrodes is made of carbon, the other of zinc. That part of each electrode remaining outside the solution is called the **terminal**, and connections to wires and circuits are made here. The acid attacks the zinc electrode and tends to dissolve it. But each zinc atom leaves two electrons behind, so it enters the solution as a positive ion. The zinc electrode thus acquires a negative charge. As more zinc ions enter solution, the electrolyte can momentarily become positively charged. Because of this, and through other chemical reactions, electrons are pulled off the carbon electrode. Thus the carbon electrode becomes positively charged. Because there is an opposite charge on the two electrodes, there is a potential difference between the two terminals. In a cell whose terminals are not connected, only a small amount of the zinc is dissolved, for as the zinc electrode becomes increasingly negative, any new positive zinc ions produced are attracted back to the electrode. Thus, a particular potential difference or voltage is maintained between the two terminals. If charge is allowed to flow between the terminals, say, through a wire (or a lightbulb),

Cells and batteries

How a simple battery works

FIGURE 18–4 Simple electric cell.

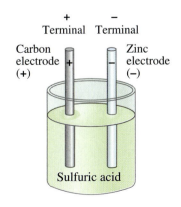

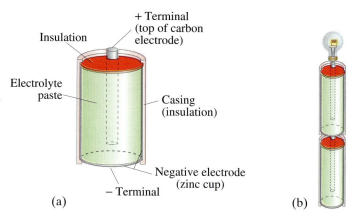

FIGURE 18–5 (a) Diagram of an ordinary dry cell (like a D-cell or AA). The cylindrical zinc cup is covered on the sides; its flat bottom is the negative terminal. (b) Two dry cells (AA type) connected in series. Note that the positive terminal of one cell pushes against the negative terminal of the other.

+ Terminal (top of carbon electrode)

Insulation

Electrolyte paste

Casing (insulation)

Negative electrode (zinc cup)

− Terminal

(a)

(b)

then more zinc can be dissolved. After a time, one or the other electrode is used up and the cell becomes "dead."

The voltage that exists between the terminals of a battery depends on what the electrodes are made of and their relative ability to be dissolved or give up electrons. When two or more cells are connected so that the positive terminal of one is connected to the negative terminal of the next, they are said to be connected in *series* and their voltages add up. Thus, the voltage between the ends of two 1.5-V flashlight batteries connected in series is 3.0 V, whereas the six 2-V cells of an automobile storage battery give 12 V. Figure 18–5 shows (a) a diagram of a common dry cell (or "flashlight battery") used in portable radios, Walkmans, flashlights, etc., and (b) shows two of them in series.

18–2 Electric Current

Electric circuit

When a continuous conducting path is connected between the terminals of a battery, we have an electric **circuit**, Fig. 18–6a. On any diagram of a circuit, as in Fig. 18–6b, we will represent a battery by the symbol

Battery

$$\dashv\!\vdash_{+\ \ -} .$$ [battery symbol]

The longer line on this symbol represents the positive terminal, and the shorter line, the negative terminal. The device powered by the battery could be a lightbulb (which is just a fine wire inside an evacuated glass bulb), a heater, a radio, or whatever. When such a circuit is formed, charge can flow through the wires of the circuit, from one terminal of the battery to the other. A flow of charge such as this is called an **electric current**.

FIGURE 18–6 (a) A simple electric circuit. (b) Schematic drawing of the same circuit.

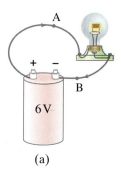

A

+   −

6 V

B

(a)

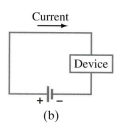

Current

Device

+   −

(b)

More precisely, the electric current in a wire is defined as the net amount of charge that passes through it per unit time at any point. Thus, the average current I is defined as

$$I = \frac{\Delta Q}{\Delta t},$$ (18–1)   *Electric current.*

where ΔQ is the amount of charge that passes through the conductor at any location during the time interval Δt. Electric current is measured in coulombs per second; this is given a special name, the **ampere** (abbreviated amp or A), after the French physicist André Ampère (1775–1836). Thus, $1\,A = 1\,C/s$. Smaller units of current are often used such as the milliampere ($1\,mA = 10^{-3}\,A$) and microampere ($1\,\mu A = 10^{-6}\,A$).   *The ampere* ($1\,A = 1\,C/s$)

In any single circuit, such as in Fig. 18–6, the current at any instant is the same at one point (say point A) as at any other point (such as B). This follows from the conservation of electric charge (charge doesn't disappear).

EXAMPLE 18–1 **Current is flow of charge.** A steady current of 2.5 A flows in a wire for 4.0 min. (a) How much charge passed through any point in the circuit? (b) How many electrons would this be?

SOLUTION (a) Since the current was 2.5 A, or 2.5 C/s, then in 4.0 minutes ($= 240$ seconds) the total charge that flowed was, from Eq. 18–1,

$$\Delta Q = I\,\Delta t$$

$$= (2.5\,C/s)(240\,s) = 600\,C.$$

(b) The charge on one electron is $1.60 \times 10^{-19}\,C$, so 600 C would consist of

$$\frac{600\,C}{1.6 \times 10^{-19}\,C/\text{electron}} = 3.8 \times 10^{21}\ \text{electrons}.$$

CONCEPTUAL EXAMPLE 18–2 **How to connect a battery.** What's wrong with each of the schemes shown in Fig. 18–7 for lighting a flashlight with a flashlight battery and a single wire?

RESPONSE (a) There is no loop for current to flow around. Charges might briefly start to flow from the battery toward the lightbulb, but there they run into a "dead end," and the flow would immediately come to a stop. (b) Now there is a loop passing to and from the lightbulb, but there is no potential difference, and so there will be nothing to push the current. (c) Nothing is wrong here. This way will light the bulb.

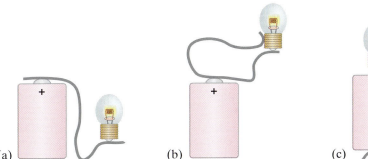

FIGURE 18–7 Example 18–2.

(a)   (b)   (c)

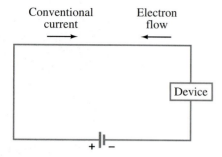

FIGURE 18–8 Conventional current from + to − is equivalent to a negative (electron) current flowing from − to +.

We saw in Chapter 16 (Section 16–3) that conductors contain many free electrons. Thus, when a conducting wire is connected to the terminals of a battery as in Fig. 18–6, it is actually the negatively charged electrons that flow in the wire. When the wire is first connected, the potential difference between the terminals of the battery sets up an electric field inside the wire[†] and parallel to it. Thus free electrons at one end of the wire are attracted into the positive terminal, and at the same time, electrons leave the negative terminal of the battery and enter the wire at the other end. There is a continuous flow of electrons through the wire that begins as soon as the wire is connected to *both* terminals. However, when the conventions of positive and negative charge were invented two centuries ago, it was assumed that positive charge flowed in a wire. Actually, for nearly all purposes, positive charge flowing in one direction is exactly equivalent to negative charge flowing in the opposite direction,[‡] as shown in Fig. 18–8. Today, we still use the historical convention of positive current flow when discussing the direction of a current. So when we speak of the current flowing in a circuit, we mean the direction positive charge would flow. This is sometimes referred to as **conventional current**. When we want to speak of the direction of electron flow, we will specifically state it is the electron current. In liquids and gases, both positive and negative charges (ions) can move.

Conventional current

18–3 Ohm's Law: Resistance and Resistors

To produce an electric current in a circuit, a difference in potential is required. One way of producing a potential difference is by a battery. It was Georg Simon Ohm (1787–1854) who established experimentally that the current in a metal wire is proportional to the potential difference V applied to its ends:

$$I \propto V.$$

Current ∝ voltage

If, for example, we connect a wire to a 6-V battery, the current flow will be twice what it would be if the wire were connected to a 3-V battery.

It is helpful to compare an electric current to the flow of water in a river or a pipe acted on by gravity. If the pipe (or river) is nearly level, the flow rate is small. But if one end is somewhat higher than the other, the flow rate—or current—is greater. The greater the difference in height, the greater the current. We saw in Chapter 17 that electric potential is analogous, in the gravitational case, to the height of a cliff; and this applies in the present case to the height through which the fluid flows. Just as an increase in height causes a greater flow of water, so a greater electric potential difference, or voltage, causes a greater electric current flow.

Water analogy

Exactly how much current flows in a wire depends not only on the voltage, but also on the resistance the wire offers to the flow of electrons. The walls of a pipe, or the banks of a river and rocks in the middle, offer resistance to the flow of current. Similarly, electrons are slowed down because of interactions with the atoms of the wire. The higher this resistance, the less the current for a given voltage V. We then define *resistance* so that

---

[†]This does not contradict what was said in Section 16–9 that in the *static* case, there can be no electric field within a conductor since otherwise the charges would move. Indeed, when there is an electric field in a conductor, charges do move, and we get an electric current.

[‡]An exception is discussed in Section 20–11.

the current is inversely proportional to the resistance. When we combine this with the above proportion, we have

$$I = \frac{V}{R},$$ (18–2)

where R is the **resistance** of a wire or other device, V is the potential difference across the device, and I is the current that flows through it. This relation (Eq. 18–2) is often written

$$V = IR,$$

OHM'S LAW

and is known as **Ohm's law**. Many physicists would say that it is not a law, but rather is the *definition of resistance*. If we want to call something Ohm's law, it would be the statement that the current through a *metal conductor* is proportional to the applied voltage, $I \propto V$. That is, R is a constant, independent of V, for metal conductors. But this relation does *not* apply generally for other substances and devices such as diodes, vacuum tubes, transistors, and so on. Thus "Ohm's law" is not a fundamental law, but rather a description of a certain class of materials (metal conductors). The habit of calling it Ohm's law is so ingrained that we won't quibble with continuing this usage, as long as we keep in mind its limitations. Materials or devices that do not follow Ohm's law are said to be *nonohmic*. See Fig. 18–9. The definition of resistance,

$$R = V/I$$

(Eq. 18–2), can be applied also to nonohmic cases, but in these cases, R would not be constant and would depend on the applied voltage.

The unit for resistance is called the **ohm** and is abbreviated Ω (Greek capital omega). Because $R = V/I$, we see that $1.0\,\Omega$ is equivalent to $1.0\,V/A$.

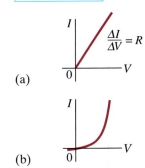

FIGURE 18–9 Graphs of current or voltage for (a) a metal conductor which obeys Ohm's law, and (b) for a nonohmic device, in this case a semiconductor diode.

The ohm $(1\,\Omega = 1\,V/A)$

FIGURE 18–10 Flashlight (Example 18–3). Note how circuit is completed along the side strip.

EXAMPLE 18–3 **Flashlight bulb resistance.** A small flashlight bulb (Fig. 18–10) draws 300 mA from its 1.5-V battery. (*a*) What is the resistance of the bulb? (*b*) If the voltage dropped to 1.2 V, how would the current change?

SOLUTION (*a*) We use Ohm's law, Eq. 18–2, and find

$$R = \frac{V}{I} = \frac{1.5\,V}{0.30\,A} = 5.0\,\Omega.$$

(*b*) If the resistance stayed constant, the current would be approximately

$$I = \frac{V}{R} = \frac{1.2\,V}{5.0\,\Omega} = 0.24\,A,$$

or a drop of 60 mA. Actually, resistance depends on temperature (Section 18–4), so this is only an approximation.

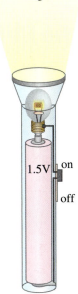

All electric devices, from heaters to lightbulbs to stereo amplifiers, offer resistance to the flow of current. The filaments of lightbulbs and electric heaters are special types of wires whose resistance results in their becoming very hot. Generally, the connecting wires have very low resistance in comparison to the resistance of the wire filaments or coils. In many circuits, particularly in electronic devices, **resistors** are used to control the *Resistors*

FIGURE 18–11 Photo of resistors (mostly).

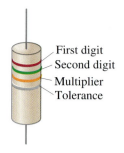

First digit
Second digit
Multiplier
Tolerance

FIGURE 18–12 The resistance value of a given resistor is written on the exterior, or may be given as a color code, as shown above and in the table: the first two colors represent the first two digits in the value of the resistance, the third color represents the power of ten that it must be multiplied by, and the fourth is the manufactured tolerance. For example, a resistor whose four colors are red, green, orange, and silver has a resistance of 25,000 Ω (25 kΩ), give or take 10 percent.

Resistor Color Code

Color	Number	Multi-plier	Toler-ance (%)
Black	0	1	
Brown	1	10^1	
Red	2	10^2	
Orange	3	10^3	
Yellow	4	10^4	
Green	5	10^5	
Blue	6	10^6	
Violet	7	10^7	
Gray	8	10^8	
White	9	10^9	
Gold		10^{-1}	5%
Silver		10^{-2}	10%
No color			20%

amount of current. Resistors have resistances from less than an ohm to millions of ohms (see Figs. 18–11 and 18–12). The two main types are "wire-wound" resistors, which consist of a coil of fine wire, and "composition" resistors, which are usually made of the semiconductor carbon.

When we draw a diagram of a circuit, we indicate a resistance with the symbol

Resistor
　　　　　　　　　⎍⋀⋀⋀⎍　.　　　　　　　　　　　[resistor symbol]

Wires whose resistance is negligible, however, are shown simply as straight lines.

18–4 Resistivity

We might expect that the resistance of a thick wire would be less than that of a thin one because a thicker wire has more area for the electrons to pass through. And you might expect the resistance to be greater if the length is greater since there would be more obstacles to electron flow. Indeed, it is found experimentally that the resistance R of a metal wire is directly proportional to its length L and inversely proportional to the cross-sectional area A. That is,

$$R = \rho \frac{L}{A}, \tag{18–3}$$

Resistivity
where ρ, the constant of proportionality, is called the **resistivity** and depends on the material used. Typical values of ρ, whose units are $\Omega \cdot m$ (see Eq. 18–3), are given for various materials in the middle column of Table 18–1. The values depend somewhat on purity, heat treatment, temperature, and other factors. Notice that silver has the lowest resistivity and is thus the best conductor (although it is expensive). Copper is not far behind, so it is clear why most wires are made of copper. Aluminum, although it has a higher resistivity, is much less dense than copper; it is thus preferable to copper in some situations, such as transmission lines, because its resistance for the same weight is less than that for copper.

TABLE 18–1   Resistivity and Temperature Coefficients (at 20°C)

Material	Resistivity, ρ ($\Omega \cdot$m)	Temperature Coefficient, α (C°)$^{-1}$
Conductors		
Silver	1.59×10^{-8}	0.0061
Copper	1.68×10^{-8}	0.0068
Gold	2.44×10^{-8}	0.0034
Aluminum	2.65×10^{-8}	0.00429
Tungsten	$5.6 \ \times 10^{-8}$	0.0045
Iron	9.71×10^{-8}	0.00651
Platinum	$10.6 \ \times 10^{-8}$	0.003927
Mercury	$98 \quad \times 10^{-8}$	0.0009
Nichrome (alloy of Ni, Fe, Cr)	$100 \quad \times 10^{-8}$	0.0004
Semiconductors[†]		
Carbon (graphite)	$(3–60) \times 10^{-5}$	-0.0005
Germanium	$(1–500) \times 10^{-3}$	-0.05
Silicon	$0.1 - 60$	-0.07
Insulators		
Glass	$10^{9} - 10^{12}$	
Hard rubber	$10^{13} - 10^{15}$	

[†]Values depend strongly on presence of even slight amounts of impurities.

EXAMPLE 18–4   Speaker wires. Suppose you want to connect your stereo to remote speakers (Fig. 18–13). (*a*) If each wire must be 20 m long, what diameter copper wire should you use to keep the resistance less than 0.10 Ω per wire? (*b*) If the current to each speaker is 4.0 A, what is the voltage drop across each wire?

SOLUTION   (*a*) We solve Eq. 18–3 for the area A and use Table 18–1:

$$A = \rho \frac{L}{R} = \frac{(1.68 \times 10^{-8}\ \Omega \cdot \text{m})(20\ \text{m})}{(0.10\ \Omega)} = 3.4 \times 10^{-6}\ \text{m}^2.$$

The cross-sectional area A of a circular wire is related to its diameter d by $A = \pi d^2 / 4$. The diameter must then be at least

$$d = \sqrt{\frac{4A}{\pi}} = 2.1 \times 10^{-3}\ \text{m} = 2.1\ \text{mm}.$$

(*b*) From Ohm's law,

$$V = IR = (4.0\ \text{A})(0.10\ \Omega) = 0.40\ \text{V}.$$

FIGURE 18–13   Example 18–4.

CONCEPTUAL EXAMPLE 18–5   Stretching changes resistance. A wire of resistance R is stretched uniformly until it is twice its original length. What happens to its resistance?

RESPONSE   If the length L doubles then the cross-sectional area A halves, so that the volume ($V = AL$) of the wire remains the same. From Eq. 18–3 we see that the resistance would increase by a factor of four ($2/\frac{1}{2} = 4$).

The resistivity of a material depends somewhat on temperature. In general, the resistance of metals increases with temperature. This is not surprising, for at higher temperatures, the atoms are moving more rapidly and are arranged in a less orderly fashion; so they might be expected to interfere more with the flow of electrons. If the temperature change is not too great, the resistivity of metals usually increases nearly linearly with temperature. That is,

$$\rho_T = \rho_0[1 + \alpha(T - T_0)] \tag{18–4}$$

where ρ_0 is the resistivity at some reference temperature T_0 (such as 0°C or 20°C), ρ_T is the resistivity at a temperature T and α is the *temperature coefficient of resistivity*. Values for α are given in Table 18–1. Note that the temperature coefficient for semiconductors can be negative. Why? It seems that at higher temperatures, some of the electrons that are not normally free in a semiconductor become free and can contribute to the current. Thus, the resistance of a semiconductor can decrease with an increase in temperature, although this is not always the case.

EXAMPLE 18–6 **Resistance thermometer.** The variation in electrical resistance with temperature can be used to make precise temperature measurements. Platinum is usually used since it is relatively free from corrosive effects and has a high melting point. To be specific, suppose at 20°C the resistance of a platinum resistance thermometer is 164.2 Ω. When placed in a particular solution, the resistance is 187.4 Ω. What is the temperature of this solution?

SOLUTION Since the resistance R is directly proportional to the resistivity ρ, we can combine Eq. 18–3 with Eq. 18–4:

$$R = R_0[1 + \alpha(T - T_0)].$$

Here $R_0 = \rho_0 L/A$ is the resistance of the wire at $T_0 = 20$°C. We solve this equation for T and find

$$T = T_0 + \frac{R - R_0}{\alpha R_0} = 20°C + \frac{187.4\ \Omega - 164.2\ \Omega}{(3.927 \times 10^{-3}(\text{C}°)^{-1})(164.2\ \Omega)} = 56.0°C.$$

More convenient for some applications is a *thermistor*, which consists of a metal oxide or semiconductor whose resistance also varies in a repeatable way with temperature. They can be made quite small and respond very quickly to temperature changes. They have another advantage in that they can be used at very high or low temperatures where gas or liquid thermometers would be useless.

The value of α in Eq. 18–4 depends on temperature, so it is important to check the temperature range of validity of any value (say, in a handbook of physical data). If the temperature range is wide, Eq. 18–4 is not adequate and terms proportional to the square and cube of the temperature are needed, but they are generally very small except when $T - T_0$ is large.

*18–5 Superconductivity

At very low temperatures, near absolute zero, the resistivity of certain metals and their compounds or alloys becomes zero as measured by the highest-precision techniques. Materials in such a state are said to be **superconducting**. This phenomenon was first observed by H. K. Onnes (1853–1926) in 1911 when he cooled mercury below 4.2 K ($-269°C$). He found that at this temperature, the resistance of mercury suddenly dropped to zero. In general, superconductors become superconducting only below a certain *transition temperature* T_C, which is usually within a few degrees of absolute zero. Current in a ring-shaped superconducting material has been observed to flow for years in the absence of a potential difference, with no measurable decrease. Measurements show that the resistivity ρ of superconductors is less than $4 \times 10^{-25}\,\Omega\cdot m$, which is over 10^{16} times smaller than that for copper, and is considered to be zero in practice.

Much research has been done on superconductivity in recent years to try to understand why it occurs,[†] and to find materials that superconduct at more reasonable temperatures to reduce the cost and inconvenience of refrigeration at the required very low temperature. Before 1986 the highest temperature at which a material was found to superconduct was 23 K, and this required liquid helium to keep the material cold. In 1987, a compound of yttrium, barium, copper, and oxygen was developed that can be superconducting at 90 K. Since this is above the temperature of liquid nitrogen, 77 K, boiling liquid nitrogen is sufficiently cold to keep the material superconducting. This was an important breakthrough since liquid nitrogen is much more easily and cheaply obtained than is the liquid helium needed for previous superconductors. Since then, superconductivity at temperatures in the vicinity of 160K have been reported, though in fragile compounds.

High-temperature superconductors

Considerable research is being done to develop high-T_C superconductors as wires that can carry currents strong enough to be practical. Most applications today use a bismuth-strontium-calcium-copper oxide, known (for short) as BSCCO. Applications of superconductivity that once seemed like science fiction may become real. A major use today of superconductors is for carrying the current in electromagnets (we shall see in Chapter 20 that electric currents produce magnetic fields). In large magnets, a great amount of energy is needed just to maintain the current, and this energy is wasted as heat.

The use of higher-temperature superconductors would enable motors and generators to be much smaller (perhaps $\frac{1}{10}$ the size today) if superconductors can be developed that can sustain large currents. Transmission of power over long distances using superconductors would similarly require much smaller and less costly transmission lines.

Superconductors could make electric cars more practical, make computers far faster than they are today, and have great potential in devices to store energy for use at peak demand. Superconductors are already being studied for use in high-speed ground transportation: the magnetic fields produced by superconducting magnets would be used to "levitate" vehicles over tracks so there is essentially no friction (Fig. 18–14). The levitation arises from the repulsive force between the magnet (on the train) and the eddy currents (Section 21–6) produced in the track below.

FIGURE 18–14 An experimental train in Japan, supported by the magnetic field produced by current in coils beneath the tracks (in the red containers).

[†]The first successful superconductivity theory was published by Bardeen, Cooper, and Schrieffer (the BCS theory) in 1957. It is based on quantum theory and cannot be explained on the basis of classical physics.

18–6 Electric Power

Electric energy is useful to us because it can be easily transformed into other forms of energy. Motors, whose operation we will examine in Chapter 20, transform electric energy into mechanical work.

In other devices such as electric heaters, stoves, toasters, and hair dryers, electric energy is transformed into thermal energy in a wire resistance known as a "heating element." And in an ordinary lightbulb, the tiny wire filament (Fig. 18–15) becomes so hot it glows; only a few percent of the energy is transformed into visible light, and the rest, over 90 percent, into thermal energy. Lightbulb filaments and heating elements in household appliances have resistances typically of a few ohms to a few hundred ohms.

Electric energy is transformed into thermal energy or light in such devices because the current is usually rather large, and there are many collisions between the moving electrons and the atoms of the wire. In each collision, part of the electron's kinetic energy is transferred to the atom with which it collides. As a result, the kinetic energy of the atoms increases and hence the temperature of the wire element increases (see Section 13–11). The increased thermal energy (internal energy) can be transferred as heat by conduction and convection to the air in a heater or to food in a pan, by radiation to toast in a toaster, or radiated as light.

To find the power transformed by an electric device recall that the energy transformed when a charge Q moves through a potential difference V is QV (Eq. 17–1). Then the power P, which is the rate energy is transformed, is

$$P = \text{power} = \frac{\text{energy transformed}}{\text{time}} = \frac{QV}{t}.$$

The charge that flows per second, Q/t, is simply the electric current I. Thus we have

Electric power (general)

$$P = IV. \qquad \text{(18–5)}$$

This general relation gives us the power transformed by any device, where I is the current passing through it and V is the potential difference across it. It also gives the power delivered by a source such as a battery. The SI unit of electric power is the same as for any kind of power, the **watt** (1 W = 1 J/s).

The rate of energy transformation in a resistance R can be written, using Ohm's law ($V = IR$), in two other ways:

$$P = IV \qquad \text{(18–6a)}$$

Electric power (in resistance R)

$$= I(IR) = I^2R \qquad \text{(18–6b)}$$

$$= \left(\frac{V}{R}\right)V = \frac{V^2}{R}. \qquad \text{(18–6c)}$$

Equations 18–6b and c apply only to resistors, whereas Eq. 18–6a, $P = IV$, applies to any device.

FIGURE 18–15
Incandescent lightbulb.

Filament

Insulator

EXAMPLE 18–7 **Headlights.** Calculate the resistance of a 40-W automobile headlight designed for 12 V (Fig. 18–16).

SOLUTION Since we are given $P = 40$ W and $V = 12$ V, we can use Eq. 18–6c and solve for R:

$$R = \frac{V^2}{P} = \frac{(12 \text{ V})^2}{(40 \text{ W})} = 3.6 \ \Omega.$$

This is the resistance when the bulb is burning brightly at 40 W. When the bulb is cold, the resistance is much lower, as we saw in Eq. 18–4. (Since the current is high when the resistance is low, lightbulbs burn out most often when first turned on.)

40-W Headlight

FIGURE 18–16 Example 18–7.

It is energy, not power, that you pay for on your electric bill. Since power is the *rate* energy is transformed, the total energy used by any device is simply its power consumption multiplied by the time it is on. If the power is in watts and the time is in seconds, the energy will be in joules since $1 \text{ W} = 1 \text{ J/s}$. Electric companies usually specify the energy with a much larger unit, the **kilowatt-hour** (kWh). One kWh = $(1000 \text{ W})(3600 \text{ s}) = 3.60 \times 10^6$ J.

You pay for energy

Kilowatt-hour (unit of energy)

EXAMPLE 18–8 **Electric heater.** An electric heater draws 15.0 A on a 120-V line. How much power does it use and how much does it cost per month (30 days) if it operates 3.0 h per day and the electric company charges 10.5 cents per kWh? (For simplicity, assume the current flows steadily in one direction.)

SOLUTION The power is

$$P = IV = (15.0 \text{ A})(120 \text{ V}) = 1800 \text{ W}$$

or 1.80 kW. To operate it for $(3.0 \text{ h/d})(30 \text{ d}) = 90$ h would cost $(1.80 \text{ kW})(90 \text{ h})(\$0.105) = \$17.$

EXAMPLE 18–9 **ESTIMATE** **Lightning bolt.** Lightning is a spectacular example of electric current in a natural phenomenon. There is much variability to lightning bolts, but a typical event can transfer 10^9 J of energy across a potential difference of perhaps 5×10^7 V during a time interval of about 0.2 s. Use this information to estimate the total amount of charge transferred, the current, and the average power over the 0.2 s.

SOLUTION From Eq. 17–1, energy = QV, so

$$Q \approx \frac{10^9 \text{ J}}{5 \times 10^7 \text{ V}} = 20 \text{ C}.$$

The current over the 0.2 s is about

$$I = \frac{Q}{t} \approx \frac{20 \text{ C}}{0.2 \text{ sec}} = 100 \text{ A}.$$

[Since most lightning bolts consist of several stages, it is possible that individual parts could carry currents much higher than this.] The average power delivered is

$$\overline{P} = \frac{\text{energy}}{\text{time}} = \frac{10^9 \text{ J}}{0.2 \text{ sec}} = 5 \times 10^9 \text{ W} = 5 \text{ GW}.$$

We can also use Eq. 18–6a:

$$P = IV = (100 \text{ A})(5 \times 10^7 \text{ V}) = 5 \text{ GW}.$$

539

18–7 Power in Household Circuits

PHYSICS APPLIED

Safety—wires getting hot

The electric wires that carry electricity to lights and other electric appliances have some resistance, although usually it is quite small. Nonetheless, if the current is large enough, the wires will heat up and produce thermal energy at a rate equal to I^2R, where R is the wire's resistance. One possible hazard is that the current-carrying wires in the wall of a building may become so hot as to start a fire. Thicker wires, of course, have less resistance (see Eq. 18–3) and thus can carry more current without becoming too hot. When a wire carries more current than is safe, it is said to be "overloaded." A building should, of course, be designed with wiring heavy enough for any expected load. To prevent overloading, fuses or circuit breakers are installed in circuits. They are basically switches (Fig. 18–17) that open the circuit when the current exceeds some particular value. A 20-A fuse or circuit breaker, for example, opens when the current passing through it exceeds 20 A. If a circuit repeatedly burns out a fuse or opens a circuit breaker, there are two possibilities: there may be too many devices drawing current in that circuit; or there is a fault somewhere, such as a "short." A short, or "short circuit," means that two wires have crossed (perhaps because the insulation has worn down) so the path of the current is shortened. The resistance of the circuit is then very small, so the current will be very large. Short circuits, of course, should be remedied immediately.

Fuses, circuit breakers, and shorts

Short circuit

FIGURE 18–17 (a) Fuses. When the current exceeds a certain value, the metallic ribbon melts and the circuit opens. Then the fuse must be replaced. (b) A circuit breaker. Electric current passes through a bimetallic strip. When the current is great enough (i.e., too great to be safe), the bimetallic strip is heated and bends so far to the left that the notch in the spring-loaded metal strip drops down over the end of the bimetallic strip; (c) the circuit then opens at the contact points (one is attached to the metal strip) and the outside switch is also flipped. As soon as the bimetallic strip cools down, it can be reset using the outside switch.

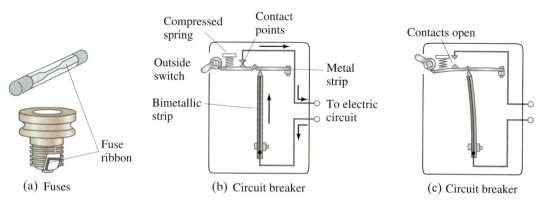

(a) Fuses   (b) Circuit breaker   (c) Circuit breaker

Household circuits are designed with the various devices connected so that each receives the standard voltage (usually 120 V in the United States) from the electric company (Fig. 18–18). (Circuits with the devices arranged as in Fig. 18–18 are called *parallel circuits*, as we will discuss more fully in the next chapter.) When a fuse blows or circuit breaker opens, the total current being drawn on that circuit should be checked.

EXAMPLE 18–10 Will a fuse blow? Determine the total current drawn by all the devices in the circuit of Fig. 18–18.

SOLUTION The circuit in Fig. 18–18 draws the following currents: the lightbulb draws $I = P/V = 100\,\text{W}/120\,\text{V} = 0.8\,\text{A}$; the heater draws $1800\,\text{W}/120\,\text{V} = 15.0\,\text{A}$; the stereo draws a maximum of $350\,\text{W}/120\,\text{V} = 2.9\,\text{A}$; and the hair dryer draws $1200\,\text{W}/120\,\text{V} = 10.0\,\text{A}$. The total current drawn, if all devices are used at the same time, is

$$0.8\,\text{A} + 15.0\,\text{A} + 2.9\,\text{A} + 10.0\,\text{A} = 28.7\,\text{A}.$$

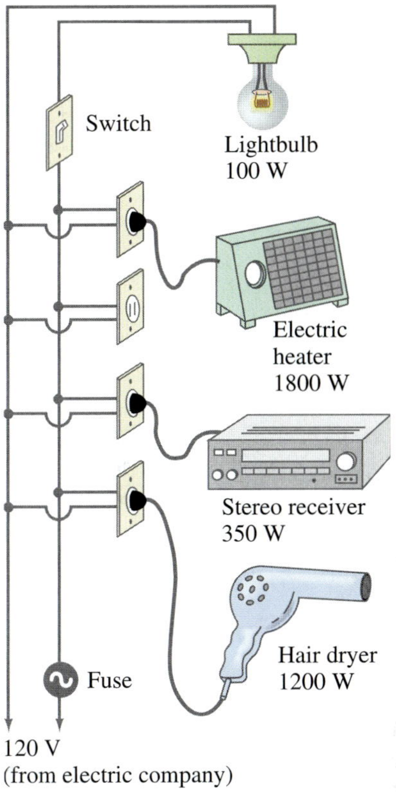

If the circuit in Fig. 18–18 has a 20-A fuse, it would be expected to blow, and we hope it will, to prevent overloaded wires from getting hot enough to start a fire. Something will have to be turned off to get this circuit below 20 A. (Houses and apartments usually have several circuits, each with its own fuse or circuit breaker; try moving one of the devices to another circuit.) If the circuit is designed for a 30-A fuse, it shouldn't blow, so if it does, a short may be the problem. (The most likely place is in the cord of one of the devices.) Proper fuse size is selected according to the wire used to supply the current; a properly rated fuse should *never* be re-placed by a higher-rated one. A fuse blowing or a circuit breaker opening is acting like a switch, making an "open circuit." By an open circuit, we mean that there is no longer a complete conducting path, so no current can flow; it is as if $R = \infty$ (infinity).

In electric circuits, heat dissipation by resistors must be considered. The physical size of a resistor is a rough indicator of the maximum per-missible power it can dissipate $(= I^2R)$ without appreciable rise in temper-ature. Common values are $\frac{1}{4}$ W, $\frac{1}{2}$ W, and 1 W; the higher the wattage, the larger the physical size.

FIGURE 18–18 Connection of household appliances.

18–8 Alternating Current

When a battery is connected to a circuit, the current flows steadily in one direction. This is called a **direct current**, or **DC**. Electric generators at elec-tric power plants, however, produce **alternating current**, or **AC**.[†] An alter-

DC and AC

[†]Although it is redundant, we sometimes say "ac current" and "ac voltage," which really mean "alternating current" and "alternating voltage."

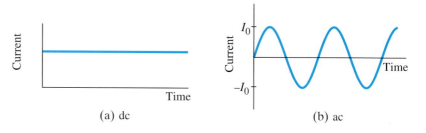

FIGURE 18–19 (a) Direct current. (b) Alternating current.

(a) dc

(b) ac

nating current reverses direction many times per second and is usually sinusoidal, as shown in Fig. 18–19. The electrons in a wire first move in one direction and then in the other. The current supplied to homes and businesses by electric companies is ac throughout virtually the entire world.

The voltage produced by an ac electric generator is sinusoidal, as we shall see in Chapter 21. The current it produces is thus sinusoidal (Fig. 18–19b). We can write the voltage as a function of time as

$$V = V_0 \sin 2\pi f t.$$

The potential V oscillates between $+V_0$ and $-V_0$. V_0 is referred to as the **peak voltage**. The frequency f is the number of complete oscillations made per second. In most areas of the United States and Canada, f is 60 Hz (the unit "hertz," as we saw in Chapter 11, means cycles per second). In some countries, 50 Hz is used.

From Ohm's law, if a voltage V exists across a resistance R, then the current I is

$$I = \frac{V}{R} = \frac{V_0}{R} \sin 2\pi f t = I_0 \sin 2\pi f t. \tag{18–7}$$

The quantity $I_0 = V_0/R$ is the **peak current**. The current is considered positive when the electrons flow in one direction and negative when they flow in the opposite direction. It is clear from Fig. 18–19b that an alternating current is as often positive as it is negative. Thus, the average current is zero. This does not mean, however, that no power is needed or that no heat is produced in a resistor. Electrons do move back and forth, and do produce heat. Indeed, the power delivered to a resistance R at any instant is

$$P = I^2 R = I_0^2 R \sin^2 2\pi f t.$$

Because the current is squared, we see that the power is always positive, Fig. 18–20. The quantity $\sin^2 2\pi f t$ varies between 0 and 1; and it is not too difficult to show that its average value is $\frac{1}{2}$ as can be seen graphically in the figure. Thus, the *average power* developed, $\overline{P}$, is

$$\overline{P} = \tfrac{1}{2} I_0^2 R.$$

Since power can also be written $P = V^2/R = (V_0^2/R) \sin^2 2\pi f t$, we also have that the average power is

$$\overline{P} = \tfrac{1}{2} \frac{V_0^2}{R}.$$

The average or mean value of the *square* of the current or voltage is thus what is important for calculating average power: $\overline{I^2} = \tfrac{1}{2} I_0^2$ and $\overline{V^2} = \tfrac{1}{2} V_0^2$. The

FIGURE 18–20 Power delivered to a resistor in an ac circuit.

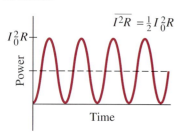

square root of each of these is the **rms** (root-mean-square) value of the current or voltage:

$$I_{rms} = \sqrt{\overline{I^2}} = \frac{I_0}{\sqrt{2}} = 0.707I_0, \tag{18-8a}$$

rms current

$$V_{rms} = \sqrt{\overline{V^2}} = \frac{V_0}{\sqrt{2}} = 0.707V_0. \tag{18-8b}$$

rms voltage

The rms values of V and I are sometimes called the "effective values." They are useful because they can be substituted directly into the power formulas, Eqs. 18–6, to get the average power:

$$\overline{P} = \tfrac{1}{2}I_0^2R = I_{rms}^2R \tag{18-9a}$$

$$\overline{P} = \tfrac{1}{2}\frac{V_0^2}{R} = \frac{V_{rms}^2}{R}. \tag{18-9b}$$

Thus, a direct current whose values of I and V equal the rms values of I and V for an alternating current will produce the same power. Hence it is usually the rms value of current that is specified or measured. For example, in the United States and Canada, standard line voltage[†] is 120 V ac. The 120 V is V_{rms}; the peak voltage V_0 is

$$V_0 = \sqrt{2}V_{rms} = 170 \text{ V}.$$

In most of Europe the rms voltage is 240 V, so the peak voltage is 340 V.

EXAMPLE 18–11 **Hair dryer.** (*a*) Calculate the resistance and the peak current in a 1000-W hair dryer (Fig. 18–21) connected to a 120-V line. (*b*) What happens if it is connected to a 240-V line in Britain?

SOLUTION (*a*) We can apply Eq. 18–6a using rms values. Then the rms current is

$$I_{rms} = \frac{\overline{P}}{V_{rms}} = \frac{1000 \text{ W}}{120 \text{ V}} = 8.33 \text{ A}.$$

Thus $I_0 = \sqrt{2}I_{rms} = 11.8$ A. From Ohm's law the resistance is

$$R = \frac{V_{rms}}{I_{rms}} = \frac{120 \text{ V}}{8.33 \text{ A}} = 14.4 \ \Omega.$$

The resistance could equally well be calculated using peak values: $R = V_0/I_0 = 170 \text{ V}/11.8 \text{ A} = 14.4 \ \Omega$.
(*b*) When connected to a 240-V line, more current would flow and the resistance would change with the increased temperature (Section 18–4). But let us make an estimate based on the same 14.4 Ω resistance. The average power delivered would be

$$\overline{P} = \frac{V_{rms}^2}{R} = \frac{(240 \text{ V})^2}{(14.4 \ \Omega)} = 4000 \text{ W}.$$

This is four times the dryer's power rating and would undoubtedly melt the heating element or the wire coils of the motor. Be sure your hair dryer (or electric shaver) has a 120/240 V switch before traveling too far from home, or carry a transformer (Section 21–7) for travelers.

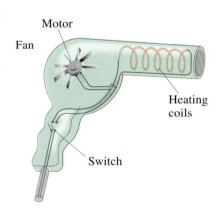

FIGURE 18–21 A hair dryer. Most of the current goes through the heating coils, a pure resistance; a small part goes to the motor to turn the fan. Example 18–11.

[†]The line voltage can vary, depending on the total load; the frequency of 60 Hz, however, remains extremely steady.

EXAMPLE 18–12 **Stereo power.** Each channel of a stereo receiver is capable of an average power output of 100 W into an 8-Ω loudspeaker (see Fig. 18–13). What is the rms voltage and rms current fed to the speaker (a) at the maximum power of 100 W, and (b) at 1.0 W?

SOLUTION We assume that the loudspeaker can be treated as a simple resistance (not quite true—see Chapter 21) with $R = 8.0\ \Omega$. (a) From Eq. 18–9, with $\overline{P} = 100$ W

$$V_{rms} = \sqrt{\overline{P}R} = \sqrt{(100\ \text{W})(8.0\ \Omega)} = 28\ \text{V},$$

and

$$I_{rms} = \sqrt{\frac{\overline{P}}{R}} = \sqrt{\frac{100\ \text{W}}{8.0\ \Omega}} = 3.5\ \text{A},$$

or

$$I_{rms} = \frac{V_{rms}}{R} = \frac{28\ \text{V}}{8.0\ \Omega} = 3.5\ \text{A}.$$

(b) At $\overline{P} = 1.0$ W,

$$V_{rms} = \sqrt{(1.0\ \text{W})(8.0\ \Omega)} = 2.8\ \text{V}$$

$$I_{rms} = \frac{2.8\ \text{V}}{8.0\ \Omega} = 0.35\ \text{A}.$$

This section has given a brief introduction to the simpler aspects of alternating currents. We will discuss ac circuits in more detail in Chapter 21. In Chapter 19 we will deal with the details of dc circuits only.

* 18–9 Microscopic View of Electric Current

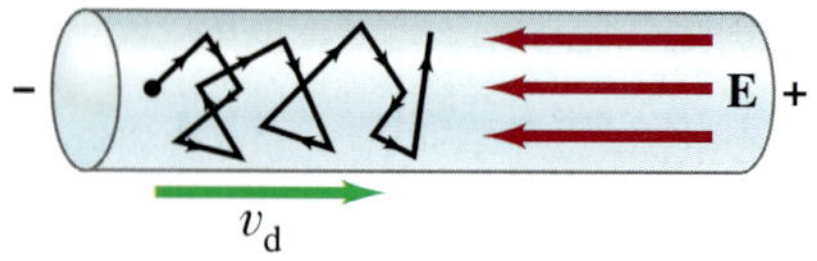

It can be useful to analyze a simple model of electric current at the microscopic level of atoms and electrons. In a conducting wire, for example, we can imagine the free electrons as moving about randomly at high speeds, bouncing off the atoms of the wire (somewhat like the molecules of a gas—Sections 13–10 and ff). When an electric field exists in the wire, Fig. 18–22, the electrons feel a force and initially begin to accelerate. But they soon reach a more or less steady average speed (due to collisions with atoms in the wire), known as their **drift speed**, v_d. The drift speed is normally very much smaller than the electrons' average random speed.

We can relate v_d to the macroscopic current I in the wire. In a time Δt, the electrons will travel a distance $l = v_d\ \Delta t$ on average. Suppose the wire has cross-sectional area A. Then in time Δt, electrons in a volume $V = Al = Av_d\ \Delta t$ will pass through the cross section A of wire, as shown in Fig. 18–23. If there are n free electrons (each of charge e) per unit volume, $(n = N/V)$ then the total charge ΔQ that passes through the area A in a time Δt is

$$\Delta Q = (\text{no. of charges, } N) \times (\text{charge per particle})$$
$$= (nV)(e) = (nAv_d\ \Delta t)(e).$$

The current I in the wire is thus

$$I = \frac{\Delta Q}{\Delta t} = neAv_d. \tag{18–10}$$

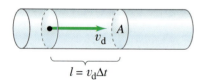

FIGURE 18–22 Electric field **E** in a wire gives electrons in random motion a drift velocity v_d.

Drift speed

FIGURE 18–23 Electrons in the volume Al will all pass through the cross-section indicated in a time Δt where $l = v_d\ \Delta t$.

Current (microscopic variables)

EXAMPLE 18–13 **Electron speeds in a wire.** A copper wire, 3.2 mm in diameter, carries a 5.0-A current. Determine (*a*) the drift speed of the free electrons, and (*b*) the rms speed of electrons assuming they behave like an ideal gas at 20°C. Assume that one electron per Cu atom is free to move[†] (the others remain bound to the atom).

SOLUTION (*a*) The cross-sectional area of the wire is

$$A = \pi r^2 = (3.14)(1.60 \times 10^{-3} \text{ m})^2 = 8.0 \times 10^{-6} \text{ m}^2.$$

Since we assume there is one free electron per atom, the density of free electrons, n, is the same as the density of Cu atoms. The atomic mass of Cu is 63.5 u (see Periodic Table inside the back cover), so 63.5 g of Cu contains one mole or 6.02×10^{23} free electrons. The density of copper (Table 10–1) is $\rho = 8.9 \times 10^3 \text{ kg/m}^3$, where $\rho = m/V$. So the number of free electrons per unit volume is

$$n = \frac{N}{V} = \frac{N}{m/\rho} = \frac{N(1 \text{ mole})}{m(1 \text{ mole})}\rho$$

$$n = \left(\frac{6.02 \times 10^{23} \text{ electrons}}{63.5 \times 10^{-3} \text{ kg}}\right)(8.9 \times 10^3 \text{ kg/m}^3) = 8.4 \times 10^{28} \text{ m}^{-3}.$$

Then, by Eq. 18–10, the drift speed is

$$v_d = \frac{I}{neA} = \frac{5.0 \text{ A}}{(8.4 \times 10^{28} \text{ m}^{-3})(1.6 \times 10^{-19} \text{ C})(8.0 \times 10^{-6} \text{ m}^2)}$$

$$= 4.7 \times 10^{-5} \text{ m/s},$$

which is only about 0.05 mm/s.

(*b*) If we model the free electrons as an ideal gas (a rough approximation), we use Eq. 13–9 to estimate the random rms speed of an electron as it darts around:

$$v_{\text{rms}} = \sqrt{\frac{3kT}{m}} = \sqrt{\frac{3(1.38 \times 10^{-23} \text{ J/K})(293 \text{ K})}{9.11 \times 10^{-31} \text{ kg}}} = 1.2 \times 10^5 \text{ m/s}.$$

Thus we see that the drift speed (average speed in the direction of the current) is very much less than the rms thermal speed of the electrons (by a factor of about 10^9). [*Note*: The result in (*b*) is an underestimate. Quantum theory calculations, and experiments, give the rms speed in copper to be about 1.6×10^6 m/s.]

The drift velocity of electrons in a wire is clearly very slow, only about 0.05 mm/s for the example above, which means it takes an electron 20×10^3 s, or $5\frac{1}{2}$ h, to travel only 1 m. This is not, of course, how fast "electricity travels": when you flip a light switch, the light—even if many meters away—goes on nearly instantaneously, for electric fields travel essentially at the speed of light (3×10^8 m/s). We can think of electrons in a wire as being like a pipe full of water: when a little water enters one end of the pipe, almost immediately some water issues forth at the other end.

Electricity's "speed"

[†]The number of free electrons in many cases can be determined by measuring the drift speed using the Hall effect (Section 20–11)—that is, the reverse of what we are calculating here.

An interesting example of the flow of electric charge is our remarkable and complex nervous system, which provides us with the means for being aware of the world, for communication within the body, and for controlling the body's muscles. Although the detailed functioning of the nervous system is still not well understood, we do have a reasonably good understanding of how messages are transmitted within the nervous system: they are electrical signals passing along the basic element of the nervous system, the *neuron*.

Neurons are living cells of unusual shape (Fig. 18–24). Attached to the main cell body are several small appendages known as *dendrites* and a long tail called the *axon*. Signals are received by the dendrites and are propagated along the axon. When a signal reaches the nerve endings, it is transmitted to the next nerve or to a muscle at a connection called a *synapse*. (Some neurons have separate cells, called Schwann cells, wrapped around their axons; they form a layered sheath called a myelin sheath and help to insulate neurons from one another.) Neurons serve in three capacities. "Sensory neurons" carry messages from the eyes, ears, skin, and other organs to the central nervous system, which consists of the brain and spinal cord. "Motor neurons" carry signals from the central nervous system to particular muscles and can signal them to contract. These two types of neuron make up the "peripheral system" as distinguished from the central nervous system. The third type of neuron is the interneuron, which transmits signals between neurons. Interneurons are in the brain and spinal column, and often are connected in an incredibly complex array.

A neuron, before transmitting an electrical signal, is in the so-called "resting state." Like nearly all living cells, neurons have a net positive charge on the outer surface of the cell membrane and a negative charge on the inner surface. This was already mentioned in Section 17–11 with regard to heart muscles and the ECG. This difference in charge, or "dipole layer," means that a potential difference exists across the cell membrane. When a neuron is not transmitting a signal, this "resting potential," which is normally stated as

$$V_{\text{inside}} - V_{\text{outside}},$$

is typically $-60\,\text{mV}$ to $-90\,\text{mV}$, depending on the type of organism. The most common ions in a cell are K^+, Na^+, and Cl^-. There are large differences in the concentrations of these ions inside and outside a cell, as indicated by the typical values given in Table 18–2. Other ions are also present, so the fluids both inside and outside the axon are electrically neutral. Because of the differences in concentration, there is a tendency for ions to diffuse across the membrane (recall diffusion, Section 13–15). However, in the resting state the cell membrane prevents any net flow of Na^+ (through a mechanism of "active pumping" of Na^+ out of the cell). But it does allow the flow of Cl^- ions and less so of K^+ ions, and it is these

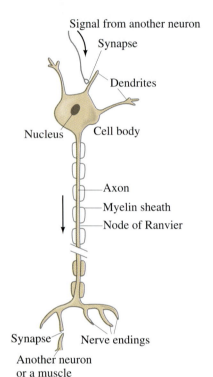

FIGURE 18–24   A neuron.

TABLE 18–2
Concentrations of Ions Inside and Outside a Typical Axon

	Concentration inside axon (mol/m³)	Concentration outside axon (mol/m³)
K^+	140	5
Na^+	15	140
Cl^-	9	125

two ions that produce the dipole charge layer on the membrane. Because there is a greater concentration of K^+ inside the cell than outside, more K^+ ions tend to diffuse outward across the membrane than diffuse inward. A K^+ ion that passes through the membrane becomes attached to the outer surface of the membrane, and leaves behind an equal negative charge that lies on the inner surface of the membrane (Fig. 18–25). The fluids themselves remain neutral. Indeed, what keeps the ions on the membrane is their attraction for each other across the membrane. Independent of this process, Cl^- ions tend to diffuse *into* the cell since their concentration outside is higher. Both K^+ and Cl^- diffusion tends to charge the interior surface of the membrane negatively and the outside positively. As charge accumulates on the membrane surface, it becomes increasingly difficult for more ions to diffuse: K^+ ions trying to move outward, for example, are repelled by the positive charge already there. Equilibrium is reached when the tendency to diffuse because of the concentration difference is just balanced by the electrical potential difference across the membrane. The greater the concentration difference, the greater the potential difference across the membrane, which, as mentioned above, is in the range -60 mV to -90 mV.

The most important aspect of a neuron is not that it has a resting potential (most cells do), but rather that it can respond to a stimulus and conduct an electrical signal along its length. A nerve can be stimulated in various ways. The stimulus could be thermal (when you touch a hot stove) or chemical (as in taste buds); it could be pressure (as on the skin or at the eardrum), or light (as in the eye); or it could be the electric stimulus of a signal coming from the brain or another neuron. In the laboratory, the stimulus is usually electrical and is applied by a tiny probe at some point on the neuron. If the stimulus exceeds some threshold, a voltage pulse will travel down the axon. This voltage pulse can be detected at a point on the axon using a voltmeter or an oscilloscope connected as in Fig. 18–26. This voltage pulse has the shape shown in Fig. 18–27, and is called an *action potential*. As can be seen, the potential increases from a resting potential of about -70 mV and becomes a positive 30 mV or 40 mV. The action potential lasts for about 1 ms and travels down an axon with a speed of 30 m/s to 150 m/s.

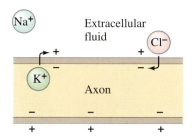

FIGURE 18–25 How a dipole layer of charge forms on a cell membrane.

FIGURE 18–26 Measuring the potential difference between the inside and outside of a nerve.

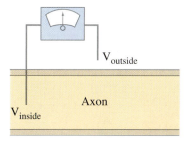

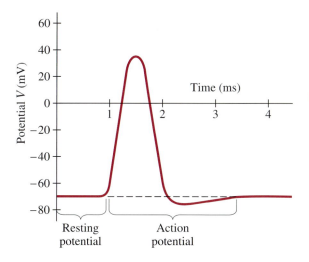

FIGURE 18–27 Action potential.

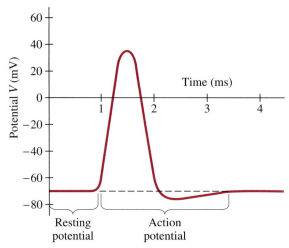

FIGURE 18–27
Action potential.

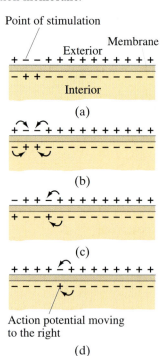

FIGURE 18–28 Propagation of an action potential along an axon membrane.

Point of stimulation

(a)

(b)

(c)

Action potential moving to the right

(d)

But what causes the action potential? Apparently, the cell membrane has the ability to alter its permeability properties. At the point where the stimulation occurs, the membrane suddenly becomes much more permeable to Na^+ than to K^+ and Cl^- ions. Thus, Na^+ ions rush into the cell and the inner surface of the wall becomes positively charged, and the potential difference quickly swings positive ($\approx +30\ mV$ in Fig. 18–27). But then the membrane suddenly returns to its original characteristics: it becomes impermeable to Na^+ and in fact pumps out Na^+ ions. The diffusion of Cl^- and K^+ ions again predominates and the original resting potential is restored ($-70\ mV$ in Fig. 18–27).

What causes the action potential to travel along the axon? The action potential occurs at the point of stimulation, as shown in Fig. 18–28a. The membrane momentarily is positive on the inside and negative on the outside at this point. Nearby charges are attracted toward this region, as shown in Fig. 18–28b. The potential in these adjacent regions then drops, causing an action potential there. Thus, as the membrane returns to normal at the original point, nearby it experiences an action potential, so the action potential moves down the axon (Figs. 18–28c and d).

You may wonder if the number of ions that pass through the membrane would significantly alter the concentrations. The answer is no; and we can show why by treating the axon as a capacitor in the following Example.

EXAMPLE 18–14 ESTIMATE Capacitance of an axon. (a) Do an order-of-magnitude estimate for the capacitance of an axon 10 cm long of radius 10 μm. The thickness of the membrane is about $10^{-8}\ m$ and the dielectric constant is about 3. (b) By what factor does the concentration of Na^+ ions in the cell change as a result of one action potential?

SOLUTION (a) The membrane of an axon resembles a cylindrically shaped parallel-plate capacitor, with opposite charges on each side. The separation of the "plates" is the thickness of the membrane, $d \approx 10^{-8}\ m$. The area A is the area of a cylinder of radius r and length l:

$$A = 2\pi rl \approx (6.28)(10^{-5}\ m)(0.1\ m) \approx 6 \times 10^{-6}\ m^2.$$

From Eq. 17–7, we have

$$C = K\epsilon_0 \frac{A}{d} \approx (3)(8.85 \times 10^{-12}\ C^2/N\cdot m^2)\frac{6 \times 10^{-6}\ m^2}{10^{-8}\ m} \approx 10^{-8}\ F.$$

(*b*) Since the voltage changes from $-70\,\mathrm{mV}$ to about $+30\,\mathrm{mV}$, the total change is about $100\,\mathrm{mV}$. The amount of charge that moves is then

$$Q = CV \approx (10^{-8}\,\mathrm{F})(0.1\,\mathrm{V}) = 10^{-9}\,\mathrm{C}.$$

Each ion carries a charge $e = 1.6 \times 10^{-19}\,\mathrm{C}$, so the number of ions that flow per action potential is $Q/e = (10^{-9}\,\mathrm{C})/(1.6 \times 10^{-19}\,\mathrm{C}) \approx 10^{10}$. The volume of our cylindrical axon is

$$V = \pi r^2 l \approx (3)(10^{-5}\,\mathrm{m})^2(0.1\,\mathrm{m}) = 3 \times 10^{-11}\,\mathrm{m}^3,$$

and the concentration of Na^+ ions inside the cell (Table 18–2) is $15\,\mathrm{mol/m^3} = 15 \times 6.02 \times 10^{23}\,\mathrm{ions/m^3} \approx 10^{25}\,\mathrm{ions/m^3}$. Thus, the cell contains $(10^{25}\,\mathrm{ions/m^3}) \times (3 \times 10^{-11}\,\mathrm{m}^3) \approx 3 \times 10^{14}\,\mathrm{Na}^+$ ions. One action potential, then, will change the concentration of Na^+ ions by at most $10^{10}/(3 \times 10^{14}) = \frac{1}{3} \times 10^{-4}$ or 1 part in 30,000. This tiny change would not be measurable.

Thus, even 1000 action potentials will not alter the concentration significantly. The sodium pump does not, therefore, have to remove Na^+ ions quickly after an action potential, but can operate slowly over time to maintain a relatively constant concentration.

The propagation of a nerve pulse as described here applies to an unmyelinated axon. Myelinated axons, on the other hand, are insulated from the extracellular fluid by the myelin sheath except at the nodes of Ranvier (see Fig. 18–24). An action potential cannot be regenerated where there is a myelin sheath. Once such a neuron is stimulated, the pulse will still travel along the membrane, but there is resistance and the pulse becomes smaller as it moves down the axon. Nonetheless, the weakened signal can still stimulate a full-fledged action potential when it reaches a node of Ranvier. Thus, the signal is repeatedly amplified at these points. Compare this to an unmyelinated neuron, in which the signal is continually amplified by repeated action potentials all along its length. This naturally requires much more energy. Development of myelinated neurons was a significant evolutionary step, for it meant reliable transmission of nerve pulses with less energy expended. And the pulses travel more quickly, since ordinary conduction is faster than the repeated production of action potentials, whose speed depends on the flow of ions across the membrane.

SUMMARY

An electric battery serves as a source of potential difference by transforming chemical energy into electric energy. A simple battery consists of two electrodes made of different metals immersed in a solution or paste known as an electrolyte.

Electric current, I, refers to the rate of flow of electric charge and is measured in **amperes** (A): 1 A equals a flow of 1 C/s past a given point.

The direction of **conventional current** flow is that of positive charge. In a wire, it is actually negatively charged electrons that move, so they flow in a direction opposite to the direction of the conventional current. Positive conventional current always flows from a high potential to a low potential.

Ohm's law states that the current in a good conductor is proportional to the potential difference applied to its two ends. The proportionality constant is called the **resistance** R of the material, so

$$V = IR.$$

The unit of resistance is the **ohm** (Ω), where $1\,\Omega = 1\,\mathrm{V/A}$. See Table 18–3.

TABLE 18–3 Summary of Units

Current	$1\,\mathrm{A} = 1\,\mathrm{C/s}$
Potential difference	$1\,\mathrm{V} = 1\,\mathrm{J/C}$
Power	$1\,\mathrm{W} = 1\,\mathrm{J/s}$
Resistance	$1\,\Omega = 1\,\mathrm{V/A}$

The resistance R of a wire is inversely proportional to its cross-sectional area A, and directly proportional to its length L and to a property of the material called its resistivity: $R = \rho L/A$. The **resistivity**, ρ, increases with temperature for metals, but for semiconductors it may decrease.

At very low temperatures certain materials become **superconducting**, which means their electrical resistance becomes zero.

The rate at which energy is transformed in a resistance R from electric to other forms of energy (such as heat and light) is equal to the product of current and voltage. That is, the **power** transformed, measured in watts, is given by

$$P = IV$$

and for resistors can be written with the help of Ohm's law as

$$P = I^2R = \frac{V^2}{R}.$$

The total electric energy transformed in any device equals the product of power and the time during which the device is operated. In SI units, energy is given in joules ($1\,\text{J} = 1\,\text{W·s}$), but electric companies use a larger unit, the **kilowatt-hour** ($1\,\text{kWh} = 3.6 \times 10^6\,\text{J}$).

Electric current can be **direct current (dc)**, in which the current is steady in one direction; or it can be **alternating current (ac)**, in which the current reverses direction at a particular frequency, typically $60\,\text{Hz}$. Alternating currents are often sinusoidal in time, $I = I_0 \sin 2\pi ft$, and are produced by an alternating voltage.

The rms values of sinusoidally alternating currents and voltages are given by

$$I_{\text{rms}} = \frac{I_0}{\sqrt{2}} \quad \text{and} \quad V_{\text{rms}} = \frac{V_0}{\sqrt{2}},$$

respectively, where I_0 and V_0 are the peak values. The power relationship, $P = IV = I^2R = V^2/R$, is valid for the average power in alternating currents when the rms values of V and I are used.

QUESTIONS

1. Car batteries can be rated in ampere-hours (A·h). What quantity is being rated?

2. When an electric cell is connected to a circuit, electrons flow away from the negative terminal in the circuit. But within the cell, electrons flow *to* the negative terminal. Explain.

3. Develop an analogy between blood circulation and an electrical circuit. Discuss what plays the role of the heart for the electric case, and so on.

4. Design a circuit in which two different switches of the type shown in Fig. 18–29 can be used to operate the same lightbulb from opposite sides of a room.

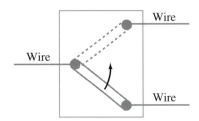

FIGURE 18–29   Question 4.

5. Can a copper wire and an aluminum wire of the same length have the same resistance? Explain.

6. If a rectangular solid made of carbon has sides a, $2a$, $3a$, how would you connect the wires from a battery so as to obtain (*a*) the least resistance, (*b*) the greatest resistance?

7. In a car, one terminal of the battery is said to be connected to "ground." Since it is not really connected to the ground, what is meant by this expression?

8. The equation $P = V^2/R$ indicates that the power dissipated in a resistor decreases if the resistance is increased, whereas the equation $P = I^2R$ implies the opposite. Is there a contradiction here? Explain.

9. What happens when a lightbulb burns out?

10. Explain why lightbulbs almost always burn out just as they are turned on and not after they have been on for some time.

11. Which draws more current, a 100-W lightbulb or a 75-W bulb? Which has the higher resistance?

12. Electric power is transferred over large distances at very high voltages. Explain how the high voltage reduces power losses in the transmission lines.

13. Why is it dangerous to replace a 15-A fuse that blows repeatedly with a 25-A fuse?

14. Electric lights operated on low-frequency ac (say, 10 Hz) flicker noticeably. Why?

15. Some lamps might have batteries connected in either of the two arrangements shown in Fig. 18–30. What would be the advantages of each scheme?

16. Driven by ac power, the same electrons pass back and forth through your reading lamp over and over again. Explain why the light stays lit instead of going out after the first pass of electrons.

17. The heating element in a toaster is made of Nichrome wire. Immediately after the toaster is turned on, is the current (I_{rms}) in the wire increasing, decreasing, or staying constant? Explain.

18. Is current used up in a resistor?

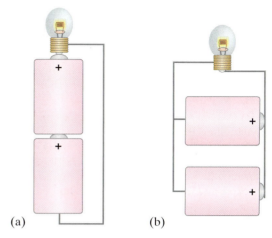

(a)   (b)

FIGURE 18–30   Question 15.

PROBLEMS

SECTIONS 18–2 AND 18–3

1. (I) A service station charges a battery using a current of 5.7 A for 7.0 h. How much charge passes through the battery?

2. (I) A current of 1.00 A flows in a wire. How many electrons are flowing past any point in the wire per second? The charge on one electron is 1.60×10^{-19} C.

3. (I) What is the current in amperes if 1000 Na$^+$ ions were to flow across a cell membrane in 6.5 μs? The charge on the sodium is the same as on an electron, but positive.

4. (I) What voltage will produce 0.25 A of current through a 3000-Ω resistor?

5. (I) What is the resistance of a toaster if 110 V produces a current of 3.1 A?

6. (II) An electrical device draws 5.50 A at 110 V. (*a*) If the voltage drops by 10 percent, what will be the current, assuming nothing else changes? (*b*) If the resistance of the device were reduced by 10 percent, what current would be drawn at 110 V?

7. (II) A 9.0-V battery is connected to a bulb whose resistance is 1.6 Ω. How many electrons leave the battery per minute?

8. (II) A hair dryer draws 9.0 A when plugged into a 120-V line. (*a*) What is its resistance? (*b*) How much charge passes through it in 15 min?

9. If a 12-V battery pushes a current of 0.50 A through a resistor, what is its resistance, and how many joules of energy does the battery lose in a minute?

10. (II) A bird stands on an electric transmission line carrying 2500 A (Fig. 18–31). The line has $2.5 \times 10^{-5}\ \Omega$ resistance per meter and the bird's feet are 4.0 cm apart. What voltage does the bird feel?

FIGURE 18–31   Problem 10.

SECTION 18–4

11. (I) What is the resistance of a 3.0-m length of copper wire 1.5 mm in diameter?

12. (I) What is the diameter of a 1.00-m length of tungsten wire whose resistance is 0.22 Ω?

13. (II) Compare the resistance of 10.0 m of aluminum wire 2.0 mm in diameter with 15.0 m of copper filament wire 2.5 mm in diameter.

14. (II) Can a 2.5-mm-diameter copper wire have the same resistance as a tungsten wire of the same length? Give numerical details.

15. (II) A certain copper wire has a resistance of 10.0 Ω. At what point along the length must the wire be cut so that the resistance of one piece is 7.0 times the resistance of the other? Calculate the resistance of each piece.

16. (II) How much would you have to raise the temperature of a copper wire (originally at 20°C) to increase its resistance by 20 percent?

17. (II) A length of aluminum wire is connected to a precision 10.00-V power supply, and a current of 0.4212 A is precisely measured at 20.0°C. The wire is placed in a new environment of unknown temperature where the measured current is 0.3618 A. What is the unknown temperature?

18. (II) Estimate at what temperature copper will have the same resistivity as tungsten does at 20°C.

19. (II) A 100-W lightbulb has a resistance of about 12 Ω when cold (20°C) and 140 Ω when on (hot). Estimate the temperature of the filament when "on" assuming an average temperature coefficient of resistivity $\alpha = 0.0060$ (C°)$^{-1}$.

20. (II) A rectangular solid made of carbon has sides lying along the x, y, and z axes, whose lengths are 1.0 cm, 2.0 cm, and 4.0 cm, respectively (Fig. 18–32). Determine the resistance for current that flows through the solid in (a) the x direction, (b) the y direction, and (c) the z direction. Assume the resistivity is $\rho = 3.0 \times 10^{-5}$ Ω·m.

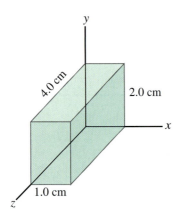

FIGURE 18–32   Problem 20.

21. (II) 10.0 m of wire consists of 5.0 m of copper followed by 5.0 m of aluminum of equal diameter (both 1.0 mm). A voltage difference of 80 V is placed across the composite wire. (a) What is the total resistance of the wire? (b) What is the current flow through the wire? (c) What are the voltages across the aluminum part and across the copper part?

22. (II) For some applications, it is important that the value of a resistance not change with temperature. For example, suppose you made a 4.70-kΩ resistor from a carbon resistor and a Nichrome wire-wound resistor connected together so the total resistance is the sum of their separate resistances. What value should each of these resistors have (at 0°C) so that the combination is temperature independent?

SECTIONS 18–6 AND 18–7

23. (I) The element of an electric oven is designed to produce 3.3 kW of heat when connected to a 240-V source. What must be the resistance of the element?

24. (I) What is the maximum power consumption of a 9.0-V portable cassette player that draws a maximum of 350 mA of current?

25. (I) What is the maximum voltage that can be applied to a 2.7-kΩ resistor rated at $\frac{1}{4}$ watt?

26. (I) A hair dryer has two settings: 600 W and 1200 W. (a) At which setting do you expect the resistance to be higher? After making a guess, determine the resistance at (b) the lower setting, (c) the higher setting.

27. (I) (a) What is the resistance and current through a 60-W lightbulb if it is connected to its proper source voltage of 120 V? (b) Repeat for a 440-W bulb.

28. (I) You buy a 60-W lightbulb in Europe, where electricity is delivered to homes at 240 V. If you use the lightbulb at home in the United States, how bright will it be relative to 60-W 120-V bulbs? Estimate how much power it will consume.

29. (II) How many kWh of energy does a 550-W toaster use in the morning if it is in operation for a total of 10 min? At a cost of 12 cents/kWh, how much would this add to your monthly electric energy bill if you made toast four mornings per week?

30. (II) At $0.110 per kWh, what does it cost to leave a 40-W porch light on day and night for a year?

31. (II) An ordinary flashlight uses 2 D-cell batteries (1.5 V each) and draws 350 mA when turned on. (a) Calculate the resistance of the bulb and the power dissipated. (b) By what factor would the power increase if 4 D-cells in series were used with the same bulb? (Neglect heating effects of the filament.) Why shouldn't you try this?

32. (II) What is the total amount of energy stored in a 12-V, 90-A·h car battery when it is fully charged?

33. (II) A transistor to be used in a circuit is rated for a maximum current of 25 mA if operated at 9.0 V. (a) What is the maximum power output of the transistor, and (b) what would be the limiting current if the voltage applied were actually only 7.0 V?

34. (II) How many 100-W lightbulbs, connected to 120 V as in Fig. 18–18, can be used without blowing a 2.5-A fuse?

35. (II) What is the efficiency of a 0.50-hp electric motor that draws 4.4 A from a 120-V line?

36. (II) A power station delivers 520 kW of power to a factory through wires of total resistance of 3.0 Ω. How much less power is wasted if the electricity is delivered at 50,000 V rather than 12,000 V?

37. (II) A 2200-W oven is hooked to a 240-V source. (a) What is the resistance of the oven? (b) How long will it take to boil 100 mL of water assuming 80 percent efficiency? (c) How much will this cost at 10 cents/kWh?

38. (III) The current in an electromagnet connected to a 240-V line is 14.5 A. At what rate must cooling water pass over the coils if the water temperature is to rise by no more than 7.50 C°?

39. (III) A small immersion heater can be used in a car to heat a cup of water for coffee. If the heater can heat 150 mL of water from 5°C to 95°C in 5.0 min, approximately how much current does it draw from the 12-V battery and what is its resistance? Assume the manufacturer's claim of 60 percent efficiency.

SECTION 18–8

40. (I) Calculate the peak current in a 2.2-kΩ resistor connected to a 120-V rms ac source.

41. (I) An ac voltage, whose peak value is 180 V, is across a 330-Ω resistor. What is the value of the rms and peak currents in the resistor?

42. (I) What is the resistance of your house when (a) everything electrical is turned off, and (b) there is a lone 75-W lightbulb burning on the porch?

43. (II) The peak value of an alternating current passing through a 1500-W device is 4.0 A. What is the rms voltage across it?

44. (II) Calculate the peak voltage and peak current through an 1800-W arc welder connected to a 450-V ac line.

45. (II) What is the maximum instantaneous power dissipated by, and maximum current passing through, a 3.0-hp pump connected to a 240-V ac power source?

46. (II) A heater coil connected to a 240-V ac line has a resistance of 34 Ω. What is the average power used? What are the maximum and minimum values of the instantaneous power?

*SECTION 18–9

* 47. (II) A 0.55-mm-diameter copper wire carries a tiny current of 2.5 μA. What is the electron drift speed in the wire?

* 48. (II) A 5.00-m length of 2.0-mm-diameter wire carries a 750-mA current when 22.0 mV is applied to its ends. If the drift velocity has been measured (by the Hall effect—Section 20–11) to be 1.7×10^{-5} m/s, determine (a) the resistance R of the wire, (b) the resistivity ρ, and (c) the number n of free electrons per unit volume.

* 49. (II) In a liquid solution, Na^+ ions in a concentration of 5.00 mol/m^3 move with a drift speed of 5.00×10^{-4} m/s, and SO_4^{2-} ions move with a speed of 2.00×10^{-4} m/s in the same direction, with the result of producing zero net current. What is the concentration of SO_4^{2-} ions?

* 50. (III) At a point high in the Earth's atmosphere, He^{2+} ions in a concentration of $2.8 \times 10^{12}/m^3$ are moving due north at a speed of 2.0×10^6 m/s. Also, an $8.0 \times 10^{11}/m^3$ concentration of O_2^- ions is moving due south at a speed of 7.2×10^6 m/s. Determine the magnitude and direction of the net current passing through unit area (A/m^2).

*SECTION 18–10

* 51. (I) What is the magnitude of the electric field across an axon membrane 1.0×10^{-8} m thick if the resting potential is -70 mV?

* 52. (II) A nerve is stimulated with an electric pulse. The action potential is detected at a point 3.40 cm down the axon 0.0052 s later. When the action potential is detected 7.20 cm from the point of stimulation, the time required is 0.0063 s. What is the speed of the electric pulse along the axon? (Why are two measurements needed instead of only one?)

* 53. (III) Estimate how much energy is required to transmit one action potential along the axon of Example 18–14. [Hint: One pulse is equivalent to charging and discharging the axon capacitance; see Section 17–9]. What minimum average power is required for 10^4 neurons transmitting 100 pulses per second?

* 54. (III) During the action potential, Na^+ ions move into the cell at a rate of about 3×10^{-7} mol/m$^2 \cdot$s. How much power must be produced by the active transport system to produce this flow against a $+30$-mV potential difference? Assume that the axon is 10 cm long and 20 μm in diameter.

GENERAL PROBLEMS

55. How many coulombs are there in 1.00 ampere-hour?

56. What is the average current drawn by a 1.0-hp 120-V motor?

57. A person accidentally leaves a car with the lights on. If each of the two front lights uses 40 W and each of the two rear lights 6 W, for a total of 92 W, how long will a fresh 12-V battery last if it is rated at 90 A·h? Assume the full 12 V appears across each bulb.

58. The heating element of a 110-V, 1500-W heater is 5.4 m long. If it is made of iron, what must its diameter be?

59. The *conductance G* of an object is defined as the reciprocal of the resistance R; that is, $G = 1/R$. The unit of conductance is a mho ($= $ohm^{-1}), which is also called the siemens (S). What is the conductance (in siemens) of an object that draws 700 mA of current at 3.0 V?

60. (a) A particular household uses a 1.8-kW heater 3.0 h/day ("on" time), four 100-W lightbulbs 6.0 h/day, a 3.0-kW electric stove element for a total of 1.4 h, and miscellaneous power amounting to 2.0 kWh/day. If electricity costs $.105 per kWh, what will be their monthly bill (30 d)? (b) How much coal (which produces 7000 kcal/kg) must be burned by a 35-percent-efficient power plant to provide the yearly needs of this household?

61. A length of wire is cut in half and the two lengths are wrapped together side by side to make a thicker wire. How does the resistance of this new combination compare to the resistance of the original wire?

62. A 1200-W hair dryer is designed for 117 V. (a) What will be the percentage change in power output if the voltage drops to 105 V? Assume no change in resistance. (b) How would the actual change in resistivity with temperature affect your answer?

63. The wiring in a house must be thick enough so it doesn't become so hot as to start a fire. What diameter must a copper wire be if it is to carry a maximum current of 35 A and produce no more than 1.6 W of heat per meter of length?

64. Suppose a current is given by the equation $I = 1.80 \sin 210t$, where I is in amperes, and t in seconds. (a) What is the frequency? (b) What is the rms value of the current? (c) If this is the current through a 42.0-Ω resistor, what is the equation that describes the voltage as a function of time?

65. In a "brownout" situation, the voltage supplied by the electric company falls. Assuming the percent drop is small, show that the power output of a given appliance falls by approximately twice that percent, assuming the resistance does not change. How much of a voltage drop does it take for a 60-W lightbulb to begin acting like a 50-W bulb?

66. A microwave oven running at 60 percent efficiency delivers 900 W of energy per second to the interior. Find: (a) the power drawn from the source; (b) the current drawn. Assume a source voltage of 120 V.

67. A 1.00-Ω wire is drawn out to 3.00 times its original length. What is its resistance now?

68. 220 V is applied to two different conductors made of the same material. One conductor is twice as long and twice as thick as the second. What is the ratio of the power transformed in the first relative to the second?

69. An electric heater is used to heat a room of volume 62 m³. Air is brought into the room at 5°C and is changed completely twice per hour. Heat loss through the walls amounts to approximately 850 kcal/h. If the air is to be maintained at 20°C, what minimum wattage must the heater have? (The specific heat of air is about 0.17 kcal/kg·C°.)

FIGURE 18–33 The EV-1 electric car. (Problem 70.)

70. The new EV-1 electric car (Fig. 18–33) makes use of storage batteries as its source of energy. Its mass is 1300 kg and it is powered by 26 batteries, each 12 V, 52 A·h. Assume that the car is driven on the level at an average speed of 40 km/h, and the average friction force is 240 N. Assume 100 percent efficiency and neglect energy used for acceleration. Note that no energy is consumed when the vehicle is stopped since the engine doesn't need to idle. (a) Determine the horsepower required. (b) After approximately how many kilometers must the batteries be recharged?

71. A 12.5-Ω resistor is made from a coil of copper wire whose total mass is 18.0 g. What is the diameter of the wire and how long is it?

72. A 100-W, 120-V lightbulb has a resistance of 12 Ω when cold (20°C) and 140 Ω when on (hot). Calculate its power consumption at (a) the instant it is turned on, and (b) after a few moments when it is hot.

73. A capacitor is often used in electronics to keep energy flowing even if there is a momentary loss of power from the electric company. What capacitance would be required for a 150-W television (plugged into a standard 120 V ac outlet) to protect it from a 0.10 s lapse in power?

* **74.** The Tevatron accelerator at Fermilab (Illinois) is designed to carry an 11 mA beam of protons traveling at very nearly the speed of light (3.0×10^8 m/s) around a ring 6300 m in circumference. How many protons are stored in the beam?

* **75.** How far does an average electron move along the wires of a 300-W toaster during an alternating current cycle? The power cord has copper wires of diameter 1.8 mm and is plugged into a standard ac outlet. [*Hint*: the maximum current in the cycle is related to the maximum drift velocity. The maximum velocity in an oscillation is related to the maximum displacement; see Section 11–3.]

Years ago when one bulb of a certain type of Christmas tree lights burned out, the whole string of lights went out. (Was this a series or a parallel circuit?) Today, if one bulb goes out, the rest stay lit. For other strings however (usually those with very tiny bulbs), if you *remove* a bulb (burned out or not), all the others go out. What kind of circuit could this be?

C H A P T E R
19

DC CIRCUITS

Electric circuits abound in today's world. They are basic parts of all electronic gear from radio and TV sets to computers and even automobiles. Scientific measurements, from physics to biology and medicine, make use of electric circuits. In Chapter 18, we discussed the basic principles of electric current. Now we will apply these principles to analyze dc circuits and to understand the operation of a number of useful instruments.[†]

When we draw a diagram for a circuit, we represent batteries, capacitors, and resistors by the symbols shown in Table 19–1. Wires whose resistance is negligible compared to other resistance in the circuit are drawn simply as straight lines. For the most part in this chapter, except in Section 19–7, we will be interested in circuits operating in their steady state—that is, we won't be looking at a circuit at the moment any change is made in it, such as when a battery or resistor is connected or disconnected, but rather a short time later when the currents have reached their steady values.

TABLE 19–1 Symbols for Circuit Elements

Symbol	Device
⊣⊢	Battery
⊣⊢	Capacitor
-⋁⋁-	Resistor
——	Wire with negligible resistance

[†]Ac circuits that contain only a voltage source and resistors can be analyzed like the dc circuits in this chapter. However, ac circuits that contain capacitors and other circuit elements are more complicated, and we discuss them in Chapter 21.

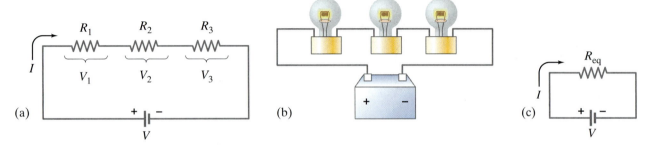

FIGURE 19–1 (a) Resistances connected in series: $R_{eq} = R_1 + R_2 + R_3$. (b) Resistances could be lightbulbs, or any other type of resistance. (c) Equivalent single resistance R_{eq} that draws the same current.

19–1 Resistors in Series and in Parallel

When two or more resistors are connected end to end as shown in Fig. 19–1, they are said to be connected in **series**. The resistors could be simple resistors as in Fig. 18–11, or they could be lightbulbs, heating elements, or other resistive devices. Any charge that passes through R_1 in Fig 19–1a will also pass through R_2 and then R_3. Hence the same current I passes through each resistor. (If it did not, this would imply that charge was accumulating at some point in the circuit, which does not happen in the steady state.) We let V represent the voltage across all three resistors. We assume all other resistance in the circuit can be ignored, and so V equals the voltage of the battery. We let V_1, V_2, and V_3 be the potential differences across each of the resistors, R_1, R_2, and R_3, respectively, as shown in Fig. 19–1a. By Ohm's law, $V_1 = IR_1$, $V_2 = IR_2$, and $V_3 = IR_3$. Because the resistors are connected end to end, energy conservation tells us that the total voltage V is equal to the sum of the voltages across each resistor:

Series circuit: voltages add; current the same in each R

$$V = V_1 + V_2 + V_3 = IR_1 + IR_2 + IR_3. \qquad \text{[series]} \quad \textbf{(19–1)}$$

[To see in more detail why this is true, note that an electric charge q passing through R loses potential energy by qV_1. In passing through R_2 and R_3, the PE decreases by qV_2 and qV_3, for a total ΔPE $= qV_1 + qV_2 + qV_2$; this sum must equal the energy given to q by the battery, qV, so that energy is conserved. Hence $qV = q(V_1 + V_2 + V_3)$, and so $V = V_1 + V_2 + V_3$, which is Eq. 19–1.]

Now let us determine the equivalent single resistance R_{eq} that would draw the same current as our combination (Fig. 19–1c). Such a single resistance R_{eq} would be related to V by

$$V = IR_{eq}.$$

We equate this expression with Eq. 19–1, $V = I(R_1 + R_2 + R_3)$, and find

Resistances in series

$$R_{eq} = R_1 + R_2 + R_3. \qquad \text{[series]} \quad \textbf{(19–2)}$$

This is, in fact, what we expect. When we put several resistances in series, the total resistance is the sum of the separate resistances. This applies to any number of resistances in series. Note that when you add more resistance to the circuit, the current will decrease. For example, if a 12-V battery is connected to a 4-Ω resistor, the current will be 3 A. But if the 12-V battery is connected to three 4-Ω resistors in series, the total resistance is 12 Ω and the current will be only 1 A.

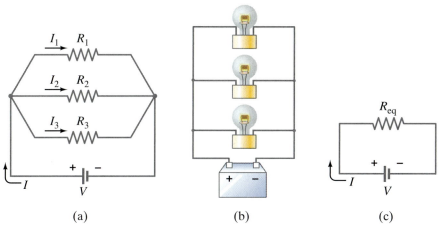

(a)   (b)   (c)

FIGURE 19–2
(a) Resistances connected in parallel:
$1/R_{eq} = 1/R_1 + 1/R_2 + 1/R_3$,
which could be (b) lightbulbs;
(c) shows the equivalent circuit with R_{eq} obtained from Eq. 19–3.

Another simple way to connect resistors is in **parallel**, so that the current from the source splits into separate branches, as shown in Fig. 19–2. The wiring in houses and buildings is arranged so all electric devices are in parallel, as we already saw in Fig. 18–18. With parallel wiring, if you disconnect one device (say R_1 in Fig. 19–2), the current to the others is not interrupted. But in a series circuit, if one device (say R_1 in Fig. 19–1) is disconnected, the current *is* stopped to all the others.

In a parallel circuit, Fig. 19–2a, the total current I that leaves the battery breaks into three branches. We let $I_1, I_2,$ and I_3 be the currents through each of the resistors, $R_1, R_2,$ and R_3, respectively. Because electric charge is conserved, the current flowing into a junction (where different wires or conductors meet) must equal the current flowing out of the junction. Thus, in Fig. 19–2a,

$$I = I_1 + I_2 + I_3. \qquad \text{[parallel]}$$

Parallel circuit: currents add; voltage the same across each R

When resistors are connected in parallel, each experiences the same voltage. (Indeed, any two points in a circuit connected by a wire of negligible resistance are at the same potential.) Hence the full voltage of the battery is applied to each resistor in Fig. 19–2a, so

$$I_1 = \frac{V}{R_1}, \qquad I_2 = \frac{V}{R_2}, \qquad \text{and} \qquad I_3 = \frac{V}{R_3}.$$

Let us now determine what single resistor R_{eq} (Fig. 19–2c) will draw the same current I as these three resistances in parallel. This equivalent resistance R_{eq} must satisfy

$$I = \frac{V}{R_{eq}}.$$

We now combine the equations above:

$$I = I_1 + I_2 + I_3,$$

$$\frac{V}{R_{eq}} = \frac{V}{R_1} + \frac{V}{R_2} + \frac{V}{R_3}.$$

When we divide out the V from each term, we have

$$\frac{1}{R_{eq}} = \frac{1}{R_1} + \frac{1}{R_2} + \frac{1}{R_3}. \qquad \text{[parallel]} \quad \textbf{(19–3)}$$

Resistances in parallel

For example, suppose you connect two 8-Ω loudspeakers to a single set of output terminals of your stereo amplifier or receiver. (Ignore the other

FIGURE 19–3 Water pipes in parallel—analogy to electric currents in parallel.

[*Optional paragraph*]

channel for a moment—our two speakers are both connected to the left channel, say.) The equivalent resistance will then be found from

$$\frac{1}{R} = \frac{1}{8\ \Omega} + \frac{1}{8\ \Omega} = \frac{2}{8\ \Omega} = \frac{1}{4\ \Omega}$$

and so $R = 4\ \Omega$. Thus the net resistance is *less* than that of each single resistance. This may at first seem surprising. But remember that when you put resistors in parallel, you are giving the current additional paths to follow. Hence the net resistance will be less.

[An analogy may help here. Consider two pipes taking in water near the top of a dam and releasing it below as shown in Fig. 19–3. The gravitational potential difference, proportional to the height h, is the same for both pipes, just as in the electrical case of parallel resistors. If both pipes are open, rather than only one, twice as much current will flow through. That is, with two equal pipes open, the net resistance to the flow of water will be reduced, by half. Note that if both pipes are closed, the dam offers infinite resistance to the flow of water. This corresponds in the electrical case to an open circuit—when no current flows—so the electrical resistance is infinite.]

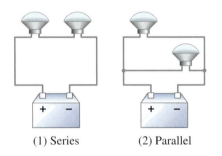

(1) Series (2) Parallel

FIGURE 19–4 Example 19–1.

CONCEPTUAL EXAMPLE 19–1 **Series or Parallel?** (*a*) The lightbulbs in Fig 19–4 are identical and have identical resistance R. Which configuration produces more light? (*b*) Which way do you think the headlights of a car are wired?

RESPONSE (*a*) The parallel combination has lower resistance ($= R/2$) than the series combination ($= 2R$). There will be more total current in configuration (2). The total power consumed, which is proportional to the light produced, is $P = IV$, so the greater current in (2) means more light.
(*b*) In parallel (2), because if one goes out, the other light can stay lit.

EXAMPLE 19–2 **Series and parallel resistors.** Two 100-Ω resistors are connected (*a*) in series, and (*b*) in parallel, to a 24.0-V battery. See Fig. 19–5. What is the current through each resistor and what is the equivalent resistance of each circuit?

FIGURE 19–5 Example 19–2.

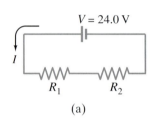

(a)

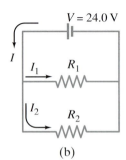

(b)

SOLUTION (*a*) Every charge, every moving electron, passes first through R_1 and then R_2. So the current I is the same in both resistors; and the potential difference across the battery, V, equals the total change in potential across the two resistors:

$$V = V_1 + V_2 = IR_1 + IR_2.$$

Hence

$$I = \frac{V}{R_1 + R_2} = \frac{24.0 \text{ V}}{100 \text{ } \Omega + 100 \text{ } \Omega} = 0.120 \text{ A}.$$

The equivalent resistance, using Eq. 19–2, is $R_{eq} = R_1 + R_2 = 200 \text{ } \Omega$. We could also get R_{eq} by thinking from the point of view of the battery: the total resistance R_{eq} must equal the battery voltage divided by the current it puts out:

$$R_{eq} = \frac{V}{I} = \frac{24.0 \text{ V}}{0.120 \text{ A}} = 200 \text{ } \Omega.$$

[Note that the voltage across R_1 is $V_1 = IR_1, = (0.120 \text{ A})(100 \text{ } \Omega) = 12.0 \text{ V}$, and across R_2 is $V_2 = IR_2 = 12.0 \text{ V}$, each being half of the battery voltage. A simple circuit like Fig. 19–5a is thus often called a simple "voltage divider."]
(*b*) Any given charge (or electron) can flow through only one or the other of the two resistors. Just as a river may break into two streams when going around an island, here too the total current I from the battery (Fig. 19–5b) equals the sum of the separate currents through the two resistors:

$$I = I_1 + I_2.$$

The potential difference across each resistor is the battery voltage $V = 24.0 \text{ V}$. Hence

$$I = I_1 + I_2 = \frac{V}{R_1} + \frac{V}{R_2} = \frac{24.0 \text{ V}}{100 \text{ } \Omega} + \frac{24.0 \text{ V}}{100 \text{ } \Omega} = 0.24 \text{ A} + 0.24 \text{ A} = 0.48 \text{ A}.$$

The equivalent resistance is

$$R_{eq} = \frac{V}{I} = \frac{24.0 \text{ V}}{0.48 \text{ A}} = 50 \text{ } \Omega.$$

We could also have obtained this result from Eq. 19–3:

$$\frac{1}{R_{eq}} = \frac{1}{100 \text{ } \Omega} + \frac{1}{100 \text{ } \Omega} = \frac{2}{100 \text{ } \Omega} = \frac{1}{50 \text{ } \Omega},$$

and so $R_{eq} = 50 \text{ } \Omega$.

Note that whenever a group of resistors is replaced by the equivalent resistance, current and voltage and power in the rest of the circuit are unaffected.

EXAMPLE 19–3 **Series with parallel in a circuit.** How much current flows from the battery shown in Fig. 19–6a?

SOLUTION The current I that flows out of the battery all passes through the 400-Ω resistor, but then it splits into I_1 and I_2 passing through the 500-Ω and 700-Ω resistors. The latter two are in parallel. This seems to be a complicated circuit. We look for simplicity, something that we already know how to treat. So let's start by finding the equivalent resistance, R_P, of the parallel resistors, 500-Ω and 700-Ω:

$$\frac{1}{R_P} = \frac{1}{500 \ \Omega} + \frac{1}{700 \ \Omega} = 0.0020 \ \Omega^{-1} + 0.0014 \ \Omega^{-1} = 0.0034 \ \Omega^{-1}.$$

This is $1/R_P$, so we must take the reciprocal to find R_P. (It is a common mistake to forget to take this reciprocal. Notice that the units of reciprocal ohms, Ω^{-1}, help to remind us of this.) Thus

$$R_P = \frac{1}{0.0034 \ \Omega^{-1}} = 290 \ \Omega.$$

This 290 Ω is the equivalent resistance of the two parallel resistors, and is in series with the 400-Ω resistor as shown in the equivalent circuit of Fig. 19–6b. To find the total equivalent resistance R_{eq}, we add the 400-Ω and 290-Ω resistances together, since they are in series, and find

$$R_{eq} = 400 \ \Omega + 290 \ \Omega = 690 \ \Omega.$$

The total current flowing from the battery is then

$$I = \frac{V}{R_{eq}} = \frac{12.0 \ \text{V}}{690 \ \Omega} = 0.017 \ \text{A} = 17 \ \text{mA}.$$

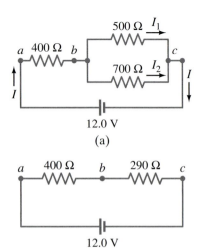

FIGURE 19–6 (a) Circuit for Examples 19–3 and 19–4. (b) Equivalent circuit, showing the equivalent resistance of 290 Ω for the two parallel resistors in (a).

EXAMPLE 19–4 **Current in one branch.** What is the current flowing through the 500-Ω resistor in Fig. 19–6a?

SOLUTION To solve this problem, we must find the voltage across the 500-Ω resistor, which is the voltage between points b and c in the diagram, and we call it V_{bc}. Once V_{bc} is known, we can apply Ohm's law to get the current. First we find the voltage across the 400-Ω resistor, V_{ab}, since we know that 17 mA passes through it; V_{ab} can be found using Ohm's law, $V = IR$:

$$V_{ab} = (0.017 \ \text{A})(400 \ \Omega) = 6.8 \ \text{V}.$$

Since the total voltage across the network of resistors is $V_{ac} = 12.0$ V, then V_{bc} must be 12.0 V − 6.8 V = 5.2 V. Then Ohm's law tells us that the current I_1 through the 500-Ω resistor is

$$I_1 = \frac{5.2 \ \text{V}}{500 \ \Omega} = 1.0 \times 10^{-2} \ \text{A} = 10 \ \text{mA}.$$

This is the answer we wanted. However, we can also calculate the current I_2 through the 700-Ω resistor since the voltage across it is also 5.2 V:

$$I_2 = \frac{5.2 \ \text{V}}{700 \ \Omega} = 7 \ \text{mA}.$$

Notice that when I_1 combines with I_2 to form the total current I (at point c in Fig. 19–6a), their sum is 10 mA + 7 mA = 17 mA. This is, of course, the total current as calculated in Example 19–3.

CONCEPTUAL EXAMPLE 19–5 Bulb brightness in a circuit. The circuit shown in Fig. 19–7 has three identical lightbulbs, each of resistance R. (*a*) When switch S is closed, how will the brightness of bulbs A and B compare with that of bulb C? (*b*) What happens when switch S is opened? Use a minimum of mathematics in your answers.

RESPONSE (*a*) With switch S closed, the current that passes through bulb C must then split into two equal parts when it reaches the junction leading to bulbs A and B. It splits into equal parts because the resistance of bulbs A and B are equal. Thus, bulbs A and B receive half the current, and will be less bright than bulb C.
(*b*) When the switch S is open, we have a simple one-loop series circuit, and we expect bulbs B and C to be equally bright. However, the equivalent resistance of this circuit ($= R + R$) is greater than that of the circuit with the switch closed. So when we open the switch, we increase the resistance and reduce the current leaving the battery. Thus, bulb C will dim when we open the switch. But bulb B gets more current when the switch is open (you may have to use some mathematics here), and so it will brighten somewhat when the switch is opened.

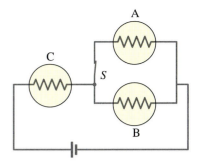

FIGURE 19–7 Example 19–5, three identical lightbulbs.

EXAMPLE 19–6 Resistor "ladder." Estimate the equivalent resistance of the "ladder" of equal 100-Ω resistors shown in Fig. 19–8a. In other words, what would an ohmmeter read if connected between points A and B?

SOLUTION It may seem that none of these resistors is in series or parallel. But there is a place to start: the three resistors on the far right are definitely in series with one another. Each has resistance $R(= 100\ \Omega)$, so those last three have a net resistance of $3R(= 300\ \Omega)$.
  Next we can see that this combination is in parallel with the next resistor to the left, as shown in the dashed box in Fig. 19–8b. The equivalent resistance of the resistors in the dashed box (b), call it R_{eq1}, is given by

$$\frac{1}{R_{eq1}} = \frac{1}{R} + \frac{1}{3R} = \frac{4}{3R}, \quad \text{so} \quad R_{eq1} = \frac{3R}{4}$$

which numerically is $300\ \Omega/4 = 75\ \Omega$. Next, this equivalent resistance of $3R/4$ is in series with the next two (Fig. 19–8c).
The resistors in the dashed box (c) are in series, and are equivalent to $2R + 3R/4 = 11R/4$ (which equals $1100\ \Omega/4 = 275\ \Omega$). Now this $11R/4$ is in parallel with the next step on the ladder (Fig. 19–8d).
The resistance in the dashed box (d) is equivalent to

$$\frac{1}{R_{eq2}} = \frac{1}{R} + \frac{4}{11R} = \frac{15}{11R}, \quad \text{so} \quad R_{eq2} = \frac{11R}{15}.$$

This in turn is in series with two more, yielding the final equivalent resistance of

$$R_{eq} = \frac{11R}{15} + R + R = \frac{41}{15}R.$$

We are given that $R = 100\ \Omega$, so $R_{eq} = 273\ \Omega$. Notice again the value of using algebra all along: we can have an answer that enables us to calculate the overall resistance regardless of the value of the individual resistor.

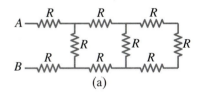

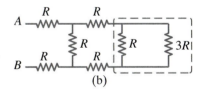

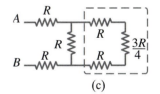

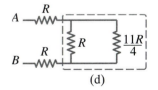

FIGURE 19–8 Example 19–6.

EMF and Terminal Voltage

emf

A device such as a battery or an electric generator that transforms one type of energy (chemical, mechanical, light, and so on) into electric energy is called a **seat** or **source** of **electromotive force** or of **emf**. (The term "electromotive force" is a misnomer since it does not refer to a "force" that is measured in newtons. Hence, to avoid confusion, we prefer to use the abbreviation, emf.) The potential difference between the terminals of such a source, when no current flows to an external circuit, is called the emf of the source. The symbol $\mathscr{E}$ is usually used for emf (don't confuse it with E for electric field).

Why battery voltage isn't constant

You may have noticed in your own experience that when a current is drawn from a battery, the voltage across its terminals drops below its rated emf. For example, if you start a car with the headlights on, you may notice the headlights dim. This happens because the starter draws a large current, and the battery voltage drops as a result. The voltage drop occurs because the chemical reactions in a battery cannot supply charge fast enough to maintain the full emf. For one thing, charge must flow (within the electrolyte) between the electrodes of the battery, and there is always some hindrance to completely free flow. Thus, a battery itself has some resistance, which is called its **internal resistance**; it is usually designated r. A real battery is then modeled as if it were a perfect emf $\mathscr{E}$ in series with a resistor r, as shown in Fig. 19–9. Since this resistance r is inside the battery, we can never separate it from the battery. The two points a and b in the diagram represent the two terminals of the battery. What we measure is the **terminal voltage** V_{ab}. When no current is drawn from the battery, the terminal voltage equals the emf, which is determined by the chemical reactions in the battery: $V_{ab} = \mathscr{E}$. However, when a current I flows from the battery, there is an internal drop in voltage equal to Ir. Thus the terminal voltage (the actual voltage delivered) is[†]

Terminal voltage

$$V_{ab} = \mathscr{E} - Ir.$$

For example, if a 12-V battery has an internal resistance of 0.1 Ω, then when 10 A flows from the battery, the terminal voltage is 12 V − (10 A)(0.1 Ω) = 11 V. The internal resistance of a battery is usually small. For example, an ordinary flashlight battery when fresh may have an internal resistance of perhaps 0.05 Ω. (However, as it ages and the electrolyte dries out, the internal resistance increases to many ohms.) Car batteries have even lower internal resistance.

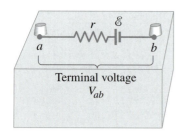

FIGURE 19–9 Diagram for an electric cell or battery.

[†]When a battery is being charged, a current is forced to pass through it (this happens in Fig. 19–14b), and we have to write

$$V_{ab} = \mathscr{E} + Ir.$$

See, for example, Problem 25 and Fig. 19–31.

EXAMPLE 19–7 **Analyzing a circuit.** A 9.0-V battery whose internal resistance r is 0.50 Ω is connected in the circuit shown in Fig. 19–10a. (a) How much current is drawn from the battery? (b) What is the terminal voltage of the battery? (c) What is the current in the 6.0-Ω resistor?

SOLUTION (a) First, we determine the equivalent resistance of the circuit. But where do we start? We note that the 4.0-Ω and 8.0-Ω resistors are in parallel, and so have an equivalent resistance R_{eq1} given by

$$\frac{1}{R_{eq1}} = \frac{1}{8.0\ \Omega} + \frac{1}{4.0\ \Omega} = \frac{3}{8.0\ \Omega};$$

so $R_{eq1} = 2.7\ \Omega$. This 2.7 Ω is in series with the 6.0-Ω resistor, as shown in Fig. 19–10b; so the net resistance of the lower arm of the circuit is

$$R_{eq2} = 6.0\ \Omega + 2.7\ \Omega = 8.7\ \Omega,$$

as shown in Fig. 19–10c. The equivalent resistance R_{eq3} of this 8.7-Ω and the 10.0-Ω resistances in parallel is given by

$$\frac{1}{R_{eq3}} = \frac{1}{10.0\ \Omega} + \frac{1}{8.7\ \Omega} = 0.21\ \Omega^{-1},$$

so $R_{eq3} = (1/0.21\ \Omega^{-1}) = 4.8\ \Omega$. This 4.8 Ω is in series with the 5.0-Ω resistor and the 0.50-Ω internal resistance of the battery (Fig. 19–10d), so the total equivalent resistance R_{eq} of the circuit is $R_{eq} = 4.8\ \Omega + 5.0\ \Omega + 0.50\ \Omega = 10.3\ \Omega$. Hence the current drawn is

$$I = \frac{\mathcal{E}}{R_{eq}} = \frac{9.0\ \text{V}}{10.3\ \Omega} = 0.87\ \text{A}.$$

(b) The terminal voltage of the battery is

$$V_{ab} = \mathcal{E} - Ir = 9.0\ \text{V} - (0.87\ \text{A})(0.50\ \Omega) = 8.6\ \text{V}.$$

(c) Now we can work back and get the current in the 6.0-Ω resistor. It must be the same as the current through the 8.7 Ω shown in Fig. 19–10c (why?). The voltage across the 8.7 Ω shown will be the emf of the battery minus the voltage drops across r and the 5.0-Ω resistor. Thus the current will be (call it I')

$$I' = \frac{9.0\ \text{V} - (0.87\ \text{A})(0.50\ \Omega + 5.0\ \Omega)}{8.7\ \Omega} = 0.48\ \text{A},$$

and this is the current through the 6.0-Ω resistor.

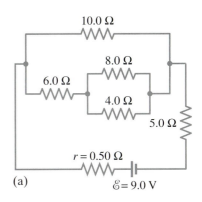

(a)

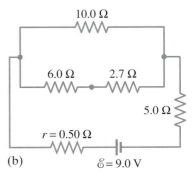

(b)

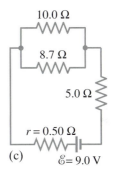

(c)

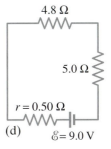

(d)

FIGURE 19–10 Circuit for Example 19–7, where r is the internal resistance of the battery.

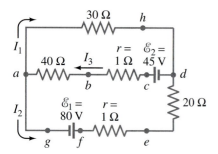

FIGURE 19–11 Currents can be calculated using Kirchhoff's rules.

19–3 Kirchhoff's Rules

In the last few examples we have been able to find the currents flowing in circuits by combining resistances in series and parallel, and using Ohm's law. This technique can be used for many circuits. However, we sometimes encounter a circuit that is too complicated for this analysis. For example, we cannot find the currents flowing in each part of the circuits shown in Fig. 19–11 simply by combining resistances as we did before.

To deal with such complicated circuits, we use Kirchhoff's rules, invented by G. R. Kirchhoff (1824–1887) in the mid-nineteenth century. There are two of them, and they are simply convenient applications of the laws of conservation of charge and energy. **Kirchhoff's first** or **junction rule** is based on the conservation of charge, and we already used it in deriving the rule for parallel resistors. It states that

Junction rule
(conservation of charge)

at any junction point, the sum of all currents entering the junction must equal the sum of all currents leaving the junction.

(That is, what goes in must come out.) For example, at the junction point a in Fig. 19–11, I_3 is entering whereas I_1 and I_2 are leaving. Thus Kirchhoff's junction rule states that $I_3 = I_1 + I_2$. We already saw an instance of this at the end of Example 19–4, where the currents passing through the 400-Ω and 700-Ω resistors were 10 mA and 7 mA, respectively, and they added at point c in Fig. 19–6a to give the outgoing current of 17 mA.

Kirchhoff's junction rule is based on the conservation of charge. Charges that enter a junction must also leave—none is lost or gained. **Kirchhoff's second** or **loop rule** is based on the conservation of energy. It states that

Loop rule
(conservation of energy)

the sum of the changes in potential around any closed path of a circuit must be zero.

To see why this should hold, consider the analogy of a roller coaster on its track. When it starts from the station, it has a particular potential energy. As it climbs the first hill, its potential energy increases and reaches a peak at the top. As it descends the other side, its potential energy decreases and reaches a local minimum at the bottom of the hill. As the roller coaster continues on its path, its potential energy goes through more changes. But when it arrives back at the starting point, it has exactly as much potential energy as it had when it started at this point. Another way of saying this is that there was as much uphill as there was downhill.

The same reasoning can be applied to an electric circuit. We will do the circuit of Fig. 19–11 shortly but first we consider the simple circuit in Fig. 19–12. We have chosen it to be the same as the equivalent circuit of Fig. 19–6b already discussed. The current flowing in this circuit is $I = (12.0\,\text{V})/(690\,\Omega) = 0.0174\,\text{A}$, as we calculated in Example 19–3, although now we calculate to three significant figures and assume all values are known to three significant figures. The positive side of the battery, point e in Fig. 19–12a, is at a high potential compared to point d at the negative side of the battery. That is, point e is like the top of a hill for a roller coaster. We can now follow the current around the circuit starting at any point we choose. Let us start at point e and follow a positive test charge completely around this circuit. As we go, we will note all changes in potential. When the test charge returns to point e, the potential there will be the same as when we started, so the total change in potential will be zero. It is helpful to plot the changes in voltage around the circuit, and we do this in Fig. 19–12b; point d is arbitrarily taken as zero. As our positive test charge goes from point e to point a, there is no change in potential since there is no source of potential nor any resistance. However, as the charge passes through the 400-Ω resistor to get to point b, there is a decrease in potential of $V = IR = (0.0174\,\text{A})(400\,\Omega) = 6.96\,\text{V}$. In effect, the positive test charge is flowing "downhill" since it is heading toward the negative terminal of the battery. This is indicated in the graph of Fig. 19–12b. The decrease in voltage between the two ends of a resistor ($= IR$) is called a **voltage drop**. Because this is a *decrease* in a voltage, we use a *negative* sign when applying Kirchhoff's loop rule; that is,

$$V_{ba} = V_b - V_a = -6.96\,\text{V}.$$

As the charge proceeds from b to c there is a further voltage drop of $(0.0174\,\text{A}) \times (290\,\Omega) = 5.04\,\text{V}$, and since this is a decrease, we write

$$V_{cb} = -5.04\,\text{V}.$$

There is no change in potential as our test charge moves from c to d. But when it moves from d, which is the negative or low potential side of the battery, to point e which is the positive terminal, the voltage *increases* by 12.0 V. That is,

$$V_{ed} = +12.0\,\text{V}.$$

The sum of all the changes in potential in going around the circuit of Fig. 19–12 is then

$$-6.96\,\text{V} - 5.04\,\text{V} + 12.0\,\text{V} = 0.$$

And this is exactly what Kirchhoff's loop rule said it would be.

We already studied the details of this circuit, but have studied it again to illustrate how the loop rule is applied. In the next Section, we will see how to use Kirchhoff's rules to determine the currents in more difficult circuits.

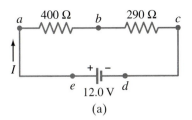

(a)

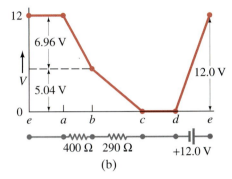

(b)

FIGURE 19–12 Changes in potential around the circuit in (a) are plotted in (b).

➡ **PROBLEM SOLVING**

Be consistent with signs

19–4 Solving Problems with Kirchhoff's Rules

When using Kirchhoff's rules, we will find it useful to designate the current in each separate branch of the given circuit by a different subscript, such as I_1, I_2, and I_3 in Fig. 19–13 (this is the same circuit as in Fig. 19–11). You do not have to know in advance in which direction these currents actually are moving. You make a guess and calculate the potentials around the circuit for that direction. If the current actually flows in the opposite direction, your answer will merely have a negative sign. This and other details of using Kirchhoff's rules will become clearer in the following Example.

➡ **PROBLEM SOLVING**

Choose current directions arbitrarily

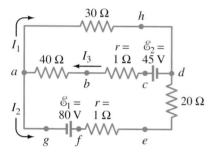

FIGURE 19–13 Currents can be calculated using Kirchhoff's rules. See Example 19–8.

EXAMPLE 19–8 **Using Kirchhoff's rules.** Calculate the currents I_1, I_2, and I_3 in each of the branches of the circuit in Fig. 19–13.

SOLUTION We choose the directions of the currents as shown in Fig. 19–13. Since (positive) current tends to move away from the positive terminal of a battery, we expect I_2 and I_3 to have the directions shown. It is hard to tell the direction of I_1 in advance, so we arbitrarily chose the direction indicated. We have three unknowns and therefore we need three equations. We first apply Kirchhoff's junction rule to the currents at point a, where I_3 enters and I_2 and I_1 leave:

$$I_3 = I_1 + I_2. \tag{a}$$

This same equation holds at point d, so we get no new information there. We now apply Kirchhoff's loop rule to two different closed loops. First we apply it to the loop $ahdcba$. From a to h we have a voltage drop $V_{ha} = -(I_1)(30\,\Omega)$. From h to d there is no change, but from d to c the potential increases by 45 V: that is, $V_{cd} = +45$ V. From c to a the voltage drops through the two resistances by an amount $V_{ac} = -(I_3)(40\,\Omega + 1\,\Omega)$. Thus we have $V_{ha} + V_{cd} + V_{ac} = 0$, or

$$-30I_1 - 41I_3 + 45 = 0, \tag{b}$$

where we have omitted the units. For our second loop, we take the complete circuit $ahdefga$. (We could have just as well taken $abcdefg$ instead.) Again we have $V_{ha} = -(I_1)(30\,\Omega)$, and $V_{dh} = 0$. But when we take our positive test charge from d to e, it actually is going uphill, against the flow of current—or at least against the *assumed* direction of the current, which is what counts in this calculation. Thus $V_{ed} = I_2(20\,\Omega)$ has a *positive* sign. Similarly, $V_{fe} = I_2(1\,\Omega)$. From f to g there is a decrease in potential of 80 V since we go from the high potential terminal of the battery to the low. Thus $V_{gf} = -80$ V. Finally, $V_{ag} = 0$, and the sum of the potentials around this loop is then

$$-30I_1 + 21I_2 - 80 = 0. \tag{c}$$

The physics is now done. The rest is algebra. We have three equations—labeled (a), (b), and (c)—in three unknowns. From Eq. (c) we have

$$I_2 = \frac{80 + 30I_1}{21} = 3.8 + 1.4I_1. \tag{d}$$

From Eq. (b) we have

$$I_3 = \frac{45 - 30I_1}{41} = 1.1 - 0.73I_1. \qquad\qquad (e)$$

We substitute these into Eq. (a) and solve for I_1:

$$I_1 = I_3 - I_2 = 1.1 - 0.73I_1 - 3.8 - 1.4I_1$$

$$I_1 = -2.7 - 2.1I_1,$$

so

$$3.1I_1 = -2.7$$

$$I_1 = -0.87 \text{ A}.$$

I_1 has magnitude 0.87 A. The negative sign indicates that its direction is actually opposite to that initially assumed and shown in Fig. 19–13. Note that the answer automatically comes out in amperes since all values were in volts and ohms. From Eq. (d) we have

$$I_2 = 3.8 + 1.4I_1 = 2.6 \text{ A},$$

and from Eq. (e)

$$I_3 = 1.1 - 0.73I_1 = 1.7 \text{ A}.$$

This completes the solution.

The unknowns in different situations are not necessarily currents. It might be that the currents are given and we have to solve for unknown resistance or voltage.

➡ **PROBLEM SOLVING** Kirchhoff's Rules

1. Label + and − for each battery. The long side of a battery symbol is +.

2. Label the current in each branch of the circuit with a symbol and an arrow (as in Fig. 19–13). The direction of the arrow can be chosen arbitrarily. If the current is actually in the opposite direction, it will come out with a minus sign in the solution.

3. Apply Kirchhoff's junction rule at each junction, and the loop rule for one or more loops. You will need as many independent equations as there are unknowns. You may write down more equations than this, but you will find that some of the equations will be redundant (that is, not be independent in the sense of providing new information). You may use $V = IR$ for each resistor, which sometimes will reduce the number of unknowns.

4. In applying the loop rule, follow each loop in one direction only (either clockwise or coun-

terclockwise—your choice). Pay careful attention to signs:

(a) For a resistor, the sign of the potential difference is negative if your chosen loop direction is the same as the chosen current direction through that resistor; the sign is positive if you are moving opposite to the chosen current direction.

(b) For a battery, the sign of the potential difference is positive if your loop direction moves from the negative terminal toward the positive; the sign will be negative if you are moving from the positive terminal toward the negative terminal.

5. Solve the equations algebraically for the unknowns. Be careful in manipulating equations not to err with signs. At the end, check your answers by plugging them into the original equations, or even by using any additional equations not used previously (either loop or junction rule equations).

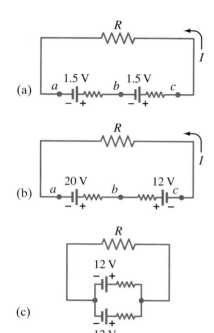

(a)

(b)

(c)

FIGURE 19–14 Batteries in series, (a) and (b), and in parallel, (c).

19–5 EMFs in Series and in Parallel; Charging a Battery

When two or more sources of emf, such as batteries, are arranged in series, the total voltage is the algebraic sum of their respective voltages. For example, if two 1.5-V flashlight batteries are connected as shown in Fig. 19–14a, the voltage V_{ca} across the light bulb, represented by the resistance R, is 3.0 V. (To be absolutely correct, we should also take into account the internal resistance of the batteries, but we assume it to be small.) On the other hand, when a 20-V and a 12-V battery are connected oppositely, as shown in Fig. 19–14b, the net voltage V_{ca} is 8 V. That is, a positive test charge moved from a to b gains in potential by 20 V, but when it passes from b to c it drops by 12 V. So the net change is 20 V − 12 V = 8 V. You might think that connecting batteries in reverse like this would be wasteful. And for most purposes that would be true. But such a reverse arrangement is precisely how a battery charger works. In Fig. 19–14b, the 20-V source is charging up the 12-V battery. Because of its greater voltage, the 20-V source is forcing charge back into the 12-V battery: electrons are being forced into its negative terminal and removed from its positive terminal. An automobile alternator keeps the car battery charged in the same way. A voltmeter placed across the terminals of a (12-V) car battery with the engine running fairly fast can tell you whether or not the alternator is charging the battery. If it is, the voltmeter reads 13 or 14 V. If the battery is not being charged, the voltage will drop below 12 V because the battery is discharging. Car batteries can be recharged, but other batteries may not be rechargeable, since the chemical reactions in many cannot be reversed. In such cases, the arrangement of Fig. 19–14b would simply waste energy.

Sources of emf can also be arranged in parallel, Fig. 19–14c. A parallel arrangement is not used to increase voltage, but rather to provide more energy when large currents are needed (such as for starting some diesel engines). Each of the cells in parallel has to produce only a fraction of the total current, so the loss due to internal resistance is less than for a single cell; and the batteries will go dead less quickly.

19–6 Circuits Containing Capacitors in Series and in Parallel

Just as resistors can be placed in series or in parallel in a circuit, so can capacitors (Chapter 17). We first consider a **parallel** connection as shown in Fig. 19–15a. If a battery of voltage V is connected to points a and b, this voltage exists across each of the capacitors. Each acquires a charge given by $Q_1 = C_1V$, $Q_2 = C_2V$, and $Q_3 = C_3V$. The total charge Q that must leave the battery is then

$$Q = Q_1 + Q_2 + Q_3 = C_1V + C_2V + C_3V.$$

A single equivalent capacitor that will hold the same charge Q at the same voltage V will have a capacitance C_{eq} given by

$$Q = C_{eq}V.$$

Combining the previous two equations, we have

$$C_{eq}V = C_1V + C_2V + C_3V,$$

or

$$C_{eq} = C_1 + C_2 + C_3. \qquad \text{[parallel]} \quad \textbf{(19–4)} \qquad \textit{Capacitors in parallel}$$

The net effect of connecting capacitors in parallel is thus to increase the capacitance. This makes sense because we are essentially increasing the area of the plates where charge can accumulate (see Eq. 17–6).

    If the capacitors are connected in **series**, as in Fig. 19–15b, a charge $+Q$ flows from the battery to one plate of C_1, and $-Q$ flows to one plate of C_3. The regions A and B between the capacitors were originally neutral, so the net charge there must still be zero. The $+Q$ on the left plate of C_1 attracts a charge of $-Q$ on the opposite plate. Because region A must have a zero net charge, there is thus $+Q$ on the left plate of C_2. The same considerations apply to the other capacitors, so we see that the charge on each capacitor is the same, namely Q. A single capacitor that could replace these three in series without affecting the circuit (that is, Q and V the same) would have a capacitance C_{eq} given by

$$Q = C_{eq}V.$$

Now the total voltage V across the three capacitors in series must equal the sum of the voltages across each capacitor:

$$V = V_1 + V_2 + V_3.$$

We also have $Q = C_1V_1$, $Q = C_2V_2$, and $Q = C_3V_3$, so we substitute for V_1, V_2, and V_3 into the last equation and get

$$\frac{Q}{C_{eq}} = \frac{Q}{C_1} + \frac{Q}{C_2} + \frac{Q}{C_3},$$

or

$$\frac{1}{C_{eq}} = \frac{1}{C_1} + \frac{1}{C_2} + \frac{1}{C_3}. \qquad \text{[series]} \quad \textbf{(19–5)} \qquad \textit{Capacitors in series}$$

Notice that the forms of the equations for capacitors in series or in parallel are just the reverse of their counterparts for resistance. That is, the formula for capacitors in series resembles the formula for resistors in parallel, and vice versa.

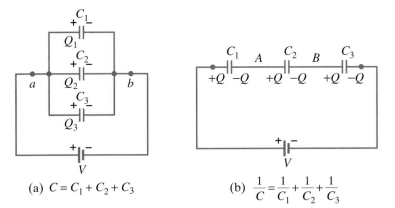

(a) $C = C_1 + C_2 + C_3$

(b) $\dfrac{1}{C} = \dfrac{1}{C_1} + \dfrac{1}{C_2} + \dfrac{1}{C_3}$

FIGURE 19–15 Capacitors (a) in parallel, (b) in series.

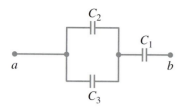

FIGURE 19–16 Example 19–9.

➡ **PROBLEM SOLVING**

Remember to take the reciprocal

EXAMPLE 19–9 **Equivalent capacitance.** Determine the capacitance of a single capacitor that will have the same effect as the combination shown in Fig. 19–16. Take $C_1 = C_2 = C_3 = C$.

SOLUTION C_2 and C_3 are connected in parallel, so they are equivalent to a single capacitor having capacitance

$$C_{23} = C_2 + C_3 = 2C.$$

C_{23} is in series with C_1, so the equivalent capacitance, C_{eq}, is given by

$$\frac{1}{C_{eq}} = \frac{1}{C_1} + \frac{1}{C_{23}} = \frac{1}{C} + \frac{1}{2C} = \frac{3}{2C}.$$

Hence $C_{eq} = \frac{2}{3}C$, which is the equivalent capacitance of the entire combination.

19–7 Circuits Containing a Resistor and a Capacitor

RC circuits

Charging the capacitor

Capacitors and resistors are often found together in a circuit. A simple example is shown in Fig. 19–17a. We now analyze this *RC circuit*. When the switch S is closed, current immediately begins to flow through the circuit. Electrons will flow out from the negative terminal of the battery, through the resistor R, and accumulate on the upper plate of the capacitor. And electrons will flow into the positive terminal of the battery, leaving a positive charge on the other plate of the capacitor. As charge accumulates on the capacitor, the potential difference across it increases, and the current is reduced until eventually the voltage across the capacitor equals the emf of the battery, $\mathcal{E}$. There is then no potential difference across the resistor, and no further current flow. The potential difference across the capacitor, which is proportional to the charge on the capacitor ($V = Q/C$, Eq. 17–5), thus increases in time, as shown in Fig. 19–17b. The actual shape of this curve is a type of exponential. It is given by the formula[†]

$$V = \mathcal{E}(1 - e^{-t/RC}),$$

where V is the voltage across the capacitor as a function of time t.

FIGURE 19–17 For the *RC* circuit shown in (a), the voltage across the capacitor increases with time, as shown in (b), after the switch S is closed.

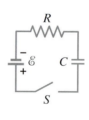

(a)

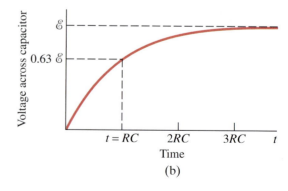

(b)

[†]The derivation is easy, using calculus. The exponential, e, has the value $e = 2.718\cdots$.

The product of the resistance R times the capacitance C, which appears in the exponent, is called the **time constant** τ of the circuit: $\tau = RC$. The time constant is a measure of how quickly the capacitor becomes charged. [The units of RC are $\Omega \cdot F = (V/A)(C/V) = C/(C/s) = s$.] Specifically, it can be shown that the product RC gives the time required for the capacitor to reach 63 percent of full voltage. This can be checked[†] using any calculator with an e^x key: $e^{-1} = 0.37$, so for $t = RC$, then $(1 - e^{-t/RC}) = (1 - e^{-1}) = (1 - 0.37) = 0.63$. In a circuit, for example, where $R = 200\,k\Omega$ and $C = 3.0\,\mu F$, the time constant is $(2.0 \times 10^5\,\Omega)(3.0 \times 10^{-6}\,F) = 0.60\,s$. If the resistance is much lower, the time constant is much smaller and the capacitor becomes charged almost instantly. This makes sense, since a lower resistance will retard the flow of charge less. All circuits contain some resistance (if only in the connecting wires), so a capacitor can never be charged instantaneously when connected to a battery.

Time constant = *RC*

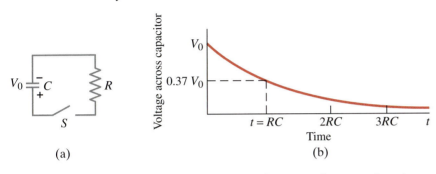

(a)

(b)

FIGURE 19–18 For the RC circuit shown in (a), the voltage V on the capacitor decreases with time, as shown in (b), after the switch S is closed. The charge on the capacitor follows the same curve since $Q \propto V$.

The circuit just discussed involved the *charging* of a capacitor by a battery through a resistance. Now let us look at another situation: when a capacitor is already charged (say to a voltage V_0), and it is allowed to *discharge* through a resistance R as shown in Fig. 19–18a. (In this case there is no battery.) When the switch S is closed, charge begins to flow through resistor R from one side of the capacitor toward the other side, until it is fully discharged. The voltage across the capacitor decreases, as shown in Fig. 19–18b. This "exponential decay" curve is given by

$$V = V_0 e^{-t/RC},$$

where V_0 is the initial voltage across the capacitor. The voltage falls 63 percent of the way to zero (to $0.37\,V_0$) in a time $\tau = RC$.

The charging and discharging in an RC circuit can be used to produce voltage pulses at a regular frequency. The charge on the capacitor increases to a particular voltage, and then discharges. A simple way of initiating the discharge is by the use of a gas-filled tube that breaks down when the voltage across it reaches a certain value V_0. After the discharge is finished, the tube no longer conducts current and the recharging process repeats itself, starting at V_0'. Figure 19–19 shows a possible circuit, and the "sawtooth" voltage it produces.

Capacitor discharges

FIGURE 19–19 (a) An RC circuit, coupled with a gas-filled tube as a switch, can produce a repeating "sawtooth" voltage, as shown in (b).

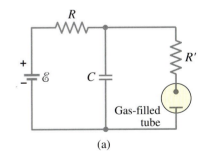

(a)

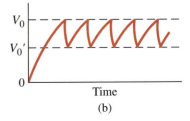

(b)

[†]More simply, since $e = 2.718\cdots$, then $e^{-1} = 1/e = 1/2.718 = 0.37$. Note that the exponential e is the inverse operation to the natural logarithm ln: $\ln(e) = 1$, and $\ln(e^x) = x$.

EXAMPLE 19–10 **An RC circuit.** If a charged capacitor, $C = 35\ \mu\text{F}$, is connected to a resistance, $R = 120\ \Omega$ (as in Fig. 19–18a), how much time will elapse until the voltage falls to 10 percent of its original (maximum) value?

SOLUTION The time constant for this circuit is given by

$$\tau = RC = (120\ \Omega)(35\ \mu\text{F}) = 4.2 \times 10^{-3}\ \text{s}.$$

After a time t the voltage across the capacitor will be

$$V = V_0\,(e^{-t/RC}).$$

We want to know the time t for which $V = (0.10)V_0$. Thus we set

$$(0.10)V_0 = V = V_0\,e^{-t/RC}$$

so

$$e^{-t/RC} = 0.10.$$

The inverse operation to the exponential e is the natural log, ln. Thus

$$\ln(e^{-t/RC}) = -\frac{t}{RC} = \ln 0.10 = -2.3.$$

Thus the elapsed time is

$$t = -2.3(RC) = (2.3)(4.2 \times 10^{-3}\ \text{s}) = 9.7 \times 10^{-3}\ \text{s}$$

or 9.7 ms.

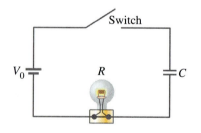

FIGURE 19–20 Example 19–11.

CONCEPTUAL EXAMPLE 19–11 **Bulb in RC circuit.** In the circuit of Fig. 19–20, the capacitor is originally uncharged. Describe the behavior of the lightbulb from the instant the switch is thrown until a long time later.

RESPONSE When the switch is first thrown, the charging current is high and the lightbulb burns brightly. As the capacitor charges, the voltage across it increases and the current is reduced causing the lightbulb to dim. As the capacitor approaches the same voltage as the battery, the current drops toward zero and the lightbulb goes out.

* 19–8 Heart Pacemakers

An interesting use of an RC circuit is the electronic pacemaker, which can make a stopped heart start beating again by applying an electric stimulus through electrodes attached to the chest. The stimulus can be repeated at the normal heartbeat rate if necessary.

The heart itself contains a *pacemaker*, which sends out tiny electric pulses at a rate of 60 to 80 per minute. These are the signals that induce the start of each heartbeat. In some forms of heart disease, the pacemaker cells fail to function properly, and the heart loses its beat. People suffering from this ailment now commonly make use of *electronic pacemakers*. These devices produce a regular voltage pulse that starts and controls the frequency of the heartbeat. The "fixed-rate" type produces signals continuously. The "demand" type operates only when the natural pacemaker fails. The electrodes are implanted in or near the heart and the circuit usually contains a capacitor and a resistor. The charge on the capacitor increases to a certain point and then discharges. Then it starts charging again. The pulsing rate depends on the values of R and C. Generally, the power source is a battery that must be replaced or recharged, depending on type. Some pacemakers obtain their energy from the heat produced by a radioactive element; the thermal energy is transformed to electricity by a thermocouple.

19–9 Electric Hazards; Leakage Currents

An electric shock can cause damage to the body or may even be fatal. The severity of a shock depends on the magnitude of the current, how long it acts, and through what part of the body it passes. A current passing through vital organs such as the heart or brain is especially serious for it can interfere with their operation. Electric current heats tissue and can cause burns. A current also stimulates nerves and muscles (whose operation, as we have seen in Sections 17–11 and 18–9, is electrical), and we feel a "shock."

Most people can "feel" a current of about 1 mA. Currents of a few mA cause pain but rarely much damage in a healthy person. However, currents above 10 mA cause severe contraction of the muscles, and a person may not be able to release the source of the current (say, a faulty appliance or wire). Death from paralysis of the respiratory system can occur. Artificial respiration, however, can often revive a victim. If a current above about 70 mA passes across the torso so that a portion passes through the heart for a second or more, the heart muscles will begin to contract irregularly and blood will not be properly pumped. This condition is called "ventricular fibrillation." If it lasts for long, death results. Strangely enough, if the current is much larger, on the order of 1 A, the damage may be less and death by heart failure may be less likely[†] under some conditions.

The seriousness of a shock depends on the effective resistance of the body. Living tissue has quite low resistance since the fluid of cells contains ions that can conduct quite well. However, the outer layers of skin, when dry, offer much resistance. The effective resistance between two points on opposite sides of the body when the skin is dry is in the range of 10^4 to $10^6 \, \Omega$. However, when the skin is wet, the resistance may be $10^3 \, \Omega$ or less. A person in good contact with the ground who touches a 120-V dc line with wet hands can suffer a current

$$I = \frac{120 \text{ V}}{1000 \, \Omega} = 120 \text{ mA}.$$

As we saw above, this could be lethal.

[†]Apparently, larger currents bring the entire heart to a standstill. Upon release of the current, the heart returns to its normal rhythm. This may not happen when fibrillation occurs since it is often hard to stop once it starts. Fibrillation may also occur as a result of a heart attack or during heart surgery. A device known as a *defibrillator* can apply a brief high current to the heart; this causes complete heart stoppage and is often followed by resumption of normal beating.

PHYSICS APPLIED

Electric shock

Beware of wet skin

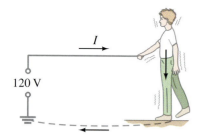

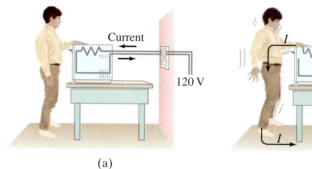

FIGURE 19–21 A person receives an electric shock when the circuit is completed.

Figure 19–21 shows how the circuit is completed when a person touches an electric wire. One side of a 120-V source is connected to ground by a wire connected to a buried conductor (say, a water pipe). Thus the current passes from the high-voltage wire, through the person, to the ground; it passes through the ground back to the other terminal of the source to complete the circuit. If the person in Fig. 19–21 stands on a good insulator—thick-soled shoes or a dry wood floor—there will be much more resistance in the circuit and consequently much less current will flow. However, if the person stands with bare feet on the ground, or is sitting in a bathtub, there is considerable danger because the resistance is much less. In a bathtub, not only are you wet, but the water is in contact with the drain pipe that leads to the ground. That is why it is strongly recommended not to touch anything electrical in such a situation.

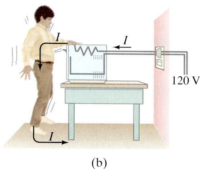

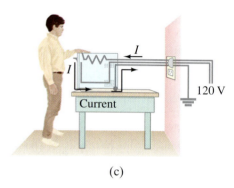

(a)   (b)   (c)

FIGURE 19–22   (a) An electric oven operating normally with a two-prong plug.
(b) Short to the case with ungrounded case: shock.
(c) Short to the case with the case grounded with third prong.

➡ PHYSICS APPLIED

Grounding and shocks

FIGURE 19–23   Human body modeled electrically as resistance and capacitor in parallel when a voltage is applied.

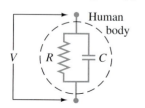

A principal danger comes from touching a bare wire whose insulation has worn off, or from a bare wire inside an appliance when you're tinkering with it. (Always unplug an electrical device before investigating its insides!) Sometimes a wire inside a device breaks or loses its insulation and comes in contact with the case. If the case is metal, it will conduct electricity. A person could then suffer a severe shock merely by touching the case, as shown in Fig. 19–22b. To prevent an accident, metal cases are supposed to be connected directly to ground, so they cannot become "hot." Then if a "hot" wire touches the grounded case, a short circuit to ground immediately occurs internally, as shown in Fig. 19–22c; most of the current passes through the low resistance ground wire rather than through the person. Furthermore, the high current immediately opens the fuse or circuit breaker in the circuit. Grounding a metal case is best done by a separate ground wire connected to the third (round) prong of a 3-prong plug; it can also be done by connecting the case to the larger prong of a so-called "polarized" 2-prong plug. Of course not only the device, but also the outlets must be wired correctly to ground.

The human body acts as if it had capacitance in parallel with its resistance (Fig. 19–23). A dc current can pass through the resistance, but not the capacitance. An ac current, like the changing currents discussed in Section 18–8 (more on this in Chapter 21), can exist also in the capacitive branch. Because of the additional path allowing current flow, the ac current for a given V_{rms} will be greater than for the same dc voltage. Thus an ac voltage is more dangerous than an equal dc voltage.

Another danger is *leakage current*, by which we mean a current along an unintended path. Leakage currents are often capacitively coupled. For example, a wire in a lamp forms a capacitor with the metal case; charges moving in one conductor attract or repel charge in the other, so there is a current. Typical electrical codes limit leakage currents to 1 mA for any device. A 1-mA leakage current is usually harmless. It can be very dangerous, however, to a hospital patient with implanted electrodes connected to ground through the apparatus. This is because the current can pass directly through the heart as compared to the usual situation where the current enters at the hands and spreads out through the body. Although 70 mA may be needed to cause heart fibrillation when entering through the hands (very little of it actually passes through the heart), as little as 0.02 mA has been known to cause fibrillation when passing directly to the heart. Thus, a "wired" patient is in considerable danger from leakage current even from as simple an act as touching a lamp.

Leakage current

∗ 19–10 DC Ammeters and Voltmeters

An **ammeter** is used to measure current, and a **voltmeter** measures potential difference or voltage. The crucial part of an analog ammeter or voltmeter, in which the reading is by a pointer on a scale (Fig. 19–24), is a *galvanometer*. The galvanometer works on the principle of the force between a magnetic field and a current-carrying coil of wire, and will be discussed in Chapter 20. For now, we merely need to know that the deflection of the needle of a galvanometer is proportional to the current flowing through it. The *full-scale current sensitivity*, I_m, of a galvanometer is the current needed to make the needle deflect full scale. If the sensitivity I_m is 50 μA, a current of 50 μA will cause the needle to go to the end of the scale. A current of 25 μA will then make it go only half way. If there is no current, the needle should be on zero, and usually there is an adjustment screw to make it so.

Many meters we see in everyday life are galvanometers connected as ammeters or voltmeters, such as some of the meters on a car's instrument panel.

A galvanometer can be used directly to measure small dc currents. For example, a galvanometer whose sensitivity I_m is 50 μA can measure currents from about 1 μA (currents smaller than this would be hard to read on the scale) up to 50 μA. To measure larger currents, a resistor is placed in parallel with the galvanometer. Thus, an ammeter (represented by the symbol •–Ⓐ–•) consists of a galvanometer (•–Ⓖ–•) in parallel with a resistor called the **shunt resistor**. ("Shunt" is a synonym for "in parallel.") This is shown in Fig. 19–25. The shunt resistance is R, and the resistance of the galvanometer coil (which carries the current) is r. The value of R is chosen according to what full-scale deflection is desired and is normally very small, giving an ammeter a very small internal resistance.

Meters

FIGURE 19–24 A multimeter used as a voltmeter.

Ammeter uses shunt resistor

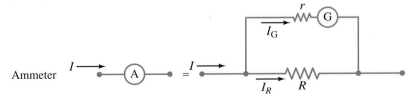

FIGURE 19–25 An ammeter is a galvanometer in parallel with a (shunt) resistor with low resistance, R.

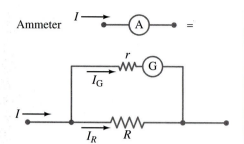

Ammeter

FIGURE 19–25 (Repeated.) An ammeter is a galvanometer in parallel with a (shunt) resistor with low resistance, R.

EXAMPLE 19–12 **Ammeter design.** Design an ammeter to read 1.0 A at full scale using a galvanometer with a full-scale sensitivity of 50 μA and a resistance $r = 30\,\Omega$. Check if the scale is linear.

SOLUTION When the total current I entering the ammeter is 1.0 A, we want the current I_G through the galvanometer to be precisely 50 μA (to give full-scale deflection). See Fig 19–25. Thus, when 1.0 A flows into the meter, we want 0.999950 A $(= I_R)$ to pass through the shunt resistor R. Since the potential difference across the shunt is the same as across the galvanometer,

$$I_R R = I_G r,$$

then

$$R = \frac{I_G r}{I_R} = \frac{(5.0 \times 10^{-5}\,\text{A})(30\,\Omega)}{(0.999950\,\text{A})} = 1.5 \times 10^{-3}\,\Omega,$$

or 0.0015 Ω. The shunt resistor must thus have a very low resistance so that most of the current passes through it.

If the current I into the meter is 0.50 A, this will produce a current to the galvanometer equal to $I_G = I_R R/r = (0.50\,\text{A})(1.5 \times 10^{-3}\,\Omega)/30\,\Omega = 25\,\mu$A, which gives a deflection half of full scale, as required.

Voltmeter uses series resistor

A voltmeter ($\bullet\text{–}\textcircled{V}\text{–}\bullet$) also consists of a galvanometer and a resistor. But the resistor R is connected in series, Fig. 19–26, and it is usually large, giving a voltmeter a high internal resistance.

FIGURE 19–26 A voltmeter is a galvanometer in series with a resistor with high resistance, R.

Voltmeter

EXAMPLE 19–13 **Voltmeter design.** Using the same galvanometer with internal resistance $r = 30\,\Omega$ and full-scale current sensitivity of 50 μA, design a voltmeter that reads from 0 to 15 V. Is the scale linear?

SOLUTION When a potential difference of 15 V exists across the terminals of our voltmeter, we want 50 μA to be passing through it so as to give a full-scale deflection. From Ohm's law we have (see Fig. 19–26)

$$15\,\text{V} = (50\,\mu\text{A})(r + R),$$

so

$$R = \frac{15\,\text{V}}{5.0 \times 10^{-5}\,\text{A}} - r = 300\,\text{k}\Omega - 30\,\Omega = 300\,\text{k}\Omega.$$

Notice that $r = 30\,\Omega$ is so small compared to the value of R that it doesn't influence the calculation significantly.

The scale will again be linear: if the voltage to be measured is 6.0 V, the current passing through the voltmeter will be $(6.0\,\text{V})/(3.0 \times 10^5\,\Omega) = 2.0 \times 10^{-5}$ A, or 20 μA. This will produce two fifths of full-scale deflection, as required $(6.0\,\text{V}/15.0\,\text{V} = 2/5)$.

The meters discussed above are for direct current. If an alternating current passed through these meters, the reading would fluctuate wildly, or possibly not move at all because the back and forth changes would be too

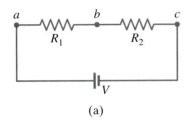

(a)

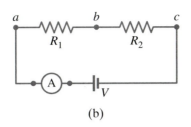

(b)

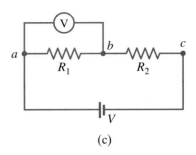

(c)

FIGURE 19–27 Measuring current and voltage.

rapid. A dc meter like those described above can be modified to measure ac with the addition of diodes (Chapter 29) which allow current to flow in one direction only. The resulting unidirectional current can be read by the meter. An ac meter can be calibrated to read rms or peak values.

Suppose you wish to determine the current I in the circuit shown in Fig. 19–27a and the voltage V across the resistor R_1. How exactly are ammeters and voltmeters connected to the circuit being measured?

How to connect meters

Because an ammeter is used to measure the current flowing in the circuit, it must be inserted directly into the circuit, in series with the other elements, as shown in Fig. 19–27b. The smaller its internal resistance, the less it will affect the circuit.

Ammeter is inserted into the circuit

A voltmeter, on the other hand, is connected in parallel with the circuit element across which the voltage is to be measured. It is used to measure the potential difference between two points and its two wire leads (connecting wires) are connected to the two points, as shown in Fig. 19–27c where the voltage across R_1 is being measured. The larger its internal resistance ($R + r$ in Fig. 19–26) the less it affects the circuit being measured.

Voltmeter is connected in parallel

Voltmeters and ammeters can have several series or shunt resistors to offer a choice of range. **Multimeters** can measure voltage, current, and resistance. Sometimes they are called VOMs (Volt-Ohm-Meter). Meters with digital readout are called digital voltmeters (DVM) or digital multimeters (DMM), Fig. 19–28.

To measure resistance, the meter must contain a battery of known voltage connected in series to a resistor (R_{ser}) and to an ammeter (consisting of a galvanometer of resistance r and a shunt resistor R_{sh}). This makes an **ohmmeter** (see Fig. 19–29, and Problem 83). The resistor whose resistance is to be measured completes the circuit. The deflection in this case is inversely proportional to the resistance. Thus, if the resistance is small the deflection is large, and vice versa. The calibration of the scale depends on the value of the series resistor. Since an ohmmeter sends a current through the device whose resistance is to be measured, it should not be used on very delicate devices that could be damaged by the current.

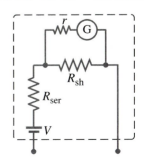

FIGURE 19–28 A digital multimeter being used to measure resistance.

FIGURE 19–29 An ohmmeter.

The **sensitivity** of a meter is generally specified on the face. It may be given as so many ohms per volt, which indicates how many ohms of resistance there are in the meter per volt of full-scale reading. For example, if the sensitivity is 30,000 Ω/V, this means that on the 10-V scale the meter has a resistance of 300,000 Ω. The full-scale current sensitivity, I_m, discussed earlier, is just the reciprocal of the sensitivity in Ω/V (note that the units Ω/V = A^{-1}). For example, a meter with sensitivity 30,000 Ω/V produces a full-scale deflection at 1.0 V when 30,000 Ω is in series with the galvanometer. Thus the current sensitivity is (1.0 V)/(3.0 × 10^4 Ω) = 33 μA in this case.

*19–11 Effects of Meter Resistance

⇒ **PHYSICS APPLIED**

Correcting for meter resistance

It is important to know the sensitivity of a meter, for in many cases the resistance of the meter can seriously affect your results. Take the following Example.

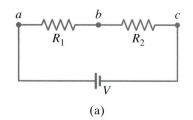

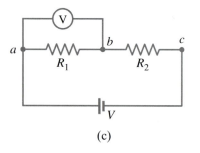

FIGURE 19–27 (a) and (c) repeated for Example 19–14.

EXAMPLE 19–14 **Voltage reading versus true voltage.** Suppose you are testing an electronic circuit which has two resistors, R_1 and R_2, each 15 kΩ, connected in series as shown in Fig. 19–27a. The battery maintains 8.0 V across them and has negligible internal resistance. A voltmeter whose sensitivity is 10,000 Ω/V is put on the 5.0-V scale. What voltage does the meter read when connected across R_1, and what error is caused by the finite resistance of the meter?

SOLUTION On the 5.0-V scale, the voltmeter has an internal resistance of $(5.0\,\text{V})(10,000\,\Omega/\text{V}) = 50,000\,\Omega$. When connected across R_1, as in Fig. 19–27c, we have this 50 kΩ in parallel with $R_1 = 15$ kΩ. The net resistance R_{eq} of these two is given by

$$\frac{1}{R_{eq}} = \frac{1}{50\,\text{k}\Omega} + \frac{1}{15\,\text{k}\Omega} = \frac{13}{150\,\text{k}\Omega},$$

so $R_{eq} = 11.5$ kΩ. This $R_{eq} = 11.5$ kΩ is in series with $R_2 = 15$ kΩ, so the total resistance of the circuit is now 26.5 kΩ. Hence the current from the battery is

$$I = \frac{8.0\,\text{V}}{26.5\,\text{k}\Omega} = 0.30\,\text{mA}.$$

Then the voltage drop across R_1, which is the same as that across the voltmeter, is $(3.0 \times 10^{-4}\,\text{A})(11.5 \times 10^3\,\Omega) = 3.5$ V. [The voltage drop across R_2 is $(3.0 \times 10^{-4}\,\text{A})(15 \times 10^3\,\Omega) = 4.5$ V, for a total of 8.0 V.] If we assume the meter is precise, it will read 3.5 V. In the normal circuit, without the meter, $R_1 = R_2$ so the voltage across R_1 is half that of the battery, or 4.0 V. Thus the voltmeter, because of its internal resistance, gives a low reading. In this case it is off by 0.5 V, or more than 10 percent.

Example 19–14 illustrates how seriously a meter can affect a circuit and give a misleading reading. If the resistance of a voltmeter is much higher than the resistance of the circuit, however, it will have little effect and its readings can be trusted, at least to the manufactured precision of the meter, which for ordinary analog meters is typically 3 to 4 percent of full-scale deflection. An ammeter also can interfere with a circuit, but the effect is minimal if its resistance is much less than that of the circuit as a whole. For both voltmeters and ammeters, the more sensitive the galvanometer the less effect it will have. A 50,000-Ω/V meter is far better than a 1,000-Ω/V meter.

Electronic voltmeters using transistors, including digital meters, have very high input resistance (usually specified in ohms) in the range 10^6 to $10^8\,\Omega$, and even higher. Hence they have very little effect on most circuits and their readings are reliable for nearly any circuit. The precision of digital meters is typically one part in 10^4 ($= 0.01$ percent) or better. State-of-the-art instruments reach a precision of one part in 10^6.

SUMMARY

When resistances are connected in **series** (end to end), the equivalent resistance is the sum of the individual resistances:

$$R_{eq} = R_1 + R_2 + \cdots .$$

When resistors are connected in **parallel**, the reciprocal of the total resistance equals the sum of the reciprocals of the individual resistances:

$$\frac{1}{R_{eq}} = \frac{1}{R_1} + \frac{1}{R_2} + \cdots .$$

In a parallel connection, the net resistance is less than any of the individual resistances.

A device that transforms one type of energy into electrical energy is called a **seat** or **source** of **emf**. A battery behaves like a source of emf in series with an **internal resistance**. The emf is the potential difference determined by the chemical reactions in the battery and equals the terminal voltage when no current is drawn. When a current is drawn, the voltage at the battery's terminals is less than its emf by an amount equal to the Ir drop across the internal resistance.

Kirchhoff's rules are helpful in determining the currents and voltages in a complex circuit. Kirchhoff's **junction rule** is based on conservation of electric charge and states that the sum of all currents entering any junction equals the sum of all currents leaving that junction. The second, or **loop rule**, is based on conservation of energy and states that the algebraic sum of the voltage changes around any closed path of the circuit must be zero.

When capacitors are connected in **parallel**, the equivalent capacitance is the sum of the individual capacitances:

$$C_{eq} = C_1 + C_2 + \cdots .$$

When capacitors are connected in **series**, the reciprocal of the equivalent capacitance equals the sum of the reciprocals of the individual capacitances:

$$\frac{1}{C_{eq}} = \frac{1}{C_1} + \frac{1}{C_2} + \cdots .$$

If an **RC circuit** containing a resistor R in series with a capacitance C is connected to a dc source of emf, the voltage across the capacitor rises gradually in time characterized by the **time constant**

$$\tau = RC.$$

This is the time it takes for the voltage to reach 63 percent of its maximum value. A capacitor discharging through a resistor is characterized by the same time constant: in a time $\tau = RC$, the voltage across the capacitor drops to 37 percent of its initial value.

Electric shocks are caused by current passing through the body. To avoid shocks, the body must not become part of a circuit by allowing different parts of the body to touch objects at different potentials. Commonly, one part of the body may be touching ground and another part a high or low potential.

QUESTIONS

1. Explain why birds can sit on power lines safely, while leaning a metal ladder up against one to fetch a stuck kite is extremely dangerous.

2. Discuss the advantages and disadvantages of Christmas tree lights connected in parallel versus those connected in series.

3. If all you have is a 120-V line, would it be possible to light several 6-V lamps without burning them out? How?

4. Two lightbulbs of resistance R_1 and R_2 ($> R_1$) are connected in series. Which is brighter? What if they are connected in parallel?

5. Describe carefully the difference between emf and potential difference.

6. Household outlets are often double outlets. Are these connected in series or parallel? How do you know?

7. With two identical lightbulbs and two identical batteries, how would you arrange the bulbs and batteries in a circuit in order to get the maximum possible total power out. (Assume that the batteries have negligible internal resistance.)

8. Explain why Kirchhoff's first (junction) rule is equivalent to conservation of electric charge.

9. Explain why Kirchhoff's second (loop) rule is a result of the conservation of energy.

10. How does the overall resistance of your room's electric circuit change when instead of having a single 60-W lightbulb on, you turn on an additional 100-W bulb?

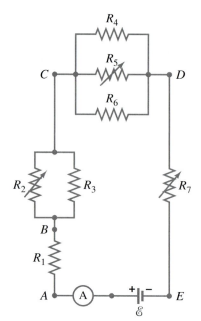

FIGURE 19–30
Question 11.

11. Given the circuit shown in Fig. 19–30 use the words "increases," "decreases," or "stays the same" to complete the following statements:
(a) If R_7 increases, the potential difference between A and E (assume no resistance in   and $\mathscr{E}$) ___.
(b) If R_7 increases, the potential difference between A and E (assume   and $\mathscr{E}$ have resistance) ___.
(c) If R_7 increases, the voltage drop across R_4 _____.
(d) If R_2 decreases, the current through R_1 _____.
(e) If R_2 decreases, the current through R_6 _____.
(f) If R_2 decreases, the current through R_3 _____.
(g) If R_5 increases, the voltage drop across R_2 _____.
(h) If R_5 increases, the voltage drop across R_4 _____.
(i) If R_2, R_5, and R_7 increase, $\mathscr{E}$ _____.

12. Why are batteries connected in series? Why in parallel? Does it matter if the batteries are nearly identical or not in either case?

13. Describe a situation in which the terminal voltage of a battery is greater than its emf.

14. The 18-V source in Fig. 19–31 is "charging" the 12-V battery. Explain how it does this.

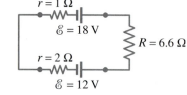

FIGURE 19–31
Question 14 and
Problem 25.

15. Explain in detail how you could measure the internal resistance of a battery.

16. Compare and discuss the formulas for resistors and for capacitors when connected in series and in parallel.

17. Suppose that three identical capacitors are connected to a battery. Will they store more energy if connected in series or in parallel?

18. Why is it more dangerous to turn on an electric appliance when you are standing outside in bare feet than when you are inside wearing shoes with thick soles?

19. Figure 19–32 is a diagram of a capacitor (or condenser) *microphone*. The changing air pressure in a sound wave causes one plate of the capacitor C to move back and forth. Explain how a current of the same frequency as the sound wave is produced.

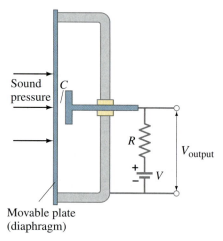

FIGURE 19–32   Diagram of a capacitor microphone. Question 19.

* 20. In an RC circuit, current flows from the battery until the capacitor is completely charged. Is the total energy supplied by the battery equal to the total energy stored by the capacitor? If not, where does the extra energy go?

* 21. What is the main difference between a voltmeter and an ammeter?

* 22. What would happen if you mistakenly used an ammeter where you needed to use a voltmeter?

* 23. Explain why an ideal ammeter would have zero resistance and an ideal voltmeter infinite resistance.

PROBLEMS

SECTION 19–1

In these problems neglect the internal resistance of a battery unless the problem refers to it.

1. (I) Four 140-Ω lightbulbs are connected in series. What is the total resistance of the circuit? What is their resistance if they are connected in parallel?

2. (I) Three 40-Ω lightbulbs and three 80-Ω lightbulbs are connected in series. (*a*) What is the total resistance of the circuit? (*b*) What is their resistance if all six are wired in parallel?

3. (I) Given only one 30-Ω and one 50-Ω resistor, list all possible values of resistance that can be obtained.

4. (I) Suppose that you have a 500-Ω, a 900-Ω, and a 1.40-$k\Omega$ resistor. What is (*a*) the maximum, and (*b*) the minimum resistance you can obtain by combining these?

5. (II) Suppose that you have a 6.0-V battery and you wish to apply a voltage of only 4.0 V. Given an unlimited supply of 1.0-Ω resistors, how could you connect them so as to make a "voltage divider" that produced a 4.0-V output for a 6.0-V input?

6. (II) Three 240-Ω resistors can be connected together in four different ways, making combinations of series and/or parallel circuits. What are these four ways and what is the net resistance in each case?

7. (II) What is the net resistance of the circuit connected to the battery in Fig. 19–33? Each resistance has $R = 2.8\ k\Omega$.

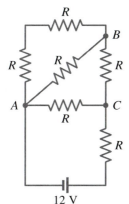

FIGURE 19–33
Problems 7 and 14.

8. (II) Eight lights are connected in series across a 110-V line. (*a*) What is the voltage across each bulb? (*b*) If the current is 0.40 A, what is the resistance of each bulb and the power dissipated in each?

9. (II) Eight lights are connected in parallel to a 110-V source by two leads of total resistance 1.5 Ω. If 240 mA flows through each bulb, what is the resistance of each, and what fraction of the total power is wasted in the leads?

10. (II) Eight 7.0-W Christmas tree lights are connected in series to each other and to a 110-V source. What is the resistance of each bulb?

11. (II) A close inspection of an electric circuit reveals that a 480-Ω resistor was inadvertently soldered in the place where a 320-Ω resistor is needed. How can this be fixed without removing anything from the existing circuit?

12. (II) Two resistors when connected in series to a 110-V line use one fourth the power that is used when they are connected in parallel. If one resistor is 2.2 kΩ, what is the resistance of the other?

13. (II) A 75-W, 110-V bulb is connected in parallel with a 40-W, 110-V bulb. What is the net resistance?

14. (II) Calculate the current through each resistor in Fig. 19–33 if each resistance $R = 2.20\ k\Omega$. What is the potential difference between points A and B?

15. (II) Consider the network of resistors shown in Fig. 19–34. Answer qualitatively: (*a*) What happens to the voltage across each resistor when the switch S is closed? (*b*) What happens to the current through each when the switch is closed? (*c*) What happens to the power output of the battery when the switch is closed? (*d*) Let $R_1 = R_2 = R_3 = R_4 = 100\ \Omega$ and $V = 45.0$ V. Determine the current through each resistor before and after closing the switch. Are your qualitative predictions confirmed?

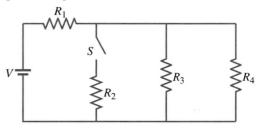

FIGURE 19–34   Problem 15.

16. (II) Three equal resistors (R) are connected to a power supply, as shown in Fig. 19–35. Qualitatively, what happens to (*a*) the voltage drop across each of these resistors, (*b*) the current flow through each, and (*c*) the terminal voltage of the battery, when the switch S is opened, after having been closed for a long time? (*d*) If the emf of the battery is 18.0 V, what is its terminal voltage when the switch is closed if the internal resistance is 0.50 Ω and $R = 5.50\ \Omega$? (*e*) What is the terminal voltage when the switch S is open?

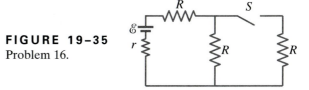

FIGURE 19–35
Problem 16.

17. (III) A 2.8-kΩ and a 2.1-kΩ resistor are connected in parallel; this combination is connected in series with a 1.8-kΩ resistor. If each resistor is rated at $\frac{1}{2}$ W, what is the maximum voltage that can be applied across the whole network?

581

18. (I) Calculate the terminal voltage for a battery with an internal resistance of $0.900\ \Omega$ and an emf of $8.50\ V$ when the battery is connected in series with (a) an $81.0\text{-}\Omega$ resistor, and (b) an $810\text{-}\Omega$ resistor.

19. (I) Four 2.0-V cells are connected in series to a $12\text{-}\Omega$ lightbulb. If the resulting current flow is $0.62\ A$, what is the internal resistance of each cell, assuming they are identical and neglecting the wires?

20. (II) A battery with an emf of $12.0\ V$ shows a terminal voltage of $11.8\ V$ when operating in a circuit with two lightbulbs rated at $3.0\ W$ (at $12.0\ V$) which are connected in parallel with it. What is the battery's internal resistance?

21. (II) A 1.5-V dry cell can be tested by connecting it to a low-resistance ammeter. It should be able to supply at least $25\ A$. What is the internal resistance of the cell in this case?

22. (II) What is the internal resistance of a 12.0-V car battery whose terminal voltage drops to $8.8\ V$ when the starter draws $60\ A$? What is the resistance of the starter?

23. (II) What is the current in the $8.0\text{-}\Omega$ resistor in Fig. 19–10a?

SECTIONS 19–3 AND 19–4

24. (II) Calculate the current in the circuit of Fig. 19–36 and show that the sum of all the voltage changes around the circuit is zero.

FIGURE 19–36
Problem 24.

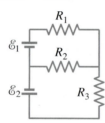

$r = 2.0\ \Omega$
$9.0\ V$
$8.0\ \Omega$
$12.0\ \Omega$

25. (II) Determine the terminal voltage of each battery in Fig. 19–31.

26. (II) What is the potential difference between points a and d in Fig. 19–13?

27. (II) What is the terminal voltage of each battery in Fig. 19–13?

28. (II) Determine the magnitudes and directions of the currents through R_1 and R_2 in Fig. 19–37.

29. (II) Repeat Problem 28, now assuming that each battery has an internal resistance $r = 1.2\ \Omega$.

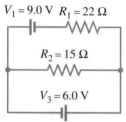

$V_1 = 9.0\ V$ $R_1 = 22\ \Omega$
$R_2 = 15\ \Omega$
$V_3 = 6.0\ V$

FIGURE 19–37
Problems 28 and 29.

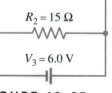

R_1
$\mathscr{E}_1$ R_2
$\mathscr{E}_2$ R_3

FIGURE 19–38
Problems 30 and 31.

582   CHAPTER 19   DC Circuits

30. (II) Determine the magnitudes and directions of the currents through each resistor shown in Fig. 19–38. The batteries have emfs of $\mathscr{E}_1 = 9.0\ V$ and $\mathscr{E}_2 = 12.0\ V$ and the resistors have values of $R_1 = 15\ \Omega$, $R_2 = 20\ \Omega$, and $R_3 = 30\ \Omega$.

31. (II) Repeat Problem 30 assuming each battery has internal resistance $r = 1.0\ \Omega$.

32. (III) Determine the current through each of the resistors in Fig. 19–39.

33. (III) If the $25\text{-}\Omega$ resistor in Fig. 19–39 were shorted out (resistance $= 0$), what then would be the current through the $10\text{-}\Omega$ resistor?

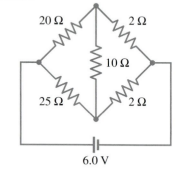

$20\ \Omega$ $2\ \Omega$
$10\ \Omega$
$25\ \Omega$ $2\ \Omega$

FIGURE 19–39
Problems 32 and 33.

$6.0\ V$

34. (III) Determine the currents I_1, I_2, and I_3 in Fig. 19–40. Assume the internal resistance of each battery is $r = 1.0\ \Omega$. What is the terminal voltage of the 6.0-V battery?

35. (III) What would the current I_1 be in Fig. 19–40 if the $12\text{-}\Omega$ resistor were shorted out ($r = 1.0\ \Omega$)?

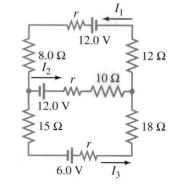

r I_1
$12.0\ V$
$8.0\ \Omega$ $12\ \Omega$
I_2 r $10\ \Omega$
$12.0\ V$
$15\ \Omega$ $18\ \Omega$
r
$6.0\ V$ I_3

FIGURE 19–40
Problems 34 and 35.

SECTION 19–5

36. (II) Suppose two batteries, with unequal emfs of $2.0\ V$ and $3.0\ V$, are connected as shown in Fig. 19–41 with unequal emfs. If each internal resistance is $r = 0.10\ \Omega$, and $R = 4.0\ \Omega$, what is the voltage across the resistor R?

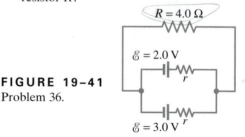

$R = 4.0\ \Omega$
$\mathscr{E} = 2.0\ V$ r
$\mathscr{E} = 3.0\ V$ r

FIGURE 19–41
Problem 36.

SECTION 19-6

37. (I) Six 3.7-μF capacitors are connected in parallel. What is the equivalent capacitance? What is their equivalent capacitance if connected in series?

38. (I) An electric circuit was accidentally constructed using a 5.0-μF capacitor instead of the required 16-μF value. What can a technician add to correct this circuit?

39. (I) The capacitance of a portion of a circuit is to be reduced from 4800 pF to 3300 pF. What capacitance can be added to the circuit to produce this effect without removing existing circuit elements? Must any existing connections be broken in the process?

40. (I) You have three capacitors, of capacitance 2000 pF, 7500 pF, and 0.0100 μF. What is the maximum and minimum capacitance that you can form from these? How do you make the connection in each case?

41. (II) A circuit contains a single 150-pF capacitor hooked across a battery. It is desired to store three times as much energy in a combination of two capacitors by adding a single capacitor to this one. How would you hook it up and what would its value be?

42. (II) (a) Determine the equivalent capacitance of the circuit shown in Fig. 19–42. (b) If $C_1 = C_2 = 2C_3 = 12.5\ \mu$F, how much charge is stored on each capacitor when $V = 45.0$ V?

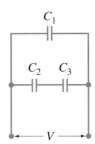

FIGURE 19–42
Problems 42 and 43.

43. (II) In Fig. 19–42, let $V = 90$ V and $C_1 = C_2 = C_3 = 8.8\ \mu$F. How much energy is stored in the capacitor network?

44. (II) A 0.40-μF and a 0.50-μF capacitor are connected in series to a 9.0-V battery. Calculate (a) the potential difference across each capacitor, and (b) the charge on each. (c) Repeat parts (a) and (b) assuming the two capacitors are in parallel.

45. (II) Suppose three parallel-plate capacitors, whose plates have areas A_1, A_2, and A_3, and separations d_1, d_2, and d_3, are connected in parallel. Show, using Eq. 17–6, that Eq. 19–4 is valid.

46. (II) An air-filled parallel-plate capacitor of capacitance C_0 is filled with two identically sized dielectric slabs of dielectric constants K_1 and K_2 as shown in Fig. 19–43. What is the new capacitance? [*Hint*: Treat this as two capacitors in combination.]

FIGURE 19–43
Problem 46.

FIGURE 19–44
Problem 47.

47. (II) An air-filled parallel-plate capacitor of capacitance C_0 is filled with two identically sized dielectric slabs of dielectric constants K_1 and K_2 as shown in Fig. 19–44. What is the new capacitance? [*Hint*: Treat this as two capacitors in combination.]

48. (III) A 200-pF capacitor is connected in series with an unknown capacitor, and as a series combination they are connected to a battery with an emf of 25.0 V. If the 200-pF capacitor stores 125 pC of charge on its plates, what is the unknown capacitance?

49. (III) A 7.0-μF and a 3.0-μF capacitor are connected in series and this combination is connected in parallel with a 4.0-μF capacitor. (a) What is the net capacitance? (b) If 24 V is applied across the whole network, calculate the voltage across each capacitor.

SECTION 19-7

50. (I) Electrocardiographs are often connected as shown in Fig. 19–45. The leads are said to be capacitively coupled. A time constant of 3.0 s is typical and allows rapid changes in potential to be accurately recorded. If $C = 3.0\ \mu$F, what value must R have?

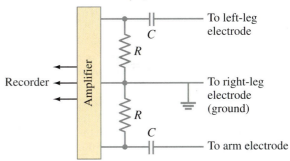

FIGURE 19–45   Problem 50.

51. (II) Suppose a 7.5-μF capacitor replaced the 10.0-Ω resistor in Fig. 19–10a. Calculate: (a) the charge on the capacitor in the steady state—that is, after the capacitor reaches its maximum charge; (b) the steady state current through the other resistors.

52. (II) In Fig. 19–17a, the total resistance is 15 kΩ, and the battery's emf is 24.0 V. If the time constant is measured to be 35 μs, calculate (a) the total capacitance of the circuit and (b) the time it takes for the voltage across the resistor to reach 16.0 V.

53. (II) The RC circuit of Fig. 19–18a has $R = 6.7$ kΩ and $C = 3.0\ \mu$F. The capacitor is at voltage V_0 at $t = 0$, when the switch is closed. How long does it take the capacitor to discharge to 1.0 percent of its initial voltage?

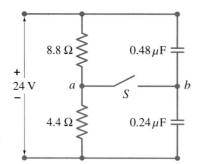

FIGURE 19–46
Problem 54.

54. (III) Two resistors and two uncharged capacitors are arranged as shown in Fig. 19–46. With a potential difference of 24 V across the combination, (a) what is the potential at point a with S open? (Let $V = 0$ at the negative terminal of the source.) (b) What is the potential at point b with the switch open? (c) When the switch is closed, what is the final potential of point b? (d) How much charge flows through the switch S after it is closed?

*SECTION 19–10

* **55.** (I) What is the resistance of a voltmeter on the 250-V scale if the meter sensitivity is 30,000 Ω/V?

* **56.** (II) An ammeter has a sensitivity of 10,000 Ω/V. What current passing through the galvanometer produces full-scale deflection?

* **57.** (II) A galvanometer has an internal resistance of 30 Ω and deflects full scale for a 50-μA current. Describe how to use this galvanometer to make (a) an ammeter to read currents up to 30 A, and (b) a voltmeter to give a full-scale deflection of 1000 V.

* **58.** (II) A galvanometer has a sensitivity of 35 kΩ/V and internal resistance 20.0 Ω. How could you make this into (a) an ammeter that reads 2.0 A full scale, or (b) a voltmeter reading 1.00 V full scale?

* **59.** (II) A milliammeter reads 10 mA full scale. It consists of a 0.20-Ω resistor in parallel with a 30-Ω galvanometer. How can you change this ammeter to a voltmeter giving a full-scale reading of 10 V without taking the ammeter apart? What will be the sensitivity (Ω/V) of your voltmeter?

*SECTION 19–11

* **60.** (II) A 45-V battery of negligible internal resistance is connected to a 37-kΩ and a 28-kΩ resistor in series. What reading will a voltmeter, of internal resistance 100 kΩ, give when used to measure the voltage across each resistor? What is the percent inaccuracy due to meter resistance for each case?

* **61.** (II) An ammeter whose internal resistance is 60 Ω reads 5.25 mA when connected in a circuit containing a battery and two resistors in series whose values are 700 Ω and 400 Ω. What is the actual current when the ammeter is absent?

* **62.** (II) A battery with $\mathcal{E} = 12.0$ V and internal resistance $r = 1.0\ \Omega$ is connected to two 9.0-kΩ resistors in series. An ammeter of internal resistance 0.50 Ω measures the current and at the same time a voltmeter with internal resistance 15 kΩ measures the voltage across one of the 9.0-kΩ resistors in the circuit. What do the ammeter and voltmeter read?

* **63.** (III) The value of an unknown resistance, R, can be measured in at least two different ways using an ammeter and a voltmeter (Fig. 19–47). The measured value of R in each case is assumed to be $R = V/I$, where V is the voltmeter reading and I is the ammeter reading. Assume that the resistance of the ammeter is 1.00 Ω and that of the voltmeter is 10.0 kΩ in each case. Calculate the measured value for R for each setup, and determine which setup gives the most accurate value for R if the actual value of R is (a) $R = 2.00\ \Omega$, (b) $R = 100\ \Omega$, or (c) $R = 5000\ \Omega$.

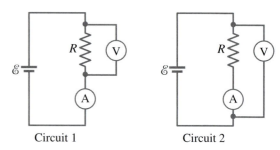

Circuit 1   Circuit 2

FIGURE 19–47 Problem 63.

* **64.** (III) Two 7.4-kΩ resistors are placed in series and connected to a battery. A voltmeter of sensitivity 1000 Ω/V is on the 3.0-V scale and reads 2.0 V when placed across either of the resistors. What is the emf of the battery? (Ignore its internal resistance.)

* **65.** (III) What internal resistance should the voltmeter have to be in error by less than 3 percent for the situation of Example 19–14?

* **66.** (III) The voltage across a 120-kΩ resistor in a circuit containing additional resistance (R_2) in series with a battery (V) is measured, by a 20,000-Ω/V meter on the 100-V scale, to be 25 V. On the 30-V scale, the reading is 23 V. What is the actual voltage in the absence of the voltmeter? What is R_2?

* **67.** (III) A 12.0-V battery (assume the internal resistance = 0) is connected to two resistors in series. A voltmeter whose internal resistance is 15.0 kΩ measures 5.5 V and 4.0 V, respectively, when connected across each of the resistors. What is the resistance of each resistor?

GENERAL PROBLEMS

68. Suppose that you wish to apply a 0.25-V potential difference between two points on the body. The resistance is about 2000 Ω, and you only have a 9.0-V battery. How can you connect up one or more resistors so that you can produce the desired voltage?

69. A three-way lightbulb can produce 50 W, 100 W, or 150 W, at 120 V. Such a bulb contains two filaments that can be connected to the 120 V individually or in parallel. Describe how the connections to the two filaments are made to give each of the three wattages. What must be the resistance of each filament?

70. Suppose you want to run some apparatus that is 95 m from an electric outlet. Each of the wires connecting your apparatus to the 120-V source has a resistance per unit length of 0.0065 Ω/m. If your apparatus draws 3.0 A, what will be the voltage drop across the connecting wires and what voltage will be applied to your apparatus?

71. Electricity can be a hazard in hospitals, particularly to patients who are connected to electrodes, such as an ECG. For example, suppose that the motor of a motorized bed shorts out to the bed frame, and the bed frame's connection to a ground has broken (or was not there in the first place). If a nurse touches the bed and the patient at the same time, she becomes a conductor and a complete circuit can be made through the patient to ground through the ECG apparatus. This is shown schematically in Fig. 19–48. Calculate the current through the patient.

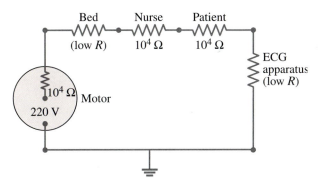

FIGURE 19–48 Problem 71.

72. How much energy must a 45-V battery expend to fully charge a 0.40-μF and a 0.60-μF capacitor when they are placed (a) in parallel, (b) in series? (c) How much charge flowed from the battery in each case?

73. A heart pacemaker is designed to operate at 72 beats/min using a 7.5-μF capacitor in a simple RC circuit. What value of resistance should be used if the pacemaker is to fire when the voltage reaches 63 percent of maximum?

74. The internal resistance of a 1.35-V mercury cell is 0.030 Ω, whereas that of a 1.5-V dry cell is 0.35 Ω. Explain why three mercury cells can more effectively power a 2-W hearing aid that requires 4.0 V than can three dry cells.

75. Suppose that a person's body resistance is 900 Ω. (a) What current passes through the body when the person accidentally is connected to 110 V? (b) If there is an alternative path to ground whose resistance is 40 Ω, what current passes through the person? (c) If the voltage source can produce at most 1.5 A, how much current passes through the person in case (b)?

76. The **Wheatstone bridge** is a circuit used to make precise measurements of resistance. The unknown resistance to be measured, R_x, is placed in the bridge circuit as shown in Fig. 19–49, and the variable resistor R_3 is adjusted until the galvanometer does not deflect when the switch is closed. For this setting, no current flows from B to D, so B and D are at the same potential. (a) Show that the unknown resistance R_x is given by

$$R_x = \frac{R_2}{R_1} R_3$$

when R_1, R_2, and R_3 are all accurately known. (b) A Wheatstone bridge is balanced when $R_1 = 630\ \Omega$, $R_2 = 972\ \Omega$, and $R_3 = 42.6\ \Omega$. What is the value of the unknown resistance in the fourth arm?

77. An unknown length of platinum wire 0.920 mm in diameter is placed as the unknown resistance in a Wheatstone bridge (see Problem 76). Arms 1 and 2 have resistance of 38.0 Ω and 46.0 Ω, respectively. Balance is achieved when R_3 is 3.48 Ω. How long is the platinum wire?

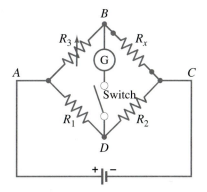

FIGURE 19–49 Wheatstone bridge. Problems 76 and 77.

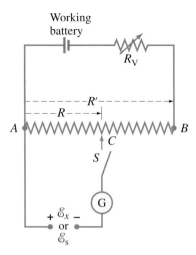

FIGURE 19–50 Potentiometer circuit. Problem 78.

FIGURE 19–51 Problems 79 and 86.

78. A **potentiometer** is a device to precisely measure potential differences or emf, using a "null" technique. In the simple potentiometer circuit shown in Fig. 19–50, R' represents the total resistance of the resistor from A to B (which could be a long uniform "slide" wire), whereas R represents the resistance of only the part from A to the movable contact at C. When the unknown emf to be measured, $\mathcal{E}_x$, is placed into the circuit as shown, the movable contact C is moved until the galvanometer G gives a null reading (i.e., zero) when the switch S is closed. The resistance between A and C for this situation we call R_x. Next, a standard emf, $\mathcal{E}_s$, which is known precisely, is inserted into the circuit in place of $\mathcal{E}_x$ and again the contact C is moved until zero current flows through the galvanometer when the switch S is closed. The resistance between A and C now is called R_s. (a) Show that the unknown emf is given by

$$\mathcal{E}_x = \left(\frac{R_x}{R_s}\right)\mathcal{E}_s$$

where R_x, R_s, and $\mathcal{E}_s$ are all precisely known. The working battery is assumed to be fresh and give a constant voltage. (b) A slide-wire potentiometer is balanced against a 1.0182-V standard cell when the slide wire is set at 25.4 cm out of a total length of 100.0 cm. For an unknown source, the setting is 45.8 cm. What is the emf of the unknown? (c) The galvanometer of a potentiometer has an internal resistance of 30 Ω and can detect a current as small as 0.015 mA. What is the minimum uncertainty possible in measuring an unknown voltage? (d) Explain the advantage of using this "null" method of measuring emf.

79. The variable capacitance of a radio tuner consists of four plates connected together placed alternately between four other plates, also connected together (Fig. 19–51). Each plate is separated from its neighbor by 2.0 mm of air. One set of plates can move so that the area of overlap varies from 2.0 cm² to 12.0 cm². (a) Are these seven capacitors connected in series or in parallel? (b) Determine the range of capacitance values.

80. A battery produces 40.8 V when 7.40 A are drawn from it and 47.3 V when 2.20 A are drawn. What is the emf and internal resistance of the battery?

81. How many $\frac{1}{2}$-W resistors, each of the same resistance, must be used to produce an equivalent 2.2-kΩ, 5-W resistor? What is the resistance of each, and how must they be connected?

82. The current through the 4.0-kΩ resistor in Fig. 19–52 is 3.50 mA. What is the terminal voltage V_{ba} of the "unknown" battery? (There are two answers. Why?) [*Hint:* Use conservation of energy or Kirchhoff's rules.]

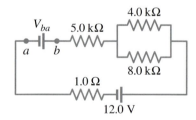

FIGURE 19–52 Problem 82.

83. One type of *ohmmeter* consists of an ammeter connected to a series resistor and a battery, Fig. 19–29. The scale is different from ammeters and voltmeters. Zero resistance corresponds to full-scale deflection (because maximum current flows from the battery), whereas infinite resistance corresponds to no deflection (no current flow). Suppose a 3.0-V battery is used, and the galvanometer has internal resistance of $25\,\Omega$ and deflects full scale for a current of $35\,\mu\text{A}$. What values of shunt resistance, R_{sh}, and series resistance, R_{ser}, are needed to make an ohmmeter that registers a midscale deflection (that is, half of maximum) for a resistance of $30\,\text{k}\Omega$? [Note: An additional series resistor (variable) is also needed so the meter can be zeroed. The zero should be checked frequently by touching the leads together, since the battery voltage can vary. Because of battery voltage variation, such ohmmeters are not precision instruments, but they are useful to obtain approximate values.]

84. (II) Some light dimmer switches use a variable resistor called a *potentiometer*, as shown in Fig. 19–53 (a variation of Fig. 19–50). The slide moves from position $x = 0$ to $x = 1$, and the resistance up to slide position x is proportional to x (the total resistance is $R_{\text{pot}} = 100\,\Omega$). What is the power expended in the lightbulb if (a) $x = 1$, (b) $x = \frac{1}{2}$, (c) $x = \frac{1}{4}$?

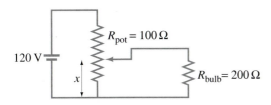

FIGURE 19–53   Problem 84.

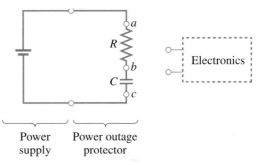

FIGURE 19–54   Problem 85.

85. (II) Electronic devices often use an *RC* circuit to protect against power outages as shown in Fig. 19–54. (a) If the device is supposed to keep the supply voltage at least 70 percent of nominal for as long as 0.20 s, how big a resistance is needed? The capacitor is $22\,\mu\text{F}$. (b) Between which two terminals should the device be connected, a and b, b and c, or a and c?

86. (II) A high-voltage supply can be constructed from a variable capacitor with interleaving plates, one of which can be rotated, as in Fig. 19–51. A version of this type of capacitor with more plates has a capacitance which can be varied from $10\,\text{pF}$ to $1\,\text{pF}$ (a) Initially, this capacitor is connected to a 10,000-volt power supply when the capacitance is 10 pF and charged. It is then disconnected from the power supply and the capacitance reduced to 1 pF by rotating the plates. What is the voltage across the capacitor now? (b) What is a major disadvantage of this as a high-voltage power supply?

Magnets produce magnetic fields, but so do electric currents. An electric current flowing in this coil of wire produces a magnetic field which causes the tiny pieces of iron (iron "filings") to align in the field. Fig. 20–4 shows iron filings in the magnetic field produced by a magnet.

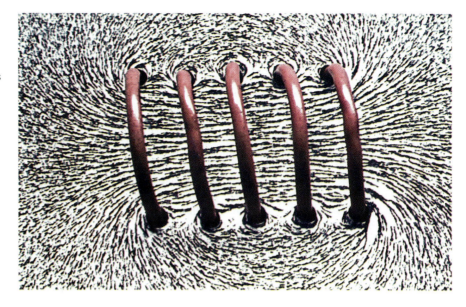

20 MAGNETISM

Today it is clear that magnetism and electricity are closely related. This relationship was not discovered, however, until the nineteenth century. The history of magnetism begins much earlier with the ancient civilizations in Asia Minor. It was in a region of Asia Minor known as Magnesia that rocks were found that would attract each other. These rocks were called "magnets" after their place of discovery.

20–1 Magnets and Magnetic Fields

A magnet will attract paper clips, nails, and other objects made of iron. Any magnet, whether it is in the shape of a bar or a horseshoe, has two ends or faces, called **poles**, which is where the magnetic effect is strongest. If a magnet is suspended from a fine thread, it is found that one pole of the magnet will always point toward the north. It is not known for sure when this fact was discovered, but it is known that the Chinese were making use of it as an aid to navigation by the eleventh century and perhaps earlier. This is, of course, the principle of a compass. A compass needle is simply a magnet that is supported at its center of gravity so it can rotate freely. That pole of a freely suspended magnet which points toward the north is called the **north pole** of the magnet. The other pole points toward the south and is called the **south pole**.

Poles of a magnet

It is a familiar fact that when two magnets are brought near one another, each exerts a force on the other. The force can be either attractive or repulsive and can be felt even when the magnets don't touch. If the north pole of one magnet is brought near the north pole of a second magnet, the force is repulsive. Similarly, if two south poles are brought close, the force is repulsive. But when a north pole is brought near a south pole, the force is attractive. These results are shown in Fig. 20–1, and are reminiscent of the force between electric charges; like poles repel, and unlike poles attract. *But do not confuse magnetic poles with electric charge.* They are not the same thing. One important difference is that a positive or negative electric charge can easily be isolated. But the isolation of a single magnetic pole seems impossible. If a bar magnet is cut in half, you do not obtain isolated north and south poles. Instead, two new magnets are produced, Fig. 20–2. If the cutting operation is repeated, more magnets are produced, each with a north and a south pole. Physicists have tried complicated means to isolate single magnetic poles (monopoles), but so far there is no firm experimental evidence for their existence.

Only iron and a few other materials such as cobalt, nickel, and gadolinium show strong magnetic effects. They are said to be **ferromagnetic** (from the Latin word *ferrum* for iron). All other materials show some slight magnetic effect, but it is extremely weak and can be detected only with delicate instruments. (We will look in more detail at ferromagnetism in Sections 20–13 and 20–15.)

We found it useful to speak of an electric field surrounding an electric charge. In the same way, we can imagine a **magnetic field** surrounding a magnet. The force one magnet exerts on another can then be described as the interaction between one magnet and the magnetic field of the other. Just as we drew electric field lines, we can also draw **magnetic field lines**. They can be drawn, as for electric field lines, so that (1) the direction of the magnetic field is tangent to a line at any point, and (2) the number of lines per unit area is proportional to the magnitude of the magnetic field.

The *direction* of the magnetic field at a given point can be defined as the direction that the north pole of a compass needle would point when placed at that point. Figure 20–3a shows how one magnetic field line around a bar magnet is found using compass needles. The magnetic field determined in this way for the field outside a bar magnet is shown in Fig. 20–3b. Notice that because of our definition, the lines always point from the north toward the south pole of a magnet (the north pole of a magnetic compass needle is attracted to the south pole of another magnet). Figure 20–4 shows how thin iron filings reveal the magnetic field lines by lining up like compass needles.

FIGURE 20–3 (a) Plotting a magnetic field line of a bar magnet. (b) Magnetic field lines outside of a bar magnet.

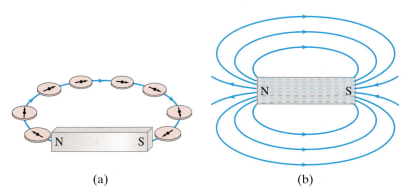

(a)   (b)

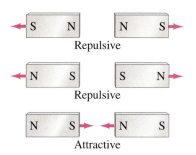

FIGURE 20–1 Like poles of a magnet repel; unlike poles attract.

Magnetic poles not found singly

FIGURE 20–2 If you break a magnet in half, you do not obtain isolated north and south poles; instead, two new magnets are produced, each with a north and a south pole.

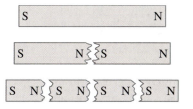

Magnetic field lines

Magnetic field lines point from north to south magnetic poles

FIGURE 20–4 Thin iron filings indicate the magnetic field lines around a bar magnet.

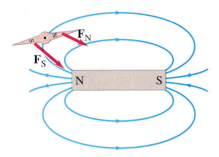

FIGURE 20–5 Forces on a compass needle that produce a torque to orient it parallel to the magnetic field lines. The net torque will be zero when the needle is parallel to the magnetic field line at that point. (Only the attractive forces are shown; try drawing in the repulsive forces and show that they produce a similar torque.)

➡ **PHYSICS APPLIED**

Use of a compass.

FIGURE 20–6 The Earth acts like a huge magnet but its magnetic poles are not at the geographic poles.

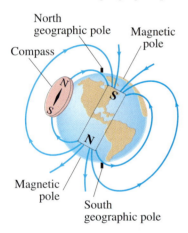

FIGURE 20–7 Using a map and compass in the wilderness. First you align the compass so the needle points away from true north (N) exactly the number of degrees of declination as stated on the (topographic) map: 15° in the case shown. Then align the map with true north, as shown, *not* with the compass needle.

We can define the magnetic field at any point as a vector, represented by the symbol **B**, whose direction is determined as discussed above, using a compass needle. The *magnitude* of **B** can be defined in terms of the torque exerted on a compass needle when it makes a certain angle with the magnetic field, as in Fig. 20–5. That is, the greater the torque, the greater the magnetic field strength. We can use this definition for now, but a more precise definition will be given in Section 20–3.

The Earth's magnetic field is shown in Fig. 20–6. The pattern of field lines is as if there were an (imaginary) bar magnet inside the Earth. Since the north pole of a compass needle points north, the magnetic pole which is in the geographic north is actually a south pole magnetically, as indicated in Fig. 20–6 by the S on the schematic bar magnet inside the Earth. (Remember that the north pole of one magnet is attracted to the south pole of a second.) Nonetheless, this pole is still often called the "north magnetic pole" simply because it is in the north. Similarly, the Earth's south magnetic pole, near the geographic south pole, is magnetically a north pole. The Earth's magnetic poles do not coincide with the geographic poles (which are on the Earth's axis of rotation). The north magnetic pole, for example, is in northern Canada, about 1300 km from the geographic north pole. This must be taken into account when using a compass (Fig. 20–7). The angular difference between magnetic north as indicated by a compass, and true (geographical) north, is called the **magnetic declination**. In the U.S. it varies from 0° to about 25°, depending on location.

Notice in Fig. 20–6 that the Earth's magnetic field is not tangent to the Earth's surface at all points. The angle that the Earth's magnetic field makes with the horizontal at any point is referred to as the **angle of dip**.

The simplest magnetic field is one that is uniform—it doesn't change from one point to another. A perfectly uniform field over a large area is not easy to produce. But the field between two flat parallel pole pieces of a magnet is nearly uniform if the area of the pole faces is large compared to their separation, as shown in Fig. 20–8. At the edges, the field "fringes" out somewhat and is no longer uniform. The parallel evenly spaced field lines in the drawing indicate that the field is uniform at points not too near the edge, much like the electric field between two parallel plates (Fig. 17–1).

FIGURE 20–8 Magnetic field between two large poles of a magnet is nearly uniform except at the edges.

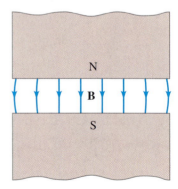

20-2 Electric Currents Produce Magnetism

During the eighteenth century, many natural philosophers sought to find a connection between electricity and magnetism. A stationary electric charge and a magnet were shown not to have any influence on each other. But in 1820, Hans Christian Oersted (1777–1851) found that when a compass needle is placed near an electric wire, the needle deflects as soon as the wire is connected to a battery and a current flows. As we have seen, a compass needle can be deflected by a magnetic field. What Oersted found was that **an electric current produces a magnetic field**. He had found a connection between electricity and magnetism.

A compass needle placed near a straight section of current-carrying wire aligns itself so it is tangent to a circle drawn around the wire, Fig. 20–9. Thus, the magnetic field lines produced by a current in a straight wire are in the form of circles with the wire at their center, Fig. 20–10a. The direction of these lines is indicated by the north pole of the compass in Fig. 20–9. There is a simple way to remember the direction of the magnetic field lines in this case. It is called a **right-hand rule**: you grasp the wire with your right hand so that your thumb points in the direction of the conventional (positive) current; then your fingers will encircle the wire in the direction of the magnetic field, Fig. 20–10b. The magnetic field lines due to a circular loop of current-carrying wire can be determined in a similar way using a compass. The result is shown in Fig. 20–11. Again the right-hand rule can be used, as shown in Fig. 20–12.

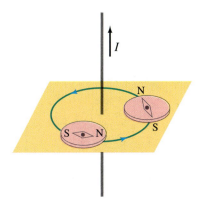

FIGURE 20–9 Deflection of a compass needle near a current-carrying wire, showing the presence and direction of the magnetic field.

FIGURE 20–10 (a) Magnetic field lines around an electric current in a straight wire. (b) Right-hand rule for remembering the direction of the magnetic field: when the thumb points in the direction of the conventional current, the fingers wrapped around the wire point in the direction of the magnetic field.

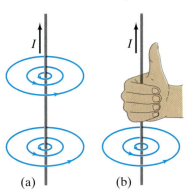

FIGURE 20–11 Magnetic field due to a circular loop of wire.

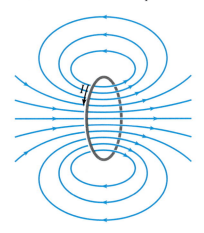

FIGURE 20–12 Right-hand rule for determining the direction of the magnetic field relative to the current.

Magnetic field

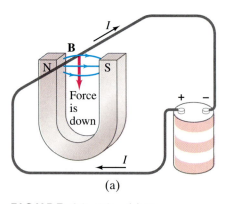

Force is up

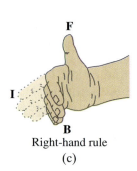

F

I

B

Right-hand rule

(a)   (b)   (c)

FIGURE 20–13 (a) Force on a current-carrying wire placed in a magnetic field **B**; (b) same, but current reversed; (c) right-hand rule for setup in (b).

Force on an Electric Current in a Magnetic Field; Definition of B

In Section 20–2 we saw that an electric current exerts a force on a magnet, such as a compass needle. By Newton's third law, we might expect the reverse to be true as well: we should expect that *a magnet exerts a force on a current-carrying wire.* Experiments indeed confirm this effect, and it too was first observed by Oersted.

Magnet exerts a force on an electric current

Let us look at the force exerted on a wire in detail. Suppose a straight wire is placed between the poles of a horseshoe magnet as shown in Fig. 20–13. When a current flows in the wire, a force is exerted on the wire. But this force is *not* toward one or the other poles of the magnet. Instead, the force is directed at right angles to the magnetic field direction. If the current is reversed in direction, the force is in the opposite direction. It is found that *the direction of the force is always perpendicular to the direction of the current and also perpendicular to the direction of the magnetic field,* **B**. This statement does not completely describe the direction, however: the force could be either up or down in Fig. 20–13b and still be perpendicular to both the current and to **B**. Experimentally, the direction of the force is given by another *right-hand rule*, as illustrated in Fig. 20–13c. First you orient your right hand so that the outstretched fingers point in the direction of the (conventional) current; from this position, bend your fingers, so that they point in the direction of the magnetic field lines (which point from the N toward the S pole outside a magnet); you may have to rotate your hand and arm about the wrist until they do point along **B** when bent, remembering that straightened fingers must point along the direction of the current first. When your hand is oriented in this way, then the extended thumb points in the direction of the force on the wire.

Right-hand rule for force on current due to **B**

FIGURE 20–14 Current-carrying wire in a magnetic field. Force on the wire is directed into the page.

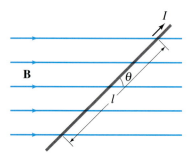

This describes the direction of the force. What about its magnitude? It is found experimentally that the magnitude of the force is directly proportional to the current I in the wire, to the length l of wire in the magnetic field (assumed uniform), and to the magnetic field B. The force also depends on the angle θ between the current direction and the magnetic field (Fig. 20–14). When the current is perpendicular to the field lines, the force is strongest. When the wire is parallel to the magnetic field lines, there is no force at all. At other angles, the force is proportional to $\sin \theta$ (Fig. 20–14).

Thus we have

$$F \propto IlB \sin \theta.$$

Up to now we have talked of magnetic field strengths in terms of the torque exerted by the field on a compass needle (Fig. 20–5). But we have not defined the magnetic field strength precisely. In fact, the magnetic field B is defined in terms of the above proportion so that the proportionality constant is precisely 1. Thus we have

$$F = IlB \sin \theta. \qquad (20\text{–}1)$$

FORCE ON ELECTRIC CURRENT IN A MAGNETIC FIELD

If the direction of the current is perpendicular to the field ($\theta = 90°$), then the force is

$$F_{max} = IlB. \qquad [\mathbf{I} \perp \mathbf{B}] \quad (20\text{–}2)$$

If the current is parallel to the field ($\theta = 0°$), the force is zero.[†]

In summary, the magnetic field vector $\mathbf{B}$ is defined as follows. The direction of $\mathbf{B}$ in a region of space is the direction that a straight section of current-carrying wire would have when placed in the field and the force on it is zero ($\theta = 0°$ in Eq. 20–1), and consistent with the right-hand rule when the wire is oriented in another direction. The magnitude of $\mathbf{B}$ is defined (from Eq. 20–2) as

Definition of magnetic field

$$B = \frac{F_{max}}{Il},$$

where F_{max} is the magnitude of the force on a straight length l of wire carrying a current I when the wire is perpendicular to $\mathbf{B}$.

The SI unit for magnetic field B is the **tesla** (T). From Eq. 20–1, it is clear that $1\,\text{T} = 1\,\text{N/A·m}$. An older name for the tesla is the "weber per meter squared" ($1\,\text{Wb/m}^2 = 1\,\text{T}$). Another unit commonly used to specify magnetic field is a cgs unit, the **gauss** (G): $1\,\text{G} = 10^{-4}\,\text{T}$. A field given in gauss should always be changed to teslas before using with other SI units. To get a "feel" for these units, we note that the magnetic field of the Earth at its surface is about $\frac{1}{2}$ G or $0.5 \times 10^{-4}\,\text{T}$. On the other hand, strong electromagnets can produce fields on the order of 2 T and superconducting magnets over 10 T.

The tesla and the gauss (units)

EXAMPLE 20–1 **Magnetic force on a current-carrying wire.** A wire carrying a 30-A current has a length $l = 12\,\text{cm}$ between the pole faces of a magnet at an angle $\theta = 60°$ (Fig. 20–14). The magnetic field is approximately uniform at 0.90 T. We ignore the field beyond the pole pieces. What is the force on the wire?

SOLUTION We use Eq. 20–1 and find that

$$F = IlB \sin \theta$$

$$= (30\ \text{A})(0.12\ \text{m})(0.90\ \text{T})(0.866) = 2.8\ \text{N}.$$

On a diagram, when we want to represent a magnetic field that is pointing out of the page (toward us) or into the page, we use $\odot$ or $\times$. The $\odot$ is meant to

---

[†]In our discussion, we have assumed that the magnetic field is uniform. If it is not, then B in Eqs. 20–1 and 20–2 is the average field over the length l of the wire. In practical cases, we can consider a wire as made up of many short segments Δl and the force on each segment is proportional to the length Δl of that segment and to the magnetic field B at that segment. The total force is the vector sum of the individual forces.

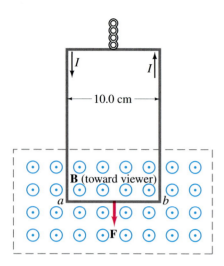

FIGURE 20–15 Measuring a magnetic field **B**. Example 20–2.

resemble the tip of an arrow pointing directly toward the reader, whereas the × or ⊗ resembles the tail of an arrow going away. (See Fig. 20–15.)

EXAMPLE 20–2 **Measuring a magnetic field.** A rectangular loop of wire hangs vertically as shown in Fig. 20–15. A magnetic field **B** is directed horizontally, perpendicular to the wire, and points out of the page at all points as represented by the symbol ⊙. The magnetic field **B** is very nearly uniform along the horizontal portion of wire *ab* (length $l = 10.0$ cm) which is near the center of a large magnet producing the field. The top portion of the wire loop is free of the field. The loop hangs from a balance which measures a downward force (in addition to the gravitational force) of $F = 3.48 \times 10^{-2}$ N when the wire carries a current $I = 0.245$ A. What is the magnitude of the magnetic field B at the center of the magnet?

SOLUTION The magnetic forces on the two vertical sections of the wire loop point to the left and right, respectively. They are equal and in opposite directions and so add up to zero. Hence, the net magnetic force on the loop is that on the horizontal section *ab* whose length is $l = 0.100$ m (and $\theta = 90°$ so $\sin \theta = 1$); thus

$$B = \frac{F}{Il} = \frac{3.48 \times 10^{-2} \text{ N}}{(0.245 \text{ A})(0.100 \text{ m})} = 1.42 \text{ T}.$$

This technique is a highly precise means of determining magnetic fields.

20–4 Force on an Electric Charge Moving in a Magnetic Field

We have seen that a current-carrying wire experiences a force when placed in a magnetic field. Since a current in a wire consists of moving electric charges, we might expect that freely moving charged particles (not in a wire) would also experience a force when passing through a magnetic field. Indeed, this is the case.

From what we already know, let us determine the force on a single moving electric charge. If N such particles of charge q pass by a given point in time t, they constitute a current $I = Nq/t$. We let t be the time for a charge q to travel a distance l in a magnetic field B; then $l = vt$, where v is the velocity of the particle. Thus, the force on these N particles is, by Eq. 20–1, $F = IlB \sin \theta = (Nq/t)(vt)B \sin \theta$. The force on *one* of the particles is found by dividing by N:

$$F = qvB \sin \theta. \tag{20–3}$$

This equation gives the magnitude of the force on a particle of charge q moving with velocity v in a magnetic field of strength B, where θ is the angle between **v** and **B**. The force is greatest when the particle moves perpendicular to **B** ($\theta = 90°$):

$$F_{max} = qvB. \qquad [\mathbf{v} \perp \mathbf{B}] \quad (20–4)$$

The force is *zero* if the particle moves *parallel* to the field lines ($\theta = 0°$). The *direction* of the force is perpendicular to the magnetic field **B** and to the velocity **v** of the particle. It is again given by a *right-hand rule*: you orient your right hand so that your outstretched fingers point along the direction of motion of the particle (**v**), and when you bend your fingers they must point

Right-hand rule

along the direction of **B**; then your thumb will point in the direction of the force. This is true only for *positively* charged particles, and will be "down" for the situation shown in Fig. 20–16. For negatively charged particles, the force is in exactly the opposite direction ("up" in Fig. 20–16).

EXAMPLE 20–3 **Magnetic force on a proton.** A proton having a speed of 5.0×10^6 m/s in a magnetic field feels a force of 8.0×10^{-14} N toward the west when it moves vertically upward. When moving horizontally in a northerly direction, it feels zero force. What is the magnitude and direction of the magnetic field in this region? (The charge on a proton is $q = +e = 1.6 \times 10^{-19}$ C.)

SOLUTION Since the proton feels no force when moving north, the field must be in a north–south direction. The right-hand rule tells us that **B** must point toward the north in order to produce a force to the west when the proton moves upward. (Your thumb points west and the outstretched fingers of your right hand point upward only when your bent fingers point north.) The magnitude of **B**, from Eq. 20–3 with $\theta = 90°$, is

$$B = \frac{F}{qv} = \frac{8.0 \times 10^{-14} \text{ N}}{(1.6 \times 10^{-19} \text{ C})(5.0 \times 10^6 \text{ m/s})} = 0.10 \text{ T}.$$

The path of a charged particle moving in a plane perpendicular to a uniform magnetic field is a circle (or the arc of a circle if the particle later passes out of the magnetic field region). See Fig. 20–17, where the magnetic field is directed *into* the paper, as represented by ×'s. An electron at point *P* is moving to the right, and the force on it at this point is downward as shown (use the right-hand rule and reverse the direction for negative charge). The electron is thus deflected downward. A moment later, say when it reaches point *Q*, the force is still perpendicular to the velocity and is in the direction shown. Since the force is always perpendicular to **v**, the magnitude of **v** does not change but the particle changes direction and moves in a circular path, with a centripetal acceleration (we demonstrate this in Example 20–4). The force is directed toward the center of this circle at all points. Note that the electron moves clockwise in Fig. 20–17. A positive particle would feel a force in the opposite direction and would thus move counterclockwise.

EXAMPLE 20–4 **Electron's path in a uniform magnetic field.** An electron travels at 2.0×10^7 m/s in a plane perpendicular to a 0.010-T magnetic field. Describe its path.

SOLUTION The electron moves at constant speed in a curved path whose radius of curvature is found using Newton's second law, $F = ma$. We have a centripetal acceleration $a = v^2/r$ (Eq. 5–1). The force is given by Eq. 20–4, $F = qvB$, so we have

$$F = ma$$

$$qvB = \frac{mv^2}{r}.$$

We solve for *r* and find

$$r = \frac{mv}{qB}.$$

Since **F** is perpendicular to **v**, the magnitude of **v** doesn't change. From this equation we see that if **B** = constant, then *r* = constant, and the curve must be a circle as we claimed above.

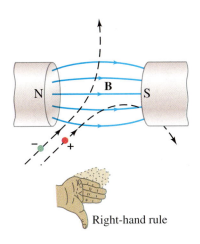

Right-hand rule

FIGURE 20–16 Force on charged particles due to a magnetic field is perpendicular to the magnetic field direction.

FIGURE 20–17 Force exerted by a uniform magnetic field on a moving charged particle (in this case, an electron) produces a circular path.

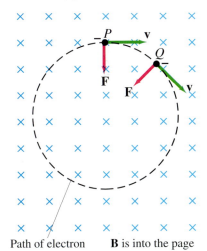

Path of electron   B is into the page

To get r, we put in the numbers:

$$r = \frac{(9.1 \times 10^{-31}\ \text{kg})(2.0 \times 10^7\ \text{m/s})}{(1.6 \times 10^{-19}\ \text{C})(0.010\ \text{T})} = 1.1 \times 10^{-2}\ \text{m},$$

or 1.1 cm.

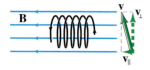

FIGURE 20-18 Example 20-5.

CONCEPTUAL EXAMPLE 20-5   **A spiral path.** What is the path of a charged particle if its velocity is *not* perpendicular to the magnetic field?

RESPONSE The velocity vector can be broken down into components parallel and perpendicular to the field. The velocity component parallel to the field lines experiences no force, and so this component remains constant. The velocity component perpendicular to the field results in circular motion about the field lines. Putting these two motions back together produces a helical (spiral) motion around the field lines as shown in Fig. 20-18.

➡ **PHYSICS APPLIED**

The aurora borealis

CONCEPTUAL EXAMPLE 20-6   **Aurora borealis.** Charged ions approach the Earth from the Sun (the "solar wind") and are drawn toward the poles, sometimes causing a phenomenon called the *aurora borealis* or "northern lights" in northern latitudes. Why toward the poles?

RESPONSE A glance at Fig. 20-19 (see also Fig. 20-18) provides the answer. Imagine a stream of charged particles approaching the Earth as shown. The velocity component perpendicular to the field for each particle becomes a circular orbit around the field lines, whereas the velocity component parallel to the field carries the particle along the field lines toward the poles. The high concentration of charged particles ionizes the air, and the recombining of electrons with atoms emits light (Chapter 27), which is the aurora, especially during periods of high sun spot activity when the solar wind is greater.

FIGURE 20-19 (a) Diagram showing a charged particle approaching the Earth which is "captured" by the magnetic field of the Earth. Such particles follow the field lines toward the poles as shown. (b) Photo of aurora borealis.

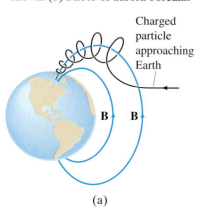

(a)

(b)

20–5 Magnetic Field Due to a Straight Wire

We saw in Section 20–2, Fig. 20–10, that the magnetic field due to the electric current in a long straight wire is such that the field lines are circles with the wire at the center (Fig. 20–20). You might expect that the field strength at a given point would be greater if the current flowing in the wire were greater; and that the field would be less at points farther from the wire. This is indeed the case. Careful experiments show that the magnetic field B at a point near a long straight wire is directly proportional to the current I in the wire and inversely proportional to the distance r from the wire:

$$B \propto \frac{I}{r}.$$

This relation is valid as long as r, the perpendicular distance to the wire, is much less than the distance to the ends of the wire (i.e., the wire is long). The proportionality constant is written[†] as $\mu_0/2\pi$; thus,

$$B = \frac{\mu_0}{2\pi}\frac{I}{r}. \qquad \text{[outside a long straight wire]} \quad \textbf{(20–5)}$$

The value of the constant μ_0, which is called the **permeability of free space**, is $\mu_0 = 4\pi \times 10^{-7}\,\text{T·m/A}$.

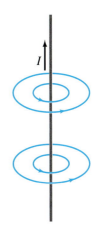

FIGURE 20–20 Same as Fig. 20–10a, magnetic field lines around a long straight wire carrying an electric current I.

Magnetic field due to current in straight wire

EXAMPLE 20–7 **Calculation of B near a wire.** A vertical electric wire in the wall of a building carries a dc current of 25 A upward. What is the magnetic field at a point 10 cm due north of this wire (Fig. 20–21)?

SOLUTION According to Eq. 20–5:

$$B = \frac{\mu_0 I}{2\pi r} = \frac{(4\pi \times 10^{-7}\,\text{T·m/A})(25\,\text{A})}{(2\pi)(0.10\,\text{m})} = 5.0 \times 10^{-5}\,\text{T},$$

or 0.50 G. By the right-hand rule (Fig. 20–10b), the field points to the west (into the page in Fig. 20–21) at this point. Since this field has about the same magnitude as Earth's, a compass would not point north but in a northwesterly direction.

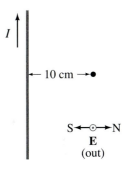

FIGURE 20–21 Example 20–7.

➡ **PHYSICS APPLIED**

A compass, near a current, may not point North

20–6 Force Between Two Parallel Wires

We have seen that a wire carrying a current produces a magnetic field (magnitude given by Eq. 20–5 for a long straight wire), and furthermore that such a wire feels a force when placed in a magnetic field (Section 20–3, Eq. 20–1). Thus, we expect that two current-carrying wires would exert a force on each other.

[†]The constant is chosen in this complicated way so that Ampère's law (Section 20–8), which is considered more fundamental, will have a simple and elegant form.

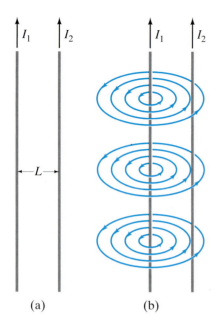

(a)   (b)

FIGURE 20–22 (a) Two parallel conductors carrying currents I_1 and I_2. (b) Magnetic field produced by I_1. (Field produced by I_2 is not shown.)

FIGURE 20–23 (a) Parallel currents in the same direction exert attractive force on each other. (b) Antiparallel currents (in opposite directions) exert repulsive force on each other.

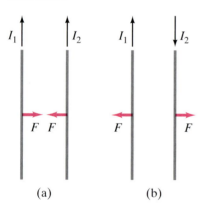

(a)   (b)

Consider two long parallel conductors separated by a distance L, as in Fig. 20–22a. They carry currents I_1 and I_2, respectively. Each current produces a magnetic field that is "felt" by the other so that each must exert a force on the other, as Ampère first pointed out. For example, the magnetic field B_1 produced by I_1 is given by Eq. 20–5. At the location of the second conductor, the magnitude of this field is

$$B_1 = \frac{\mu_0}{2\pi} \frac{I_1}{L}.$$

See Fig. 20–22b where the field due *only* to I_1 is shown. According to Eq. 20–2, the force F per unit length l on the conductor carrying current I_2 is

$$\frac{F}{l} = I_2 B_1.$$

Note that the force on I_2 is due only to the field produced by I_1. Of course I_2 also produces a field, but it does not exert a force on itself. We substitute in the above formula for B_1 and find

$$\frac{F}{l} = \frac{\mu_0}{2\pi} \frac{I_1 I_2}{L}. \qquad (20\text{–}6)$$

If we use the right-hand rule of Fig. 20–10b, we see that the lines of B_1 are as shown in Fig. 20–22b. Then using the right-hand rule of Fig. 20–13c, we see that the force exerted on I_2 will be to the left in the figure. That is, I_1 exerts an attractive force on I_2 (Fig. 20–23a). This is true as long as the currents are in the same direction. If I_2 is in the opposite direction, the right-hand rule indicates that the force is in the opposite direction. That is, I_1 exerts a repulsive force on I_2 (Fig. 20–23b). Reasoning similar to that above shows that the magnetic field produced by I_2 exerts an equal but opposite force on I_1. We expect this to be true also, of course, from Newton's third law. Thus, as shown in Fig. 20–23, parallel currents in the same directions attract each other, whereas parallel currents in opposite directions repel.

EXAMPLE 20–8 Force between two current carrying wires. The two wires of a 2.0-m-long appliance cord are 3.0 mm apart and carry a current of 8.0 A dc. Calculate the force between these wires.

SOLUTION Equation 20–6 gives us

$$F = \frac{(2.0 \times 10^{-7}\,\text{T·m/A})(8.0\,\text{A})^2(2.0\,\text{m})}{(3.0 \times 10^{-3}\,\text{m})} = 8.5 \times 10^{-3}\,\text{N},$$

where we have written $\mu_0/2\pi = 2.0 \times 10^{-7}\,\text{T·m/A}$. Since the currents are in opposite directions, the force would tend to spread them apart.

EXAMPLE 20–9 **Suspending a current with a current.** A horizontal wire carries a current $I_1 = 80$ A dc. A second parallel wire 20 cm below it (Fig. 20–24) must carry how much current I_2 so that it doesn't fall due to gravity? The lower wire has a mass of 0.12 g per meter of length.

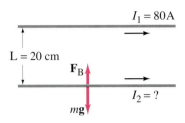

SOLUTION The force of gravity on the lower wire is downward and per each meter of length has magnitude

$$\frac{F}{l} = \frac{mg}{l} = \frac{(0.12 \times 10^{-3}\,\text{kg})(9.8\,\text{m/s}^2)}{1.0\,\text{m}} = 1.18 \times 10^{-3}\,\text{N/m}.$$

The magnetic force on wire 2 must be upward (hence I_2 must have the same direction as I_1) and with $L = 0.20$ m and $I_1 = 80$ A has magnitude

$$\frac{F}{l} = \frac{\mu_0}{2\pi}\frac{I_1 I_2}{L}$$

FIGURE 20–24 Example 20–9.

We solve for I_2 and find

$$I_2 = \frac{2\pi L}{\mu_0 I_1}\left(\frac{F}{l}\right) = \frac{2\pi (0.20\,\text{m})}{(4\pi \times 10^{-7}\,\text{T·m/A})(80\,\text{A})}(1.18 \times 10^{-3}\,\text{N/m}) = 15\,\text{A}.$$

* 20–7 Definition of the Ampere and the Coulomb

You may have wondered how the constant μ_0 in Eq. 20–5 could be exactly $4\pi \times 10^{-7}$ T·m/A. Here is how it happened. With an older definition of the ampere, μ_0 was measured experimentally to be very close to this value. Today, however, μ_0 is *defined* to be exactly $4\pi \times 10^{-7}$ T·m/A. This, of course, could not be done if the ampere were defined independently. The ampere, the unit of current, is now defined in terms of the magnetic field B it produces using the defined value of μ_0.

In particular, we use the force between two parallel current-carrying wires, Eq. 20–6, to define the ampere precisely. If $I_1 = I_2 = 1$ A exactly, and the two wires are exactly 1 m apart, then

$$\frac{F}{l} = \frac{(4\pi \times 10^{-7}\,\text{T·m/A})}{(2\pi)}\frac{(1\,\text{A})(1\,\text{A})}{(1\,\text{m})} = 2 \times 10^{-7}\,\text{N/m}.$$

Thus, *one* **ampere** *is defined as that current flowing in each of two long parallel conductors 1 m apart, which results in a force of exactly 2×10^{-7} N/m of length of each conductor.*

Definitions of ampere and coulomb

This is the precise definition of the ampere. The **coulomb** is then defined as being *exactly* one ampere-second: $1\,\text{C} = 1\,\text{A·s}$. The value of k or ϵ_0 in Coulomb's law (Section 16–5) is obtained from experiment.

* 20–8 Ampère's Law

In Section 20–5, we saw that Eq. 20–5 gives the relation between the current in a long straight wire and the magnetic field it produces. This equation is valid only for a long straight wire. The following important question arises: Is there a general relation between a current in a wire of whatever shape and the magnetic field around it? The answer is yes: the French scientist André Marie Ampère (1775–1836) proposed such a relation shortly after Oersted's discovery. Consider any (arbitrary) closed path around a current, as shown in Fig. 20–25, and imagine this path as being made up of short segments each of length Δl. First, we take the product of the length of each segment times the component of **B** parallel to that segment. If we now sum all these terms, according to Ampère, the result will be equal to μ_0 times the net current I that passes through the surface enclosed by the path. This is known as **Ampère's law** and can be written mathematically as

$$\sum B_{\parallel}\, \Delta l = \mu_0 I. \qquad \textbf{(20–7)}$$

The symbol Σ means "the sum of" and $B_{\parallel}$ means the component of **B** parallel to that particular Δl. The lengths Δl are chosen so that $B_{\parallel}$ is essentially constant on each length. The sum must be made over a closed path; and I is the net current passing through the surface bounded by this closed path.

We can check Ampère's law by applying it to the simple case of a long straight wire carrying a current I, which we have already examined and which served as an inspiration for Ampère himself. Suppose that we want to find the magnitude of B at point A, a distance r from the wire in Fig. 20–26. We know that the magnetic field lines are circles with the wire at their center. We then choose a path to be used in Eq. 20–7: we choose a circle of radius r (the choice of path is ours—so we choose one that will be convenient). We choose this circular path because at any point on this path, **B** will be tangent to this circle. Thus, for any short segment of the circle (Fig. 20–26), **B** will be parallel to that segment, so $B_{\parallel} = B$. Suppose that we break the circular path down into 100 segments.[†] Then Ampère's law states that

$$(B\, \Delta l)_1 + (B\, \Delta l)_2 + (B\, \Delta l)_3 + \cdots + (B\, \Delta l)_{100} = \mu_0 I.$$

The dots represent all the terms we did not write down. Since all the segments are the same distance from the wire, we expect B to be the same at each segment. We can then factor out B from the sum:

$$B(\Delta l_1 + \Delta l_2 + \Delta l_3 + \cdots + \Delta l_{100}) = \mu_0 I.$$

The sum of the segment lengths is just the circumference of the circle, $2\pi r$. Thus we have

$$B(2\pi r) = \mu_0 I,$$

or

$$B = \frac{\mu_0 I}{2\pi r}.$$

This is just Eq. 20–5 for the field near a long straight wire, as discussed earlier.

Ampère's law thus works for this simple case. A great many experiments indicate that Ampère's law is valid in general. However, it can be used to calculate the magnetic field mainly for simple cases. Its importance is that it relates the magnetic field to the current in a direct and mathematically elegant way. Ampère's law is thus considered one of the basic laws of electricity and magnetism. It is valid for any situation where the currents and fields are not changing in time.

We now can see why the constant in Eq. 20–5 is written $\mu_0/2\pi$. This is done so that only μ_0 appears in Eq. 20–7 (rather than, say, $2\pi k$ if we had used k in Eq. 20–5). In this way, the more fundamental equation, Ampère's law, has the simpler form.

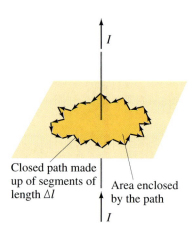

FIGURE 20–25 Arbitrary path enclosing a current, for Ampère's law. The path is broken down into segments of equal length Δl.

Closed path made up of segments of length Δl

Area enclosed by the path

FIGURE 20–26 Circular path of radius r.

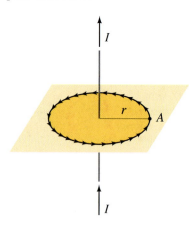

[†]Actually, Ampère's law is precisely accurate when there is an infinite number of infinitesimally short segments, but that leads into calculus.

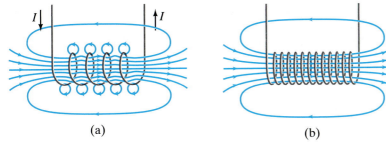

(a)

(b)

We now use Ampère's law to calculate the magnetic field inside a long coil of wire with many loops, as shown in Fig. 20–27, which is known as a **solenoid**. Each coil produces a magnetic field as shown in Fig. 20–11, and the total field inside the solenoid will be the sum of the fields due to each current loop as shown in Fig. 20–27a. If the coils of the solenoid are very closely spaced, the field inside will be essentially parallel to the axis except at the ends, as shown in Fig. 20–27b. For applying Ampère's law, we choose the path $abcd$ shown in Fig. 20–28, far from either end. We will consider this path as made up of four segments, the sides of the rectangle: ab, bc, cd, da. Then the left side of Eq. 20–7 becomes

$$(B_\parallel \Delta l)_{ab} + (B_\parallel \Delta l)_{bc} + (B_\parallel \Delta l)_{cd} + (B_\parallel \Delta l)_{da}.$$

The first term in this sum will be very small since the field outside the solenoid is so small as to be negligible compared to the field inside (the same number of lines inside the solenoid spread throughout space outside). Thus the first term will be zero. Furthermore, **B** is perpendicular to the segments bc and da, so these terms are zero, too. Therefore, the left side of Eq. 20–7 is simply $(B_\parallel \Delta l)_{cd} = Bl$, where B is the field inside the solenoid, and l is the length cd. Now we determine the current enclosed by our chosen rectangular loop, to use for the right side of Eq. 20–7. If a current I flows in the wire of the solenoid, the total current enclosed by our path $abcd$ is NI, where N is the number of loops our path encircles (five in Fig. 20–28). Thus Ampère's law gives us

$$Bl = \mu_0 NI.$$

If we let $n = N/l$ be the *number of loops per unit length*, then

$$B = \mu_0 nI. \qquad \text{[solenoid]} \quad \textbf{(20–8)}$$

This is the magnitude of the magnetic field within a solenoid. Note that B depends only on the number of loops per unit length, n, and the current I. The field does not depend on the position within the solenoid, so B is uniform. This is strictly true only for an infinite solenoid, but it is a good approximation for real ones for points not close to the ends.

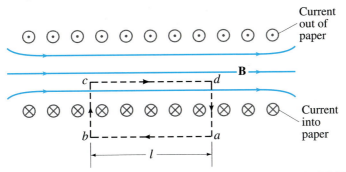

Current out of paper

c

d

B

Current into paper

b

a

l

EXAMPLE 20–10 **Field inside a solenoid.** A thin 10-cm-long solenoid has a total of 400 turns of wire and carries a current of 2.0 A. Calculate the field inside near the center.

SOLUTION The number of turns per unit length is $n = 400/0.10\,\text{m} = 4.0 \times 10^3\,\text{m}^{-1}$. From Eq. 20–8:

$$B = \mu_0 nI = (12.57 \times 10^{-7}\,\text{T·m/A})(4.0 \times 10^3\,\text{m}^{-1})(2.0\,\text{A})$$
$$= 1.0 \times 10^{-2}\,\text{T}.$$

➡ **PHYSICS APPLIED**

Coaxial cable
(shielding)

CONCEPTUAL EXAMPLE 20–11 **Coaxial cable.** A *coaxial cable* is a single wire surrounded by a cylindrical metallic braid, as shown in Fig. 20–29. The two conductors are separated by an insulator. The central wire carries current to the other end of the cable, and the outer braid carries the return current and is usually considered ground. Describe the magnetic field (*a*) in the space between the conductors, and (*b*) outside the cable.

RESPONSE (*a*) In the space between the conductors, we can apply Ampère's law for a circular path around the center wire, just as we did for the case shown in Fig. 20–26 and the magnitude is as given by Eq. 20–5. The current in the outer conductor has no bearing on this result. (Ampère's law uses only the current enclosed *inside* the path; as long as the currents outside the path don't affect the symmetry of the field, they do not contribute to the field along the path at all).
(*b*) Outside the cable, we can draw a similar circular path, for we expect the field to have the same circular symmetry. Now, however, there are two currents enclosed by the path, and they add up to zero. The field outside the cable is zero.

The nice feature of coaxial cables is that they are self-shielding: no stray magnetic fields escape outside the cable. The outer cylindrical conductor also shields external electric fields from coming in (see also Example 16–9). This makes them ideal for carrying signals near sensitive equipment. Audiophiles use coaxial cables between stereo equipment components and even to the loudspeakers.

FIGURE 20–29 Coaxial cable. Example 20–11.

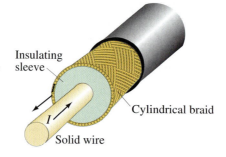

Insulating sleeve

Cylindrical braid

Solid wire

I

*20-9 Torque on a Current Loop; Magnetic Moment

When an electric current flows in a closed loop of wire placed in a magnetic field, as shown in Fig. 20–30, the magnetic force on the current can produce a torque. This is the basic principle behind a number of important practical devices, including meters and motors. (We discuss these applications in the next Section.) The interaction between a current and a magnetic field is important in other areas as well, including atomic physics.

When current flows through the loop in Fig. 20–30a, whose face we assume is parallel to **B** and is rectangular, the magnetic field exerts a force on both vertical sections of wire as shown, **F₁** and **F₂** (see also top view, Fig. 20–30b). Notice that, by the right-hand rule (Fig. 20–13c), the direction of the force on the upward current on the left is in the opposite direction from the equal magnitude force **F₂** on the descending current on the right. These forces give rise to a net torque that tends to rotate the coil about its vertical axis.

Let us calculate the magnitude of this torque. From Eq. 20–2, the force $F = IaB$, where a is the length of the vertical arm of the coil. The lever arm for each force is $b/2$, where b is the width of the coil and the "axis" is at the midpoint. The total torque is the sum of the torques due to each of the forces, so

$$\tau = IaB\frac{b}{2} + IaB\frac{b}{2} = IabB = IAB,$$

where $A = ab$ is the area of the coil. If the coil consists of N loops of wire, the current is then NI, so the torque becomes

$$\tau = NIAB. \tag{20-9a}$$

If the coil makes an angle θ with the magnetic field, as shown in Fig. 20–30c, the forces are unchanged, but each lever arm is reduced from $\frac{1}{2}b$ to $\frac{1}{2}b\sin\theta$. Note that the angle θ is chosen to be the angle between **B** and the perpendicular to the face of the coil, Fig. 20–30c. So the torque becomes

$$\tau = NIAB\sin\theta. \tag{20-9b}$$

This formula, derived here for a rectangular coil, is valid for any shape of flat coil.

The quantity NIA is called the **magnetic dipole moment** of the coil and is considered a vector:

$$\mathbf{M} = NI\mathbf{A}, \tag{20-10}$$

where the direction of **A** (and therefore of **M**) is *perpendicular* to the plane of the coil (the black arrow in Fig. 20–30c).

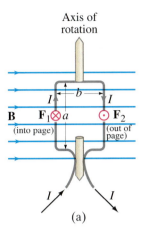

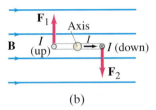

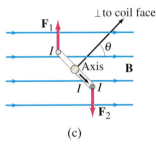

FIGURE 20-30 Calculating the torque on a current loop in a magnetic field **B**. (a) Loop face parallel to **B** field lines; (b) Top view; (c) Loop makes an angle to **B**, reducing the torque since the lever arm is reduced.

EXAMPLE 20–12 **Torque on a coil.** A circular coil of wire has a diameter of 20.0 cm and contains 10 loops. The current in each loop is 3.00 A, and the coil is placed in a 2.00-T magnetic field. Determine the maximum and minimum torque exerted on the coil by the field.

SOLUTION Equation 20–9 is valid for any shape of coil, including circular, where the area is

$$A = \pi r^2 = \pi (0.100 \text{ m})^2 = 3.14 \times 10^{-2} \text{ m}^2.$$

The maximum torque occurs when the coil's face is parallel to the magnetic field, so $\theta = 90°$ in Fig. 20–30c, and $\sin \theta = 1$ in Eq. 20–9b:

$$\tau = NIAB \sin \theta = (10)(3.00 \text{ A})(3.14 \times 10^{-2} \text{ m}^2)(2.00 \text{ T})(1) = 1.88 \text{ N·m}.$$

The minimum torque occurs if $\sin \theta = 0$, for which $\theta = 0°$, and then $\tau = 0$ from Eq. 20–9b.

* 20–10 Applications: Galvanometers, Motors, Loudspeakers

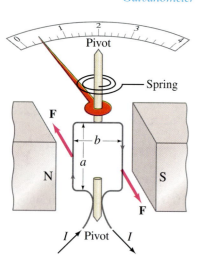

FIGURE 20–31 Galvanometer.

FIGURE 20–32 Galvanometer coil wrapped on an iron core.

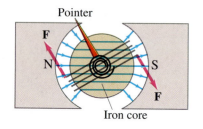

The basic component of most meters, including ammeters, voltmeters, and ohmmeters, is a galvanometer. We have already seen how these meters are designed (Section 19–10), and now we can examine how the crucial element, a galvanometer, itself works. As shown in Fig. 20–31, a **galvanometer** consists of a coil of wire (with attached pointer) suspended in the magnetic field of a permanent magnet. When current flows through the loop of wire, which is usually rectangular, the magnetic field exerts a torque on the loop, as given by Eq. 20–9b, $\tau = NIAB \sin \theta$. This torque is opposed by a spring which exerts a torque τ_s approximately proportional to the angle ϕ through which it is turned (Hooke's law). That is,

$$\tau_s = k\phi,$$

where k is the stiffness constant of the spring. Thus the coil and the attached pointer will rotate only to the point where the spring torque balances the torque due to the magnetic field. From Eq. 20–9b we then have $k\phi = NIAB \sin \theta$ or

$$\phi = \frac{NIAB \sin \theta}{k}.$$

Thus the deflection of the pointer, ϕ, is directly proportional to the current I flowing in the coil. But it also depends on the angle θ the coil makes with **B**. For a useful meter we need ϕ to depend only on I, independent of θ. To solve this problem, curved pole pieces are used and the galvanometer coil is wrapped around a cylindrical iron core as shown in Fig. 20–32. The iron tends to concentrate the magnetic field lines so that **B** always points parallel to the face of the coil at the wire outside the core. The force is then always perpendicular to the face of the coil and the torque will not vary with angle. Thus ϕ will be proportional to I, as required.

A **chart recorder**, in which a pen graphs a signal such as an ECG (electrocardiogram–Section 17–11) on a moving roll of paper, is basically

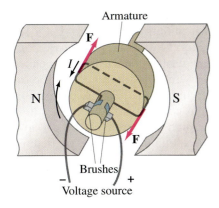

FIGURE 20–33 Diagram of a simple dc motor.

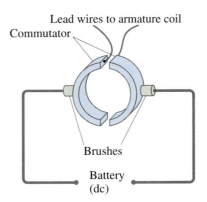

FIGURE 20–34 The commutator-brush arrangement in a dc motor assures alternation of the current in the armature to keep rotation continuous. The commutators are attached to the motor shaft and turn with it, whereas the brushes remain stationary.

a galvanometer. The pen is attached to an arm, which is connected to the galvanometer coil. The instrument could record either voltage or current, just as any galvanometer can be connected as a voltmeter or ammeter.

An **electric motor** changes electric energy into (rotational) mechanical energy. A motor works on the same principle as a galvanometer, except that there is no spring so the coil can rotate continuously in one direction. The coil is larger and is mounted on a large cylinder called the **rotor** or **armature**, Fig. 20–33. Actually, there are several coils, although only one is indicated in the figure. The armature is mounted on a shaft or axle. At the moment shown in Fig. 20–33, the magnetic field exerts forces on the current in the loop as shown. However, when the coil, which is rotating clockwise in Fig. 20–33, passes beyond the vertical position the forces would then act to return the coil back to vertical if the current remained the same. But if the current could somehow be reversed at that critical moment, the forces would reverse, and the coil would continue rotating in the same direction. Thus, alternation of the current is necessary if a motor is to turn continuously in one direction. This can be achieved in a **dc motor** with the use of **commutators** and **brushes**: as shown in Fig. 20–34, the brushes are stationary contacts that rub against the conducting commutators mounted on the motor shaft. At every half revolution, each commutator changes its connection to the other brush. Thus the current in the coil reverses every half revolution as required for continuous rotation. Most motors contain several coils, called "windings," each located in a different place on the armature, Fig. 20–35. Current flows through each coil only during a small part of a revolution, at the time when its orientation results in the maximum torque. In this way, a motor produces a much steadier torque than can be obtained from a single coil. An **ac motor**, with ac current as input, can work without commutators since the current itself alternates. Many motors use wire coils to produce the magnetic field (electromagnets) instead of a permanent magnet. Indeed the design of most practical motors is more complex than described here, but the general principles remain the same.

Electric motor

➡ **PHYSICS APPLIED**

DC motor

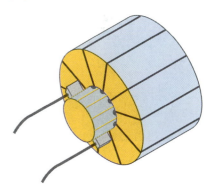

FIGURE 20–35 Motor with many windings.

➡ **PHYSICS APPLIED**

AC motor

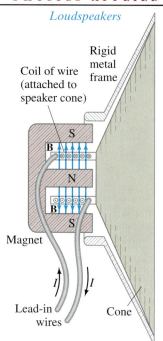

Coil of wire
(attached to
speaker cone)

Rigid
metal
frame

Magnet

Lead-in
wires

Cone

FIGURE 20–36
Loudspeaker.

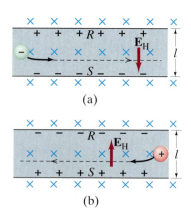

(a)

(b)

FIGURE 20–37 The Hall effect. (a) Negative charges moving to the right as the current. (b) Positive charges moving to the left as the current.

A **loudspeaker** also works on the principle that a magnet exerts a force on a current-carrying wire. The electrical output of a radio or TV set is connected to the wire leads of the speaker. The speaker leads are connected internally to a coil of wire, which is itself attached to the speaker cone, Fig. 20–36. The speaker cone is usually made of stiffened cardboard and is mounted so that it can move back and forth freely. A permanent magnet is mounted directly in line with the coil of wire. When the alternating current of an audio signal flows through the wire coil, the coil and the attached speaker cone experience a force due to the magnetic field of the magnet. As the current alternates at the frequency of the audio signal, the speaker cone moves back and forth at the same frequency, causing alternate compressions and rarefactions of the adjacent air, and sound waves are produced. A speaker thus changes electrical energy into sound energy, and the frequencies and intensities of the emitted sound waves can be an accurate reproduction of the electrical input.

* 20–11 The Hall Effect

When a current-carrying conductor is held firmly in a magnetic field, the field exerts a sideways force on the charges moving in the conductor. For example, if electrons move to the right in the rectangular conductor shown in Fig. 20–37a, the inward magnetic field will exert a downward force on the electrons $F_B = ev_d B$, where v_d is the drift velocity of the electrons (Section 18–9). So the electrons will tend to move nearer face S than face R. There will thus be a potential difference between faces R and S of the conductor. This potential difference builds up until the electric field $\mathbf{E}_H$ it produces exerts a force, $e\mathbf{E}_H$, on the moving charges that is equal and opposite to the magnetic force. This effect is called the **Hall effect** after E. H. Hall, who discovered it in 1879. The difference of potential produced is called the **Hall emf**.

The electric field due to the separation of charge is called the *Hall field*, $\mathbf{E}_H$, and points downward in Fig. 20–37a, as shown. In equilibrium, the force due to this electric field is balanced by the magnetic force $ev_d B$, so

$$eE_H = ev_d B.$$

Hence $E_H = v_d B$. The Hall emf is then (assuming the conductor is long and thin so E_H is uniform)

$$\mathcal{E}_H = E_H l = v_d B l, \tag{20–11}$$

where l is the width of the conductor.

A current of negative charges moving to the right is equivalent to positive charges moving to the left, at least for most purposes. But the Hall effect can distinguish these two. As can be seen in Fig. 20–37b, positive particles moving to the left are deflected downward, so that the bottom surface is positive relative to the top surface. This is the reverse of part (a). Indeed, the direction of the emf in the Hall effect first revealed that it is negative particles that move in metal conductors. In some semiconductors, however, the Hall effect reveals that the carriers of current are positive (more on this in Chapter 29).

The magnitude of the Hall emf is proportional to the strength of the magnetic field. The Hall effect can thus be used to measure magnetic field strengths. First the conductor, called a *Hall probe*, is calibrated with known magnetic fields. Then, for the same current, its emf output will be a measure of B. Hall probes can be made very small and are convenient and accurate to use.

The Hall effect can also be used to measure the drift velocity of charge carriers when the external magnetic field B is known. Such a measurement also allows us to determine the density of charge carriers in the material.

EXAMPLE 20–13 **Drift velocity using the Hall effect.** A long copper strip 1.8 cm wide and 1.0 mm thick is placed in a 1.2-T magnetic field as in Fig. 20–37a. When a steady current of 15 A passes through it, the Hall emf is measured to be 1.02 μV. Determine the drift velocity of the electrons and the density of free (conducting) electrons (number per unit volume) in the copper.

SOLUTION The drift velocity (Eq. 20–11) is

$$v_d = \frac{\mathscr{E}_H}{Bl} = \frac{1.02 \times 10^{-6} \text{ V}}{(1.2 \text{ T})(1.8 \times 10^{-2} \text{ m})} = 4.7 \times 10^{-5} \text{ m/s.}$$

The density of charge carriers n is obtained from Eq. 18–10, $I = nev_dA$, where A is the cross-sectional area through which the current I flows. Then

$$n = \frac{I}{ev_d A} = \frac{15 \text{ A}}{(1.6 \times 10^{-19} \text{ C})(4.7 \times 10^{-5} \text{ m/s})(1.8 \times 10^{-2} \text{ m})(1.0 \times 10^{-3} \text{ m})}$$
$$= 11 \times 10^{28} \text{ m}^{-3}.$$

This value for the density of free electrons in copper, $n = 11 \times 10^{28}$ per m³, is the experimentally measured value. It represents *more* than one free electron per atom, which as we saw in Example 18–13 is $8.4 \times 10^{28} \text{ m}^{-3}$.

* 20–12 Mass Spectrometer

Various methods were developed in the early part of this century to measure the masses of atoms. One of the most accurate was the **mass spectrometer**[†] of Fig. 20–38. Ions are produced by heating, or by an electric current, in the source S. Those that pass through slit S_1 enter a region where there are both electric and magnetic fields: the magnetic field points out of the page in Fig. 20–38, and the electric field points up (from the $+$ plate toward the $-$ plate). Ions will follow a straight-line path in this region, as shown, if the electric force qE (upward on a positive ion) is just balanced by the magnetic force qvB (downward on a positive ion): that is, if

$$qE = qvB$$

or

$$v = \frac{E}{B}.$$

In other words, those ions (and only those) whose speed is $v = E/B$ will pass through undeflected and emerge through slit S_2. (This arrangement is called a *velocity selector*.) In the second region, after S_2, there is only a

FIGURE 20–38 Bainbridge mass spectrometer. The magnetic fields B and B' point out of the paper (indicated by the dots).

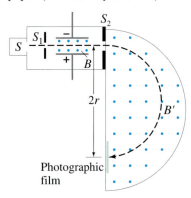

[†]The term *mass spectrograph* is also used.

magnetic field B' so the ions follow a circular path. The radius of their path can be measured because the ions darken the photographic plate where they strike. Since $qvB' = mv^2/r$ and $v = E/B$, then we have

$$m = \frac{qB'r}{v} = \frac{qBB'r}{E}.$$

All the quantities on the right can be measured, and thus m can be determined. Note that for ions of the same charge, the mass of each is proportional to the radius of its path.

The masses of many atoms were measured in this way. When a pure substance was used, it was sometimes found that two or more closely spaced marks would appear on the film. For example, neon produced two marks whose radii corresponded to atoms of mass 20 and 22 atomic mass units (u). Impurities were ruled out and it was concluded that there must be two types of neon with different masses. These different forms were called **isotopes.** It was soon found that most elements are mixtures of isotopes. We shall see in Chapter 30 that the difference in mass is due to different numbers of neutrons.

Mass spectrometers can be used to separate not only different elements and isotopes, but different molecules as well. They are used in physics and chemistry, and in biological and biomedical laboratories.

EXAMPLE 20–14 **Mass spectrometry.** Carbon atoms of atomic mass 12.0 u are found to be mixed with another, unknown, element. In a mass spectrometer, the carbon traverses a path of radius 22.4 cm and the unknown's path has a 26.2 cm radius. What is the unknown element? Assume they have the same charge.

SOLUTION Since mass is proportional to the radius, we have

$$\frac{m_x}{m_C} = \frac{26.2 \text{ cm}}{22.4 \text{ cm}} = 1.17.$$

Thus $m_x = 1.17 \times 12.0 \text{ u} = 14.0 \text{ u}$. The other element is probably nitrogen (see the periodic table, inside the back cover). However, it could also be an isotope of carbon or oxygen. Further physical or chemical analysis would be needed.

*20–13 Ferromagnetism; Domains

We saw in Section 20–1 that iron (and a few other materials) can be made into strong magnets. These materials are said to be **ferromagnetic.** We now look more deeply into the sources of ferromagnetism.

A bar magnet, with its two opposite poles at either end, resembles an electric dipole (equal-magnitude positive and negative charges separated by a distance). Indeed, a bar magnet is sometimes referred to as a "magnetic dipole." There are opposite "poles" separated by a distance. And the magnetic field lines of a bar magnet form a pattern much like that for the electric field of an electric dipole: compare Fig. 16–29a with Fig. 20–3b.

Domains in iron   Microscopic examination reveals that a magnet is actually made up of tiny regions known as **domains,** which are at most about 1 mm in length or width. Each domain behaves like a tiny magnet with a north and a south pole. In an unmagnetized piece of iron, these domains are arranged randomly, as shown in Fig. 20–39a. The magnetic effects of the domains cancel each other out, so this piece of iron is not a magnet. In a magnet, the domains are preferentially aligned in one direction as shown in Fig. 20–39b (downward in this case). A magnet can be made from an unmagnetized piece of iron by placing it in a strong magnetic field. (You can make a needle magnetic, for example, by stroking it with one pole of a strong magnet.) Careful observations show in this case that the magnetization of domains may actually rotate slightly so as to be more nearly parallel to the external field. Or, more commonly, the borders of domains move so that those domains whose magnetic orientation is parallel to the external field grow in size at the expense of other domains. This can be seen by comparing Figs. 20–39a and b. This explains how a magnet can pick up unmagnetized pieces of iron like paper clips or bobby pins. The magnet's field causes a slight alignment of the domains in the unmagnetized object so that the object becomes a temporary magnet with its north pole facing the south pole of the permanent magnet, and vice versa; thus, attraction results. In the same way, elongated iron filings will arrange themselves in a magnetic field just as a compass needle does, and will reveal the shape of the magnetic field, Fig. 20–40.

An iron magnet can remain magnetized for a long time, and thus it is referred to as a "permanent magnet." However, if you drop a magnet on the floor or strike it with a hammer, you may jar the domains into randomness. The magnet can thus lose some or all of its magnetism. Heating a magnet too can cause a loss of magnetism, for raising the temperature increases the random thermal motion of the atoms which tends to randomize the domains. Above a certain temperature known as the **Curie temperature** (1043 K for iron), a magnet cannot be made at all.[†]

There is a striking similarity between the fields produced by a bar magnet and by a loop of electric current or a solenoid (compare Fig. 20–3b with Figs. 20–11 and 20–27). This suggests that the magnetic field produced by a current may have something to do with ferromagnetism, an idea proposed by Ampère in the nineteenth century. According to modern atomic theory, the atoms that make up any material can be roughly visualized as containing electrons that orbit around a central nucleus. Since the electrons are charged, they constitute an electric current and therefore produce a magnetic field. But if there is no external field, the electron orbits in different atoms are arranged randomly, so the magnetic effects due to the many orbits in all the atoms in a material cancel out. However, electrons produce an additional magnetic field, almost as if they and their electric charge were spinning about their own axes. It is the magnetic field due to electron spin[‡] that is believed to produce ferromagnetism.

FIGURE 20–39 (a) An unmagnetized piece of iron is made up of domains that are randomly arranged. Each domain is like a tiny magnet; the arrows represent the magnetization direction, with the arrowhead being the N pole. (b) In a magnet, the domains are preferentially aligned in one direction, and may be altered in size by the magnetization process.

FIGURE 20–40 Iron filings line up along magnetic field lines.

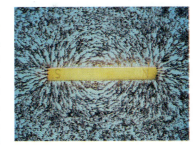

[†]Iron, nickel, cobalt, gadolinium, and certain alloys are ferromagnetic at room temperature; several other elements and alloys have low Curie temperature and thus are ferromagnetic only at low temperatures.

[‡]The name "spin" comes from the early suggestion that the additional magnetic field arises from the electron "spinning" on its axis (as well as "orbiting" the nucleus) and this additional motion of the charge was supposed to produce the extra field. However this view of a spinning electron is oversimplified: see Chapter 28.

In most materials, the magnetic fields due to electron spin cancel out. But in iron and other ferromagnetic materials, a complicated cooperative mechanism seems to operate. The result is that the electrons contributing to the ferromagnetism in a domain "spin" in the same direction. Thus the tiny magnetic fields due to each of the electrons add up to give the magnetic field of a domain. And when the domains are aligned, as we have seen, a strong magnet results.

It is believed possible today that *all* magnetic fields are caused by electric currents. This would explain why no single magnetic pole has ever been found: there is no way to divide up a current and obtain a single magnetic pole. Of course if an isolated pole is found, we will have to alter the idea that all magnetic fields are produced by currents.

The lack of single magnetic poles means that magnetic field lines form closed loops, unlike electric field lines which begin on positive charges and end on negative charges.

* 20–14 Electromagnets and Solenoids

A long coil of wire consisting of many loops of wire, as discussed in Section 20–8, is called a solenoid. The magnetic field within a solenoid can be fairly large since it will be the sum of the fields due to the current in each loop (see Fig. 20–41). The solenoid acts like a magnet; one end can be considered the north pole and the other the south pole, depending on the direction of the current in the loops (use the right-hand rule). Since the magnetic field lines leave the north pole of a magnet, the north pole of the solenoid in Fig. 20–41 is on the right.

If a piece of iron is placed inside a solenoid, the magnetic field is increased greatly because the domains of the iron are aligned by the magnetic field produced by the current. The resulting magnetic field is the sum of that due to the current and that due to the iron, and can be hundreds or thousands of times that due to the current alone (see Section 20–15). This arrangement is called an **electromagnet**. The iron used in electromagnets acquires and loses its magnetism quite readily when the current is turned on or off, and so is referred to as "soft iron." (It is "soft" only in a magnetic sense.) Iron that holds its magnetism even when there is no externally applied field is called "hard iron." Hard iron is used in permanent magnets. Soft iron is usually used in electromagnets so that the field can be turned on and off readily. Whether iron is hard or soft depends on heat treatment and other factors.

Electromagnets find use in many practical applications, from use in motors and generators to producing large magnetic fields for research. Because the current flows continuously, a great deal of waste heat (I^2R power) is often produced. Cooling coils, which are tubes carrying water, must be used to absorb the heat in bigger installations. For some applications, superconducting magnets are coming into use. The current-carrying wires are made of superconducting material (Section 18–5) kept below the transition temperature. No electric power is needed to maintain large current, which means large savings of electricity. Of course, energy is needed to keep the superconducting coils at the necessary low temperature.

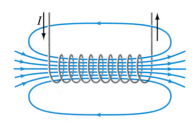

FIGURE 20–41 Magnetic field of a solenoid. The north pole of this solenoid, thought of as a magnet, is on the right, and the south pole is on the left.

PHYSICS APPLIED

Electromagnets and solenoids

Another useful device consists of a solenoid into which a rod of iron is partially inserted. This combination is also referred to as a solenoid. One simple use is as a doorbell (Fig. 20–42). When the circuit is closed by pushing the button, the coil effectively becomes a magnet and exerts a force on the iron rod. The rod is pulled into the coil and strikes the bell. A larger solenoid is used in the starters of cars; when you engage the starter, you are closing a circuit that not only turns the starter motor, but activates a solenoid that first moves the starter into direct contact with the engine. Solenoids are used as switches in many other devices, such as tape recorders. They have the advantage of moving mechanical parts quickly and accurately.

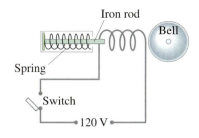

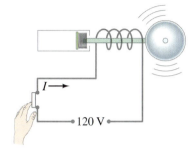

FIGURE 20–42 Solenoid used as a doorbell.

* **20–15** **Magnetic Fields in Magnetic Materials; Hysteresis**

The field of a long solenoid is directly proportional to the current. Indeed, Eq. 20–8 tells us that the field B_0 inside a solenoid is given by

$$B_0 = \mu_0 n I.$$

This is valid if there is only air inside the coil. If we put a piece of iron or other ferromagnetic material inside the solenoid, the field will be greatly increased, often by hundreds or thousands of times. This occurs because the domains in the iron become preferentially aligned by the external field. The resulting magnetic field is the sum of that due to the current and that due to the iron. It is sometimes convenient to write the total field in this case as a sum of two terms:

$$\mathbf{B} = \mathbf{B}_0 + \mathbf{B}_M. \qquad (20\text{–}12)$$

Here, $\mathbf{B}_0$ refers to the field due only to the current in the wire (the "external field"). It is the field that would be present in the absence of a ferromagnetic material. Then $\mathbf{B}_M$ represents the additional field due to the ferromagnetic material itself; often $\mathbf{B}_M \gg \mathbf{B}_0$.

The total field inside a solenoid in such a case can also be written by replacing the constant μ_0 in Eq. 20–8 by another constant, μ, characteristic of the material inside the coil:

$$B = \mu n I; \qquad (20\text{–}13)$$

μ is called the **magnetic permeability** of the material. For ferromagnetic materials, μ is much greater than μ_0. For all other materials, its value is very close to μ_0.[†] The value of μ, however, is not constant for ferromagnetic materials; it depends on the value of the external field B_0, as the following experiment shows.

[†]All materials are slightly magnetic. Nonferromagnetic materials fall into two principal classes: **paramagnetic**, in which μ is very slightly larger than μ_0; and **diamagnetic**, in which μ is very slightly less than μ_0. Paramagnetic materials apparently contain atoms that have a net magnetic dipole moment due to orbiting electrons, and these become slightly aligned with an external field just as the galvanometer coil in Fig. 20–31 experiences a torque that tends to align it. Atoms of diamagnetic materials have no net dipole moment. However, in the presence of an external field, electrons revolving in one direction are caused to increase in speed slightly, whereas those revolving in the opposite direction are reduced in speed. The result is a slight net magnetic effect that actually opposes the external field.

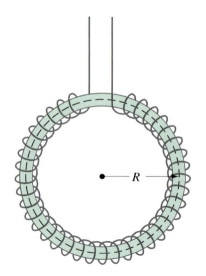

FIGURE 20–43 Iron-core torus.

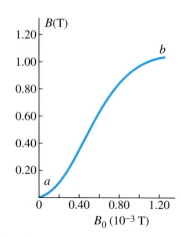

FIGURE 20–44 Total magnetic field B in an iron-core torus as a function of the external field B_0 (B_0 is caused by the current I in the coil).

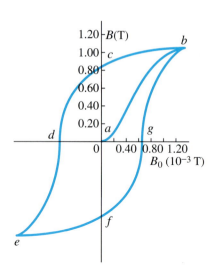

FIGURE 20–45 Hysteresis curve.

Hysteresis

Measurements on magnetic materials are generally done using a torus, which is essentially a long solenoid bent into the shape of a circle (Fig. 20–43), so that practically all the lines of **B** remain within the torus. Suppose the torus has an iron core that is initially unmagnetized and there is no current in the windings of the torus. Then the current I is slowly increased, and B_0 increases linearly with I. The total field B also increases, but follows the curved line shown in the graph of Fig. 20–44. (Note the different scales: $B \gg B_0$.) Initially, point a, the domains (Section 20–13) are randomly oriented. As B_0 increases, the domains become more and more aligned until at point b, nearly all are aligned. The iron is said to be approaching **saturation**. Point b is typically 70 percent of full saturation. (If B_0 is increased further, the curve continues to rise very slowly, and reaches 98 percent saturation only when B_0 reaches a value about a thousandfold above that at point b; the last few domains are very difficult to align.) Next, suppose the external field B_0 is reduced by decreasing the current in the coils. As the current is reduced to zero, point c in Fig. 20–45, the domains do not become completely random. Some permanent magnetism remains. If the current is then reversed in direction, enough domains can be turned around so $B = 0$ (point d). As the reverse current is increased further, the iron approaches saturation in the opposite direction (point e). Finally, if the current is again reduced to zero and then increased in the original direction, the total field follows the path $efgb$, again approaching saturation at point b.

Notice that the field did not pass through the origin (point a) in this cycle. The fact that the curves do not retrace themselves on the same path is called **hysteresis**. The curve $bcdefgb$ is called a **hysteresis loop**. In such a cycle, much energy is transformed to thermal energy (friction) due to realigning of the domains. It can be shown that the energy dissipated in this way is proportional to the area of the hysteresis loop.

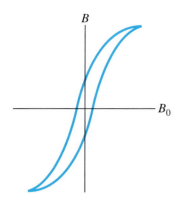

FIGURE 20–46 Hysteresis curve for soft iron.

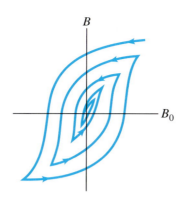

FIGURE 20–47 Successive hysteresis loops during demagnetization.

At points c and f, the iron core is magnetized even though there is no current in the coils. These points correspond to a permanent magnet. For a permanent magnet, it is desired that ac and af be as large as possible. Materials for which this is true are said to have high **retentivity**, and may be referred to as "hard." On the other hand, a hysteresis curve such as that in Fig. 20–46 occurs for so-called "soft iron" (soft from a magnetic point of view). This is preferred for *electromagnets* (Section 20–14) since the field can be more readily switched off, and the field can be reversed with less loss of energy.

A ferromagnetic material can be demagnetized—that is, made unmagnetized. This can be done by reversing the magnetizing current repeatedly while decreasing its magnitude. This results in the curve of Fig. 20–47. The heads of a tape recorder are demagnetized in this way. The alternating magnetic field acting at the heads due to a demagnetizer is strong when the demagnetizer is placed near the heads and decreases as it is moved slowly away. (Cassette tapes themselves can be erased and ruined by a magnetic field.)

Demagnetizing

Magnetic fields are somewhat analogous to the electric fields of Chapter 16, but there are several important differences to recall:

1. The force experienced by a charged particle moving in a magnetic field is *perpendicular* to the direction of the magnetic field (and to the direction of the velocity of the particle), whereas the force exerted by an electric field is *parallel* to the direction of the field (and unaffected by the velocity of the particle).

2. The *right-hand rule*, in its many forms, is intended to help you determine the directions of magnetic field, and the forces they exert, and/or the directions of electric current or charged particle velocity. The right-hand rules are specifically designed to deal with the "perpendicular" nature of these quantities.

3. Note that the equations in this chapter are generally not printed as vector equations, but involve magnitudes only. The right-hand rule is to be used to find directions of vector quantities.

SUMMARY

A magnet has two **poles**, north and south. The north pole is that end which points toward the north when the magnet is freely suspended. Unlike poles of two magnets attract each other, whereas like poles repel.

We can imagine that a **magnetic field** surrounds every magnet. The SI unit for magnetic field is the **tesla** (T). The force one magnet exerts on another is said to be an interaction between one magnet and the magnetic field produced by the other.

Electric currents produce magnetic fields. For example, the lines of magnetic field due to a current in a straight wire form circles around the wire and the field exerts a force on magnets.

The magnitude of the magnetic field a distance r from a long straight wire carrying a current I is given by

$$B = \frac{\mu_0}{2\pi} \frac{I}{r}.$$

A magnetic field exerts a force on an electric current. For a straight wire of length l carrying a current I, the force has magnitude

$$F = IlB \sin \theta,$$

where θ is the angle between the magnetic field of strength B and the wire. The direction of the force is perpendicular to the wire and to the magnetic field, and is given by the right-hand rule.

Similarly, a magnetic field exerts a force on a charge q moving with velocity v of magnitude

$$F = qvB \sin \theta,$$

where θ is the angle between **v** and **B**. The direction of **F** is perpendicular to **v** and to **B**. The path of a charged particle moving perpendicular to a uniform magnetic field is a circle.

The force exerted on a current-carrying wire by a magnetic field is the basis for operation of many devices, such as meters, motors, and loudspeakers.

QUESTIONS

1. A compass needle is not always balanced parallel to the Earth's surface but one end may dip downward. Explain.

2. Draw the magnetic field lines around a straight section of wire carrying a current horizontally to the left.

3. In what direction are the magnetic field lines surrounding a straight wire carrying a current that is moving directly toward you?

4. The magnetic field due to current in wires in your home can affect a compass. Discuss the problem in terms of currents, including if they are ac or dc.

5. What kind of field or fields surround a moving electric charge?

6. Will a magnet attract any metallic object, or only those made of iron? (Try it and see.) Why is this so?

7. Two iron bars attract each other no matter which ends are placed close together. Are both magnets? Explain.

*8. Note that the pattern of magnetic field lines surrounding a bar magnet is similar to that of the electric field around an electric dipole. From this fact, predict how the magnetic field will change with distance (*a*) when near one pole of a very long bar magnet, and (*b*) when far from a magnet as a whole.

9. Suppose you have three iron rods, two of which are magnetized but the third is not. How would you determine which two are the magnets without using any additional objects?

10. How can you make a compass without using iron or other ferromagnetic material?

11. A horseshoe magnet is held vertically with the north pole on the left and south pole on the right. A wire passes between the poles, equidistant from them, and carries a current directly away from you. In what direction is the force on the wire?

12. Can you set a resting electron into motion with a magnetic field? With an electric field?

13. A charged particle is moving in a circle under the influence of a uniform magnetic field. If an electric field that points in the same direction as the magnetic field is turned on, describe the path the charged particle will take.

14. Each of the right-hand rules you learned in this chapter can be changed to *left-hand rules* if you are specifying the direction of movement of *negative* particles, such as electrons in a wire. Show, for each right-hand rule, that the same operations using the left hand give the same results if the direction of charge flow is for negative charges.

15. A charged particle moves in a straight line through a particular region of space. Could there be a nonzero magnetic field in this region? If so, give two possible situations.

16. If a moving charged particle is deflected sideways in some region of space, can we conclude for certain that **B** $\neq$ 0 in that region?

17. If a negatively charged particle enters a region of uniform magnetic field which is perpendicular to the particle's velocity, will the kinetic energy of the particle increase, decrease, or stay the same. Explain your answer. (Neglect gravity.)

18. In Fig. 20–48, charged particles move in the vicinity of a current carrying wire. For each charged particle the arrow indicates the direction of motion of the particle and the + or − indicates the sign of the charge. For each of the particles, indicate the direction of the magnetic force due to the magnetic field produced by the wire.

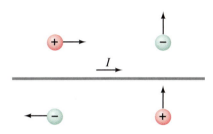

FIGURE 20–48   Question 18.

19. Explain why a strong magnet held near a television screen causes the picture to become distorted. Also, explain why the picture sometimes goes completely black where the field is the strongest.

20. In a particular region of space there is a uniform magnetic field **B**. Outside this region, $B = 0$. Can you inject an electron into the field perpendicularly so it will move in a closed circular path in the field?

21. How could you tell whether moving electrons in a certain region of space are being deflected by an electric field or by a magnetic field (or by both)?

22. A beam of electrons is directed perpendicularly toward a horizontal wire carrying a current from left to right. In what direction are the electrons deflected?

23. Two long wires carrying equal currents I are at right angles to each other, but don't quite touch. Describe the magnetic force one exerts on the other.

24. A horizontal current-carrying wire, free to move, is suspended directly above a second, parallel, current-carrying wire. (*a*) In what direction is the current in the lower wire? (*b*) Can the upper wire be held in stable equilibrium due to the magnetic force of the lower wire? Explain.

25 What factors determine the sensitivity of a galvanometer?

26. A rectangular piece of semiconductor is inserted in a magnetic field and a battery is connected to its ends as shown in Fig. 20–49. When a sensitive voltmeter is connected between points *a* and *b*, it is found that point *a* is at a higher potential than *b*. What is the sign of the charge carriers in this semiconductor material?

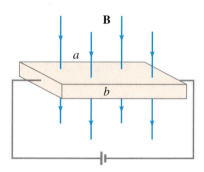

FIGURE 20–49   Question 26.

* **27.** Two ions have the same mass, but one is singly ionized and the other is doubly ionized. How will their positions on the film of the mass spectrograph of Fig. 20–38 differ?

* **28.** Why will either pole of a magnet attract an unmagnetized piece of iron?

* **29.** An unmagnetized nail will not attract an unmagnetized paper clip. However, if one end of the nail is in contact with a magnet, the other end *will* attract a paper clip. Explain.

* **30.** Another type of magnetic switch similar to a solenoid is a **relay**. A relay is an electromagnet (the iron rod inside the coil does not move) that, when activated, attracts a piece of soft iron on a pivot. Design a relay (*a*) to make a doorbell, and (*b*) to close an electrical switch. A relay is used in the latter case when you need to switch on a circuit carrying a very large current but you do not want that large current flowing through the main switch. For example, the starter switch of a car is connected to a relay so that the large currents needed for the starter do not pass to the dashboard switch.

PROBLEMS

SECTIONS 20–3 AND 20–4

1. (I) (*a*) What is the force per meter on a wire carrying a 9.80-A current when perpendicular to a 0.80-T magnetic field? (*b*) What if the angle between the wire and field is 45.0°?

2. (I) A 1.5-m length of wire carrying 6.5 A of current is oriented horizontally. At that point on the Earth's surface, the dip angle of the Earth's magnetic field makes an angle of 40° to the wire. Estimate the magnetic force on the wire due to the Earth's magnetic field of 5.5×10^{-5} T at this point.

3. (I) How much current is flowing in a wire 4.20 m long if the maximum force on it is 0.900 N when placed in a uniform 0.0800-T field?

4. (I) The force on a wire carrying 25.0 A is a maximum of 4.14 N when placed between the pole faces of a magnet. If the pole faces are 22.0 cm in diameter, what is the approximate strength of the magnetic field?

5. (I) Determine the magnitude and direction of the force on an electron traveling 3.58×10^6 m/s horizontally to the west in a vertically upward magnetic field of strength 1.30 T.

6. (I) Describe the path of an electron that is projected vertically upward with a speed of 1.80×10^6 m/s into a uniform magnetic field of 0.250 T that is directed away from the observer.

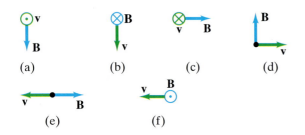

FIGURE 20–50 Problem 7.

7. (I) Find the direction of the force on a negative charge for each diagram shown in Fig. 20–50, where **v** is the velocity of the charge and **B** is the direction of the magnetic field. ($\otimes$ means the vector points inward. $\odot$ means it points outward, toward the viewer.)

8. (I) Determine the direction of **B** for each case in Fig. 20–51, where **F** represents the force on a positively charged particle moving with velocity **v**.

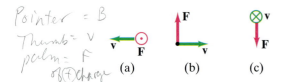

FIGURE 20–51 Problem 8.

9. (I) Alpha particles of charge $q = +2e$ and mass $m = 6.6 \times 10^{-27}$ kg are emitted from a radioactive source at a speed of 1.6×10^7 m/s. What magnetic field strength would be required to bend these into a circular path of radius $r = 0.25$ m?

10. (II) An electron experiences the greatest force as it travels 1.8×10^6 m/s in a magnetic field when it is moving southward. The force is upward and of magnitude 2.2×10^{-12} N. What is the magnitude and direction of the magnetic field?

11. (II) The magnetic force per meter on a wire is measured to be only 45 percent of its maximum possible value. Sketch the relationship of the wire and the field if the force were a maximum, and sketch the relationship as it actually is, calculating the angle between the wire and the magnetic field.

12. (II) The force on a wire is a maximum of 5.30 N when placed between the pole faces of a magnet. The current flows horizontally to the right and the magnetic field is vertical. The wire is observed to "jump" toward the observer when the current is turned on. (a) What type of magnetic pole is the top pole face? (b) If the pole faces have a diameter of 10.0 cm, estimate the current in the wire if the field is 0.15 T. (c) If the wire is tipped so that it now makes an angle of 10° with the horizontal, what force will it now feel?

13. (II) A proton moves in a circular path perpendicular to a 1.15-T magnetic field. The radius of its path is 8.40 mm. Calculate the energy of the proton in eV.

14. (II) For a particle of mass m and charge q moving in a circular path in a magnetic field B, show that its kinetic energy is proportional to r^2, the square of the radius of curvature of its path.

15. (II) A particle of charge q moves in a circular path of radius r in a uniform magnetic field B. Show that its momentum is $p = qBr$.

16. (II) For a particle of mass m and charge q moving in a circular orbit in a uniform magnetic field B, show that its angular momentum is given by $L = qBr^2$.

17. (II) A sort of "projectile launcher" is shown in Fig. 20–52. A large current moves in a closed loop composed of fixed rails, a power supply, and a very light, almost frictionless bar touching the rails. A magnetic field is perpendicular to the plane of the circuit. If the bar has a length of 20 cm, a mass of 1.5 g, and is placed in a field of 1.7 T, what constant current flow is needed in order for it to accelerate to 30 m/s in a distance of 1.0 m? In what direction must the field point?

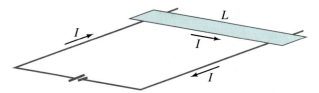

FIGURE 20–52 Problem 17.

18. (III) A 3.80-g bullet moves with a speed of 180 m/s perpendicular to the Earth's magnetic field of 5.00×10^{-5} T. If the bullet possesses a net charge of 8.10×10^{-9} C, by what distance will it be deflected from its path due to the magnetic field after it has traveled 1.00 km?

SECTIONS 20–5 AND 20–6

19. (I) Jumper cables used to start a stalled vehicle often carry a 15-A current. How strong is the magnetic field 15 cm away? What percentage of the Earth's magnetic field is this?

20. (I) If a magnetic field no larger than that of the Earth (0.55×10^{-4} T) is to be allowed 30 cm from an electrical wire, what is the maximum current the wire can carry?

21. (I) What is the magnitude and direction of the force between two parallel wires 45 m long and 6.0 cm apart, each carrying 35 A in the same direction?

22. (I) A vertical straight wire carrying an upward 12-A current exerts an attractive force per unit length of 8.8×10^{-4} N/m on a second parallel wire 7.0 cm away. What current (magnitude and direction) flows in the second wire?

23. (II) What is the maximum current that a wire can carry if an experimenter is performing an experiment 1.0 m away that deals with the Earth's magnetic field, which she wishes to measure to ± 1 percent?

24. (II) What is the acceleration (in g's) of a 175-g model airplane charged to 18.0 C and traveling at 1.8 m/s as it passes within 8.6 cm of a wire, nearly parallel to its path, carrying a 30-A current?

25. (II) A horizontal compass is placed 20 cm due south from a straight vertical wire carrying a 30-A current downward. In what direction does the compass needle point at this location? Assume the horizontal component of the Earth's field at this point is 0.45×10^{-4} T and the magnetic declination is 0°.

26. (II) A long horizontal wire carries 12.0 A of current due north. What is the net magnetic field 20.0 cm due west of the wire if the Earth's field there points downward, 40° below the horizontal, and has magnitude 5.0×10^{-5} T?

27. (II) A stream of protons passes a given point in space at a rate of 10^9 protons/s. What magnetic field do they produce 2.0 m from the beam?

28. (II) Determine the magnetic field midway between two long straight wires 2.0 cm apart in terms of the current I in one when the other carries 15 A. Assume these currents are (*a*) in the same direction, and (*b*) in opposite directions.

29. (II) A long pair of wires serves to conduct 25.0 A of dc current to (and from) an instrument. If the wires are of negligible diameter but are 2.0 mm apart, what is the magnetic field 10.0 cm from their midpoint, in their plane (Fig. 20–53)? Compare to the magnetic field of the Earth.

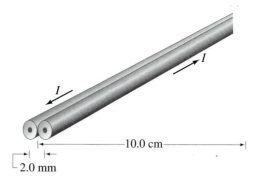

FIGURE 20–53   Problem 29.

30. (II) A compass needle points 20° E of N outdoors. However, when it is placed 8.0 cm to the east of a vertical wire inside a building, it points 55° E of N. What is the magnitude and direction of the current in the wire? The Earth's field there is 0.50×10^{-4} T and is horizontal.

31. (II) Three long parallel wires are 38.0 cm from one another. (Looking along them, they are at three corners of an equilateral triangle.) The current in each wire is 8.00 A, but that in wire A is opposite to that in wires B and C (Fig. 20–54). Determine the magnetic force per unit length on each wire due to the other two.

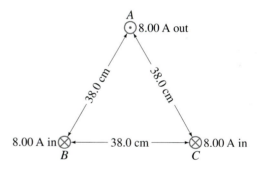

FIGURE 20–54   Problems 31 and 65.

32. (II) The magnetic field near the center of a single circular loop of radius r, carrying current I, is given by:

$$B = \frac{\mu_0 I}{2r}.$$

Assume the planetary model for the hydrogen atom, in which a single electron makes a circular orbit of radius 5.3×10^{-11} m about the nucleus. What magnitude of magnetic field would the orbiting electron produce at the nucleus?

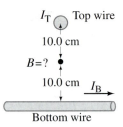

FIGURE 20-55 Problem 33.

33. (II) Two long wires are oriented so that they are perpendicular to each other, and at their closest, they are 20.0 cm apart (Fig. 20–55). What is the magnitude of the magnetic field at a point midway between them if the top one carries a current of 20.0 A and the bottom one carries 5.0 A?

34. (II) A long horizontal wire carries a current of 48 A. A second wire, made of 2.5-mm-diameter copper wire and parallel to the first but 15 cm below it, is held in suspension magnetically (Fig. 20–56). (a) What is the magnitude and direction of the current in the lower wire? (b) Is the lower wire in stable equilibrium? (c) Repeat parts (a) and (b) if the second wire is suspended 15 cm *above* the first due to the latter's field.

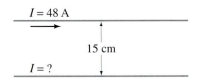

FIGURE 20-56 Problem 34.

* 35. (III) Two long parallel wires 6.00 cm apart carry 16.5-A currents in the same direction. Determine the magnetic field strength at a point 12.0 cm from one wire and 13.0 cm from the other. [*Hint*: Make a drawing in a plane containing the field lines, and recall the rules for vector addition.]

SECTION 20-8

* 36. (I) A 30.0-cm long solenoid 1.25 cm in diameter is to produce a field of 0.385 T at its center. How much current should the solenoid carry if it has 1000 turns of the wire?

* 37. (II) You have 1.0 kg of copper and want to make a practical solenoid that produces the greatest possible magnetic field. Should you make your copper wire long and thin, short and fat, or something else? Consider other variables, such as solenoid diameter, length, and so on.

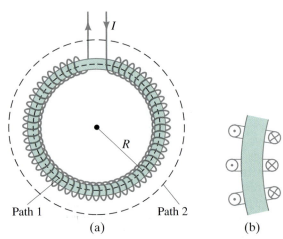

FIGURE 20-57 Problem 38. (a) A torus. (b) A section of the torus showing direction of the current for three loops: ⊙ means current toward viewer, and ⊗ means current away from viewer.

* 38. (II) A torus is a solenoid in the shape of a circle (Fig. 20–57). Use Ampère's law along the circular path, shown dashed in Fig. 20–57a, to determine that the magnetic field (a) inside the torus is $B = \mu_0 NI/2\pi R$, where N is the total number of turns, and (b) outside the torus is $B = 0$. (c) Is the field inside a torus uniform like a solenoid's? If not, how does it vary?

* 39. (II) Use Ampère's law to show that a uniform magnetic field, such as between the pole pieces of a magnet, Fig. 20–8, cannot drop abruptly to zero outside the magnet. [*Hint*: Take as your path a rectangle with one vertical side inside the field and one vertical side completely outside the field.]

* 40. (III) A current I, flowing in a long solid cylindrical wire of radius r_0, is uniform across the cross section (Fig. 20–58). (a) Use Ampère's law to show that the magnetic field inside the conductor at a distance r from the center of the conductor is

$$B = \frac{\mu_0 Ir}{2\pi r_0^2}.$$

Assume that the field lines are circles, just as they are outside the conductor. (b) Show that at the surface of the wire this agrees with the answer for the magnetic field outside of a long wire. (c) Where is the magnetic field a maximum, and what is its maximum value for a 1.0-mm-diameter wire carrying 15.0 A dc? (d) At what distance from the surface would the field be 10 percent of its maximum? [*Hint*: Make a plot of the magnetic field strength as a function of the distance perpendicularly out from the central axis of the wire.]

FIGURE 20-58
Problem 40.

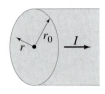

* **41.** (I) A galvanometer needle deflects full scale for a 63.0-μA current. What current will give full-scale deflection if the magnetic field weakens to 0.860 of its original value?

* **42.** (I) If the restoring spring of a galvanometer weakens by 20 percent over the years, what current will give full-scale deflection if it originally required 36 μA?

* **43.** (I) If the current to a motor drops by 15 percent, by what factor does the output torque change?

* **44.** (I) A single square loop of wire 22.0 cm on a side is placed with its face parallel to the magnetic field between the pole pieces of a large magnet. When 6.30 A flows in the coil, the torque on it is 0.325 m·N. What is the magnetic field strength?

* **45.** (II) Show that the magnetic dipole moment M of an electron orbiting the proton nucleus of a hydrogen atom is related to the orbital momentum L of the electron by

$$M = \frac{e}{2m} L.$$

* **46.** (II) A circular coil 18.0 cm in diameter and containing eleven loops lies flat on the ground. The Earth's magnetic field at this location has magnitude 5.50×10^{-5} T and points into the Earth at an angle of 56.0° below a line pointing due north. If a 7.70-A counterclockwise current passes through the coil, (a) determine the torque on the coil, and (b) which edge of the coil rises up, north, east, south, or west?

* **47.** (II) A rectangular sample of a metal is 3.0 cm wide and 500 μm thick. When it carries a 30-A current and is placed in a 0.80-T magnetic field it produces a 6.5-μV Hall emf. Determine: (a) the Hall field in the conductor; (b) the drift speed of the conduction electrons; (c) the density of free electrons in the metal.

* **48.** (II) In a probe that uses the Hall effect to measure magnetic fields, a 12.0-A current passes through a 1.50-cm-wide 1.00-mm-thick strip of sodium metal. If the Hall emf is 2.42 μV, what is the magnitude of the magnetic field (take it perpendicular to the flat face of the strip)? Assume one free electron per atom of Na, and take its specific gravity to be 0.971.

* **49.** (II) The Hall effect can be used to measure blood flow rate because the blood contains ions that constitute an electric current. (a) Does the sign of the ions influence the emf? (b) Determine the flow velocity in an artery 3.3 mm in diameter if the measured emf is 0.10 mV and B is 0.070 T. (In actual practice, an alternating magnetic field is used.)

* **50.** (I) Protons move in a circle of radius 5.10 cm in a 0.566-T magnetic field. What value of electric field could make their paths straight? In what direction must it point?

* **51.** (I) In a mass spectrometer, germanium atoms have radii of curvature equal to 21.0, 21.6, 21.9, 22.2, and 22.8 cm. The largest radius corresponds to an atomic mass of 76 u. What are the atomic masses of the other isotopes?

* **52.** (II) Suppose the electric field between the electric plates in the mass spectrometer of Fig. 20–38 is 2.48×10^4 V/m and the magnetic fields $B = B' = 0.68$ T. The source contains carbon isotopes of mass numbers 12, 13, and 14 from a long-dead piece of a tree. (To estimate atomic masses, multiply by 1.67×10^{-27} kg.) How far apart are the lines formed by the singly charged ions of each type on the photographic film? What if the ions were doubly charged?

* **53.** (II) (a) What value of magnetic field would make a beam of electrons, traveling to the right at a speed of 4.8×10^6 m/s, go undeflected through a region where there is a uniform electric field of 10,000 V/m pointing vertically up? (b) What is the direction of the magnetic field if it is known to be perpendicular to the electric field? (c) What is the frequency of the circular orbit of the electrons if the electric field is turned off?

* **54.** (II) A mass spectrometer is being used to monitor air pollutants. It is difficult, however, to separate molecules with nearly equal mass such as CO (28.0106 u) and N_2 (28.0134 u). How large a radius of curvature must a spectrometer have if these two molecules are to be separated on the film by 0.50 mm?

* **55.** (II) One form of mass spectrometer accelerates ions by a voltage V before they enter a magnetic field B. The ions are assumed to start from rest. Show that the mass of an ion is $m = qB^2R^2/2V$, where R is the radius of the ions' path in the magnetic field and q is their charge.

* **56.** (II) An iron-core solenoid is 36 cm long, 1.5 cm in diameter, and has 600 turns of wire. A magnetic field of 1.8 T is produced when 40 A flows in the wire. What is the permeability μ at this high field strength?

GENERAL PROBLEMS

57. Protons with momentum 4.8×10^{-16} kg·m/s are magnetically steered clockwise in a circular path 2.0 km in diameter at Fermi National Accelerator Laboratory in Illinois. What is the magnitude and direction of the field in the magnets surrounding the beam pipe?

58. A rectangular loop of wire is sitting next to a straight wire, as shown in Fig. 20–59. There is a current of 2.5 A in both wires. What is the magnitude and direction of the net force on the loop?

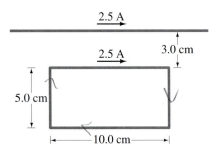

FIGURE 20–59 Problem 58.

59. A proton and an electron have the same kinetic energy upon entering a region of constant magnetic field. What is the ratio of the radii of their circular paths?

60. Near the equator, the Earth's magnetic field points almost horizontally to the north and has magnitude $B = 0.50 \times 10^{-4}$ T. What should be the magnitude and direction for the velocity of an electron if its weight is to be exactly balanced by the magnetic force?

61. Calculate the force on an airplane which has acquired a net charge of 155 C and moves with a speed of 120 m/s perpendicular to the Earth's magnetic field of 5.0×10^{-5} T.

62. The power cable for an electric trolley (Fig. 20–60) carries a horizontal current of 330 A toward the east. The Earth's magnetic field has a strength 5.0×10^{-5} T and makes an angle of dip of 22° at this location. Calculate the magnitude and direction of the magnetic force on a 10-m length of this cable.

63. A doubly charged helium atom, whose mass is 6.6×10^{-27} kg, is accelerated by a voltage of 2400 V. (*a*) What will be its radius of curvature in a uniform 0.240-T field? (*b*) What is its period of revolution?

64. A straight 1.00-mm-diameter copper wire can just "float" horizontally in air because of the force of the Earth's magnetic field **B** which is horizontal and of magnitude 5.00×10^{-5} T. What current does the wire carry?

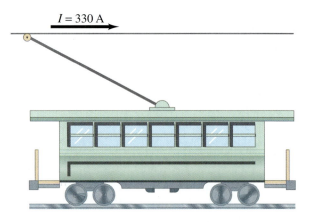

FIGURE 20–60 Problem 62.

65. In Fig. 20–54 the top wire is 2.00-mm-diameter copper wire and is suspended in air due to the two magnetic forces from the bottom two wires. The current flow through the two bottom wires is 20.0 A in each. Calculate the required current flow in the suspended wire.

66. Two stiff parallel wires a distance *l* apart in a horizontal plane act as rails to support a light metal rod of mass *m* (perpendicular to each rail), Fig. 20–61. A magnetic field **B**, directed vertically upward (outward in diagram), acts throughout. At $t = 0$, wires connected to the rails are connected to a constant current source and a current *I* begins to flow through the system. Determine the speed of the rod as a function of time (*a*) assuming no friction between the rod and the rails, and (*b*) if the coefficient of friction is μ_k. (*c*) In which direction does the rod move, east or west, if the current through it heads north?

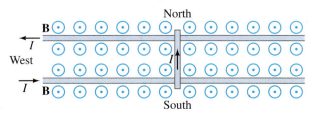

FIGURE 20–61 Looking down on a rod sliding on rails. Problem 66.

67. Estimate the approximate maximum deflection of the electron beam near the center of a TV screen due to the Earth's 5.0×10^{-5} T field. Assume the screen is 20 cm from the electron gun where the electrons are accelerated (*a*) by 2.0 kV, or (*b*) by 30 kV. Note that in color TV sets, the beam must be directed accurately to within less than 1 mm in order to strike the correct phosphor. Because the Earth's field is significant here, mu-metal shields are used to reduce the Earth's field in the CRT. (See Section 17–10.)

68. An electron enters a large solenoid at a 7.0° angle to the axis. If the field is a uniform 3.3×10^{-2} T, determine the radius and pitch (distance between loops) of the electron's helical path if its speed is 1.8×10^7 m/s.

69. The cyclotron (Fig. 20–62) is a device used to accelerate elementary particles such as protons to high speeds. Particles starting at point A with some initial velocity travel in circular orbits in the magnetic field B. The particles are accelerated to higher speeds each time they pass in the gap between the metal "dees," where there is an electric field E. (There is no electric field within the cavity of the metal dees.) The electric field changes direction each half-cycle, owing to an ac voltage $V = V_0 \sin 2\pi f t$, so that the particles are increased in speed at each passage through the gap. (*a*) Show that the frequency f of the voltage must be $f = Bq/2\pi m$, where q is the charge on the particles and m their mass. (*b*) Show that the kinetic energy of the particles increases by $2qV_0$ each revolution, assuming that the gap is small. (*c*) If the radius of the cyclotron is 2.0 m and the magnetic field strength is 0.50 T, what will be the maximum kinetic energy of accelerated protons in MeV? (*d*) How is a cyclotron like a swing?

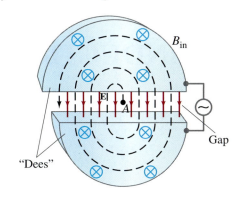

FIGURE 20–62 A cyclotron. Problem 69.

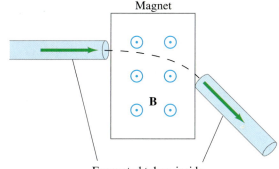

Evacuated tubes, inside of which the protons move with velocity indicated by the green arrows.

FIGURE 20–63 Problem 70.

70. Magnetic fields are very useful in particle accelerators for "beam steering"; that is, the magnetic fields can be used to change the beam's direction without altering its speed (Fig. 20–63). Show how this works with a beam of protons. What happens to protons that are not moving with the speed that the magnetic field is designed for? If the field extends over a region 5.0 cm wide and has a magnitude of 0.33 T, by approximately what angle will a beam of protons traveling at 1.0×10^7 m/s be bent?

71. A square loop of aluminum wire is 20.0 cm on a side. It is to carry 25.0 A and rotate in a 1.65-T magnetic field. (*a*) Determine the minimum diameter of the wire so that it will not fracture from tension or shear. Assume a safety factor of 10. (See Table 9–2.) (*b*) What is the resistance of a single loop of this wire?

72. The magnetic field B at the center of a circular coil of wire carrying a current I is

$$B = \frac{\mu_0 N I}{2r},$$

where N is the number of loops in the coil and r is its radius. Suppose that an electromagnet uses a coil 1.2 m in diameter made from square copper wire 1.6 mm on a side. The power supply produces 120 V at a maximum power output of 4.0 kW. (*a*) How many turns are needed to run the power supply at maximum power? (*b*) What is the magnetic field strength at the center of the coil? (*c*) If you use a greater number of turns and this same power supply (so the voltage remains at 120 V), will a greater magnetic field strength result? Explain.

Transmission lines carry electric power over great distances, at very high voltage for greater efficiency. To reduce high voltage to usable voltage, transformers are used, whose operation depends on electromagnetic induction. Induction is also the basis for electric generators, which produce the electric power in the first place.

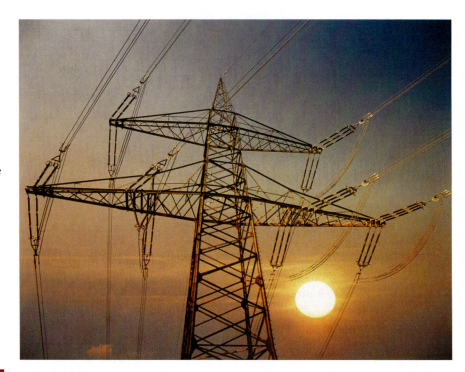

21 ELECTROMAGNETIC INDUCTION AND FARADAY'S LAW; AC CIRCUITS

I n Chapter 20, we discussed two ways in which electricity and magnetism are related: (1) an electric current produces a magnetic field; and (2) a magnetic field exerts a force on an electric current or moving electric charge. These discoveries were made in 1820–1821. Scientists then began to wonder: if electric currents produce a magnetic field, is it possible that a magnetic field can produce an electric current? Ten years later the American Joseph Henry (1797–1878) and the Englishman Michael Faraday (1791–1867) independently found that it was possible. Henry actually made the discovery first. But Faraday published his results earlier and investigated the subject in more detail. We now discuss this phenomenon and some of its world-changing applications.

21–1 Induced EMF

In his attempt to produce an electric current from a magnetic field, Faraday used an apparatus like that shown in Fig. 21–1. A coil of wire, X, was connected to a battery. The current that flowed through X produced a magnetic field that was intensified by the iron core. Faraday hoped that by using a

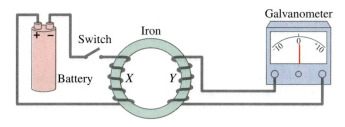

FIGURE 21-1 Faraday's experiment to induce an emf.

strong enough battery, a steady current in X would produce a great enough magnetic field to produce a current in a second coil Y. This second circuit, Y, contained a galvanometer to detect any current but contained no battery. He met no success with steady currents. But the long-sought effect was finally observed when Faraday saw the galvanometer in circuit Y deflect strongly at the moment he closed the switch in circuit X. And the galvanometer deflected strongly in the opposite direction when he opened the switch. A *steady* current in X had produced *no* current in Y. Only when the current in X was starting or stopping was a current produced in Y.

Constant **B** *induces no emf*

Faraday concluded that although a steady magnetic field produces no current, a *changing* magnetic field can produce an electric current! Such a current is called an **induced current**. When the magnetic field through coil Y changes, a current flows as if there were a source of emf in the circuit. We therefore say that an

induced emf is produced by a changing magnetic field.

Faraday did further experiments on **electromagnetic induction**, as this phenomenon is called. For example, Fig. 21–2 shows that if a magnet is moved quickly into a coil of wire, a current is induced in the wire. If the magnet is quickly removed, a current is induced in the opposite direction. Furthermore, if the magnet is held steady and the coil of wire is moved toward or away from the magnet, again an emf is induced and a current flows. Motion or change is required to induce an emf. It doesn't matter whether the magnet or the coil moves.

Changing **B** *induces an emf*

FIGURE 21-2 (a) A current is induced when a magnet is moved toward a coil. (b) The induced current is opposite when the magnet is moved away from the coil. Note that the galvanometer zero is at the center of the scale and the needle deflects left or right, depending on the direction of the current. In (c) no current is induced if the magnet does not move relative to the coil.

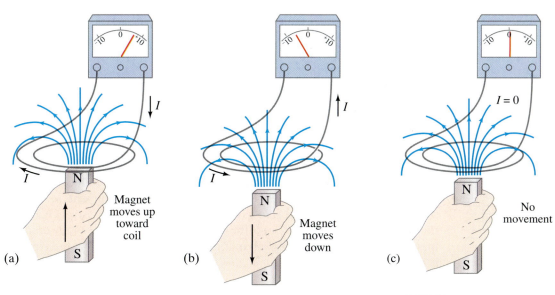

Faraday investigated quantitatively what factors influence the magnitude of the emf induced. He found first of all that it depends on time: the more rapidly the magnetic field changes, the greater the induced emf. But the emf is not simply proportional to the rate of change of the magnetic field, **B**. Rather it is proportional to the rate of change of the **magnetic flux**, Φ_B, passing through the loop of area A, which is defined as

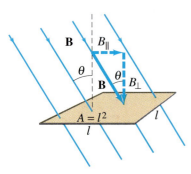

Magnetic flux defined

$$\Phi_B = B_\perp A = BA \cos\theta. \qquad (21\text{–}1)$$

Here $B_\perp$ is the component of the magnetic field **B** perpendicular to the face of the coil, and θ is the angle between **B** and a line drawn perpendicular to the face of the coil. These quantities are shown in Fig. 21–3 for a square coil of side l whose area $A = l^2$. When the face of the coil is parallel to **B**, $\theta = 90°$ and $\Phi_B = 0$. When **B** is perpendicular to the coil, $\theta = 0°$ and

$$\Phi_B = BA. \qquad [\mathbf{B} \perp \text{coil face}]$$

As we saw earlier, the lines of **B** (like lines of **E**) can be drawn such that the number of lines per unit area is proportional to the field strength. Then the flux Φ_B can be thought of as being proportional to the *total number of lines passing through the coil*. This is illustrated in Fig. 21–4, where the coil is viewed from the side (on edge). For $\theta = 90°$, no lines pass through the coil and $\Phi_B = 0$, whereas Φ_B is a maximum when $\theta = 0°$. The unit of magnetic flux is the tesla-meter^2; this is called a **weber**: $1\,\text{Wb} = 1\,\text{T·m}^2$.

With this definition of the flux, we can now write down the results of Faraday's investigations. If the flux through N loops of wire changes by an amount $\Delta\Phi_B$ during a time Δt, the average induced emf during this time is

FIGURE 21–3 Determining the flux through a flat loop of wire. This loop is square, of side l and area $A = l^2$.

FARADAY'S LAW
OF INDUCTION

$$\mathcal{E} = -N\frac{\Delta\Phi_B}{\Delta t}. \qquad (21\text{–}2)$$

This fundamental result is known as **Faraday's law of induction**, and is one of the basic laws of electromagnetism.

The minus sign in Eq. 21–2 is placed there to remind us in which direction the induced emf acts. Experiments show that

Lenz's law

> **an induced emf always gives rise to a current whose magnetic field opposes the original change in flux.**

This is known as **Lenz's law**. Let us apply it to the case of relative motion between a magnet and a coil, Fig. 21–2. The changing flux induces an emf, which produces a current in the coil. And this induced current produces its own magnetic field. In Fig. 21–2a the distance between the coil and the magnet decreases. So the magnetic field, and therefore the flux, through the coil increases. The magnetic field of the magnet points upward. To oppose this upward increase, the field produced by the induced current points *downward*. Thus, Lenz's law tells us that the current moves as shown (use the right-hand rule). In Fig. 21–2b, the flux *decreases* (because the magnet is moved away), so the induced current produces an *upward* magnetic field that is "trying" to maintain the status quo. Thus the current is as shown.

FIGURE 21–4 Magnetic flux Φ_B is proportional to the number of lines of **B** that pass through the loop.

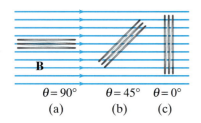

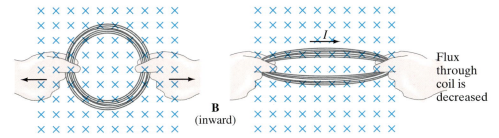

FIGURE 21–5 A current can be induced by changing the area of the coil. In both this case and that of Fig. 21–6, the flux through the coil is reduced. Here the brief induced current acts in the direction shown so as to try to maintain the original flux ($\Phi = BA$) by producing its own magnetic field into the page. That is, as the area A decreases, the current acts to increase B in the original (inward) direction.

Let us consider what would happen if Lenz's law were not true, but were just the reverse. The induced current in this imaginary situation would produce a flux in the same direction as the original change. This greater change in flux would produce an even larger current followed by a still greater change in flux, and so on. The current would continue to grow indefinitely, producing power ($= I^2R$) even after the original stimulus ended. This would violate the conservation of energy. Such "perpetual motion" devices do not exist. Thus, Lenz's law as stated above (and not its opposite) is consistent with the law of conservation of energy.

It is important to note that an emf is induced whenever there is a change in flux. Since magnetic flux $\Phi_B = BA\cos\theta$, we see that an emf can be induced in three ways: (1) by a changing magnetic field B; (2) by changing the area of the loop in the field; or (3) by changing the loop's orientation θ with respect to the field. Figures 21–1 and 21–2 illustrated case 1. Examples of cases 2 and 3 are illustrated in Figs. 21–5 and 21–6, respectively.

CONCEPTUAL EXAMPLE 21–1 | **Induction stove.** Some modern stove burners are based on induction. That is, an ac current passes around a coil that is the "burner" (a burner that never gets hot). Why will it heat a metal pan but not a glass container?

RESPONSE The ac current sets up a changing magnetic field that passes through the pan bottom. This changing field induces a current through the pan bottom, and since the pan offers resistance, electric energy is transformed to heat, heating the pot and its contents. A glass container offers very high resistance so very little current is induced and very little energy transferred. Recall Eq. 18–6c, $P = V^2/R$.

➡ **PHYSICS APPLIED**

Induction stove

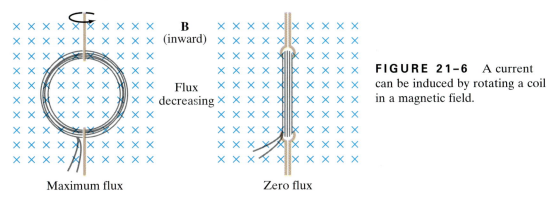

Maximum flux   Zero flux

FIGURE 21–6 A current can be induced by rotating a coil in a magnetic field.

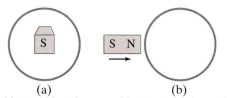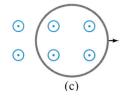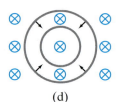

(a)	(b)	(c)	(d)	(e)
N magnetic pole moving toward coil from above the page	N magnetic pole moving toward the coil in the plane of the page	Pulling the coil to the right out of a magnetic field that points out of the page	Shrinking a coil in a magnetic field pointing into the page	Rotating the coil about the vertical diameter by pulling the left side toward the reader and pushing the right side away from the reader in a magnetic field that points from right to left in the plane of the page

FIGURE 21–7 Example 21–2.

CONCEPTUAL EXAMPLE 21–2 **Practice with Lenz's law.** In which direction is the current induced in the coil for each situation in Fig. 21–7?

RESPONSE (*a*) Magnetic field lines point out from the N pole of a magnet, so as the magnet moves down toward the coil, the field points into the page and is getting stronger. The current will be induced in the counterclockwise direction to produce a field **B** *out* of the page so that its own flux counteracts the externally imposed change.
(*b*) The field is in the plane of the page, so the flux through the coil is zero throughout the process; hence there is no change in magnetic flux with time, and there will be no induced emf or current in the coil.
(*c*) Initially, the magnetic flux pointing out of the page passes through the coil. If you remove the coil, the induced current will be in a direction to make up the deficiency: the current flow will be counterclockwise to produce an outward (toward the reader) magnetic field.
(*d*) The flux is into the page and the coil area shrinks so the flux will decrease; hence the induced current will be clockwise to try to produce its own flux into the page to make up for the flux decrease.
(*e*) Initially there is no flux through the coil (why?). When you start to rotate the coil, the flux begins passing through the coil increasing to the left. To counteract this, the coil will have current induced in a counterclockwise direction so as to produce its own flux to the right.

EXAMPLE 21–3 **Pulling a coil from a magnetic field.** A square coil of side 5.0 cm contains 100 loops and is positioned perpendicular to a uniform 0.60-T magnetic field, as shown in Fig. 21–8. It is quickly and uniformly pulled from the field (moving perpendicular to **B**) to a region where B drops abruptly to zero. It takes 0.10 s for the whole coil to reach the field-free region. Find (*a*) the change in flux through the coil, (*b*) the emf and current induced, and (*c*) how much energy is dissipated in the coil if its resistance is 100 Ω. (*d*) What was the average force required?

SOLUTION (*a*) First we find how the magnetic flux, $\Phi_B = BA$, changes during the time interval $\Delta t = 0.10$ s. The area of the coil is $A = (0.050\text{ m})^2 = 2.5 \times 10^{-3}\text{ m}^2$. The flux is initially $\Phi_B = BA = (0.60\text{ T})(2.5 \times 10^{-3}\text{ m}^2) = 1.5 \times 10^{-3}$ Wb. After 0.10 s, the flux is zero. Hence the change in flux is

$$\Delta\Phi_B = 0 - 1.5 \times 10^{-3}\text{ Wb} = -1.5 \times 10^{-3}\text{ Wb}.$$

(*b*) The rate of change of flux is constant during the 0.10 s, so the emf induced (Eq. 21–2) during this period is

$$\mathscr{E} = -(100)\frac{(0 - 1.5 \times 10^{-3}\text{ Wb})}{(0.10\text{ s})} = 1.5\text{ V}.$$

The current is

$$I = \frac{\mathscr{E}}{R} = \frac{1.5\text{ V}}{100\text{ }\Omega} = 15\text{ mA}.$$

(c) The total energy dissipated is

$$E = Pt = I^2Rt = (1.5 \times 10^{-2}\text{ A})^2(100\text{ }\Omega)(0.10\text{ s}) = 2.3 \times 10^{-3}\text{ J}.$$

(d) From the conservation of energy principle, the result in (c) is equal to the work W needed to pull the coil out of the field. Since $W = \bar{F}d$, the average force is

$$\bar{F} = \frac{W}{d} = \frac{2.3 \times 10^{-3}\text{ J}}{5.0 \times 10^{-2}\text{ m}} = 0.046\text{ N},$$

where $d = 5.0$ cm because there is no flux change (hence no force) until one edge of the coil leaves the field.

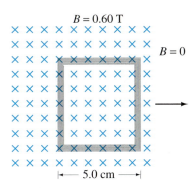

FIGURE 21–8 Example 21–3. The square coil in a magnetic field $B = 0.60$ T is pulled abruptly to the right to a region where $B = 0$.

21–3 EMF Induced in a Moving Conductor

Another way to induce an emf is shown in Fig. 21–9, and this situation helps illuminate the nature of the induced emf. Assume that a uniform magnetic field **B** is perpendicular to the area bounded by the U-shaped conductor and the movable rod resting on it. If the rod is made to move at a speed v, it travels a distance $\Delta x = v\,\Delta t$ in a time Δt. Therefore, the area of the loop increases by an amount $\Delta A = l\,\Delta x = lv\,\Delta t$ in a time Δt. By Faraday's law, there is an induced emf $\mathscr{E}$ whose magnitude is given by

$$\mathscr{E} = \frac{\Delta\Phi_B}{\Delta t} = \frac{B\,\Delta A}{\Delta t} = \frac{Blv\,\Delta t}{\Delta t} = Blv. \qquad \text{(21–3)}$$

This equation is valid as long as B, l, and v are mutually perpendicular. (If they are not, we use only the components of each that are mutually perpendicular.) An emf induced in this way is sometimes called *motional emf*.

Motional emf

We can also obtain Eq. 21–3 without using Faraday's law. We saw in Chapter 20 that a charged particle moving perpendicular to a magnetic field B with speed v experiences a force $F = qvB$. When the rod of Fig. 21–9 moves to the right with speed v, the electrons in the rod move with this same speed. Therefore, each feels a force $F = qvB$, which acts upward in the figure. If the rod were not in contact with the U-shaped conductor, electrons would collect at the upper end of the rod, leaving the lower end positive. There must thus be an induced emf. If the rod does slide on the U-shaped conductor, the electrons will flow into it. There will then be a clockwise (conventional) current flowing in the loop. To calculate the emf, we determine the work W needed to move a charge q from one end of the rod to the other against this potential difference: $W = \text{force} \times \text{distance} = (qvB)(l)$. The emf equals the work done per unit charge, so $\mathscr{E} = W/q = qvBl/q = Blv$, just as above.[†]

FIGURE 21–9 A conducting rod is moved to the right on a U-shaped conductor in a uniform magnetic field **B** that points out of the paper.

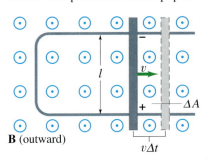

[†]This argument, which is basically the same as for the Hall effect, explains this one way of inducing an emf. It does not explain the general case of electromagnetic induction, however.

FIGURE 21–10 Example 21–4.

EXAMPLE 21–4 **Does a moving airplane develop a dangerous emf?** An airplane travels 1000 km/h in a region where the Earth's field is 5.0×10^{-5} T and is nearly vertical (Fig. 21–10). What is the potential difference induced between the wing tips that are 70 m apart?

SOLUTION Since $v = 1000$ km/h $= 280$ m/s, and $\mathbf{v} \perp \mathbf{B}$, we have

$$\mathscr{E} = Blv = (5.0 \times 10^{-5} \text{ T})(70 \text{ m})(280 \text{ m/s}) = 1.0 \text{ V}.$$

Not much to worry about.

21–4 Changing Magnetic Flux Produces an Electric Field

As we just discussed, the electrons in the moving conductor of Fig. 21–9 feel a force. This implies that there is an electric field in the conductor. Since electric field is defined as the force per unit charge, $E = F/q$, the effective field E in the rod must be (since $F = qvB$)

Electric field is produced by a changing magnetic flux

$$E = \frac{F}{q} = \frac{qvB}{q} = vB. \tag{21–4}$$

In the situation where a changing magnetic field (rather than a moving conductor) induces an emf (as, for example, in Fig. 21–2), there also is an induced current. And again this implies that there is an electric field in the wire. Thus we come to the important conclusion that

a changing magnetic flux produces an electric field.

This applies not only to wires and other conductors, but is a general result that applies to any region in space: an electric field will be produced at any point in space where there is a changing magnetic field.

→ **PHYSICS APPLIED**

Blood flow measurement

FIGURE 21–11 Measurement of blood velocity from the induced emf.

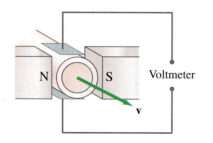

EXAMPLE 21–5 **Electromagnetic blood-flow measurement.** The rate of blood flow can be measured using the apparatus shown in Fig. 21–11, since blood contains charged ions. Suppose that the blood vessel is 2.0 mm in diameter, the magnetic field is 0.080 T, and the measured emf is 0.10 mV. What is the flow velocity of the blood?

SOLUTION We solve for v in Eq. 21–3, and we find that

$$v = \frac{\mathscr{E}}{Bl} = \frac{(1.0 \times 10^{-4} \text{ V})}{(0.080 \text{ T})(2.0 \times 10^{-3} \text{ m})} = 0.63 \text{ m/s}.$$

(In actual practice, an alternating current is used to produce an alternating magnetic field. The induced emf is then alternating.)

21–5 Electric Generators

Probably the most important practical result of Faraday's great discovery was the development of the **electric generator** or **dynamo**. A generator transforms mechanical energy into electric energy. This is just the opposite of what a motor does. Indeed, a generator is basically the inverse of a motor.[†] A simplified diagram of an **ac generator** is shown in Fig. 21–12. A generator consists of many coils of wire (only one is shown) wound on an armature that can rotate in a magnetic field. The axle is turned by some mechanical means (falling water, car motor belt), and an emf is induced in the rotating coil. An electric current is thus the *output* of a generator. In Fig. 21–12 the right-hand rule tells us that, with the armature rotating counterclockwise, the (conventional) current in the wire labeled *a* on the armature is outward; therefore it is outward at brush *a*. (Each brush presses against a continuous slip ring.) After one-half revolution, wire *a* will be where wire *b* is now in the drawing, and the current then at brush *a* will be inward. Thus the current produced is alternating. Let us look at this in more detail.

In Fig. 21–13, the loop is being made to rotate clockwise in a uniform magnetic field **B**. The velocity of the two lengths *ab* and *cd* at this instant are shown. Although the sections of wire *bc* and *da* are moving, the force on electrons in these sections is toward the side of the wire, not along its length. The emf generated is thus due only to the force on charges in the sections *ab* and *cd*. From the right-hand rule, we see that the direction of the induced current in *ab* is from *a* toward *b*. And in the lower section, it is from *c* to *d*; so the flow is continuous in the loop. The magnitude of the emf generated in *ab* is given by Eq. 21–3, except that we must take the component of the velocity perpendicular to *B*:

$$\mathscr{E} = Blv_{\perp},$$

where *l* is the length of *ab*. From the diagram we can see that $v_{\perp} = v \sin\theta$, where θ is the angle the face of the loop makes with the vertical. The emf induced in *cd* has the same magnitude and is in the same direction. Therefore they add, and the total emf is

$$\mathscr{E} = 2NBlv \sin\theta,$$

where we have multiplied by *N*, the number of loops in the coil (if there is more than one). If the coil is rotating with constant angular velocity ω, then the angle $\theta = \omega t$. We also have from the angular equations (Chapter 8) that $v = \omega r = \omega(h/2)$, where *h* is the length of *bc* or *ad*. Thus $\mathscr{E} = 2NBl\omega(h/2)\sin\omega t$, or

$$\mathscr{E} = NBA\omega \sin\omega t, \tag{21–5}$$

where $A = lh$ is the area of the loop. This equation holds for any shape coil, not just for a rectangle as derived. Thus, the output emf of the generator is sinusoidally alternating (Fig. 21–14 and Section 18–8). Since ω is expressed in radians per second, we can write $\omega = 2\pi f$, where *f* is the frequency.

PHYSICS APPLIED

AC generator

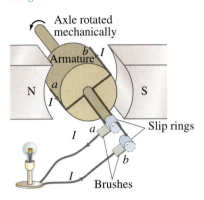

FIGURE 21–12 An ac generator.

FIGURE 21–13 The emf is induced in the segments *ab* and *cd*, whose velocity components perpendicular to the field **B** are $v \sin\theta$.

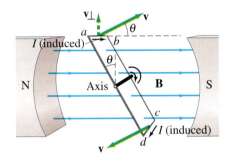

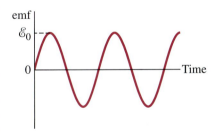

FIGURE 21–14 An ac generator produces an alternating current. The output emf $\mathscr{E} = \mathscr{E}_0 \sin\omega t$, where $\mathscr{E}_0 = NAB\omega$ (Eq. 21–5).

FIGURE 21–15 Water-driven generators at the base of Boulder Dam, Nevada.

Over 99 percent of the electricity used in the United States is produced from generators (Fig. 21–15). The frequency f is 60 Hz for general use in the United States and Canada, although 50 Hz is used in many countries. In electric power generating plants, the armature is mounted on a heavy axle connected to a turbine, which is the modern equivalent of a waterwheel. Water falling over a dam can turn the turbine at a hydroelectric plant. Most of the power generated at present in the United States, however, is done at steam plants, where the burning of fossil fuels (coal, oil, natural gas) boils water to produce high-pressure steam that turns the turbines. Likewise, at nuclear power plants, the nuclear energy released is used to produce steam to turn turbines. Thus, a heat engine (Chapter 15) connected to a generator is the principal means of generating electric power.

The frequency of 60 Hz is maintained very precisely by power companies, and in doing problems, we will assume it is at least as precise as other numbers given.

EXAMPLE 21–6 **An ac generator.** The armature of a 60-Hz ac generator rotates in a 0.15-T magnetic field. If the area of the coil is $2.0 \times 10^{-2}\,\mathrm{m^2}$, how many loops must the coil contain if the peak output is to be $\mathscr{E}_0 = 170\,\mathrm{V}$?

SOLUTION From Eq. 21–5, we see that the maximum emf is $\mathscr{E}_0 = NBA\omega$. Since $\omega = 2\pi f = (6.28)(60\,\mathrm{s^{-1}}) = 377\,\mathrm{s^{-1}}$, we have

$$N = \frac{\mathscr{E}_0}{BA\omega} = \frac{170\,\mathrm{V}}{(0.15\,\mathrm{T})(2.0 \times 10^{-2}\,\mathrm{m^2})(377\,\mathrm{s^{-1}})} = 150\ \text{turns}.$$

A **dc generator** is much like an ac generator, except the slip rings are replaced by split-ring commutators, Fig. 21–16a, just as in a dc motor. The output of such a generator is as shown and can be smoothed out by placing a capacitor in parallel with the output (Section 19–7). More common is the use of many armature windings, as in Fig. 21–16b, which produces a smoother output.

In the past, automobiles used dc generators. More common now, however, are ac generators or **alternators**, which avoid the problems of wear and electrical arcing (sparks) across the split-ring commutators of dc generators. Alternators differ from the generators discussed above in the following way. In an alternator, current from the battery produces a magnetic field in an electromagnet, called the *rotor*, which is made to rotate by a belt from the engine. Surrounding the rotating rotor are a set of stationary

FIGURE 21–16 (a) A dc generator with one set of commutators, and (b) a dc generator with many sets of commutators and windings.

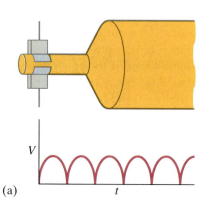

(a)

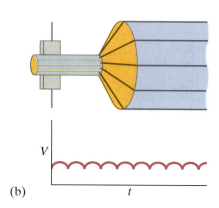

(b)

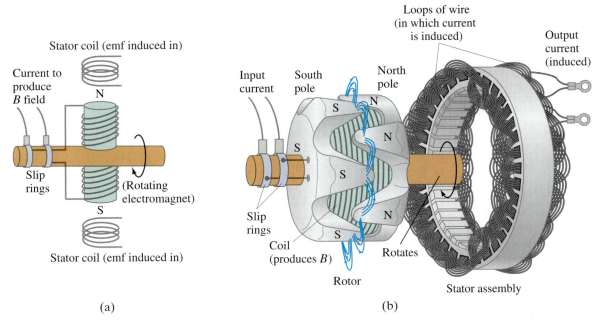

FIGURE 21–17 (a) Schematic (simplified) diagram of an alternator. The input electromagnet current to the rotor is connected through continuous slip rings. Sometimes the rotor electromagnet is replaced by a permanent magnet. (b) Actual shape of an alternator. The rotor is made to turn by a belt from the engine. The current in the wire coil of the rotor produces a magnetic field inside it on its axis that points horizontally from left to right, thus making north and south poles of the plates attached at either end. These end plates are made with triangular fingers that are bent over the coil—hence there are alternating N and S poles quite close to one another, with magnetic field lines between them as shown by the blue lines. As the rotor turns, these field lines pass through the fixed stator coils (shown on the right for clarity, but in operation the rotor rotates within the stator), inducing a current in them, which is the output.

coils called the *stator*, Fig. 21–17. The magnetic field of the rotor passes through the stator coils and, since the rotor is rotating, the field through the fixed stator coils is changing. Hence an alternating current is induced in the stator coils, which is the output. This ac output is changed to dc for charging the battery by the use of semiconductor diodes, which allow current flow in one direction only (see Section 29–8).

Output emf of alternator

* 21–6 Counter EMF and Torque; Eddy Currents

A motor turns and produces mechanical energy when a current is made to flow in it. From our description in Section 20–10 of a simple dc motor, you might expect that the armature would accelerate indefinitely due to the torque on it. However, as the armature of the motor turns, the magnetic flux through the coil changes and an emf is generated. This induced emf acts to oppose the motion (Lenz's law) and is called the **back emf** or **counter emf**. The greater the speed of the motor, the greater the counter emf. A motor normally turns and does work on something, but if there were no load, the motor's speed would increase until the counter emf equaled the input voltage. In the normal situation, when there is a mechanical load, the speed of the motor is limited also by the load. The counter emf will then be less than the external voltage. The greater the mechanical load, the slower the motor rotates and the lower is the counter emf ($\mathscr{E} \propto \omega$, Eq. 21–5).

Back emf

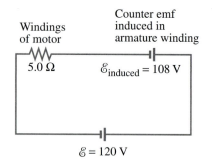

Counter emf induced in armature winding

Windings of motor

5.0 Ω $\mathcal{E}_{induced}$ = 108 V

$\mathcal{E}$ = 120 V

FIGURE 21–18 Circuit of a motor showing induced counter emf.

Effect of back emf on current

EXAMPLE 21–7 **Counter emf in a motor.** The armature windings of a dc motor have a resistance of 5.0 Ω. The motor is connected to a 120-V line, and when the motor reaches full speed against its normal load, the counter emf is 108 V. Calculate (*a*) the current into the motor when it is just starting up, and (*b*) the current when it reaches full speed.

SOLUTION (*a*) Initially, the motor is not turning (or turning very slowly), so there is no induced counter emf. Hence, from Ohm's law, the current is

$$ I = \frac{V}{R} = \frac{120\ V}{5.0\ \Omega} = 24\ A. $$

(*b*) At full speed, the counter emf is a source of emf that opposes the exterior source. We represent this counter emf as a battery in the equivalent circuit shown in Fig. 21–18. In this case, Ohm's law (or Kirchhoff's rule) gives

$$ 120\ V - 108\ V = I\,(5.0\ \Omega). $$

Therefore

$$ I = 12\ V/5.0\ \Omega = 2.4\ A. $$

This Example illustrates the fact that the current is very high when a motor first starts up. This is why the lights in your house may dim when the motor of the refrigerator (or other large motor) starts up. The large initial current causes the voltage at the outlets to drop (the house wiring has resistance, so there is some voltage drop across it when large currents are drawn).

➡ **PHYSICS APPLIED**

Burning out a motor

CONCEPTUAL EXAMPLE 21–8 **Motor overload.** When using an appliance such as a blender, electric drill, or sewing machine, if the appliance is overloaded or jammed so that the motor slows appreciably or stops while the power is still connected, the device can burn out and be ruined. Explain why this happens.

RESPONSE The motors are designed to run at a certain speed for a given applied voltage and the designer must take the expected counter emf into account. If the rotation speed is reduced, the counter emf will not be as high as expected ($\mathcal{E} \propto \omega$, Eq. 21–5), and the current will increase, and may become large enough that the windings of the motor heat up to the point of ruining the motor.

In a generator, the situation is the reverse of that for a motor. As we saw, the mechanical turning of the armature induces an emf in the loops, which is the output. If the generator is not connected to an external circuit, the emf exists at the terminals but no current flows. In this case, it takes little effort to turn the armature. But if the generator *is* connected to a device that draws current, then a current flows in the coils of the armature. Because this current-carrying coil is in a magnetic field, there will be a torque exerted on it (as in a motor), and this torque opposes the motion (use the right-hand rule for the force on a wire, in Fig. 21–12). This is called a **counter torque**. The greater the electrical load—that is, the more current that is drawn—the greater will be the counter torque. Hence the external applied torque will have to be greater to keep the generator turning.

Counter torque

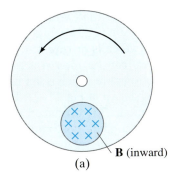
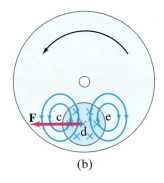

B (inward)

(a)   (b)

FIGURE 21–19 Production of eddy currents in a rotating wheel.

This of course makes sense from the conservation-of-energy principle. More mechanical-energy input is needed to produce more electrical-energy output.

Induced currents are not always confined to well-defined paths such as in wires. Consider, for example, the rotating metal wheel in Fig. 21–19a. A magnetic field is applied to a limited area as shown and points into the paper. The section of wheel in the magnetic field has an emf induced in it because the conductor is moving (carrying electrons with it). The flow of (conventional) current is upward in the region of the magnetic field (Fig. 21–19b), and the current follows a downward return path outside that region. Why? According to Lenz's law, the induced currents oppose the change that causes them. Consider the part of the wheel labeled c in Fig. 21–19b, where the magnetic field is zero but is just about to enter a region where **B** points into the page. To oppose this change, the induced current is counterclockwise to produce a field pointing out of the page (right-hand rule). Similarly region d is about to move to e, where **B** is zero; hence the current is clockwise to produce an inward field opposed to this change. These currents are referred to as **eddy currents** and can be present in any conductor that is moving across a magnetic field or through which the magnetic flux is changing. In Fig. 21–19, the magnetic field exerts a force **F** on the induced currents that opposes (use the right-hand rule) the rotational motion. Eddy currents can be used in this way as a smooth braking device on, say, a rapid-transit car. In order to stop the car, an electromagnet can be turned on that applies its field either to the wheels or to the moving steel rail below. Eddy currents can also be used to dampen (reduce) the oscillation of a vibrating system. A common example is in a galvanometer, where induced eddy currents keep the needle from overshooting or oscillating violently. Eddy currents, however, can be a problem. For example, eddy currents induced in the armature of a motor or generator produce heat ($P = I\mathscr{E}$) and waste energy. To reduce the eddy currents, the armatures are *laminated*; that is, they are made of very thin sheets of iron that are well insulated from one another. (See Fig. 21–21 in the next Section.) Thus the total path length of the eddy currents is confined to each slab, which increases the total resistance; hence the current is less and there is less wasted energy.

Eddy currents

➡ **PHYSICS APPLIED**

Brakes and damping

21–7 Transformers; Transmission of Power

A transformer is a device for increasing or decreasing an ac voltage. Transformers are found everywhere: in TV sets to give the high voltage needed for the picture tube, in converters for plugging in a portable "Walkman,"

➡ **PHYSICS APPLIED**

Transformers

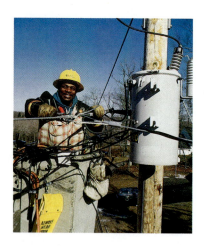

FIGURE 21–20 Repairing a transformer on a utility pole.

on utility poles (Fig. 21–20) to reduce the high voltage from the electric company to that usable in houses (110 V or 220 V), and in many other applications. A **transformer** consists of two coils of wire known as the **primary** and **secondary** coils. The two coils can be interwoven (with insulated wire); or they can be linked by a soft iron core which is laminated to prevent eddy-current losses (Section 21–6), as shown in Fig. 21–21. Transformers are designed so that (nearly) all the magnetic flux produced by the current in the primary also passes through the secondary coil, and we assume this is true in what follows. We also assume that energy losses in the resistance of the coils and hysteresis in the iron can be ignored—a good approximation for real transformers, which are often better than 99 percent efficient.

When an ac voltage is applied to the primary, the changing magnetic field it produces will induce an ac voltage of the same frequency in the secondary. However, the voltage will be different according to the number of loops in each coil. From Faraday's law, the voltage or emf induced in the secondary is

$$V_S = N_S \frac{\Delta \Phi_B}{\Delta t},$$

where N_S is the number of turns in the secondary coil, and $\Delta \Phi_B / \Delta t$ is the rate at which the magnetic flux changes. The input primary voltage, V_P, is also related to the rate at which the flux changes

$$V_P = N_P \frac{\Delta \Phi_B}{\Delta t},$$

where N_P is the number of turns in the primary coil.[†] We divide these two equations, assuming little or no flux is lost, to find

Transformer equation

$$\frac{V_S}{V_P} = \frac{N_S}{N_P}. \tag{21–6}$$

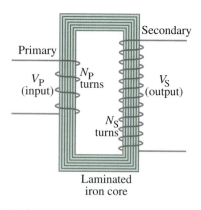

FIGURE 21–21 Step-up transformer ($N_P = 4$, $N_S = 12$).

Transformer equation II

This *transformer equation* tells how the secondary (output) voltage is related to the primary (input) voltage; V_S and V_P in Eq. 21–6 can be the rms values for both, or peak values for both.

If N_S is greater than N_P, we have a **step-up transformer**. The secondary voltage is greater than the primary voltage. For example, if the secondary has twice as many turns as the primary, then the secondary voltage will be twice that of the primary. If N_S is less than N_P, we have a **step-down transformer**.

Although ac voltage can be increased (or decreased) with a transformer, we don't get something for nothing. Energy conservation tells us that the power output can be no greater than the power input. A well-designed transformer can be greater than 99 percent efficient, so little energy is lost to heat. The power input thus essentially equals the power output. Since power $P = VI$ (Eq. 18–5), we have

$$V_P I_P = V_S I_S,$$

or

$$\frac{I_S}{I_P} = \frac{N_P}{N_S}. \tag{21–7}$$

[†]This follows because the changing flux produces a counter emf, $N_P \Delta \Phi_B / \Delta t$ in the primary that exactly balances the applied voltage V_P if the resistance of the primary can be ignored (Kirchhoff's rules).

EXAMPLE 21–9 **Portable radio transformer.** A transformer for home use of a portable radio reduces 120-V ac to 9.0-V ac. (Such a device also contains diodes to change the 9.0-V ac to dc. See Chapter 29.) The secondary contains 30 turns and the radio draws 400 mA. Calculate: (*a*) the number of turns in the primary; (*b*) the current in the primary; and (*c*) the power transformed.

SOLUTION (*a*) This is a step-down transformer, and from Eq. 21–6 we have

$$N_P = N_S \frac{V_P}{V_S} = \frac{(30)(120 \text{ V})}{(9.0 \text{ V})} = 400 \text{ turns.}$$

(*b*) From Eq. 21–7:

$$I_P = I_S \frac{N_S}{N_P} = (0.40 \text{ A})\left(\frac{30}{400}\right) = 0.030 \text{ A.}$$

(*c*) The power transformed is

$$P = I_S V_S = (9.0 \text{ V})(0.40 \text{ A}) = 3.6 \text{ W,}$$

which is, assuming 100 percent efficiency, the same as the power in the primary, $P = (120 \text{ V})(0.030 \text{ A}) = 3.6 \text{ W.}$

It is important to recognize that a transformer operates only on ac. A dc current in the primary does not produce a changing flux and therefore induces no emf in the secondary. However, if a dc voltage is applied to the primary through a switch, at the instant the switch is opened or closed there will be an induced current in the secondary. For example, if the dc is turned on and off as shown in Fig. 21–22a, the voltage induced in the secondary is as shown in Fig. 21–22b. Notice that the secondary voltage drops to zero when the dc voltage is steady.

Transformers play an important role in the transmission of electricity. Power plants are often situated some distance from metropolitan areas. Hydroelectric plants are located at a dam site and nuclear plants need much cooling water. Fossil-fuel plants too are often situated far from a city because of lack of availability of land or to avoid contributing to air pollution. In any case, electricity must often be transmitted over long distances (Fig. 21–23). There is always some power loss in the transmission lines, and this loss can be minimized if the power is transmitted at high voltage, using transformers, as the following Example shows.

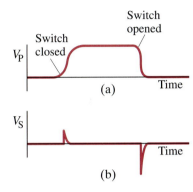

FIGURE 21–22 A dc voltage turned on and off as shown in (a) produces voltage pulses in the secondary (b). Voltage scales in (a) and (b) are not necessarily the same.

➡ **PHYSICS APPLIED**

Transformers help power transmission

FIGURE 21–23 The transmission of electric power from power plants to homes makes use of transformers at various stages.

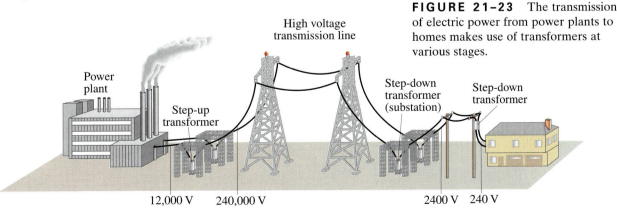

EXAMPLE 21–10 **Transmission lines.** An average of 120 kW of electric power is sent to a small town from a power plant 10 km away. The transmission lines have a total resistance of 0.40 Ω. Calculate the power loss if the power is transmitted at (*a*) 240 V and (*b*) 24,000 V.

SOLUTION For each case we determine the current I in the lines, and then find the power loss from $P = I^2R$. (*a*) If 120 kW is sent at 240 V, the total current will be

$$I = \frac{P}{V} = \frac{1.2 \times 10^5 \text{ W}}{2.4 \times 10^2 \text{ V}} = 500 \text{ A}.$$

The power loss in the lines, P_L, is then

$$P_L = I^2R = (500 \text{ A})^2(0.40 \text{ }\Omega) = 100 \text{ kW}.$$

Thus, over 80 percent of all the power would be wasted as heat in the power lines!
(*b*) When $V = 24{,}000$ V,

$$I = \frac{P}{V} = \frac{1.2 \times 10^5 \text{ W}}{2.4 \times 10^4 \text{ V}} = 5.0 \text{ A}.$$

The power loss is then

$$P_L = I^2R = (5.0 \text{ A})^2(0.40 \text{ }\Omega) = 10 \text{ W},$$

which is less than $\frac{1}{100}$ of 1 percent.

It should be clear that the greater the voltage, the less the current and thus the less power is wasted in the transmission lines. It is for this reason that power is usually transmitted at very high voltages, as high as 700 kV.

Power is generated at somewhat lower voltages than this, and the voltage in homes and factories is also much lower. The great advantage of ac, and a major reason it is in nearly universal use, is that the voltage can easily be stepped up and down by a transformer. The output voltage of an electric generating plant is stepped up prior to transmission. Upon arrival in a city, it is stepped down in stages at electric substations prior to distribution. The voltage in lines along city streets is typically 2400 V and is stepped down to 240 V or 120 V for home use by transformers (Fig. 21–20).

Direct current transmission has gained in popularity recently. Although changing voltage with dc is more difficult and expensive, it offers some advantages over ac. A few of these are as follows. Alternating current produces alternating magnetic fields which induce current in nearby wires and so reduce transmitted power; this is absent in dc. It is possible to transmit dc at a higher average voltage than ac since for dc, the rms value equals the peak; and breakdown of insulation or of air is determined by the peak voltage.

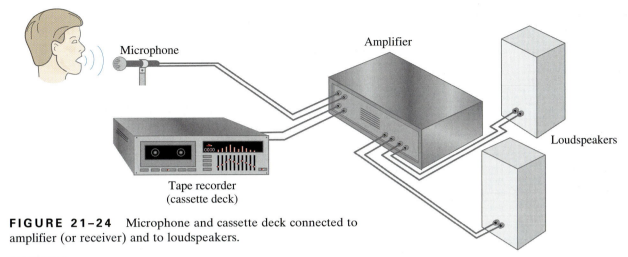

FIGURE 21–24 Microphone and cassette deck connected to amplifier (or receiver) and to loudspeakers.

*21–8 Applications of Induction: Sound Systems, Computer Memory, and the Seismograph

Figure 21–24 shows several components of a sound system. Two of these components, the microphone and the tape recorder, make use of the principle of electromagnetic induction. [The output of each goes to an amplifier that amplifies (increases) the signal and sends it to a loudspeaker to be heard; alternately, the microphone signal could go directly to the tape recorder for recording.]

There are various types of *microphones*, and many operate on the principle of induction. In one form, a microphone is just the inverse of a loudspeaker (Section 20–10). A small coil connected to a membrane is suspended close to a small permanent magnet, as shown in Fig. 21–25. The coil moves in the magnetic field when sound waves strike the membrane. The frequency of the induced emf will be just that of the impinging sound waves, and this emf is the "signal" that can be amplified and sent to loudspeakers, or sent to a tape recorder to be recorded on tape. In a "ribbon" microphone, a thin metal ribbon is suspended between the poles of a permanent magnet. The ribbon vibrates in response to sound waves, and the emf induced in the ribbon is proportional to its velocity.

Tape recording and tape playback is done by *tape heads* inside a tape recorder (or cassette deck). Recording tape for use in audio and video tape recorders contains a thin layer of magnetic oxide on a thin plastic tape. During recording, the audio or video signal voltage is sent to the recording head, which acts as a tiny electromagnet (Fig. 21–26) that magnetizes the tiny section of tape passing over the narrow gap in the head at each instant. In playback, the changing magnetism of the moving tape at the gap causes corresponding changes in the magnetic field within the soft-iron head, which in turn induces an emf in the coil (Faraday's law). This induced emf is the output signal that can be amplified and sent to a loudspeaker (or, in the case of a video signal, to the picture tube). In audio and video recorders, the signals are usually *analog*—they vary continuously in amplitude over time. The variation in degree of magnetization of the tape at any point reflects the variation in amplitude of the audio or video signal.

➡ PHYSICS APPLIED

Microphones

➡ PHYSICS APPLIED

Tape recording

FIGURE 21–25 Diagram of microphone that works by induction.

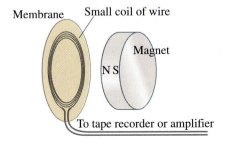

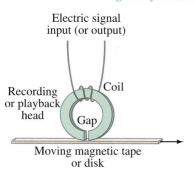

Electric signal
input (or output)

Coil

Recording
or playback
head

Gap

Moving magnetic tape
or disk

FIGURE 21–26 Recording and/or playback head for tape or disk. In recording (or "writing"), the electric input signal to the head, which acts as an electromagnet, magnetizes the passing tape or disk. In playback (or "reading"), the changing magnetic field of the passing tape or disk induces a changing magnetic field in the head, which in turn induces in the coil an emf that is the output signal.

Digital information, such as used on computer disks (floppy disks or hard disks) or on magnetic computer tape and some types of digital tape recorders, is read and written using heads that are basically the same as just described (Fig. 21–26). The essential difference is in the signals, which are not analog, but are digital, and in particular binary, meaning that only two values are possible for each predetermined space on the tape or disk. The two possible values are usually referred to as 1 and 0. The signal voltage does not vary continuously but rather takes on only two values, say $+5$ volts and 0 volts, corresponding to the 1 or 0. Thus, information is carried as a series of "bits," each of which can have only one of two values, 1 or 0.

In another field, geophysics, an important device based on electromagnetic induction is one type of *seismograph* or *geophone*. A seismograph is placed in direct contact with the Earth and converts the motion of the Earth—whether due to an earthquake or to an explosion (such as for mineral prospecting or for detecting a bomb test)—into an electrical signal. A seismograph contains a magnet and a coil of wire, one of which is fixed rigidly to the case, which moves as the Earth does where it is planted. The other element is inertial and is suspended from the case by a spring. In the type shown in Fig. 21–27, the coil moves with the Earth, and the relative motion of the magnet and coil produces an induced emf in the coil, which is the output of the device. In many geophones, the coil is inertial and the magnet moves with the Earth.

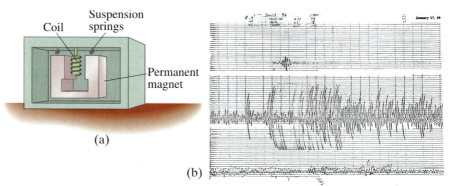

Coil

Suspension
springs

Permanent
magnet

(a)

(b)

FIGURE 21–27 (a) A seismograph or geophone. (b) A seismograph reading (Northridge, California earthquake, January 17, 1994).

* 21–9 Inductance

Mutual inductance. If two coils of wire are placed near one another, as in Fig. 21–28, a changing current in one will induce an emf in the other. According to Faraday's law, the emf $\mathscr{E}_2$ induced in coil 2 is proportional to the rate of change of flux passing through it. Since the flux is proportional to the current flowing in coil 1, $\mathscr{E}_2$ must be proportional to the rate of change of the current in coil 1, $\Delta I_1/\Delta t$. Thus we can write

Mutual inductance

$$\mathscr{E}_2 = -M\,\frac{\Delta I_1}{\Delta t}, \tag{21–8a}$$

where the constant of proportionality, M, is called the **mutual inductance**. (The minus sign is because of Lenz's law.) Mutual inductance has units of $\text{V·s/A} = \Omega\text{·s}$, which is called the **henry** (H), after Joseph Henry: $1\text{ H} = 1\ \Omega\text{·s}$. The value of M depends on whether iron is present or not, and on the geometry of the coil configuration: that is, on the size of the coils, on the number of turns, and on their separation. If we look at the inverse situation—a changing current in coil 2 inducing an emf in coil 1—the proportionality constant, M, turns out to have the same value,

$$\mathscr{E}_1 = -M\frac{\Delta I_2}{\Delta t}.$$ **(21–8b)**

A transformer is an example of mutual inductance in which the coupling is maximized so that nearly all flux lines pass through both coils. However, mutual inductance has other uses as well. For example, some pacemakers, which are used to maintain blood flow in heart patients (Section 19–8), are powered externally. Power in an external coil is transmitted via mutual inductance to a second coil in the pacemaker at the heart. This has the advantage over battery-powered pacemakers in that surgery is not needed to replace a battery when it wears out.

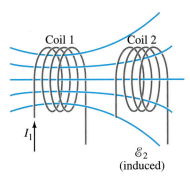

FIGURE 21–28 A changing current in one coil will induce a current in the second coil.

Self-inductance. The concept of inductance applies also to an isolated single coil. When a changing current passes through a coil or solenoid, a changing magnetic flux is produced inside the coil, and this in turn induces an emf. This induced emf opposes the change in flux (Lenz's law); it is much like the back emf generated in a motor. For example, if the current through the coil is increasing, the increasing magnetic flux induces an emf that opposes the original current and tends to retard its increase. If the current is decreasing in the coil, the decreasing flux induces an emf in the same direction as the current, tending to maintain the original current. In either case, the induced emf $\mathscr{E}$ is proportional to the rate of change in current (and is in the direction opposed to the change):

$$\mathscr{E} = -L\frac{\Delta I}{\Delta t}.$$ **(21–9)**

Self-inductance
(induced emf for an inductor)

The constant of proportionality L is called the **self-inductance**, or simply the **inductance** of the coil. It, too, is measured in henrys. The magnitude of L depends on the geometry (size and shape) and on whether an iron core is present or not.

An ac circuit always contains some inductance, but often it is quite small unless the circuit contains a coil of many turns. A coil that has significant self-inductance L is called an **inductor** or a **choke coil**. It is shown on circuit diagrams by the symbol

—oooo— .   [inductor symbol]

Inductors

It can serve a useful purpose in certain circuits. Often, inductance is to be avoided in a circuit. Precision resistors are normally wire-wound and thus would have inductance as well as resistance. The inductance can be minimized by winding the insulated wire back on itself so that the current going in the two directions cancels and little magnetic flux is produced; this is called a "noninductive winding."

If an inductor has negligible resistance, it is the inductance, or the back emf, that controls the current. If a source of alternating voltage is applied to the coil, this applied voltage will just be balanced by the induced emf of the coil as given by Eq. 21–9. Thus we can see from Eq. 21–9 that, for a given $\mathscr{E}$, if the inductance L is large, the change in the current—and therefore the current itself—will be small. The greater the inductance, the less the current. An inductance thus acts something like a resistance to impede the flow of alternating current. We use the term **impedance** for this quality of an inductor. We shall discuss impedance more fully in Sections 21–12 to 21–15, and we shall see that it depends not only on L, but also on the frequency. Here we mention one example of its importance. The resistance of the primary in a transformer is usually quite small, perhaps less than $1\,\Omega$. If resistance alone limited the current, tremendous currents would flow when *DC can burn out* a high voltage was applied. Indeed, a dc voltage applied to a transformer *a transformer* can burn it out. It is the impedance of the coil to an alternating current (or its "back" emf) that limits the current to a reasonable value.

EXAMPLE 21–11 **Solenoid inductance.** (*a*) Determine a formula for the self-inductance L of a solenoid (a long coil—see Fig. 20–27b) containing N turns of wire in its length l and whose cross-sectional area is A. (*b*) Calculate the value of L if $N = 100$, $l = 5.0\,\text{cm}$, $A = 0.30\,\text{cm}^2$, and the solenoid is air-filled. (*c*) Calculate L if the solenoid has an iron core with $\mu = 4000\,\mu_0$.

Calculating self-inductance **SOLUTION** (*a*) According to Eq. 20–8, the magnetic field inside a *of a coil* solenoid is $B = \mu_0 nI$, where $n = N/l$. From Eqs. 21–2 and 21–9, we have $\mathcal{E} = -N(\Delta\Phi_B/\Delta t) = -L(\Delta I/\Delta t)$. Thus, $L = N(\Delta\Phi_B/\Delta I)$. Since $\Phi_B = BA = \mu_0 NIA/l$, then any change in I causes a change in flux $\Delta\Phi_B = \mu_0 NA\,\Delta I/l$. Thus

$$L = N\frac{\Delta\Phi_B}{\Delta I} = \frac{\mu_0 N^2 A}{l}.$$

(*b*) Since $\mu_0 = 4\pi \times 10^{-7}\,\text{T·m/A}$,

$$L = \frac{(4\pi \times 10^{-7}\,\text{T·m/A})(100)^2(3.0 \times 10^{-5}\,\text{m}^2)}{(5.0 \times 10^{-2}\,\text{m})} = 7.5\,\mu\text{H}.$$

(*c*) Here we replace μ_0 by $\mu = 4000\mu_0$, so L will be 4000 times larger: $L = 0.030\,\text{H} = 30\,\text{mH}$.

*21–10 Energy Stored in a Magnetic Field

In Section 17–9 we saw that the energy stored in a capacitor is equal to $\frac{1}{2}CV^2$. By using a similar argument, we can show that the energy U stored in an inductance L, carrying a current I, is

$$U = \text{energy} = \tfrac{1}{2}LI^2.$$

Just as the energy stored in a capacitor can be considered to reside in the electric field between its plates, so the energy in an inductor can be considered to be stored in its magnetic field.

To write the energy in terms of the magnetic field, let us use the result of Example 21–11 that the inductance of a solenoid is $L = \mu_0 N^2 A/l$. Now the magnetic field B in a solenoid is related to the current I (see Eq. 20–8) by $B = \mu_0 NI/l$. Thus, $I = Bl/\mu_0 N$, and

$$U = \text{energy} = \tfrac{1}{2}LI^2 = \tfrac{1}{2}\left(\frac{\mu_0 N^2 A}{l}\right)\left(\frac{Bl}{\mu_0 N}\right)^2$$

$$= \tfrac{1}{2}\frac{B^2}{\mu_0}Al.$$

We can think of this energy as residing in the volume enclosed by the windings, which is Al. Then the energy per unit volume, or **energy density**, is

$$u = \text{energy density} = \frac{1}{2}\frac{B^2}{\mu_0}. \qquad \text{(21–10)}$$

Energy density in magnetic field

This formula, which was derived for the special case of a solenoid, can be shown to be valid for any region of space where a magnetic field exists. If a ferromagnetic material is present, μ_0 is replaced by μ. This equation is analogous to that for an electric field, $\frac{1}{2}\epsilon_0 E^2$, Section 17–9.

* 21–11 *LR* Circuit

Any inductor will have some resistance. We represent this situation by drawing the inductance L and the resistance R separately, as in Fig. 21–29. The resistance R could also include a separate resistor connected in series. Now we ask, what happens when a dc source is connected in series to such an *LR* circuit? At the instant the switch connecting the battery is closed, the current starts to flow. It is, of course, opposed by the induced emf in the inductor. However, as soon as current starts to flow, there is a voltage drop across the resistance. Hence, the voltage drop across the inductance is reduced and there is then less impedance to the current flow from the inductance. The current thus rises gradually, as shown in Fig. 21–30a, and approaches the steady value $I_{max} = V/R$ when all the voltage drop is across the resistance. The actual shape of the curve for I as a function of time is

$$I = \left(\frac{V}{R}\right)\left(1 - e^{-t/\tau}\right),$$

where e is the exponential function (see Section 19–7) and $\tau = L/R$ is called the **time constant** of the circuit. When $t = \tau$, then we have $(1 - e^{-1}) = 0.63$ so we see that τ is the time required for the current to reach $0.63 I_{max}$.

If the battery is suddenly removed from the circuit (dashed line in Fig. 21–29), the current drops off as shown in Fig. 21–30b. This is an exponential decay curve given by $I = I_{max}e^{-t/\tau}$. The time constant τ is the time for the current to drop to 37 percent of the original value, and again equals L/R.

These graphs show that there is always some "reaction time" when an electromagnet, for example, is turned on or off. We also see that an *LR* circuit has properties similar to an *RC* circuit (Section 19–7). Unlike the capacitor case, however, the time constant here is *inversely* proportional to R.

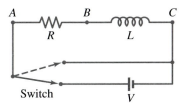

FIGURE 21–29 *LR* circuit.

FIGURE 21–30 (a) Growth of current in an *LR* circuit when connected to a battery; (b) decay of current when the *LR* circuit is shorted out (battery is out of the circuit).

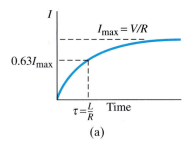

(a)

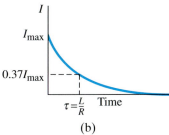

(b)

EXAMPLE 21–12 **Solenoid time constant.** A solenoid has an inductance of 87.5 mH and a resistance of 0.250 Ω. Find (a) the time constant for this circuit, and (b) how long it would take for the current to go from zero to half its final (maximum) value when connected to a battery of voltage V.

SOLUTION (a) By definition, $\tau = L/R$, so

$$\tau = \frac{L}{R} = \frac{87.5 \times 10^{-3}\,\text{H}}{0.250\,\Omega} = 0.350\,\text{s}.$$

(*b*) $I = (V/R)(1 - e^{-t/\tau})$ and we wish to find t such that $I = \frac{1}{2}(V/R)$ where $V/R = I_{max}$. So

$$\frac{1}{2}\frac{V}{R} = \frac{V}{R}(1 - e^{-t/\tau})$$

$$\frac{1}{2} = 1 - e^{-t/\tau}$$

$$e^{-t/\tau} = \frac{1}{2}$$

$$e^{t/\tau} = 2.$$

The inverse operation to the exponential e is the natural log, ln. Thus

$$\ln(e^{t/\tau}) = \frac{t}{\tau} = \ln 2,$$

so

$$t = \tau \ln 2 = (0.350 \text{ s})(0.693) = 0.243 \text{ s}.$$

* 21–12 AC Circuits and Impedance

We have previously discussed circuits that contain combinations of resistor, capacitor, and inductor, but only when they are connected to a dc source of emf or to no source (as in the discharge of a capacitor in an RC circuit). Now we discuss these circuit elements when they are connected to a source of alternating emf.

First we examine, one at a time, how a resistor, a capacitor, and an inductor behave when connected to a source of alternating emf, represented by the symbol $\bullet\!-\!\circledcirc\!-\!\bullet$, which produces a sinusoidal voltage of frequency f. We assume in each case that the emf gives rise to a current

$$I = I_0 \cos 2\pi ft,$$

where t is time and I_0 is the peak current. We must remember (Section 18–8) that $V_{rms} = V_0/\sqrt{2}$ and $I_{rms} = I_0/\sqrt{2}$ (Eq. 18–8).

Resistor. When an ac source is connected to a resistor as in Fig. 21–31a, the current increases and decreases with the alternating emf according to Ohm's law, $I = V/R$. Figure 21–31b shows the voltage (red curve) and the current (blue curve). Because the current is zero when the voltage is zero and the current reaches a peak when the voltage does, we say that the current and voltage are **in phase**. Energy is transformed into heat (Section 18–8), at an average rate $\overline{P} = \overline{IV} = I_{rms}^2 R = V_{rms}^2/R$.

Inductor. In Fig. 21–32a an inductor of inductance L, represented by the symbol $-\!\widehat{\text{ooo}}\!-$, is connected to the ac source. We ignore any resistance it might have (it is usually small). The voltage applied to the inductor will be equal to the "back" emf generated in the inductor by the changing current as given by Eq. 21–9. This is because the sum of the emfs around any closed circuit must be zero, as Kirchhoff's rule tells us. Thus

$$V - L\frac{\Delta I}{\Delta t} = 0 \qquad \text{or} \qquad V = L\frac{\Delta I}{\Delta t},$$

where V is the sinusoidally varying voltage of the source and $L\,\Delta I/\Delta t$ is the voltage induced in the inductor. According to this equation, I is in-

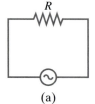

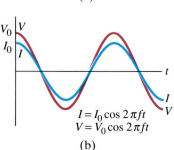

(a)

(b)

FIGURE 21–31 Resistor connected to an ac source. Current is in phase with the voltage.

Resistor: current and voltage are in phase

In figure (b): $I = I_0 \cos 2\pi ft$, $V = V_0 \cos 2\pi ft$

creasing most rapidly when V has its maximum value, $V = V_0$. And I will be decreasing most rapidly when $V = -V_0$. These two instants correspond to points d and b on the graph of voltage versus time in Fig. 21–32b. At points a and c, $V = 0$. The equation above tells us that $\Delta I/\Delta t = 0$ at these instants, so I is not changing and these points correspond to the maximum and minimum values of the current I (the slope of I versus t is zero), as confirmed by points a and c on the graph. By going point by point in this manner, the curve of I versus t as compared to that for V versus t can be constructed, and they are shown by the blue and red lines, respectively, in Fig. 21–32b. Notice that the current reaches its peaks (and troughs) one quarter of a cycle after the voltage does. We say that

in an inductor, the current lags the voltage by 90°.

Remember that 360° corresponds to a full cycle, so 90° is a quarter cycle. Alternatively, we can say that the voltage leads the current by 90°. Since we originally chose $I = I_0 \cos 2\pi ft$, we see from the graph that V must vary as $V = -V_0 \sin 2\pi ft$.

Because the current and voltage are out of phase by 90°, no energy is transformed in an inductor on the average; and no energy is dissipated as thermal energy. This can be seen as follows from Fig. 21–32b. From point c to d, the voltage is increasing from zero to its maximum. The current, however, is in the opposite direction to the voltage and is approaching zero. The average power over this interval, VI, is negative. From d to e, however, both V and I are positive so VI is positive; this contribution just balances the negative contribution of the previous quarter cycle. Similar considerations apply to the rest of the cycle. Thus, the average power transformed over one or many cycles is zero. We can see that energy from the source passes into the magnetic field of the inductor, where it is stored temporarily. Then the field decreases and the energy is transferred back to the source. None is dissipated in this process. Compare this to a resistor where the current is always in the same direction as the voltage and energy is transferred out of the source and never back into it. (The product VI is never negative.) This energy is not stored in the resistor, but is transformed to thermal energy.

As mentioned in Section 21–9, the back emf of an inductor impedes the flow of an ac current. Indeed, it is found that the magnitude of the current in an inductor is directly proportional to the applied ac voltage at a given frequency. We can therefore write an equation for a pure inductor (no resistance) that is the equivalent of Ohm's law:

$$V = IX_L, \qquad\qquad \textbf{(21–11a)}$$

where X_L is called the **inductive reactance**, or **impedance**, of the inductor. Normally, we use the term "reactance" to refer solely to the inductive properties. We then reserve the term "impedance" to include the total "impeding" qualities of the coil—its inductance plus any resistance it may have (more on this in the next section). In the absence of resistance, the impedance is the same as the reactance.

The quantities V and I in Eq. 21–11a can refer either to rms for both, or to peak values for both. Note, however, that although this equation can relate the peak values, the peak current and voltage are not reached at the same time; so Eq. 21–11a is *not* valid at a particular instant, as is the case for Ohm's law.

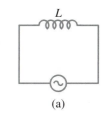

(a)

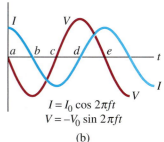

$$I = I_0 \cos 2\pi ft$$
$$V = -V_0 \sin 2\pi ft$$

(b)

FIGURE 21–32 Inductor connected to an ac source. Current lags voltage by a quarter cycle, or 90°.

From the fact that $V = L\,\Delta I/\Delta t$, we see that the larger L is, the less will be the change in current ΔI during the time Δt. Hence I itself will be smaller at any instant for a given frequency, and we expect $X_L \propto L$. The reactance X_L also depends on the frequency. The greater the frequency, the more rapidly the magnetic flux changes in the inductor. If the emf induced by this field is to remain equal to the source emf, as it must, the magnitude of the current must then be less. Hence the greater the frequency, the greater is the reactance X_L, and we expect $X_L \propto fL$. This is also consistent with the fact that if the frequency f is zero (so the current is dc), there is no back emf and no impedance to the flow of charge. Careful calculation (using calculus), as well as experiment, shows that the constant of proportionality is 2π. Thus

$$X_L = 2\pi fL. \tag{21-11b}$$

EXAMPLE 21–13 **Reactance of a coil.** A coil has a resistance $R = 1.00\,\Omega$ and an inductance of 0.300 H. Determine the current in the coil if: (*a*) 120 V dc is applied to it; (*b*) 120 V ac (rms) at 60.0 Hz is applied.

SOLUTION (*a*) There is no inductive impedance ($X_L = 0$ since $f = 0$), so we apply Ohm's law for the resistance:

$$I = \frac{V}{R} = \frac{120\text{ V}}{1.00\,\Omega} = 120\text{ A.}$$

(*b*) The inductive reactance in this case is

$$X_L = 2\pi fL = (6.28)(60.0\text{ s}^{-1})(0.300\text{ H}) = 113\,\Omega.$$

In comparison to this, the resistance can be ignored. Thus,

$$I_{\text{rms}} = \frac{V_{\text{rms}}}{X_L} = \frac{120\text{ V}}{113\,\Omega} = 1.06\text{ A.}$$

[It might be tempting to say that the total impedance is $113\,\Omega + 1\,\Omega = 114\,\Omega$. This might imply that about 1 percent of the voltage drop is across the resistor, or about 1 V; and that across the inductance is 119 V. Although the 1 V_{rms} across the resistor is correct, the other statements are not true because of the alteration in phase in an inductor. This will be discussed in the next Section.]

Capacitor. When a capacitor is connected to a battery, the capacitor plates quickly acquire equal and opposite charges; but no steady current flows in the circuit. A capacitor prevents the flow of a dc current (as long as the capacitor is not "leaky," in which case a leakage current would exist across the gap). However, if a capacitor is connected to an alternating source of voltage, as in Fig. 21–33a, an alternating current will flow continuously. This can happen because when the ac voltage is first turned on, charge begins to flow so that one plate acquires a negative charge and the other a positive charge. But when the voltage reverses itself, the charges flow in the opposite direction. Thus, for an alternating applied voltage, an ac current is present in the circuit continuously.

Let us look at this process in more detail. First, we recall that the applied voltage must equal the voltage across the capacitor: $V = Q/C$, where C is the capacitance and Q the charge on the plates. Thus the charge Q on the plates follows the voltage; when the voltage is zero, the charge is zero; when the voltage is a maximum, the charge is a maximum. But what about the current I? At point a in Fig. 21–33b, when the voltage starts increasing, the charge on the plates is zero. Thus charge flows readily toward the plates and the current is large. As the voltage approaches its maximum of V_0 (point b), the charge that has accumulated on the plates tends to prevent more charge from flowing, so the current drops to zero at point b. The charge that has accumulated now starts to flow off the plates and the magnitude of current again increases (blue curve), but in the opposite direction; it reaches a maximum (negatively) when the voltage is at point c. Thus the current follows the blue curve in Fig. 21–33b. Like an inductor, the voltage and current are out of phase by 90°. But for a capacitor, the current reaches its peaks $\frac{1}{4}$ cycle before the voltage does, so we say that the

current leads the voltage by 90°.

(Or the voltage lags the current by 90°.) This is the opposite of what happens for an inductor. Again we have chosen $I = I_0 \cos 2\pi f t$ and we see from the graph that $V = V_0 \sin 2\pi f t$.

Because the current and voltage are out of phase, the average power dissipated is zero, just as for an inductor. Energy from the source is fed to the capacitor, where it is stored in the electric field between its plates. As the field decreases, the energy returns to the source. Thus, in an ac circuit, *only a resistance will dissipate energy* to thermal energy.

A relationship between the applied voltage and the current in a capacitor can be written just as for an inductance:

$$V = IX_C, \qquad \text{(21–12a)}$$

where X_C is the **capacitive reactance** (or **impedance**) of the capacitor. This equation relates the rms or peak values for the voltage and current but is not valid at a particular instant because I and V are out of phase. X_C depends on both the capacitance C and the frequency f. The larger the capacitance, the more charge it can handle, so the less it will retard the flow of an alternating current. Hence, we expect X_C will be inversely proportional to C. It is also inversely proportional to the frequency f, since, when the frequency is higher, there is less time per cycle for the charge to build up on the plates and impede the flow. Again there is a factor of 2π, and

$$X_C = \frac{1}{2\pi f C}. \qquad \text{(21–12b)}$$

Notice that for dc conditions, $f = 0$ and X_C becomes infinite. This is as it should be, since a capacitor does not pass dc current.

It is interesting to note that the reactance of an inductor increases with frequency, but that of a capacitor decreases with frequency.

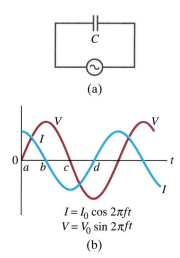

(a)

$$I = I_0 \cos 2\pi f t$$
$$V = V_0 \sin 2\pi f t$$

(b)

FIGURE 21–33 Capacitor connected to an ac source. Current leads voltage by a quarter cycle, or 90°.

Capacitor: current leads voltage

Only R (not C or L) dissipates energy

EXAMPLE 21–14 **Capacitor reactance.** What are the peak and rms currents in the circuit of Fig. 21–33a if $C = 1.0\,\mu\text{F}$ and $V_{\text{rms}} = 120\,\text{V}$? Calculate for (a) $f = 60\,\text{Hz}$, and then for (b) $f = 6.0 \times 10^5\,\text{Hz}$.

SOLUTION (a) $V_0 = \sqrt{2}\,V_{\text{rms}} = 170\,\text{V}$. Then

$$X_C = \frac{1}{2\pi f C} = \frac{1}{(6.28)(60\,\text{s}^{-1})(1.0 \times 10^{-6}\,\text{F})} = 2.7\,\text{k}\Omega.$$

Thus

$$I_0 = \frac{V_0}{X_C} = \frac{170\,\text{V}}{2.7 \times 10^3\,\Omega} = 63\,\text{mA},$$

$$I_{\text{rms}} = \frac{V_{\text{rms}}}{X_C} = \frac{120\,\text{V}}{2.7 \times 10^3\,\Omega} = 44\,\text{mA}.$$

(b) For $f = 6.0 \times 10^5\,\text{Hz}$, X_C will be $0.27\,\Omega$, $I_0 = 630\,\text{A}$, and $I_{\text{rms}} = 440\,\text{A}$. The dependence on f is dramatic.

Capacitors are used for a variety of purposes, some of which have already been described. Two other applications are illustrated in Fig. 21–34. In Fig. 21–34a, circuit A is said to be capacitively coupled to circuit B. The purpose of the capacitor is to prevent a dc voltage from passing from A to B but allowing an ac signal to pass relatively unimpeded. If C is sufficiently large, the ac signal will not be significantly attenuated, whereas dc is filtered out. The capacitor in Fig. 21–34b also passes ac but not dc. In this case, a dc voltage can be maintained between circuits A and B. If the capacitance C is large enough, the capacitor offers little impedance to an ac signal leaving A. Such a signal then passes to ground instead of into B. Thus the capacitor in Fig. 21–34b acts like a *filter* when a constant dc voltage is required; any sharp variation in voltage will pass to ground instead of into circuit B. Capacitors used in these two ways are very common in circuits.

➡ **PHYSICS APPLIED**

Capacitors as filters

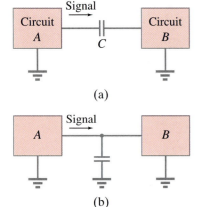

(a)

(b)

FIGURE 21–34 Two common uses for a capacitor.

* 21–13 *LRC* Series AC Circuit; Problem Solving

We now examine a circuit containing all three elements in series, a resistor R, an inductor L, and a capacitor C (Fig. 21–35). If a given circuit contains only two of these elements, we can still use the results of this section by setting $R = 0$, $L = 0$, or $C = \infty$ (infinity) as needed. We let V_R, V_L, and V_C represent the voltage across each element at a given instant in time; and V_{R0}, V_{L0}, and V_{C0} represent the *maximum* (peak) values of these voltages. The voltage across each of the elements will follow the phase relations we discussed in the last section. That is, V_R will be in phase with the current, V_L will lead the current by 90°, and V_C will lag behind the current by 90°. Also, at any instant the total voltage V supplied by the source will be $V = V_R + V_L + V_C$. However, because the various voltages are not in phase, the rms voltages (which is what ac voltmeters usually measure) will not simply add up to give the rms voltage of the source. And V_0 will *not* equal $V_{R0} + V_{L0} + V_{C0}$. Let us now examine the circuit in detail. What we would like to find in particular is the impedance of the circuit as a whole, the rms current that flows through the circuit, and the phase difference between the source voltage and the current.

First we note that the current at any instant must be the same at all points in the circuit. Thus, *the currents in each element are in phase with each other, although the voltages are not.* We choose our origin in time ($t = 0$) so that the current I at any time t is $I = I_0 \cos 2\pi f t$.

FIGURE 21–35 An *LRC* circuit.

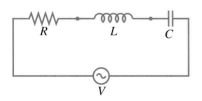

It is convenient to analyze an *LRC* circuit using a **phasor diagram**. Arrows (acting like vectors) are drawn in an *xy* coordinate system to represent each voltage. (These "vectors" are not "real." Phasors are just a useful analytical tool.) *The length of each arrow represents the magnitude of the peak voltage across each element:*

$$V_{R0} = I_0 R, \quad V_{L0} = I_0 X_L, \quad \text{and} \quad V_{C0} = I_0 X_C.$$

V_{R0} is in phase with the current and is initially ($t = 0$) drawn along the positive x axis, as is the current. Since V_{L0} leads the current by 90°, it leads V_{R0} by 90°, so is initially drawn along the positive y axis. V_{C0} lags the current by 90°, so lags V_{R0} by 90°; hence, V_{C0} is drawn initially along the negative y axis. Such a diagram is shown in Fig. 21–36a. If we let the vector diagram rotate counterclockwise at frequency f, we get the diagram shown in Fig. 21–36b; after a time, t, each arrow has rotated through an angle $2\pi ft$. Then the *projections of each arrow on the x axis represent the voltages across each element at the instant t* (see Fig. 21–36c). For example, the projection of V_{R0} on the x axis is $V_{R0} \cos 2\pi ft$ (as in Fig. 21–31); and the projections of V_{L0} and V_{C0} on the x axis are $-V_{L0} \sin 2\pi ft$ and $V_{C0} \sin 2\pi ft$, respectively, as in Figs. 21–32b and 21–33b. Maintaining the 90° angle between each arrow ensures the correct phase relations. Although these results show the validity of a phasor diagram, what we are really interested in is how to add the voltages.

The sum of the projections of the three vectors on the x axis is equal to the projection of their sum. But the sum of the projections represents the instantaneous voltage across the whole circuit, V (equal to the source voltage). Therefore, the vector sum of these vectors will be the vector that represents the peak source voltage, V_0. This is shown in Fig. 21–37, where it is seen that V_0 makes an angle ϕ with I_0 and V_{R0}. As time passes, V_0 rotates with the other vectors, so the instantaneous voltage V (projection of V_0 on the x axis) is (see Fig. 21–37)

$$V = V_0 \cos (2\pi ft + \phi).$$

The voltage V across the whole circuit must, of course, equal the source voltage (Fig. 21–35). Thus we see that the voltage from the source is out of phase[†] with the current by an angle ϕ.

From this analysis we can now draw some useful conclusions. First, we determine the total **impedance** Z of the circuit, which is defined by the relation

$$V_{\text{rms}} = I_{\text{rms}} Z, \quad \text{or} \quad V_0 = I_0 Z. \tag{21–13}$$

From Fig. 21–37 we see, using the Pythagorean theorem (V_0 is the hypotenuse of a right triangle), that

$$
\begin{aligned}
V_0 &= \sqrt{V_{R0}^2 + (V_{L0} - V_{C0})^2} \\
&= \sqrt{I_0^2 R^2 + (I_0 X_L - I_0 X_C)^2} \\
&= I_0 \sqrt{R^2 + (X_L - X_C)^2}.
\end{aligned}
$$

Thus, from Eq. 21–13, and then Eqs. 21–11b and 21–12b,

$$Z = \sqrt{R^2 + (X_L - X_C)^2} \tag{21–14a}$$

$$= \sqrt{R^2 + (2\pi fL - 1/2\pi fC)^2}. \tag{21–14b}$$

This gives the total impedance of the circuit. Also from Fig. 21–37, we can

[†]As a check, note that if $R = X_C = 0$, then $\phi = 90°$, and V_0 would lead the current by 90°, as it must for an inductor alone. Similarly, if $R = L = 0$, $\phi = -90°$ and V_0 would lag the current by 90°, as it must for a capacitor alone.

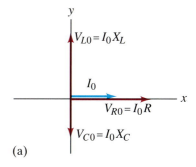

(a)

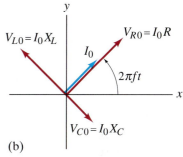

(b)

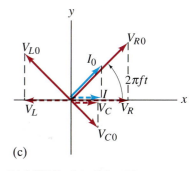

(c)

FIGURE 21–36 Phasor diagram for a series *LRC* circuit.

Impedance

FIGURE 21–37 Phasor diagram for a series *LRC* circuit showing the sum vector, V_0.

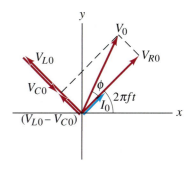

find the phase angle ϕ:

$$\tan \phi = \frac{V_{L0} - V_{C0}}{V_{R0}} = \frac{X_L - X_C}{R} \qquad \textbf{(21-15a)}$$

or

$$\cos \phi = \frac{V_{R0}}{V_0} = \frac{I_0 R}{I_0 Z} = \frac{R}{Z}. \qquad \textbf{(21-15b)}$$

Finally, we can determine the power dissipated in the circuit. We saw earlier that power is only dissipated by a resistance; none is dissipated by inductance or capacitance. Therefore, the average power $\overline{P} = I_{rms}^2 R$. But from Eq. 21–15b, $R = Z \cos \phi$. Therefore,

$$\overline{P} = I_{rms}^2 Z \cos \phi$$

$$= I_{rms} V_{rms} \cos \phi. \qquad \textbf{(21-16)}$$

The factor $\cos \phi$ is referred to as the **power factor** of the circuit. For a pure resistor, $\cos \phi = 1$ and $\overline{P} = I_{rms} V_{rms}$. For a capacitor or inductor, $\phi = -90°$ or $+90°$, respectively, so $\cos \phi = 0$ and no power is dissipated.

The test of this analysis is, of course, in experiment; and experiment is in full agreement with these results.

EXAMPLE 21–15 **LRC Circuit.** Suppose that $R = 25.0 \, \Omega$, $L = 30.0 \, \text{mH}$, and $C = 12.0 \, \mu\text{F}$ in Fig. 21–35, and that they are connected to a 90.0-V ac (rms) 500-Hz source. Calculate (a) the current in the circuit, (b) the voltmeter readings (rms) across each element, (c) the phase angle ϕ, and (d) the power dissipated in the circuit.

SOLUTION (a) First, we find the individual impedances at $f = 500 \, \text{s}^{-1}$:

$$X_L = 2\pi f L = 94.2 \, \Omega,$$

$$X_C = \frac{1}{2\pi f C} = 26.5 \, \Omega.$$

Then

$$Z = \sqrt{R^2 + (X_L - X_C)^2}$$
$$= \sqrt{(25.0 \, \Omega)^2 + (94.2 \, \Omega - 26.5 \, \Omega)^2} = 72.2 \, \Omega.$$

From Eq. 21–13,

$$I_{rms} = \frac{V_{rms}}{Z} = \frac{90.0 \, \text{V}}{72.2 \, \Omega} = 1.25 \, \text{A}.$$

(b) The rms voltage across each element is

$$(V_R)_{rms} = I_{rms} R = (1.25 \, \text{A})(25.0 \, \Omega) = 31.2 \, \text{V}$$

$$(V_L)_{rms} = I_{rms} X_L = (1.25 \, \text{A})(94.2 \, \Omega) = 118 \, \text{V}$$

$$(V_C)_{rms} = I_{rms} X_C = (1.25 \, \text{A})(26.5 \, \Omega) = 33.1 \, \text{V}.$$

Notice that these do *not* add up to give the source voltage, 90.0 V (rms). Indeed, the rms voltage across the inductance *exceeds* the source voltage. This can happen because the different voltages are out of phase with each other, and at any instant one voltage can be negative, to compensate for a large positive voltage of another. The rms voltages, however, are always positive by definition. Although the rms voltages do not have to add up to the source voltage, the instantaneous voltages at any time do add up, of course.

(c) Since

$$\tan \phi = \frac{X_L - X_C}{R} = \frac{94.2\ \Omega - 26.5\ \Omega}{25.0\ \Omega} = 2.71,$$

then $\phi = 69.7°$.

(d) $\overline{P} = I_{rms}V_{rms} \cos \phi = (1.25\ \text{A})(90.0\ \text{V})(25.0\ \Omega/72.2\ \Omega) = 39.0\ \text{W}$.

* 21–14 Resonance in AC Circuits; Oscillators

The rms current in an LRC series circuit is given by (see Eqs. 21–13 and 21–14):

$$I_{rms} = \frac{V_{rms}}{Z} = \frac{V_{rms}}{\sqrt{R^2 + \left(2\pi fL - \dfrac{1}{2\pi fC}\right)^2}}. \qquad \textbf{(21–17)}$$

Because the impedance of inductors and capacitors depends on the frequency f of the source, the current in an LRC circuit will depend on frequency. From Eq. 21–17 we can see that the current will be maximum at a frequency such that

$$2\pi fL - \frac{1}{2\pi fC} = 0.$$

We solve this for f, and call the solution f_0:

$$f_0 = \frac{1}{2\pi}\sqrt{\frac{1}{LC}}. \qquad \textbf{(21–18)}$$

Resonant frequency

This is the **resonant frequency** of the circuit. At this frequency, $X_C = X_L$, so the impedance is purely resistive and $\cos \phi = 1$. A graph of I_{rms} versus f is shown in Fig. 21–38 for particular values of R, L, and C. For smaller R compared to X_L and X_C, the resonance peak will be higher and sharper.

When R is very small, we speak of an ***LC circuit***. The energy in an LC circuit oscillates, at frequency f_0, between the inductor and the capacitor, with some being dissipated in R (some resistance is unavoidable). To see this in more detail, consider a perfect LC circuit in which $R = 0$. Assume at $t = 0$ that the capacitor C is charged and the switch is closed (Fig. 21–39). The capacitor immediately begins to discharge. As it does so, the current I through the inductor increases. At every instant, the potential difference across the capacitor, $V = Q/C$ (where Q is the charge on the capacitor at that instant), must equal the potential difference across the inductor, which is equal to its emf, $-L(\Delta I/\Delta t)$. At the instant when the charge on the capacitor reaches zero ($Q = 0$), the current I in the inductor has reached its maximum value, but at this instant I is not changing ($-L\,\Delta I/\Delta t = Q/C = 0$). At this moment, the magnetic field B in the inductor is also a maximum. The current next begins to decrease as the flowing charge starts to accumulate on the opposite plate of the capacitor. When the current has dropped to zero, the capacitor has again

LC circuit

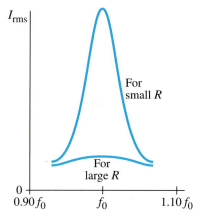

I_{rms}

For small R

For large R

0

0.90 f_0 f_0 1.10 f_0

FIGURE 21–38 Current in an LRC circuit as a function of frequency, showing resonance peak at $f = f_0 = (1/2\pi)\sqrt{1/LC}$.

FIGURE 21–39 A pure LC circuit.

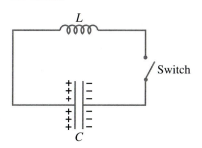

L

Switch

C

attained its maximum charge. The capacitor then begins to discharge again, with the current now flowing in the opposite direction. This process of the charge flowing back and forth from one plate of the capacitor to the other, through the inductor, continues to repeat itself. This is called an ***LC* oscillation** or an **electromagnetic oscillation**. Not only does the charge oscillate back and forth, but so does the energy, which oscillates between being stored in the electric field of the capacitor and in the magnetic field of the inductor.

EM oscillations

Electric resonance is used in many electronic devices. Radio and television sets, for example, use resonant circuits for tuning in a station. Many frequencies reach the circuit from the antenna, but a significant current flows only for frequencies at or near the resonant frequency. Either L or C is variable so that different stations can be tuned in (more on this in Chapter 22). *LC* circuits are also used in **oscillators**, which are devices that put out an oscillating signal of particular frequency. Since some resistance is always present, electrical oscillators generally need a periodic input of power to compensate for the energy converted to thermal energy in the resistance.

Electrical resonance is analogous to mechanical resonance, which we discussed in Chapter 11 (see Fig. 11–18). The energy transferred to the system is a maximum at resonance whether it is electrical resonance, the oscillation of a spring, or pushing a child on a swing (Section 11–6). That this is true in the electrical case can be seen from Eqs. 21–14, 21–15, and 21–16. At resonance, $Z = R$, $\cos\phi = 1$, and I_{rms} is a maximum. For constant V_{rms}, the power is then a maximum at resonance. A graph of power versus frequency looks much like that for the current (Fig. 21–38).

An *LRC* circuit can have the elements arranged in parallel instead of in series. Resonance will occur in this case, too, but the analysis of such circuits is more involved.

* 21–15 Impedance Matching

It is common to connect one electric circuit to a second circuit. For example, a TV antenna is connected to a TV set; an FM tuner is connected to an amplifier; the output of an amplifier is connected to a speaker; electrodes for an ECG or EEG (electrocardiogram and electroencephalogram—electrical traces of heart and brain signals) are connected to an amplifier or a recorder. In many cases it is important that the maximum power be transferred from one to the other, with a minimum of loss. This can be achieved when the output impedance of the one device matches the input impedance of the second.

To show why this is true, we consider simple circuits that contain only resistance. In Fig. 21–40, the source in circuit 1 could represent a power supply, the output of an amplifier, or the tiny signal from an antenna, a laboratory probe, or electrodes. R_1 represents the resistance of this device and includes the internal resistance of the source. R_1 is called the output impedance (or resistance) of circuit 1. The output of circuit 1 is across the terminals a and b, which are connected to the input of circuit 2. Circuit 2 may be very complicated. By combining the various resistors, we can find an equivalent resistance. This is represented by R_2, the "input resistance" (or input impedance) of circuit 2.

FIGURE 21–40 Output of the circuit on the left is input to the circuit on the right.

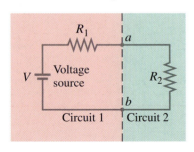

The power delivered to circuit 2 is $P = I^2 R_2$, where $I = V/(R_1 + R_2)$. Thus

$$P = I^2 R_2 = \frac{V^2 R_2}{(R_1 + R_2)^2}.$$

We divide the top and bottom of the right side by R_1 and find

$$P = \frac{V^2}{R_1} \frac{\left(\dfrac{R_2}{R_1}\right)}{\left(1 + \dfrac{R_2}{R_1}\right)^2}. \tag{21–19}$$

The question is, if the resistance of the source is R_1, what value should R_2 have so that the maximum power is transferred to circuit 2? To determine this, we plot a graph of P versus (R_2/R_1). This is shown in Fig. 21–41, where representative values are given in the table. For example, for $R_2/R_1 = 1$, Eq. 21–19 gives $P = V^2/4R_1$; for $R_2/R_1 = 3$, $P = 3V^2/16R_1 = 0.19V^2/R_1$; and so on. As can be seen from the graph, P is a maximum when $R_2 = R_1$.

Thus, the maximum power is transmitted when the *output impedance* of one device *equals the input impedance* of the second. This is called **impedance matching**.

In an ac circuit that contains capacitors and inductors, the different phases are important and the analysis is more complicated. However, the same result holds: To maximize power transfer it is important to match impedances ($Z_2 = Z_1$). In addition, one must be aware that it is possible to seriously distort a signal. For example, when a second circuit is connected, it may put the first circuit into resonance, or take it out of resonance for a certain frequency.

Without proper consideration of the impedances involved, one can make measurements that are completely meaningless. These considerations are normally examined by engineers when designing an integrated set of apparatus. It has happened that researchers have connected several components to one another without regard for impedance matching, and made a "new discovery" that later was found, embarrassingly, to be due to impedance mismatch rather than the natural phenomenon they had thought.

In some cases, a transformer is used to alter an impedance, so it can be matched to that of a second circuit. If Z_S is the secondary impedance and Z_P the primary impedance, then $V_S = I_S Z_S$ and $V_P = I_P Z_P$ (I and V are either peak or rms values of current and voltage). Hence,

$$\frac{Z_P}{Z_S} = \frac{V_P I_S}{V_S I_P} = \left(\frac{N_P}{N_S}\right)^2,$$

where we have used Eqs. 21–6 and 21–7 for a transformer. Thus the impedance can be changed with a transformer. A transformer is used for this purpose in some audio amplifiers, which may have several taps corresponding to $4\,\Omega$, $8\,\Omega$, and $16\,\Omega$ so the output can be matched to the impedance of any loudspeaker.

Some instruments, such as oscilloscopes, require only a signal voltage but very little power. Maximum power transfer is then not important and such instruments can have a high input impedance. This has the advantage that the instrument draws very little current and disturbs the original circuit as little as possible. This is often desirable in laboratory experiments.

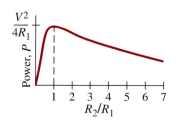

$\dfrac{R_2}{R_1}$	P
0	0
0.5	$0.22\ V^2/R_1$
1.0	$0.25\ V^2/R_1$
2.0	$0.22\ V^2/R_1$
5.0	$0.14\ V^2/R_1$
100	$0.01\ V^2/R_1$

FIGURE 21–41 Power transferred is at its maximum when $R_2 = R_1$.

Impedance matching

🡒 **PHYSICS APPLIED**

Impedance mismatch errors

SUMMARY

The **magnetic flux** passing through a loop is equal to the product of the area of the loop times the perpendicular component of the magnetic field strength: $\Phi_B = B_\perp A = BA\cos\theta$.

If the magnetic flux through a coil of wire changes in time, an emf is induced in the coil; the magnitude of the induced emf equals the time rate of change of the magnetic flux through the loop times the number N of loops in the coil:

$$\mathcal{E} = -N\frac{\Delta\phi_B}{\Delta t}.$$

This is **Faraday's law of induction**.

The induced emf produces a current whose magnetic field opposes the original change in flux (**Lenz's law**).

Faraday's law also tells us that a changing magnetic field produces an electric field; and that a straight wire of length l moving with speed v perpendicular to a magnetic field of strength B has an emf induced between its ends equal to:

$$\mathcal{E} = Blv.$$

An electric **generator** changes mechanical energy into electrical energy. Its operation is based on Faraday's law: a coil of wire is made to rotate uniformly by mechanical means in a magnetic field, and the changing flux through the coil induces a sinusoidal current, which is the output of the generator.

A motor, which operates in the reverse of a generator, acts like a generator in that a **counter emf** is induced in its rotating coil; since this counter emf opposes the input voltage, it can act to limit the current in a motor coil.

Similarly, a generator acts somewhat like a motor in that a **counter torque** acts on its rotating coil.

A **transformer**, which is a device to change the magnitude of an ac voltage, consists of a primary and a secondary coil. The changing flux due to an ac voltage in the primary induces an ac voltage in the secondary. In a 100 percent efficient transformer, the ratio of output to input voltages (V_S/V_P) equals the ratio of the number of turns N_S in the secondary to the number N_P in the primary:

$$\frac{V_S}{V_P} = \frac{N_S}{N_P}.$$

The ratio of secondary to primary current is in the inverse ratio of turns:

$$\frac{I_S}{I_P} = \frac{N_P}{N_S}.$$

QUESTIONS

1. What would be the advantage, in Faraday's experiments (Fig. 21–1), of using coils with many turns?

2. What is the difference between magnetic flux and magnetic field? Discuss in detail.

3. Suppose you are holding a circular piece of wire and suddenly thrust a magnet, south pole first, toward the center of the circle. Is a current induced in the wire? Is a current induced when the magnet is held steady within the loop? Is a current induced when you withdraw the magnet? In each case, if your answer is yes, specify the direction.

4. In what direction will the current flow in Fig. 21–9 if the rod moves to the left, which decreases the area of the loop to the left?

5. Two loops of wire are moving in the vicinity of a very long straight wire carrying a steady current as shown in Fig. 21–42. Find the direction of the induced current in each loop.

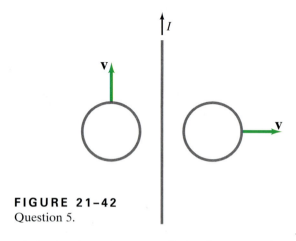

FIGURE 21–42
Question 5.

6. In situations where a small signal must travel over a distance, a "shielded cable" is used in which the signal wire is surrounded by an insulator and then enclosed by a cylindrical conductor. Why is a "shield" necessary?

7. What is the advantage of placing the two insulated electric wires carrying ac close together or even twisted about each other?

8. In some early automobiles, the starter motor doubled as a generator to keep the battery charged once the car was started. Explain how this might work.

9. Explain why, exactly, the lights may dim briefly when a refrigerator motor starts up. When an electric heater is turned on, the lights may stay dimmed as long as it is on. Explain the difference.

10. Explain what is meant by the statement "a motor acts as a motor and generator at the same time." Can the same be said for a generator?

11. Use Fig. 21–12 and the right-hand rules to show why the counter torque in a generator *opposes* the motion.

12. Will an eddy current brake (Fig. 21–19) work on a copper or aluminum wheel, or must it be ferromagnetic?

13. It has been proposed that eddy currents be used to help sort solid waste for recycling. The waste is first ground into tiny pieces and iron removed with a dc magnet. The waste then is allowed to slide down an incline over permanent magnets. How will this aid in the separation of nonferrous metals (Al, Cu, Pb, brass) from nonmetallic materials?

14. The pivoted metal bar with slots in Fig. 21–43 falls much more quickly through a magnetic field than does a solid bar. Explain in detail.

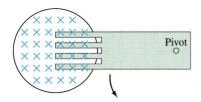

FIGURE 21–43   Question 14.

15. If an aluminum sheet is held between the poles of a large bar magnet, it requires some force to pull it out of the magnetic field even though the sheet is not ferromagnetic and does not touch the pole faces. Explain.

16. A bar magnet falling inside a vertical metal tube reaches a terminal velocity even if the tube is evacuated so that there is no air resistance. Explain.

17. A metal bar, pivoted at one end, oscillates freely in the absence of a magnetic field; but in a magnetic field, its oscillations are quickly damped out. Explain. (This *magnetic damping* is used in a number of practical devices.)

18. An enclosed transformer has four wire leads coming from it. How could you determine the ratio of turns on the two coils without taking the transformer apart? How would you know which wires paired with which?

19. The use of higher-voltage lines in homes, say 600 V or 1200 V, would reduce energy waste. Why are they not used?

* 20. Since a magnetic microphone is basically like a loudspeaker (Section 20–10) actually serve as a microphone? That is, could you speak into a loudspeaker and obtain an output signal that could be amplified? Explain. Discuss, in light of your response, how a microphone and loudspeaker differ in construction.

* 21. The primary of a transformer on a telephone pole has a resistance of $0.10\,\Omega$ and the input voltage is 2400 V ac. Can you estimate the current that will flow? Will it be 24,000 A? Explain.

* 22. A transformer designed for a 120-V ac input will often "burn out" if connected to a 120-V dc source. Explain. [*Hint*: The resistance of the primary coil is usually very low.]

* 23. How would you arrange two flat circular coils so that their mutual inductance was (*a*) greatest, (*b*) least (without separating them by a great distance)?

* 24. If you are given a fixed length of wire, how would you shape it to obtain the greatest self-inductance? The least?

* 25. Does the emf of the battery in Fig. 21–29a affect the time needed for the *LR* circuit to reach (*a*) a given fraction of its maximum possible current, (*b*) a given value of current?

* 26. Can you tell whether the current in an *LRC* circuit leads or lags the applied voltage from a knowledge of the power factor, cos ϕ?

* 27. Under what conditions is the impedance in an *LRC* circuit a minimum?

* 28. An *LC* resonance circuit is often called an *oscillator* circuit. What is it that oscillates?

* 29. Compare the oscillations of an *LC* circuit to the vibration of a mass *m* on a spring. What do *L* and *C* correspond to in the mechanical system?

P R O B L E M S

SECTIONS 21–1 TO 21–4

1. (I) A 9.2-cm-diameter loop of wire is initially oriented perpendicular to a 1.5-T magnetic field. It is rotated so that its plane is parallel to the field direction in 0.20 s. What is the average induced emf in the loop?

2. (I) A 16-cm-diameter circular loop of wire is in a 1.10-T magnetic field. It is removed from the field in 0.15 s. What is the average induced emf?

FIGURE 21–44
Problem 3.

3. (I) The rectangular loop shown in Fig. 21–44 is pushed into the magnetic field which points inward as shown. In what direction is the induced current?

4. (I) The north pole of the magnet in Fig. 21–45 is being inserted into the coil. In which direction is the induced current flowing through the resistor R?

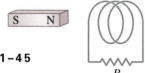

FIGURE 21–45
Problem 4.

5. (I) The magnetic flux through a coil of wire containing two loops changes from -30 Wb to $+38$ Wb in 0.42 s. What is the emf induced in the coil?

6. (I) A 7.2-cm-diameter wire coil is initially oriented so that its plane is perpendicular to a magnetic field of 0.63 T pointing up. During the course of 0.15 s, the field is changed to one of 0.25 T pointing down. What is the average induced emf in the coil?

7. (II) (a) If the resistance of the resistor in Fig. 21–46 is slowly increased, what is the direction of the current induced in the small circular loop inside the larger loop? (b) What would it be if the small loop were placed outside the larger one, to the left?

7b?
how to know
flux out of page
when moved left

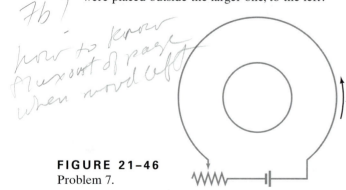

FIGURE 21–46
Problem 7.

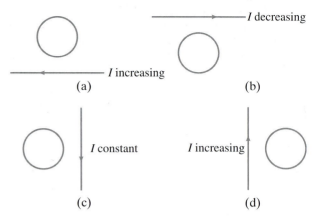

I decreasing

(a) *I increasing*

(b)

(c) *I constant*

(d) *I increasing*

FIGURE 21–47   Problem 8.

8. (II) What is the direction of the induced current in the circular loop due to the current shown in each part of Fig. 21–47?

9. (II) If the solenoid in Fig. 21–48 is being pulled away from the loop shown, in what direction is the induced current in the loop?

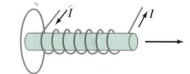

FIGURE 21–48
Problem 9.

10. (II) The magnetic field perpendicular to a circular loop of wire 20 cm in diameter is changed from $+0.52$ T to -0.45 T in 180 ms, where $+$ means the field points away from an observer and $-$ toward the observer. (a) Calculate the induced emf. (b) In what direction does the induced current flow?

11. (II) The moving rod in Fig. 21–9 is 12.0 cm long and moves with a speed of 15.0 cm/s. If the magnetic field is 0.800 T, calculate (a) the emf developed, and (b) the electric field in the rod.

12. (II) A circular loop in the plane of the paper lies in a 0.75 T magnetic field pointing into the paper. If the loop's diameter changes from 20.0 cm to 6.0 cm in 0.50 s, (a) what is the direction of the induced current, (b) what is the magnitude of the average induced emf, and (c) if the coil resistance is 2.5 Ω, what is the average induced current?

13. (II) The moving rod in Fig. 21–9 is 13.2 cm long and generates an emf of 100 mV while moving in a 0.90-T magnetic field. (a) What speed is it moving at? (b) What is the electric field in the rod?

14. (II) In Fig. 21–9, the rod moves with a speed of 1.9 m/s, is 30.0 cm long, and has a resistance of 2.5 Ω. The magnetic field is 0.75 T, and the resistance of the U-shaped conductor is 25.0 Ω at a given instant. Calculate (a) the induced emf, (b) the current flowing in the circuit, and (c) the external force necessary to ensure that the rod is moving at constant velocity at that instant.

15. (II) A single rectangular loop of wire with dimensions shown in Fig. 21–49 is situated so that part is inside a region of uniform magnetic field of 0.450 T and part is outside the field. The total resistance of the loop is 0.230 Ω. Calculate the force required to pull the loop from the field (to the right) at a constant velocity of 3.40 m/s. Neglect gravity.

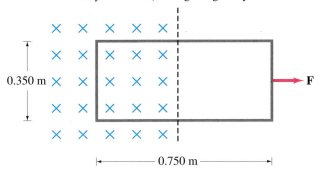

FIGURE 21–49   Problem 15.

16. (II) A 31.0-cm-diameter coil consists of 20 turns of circular copper wire 2.6 mm in diameter. A uniform magnetic field, perpendicular to the plane of the coil, changes at a rate of 8.65×10^{-3} T/s. Determine (a) the current in the loop, and (b) the rate at which thermal energy is produced.

17. (III) If the U-shaped conductor in Fig. 21–9 has resistivity ρ, whereas that of the moving rod is negligible, derive a formula for the current I as a function of time. Assume the rod has length l, starts at the bottom of the U at $t = 0$, and moves with uniform speed v in the magnetic field B. The cross-sectional area of the rod and all parts of the U is A.

18. (III) The magnetic field perpendicular to a single 13.2-cm-diameter circular loop of copper wire decreases uniformly from 0.750 T to zero. If the wire is 2.25 mm in diameter, how much charge moves past a point in the coil during this operation?

SECTION 21–5

19. (I) The generator of a car idling at 1000-rpm produces 12.4 V. What will the output be at a rotation speed of 2500 rpm assuming nothing else changes?

20. (I) A simple generator is used to generate a peak output voltage of 24.0 V. The square armature consists of windings that are 6.0 cm on a side and rotates in a field of 0.420 T at a rate of 60 rev/s. How many loops of wire should be wound on the square armature?

21. (I) Show that the rms output of an ac generator is $V_{rms} = NAB\omega/\sqrt{2}$.

22. (II) A simple generator has a 720-loop square coil 21.0 cm on a side. How fast must it turn in a 0.650-T field to produce a 120-V peak output?

23. (II) A 450-loop circular armature coil with a diameter of 10.0 cm rotates at 60 rev/s in a uniform magnetic field of strength 0.75 T. What is the rms voltage output of the generator? What would you do to the rotation frequency in order to double the rms voltage output?

*SECTION 21–6

* **24.** (I) A motor has an armature resistance of 3.75 Ω. If it draws 9.20 A when running at full speed and connected to a 120-V line, how large is the counter emf?

* **25.** (I) The counter emf in a motor is 72 V when operating at 1800 rpm. What would be the counter emf at 2500 rpm if the magnetic field is unchanged?

* **26.** (II) The counter emf in a motor is 100 V when the motor is operating at 1000 rpm. How would you change the motor's magnetic field if you wanted to reduce the counter emf to 65 V when the motor was running at 2500 rpm?

* **27.** (II) What will be the current in the motor of Example 21–7 if the load causes it to run at half speed?

* **28.** (II) The magnetic field of a "shunt-wound" dc motor is produced by field coils placed in parallel to the armature coils. Suppose that the field coils have a resistance of 66.0 Ω and the armature coils 5.00 Ω. The back emf at full speed is 105 V when the motor is connected to a 115-V line. (a) Draw the equivalent circuit for the situations when the motor is just starting and when it is running full speed. (b) What is the total current drawn by the motor at start up? (c) What is the total current drawn when the motor runs at full speed?

29. (II) A dc generator is rated at 10 kW, 200 V, and 50 A when it rotates at 1000 rpm. The resistance of the armature windings is 0.40Ω. (a) Calculate the "no-load" voltage at 1000 rpm (when there is no circuit hooked up to the generator). (b) Calculate the full-load voltage (i.e. at 50 A) when the generator is run at 800 rpm. Assume that the magnitude of the magnetic field remains constant.

SECTION 21–7

30. (I) A transformer is designed to change 120 V into 10,000 V, and there are 5000 turns in the primary. How many turns are in the secondary, assuming 100 percent efficiency?

31. (I) A transformer has 420 turns in the primary and 120 in the secondary. What kind of transformer is this and, assuming 100 percent efficiency, by what factor does it change the voltage? By what factor does it change the current?

32. (I) A step-up transformer increases 16 V to 120 V. What is the current in the secondary as compared to the primary? Assume 100 percent efficiency.

33. (I) Neon signs require 12 kV for their operation. To operate from a 220-V line, what must be the ratio of secondary to primary turns of the transformer? What would the voltage output be if the transformer were connected backward?

34. (II) A model-train transformer plugs into 120 V ac, and draws 0.75 A while supplying 15 A to the train. (*a*) What voltage is present across the tracks? (*b*) Is the transformer step-up or step-down?

35. (II) The output voltage of a 100-W transformer is 12 V and the input current is 20 A. (*a*) Is this a step-up or a step-down transformer? (*b*) By what factor is the voltage multiplied?

36. (II) High-intensity desk lamps are rated at 40 W but require only 12 V. They contain a transformer that converts 120 V household voltage. (*a*) Is the transformer step-up or step-down? (*b*) What is the current in the secondary when the lamp is on? (*c*) What is the current in the primary? (*d*) What is the resistance of the bulb when on?

37. (II) A transformer has 330 primary turns and 1240 secondary turns. The input voltage is 120 V and the output current is 15.0 A. What is the output voltage and input current assuming 100 percent efficiency?

38. (II) If 30 MW of power at 45 kV (rms) arrives at a town from a generator via 4.0-Ω transmission lines, calculate (*a*) the emf at the generator end of the lines, and (*b*) the fraction of the power generated that is lost in the lines.

39. (II) Show that the power loss in transmission lines, P_L, is given by $P_L = (P_T)^2 R_L/V^2$, where P_T is the power transmitted to the user, V is the delivered voltage, and R_L is the resistance of the power lines.

40. (II) If 50 kW is to be transmitted over two 0.100-Ω lines, estimate how much power is saved if the voltage is stepped up from 120 V to 1200 V and then down again, rather than simply transmitting at 120 V. Assume the transformers are each 99 percent efficient.

41. (III) Design a dc transmission line that can transmit 300 MW of electricity 200 km with only a 2 percent loss. The wires are to be made of aluminum and the voltage is 600 kV.

*SECTION 21–9

* 42. (I) If the current in a 120-mH coil changes steadily from 25.0 A to 10.0 A in 350 ms, what is the direction and magnitude of the induced emf?

* 43. (I) What is the inductance L of a 0.60-m-long air-filled coil 2.9 cm in diameter containing 10,000 loops?

* 44. (I) What is the inductance of a coil if it produces an emf of 8.50 V when the current in it changes from -28.0 mA to $+31.0$ mA in 42.0 ms?

* 45. (I) How many turns of wire would be required to make a 100-mH inductance out of a 30.0-cm-long air-filled coil with a diameter of 5.2 cm?

* 46. (II) An air-filled cylindrical inductor has 3000 turns, and it is 2.5 cm in diameter and 28.2 cm long. What is its inductance? How many turns would you need to generate the same inductance if the core were iron-filled instead? Assume the magnetic permeability of iron is about 1000 times that of free space.

* 47. (II) A coil has 2.25-Ω resistance and 440-mH inductance. If the current is 3.00 A and is increasing at a rate of 3.50 A/s, what is the potential difference across the coil at this moment?

* 48. (II) The wire of a tightly wound solenoid is unwound and used to make another tightly wound solenoid of twice the diameter. By what factor does the inductance change?

* 49. (II) (*a*) Show that the self-inductance L of a torus (Figs. 20–43 and 20–57) of radius R containing N loops each of radius r is

$$L \approx \frac{\mu_0 N^2 r^2}{2R}$$

if $R \gg r$. Assume the field is uniform inside the torus; is this actually true? Is this result consistent with L for a solenoid? Should it be? (*b*) Calculate the inductance L of a large toroid if the diameter of the coils is 1.5 cm and the diameter of the whole ring is 40 cm. Assume the field inside the toroid is uniform. There are a total of 1000 loops of wire.

* 50. (II) (*a*) Ignoring any mutual inductance, what is the equivalent inductance of two inductors connected in series? (*b*) What if they are connected in parallel? (*c*) How does their mutual inductance (their geometrical relationship to each other) affect the results?

* 51. (II) A long thin solenoid of length l and cross-sectional area A contains N_1 closely packed turns of wire. Wrapped tightly around it is an insulated coil of N_2 turns, Fig. 21–50. Assume all the flux from coil 1 (the solenoid) passes through coil 2, and calculate the mutual inductance.

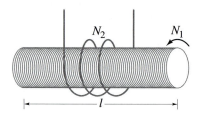

FIGURE 21–50 Problem 51.

* 52. (III) A 30-cm-long coil with 1500 loops is wound on an iron core ($\mu = 3000\,\mu_0$) along with a second coil of 800 loops. The loops of each coil have a radius of 2.0 cm. If the current in the first coil drops uniformly from 3.0 A to zero in 8.0 ms, determine (a) the emf induced in the second coil, and (b) the mutual inductance M.

* 53. (III) The potential difference across a given coil is 22.5 V at an instant when the current is 860 mA and is increasing at a rate of 3.40 A/s. At a later instant, the potential difference is 16.2 V whereas the current is 700 mA and is decreasing at a rate of 1.80 A/s. Determine the inductance and resistance of the coil.

*SECTION 21–10

* 54. (I) The magnetic field inside an air-filled solenoid 36 cm long and 2.0 cm in diameter is 0.80 T. Approximately how much energy is stored in this field?

* 55. (II) At a given instant the current through an inductor is 50.0 mA and is increasing at the rate of 100 mA/s. What is the initial energy stored in the inductor if the inductance is known to be 60.0 mH, and how long does it take for the energy to increase by a factor of 10 from the initial value?

* 56. (II) Assuming the Earth's magnetic field averages about 0.50×10^{-4} T near the surface of the Earth, estimate the total energy stored in this field in the first 10 km above the Earth's surface.

*SECTION 21–11

* 57. (II) It takes 7.20 ms for the current in an LR circuit to reach 80 percent of its maximum value. Determine (a) the time constant of the circuit, and (b) the inductance of the circuit if $R = 250\,\Omega$.

* 58. (II) Determine $\Delta I/\Delta t$ at $t = 0$ (when the battery is connected) for the LR circuit of Fig. 21–29 and show that if I continued to increase at this rate, it would reach its maximum value in one time constant.

* 59. (II) After how many time constants does the current in Fig. 21–29 reach within (a) 10 percent, (b) 1.0 percent, and (c) 0.1 percent of its maximum value?

* 60. (III) Two tightly wound solenoids have the same length and circular cross-sectional area. But solenoid 1 uses wire that is half as thick as solenoid 2. (a) What is the ratio of their inductances? (b) What is the ratio of their inductive time constants (assuming no other resistance in the circuits)?

*SECTION 21–12

* 61. (I) At what frequency will a 160-mH inductor have a reactance of 1.5 kΩ?

* 62. (I) A 9.20-μF capacitor is measured to have a reactance of 250 Ω. At what frequency is it being driven?

* 63. (I) Plot a graph of the impedance of a 1.0-μF capacitor as a function of frequency from 10 to 1000 Hz.

* 64. (I) Plot a graph of the impedance of a 1.0-mH inductor as a function of frequency from 100 to 10,000 Hz.

* 65. (II) Calculate the impedance of, and rms current in, a 160-mH radio coil connected to a 240-V (rms) 10.0-kHz ac line. Ignore resistance.

* 66. (II) An inductance coil operates at 240 V and 60 Hz. It draws 12.8 A. What is the coil's inductance?

* 67. (II) (a) What is the impedance of a well-insulated 0.030-μF capacitor connected to a 2.0-kV (rms) 700-Hz line? (b) What will be the peak value of the current?

* 68. (II) A capacitor is placed in parallel across a load, as in Fig. 21–34b, to filter out stray high-frequency signals, but to allow ordinary 60-Hz ac to pass through with little loss. Suppose that circuit B in the figure is a resistance $R = 300\,\Omega$ connected to ground, and that $C = 0.60\,\mu$F. What percent of the incoming current will pass through C rather than R if (a) it is 60 Hz and (b) it is 60,000 Hz?

* 69. (II) Suppose that circuit B in Fig. 21–34a is a resistance $R = 500\,\Omega$, connected to ground, and the capacitance $C = 2.0\,\mu$F. Will this capacitor act to eliminate 60 Hz ac but pass a high-frequency signal of frequency 60,000 Hz? To check this, determine the voltage drop across R for a 50-mV signal of frequency (a) 60 Hz, and (b) 60,000 Hz.

*SECTION 21–13

* 70. (I) A 30-kΩ resistor is in series with a 0.50-H inductor and an ac source. Calculate the impedance of the circuit if the source frequency is (a) 60 Hz, and (b) 3.0×10^4 Hz.

* 71. (I) A 2.5-kΩ resistor and a 4.0-μF capacitor are connected in series to an ac source. Calculate the impedance of the circuit if the source frequency is (a) 100 Hz, and (b) 10,000 Hz.

* 72. (I) For a 120-V rms 60-Hz voltage, a current of 70 mA passing through the body for 1.0 s could be lethal. What would be the impedance of the body for this to occur?

* 73. (II) A 2.5-kΩ resistor in series with a 420-mH inductor is driven by an ac power supply. At what frequency is the impedance double that of the impedance at 60 Hz?

* 74. (II) (a) What is the rms current in an RC circuit if $R = 28.8$ kΩ, $C = 0.80\,\mu$F, and the rms applied voltage is 120 V at 60 Hz? (b) What is the phase angle between voltage and current? (c) What is the power dissipated by the circuit? (d) What are the voltmeter readings across R and C?

* **75.** (II) (*a*) What is the rms current in an *RL* circuit when a 60-Hz 120-V rms ac voltage is applied, where $R = 1.80\,\text{k}\Omega$, and $L = 900\,\text{mH}$? (*b*) What is the phase angle between voltage and current? (*c*) How much power is dissipated? (*d*) What are the rms voltage readings across *R* and *L*?

* **76.** (II) What is the total impedance, phase angle, and rms current in an *LRC* circuit connected to a 10.0-kHz, 300-V (rms) source if $L = 22.0\,\text{mH}$, $R = 8.70\,\text{k}\Omega$, and $C = 5000\,\text{pF}$?

* **77.** (II) What is the resistance of a coil if its impedance is 35 Ω and its reactance is 30 Ω?

* **78.** (II) A voltage $V = 4.8 \sin 754t$ is applied to an *LRC* circuit. If $L = 3.0\,\text{mH}$, $R = 1.40\,\text{k}\Omega$, and $C = 3.0\,\mu\text{F}$, how much power is dissipated in the circuit?

* **79.** (II) A circuit consists of a 250-Ω resistor in series with a 40.0-mH inductor and a 50.0-V ac generator. The power dissipated by the circuit is 9.50 W. What is the frequency of the generator?

* **80.** (II) Show that for the *LRC* circuit of Fig. 21–35, if we have $I = I_0 \cos \omega t$, then

$$V_R = I_0 R \cos \omega t,$$
$$V_L = I_0 \omega L \cos (\omega t + \pi/2),$$

and

$$V_C = (I_0/\omega C) \cos (\omega t - \pi/2),$$

where $\omega = 2\pi f$.

*SECTION 21–14

* **81.** (I) A 3500-pF capacitor is connected to a 50-μH coil of resistance 3.0 Ω. What is the resonant frequency of this circuit?

* **82.** (I) The variable capacitor in the tuner of an AM radio has a capacitance of 2800 pF when the radio is tuned to a station at 580 kHz. (*a*) What must be the capacitance for a station at 1600 kHz? (*b*) What is the inductance (assumed constant)?

* **83.** (II) An *LRC* circuit has $L = 4.8\,\text{mH}$ and $R = 4.4\,\Omega$. (*a*) What value must *C* have to produce resonance at 3600 Hz? (*b*) What will be the maximum current at resonance if the peak external voltage is 50 V?

* **84.** (II) A 3000-pF capacitor is charged to 120 V and then quickly connected to an inductor. The frequency of oscillation is observed to be 20 kHz. Determine (*a*) the inductance, (*b*) the peak value of the current, and (*c*) the maximum energy stored in the magnetic field of the inductor.

*SECTION 21–15

* **85.** (I) An audio amplifier has output connections for 4 Ω, 8 Ω, and 16 Ω. If two 8-Ω speakers are to be connected in parallel, to which output terminals should they be connected?

* **86.** (I) The output of an amplifier has an impedance of 30 kΩ. It is to be connected to an 8.0-Ω loudspeaker through a transformer. What should be the turns ratio of the transformer?

GENERAL PROBLEMS

87. Suppose you are looking along a line through the centers of two circular (but separate) wire loops, one behind the other. A battery is suddenly connected to the front loop, establishing a clockwise current. (*a*) Will a current be induced in the second loop? (*b*) If so, when does this current start? (*c*) When does it stop? (*d*) In what direction is this current? (*e*) Is there a force between the two loops? (*f*) If so, in what direction?

88. Suppose you are looking at two current loops in the plane of the page as shown in Fig. 21–51. When the switch is thrown in the left-hand coil, (*a*) what is the direction of the induced current in the other loop? (*b*) What is the situation after a "long" time? (*c*) What is the direction of the induced current in the second loop if the second loop is quickly pulled horizontally to the right?

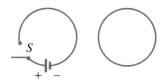

FIGURE 21–51   Problem 88.

89. A square loop 24.0 cm on a side has a resistance of 6.50 Ω. It is initially in a 0.755-T magnetic field, with its plane perpendicular to **B**, but is removed from the field in 40.0 ms. Calculate the electric energy dissipated in this process.

90. Two conducting rails 30 cm apart rest on a 5.0° ramp. They are joined at the bottom by a 0.60 Ω resistor, and at the top a copper bar of mass 0.040 kg is laid across the rails. The whole apparatus is immersed in a vertical 0.55 T field. What is the terminal (steady) velocity of the bar as it slides frictionlessly down the rails?

91. A **search coil** for measuring B (also called a *flip coil*) is a small coil with N turns, each of cross-sectional area A. It is connected to a so-called **ballistic galvanometer**, which is a device to measure the total charge Q that passes through it in a short time. The flip coil is placed in the magnetic field to be measured with its face perpendicular to the field. It is then quickly rotated $180°$. Show that the total charge Q that flows in the induced current during this short "flip" time is proportional to the magnetic field B; in particular, show that B is given by

$$B = \frac{QR}{2NA}$$

where R is the total resistance of the circuit, including that of the coil and that of the ballistic galvanometer which measures the charge Q.

92. (a) Show that the power $P = Fv$ needed to move the conducting rod to the right in Fig. 21–9 is equal to $B^2l^2v^2/R$, where R is the total resistance of the circuit. (b) Show that this equals the power dissipated in the resistance, I^2R.

93. The primary windings of a transformer, which has an 80% efficiency, are connected to 110 V ac. The secondary windings are connected across a $2.4\,\Omega$, 58 W lightbulb. (a) Calculate the current through the primary windings of the transformer. (b) Calculate the ratio of the number of primary windings of the transformer to the number of secondary windings of the transformer.

94. A pair of power transmission lines each have a 0.80-Ω resistance and carry 700 A over 9.0 km. If the rms input voltage is 42 kV, calculate (a) the voltage at the other end, (b) the power input, (c) power loss in the lines, and (d) the power output.

95. Calculate the peak output voltage of a simple generator whose square armature windings are 6.60 cm on a side if the armature contains 125 loops and rotates in a field of 0.200 T at a rate of 120 rev/s.

96. A small electric car overcomes a 250-N friction force when traveling 30 km/h. The electric motor is powered by ten 12-V batteries connected in series and is coupled directly to the wheels whose diameters are 50 cm. The 300 armature coils are rectangular, 10 cm by 15 cm, and rotate in a 0.60-T magnetic field. (a) How much current does the motor draw to produce the required torque? (b) What is the back emf? (c) How much power is dissipated in the coils? (d) What percent of the input power is used to drive the car?

* **97.** What is the inductance L of the primary of a transformer whose input is 220 V at 60 Hz and the current drawn is 5.8 A? Assume no current in the secondary.

* **98.** A 230-mH coil, whose resistance is $18.5\,\Omega$, is connected to a capacitor C and a 3360-Hz source voltage. If the current and voltage are to be in phase, what value must C have?

* **99.** A circuit contains two elements, but it is not known if they are L, R, or C. The current in this circuit when connected to a 120-V 60-Hz source is 5.6 A and lags the voltage by $50°$. What are the two elements and what are their values?

* **100.** A resonant circuit using a 220-pF capacitor is to resonate at 48.0 MHz. The air-core inductor is to be a solenoid with closely packed coils made from 14.0 m of insulated wire 1.1 mm in diameter. How many loops will the inductor contain?

* **101.** An inductance coil draws 2.5 A dc when connected to a 36-V battery. When connected to a 60-Hz 120-V (rms) source, the current drawn is 3.8 A (rms). Determine the inductance and resistance of the coil.

* **102.** The **Q factor** of a resonance circuit is defined as the ratio of the voltage across the capacitor (or inductor) to the voltage across the resistor, at resonance. The larger the Q factor, the sharper the resonance curve will be and the sharper the tuning. (a) Show that the Q factor is given by the equation $Q = (1/R)\sqrt{L/C}$. (b) At a resonant frequency $f_0 = 1.0$ MHz, what must be the value of L and R to produce a Q factor of 1000? Assume that $C = 0.010\,\mu\text{F}$.

* **103.** In a series LRC circuit, the inductance is 20 mH, the capacitance is 50 nF, and the resistance is $200\,\Omega$. At what frequencies is the power factor equal to 0.17?

Electromagnetic waves, such as radio and TV signals, can be sent, and received, by means of antennas like these. All EM waves, at whatever frequency, travel at the speed of light. And light itself is an EM wave.

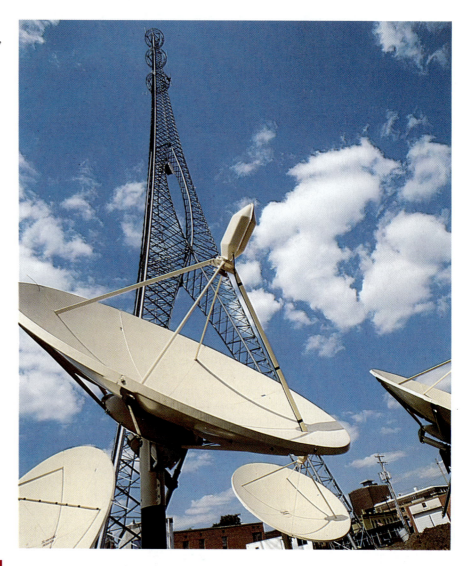

22 ELECTROMAGNETIC WAVES

T he culmination of electromagnetic theory in the nineteenth century was the prediction, and the experimental verification, that waves of electromagnetic fields could travel through space. This achievement opened a whole new world of communication—first the wireless telegraph, then radio and television. And it yielded the spectacular prediction that light is an electromagnetic wave.

The theoretical prediction of electromagnetic waves was the work of the Scottish physicist James Clerk Maxwell (1831–1879; Fig. 22–1), who unified, in one magnificent theory, all the phenomena of electricity and magnetism.

22–1 Changing Electric Fields Produce Magnetic Fields; Maxwell's Equations

The development of electromagnetic theory in the early part of the nineteenth century by Oersted, Ampère, and others was not actually done in terms of electric and magnetic fields. The idea of the field was introduced somewhat later by Faraday, and was not generally used until Maxwell showed that all electric and magnetic phenomena could be described using only four equations involving electric and magnetic fields. These equations, known as **Maxwell's equations**, are the basic equations for all electromagnetism. They are fundamental in the same sense that Newton's three laws of motion and the law of universal gravitation are for mechanics. In a sense, they are even more fundamental, since they are consistent with the theory of relativity (Chapter 26), whereas Newton's laws are not. Because all of electromagnetism is contained in this set of four equations, Maxwell's equations are considered one of the great triumphs of the human mind.

Although we will not present Maxwell's equations in mathematical form since they involve calculus, we will summarize them here in words. They are: (1) a generalized form of Coulomb's law known as Gauss's law (discussed in Appendix D) that relates electric field to its sources, electric charge; (2) a similar law for the magnetic field, except that since there seem to be no single magnetic poles (monopoles), there will be no single magnetic charges ($q_B = 0$, let's say), and so magnetic field lines are always continuous—they do not begin or end (as electric field lines do, on charges); (3) an electric field is produced by a changing magnetic field; (4) a magnetic field is produced by an electric current, or by a changing electric field.

Maxwell's equations

Law number (3) is Faraday's law (see Chapter 21, especially Section 21–4). The first part of (4), that a magnetic field is produced by an electric current, was discovered by Oersted, and the mathematical relation is given by Ampère's law (Section 20–8). But the second part of (4) is an entirely new aspect predicted by Maxwell. Maxwell argued that if a changing magnetic field produces an electric field, as given by Faraday's law, then the reverse might be true as well: **a changing electric field will produce a magnetic field**. This was a *hypothesis* by Maxwell, based on the idea of symmetry in nature. Indeed, the size of the effect in most cases is so small that Maxwell recognized it would be difficult to detect it experimentally. We look at this hypothesis in a little detail in the next (optional) Section.

Changing **E** *produces* **B**

FIGURE 22–1
James Clerk Maxwell.

*22–2 Maxwell's Fourth Equation; Displacement Current

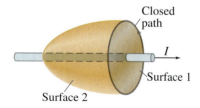

FIGURE 22–2 Ampère's law applied to two different surfaces bounded by the same closed path.

Maxwell used an indirect argument to back up his idea that a changing electric field would produce a magnetic field. In a simplified fashion it goes something like this. According to Ampère's law (Section 20–8), $\Sigma B_\parallel \Delta l = \mu_0 I$. That is, you divide any closed path you choose into short segments Δl, multiply each segment by the parallel component of the magnetic field B at that segment, and then sum all these products over the complete closed path; this sum will then equal μ_0 times the net current I that passes through the surface bounded by the path. When we applied Ampère's law to the field around a straight wire, we imagined the current as passing through the circle enclosed by our circular loop. This would be the flat surface 1 in Fig. 22–2. However, we could just as well use the sack-shaped surface 2 in the figure, since the same current I passes through it. This naturally implies that the current passing through any surface bounded by the closed path must be the same. Thus the current flowing into the volume enclosed by surfaces 1 and 2 together equals the current that flows out of this volume. This is basically Kirchhoff's point rule, and tells us that the rate at which charge enters the volume equals the rate at which it leaves.

Now consider the closed path for the situation of Fig. 22–3a, where a capacitor is being discharged. Ampère's law works for surface 1, but it does not work for surface 2, since no current passes through surface 2. There is a magnetic field around the wire, so the left side of Ampère's law is not zero; yet for surface 2, the right side *is* zero, since $I = 0$. We seem to have a contradiction of Ampère's law. Similar things can be said for the closed path indicated in Fig. 22–3b, which surrounds the region of the electric field of the plates; there is a current passing through surface 1, but not through surface 2, and there is a magnetic field around the closed path. There is a magnetic field present in Fig. 22–3, however, only if charge is flowing to or away from the capacitor plates. In this case, the electric field between the plates is changing in time. Maxwell resolved the problem of no current through surface 2 in Figs. 22–3a and b by stating that the

FIGURE 22–3 A capacitor discharging. No conduction current passes through surface 2 in either (a) or (b). An extra term is needed in Ampère's law.

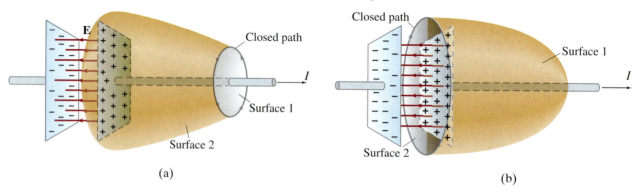

(a)

(b)

changing electric field between the plates is *equivalent to* an electric current. He called this a **displacement current**, I_D. (The name was based on an old discarded theory, and is not especially illuminating today.) An ordinary current is then called a **conduction current**, I_C. Ampère's law, as generalized by Maxwell, becomes

Displacement current defined

$$\Sigma B_{\parallel} \Delta l = \mu_0 (I_C + I_D).$$

Ampère's law will now apply even for surface 2 in Figs. 22–3a and b, where the displacement current I_D refers to the changing electric field.[†]

We can write I_D in terms of the changing electric field. The charge Q on a capacitor of capacitance C is $Q = CV$, where V is the potential difference between the plates. Also recall that $V = Ed$, where d is the (small) separation of the plates and E is the electric field strength between them; we ignore any fringing of the field. Also, for a parallel-plate capacitor, $C = \epsilon_0 A/d$, where A is the area of each plate (see Chapter 17). We combine these to obtain

$$Q = CV = \left(\epsilon_0 \frac{A}{d} \right)(Ed) = \epsilon_0 A E.$$

Now if the charge Q on the plate changes at a rate $\Delta Q / \Delta t$, the electric field changes at a proportional rate. Then, from the above expression,

$$\frac{\Delta Q}{\Delta t} = \epsilon_0 A \frac{\Delta E}{\Delta t}.$$

But $\Delta Q / \Delta t$ is also the current flowing into or out of the capacitor. If the current flow into the capacitor, $\Delta Q / \Delta t$, is set equal to the displacement current I_D between the plates, then

$$I_D = \frac{\Delta Q}{\Delta t} = \epsilon_0 A \frac{\Delta E}{\Delta t},$$

or

$$I_D = \epsilon_0 \frac{\Delta \Phi_E}{\Delta t},$$

where $\Phi_E = EA$ is the **electric flux**, defined in analogy to magnetic flux (Section 21–2). Then, Ampère's law becomes

$$\Sigma B_{\parallel} \Delta l = \mu_0 I_C + \mu_0 \epsilon_0 \frac{\Delta \Phi_E}{\Delta t}. \qquad \textbf{(22–1)}$$

Ampère's law (generalized)

This equation embodies Maxwell's idea that a magnetic field can be caused not only by a normal electric current, but also by a changing electric field or changing electric flux. Eq. 22–1 is essentially Maxwell's fourth equation.[‡]

[†]Notice that this interpretation of the changing electric field as a displacement current fits in well with our discussion in Chapter 21 where we saw that an alternating current passes through a capacitor. It also means that Kirchhoff's point rule will be valid even at a capacitor plate: when a capacitor is being charged, conduction current flows into the capacitor plate, but no conduction current flows out of the plate. Instead, a "displacement current" flows out of one plate toward the other plate.

[‡]Actually, there is also a third term on the right for the case when a magnetic field is produced by magnetized materials; but we assume in what follows that no magnets are present.

Production of Electromagnetic Waves

According to Maxwell (as mentioned at the end of Section 22–1, and expanded upon in Section 22–2), a magnetic field will be produced in empty space if there is a changing electric field. From this, Maxwell derived another startling conclusion. If a changing magnetic field produces an electric field, that electric field is itself changing. This changing electric field will, in turn, produce a magnetic field, which will be changing and so will produce a changing electric field; and so on. When Maxwell manipulated his equations, he found that the net result of these interacting changing fields was to produce a *wave* of electric and magnetic fields that can actually propagate (travel) through space. We now examine, in a simplified way, how such **electromagnetic waves** can be produced.

How EM waves are produced

Consider two conducting rods that will serve as an "antenna" (Fig. 22–4a). Suppose that these two rods are connected by a switch to the opposite terminals of a battery. As soon as the switch is closed, the upper rod quickly becomes positively charged and the lower one negatively charged. Electric field lines are formed as indicated by the lines in Fig. 22–4b. While the charges are flowing, a current exists, whose direction is indicated by the arrows. A magnetic field is therefore produced near the antenna. The magnetic field lines encircle the wires and therefore, in the plane of the page, **B** points into the page ($\otimes$) on the right and out of the page ($\odot$) on the left. Now we ask, how far out do these electric and magnetic fields extend? In the static case, the fields extend outward indefinitely far. However, when the switch in Fig. 22–4 is closed, the fields quickly appear nearby, but it takes time for them to reach distant points. Both electric and magnetic fields store energy, and this energy cannot be transferred to distant points at infinite speed.

Now we look at the situation of Fig. 22–5 where our antenna is connected to an ac generator. In Fig. 22–5a, the connection has just been completed. Charge starts building up and fields form just as in Fig. 22–4. The + and − signs in Fig. 22–5a indicate the net charge on each rod. The arrows indicate the direction of the current. The electric field is represented by lines in the plane of the page; and the magnetic field, according to the right-hand rule, is into ($\otimes$) or out of ($\odot$) the page. In Fig. 22–5b, the emf of the ac generator has reversed in direction; the current is reversed and the new magnetic field is in the opposite direction. Because the new fields have changed direction, the old lines fold back to connect up to some of the new lines and form closed loops as shown.[†] The old fields, however, don't suddenly disappear; they are on their way to distant points. Indeed, because a changing magnetic field produces an electric field, and a changing electric field produces a magnetic field, this combination of changing electric and magnetic fields moving outward is self-supporting, no longer depending on the antenna

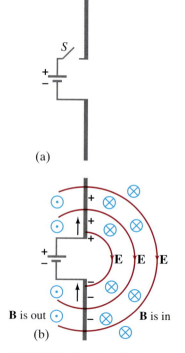

(a)

(b)

B is out $\odot$ **B** is in

FIGURE 22–4   Fields produced by charge flowing into conductors. It takes time for the **E** and **B** fields to travel outward to distant points.

[†]We are considering waves traveling through empty space, so there are no charges for lines of **E** to start or stop on, so they form closed loops. Magnetic field lines always form closed loops since there are no single (separate) magnetic poles (as far as we know).

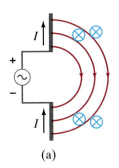 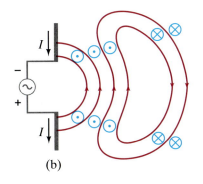

(a)   (b)

FIGURE 22–5 Sequence showing electric and magnetic fields that spread outward from oscillating charges on two conductors connected to an ac source (see the text).

charges. The fields not far from the antenna, referred to as the *near field*, become quite complicated, but we are not so interested in them. We are instead mainly interested in the fields far from the antenna (they are generally what we detect), which we refer to as the **radiation field**. The electric field lines form loops, as shown in Fig. 22–6, and continue moving outward. The magnetic field lines also form closed loops, but are not shown since they are perpendicular to the page. Although the lines are shown only on the right of the source, fields also travel in other directions. (The field strengths are greatest in directions perpendicular to the oscillating charges; and they drop to zero along the direction of oscillation—above and below the antenna in Fig. 22–6.)

The magnitudes of both **E** and **B** in the radiation field are found to decrease with distance as $1/r$. (Compare this to the static electric field given by Coulomb's law where **E** decreases as $1/r^2$.) The energy carried by the electromagnetic wave is proportional (as for any wave, Chapter 11) to the square of the amplitude, E^2 or B^2, as will be discussed further in Section 22–7, so the intensity of the wave decreases as $1/r^2$.

Several things about the radiation field can be noted from Fig. 22–6. First, *the electric and magnetic fields at any point are perpendicular to each other, and to the direction of motion*. Second, we can see that the fields alternate in direction (**B** is into the page at some points and out of the page at others; similarly for **E**). Thus, the field strengths vary from a maximum in one direction, to zero, to a maximum in the other direction. The electric and magnetic fields are "in phase"; that is, they each are zero at the same points and reach their maximum at the same points in space.

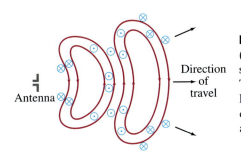

Direction
of
travel

Antenna

FIGURE 22–6 The radiation fields (far from the antenna) produced by a sinusoidal signal on a dipole antenna. The closed loops represent electric field lines. The magnetic field lines, perpendicular to the page and represented by ⊗ and ⊙, also form closed loops.

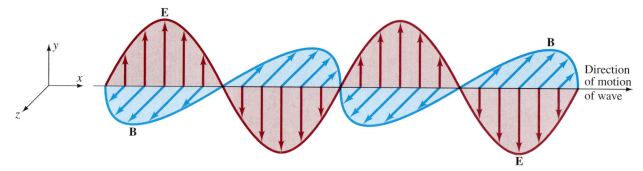

FIGURE 22–7 Electric and magnetic field strengths in an electromagnetic wave. **E** and **B** are at right angles to each other. The entire pattern moves in a direction perpendicular to both **E** and **B**.

If the source emf varies sinusoidally, then the electric and magnetic field strengths in the radiation field will also vary sinusoidally. The sinusoidal character of the waves is diagrammed in Fig. 22–7, which shows the field *strengths* plotted as a function of position. Notice that **B** and **E** are perpendicular to each other and to the direction of travel.

We call these waves electromagnetic (EM) waves. They are *transverse* waves and resemble other types of waves (Chapter 11). However, EM waves are always waves of *fields*, not of matter, as are waves on water or a rope. Because they are fields, EM waves can propagate in empty space.

EM waves are produced by accelerating electric charges

We have seen in the above analysis that EM waves are produced by electric charges that are oscillating, and hence are undergoing acceleration. In fact, we can say in general that **accelerating electric charges give rise to electromagnetic waves**.

In the next (optional) Section, we will derive a formula for the speed of EM waves:

Speed of EM waves

$$v = \frac{1}{\sqrt{\epsilon_0 \mu_0}}. \qquad (22\text{–}2)$$

which Maxwell himself derived. When Maxwell put in the values for ϵ_0 and μ_0 he found

$$v = \frac{1}{\sqrt{\epsilon_0 \mu_0}}$$

$$= \frac{1}{\sqrt{(8.85 \times 10^{-12}\ \text{C}^2/\text{N} \cdot \text{m}^2)(4\pi \times 10^{-7}\ \text{N} \cdot \text{s}^2/\text{C}^2)}}$$

$$= 3.00 \times 10^8\ \text{m/s},$$

which is equal to the measured speed of light.

* 22–4 Calculation of the Speed of Electromagnetic Waves

Maxwell's prediction that EM waves should exist was remarkable. Equally remarkable was the speed at which they were predicted to travel. We now calculate this speed.

We shall consider a region far from the source, so that the wave fronts (the field lines in Fig. 22–6) are essentially flat over a reasonable area. They are then called **plane waves**, meaning that at any instant, **E** and **B** are uniform over a plane perpendicular to the direction of propagation. Let us assume that, in a particular coordinate system, the wave is traveling in the

x direction with speed v; **E** is parallel to the y axis, and **B** is parallel to the z axis, as shown in Fig. 22–7.

Now let us apply Faraday's law to an imaginary rectangle $abcd$ placed in the xy plane as shown with gray shading in Fig. 22–8. We assume that the side ab is in a region where the wave has not yet reached, so $E = B = 0$ there. As the wave moves, its magnetic flux through our rectangular loop changes. Instead of showing the motion of the wave in Fig. 22–8, we show rather the relative position of the rectangle (it's easier to visualize). After a short time Δt, the wave moves to the right a distance $\Delta x = v\,\Delta t$, where v is the velocity of the wave; or, equivalently, the rectangular loop moves $\Delta x = v\,\Delta t$ to the left. Rectangle $a'b'c'd'$ represents the new position of the rectangle relative to the wave. The change in magnetic flux $\Delta\Phi_B$ through the loop during the time Δt is just the flux that passes through the small rectangle $dd'c'c$. Thus, $\Delta\Phi_B = B\,\Delta A = By_0\,\Delta x = By_0 v\,\Delta t$, where y_0 is the width ab or cd, and $\Delta A = y_0\,\Delta x = y_0 v\,\Delta t$ is the area of $dd'c'c$; of course B represents the magnetic field passing through this area (represented by the heavy arrow in Fig. 22–8). We assume Δx is small so that B is essentially constant over this small area. According to Faraday's law, then, the emf induced around the loop $abcd$ equals the rate of change of magnetic flux:

$$\mathscr{E} = \frac{\Delta\Phi_B}{\Delta t} = \frac{B\,\Delta A}{\Delta t} = \frac{By_0 v\,\Delta t}{\Delta t} = By_0 v.$$

(We omitted the minus sign for convenience.) The emf around the loop $abcd$ is the sum of the emfs in each straight section:

$$\mathscr{E} = \mathscr{E}_{ab} + \mathscr{E}_{bc} + \mathscr{E}_{cd} + \mathscr{E}_{da}.$$

Now the emf is the work done per unit charge: $\mathscr{E} = W/q = Fd/q = Ed$, where F is the force exerted over the distance d, and the electric field E is the force per unit charge, $E = F/q$. (The result, $\mathscr{E} = Ed$, is the same as Eq. 17–2.) The term $\mathscr{E}_{ab}$ is zero since $E = 0$ in this region; and $\mathscr{E}_{bc} = \mathscr{E}_{da} = 0$, since **E** is perpendicular to the path (Fig. 22–8). Thus

$$\mathscr{E} - \mathscr{E}_{cd} = Ey_0,$$

where E is the magnitude of the electric field along cd (or $c'd'$). From Faraday's law as we wrote it above ($\mathscr{E} = By_0 v$), we thus have

$$Ey_0 = By_0 v$$

$$E = vB. \tag{22–3}$$

Thus we see that at any point in space, E and B are in the ratio $E/B = v$, the velocity of the wave.

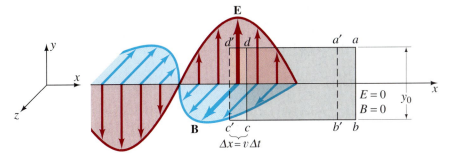

FIGURE 22–8 In a time Δt, the rectangle in the xy plane (actually stationary) moves a distance $\Delta x = v\,\Delta t$ relative to the wave (the wave moves to the right with speed v).

Now we consider a rectangle in the xz plane, as shown (gray) in Fig. 22–9. Again we show the rectangle moving to the left a distance $\Delta x = v\,\Delta t$ relative to the wave, although actually the rectangle is stationary and the wave is moving to the right; $abcd$ is the position of the rectangle initially and $a'b'c'd'$ is its position after a time Δt. There is a changing electric flux through this rectangular loop equal to the electric field E (heavy arrow) times the increasing area $\Delta A = z_0\Delta x = z_0 v\,\Delta t$ (where z_0 is the width $ab = cd$ of the rectangle). According to Ampère's law, Eq. 22–1 with $I_C = 0$ since there are no conduction currents, we have

$$\Sigma B_\parallel \,\Delta l = \mu_0\epsilon_0 \frac{\Delta\Phi_E}{\Delta t}$$

$$= \mu_0\epsilon_0 \frac{(E)(z_0 v\,\Delta t)}{\Delta t} = \mu_0\epsilon_0 E z_0 v.$$

The sum of $B_\parallel \Delta l$ on the sides ab, bc, and da are all zero, because either $B = 0$ or $\mathbf{B}$ is perpendicular to these sides. But along side cd, the contribution is Bz_0, where B is the magnetic field (heavy arrow) parallel to cd. Thus,

$$Bz_0 = \mu_0\epsilon_0 E z_0 v$$

so

$$B = \mu_0\epsilon_0 vE.$$

We combine this with Eq. 22–3 and find that

$$B = \mu_0\epsilon_0 v(vB) = \mu_0\epsilon_0 v^2 B.$$

We cancel B on both sides and solve for v:

$$v = \frac{1}{\sqrt{\epsilon_0\mu_0}}$$

which is just Eq. 22–2 that we quoted before and have now derived.

FIGURE 22–9 Rectangle in the xz plane moves a distance $\Delta x = v\,\Delta t$ relative to a wave traveling to the right.

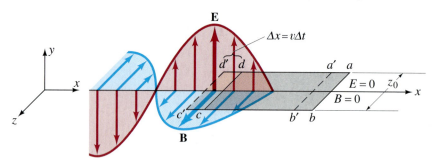

When we put in the values for ϵ_0 and μ_0, we find that

$$v = \frac{1}{\sqrt{\epsilon_0\mu_0}} = \frac{1}{\sqrt{(8.85 \times 10^{-12}\ \text{C}^2/\text{N·m}^2)(4\pi \times 10^{-7}\ \text{N·s}^2/\text{C}^2)}}$$

$$= 3.00 \times 10^8\ \text{m/s}.$$

Speed of EM waves is speed of light

This result, that the speed of EM waves is precisely equal to the measured speed of light, is truly remarkable.

22–5 Light as an Electromagnetic Wave and the Electromagnetic Spectrum

The calculation at the end of the last section gives the result that Maxwell himself determined: that the speed of EM waves is $3.00 \times 10^8\ \text{m/s}$, the same as the measured speed of light.

Light had been shown some 60 years previously to behave like a wave (we'll discuss this in Chapter 24). But nobody knew what kind of wave it was—that is, what is it that is oscillating in a light wave? Maxwell, on the basis of the calculated speed of EM waves, argued that light must be an electromagnetic wave. This idea soon came to be generally accepted by scientists, but not fully until after EM waves were experimentally detected. EM waves were first generated and detected experimentally by Heinrich Hertz (1857–1894) in 1887, eight years after Maxwell's death. Hertz used a spark-gap apparatus in which charge was made to rush back and forth for a short time, generating waves whose frequency was about $10^9\ \text{Hz}$. He detected them some distance away using a loop of wire in which an emf was produced when a changing magnetic field passed through. These waves were later shown to travel at the speed of light, $3.00 \times 10^8\ \text{m/s}$, and to exhibit all the characteristics of light such as reflection, refraction, and interference. The only difference was that they were not visible. Hertz's experiment was a strong confirmation of Maxwell's theory.

The wavelengths of visible light were measured in the first decade of the nineteenth century, long before anyone imagined that light was an electromagnetic wave. The wavelengths were found to lie between $4.0 \times 10^{-7}\ \text{m}$ and $7.5 \times 10^{-7}\ \text{m}$; or 400 nm to 750 nm (1 nm $= 10^{-9}\ \text{m}$). The frequencies of visible light can be found using Eq. 11–12, which we rewrite here:

$$f\lambda = c, \qquad\qquad (22\text{–}4)$$

where f and λ are the frequency and wavelength, respectively, of the wave. Here, c is the speed of light, $3.00 \times 10^8\ \text{m/s}$; it gets the special symbol c because of its universality for all EM waves in free space. Equation 22–4 tells us that the frequencies of visible light are between $4.0 \times 10^{14}\ \text{Hz}$ and $7.5 \times 10^{14}\ \text{Hz}$. (Recall that 1 Hz $= 1$ cycle per second $= 1\ \text{s}^{-1}$.)

c is symbol for speed of light

But visible light is only one kind of EM wave. As we have seen, Hertz produced EM waves of much lower frequency, about 10^9 Hz. These are called **radio waves**, since frequencies in this range are used today to transmit radio and TV signals. Electromagnetic waves, or EM radiation as we sometimes call it, have been produced or detected over a wide range of frequencies. They are usually categorized as shown in Fig. 22–10. This is known as the **electromagnetic spectrum**.

EM spectrum

Radio waves and microwaves can be produced in the laboratory using electronic equipment, as we saw in Fig. 22–5. Higher-frequency waves are very difficult to produce electronically. These and other types of EM waves are produced in natural processes, as emission from atoms, molecules, and nuclei (more on this in later chapters). Generally, EM waves are produced by the acceleration of electrons or other charged particles, such as electrons accelerating in the antenna of Fig. 22–5. Another example is X-rays, which are produced (see Chapters 25 and 28) when fast-moving electrons are rapidly decelerated upon striking a metal target. Even the visible light emitted by an ordinary incandescent light is due to electrons undergoing acceleration within the hot filament. We will meet various types of EM waves later. However, it is worth mentioning here that infrared (IR) radiation (EM waves whose frequency is just less than that of visible light) is mainly responsible for the heating effect of the Sun. The Sun emits not only visible light but substantial amounts of IR and UV (ultraviolet) as well. The molecules of our skin tend to "resonate" at infrared frequencies, so it is these that are preferentially absorbed and thus warm us up. We humans experience EM waves differently depending on their wavelengths: Our eyes detect wavelengths between 4 and 7×10^{-7} m (visible light), whereas our skin detects longer wavelengths. Many EM wavelengths we don't detect directly at all.

FIGURE 22–10 Electromagnetic spectrum.

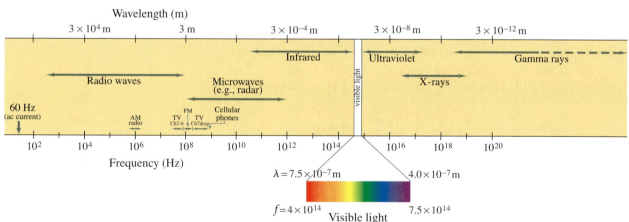

EXAMPLE 22–1 **Wavelengths of EM waves.** Calculate the wavelength: (*a*) of a 60-Hz EM wave, (*b*) of a 93.3-MHz FM radio wave, and (*c*) of a beam of visible red light from a laser at frequency 4.74×10^{14} Hz.

SOLUTION (*a*) Since $c = \lambda f$,

$$\lambda = \frac{c}{f} = \frac{3.0 \times 10^8 \text{ m/s}}{60 \text{ s}^{-1}} = 5.0 \times 10^6 \text{ m},$$

or 5000 km. 60 Hz is the frequency of ac current in the United States, and, as we see here, one wavelength stretches all the way across the country.

(*b*)
$$\lambda = \frac{3.00 \times 10^8 \text{ m/s}}{93.3 \times 10^6 \text{ s}^{-1}} = 3.22 \text{ m}.$$

The length of an FM antenna is about half this ($\frac{1}{2}\lambda$).

(*c*)
$$\lambda = \frac{3.00 \times 10^8 \text{ m/s}}{4.74 \times 10^{14} \text{ s}^{-1}} = 6.33 \times 10^{-7} \text{ m} \ (= 633 \text{ nm}).$$

Electromagnetic waves can travel along transmission lines as well as in empty space. When a source of emf is connected up to a transmission line—be it two parallel wires or a coaxial cable (Fig. 22–11)—the electric field within the wire is not set up immediately at all points along the wires. This is based on the same argument we used in Section 22–3 with reference to Fig. 22–5. Indeed, it can be shown that if the wires are separated by air, the electrical signal travels along the wires at the speed $c = 3.0 \times 10^8$ m/s. For example, when you flip a light switch, the light actually goes on a tiny fraction of a second later. If the wires are in a medium whose electric permittivity is ϵ and magnetic permeability is μ, the speed is not given by Eq. 22–2, but by

$$v = 1/\sqrt{\epsilon \mu}.$$

EXAMPLE 22–2 **Voice speed through the wires.** When you speak on the telephone from Los Angeles to a friend in New York 4000 km away, how long does it take your voice to travel?

SOLUTION Since speed = distance/time, then time = distance/speed = $(4.0 \times 10^6 \text{ m})/(3.0 \times 10^8 \text{ m/s}) = 1.3 \times 10^{-2}$ s, or about $\frac{1}{100}$ s.

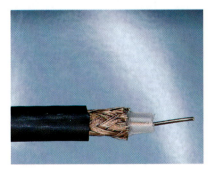

FIGURE 22–11
Coaxial cable.

22-6 Measuring the Speed of Light

The first serious attempt to actually measure the speed of light was done by Galileo by trying to measure the time required for light to travel a known distance between two hilltops. He stationed an assistant on one hilltop, and himself on another, and ordered the assistant to lift the cover from a lamp the instant he saw a flash from Galileo's lamp. Galileo measured the time between the flash of his lamp and when he received the light from his assistant's lamp. The time was so short that Galileo concluded it merely represented human reaction time, and that the speed of light must be extremely high.

The first successful determination that the speed of light is finite was made by the Danish astronomer Ole Roemer (1644–1710). Roemer had noted that the carefully measured period of Io, one of Jupiter's moons (an average period of 42.5 h to make one complete orbit around Jupiter), varied slightly, depending on the relative motion of Earth and Jupiter. When Earth was moving away from Jupiter, the period of the moon was slightly longer, and when Earth was moving toward Jupiter, the period was slightly shorter. He attributed this variation to the extra time needed for light to travel the increasing distance to Earth when Earth is receding, or to the shorter travel time for the decreasing distance when the two planets are approaching one another. Roemer concluded that the speed of light—though great—is finite.

Michelson measures c

Since then a number of techniques have been used to measure the speed of light. Among the most important were those carried out by the American Albert A. Michelson (1852–1931). Michelson used the rotating mirror apparatus diagrammed in Fig. 22–12 for a series of high-precision experiments carried out from 1880 to the 1920s. Light from a source was directed at one face of a rotating eight-sided mirror. The reflected light traveled to a stationary mirror a large distance away and back again as shown. If the rotating mirror was turning at just the right rate, the returning beam of light would reflect from one face of the mirror into a small telescope through which the observer looked. If the speed of rotation was only slightly different, the beam would be deflected to one side and would not be seen by the observer. From the required speed of the rotating mirror and the known distance to the stationary mirror, the speed of light could be calculated. In the 1920s, Michelson

FIGURE 22–12
Michelson's speed-of-light apparatus (not to scale).

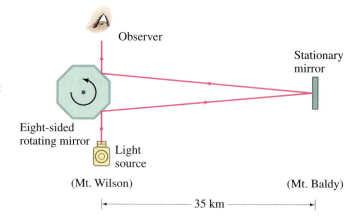

set up the rotating mirror on the top of Mt. Wilson in southern California and the stationary mirror on Mt. Baldy (Mt. San Antonio) 35 km away. He later measured the speed of light in vacuum using a long evacuated tube.

The accepted value today for the speed of light, c, in vacuum is

$$c = 2.99792458 \times 10^8 \text{ m/s.}$$

We usually round this off to

$$c = 3.00 \times 10^8 \text{ m/s}$$

when extremely precise results are not required. In air, the speed is only slightly less.

* 22–7 Energy in EM Waves

Electromagnetic waves carry energy from one region of space to another. This energy is associated with the moving electric and magnetic fields. In Section 17–9, we saw that the energy density (J/m^3) stored in an electric field E is $u = \frac{1}{2}\epsilon_0 E^2$, where u is the energy per unit volume. The energy stored in a magnetic field B, as we discussed in Section 21–10 (Eq. 21–10), is given by $u = \frac{1}{2}B^2/\mu_0$. Thus, the total energy stored per unit volume in a region of space where there is an electromagnetic wave is

$$u = \frac{1}{2}\,\epsilon_0 E^2 + \frac{1}{2}\frac{B^2}{\mu_0}. \tag{22–5}$$

In this equation, E and B represent the electric and magnetic field strengths of the wave at any instant in a small region of space. We can write Eq. 22–5 in terms of the E field only, since from Eq. 22–2 we have $\sqrt{\epsilon_0\mu_0} = 1/c$, and from Eq. 22–3, $B = E/c$. We insert these into Eq. 22–5 to obtain

$$u = \frac{1}{2}\,\epsilon_0 E^2 + \frac{1}{2}\frac{\epsilon_0\mu_0 E^2}{\mu_0}$$

$$= \epsilon_0 E^2. \tag{22–6a}$$

Notice that the energy density associated with the B field is equal to that associated with the E field, so each contributes half to the total energy. We can also write the energy density in terms of the B field only:

$$u = \epsilon_0 E^2 = \epsilon_0 c^2 B^2 = \frac{\epsilon_0 B^2}{\epsilon_0 \mu_0},$$

so

$$u = \frac{B^2}{\mu_0}. \tag{22–6b}$$

Or we can write u in one term containing both E and B:

$$u = \epsilon_0 E^2 = \epsilon_0 EcB = \frac{\epsilon_0 EB}{\sqrt{\epsilon_0\mu_0}},$$

or

$$u = \sqrt{\frac{\epsilon_0}{\mu_0}}\,EB. \tag{22–6c}$$

Equations 22–6 give the energy density in any region of space at any instant.

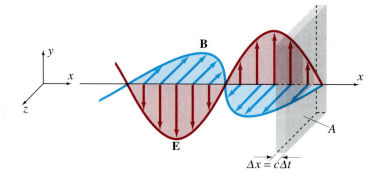

FIGURE 22–13
Electromagnetic wave carrying energy through area A.

Now let us determine the energy that is transported by the wave per unit time per unit area perpendicular to the wave direction. Let us imagine that the wave is passing through an area A perpendicular to the x axis, as shown in Fig. 22–13. In a short time Δt, the wave moves to the right a distance $\Delta x = c\,\Delta t$. The energy that has passed through the area A in the time Δt is the energy that now occupies the volume $\Delta V = A\,\Delta x = Ac\,\Delta t$. The energy density u is $u = \epsilon_0 E^2$, where E is the electric field in this volume at the given instant. So the energy ΔU contained in this volume is the energy density u times the volume: $\Delta U = u\,\Delta V = (\epsilon_0 E^2)(Ac\,\Delta t)$. Therefore, the energy crossing the area A per time Δt, which we designate[†] S, is

$$S = \frac{\Delta U}{A\,\Delta t} = \epsilon_0 c E^2,$$

and is measured in watts per square meter (W/m²). Since $E = cB$ and $c = 1/\sqrt{\epsilon_0 \mu_0}$, (Eqs. 22–3 and 22–2), this can also be written

Rate energy is transported by EM waves

$$S = \epsilon_0 c E^2 = \frac{cB^2}{\mu_0} = \frac{EB}{\mu_0}. \tag{22–7}$$

Equation 22–7 gives the energy transported per unit area per unit time at any *instant*. We often want to know the *average* over an extended period of time. If E and B are sinusoidal, then $\overline{E^2} = E_0^2/2$, just as for electric currents and voltages (Section 18–8), where E_0 is the *maximum* value of E. Thus we can write

$$\overline{S} = \frac{1}{2}\epsilon_0 c E_0^2 = \frac{1}{2}\frac{c}{\mu_0} B_0^2 = \frac{E_0 B_0}{2\mu_0}, \tag{22–8}$$

where B_0 is the maximum value of B. We can also write

$$\overline{S} = \frac{E_{\text{rms}} B_{\text{rms}}}{\mu_0}$$

where E_{rms} and B_{rms} are the rms values ($E_{\text{rms}} = \sqrt{\overline{E^2}}$, $B_{\text{rms}} = \sqrt{\overline{B_{\text{rms}}^2}}$).

[†]The quantity S is called the *Poynting vector*. Its direction is that in which the energy is being transported, which is the direction the wave is traveling.

EXAMPLE 22–3 **E and B from the Sun.** Radiation from the Sun reaches the Earth (above the atmosphere) at a rate of about 1350 J/s·m². Assume that this is a single EM wave and calculate the maximum values of E and B.

SOLUTION Since $\bar{S} = 1350\ \text{J/s·m}^2 = \epsilon_0 c E_0^2/2$, then

$$E_0 = \sqrt{\frac{2\bar{S}}{\epsilon_0 c}}$$

$$= \sqrt{\frac{2(1350\ \text{J/s·m}^2)}{(8.85 \times 10^{-12}\ \text{C}^2/\text{N·m}^2)(3.0 \times 10^8\ \text{m/s})}}$$

$$= 1.01 \times 10^3\ \text{V/m}.$$

From Eq. 22–3, $B = E/c$, so

$$B_0 = \frac{E_0}{c} = \frac{1.01 \times 10^3\ \text{V/m}}{3.00 \times 10^8\ \text{m/s}} = 3.37 \times 10^{-6}\ \text{T}.$$

This Example illustrates that B has a small numerical value compared to E. This is because of the different units for E and B and the way these units are defined. But, as we saw earlier, B contributes the same energy to the wave as E does.

EXAMPLE 22–4 **Energy in E and B.** (a) Show that Eq. 22–8 connecting the power flux (S has units W/m²) and the electric field works out unit-wise in the metric system. (b) Calculate the energy density due to electric and magnetic contributions in Example 22–3.

SOLUTION (a) The equation reads $\bar{S} = \frac{1}{2}\epsilon_0 c E_0^2$ so the units break down as follows:

$$\epsilon_0 c E_0^2\ \text{units} \Rightarrow \frac{\text{C}^2}{\text{N·m}^2} \times \frac{\text{m}}{\text{s}} \times \frac{\text{N}^2}{\text{C}^2} = \frac{(\text{N·m})}{\text{m}^2 \cdot \text{s}} = \frac{\text{J/s}}{\text{m}^2} = \frac{\text{W}}{\text{m}^2} \Rightarrow \text{OK}.$$

(b) The electric energy density is given by

$$u_E = \frac{1}{2}\epsilon_0 E_0^2$$

$$= \frac{1}{2}(8.85 \times 10^{-12}\ \text{C}^2/\text{N·m}^2)(1.01 \times 10^3\ \text{V/m})^2$$

$$= 4.5 \times 10^{-6}\ \frac{\text{J}}{\text{m}^3}.$$

The magnetic field energy density for the field value calculated in the previous Example is

$$u_B = \frac{B_0^2}{2\mu_0} = \frac{(3.37 \times 10^{-6}\ \text{T})^2}{2(4\pi \times 10^{-7}\ \text{T·m/A})} = 4.5 \times 10^{-6}\ \frac{\text{J}}{\text{m}^3}.$$

They are equal as expected. The total energy density in the EM wave is $u = u_E + u_B = 9.0 \times 10^{-6}\ \text{J/m}^3$, with equal contributions from electric and magnetic fields.

Electromagnetic waves offer the possibility of transmitting information over long distances. Among the first to realize this and put it in practice was Guglielmo Marconi (1874–1937), who, in the 1890s invented and developed the wireless telegraph. With it, messages could be sent hundreds of kilometers at the speed of light without the use of wires. The first signals were merely long and short pulses that could be translated into words by a code, such as the "dots" and "dashes" of the Morse code. The next decade saw the development of vacuum tubes. Out of this early work radio and television were born. We now discuss briefly (1) how radio and TV signals are transmitted, and (2) how they are received at home.

Transmission of radio waves

The process by which a radio station transmits information (words and music) is outlined in Fig. 22–14. The audio (sound) information is changed into an electrical signal of the same frequencies by, say, a microphone or tape recorder head. This electrical signal is called an audio-frequency (AF) signal, since the frequencies are in the audio range (20 to 20,000 Hz). The signal is amplified[†] electronically and is then mixed with a radio-frequency (RF) signal. The RF frequency is determined by the values of L and C in a resonant LCR circuit (Section 21–14) and is chosen to

Carrier frequency

produce a particular frequency for each station, called its **carrier frequency**. AM radio stations have carrier frequencies from about 530 kHz to 1600 kHz. For example, "710 on your dial" means a station whose carrier frequency is 710 kHz. FM radio stations have much higher carrier frequencies, between 88 MHz and 108 MHz. The carrier frequencies for TV stations in the United States lie between 54 and 88 MHz for channels 2 through 6, and between 174 and 216 MHz for channels 7 through 13; UHF (ultra-high-frequency) stations have even higher carrier frequencies, between 470 and 890 MHz.

FIGURE 22–14 Block diagram of a radio transmitter.

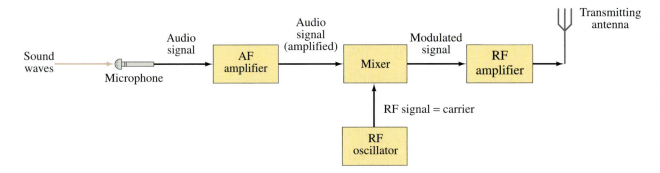

[†]How amplifiers work is discussed in Section 29–9.

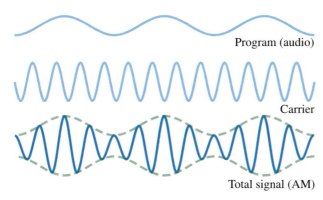

FIGURE 22–15 In amplitude modulation (AM), the amplitude of the carrier signal is made to vary in proportion to the audio signal's amiplitude.

The mixing of the audio and carrier frequencies is done in two ways. In **amplitude modulation** (AM), the amplitude of the higher frequency carrier wave is made to vary so it follows the audio signal as shown in Fig. 22–15. It is called "amplitude modulation" because the amplitude of the carrier is altered ("modulate" means to change or alter). In **frequency modulation** (FM), the *frequency* of the carrier wave is made to change in proportion to the audio signal, as shown in Fig. 22–16.

The mixed signal is amplified further (since the signal contains radio frequencies, this is called an RF amplifier) and is then sent to the antenna, where the complex mixture of frequencies is sent out in the form of EM waves.

A television transmitter works in a similar way, using frequency modulation, except that both audio and video signals are mixed with carrier frequencies.

FIGURE 22–16 In frequency modulation (FM), the frequency of the carrier signal is made to change in proportion to the audio signal's amplitude. This method is used by FM radio and television.

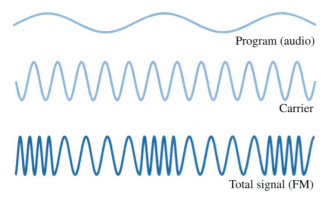

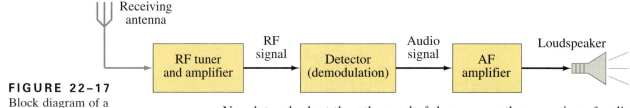

FIGURE 22–17
Block diagram of a simple radio receiver.

➥ **PHYSICS APPLIED**

Radio and TV receivers

Antennas

FIGURE 22–18 Antennas. (a) Electric field of EM wave produces a current in an antenna consisting of straight wire or rods. (b) Changing magnetic field induces an emf and current in a loop antenna.

Now let us look at the other end of the process, the reception of radio and TV programs at home. A simple radio receiver is diagrammed in Fig. 22–17. The EM waves sent out by all stations are received by the antenna. One kind of antenna consists of one or more conducting rods; the electric field in the EM waves exerts a force on the electrons in the conductor, causing them to move back and forth at the frequencies of the waves (Fig. 22–18a). A second type of antenna consists of a tubular coil of wire, often found in AM radios, or the simple loop of a UHF television antenna. These antennas detect the magnetic field of the wave, for the changing B field induces an emf in the coil (Fig. 22–18b). The signal the antenna detects and sends to the receiver is very small and contains frequencies from many different stations. The receiver selects out a particular RF frequency (actually a narrow range of frequencies) corresponding to a particular station using a resonant LC circuit (Section 21–14) with a vari-

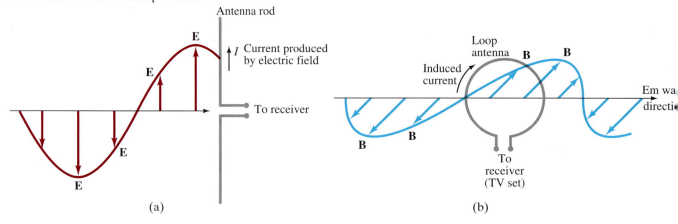

(a)

(b)

able capacitor or inductor. A simple example is shown in Fig. 22–19. A particular station is "tuned-in" by adjusting L and C so that the resonant frequency of the circuit equals that of the station's carrier frequency. The RF signal may be amplified both before and after the tuning is done. The signal, containing both audio and carrier frequencies, next goes to the *detector* (Fig. 22–17) where "demodulation" takes place—that is, the RF carrier frequency is separated from the audio signal. The audio signal is then amplified and sent to a loudspeaker or headphones.

Modern receivers have more stages than those shown. Various means are used to increase the sensitivity and selectivity (ability to detect weak signals and distinguish them from other stations), and to minimize distortion of the original signal.[†]

FIGURE 22–19 Simple tuning stage of a radio.

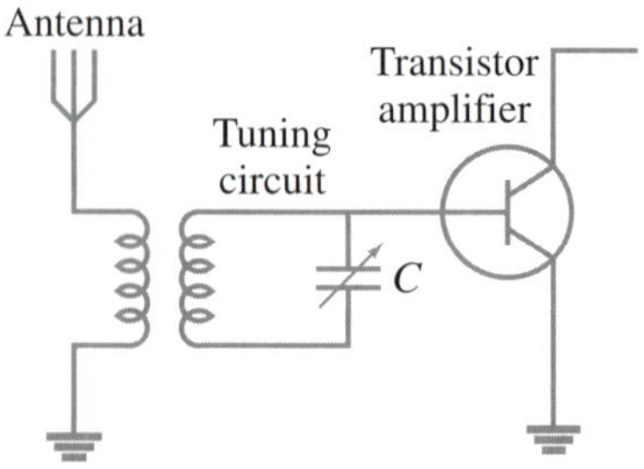

[†]For *FM stereo broadcasting*, two signals are carried by the carrier wave. One of these contains frequencies up to about 15 kHz, which includes most audio frequencies. The other signal includes the same range of frequencies, but 19 kHz is added to it. A stereo receiver subtracts this 19,000-Hz signal and distributes the two signals to the left and right channels. The first signal actually consists of the sum of left and right channels ($L + R$), so mono radios detect all the sound. The second signal is the difference between left and right ($L - R$). Hence the receiver must add and subtract the two signals to get pure left and right signals for each channel.

A television receiver does similar things to both the audio and the video signals. The audio signal goes finally to the loudspeaker, and the video signal to the picture tube, a *cathode ray tube* (CRT) whose operation was discussed in Section 17–10.

EXAMPLE 22–5 **Tuning a station.** An FM radio station transmits at 100 MHz. Calculate (*a*) its wavelength, and (*b*) the value of the capacitance in the tuning circuit if $L = 0.40 \, \mu H$.

SOLUTION (*a*) The carrier frequency is $f = 100 \, MHz = 1.0 \times 10^8 \, s^{-1}$. From Eq. 22–4, $\lambda = c/f = (3.0 \times 10^8 \, m/s)/(1.0 \times 10^8 \, s^{-1}) = 3.0 \, m$. The wavelengths of other FM signals (88 to 108 MHz) are close to this. FM antennas are typically 1.5 m long, or about a half wavelength. This length is chosen so that the antenna reacts in a resonant fashion and thus is more sensitive.

(*b*) According to Eq. 21–18, the resonant frequency is $f_0 = 1/(2\pi\sqrt{LC})$. Therefore,

$$C = \frac{1}{4\pi^2 f_0^2 L} = \frac{1}{4(3.14)^2 (1.0 \times 10^8 \, s^{-1})^2 (4.0 \times 10^{-7} \, H)}$$

$$= 6.3 \, pF.$$

Of course, the capacitor or inductor is variable, so other stations can be selected too.

The various regions of the radio-wave spectrum are assigned by governmental agencies to various purposes. Besides those mentioned above, there are "bands" assigned for use by ships, airplanes, police, military, amateurs, satellites and space, and radar. Cellular telephones, for example, occupy a band from 824 MHz to 894 MHz.

SUMMARY

James Clerk Maxwell synthesized an elegant theory in which all electric and magnetic phenomena could be described using four equations, now called **Maxwell's equations**. They are based on earlier ideas, but Maxwell added one more—that a changing electric field produces a magnetic field.

Maxwell's theory predicted that transverse **electromagnetic** (EM) **waves** would be produced by accelerating electric charges, and these waves would propagate through space at the speed of light, given by the formula

$$c = \frac{1}{\sqrt{\epsilon_0 \mu_0}}.$$

The oscillating electric and magnetic fields in an EM wave are perpendicular to each other and to the direction of propagation.

The wavelength λ and frequency f of EM waves are related to their speed c by

$$c = \lambda f,$$

just as for other waves.

After EM waves were experimentally detected in the late 1800s, the idea that light is an EM wave (of very high frequency) became generally accepted. The **electromagnetic spectrum** includes EM waves of a wide variety of wavelengths, from microwaves and radio waves to visible light to X-rays and γ-rays, all of which travel through space at a speed $c = 3.0 \times 10^8 \, m/s$.

QUESTIONS

1. The electric field in an EM wave traveling north oscillates in an east–west plane. Describe the direction of the magnetic field vector in this wave.

2. Is sound an electromagnetic wave? If not, what kind of wave is it?

3. Can EM waves travel through a perfect vacuum? Can sound waves?

4. How are light and sound alike? How are they different?

5. Are the wavelengths of radio and television signals longer or shorter than those detectable by the human eye?

6. What does the result of Example 22–1 tell you about the phase of a 60-Hz ac current that starts at a power plant as compared to its phase at a house 200 km away?

7. When you connect two loudspeakers to the output of a stereo amplifier, should you be sure the lead wires are equal in length so that there will not be a time lag between speakers? Explain.

8. In the electromagnetic spectrum, what type of EM wave would have a wavelength of 10^3 km? 1 km? 1 m? 1 cm? 1 mm? 1 μm?

* 9. A lost person may signal by flashing a flashlight on and off using Morse code. This is actually a modulated EM wave. Is it AM or FM? What is the frequency of the carrier, approximately?

* 10. Can two radio or TV stations broadcast on the same carrier frequency? Explain.

* 11. If a radio transmitter has a vertical antenna, should a receiver's antenna (rod type) be vertical or horizontal to obtain best reception?

* 12. The carrier frequencies of FM broadcasts are much higher than for AM broadcasts. On the basis of what you learned about diffraction in Chapter 11, explain why AM signals can be detected more readily than FM signals behind low hills or buildings.

13. Discuss how cordless telephones make use of EM waves. What about cellular telephones?

PROBLEMS

*SECTION 22–2

* 1. (I) Calculate the displacement current I_D between the square plates, 2.8 cm on a side, of a capacitor if the electric field is changing at a rate of 2.0×10^6 V/m·s.

* 2. (I) Calculate the displacement current between the round plates of a capacitor, 6.0 cm in diameter, if the plates are spaced 1.3 mm apart and the voltage across them is changing at a rate of 120 V/s.

* 3. (II) At a given instant, a 3.8-A current flows in the wires connected to a parallel-plate capacitor. What is the rate at which the electric field is changing between the plates if the square plates are 1.90 cm on a side?

* 4. (II) A 1200-nF capacitor with circular parallel plates 1.0 cm in diameter is accumulating charge at the rate of 25.0 mC/s at some instant in time. What will be the induced magnetic field strength 10.0 cm radially outward from the center of the plates? What will be the value of the field strength after the capacitor is fully charged?

* 5. (II) Show that the displacement current through a parallel-plate capacitor can be written $I_D = C\,\Delta V/\Delta t$, where V is the voltage across the capacitor at any instant.

* 6. (III) The electric field between two parallel circular capacitor plates (capacitance C) changes at a rate $\Delta E/\Delta t$. (a) If the radius of the plates is R, show that the magnetic field B a distance r from the center of the plates, if $r \leq R$, is $B = \frac{1}{2}\mu_0\epsilon_0 r\ (\Delta E/\Delta t)$; and outside the plates, $r \geq R$, show that $B = (\mu_0\epsilon_0 R^2/2r)(\Delta E/\Delta t)$. (b) Show that the magnetic field beyond the edges of the capacitor plates is given by

$$B = \frac{\mu_0 I_D}{2\pi r},$$

where I_D is the displacement current. (c) This is the same formula for the field outside a straight wire; explain the similarity.

SECTION 22–4

* 7. (I) If the magnetic field in a traveling EM wave has a peak value of 17.5 nT, what is the peak value of the electric field strength?

* 8. (I) If the electric field in an EM wave has a peak of 0.43×10^{-4} V/m, what is the peak value of the magnetic field strength?

* 9. (I) In an EM wave traveling west, the B field oscillates vertically and has a frequency of 80.0 kHz and an rms strength of 9.75×10^{-9} T. What is the frequency and rms strength of the electric field and what is its direction?

SECTIONS 22–5 AND 22–6

10. (I) What is the wavelength of a 27.75×10^9 Hz radar signal?

11. (I) What is the frequency of an X-ray with wavelength 0.10 nm?

12. (I) What is the frequency of a microwave whose wavelength is 1.50 cm?

13. (I) An EM wave has a wavelength of 750 nm. What is its frequency, and what kind of light would we call it?

14. (I) An EM wave has frequency 9.56×10^{14} Hz. What is its wavelength, and how would we classify it?

15. (I) How long does it take light to reach us from the Sun, 1.50×10^8 km away?

16. (I) Our nearest star (other than the Sun) is 4.2 light-years away. That is, it takes 4.2 years for the light it emits to reach Earth. How far away is it in meters?

17. (I) A light-year is a measure of distance (not time). How many meters does light travel in a year?

18. (II) How long would it take a message sent as radio waves from Earth to reach (a) Mars, (b) a spacecraft near Saturn?

19. (II) What is the minimum angular speed at which Michelson's eight-sided mirror would have had to rotate in order that light would be reflected into an observer's eye by succeeding mirror faces (Fig. 22–12)?

20. (II) Who will hear the voice of a singer first—a person in the balcony 50 m away from the stage (Fig. 22–20), or a person 3000 km away at home whose ear is next to the radio? How much sooner? Assume that the microphone is a few centimeters from the singer and the temperature is 20°C.

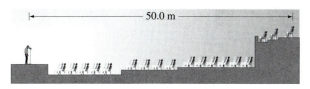

FIGURE 22–20 Problem 20.

21. (II) Pulsed lasers used for science and medicine produce very short bursts of electromagnetic energy. If the laser light wavelength is 1062 nm (this corresponds to a Neodymium-YAG laser), and the pulse lasts for 30 picoseconds, how many wavelengths are found within the laser pulse? How short would the pulse need to be to fit only one wavelength?

*SECTION 22–7

* 22. (I) The **E** field in an EM wave has a peak of 26.5 mV/m. What is the average rate at which this wave carries energy across unit area per unit time?

* 23. (II) The magnetic field in a traveling EM wave has an rms strength of 22.5 nT. How long does it take to deliver 135 J of energy to 1.00 cm^2 of a wall that it hits perpendicularly?

* 24. (II) How much energy is transported across a 1.00-cm^2 area per hour by an EM wave whose E field has an rms strength of 18.6 mV/m?

* 25. (II) A spherically spreading EM wave comes from a 1000-W source. At a distance of 10.0 m, what is the average power crossing unit area, and what is the rms value of the electric field?

* 26. (II) What is the energy contained in a 1.00-m^3 volume near the Earth's surface due to radiant energy from the Sun? See Example 22–3.

* 27. (II) A 12.8-mW laser puts out a narrow beam 2.00 mm in diameter. What are the average (rms) values of E and B in the beam?

* 28. (II) Estimate the average power output of the Sun, given that about 1350 W/m^2 reaches the upper atmosphere of the Earth.

* 29. (II) A high-energy pulsed laser emits a 1.0-ns-long pulse of average power 2.5×10^{11} W. The beam is 2.2×10^{-3} m in radius. Determine (a) the energy delivered in each pulse, and (b) the rms value of the electric field.

*SECTION 22–8

* 30. (I) The variable capacitor in the tuner of an AM radio has a capacitance of 2400 pF when the radio is tuned to a station at 550 kHz. What must the capacitance be for a station at the other end of the dial, 1550 kHz?

* 31. (I) The oscillator of a 96.1-MHz FM station has an inductance of 1.8 μH. What value must the capacitance be?

* 32. (II) A certain FM radio tuning circuit has a fixed capacitor $C = 840$ pF. Tuning is done by a variable inductance. What range of values must the inductance have to tune stations from 88 MHz to 108 MHz?

* 33. (II) An amateur radio operator wishes to build a receiver that can tune a range from 14.0 MHz to 15.0 MHz. A variable capacitor has a minimum capacitance of 92 pF. (a) What is the required value of the inductance? (b) What is the maximum capacitance used on the variable capacitor?

* 34. (II) A 1.40-m-long FM antenna is oriented parallel to the electric field of an EM wave. How large must the electric field be to produce a 1.00-mV (rms) voltage between the ends of the antenna? What is the rate of energy transport per square meter?

35. What is the wavelength of an AM station at 680 kHz?

36. An FM station broadcasts at 100.7 MHz. What is the wavelength of this wave?

37. Compare 940 on the AM dial to 94 on the FM dial. Which has the longer wavelength, and by what factor is it larger?

38. Television broadcast frequencies range between 54.0 MHz for Channel 2 up to 806 MHz for Channel 69. What are the wavelengths for these channels?

39. If the Sun were to disappear or somehow radically change its output, how long would it take for us on Earth to learn about it?

40. (a) How long did it take for a message sent from Earth to reach the first astronauts on the Moon? (b) How long will it take for a message from Earth to reach the first astronauts who arrive on Mars; assume Mars is at its closest approach to Earth (78×10^6 km)?

41. A radio voice signal from the Apollo crew on the Moon is beamed to a listening crowd from a radio speaker. If you are standing 50 m from the loudspeaker, what is the total time lag between when you hear the sound and when the sound left the Moon?

* 42. A point source emits light energy uniformly in all directions at an average rate P_0 with a single frequency f. Show that the peak electric field in the wave is given by

$$E_0 = \sqrt{\frac{\mu_0 c P_0}{2\pi r^2}}.$$

* 43. What are E_0 and B_0 2.00 m from a 100-W light source? Assume the bulb emits radiation of a single frequency uniformly in all directions.

* 44. Estimate the rms electric field in the sunlight that hits Mars, knowing that the Earth receives about 1350 W/m^2 and that Mars is 1.52 times farther away from the Sun (on average) than is the Earth.

* 45. At a given instant in time, a traveling EM wave is noted to have its maximum magnetic field pointing west and its maximum electric field pointing south. In which direction is the wave traveling? If the rate of energy flow is 500 W/m^2, what are the maximum values for the two fields?

* 46. Light is emitted from an ordinary lightbulb filament in wave-train bursts of about 10^{-8} s in duration. What is the length in space of such wave trains?

* 47. Suppose a 50-kW radio station emits EM waves uniformly in all directions. (a) How much energy per second crosses a 1.0-m^2 area 100 m from the transmitting antenna? (b) What is the rms magnitude of the **E** field at this point, assuming the station is operating at full power? (c) What is the voltage induced in a 1.0-m-long vertical car antenna?

* 48. Repeat Problem 47 for a distance of 100 km from the station.

* 49. Referring to Problem 47, what is the maximum power level of the radio station so as to avoid electrical breakdown of air at a distance of 1.0 m from the antenna? Assume the antenna is a point source. Air breaks down in an electric field of about 3×10^6 V/m. [*Hint*: See Problem 42.]

* 50. How large an emf (rms) will be generated in an antenna that consists of a 480-loop circular coil of wire 1.7 cm in diameter if the EM wave has a frequency of 810 kHz and is transporting energy at an average rate of 1.0×10^{-4} W/m^2 at the antenna? [*Hint*: You can use Eq. 21–5, since it could be applied to an observer moving with the coil so that the magnetic field is oscillating with the frequency $f = \omega/2\pi$.]

* 51. The variable capacitance of a radio tuner consists of six plates connected together placed alternately between six other plates, also connected together (Fig. 22–21). Each plate is separated from its neighbor by 1.1 mm of air. One set of plates can move so that the area of overlap varies from 1.0 cm^2 to 9.0 cm^2. (a) Are these capacitors connected in series or in parallel? (b) Determine the range of capacitance values. (c) What value of inductor is needed if the radio is to tune AM stations from 550 kHz to 1600 kHz?

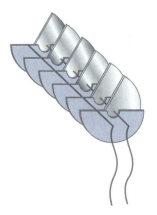

FIGURE 22–21 Problem 51.

Reflection from still water, as from a glass mirror, can be analyzed using the ray model of light.

Is this picture right side up? How can you tell? What are the clues? (Just for fun, this picture is, in fact, actually upside down. See Example 23–2.)

LIGHT: GEOMETRIC OPTICS

The sense of sight is extremely important to us, for it provides us with a large part of our information about the world. How do we see? What is the something called *light* that enters our eyes and causes the sensation of sight? How does light behave so that we can see everything that we do? The subject of light will occupy us for the next three chapters, and we will return to it in later chapters.

We see an object in one of two ways: (1) the object may be a source of light, such as a lightbulb, a flame, or a star, in which case we see the light emitted directly from the source; or (2), more commonly, we see an object by light reflected from it. In the latter case, the light may have originated from the sun, artificial lights, or a campfire. An understanding of how bodies *emit* light was not achieved until the 1920s, and this will be discussed in Chapter 27. How light is *reflected* from objects was understood much earlier, and we will discuss this in Section 23–2.

23–1 The Ray Model of Light

A great deal of evidence suggests that *light travels in straight lines* under a wide variety of circumstances. For example, a point source of light like the sun casts distinct shadows, and the beam of a flashlight appears to be a straight line. In fact, we infer the positions of objects in our environment by assuming that light moves from the object to our eyes in straight-line paths. Our whole orientation to the physical world is based on this assumption.

This reasonable assumption has led to the **ray model** of light. This model assumes that light travels in straight-line paths called light **rays**. Actually, a ray is an idealization; it is meant to represent an extremely narrow beam of light. When we see an object, according to the ray model, light reaches our eyes from each point on the object; although light rays leave each point in many different directions, normally only a small bundle of these rays can enter an observer's eye, as shown in Fig. 23–1. If the person's head moves to one side, a different bundle of rays will enter the eye from each point.

We saw in Chapter 22 that light can be considered as an electromagnetic wave. Although the ray model of light does not deal with this aspect of light (we discuss the wave nature of light in Chapter 24), the ray model has been very successful in describing many aspects of light such as reflection, refraction, and the formation of images by mirrors and lenses. Because these explanations involve straight-line rays at various angles, this subject is referred to as **geometric optics**.

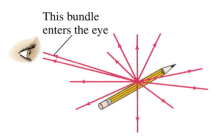

Light rays

This bundle enters the eye

FIGURE 23–1 Light rays come from each single point on an object. A small bundle of rays leaving one point is shown entering a person's eye.

23–2 Reflection; Image Formation by a Plane Mirror

When light strikes the surface of an object, some of the light is reflected. The rest is either absorbed by the object (and transformed to thermal energy) or, if the object is transparent like glass or water, part of it is transmitted through. For a very shiny object such as a silvered mirror, over 95 percent of the light may be reflected.

When a narrow beam of light strikes a flat surface (Fig. 23–2) we define the **angle of incidence**, θ_i, to be the angle an incident ray makes with the normal to the surface ("normal" means perpendicular) and the **angle of reflection**, θ_r, to be the angle the reflected ray makes with the normal. For flat surfaces, it is found that the *incident and reflected rays lie in the same plane with the normal to the surface*, and that

Angle of incidence and Angle of reflection (angle with the normal)

the angle of incidence equals the angle of reflection.

Law of reflection

This is the **law of reflection** and is indicated in Fig. 23–2 (most clearly in (b) which is a side view). It was known to the ancient Greeks, and you can confirm it yourself by shining a narrow flashlight beam at a mirror in a darkened room.

FIGURE 23–2 Law of reflection. (a) Shows an incident ray being reflected at the top of a flat surface; (b) shows a side or "end-on" view, which we will usually use because of its clarity.

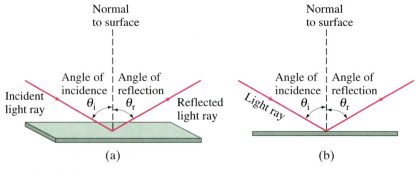

Normal to surface

Angle of incidence θ_i | Angle of reflection θ_r

Incident light ray

Reflected light ray

Normal to surface

Angle of incidence θ_i | Angle of reflection θ_r

Light ray

(a)　　　　　(b)

When light is incident upon a rough surface, even microscopically rough such as this page, it is reflected in many directions, Fig. 23–3. This is called **diffuse reflection**. The law of reflection still holds, however, at each small section of the surface. Because of diffuse reflection in all directions, an ordinary object can be seen from many different angles. When you move your head to the side, different reflected rays reach your eye from each point on the object (such as this page), Fig. 23–4a. Let us compare diffuse reflection to reflection from a mirror, which is known as *specular* reflection ("speculum" is Latin for mirror). When a narrow beam of light is shone on a mirror, the light will not reach your eye unless it is placed at just the right place where the law of reflection is satisfied, as shown in Fig. 23–4b. This is what gives rise to the unusual properties of mirrors. (Galileo, using similar arguments, showed that the Moon must have a rough surface rather than a highly polished surface like a mirror, as some people thought.)

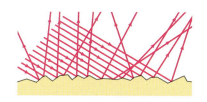

FIGURE 23–3 Diffuse reflection from a rough surface.

Mirrors (specular reflection)

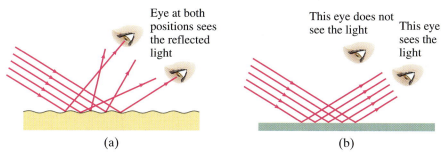

Eye at both positions sees the reflected light

This eye does not see the light

This eye sees the light

(a)

(b)

FIGURE 23–4 A beam of light from a flashlight shines on (a) white paper, and (b) a mirror. In part (a), you can see the white light reflected at various points because of diffuse reflection. But in part (b), you see the reflected light only when your eye is placed correctly ($\theta_r = \theta_i$); this is known as specular reflection.

When you look straight in a mirror, you see what appears to be yourself as well as various objects around and behind you, Fig. 23–5. Your face and the other objects look as if they are in front of you, beyond the mirror; but, of course, they are not. What you see in the mirror is an **image** of the objects.

Image

FIGURE 23–5 When you look in a mirror, you see an image of yourself and objects around you. Note that you don't see yourself as others see you, because left and right are reversed in the image.

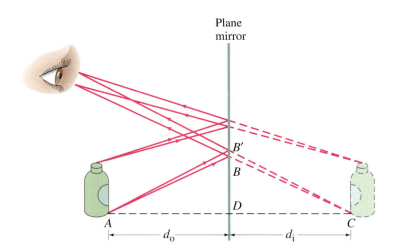

Plane
mirror

B'

B

D

A $\longleftarrow d_o \longrightarrow$ $\longleftarrow d_i \longrightarrow$ C

FIGURE 23–6 Formation of a virtual image by a plane mirror.

Ray model of how images are formed

Figure 23–6 shows how an image is formed by a plane mirror (that is, flat), according to the ray model. We are viewing the mirror, on edge, in the diagram of Fig. 23–6, and the rays are shown reflecting from the front surface. (Good mirrors are generally made by putting a highly reflective metallic coating on one surface of a very flat piece of glass.) Rays from two different points on an object are shown in Fig. 23–6: rays leaving from a point on the top of the bottle, and from a point on the bottom. Rays leave each point on the object going in many directions, but only those that enclose the bundle of rays that reach the eye from the two points are shown. The diverging rays that enter the eye *appear* to come from behind the mirror as shown by the dashed lines. That is, our eyes and brain interpret any rays that enter an eye as having traveled a straight-line path. The point from which each bundle of rays seems to come is one point on the image. For each point on the object, there is a corresponding image point. Let us concentrate on the two rays that leave point A on the object and strike the mirror at points B and B'. The angles ADB and CDB are right angles; and angles ABD and CBD are equal because of the law of reflection. Therefore, the two triangles ABD and CBD are congruent, and the length $AD = CD$. Thus the image appears as far behind the mirror as the object is in front: the **image distance**, d_i (distance from mirror to image, Fig. 23–6), equals the **object distance**, d_o. From the geometry, we also see that the height of the image is the same as that of the object.

The light rays do not actually pass through the image location itself. It merely *seems* as though the light is coming from the image because our brains interpret any light entering our eyes as having come in a straight line path from in front of us. Because the rays do not actually pass through the image, the image would not appear on paper or film placed at the location of the image. Therefore, it is called a **virtual image**. This is to distinguish it from a **real image** in which the light does pass through the image and which therefore could appear on paper or film placed at the image position. We will see that curved mirrors and lenses can form real images. A movie projector lens, for example, produces a real image that is visible on the screen.

Real and virtual images

| CONCEPTUAL EXAMPLE 23–1 | How tall must a full-length mirror be? |

A woman 1.60 m tall stands in front of a vertical plane mirror. What is the minimum height of the mirror, and how high must its lower edge be above the floor, if she is to be able to see her whole body? (Assume her eyes are 10 cm below the top of her head.)

RESPONSE The situation is diagrammed in Fig. 23–7. First consider the ray from her foot, AB, which upon reflection becomes BE and enters the eye E. The light from point A (her foot) enters the eye after reflecting at B; so the mirror needs to extend no lower than B. Because the angle of reflection equals the angle of incidence, the height BD is half of the height AE. Because $AE = 1.60\,\text{m} - 0.10\,\text{m} = 1.50\,\text{m}$, then $BD = 0.75\,\text{m}$. Similarly, if the woman is to see the top of her head, the top edge of the mirror only needs to reach point F, which is 5 cm below the top of her head (half of $GE = 10\,\text{cm}$). Thus, $DF = 1.55\,\text{m}$, and the mirror need have a vertical height of only $(1.55\,\text{m} - 0.75\,\text{m}) = 0.80\,\text{m}$. And the mirror's bottom edge must be 0.75 m above the floor. In general, a mirror need be only half as tall as a person for that person to see all of himself or herself. Does this result depend on the person's distance from the mirror? (The answer is no—try it and see, it's fun.)

➡ **PHYSICS APPLIED**

How tall a mirror is needed to see your whole reflection?

FIGURE 23–7 Seeing oneself in a mirror. Example 23–1.

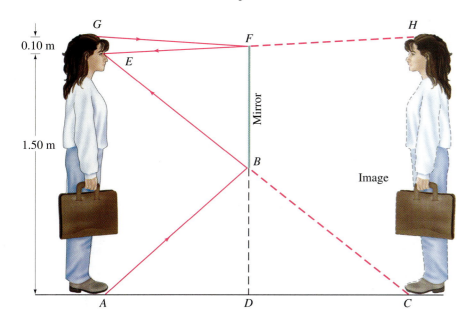

CONCEPTUAL EXAMPLE 23–2 **Is the photo upside down?** Close examination of the photograph on the first page of this chapter reveals that in the top portion, the image of the sun is seen clearly, whereas in the lower portion, the image of the sun is partially blocked by the tree branches. Why isn't the reflection in the water an exact replica of the the real scene? Illustrate your answer by drawing a sketch of this situation, showing the sun, the camera, the branch, and two rays going from the sun to the camera (one direct and one reflected). Does your sketch tell you if the photograph is right side up?

RESPONSE The caption on page 683 claims the photo is upside down. Let's try to confirm that, thus assuming that the sun blocked by the tree is actually the direct view, and the full view of the sun is the reflection. Thus, as diagrammed in Fig. 23–8, the ray which reflects off the water and into the camera travels at an angle below the branch, whereas the ray that travels directly to the camera passes through the branches. Try to draw a diagram assuming the photo is right side up (thus assuming that the image of the sun in the reflection is higher above the horizon than it is as viewed directly). It won't work.

Also, what about the people in the photo? Try to draw a diagram showing why they don't appear in the reflection. [*Hint:* assume they are not sitting on the edge of poolside, but back from the edge a bit.] Then try to draw a diagram of the reverse (i.e., assume the photo is right side up so the people are visible in the reflection but not directly).

In general, note that reflected images are not perfect replicas when different planes (distances) are involved.

FIGURE 23–8
Example 23–2

Branches

Direct ray

Camera or eye

Sun

Reflected ray

Water

23–3 Formation of Images by Spherical Mirrors

Reflecting surfaces do not have to be flat. The most common *curved* mirrors are *spherical*, which means they form a section of a sphere. A spherical mirror is called **convex** if the reflection takes place on the outer surface of the spherical shape so that the center of the mirror surface bulges out toward the viewer (Fig. 23–9a). A mirror is called **concave** if the reflecting surface is

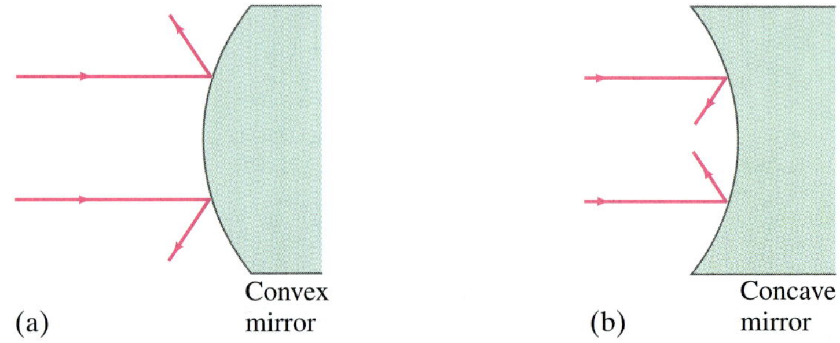

Convex and concave mirrors

FIGURE 23–9 Mirrors with convex and concave spherical surfaces.

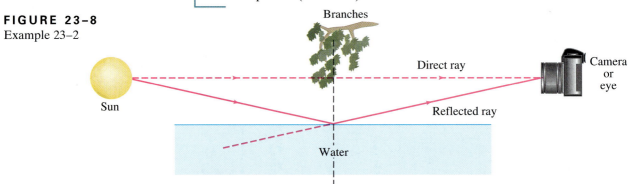

(a) Convex mirror

(b) Concave mirror

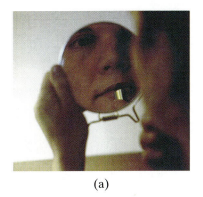

(a)

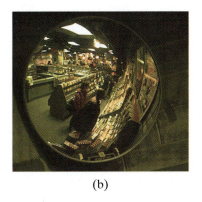

(b)

FIGURE 23–10 (a) A concave makeup mirror gives a magnified image. (b) A convex mirror in a store demagnifies and so includes a wide field of view.

on the inner surface of the sphere so that the center of the mirror sinks away from the viewer (like a "cave") (Fig. 23–9b). Concave mirrors are used as shaving or makeup mirrors (Fig. 23–10a), and convex mirrors are sometimes used on cars and trucks (rearview mirrors) and in shops (to watch for thieves), because they take in a wide field of view (Fig. 23–10b).

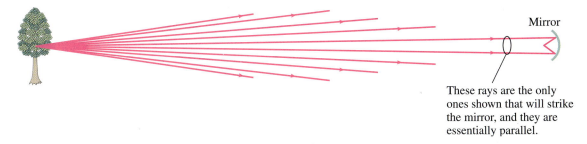

Mirror

These rays are the only ones shown that will strike the mirror, and they are essentially parallel.

FIGURE 23–11 If the object's distance is large compared to the size of the mirror (or lens), the rays are nearly parallel. They are parallel for an object at infinity (the symbol for infinity is ∞).

To see how spherical mirrors form images, we first consider an object that is very far from a concave mirror. For a distant object, as shown in Fig. 23–11, the rays from each point on the object that reach the mirror will be nearly parallel. *For an object infinitely far away* (the Sun and stars approach this), *the rays would be precisely parallel*. Now consider such parallel rays falling on a concave mirror as in Fig. 23–12. The law of reflection holds for each of these rays at the point each strikes the mirror. As can be seen, they are not all brought to a single point. In order to form a sharp image, the rays must come to a point. Thus a spherical mirror will

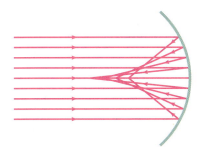

FIGURE 23–12 Parallel rays striking a concave spherical mirror do not focus at precisely a single point.

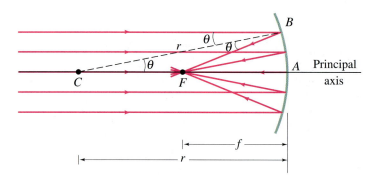

FIGURE 23–13 Rays parallel to the principal axis of a spherical mirror come to a focus at F, called the focal point, as long as the mirror is small in width as compared to its radius of curvature, r.

Small angle approximation

not make as sharp an image as a plane mirror will. However, if the mirror is small compared to its radius of curvature, so that the reflected rays make only a *small angle* upon reflection, then the rays will cross each other at very nearly a single point, or **focus**, as shown in Fig. 23–13. In the case shown, the rays are parallel to the **principal axis**, which is defined as the straight line perpendicular to the curved surface at its center (line CA in the diagram). The point F, where rays parallel to the principal axis come

Focal point
Focal length

to a focus, is called the **focal point** of the mirror. The distance between F and the center of the mirror, length FA, is called the **focal length**, f, of the mirror. Another way of defining the focal point is to say that it is the *image point for an object infinitely far away* along the principal axis. The image of the Sun, for example, would be at F.

Now we will show, for a mirror whose reflecting surface is small compared to the radius of curvature, that the rays do indeed meet at a common point, F, and we will also calculate the focal length f. We consider a ray that strikes the mirror at B in Fig. 23–13. The point C is the center of curvature of the mirror (the center of the sphere of which the mirror is a part). So the dashed line CB is equal to r, the radius of curvature, and CB is normal to the mirror's surface at B. The incoming ray that hits the mirror at B makes an angle θ with this normal, and hence the reflected ray, BF, also makes an angle θ with the normal. Note also from the geometry that angle BCF is also θ as shown. The triangle CBF is isosceles because two of its angles are equal. Thus length $CF = BF$. We assume the mirror has a width or diameter that is small compared to its radius of curvature, so the angles are small, and the length FB is nearly equal to length FA. In this approximation, $FA = FC$. But $FA = f$, the focal length, and $CA = 2\,FA = r$. Thus the focal length is half the radius of curvature:

Focal length of mirror

$$f = \frac{r}{2}. \qquad\qquad \textbf{(23–1)}$$

This argument assumed only that the angle θ was small, so the same result applies for all the other rays. And, we have also shown that all the rays pass through the same point F in this approximation of a mirror small compared to its radius of curvature.

We saw that, for an object at infinity, the image is located at the focal point of a concave spherical mirror, where $f = r/2$. But where does the image lie for an object not at infinity? First consider the object shown in Fig. 23–14, which is placed on point O between F and C. Let us determine where the image will be for a given point O' on the object. To do this, we can draw several rays and make sure these reflect from the mirror such that the reflection angle equals the incidence angle. This can involve much work and our task is simplified if we deal with three rays that are particularly easy to draw. These are the rays labeled 1, 2, and 3 in Fig. 23–14 and we draw them for object point O' as follows:

Ray 1 is drawn parallel to the axis; therefore it must pass along a line through F after reflection (as we see in Fig. 23–13; see also Fig. 23–14a).

Ray 2 is drawn through F; therefore it must reflect so it is parallel to the axis (Fig. 23–14b).

Ray 3 is chosen to be perpendicular to the mirror, and so is drawn so that it passes through C, the center of curvature; it is along a radius of the spherical surface, and because it is perpendicular to the mirror it will be reflected back on itself (Fig. 23–14c).

The point at which these rays cross is the image point I'. All other rays from the same object point will pass through this image point. To find the image point for any object point, only these three particular rays need to be used. Actually, only two of these rays are needed, but the third serves as a check.

We have shown the image point in Fig. 23–14 only for a single point on the object. Other points on the object are imaged nearby, so a complete image of the object is formed, as shown by the dashed outline. Because the light actually passes through the image itself, this is a real image that will appear on a piece of paper or film placed there. This can be compared to the virtual image formed by a plane mirror (the light does not actually pass through that image, Fig. 23–6).

The image in Fig. 23–14 can be seen by the eye when the eye is placed to the left of the image so that some of the rays diverging from each point on the image (as point I') can enter the eye as shown in part (c). (See also Figs. 23–1 and 23–6.)

Real image

FIGURE 23–14 Rays leave point O' on the object (an arrow). Shown are the three most useful rays for determining where the image I' is formed. [Note that our mirror's height is not small compared to f, so our diagram will not give the precise position of the image.]

(a) Ray 1 goes out from O' parallel to the axis and reflects through F.

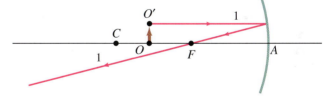

(b) Ray 2 goes through F and then reflects back parallel to the axis.

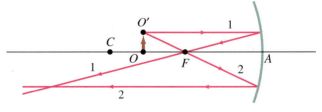

(c) Ray 3 heads out perpendicular to mirror and then reflects back on itself and goes through C (center of curvature).

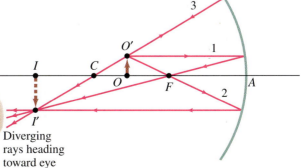

Diverging rays heading toward eye

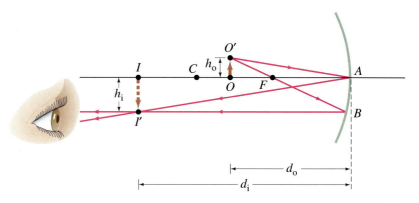

FIGURE 23–15 Diagram for deriving the mirror equation. For the derivation, we assume the mirror size is small compared to its radius of curvature.

Image points can always be determined (at least roughly) by drawing the three rays as described above, although high accuracy is hard to obtain. For one thing it is difficult to draw, maintaining small angles for the rays as we assumed (mirror small compared to f so all rays come to a focus). However, it is possible to derive an equation that gives the image distance if the object distance and radius of curvature of the mirror are known. To do this, we refer to Fig. 23–15. The distance of the object from the center of the mirror, called the **object distance**, is labeled d_o. The **image distance** is labeled d_i. The height of the object OO' is called h_o and the height of the image, $I'I$, is h_i. Two rays are shown, $O'FBI'$ (same as ray 2 in the previous figure) and $O'AI'$. The ray $O'AI'$ obeys the law of reflection, of course, so the two right triangles $O'AO$ and $I'AI$ are similar. Therefore, we have

$$\frac{h_o}{h_i} = \frac{d_o}{d_i}.$$

For the other ray, $O'FBI'$, the triangles $O'FO$ and AFB are also similar since the length $AB = h_i$ (in our approximation of a mirror that is small compared to its radius) and $FA = f$, the focal length of the mirror. Therefore,

$$\frac{h_o}{h_i} = \frac{OF}{FA} = \frac{d_o - f}{f}.$$

The left sides of the two preceding expressions are the same, so we can equate the right sides:

$$\frac{d_o}{d_i} = \frac{d_o - f}{f}.$$

We now divide both sides by d_o and rearrange to obtain

Mirror equation

$$\frac{1}{d_o} + \frac{1}{d_i} = \frac{1}{f}. \qquad \textbf{(23–2)}$$

This is the equation we were seeking. It is called the **mirror equation** and relates the object and image distances to the focal length f (where $f = r/2$).

The **lateral magnification**, m, of a mirror is defined as the height of the image divided by the height of the object. From our first set of similar triangles above, we can write:

Lateral magnification for curved mirror

$$m = \frac{h_i}{h_o} = -\frac{d_i}{d_o}, \qquad \textbf{(23–3)}$$

where the minus sign is inserted as a convention.

For consistency, we must be careful about the signs of all quantities in Eqs. 23–2 and 23–3. The sign conventions we use are: the image height h_i is positive if the image is upright, and negative if inverted, relative to the object (h_o is always taken as positive); d_i and d_o are positive if image and object are on the reflecting side of the mirror (as in Fig. 23–15), but if either image or object is behind the mirror, the corresponding distance is negative (an example can be seen in Fig. 23–16, Example 23–5). Thus the magnification (Eq. 23–3) is positive for an upright image and negative for an inverted image. We will summarize sign conventions more fully after discussing convex mirrors later in this Section.

EXAMPLE 23–3 **Image in a concave mirror.** A 1.50-cm-high diamond ring is placed 20.0 cm from a concave mirror whose radius of curvature is 30.0 cm. Determine (a) the position of the image, and (b) its size.

SOLUTION The focal length $f = r/2 = 15.0$ cm. The ray diagram is basically like that shown in Fig. 23–14 and Fig. 23–15, since the object is between F and C; a precise ray diagram would have (referring to either Fig. 23–14 or Fig. 23–15): $CA = 30.0$ cm, $FA = 15.0$ cm, and OA (object distance) = 20.0 cm. (a) Since $d_o = 20.0$ cm, we have from Eq. 23–2 that

$$\frac{1}{d_i} = \frac{1}{f} - \frac{1}{d_o} = \frac{1}{15.0 \text{ cm}} - \frac{1}{20.0 \text{ cm}} = 0.0167 \text{ cm}^{-1}.$$

So $d_i = 1/0.0167 \text{ cm}^{-1} = 60.0$ cm. The image is 60.0 cm from the mirror on the same side as the object.
(b) From Eq. 23–3, the lateral magnification is

$$m = -\frac{60.0 \text{ cm}}{20.0 \text{ cm}} = -3.00.$$

Therefore the image height is

$$h_i = mh_o = (-3.00)(1.5 \text{ cm}) = -4.5 \text{ cm}.$$

The minus sign tells us the image is inverted, as it is in Figs. 23–14 and 23–15.

➡ **PROBLEM SOLVING**

Remember to take the reciprocal

CONCEPTUAL EXAMPLE 23–4 **Reversible rays.** If the object in Example 23–3 is placed instead where the image is in Fig. 23–15, where will the new image be?

RESPONSE The mirror equation is symmetric in d_o and d_i. Thus the new image will be where the old object was. Indeed, in Fig. 23–15 we need only reverse the direction of the rays to get our new situation.

EXAMPLE 23–5 **Object closer to concave mirror.** A 1.00-cm-high object is placed 10.0 cm from a concave mirror whose radius of curvature is 30.0 cm. (a) Draw a ray diagram to locate (approximately) the position of the image. (b) Determine the position of the image and the magnification analytically.

SOLUTION (a) Since $f = r/2 = 15.0$ cm, the object is between the mirror and the focal point. We draw the three rays as described earlier (Fig. 23–14) and this is shown in Fig. 23–16. Ray 1 leaves the tip of our object heading toward the mirror parallel to the axis, and reflects through F. Ray 2 cannot head toward F because it would not strike the mirror; so ray 2 must head as if it started at F (dashed line) and heads to the mirror and then is reflected parallel to the principal axis. Ray 3 is perpendicular to the mirror, as before. The rays reflected from the mirror diverge and so never meet at a point. They appear, however, to be coming from a point behind the mirror. This point is the image which is thus behind the mirror and *virtual*. (Why?)

(b) We use Eq. 23–2 to find d_i when $d_o = 10.0$ cm:

$$\frac{1}{d_i} = \frac{1}{15.0 \text{ cm}} - \frac{1}{10.0 \text{ cm}} = \frac{2 - 3}{30.0 \text{ cm}} = -\frac{1}{30.0 \text{ cm}}.$$

Therefore, $d_i = -30.0$ cm. The minus sign means the image is behind the mirror. The lateral magnification is $m = -d_i/d_o = -(-30.0 \text{ cm})/(10.0 \text{ cm}) = +3.00$. So the image is 3.00 times larger than the object; the plus sign indicates that the image is upright (which is consistent with the ray diagram, Fig. 23–16).

FIGURE 23–16 Object placed within the focal point F. The image is *behind* the mirror and is *virtual*, Example 23–5. [Note that the vertical scale (height of object = 1.0 cm) is different from the horizontal ($OA = 10.0$ cm) for ease of drawing, and this will affect the precision of the drawing.]

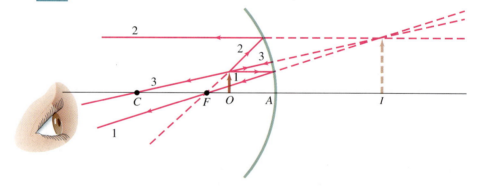

These Examples show that a spherical mirror can produce a magnified image, one that is larger than the object. (There is an old story—maybe fable is a better word—that Julius Caesar spied on the British forces by setting up a curved mirror on the coast of France. Is this reasonable?)

It is useful to compare Figs. 23–14 and 23–16. We can see that if the object is within the focal point, as in Fig. 23–16, the image is virtual, upright, and magnified. This is how a shaving or makeup mirror is used—you must place your head within the focal point if you are to see yourself right side up (Fig. 23–10a). If the object is *beyond* the focal point, as in Fig. 23–14, the image is real and inverted (upside down—and hard to use!). Whether the magnification is greater or less than 1.0 in the latter case depends on the position of the object relative to the center of curvature, point C.

➡ PHYSICS APPLIED

Shaving/makeup mirror

Analysis for convex mirrors

The analysis used for concave mirrors can be applied to **convex** mirrors. Even the mirror equation (Eq. 23–2) holds for a convex mirror, although the quantities involved must be carefully defined. Figure 23–17a shows parallel rays falling on a convex mirror. Again spherical aberration

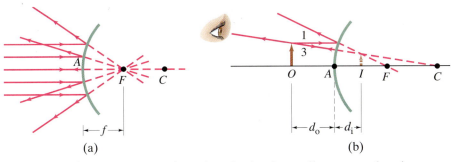

FIGURE 23–17 Convex mirror: (a) the focal point is at F, behind the mirror; (b) the image I of the object at O is virtual, upright, and smaller than the object. [Not to scale for Example 23–6.]

will be present, but we assume the mirror's size is small compared to its radius of curvature. The reflected rays diverge, but seem to come from point F behind the mirror. This is the **focal point**, and its distance from the center of the mirror is the **focal length**, f. It is easy to show that again $f = r/2$. We see that an object at infinity produces a virtual image in a convex mirror. Indeed, no matter where the object is placed on the reflecting side of a convex mirror, the image will be virtual and erect, as indicated in Fig. 23–17b. To find the image we draw rays 1 and 3 according to the rules used before on the concave mirror, as shown in Fig. 23–17b.

The mirror equation, Eq. 23–2, holds for convex mirrors but the focal length f must be considered negative, as must the radius of curvature. The proof is left as a problem. It is also left as a problem to show that Eq. 23–3 for the magnification is also valid.

EXAMPLE 23–6 **Convex rearview mirror.** A convex rearview car mirror has a radius of curvature of 40.0 cm. Determine the location of the image and its magnification for an object 10.0 m from the mirror.

➡ **PHYSICS APPLIED**

Convex rear-view mirror

SOLUTION The ray diagram will be like Fig. 23–17, but the large object distance ($d_o = 10.0$ m) makes a precise drawing difficult; Fig. 23–17 is not to scale. We have a convex mirror, so r is negative by convention. Specifically, $r = -40.0$ cm, so $f = -20.0$ cm $= -0.200$ m. The mirror equation gives

$$\frac{1}{d_i} = \frac{1}{f} - \frac{1}{d_o} = -\frac{1}{0.200 \text{ m}} - \frac{1}{10.0 \text{ m}} = -\frac{51.0}{10.0 \text{ m}}.$$

So $d_i = -10.0$ m$/51.0 = -0.196$ m, or 19.6 cm behind the mirror. The lateral magnification is $m = -d_i/d_o = -(-0.196 \text{ m})/(10.0 \text{ m}) = 0.0196$ or $1/51$. So the upright image is smaller than the object by a factor of 51.

➡ **PROBLEM SOLVING** **Spherical Mirrors**

1. Always draw a ray diagram even though you are going to make an analytic calculation—the diagram serves as a check, even if not precise. Draw at least two, and preferably three, of the easy-to-draw rays as described in Fig. 23–14. Generally draw the rays starting on a point on the object to the left of the mirror and moving to the right.

2. Use Eqs. 23–2 and 23–3; it is crucially important to follow the sign conventions.

3. **Sign Conventions**
   (a) When the object, image, or focal point is on the reflecting side of the mirror (on the left in all our drawings), the corresponding distance is considered positive. If any of these points is behind the mirror (on the right) the corresponding distance is negative.[†]
   (b) The image height h_i is positive if the image is upright, and negative if inverted, relative to the object (h_o is always taken as positive).

[†]Object distances are positive for material objects, but can be negative in systems with more than one mirror or lens—see Section 23–10.

23–4 Index of Refraction

We saw in Chapter 22 that the speed of light in vacuum is

$$c = 2.99792458 \times 10^8 \text{ m/s},$$

which we usually round off to

$$c = 3.00 \times 10^8 \text{ m/s}.$$

This speed applies to all electromagnetic waves, including visible light.

In air the speed is only very slightly less. In other transparent materials, such as glass and water, the speed is always less than that in vacuum. For example, in water light travels at about $\frac{3}{4}c$. The ratio of the speed of light in vacuum to the speed v in a given material is called the **index of refraction**, n, of that material:

Index of refraction

$$n = \frac{c}{v}. \tag{23–4}$$

The index of refraction is never less than 1 (that is, $n \geq 1$), and its value for various materials is given in Table 23–1. (As we shall see later, n varies somewhat with the wavelength of the light—except in vacuum—so a particular wavelength is specified, that of yellow light with wavelength $\lambda = 589$ nm.)

TABLE 23–1
Indices of Refraction[†]

Medium	$n = c/v$
Vacuum	1.0000
Air (at STP)	1.0003
Water	1.33
Ethyl alcohol	1.36
Glass	
Fused quartz	1.46
Crown glass	1.52
Light flint	1.58
Lucite or Plexiglas	1.51
Sodium chloride	1.53
Diamond	2.42

[†]$\lambda = 589$ nm

EXAMPLE 23–7 Light's speed in diamond. Calculate the speed of light in diamond.

SOLUTION From Table 23–1, $n = 2.42$ for diamond, so the speed of light traveling inside a diamond is

$$v = \frac{c}{n} = \frac{c}{2.42} = 0.413c$$

or

$$v = \frac{3.00 \times 10^8 \text{ m/s}}{2.42} = 1.24 \times 10^8 \text{ m/s}.$$

That light travels more slowly in matter than in vacuum can be explained at the atomic level as being due to the absorption and reemission of light by atoms and molecules of the material.

23–5 Refraction: Snell's Law

When light passes from one medium into another, part of the incident light is reflected at the boundary. The remainder passes into the new medium. If a ray of light is incident at an angle to the surface (other than perpendicular), the ray is bent as it enters the new medium. This bending is called **refraction**. Figure 23–18a shows a ray passing from air into water. The angle θ_1 is the angle of incidence and θ_2 is the **angle of refraction**. Notice that the ray bends toward the normal when entering the water. This is always the case when the ray enters a medium where the speed of light is *less*. If light travels from one

Angle of refraction

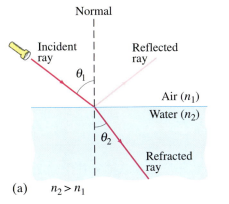

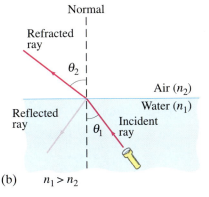

FIGURE 23–18 Refraction.

(a) $n_2 > n_1$ (b) $n_1 > n_2$

medium into a second where its speed is *greater*, the ray bends away from the normal; this is shown in Fig. 23–18b for a ray traveling from water to air.

Refraction is responsible for a number of common optical illusions. For example, a person standing in waist-deep water appears to have shortened legs. As shown in Fig. 23–19, the rays leaving the person's foot are bent at the surface. The observer's eye (and brain) assumes the rays to have traveled a straight-line path, and so the feet appear to be higher than they really are. Similarly, when you put a pencil in water, it appears to be bent (Fig. 23–20).

The angle of refraction depends on the speed of light in the two media and on the incident angle. An analytical relation between θ_1 and θ_2 was arrived at experimentally about 1621 by Willebrord Snell (1591–1626). It is known as **Snell's law** and is written:

$$n_1 \sin \theta_1 = n_2 \sin \theta_2; \qquad \textbf{(23–5)}$$

θ_1 is the angle of incidence and θ_2 is the angle of refraction (both measured with respect to a line perpendicular to the surface between the two media, as shown in Fig. 23–18); n_1 and n_2 are the respective indices of refraction in the materials. The incident and refracted rays lie in the same plane, which also includes the perpendicular to the surface. Snell's law is the basic **law of refraction**.

It is clear from Snell's law that if $n_2 > n_1$, then $\theta_2 < \theta_1$; that is, if light enters a medium where n is greater (and its speed less), then the ray is bent toward the normal. And if $n_2 < n_1$, then $\theta_2 > \theta_1$, so the ray bends away from the normal. This is what we saw in Fig. 23–18.

➡ **PHYSICS APPLIED**

Optical illusions

Snell's law
(law of refraction)

FIGURE 23–19 Ray diagram showing why a person's legs look shorter when standing in waist-deep water: the path of light traveling from the bather's foot to the observer's eye bends at the water's surface, and our brain interprets the light as having traveled in a straight line, from higher up (dashed line).

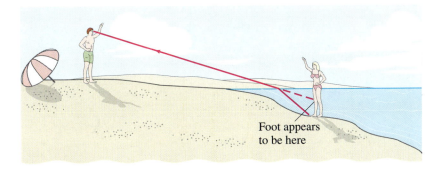

Foot appears to be here

FIGURE 23–20 A pencil in water looks bent even when it isn't.

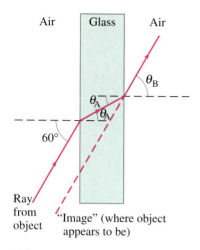

Air   Glass   Air

θ_B

θ_A

θ_A

60°

Ray from object

"Image" (where object appears to be)

FIGURE 23–21 Light passing through a piece of glass (Example 23–8).

EXAMPLE 23–8 **Refraction through flat glass.** Light strikes a flat piece of glass at an incident angle of 60°, as shown in Fig. 23–21. If the index of refraction of the glass is 1.50, (*a*) what is the angle of refraction θ_A in the glass; (*b*) what is the angle θ_B at which the ray emerges from the glass?

SOLUTION (*a*) We assume the incident ray is in air, so $n_1 = 1.00$ and $n_2 = 1.50$. Then, from Eq. 23–5 we have

$$\sin \theta_A = \frac{1.00}{1.50} \sin 60° = 0.577,$$

so $\theta_A = 35.2°$.
(*b*) Since the faces of the glass are parallel, the incident angle in this case is just θ_A, so $\sin \theta_A = 0.577$. This time $n_1 = 1.50$ and $n_2 = 1.00$. Thus, $\theta_B (= \theta_2)$ is

$$\sin \theta_B = \frac{1.50}{1.00} \sin \theta_A = 0.866,$$

and $\theta_B = 60.0°$. The direction of the beam is thus unchanged by passing through a flat piece of glass. It should be clear that this works for any angle of incidence. The ray is displaced slightly to one side, however. You can observe this by looking through a piece of glass (near its edge) at some object and then moving your head to the side so that you see the object directly.

FIGURE 23–22 Example 23–9.

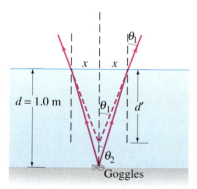

θ_1

x | x

$d = 1.0$ m

θ_1

d'

θ_2

Goggles

EXAMPLE 23–9 **Apparent depth of a pool.** A swimmer has dropped her goggles in the shallow end of a pool, marked as 1.0 m deep. But the goggles don't look that deep. Why? How deep do the goggles appear to be when you look straight down into the water?

SOLUTION The ray diagram of Fig. 23–22 shows why the water seems less deep than it actually is. Rays traveling upward from the goggles on the bottom are refracted *away* from the normal as they exit the water. The rays appear to be diverging from a point higher in the water. To calculate the apparent depth d', given a real depth $d = 1.0$ m, we use Snell's law with $n_1 = 1$ for air and $n_2 = 1.33$ for water:

$$\sin \theta_1 = n_2 \sin \theta_2.$$

We are considering only small angles, so $\sin \theta \approx \tan \theta \approx \theta$, with θ in radians. So Snell's law becomes

$$\theta_1 = n_2 \theta_2.$$

From Fig. 23–22, we see that

$$\theta_1 = \tan \theta_1 = \frac{x}{d'} \qquad \text{and} \qquad \theta_2 = \tan \theta_2 = \frac{x}{d}.$$

Putting these into Snell's law, $\theta_1 = n_2 \theta_2$, we get

$$\frac{x}{d'} = n_2 \frac{x}{d}$$

or

$$d' = \frac{d}{n_2} = \frac{1.0 \text{ m}}{1.33} = 0.75 \text{ m}.$$

23–6 Total Internal Reflection; Fiber Optics

When light passes from one material into a second material where the index of refraction is less (say, from water into air), the light bends away from the normal, as for ray J in Fig. 23–23. At a particular incident angle, the angle of refraction will be 90°, and the refracted ray would skim the surface (ray K) in this case. The incident angle at which this occurs is called the **critical angle**, θ_C. From Snell's law, θ_C is given by

$$\sin \theta_C = \frac{n_2}{n_1} \sin 90° = \frac{n_2}{n_1}. \qquad \textbf{(23–6)} \qquad \textit{Critical angle}$$

For any incident angle less than θ_C there will be a refracted ray, although part of the light will also be reflected at the boundary. However, for incident angles greater than θ_C, Snell's law would tell us that $\sin \theta_2$ is greater than 1.00. Yet the sine of an angle can never be greater than 1.00. In this case there is no refracted ray at all, and *all of the light is reflected*, as for ray L in Fig. 23–23. This effect is called **total internal reflection**. But note that total internal reflection can occur only when light strikes a boundary where the medium beyond has a lower index of refraction.

Total internal reflection

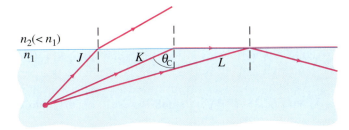

FIGURE 23–23 Since $n_2 < n_1$, light rays are totally internally reflected if $\theta > \theta_C$ as for ray L. If $\theta < \theta_C$, as for ray J, only a part of the light is reflected (this part is not shown), and the rest is refracted.

CONCEPTUAL EXAMPLE 23–10 | **View up from under water.** Describe what a person would see who looked up at the world from beneath the perfectly smooth surface of a lake or swimming pool.

RESPONSE   For an air–water interface, the critical angle is given by

$$\sin \theta_C = \frac{1.00}{1.33} = 0.750.$$

Therefore, $\theta_C = 49°$. Thus the person would see the outside world compressed into a circle whose edge makes a 49° angle with the vertical. Beyond this angle, the person would see reflections from the sides and bottom of the pool or lake (Fig. 23–24).

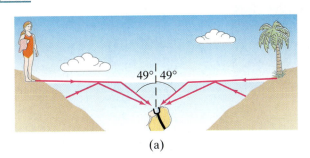

(a)

(b)

FIGURE 23–24 View looking upward from beneath the water (the surface of the water must be very smooth).

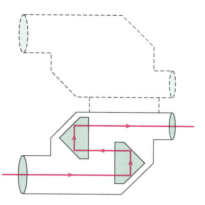

FIGURE 23–25 Total internal reflection of light by prisms in binoculars.

Many optical instruments, such as binoculars, use total internal reflection within a prism to reflect light. The advantage is that very nearly 100 percent of the light is reflected, whereas even the best mirrors reflect somewhat less than 100 percent. Thus the image is brighter. For glass with $n = 1.50$, $\theta_C = 41.8°$. Therefore, 45° prisms will reflect all the light internally, if oriented as shown in the binoculars of Fig. 23–25.

Total internal reflection is the principle behind **fiber optics**. Glass and plastic fibers as thin as a few micrometers in diameter can now be made. A bundle of such tiny fibers is called a **light pipe** or cable, and light can be transmitted along it with almost no loss because of total internal reflection. Figure 23–26 shows how light traveling down a thin fiber makes only glancing collisions with the walls so that total internal reflection occurs. Even if the light pipe is bent into a complicated shape, the critical angle still won't be exceeded, so light is transmitted practically undiminished to the other end (see Fig. 23–27). Very small losses do occur, mainly by reflection at the ends and absorption within the fiber.

Fiber optics

FIGURE 23–26 Light reflected totally at the interior surface of a glass or transparent plastic fiber.

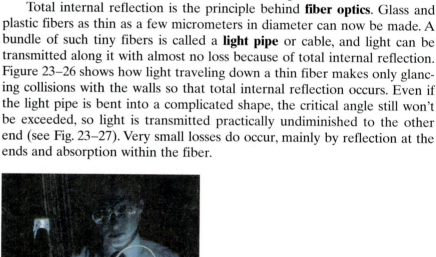

FIGURE 23–27 Total internal reflection within the tiny fibers of this light pipe makes it possible to transmit light in complex paths with minimal loss.

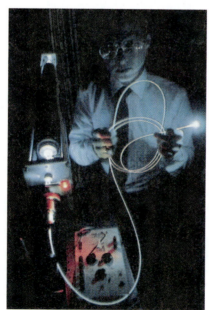

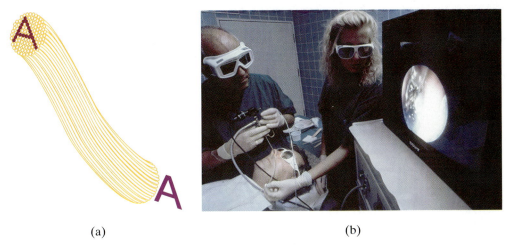

(a)   (b)

FIGURE 23-28 (a) How a fiber-optic image is made. (b) Example of a fiber-optic endoscope inserted through the nose, and the image seen.

Important applications of fiber-optic cables are in telecommunications and medicine. They are being used to transmit telephone calls, video signals, and computer data. The signal is a modulated light beam (a light beam whose intensity can be varied) and is transmitted at a much higher rate and with less loss and less interference than an electrical signal in a copper wire. The sophisticated use of fiber optics to transmit a clear picture, is particularly useful in medicine, Fig. 23–28. For example, a patient's lungs can be examined by inserting a light pipe known as a bronchoscope through the mouth and down the bronchial tube. Light is sent down an outer set of fibers to illuminate the lungs. The reflected light returns up a central core set of fibers. Light directly in front of each fiber travels up that fiber. At the opposite end, a viewer sees a series of bright and dark spots, much like a TV screen—that is, a picture of what lies at the opposite end.[†] The image may be viewed directly or on a TV monitor or film. The fibers must be optically insulated from one another, usually by a thin coating of material whose refractive index is less than that of the fiber. The fibers must be arranged precisely parallel to one another if the picture is to be clear. The more fibers there are, and the smaller they are, the more detailed the picture. Such instruments, including bronchoscopes, colonoscopes (for viewing the colon) and endoscopes (stomach or other organs) are extremely useful for observing hard-to-reach places for surgery or searching for lesions without surgery. The instrument may also be equipped with a mechanism for pinching off tissue samples for biopsies or removal.

➡ PHYSICS APPLIED
Fiber optics in telecommunications

➡ PHYSICS APPLIED
Medicine—endoscopes

[†]Lenses are used at each end: at the object end to bring the rays in parallel, and at the viewing end as a telescope.

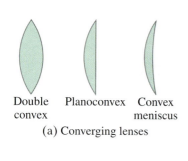

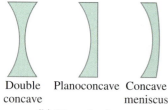

Double convex Planoconvex Convex meniscus

(a) Converging lenses

Double concave Planoconcave Concave meniscus

(b) Diverging lenses

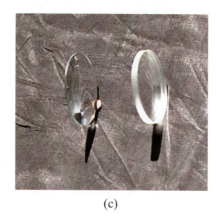

(c)

(d)

FIGURE 23–29
(a) Converging and (b) diverging lenses, shown in cross section. (c) Photo of a converging lens (on the left) and a diverging lens. (d) Converging lenses (above), and diverging lenses, lying flat, and raised off the paper to form images.

23–7 Thin Lenses; Ray Tracing

The most important simple optical device is no doubt the thin lens. The development of optical devices using lenses dates to the sixteenth and seventeenth centuries, although the earliest record of eyeglasses dates from the late thirteenth century.[†] Today we find lenses in eyeglasses, cameras, magnifying glasses, telescopes, binoculars, microscopes, and medical instruments. A thin lens is usually circular, and its two faces are portions of a sphere. (Although cylindrical surfaces are also possible, we will concentrate on spherical.) The two faces can be concave, convex, or plane; several types are shown in Fig. 23–29, in cross section. The importance of lenses is that they form images of objects, as shown in Fig. 23–30.

[†]Rounded gemstones used as magnifiers probably date from much earlier.

FIGURE 23–30
Converging lens (in holder) forms an image (large "F" on screen at right) of a bright object (illuminated "F" at the left).

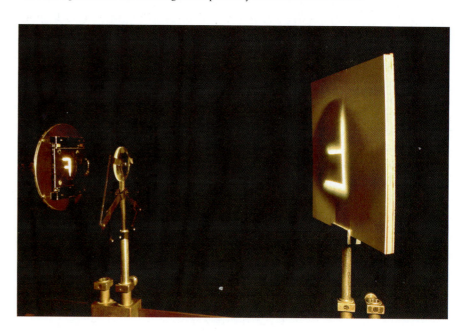

Consider the rays parallel to the axis of the double convex lens which is shown in cross section in Fig. 23–31a. We assume the lens is made of glass or transparent plastic, so its index of refraction is greater than that of the air outside. The **axis** of a lens is a straight line passing through the very center of the lens and perpendicular to its two surfaces (Fig. 23–31). From Snell's law, we can see that each ray in Fig. 23–31a is bent toward the axis at both lens surfaces (note the dashed lines indicating the normals to each surface for the top ray). If rays parallel to the axis fall on a thin lens, they will be focused to a point called the **focal point**, F. This will not be precisely true for a lens with spherical surfaces. But it will be very nearly true—that is, parallel rays will be focused to a tiny region that is nearly a point—if the diameter of the lens is small compared to the radii of curvature of the two lens surfaces. This criterion is satisfied by a **thin lens**, one that is very thin compared to its diameter, and we consider only thin lenses here.

The rays from a point on a distant object are essentially parallel—see Fig. 23–11. Therefore we can also say that *the focal point is the image point for an object at infinity on the principal axis*. Thus, the focal point of a lens can be found by locating the point where the Sun's rays (or those of some other distant object) are brought to a sharp image, Fig. 23–32. The distance of the focal point from the center of the lens is called the **focal length**, f. A lens can be turned around so that light can pass through it from the opposite side. The focal length is the same on both sides, as we shall see later, even if the curvatures of the two lens surfaces are different. If parallel rays fall on a lens at an angle, as in Fig. 23–31b, they focus at a point F_a. The plane in which all points such as F and F_a fall is called the **focal plane** of the lens.

Any lens[†] that is thicker in the center than at the edges will make parallel rays converge to a point, and is called a **converging lens** (see Fig. 23–29a). Lenses that are thinner in the center than at the edges (Fig. 23–29b) are called **diverging lenses** because they make parallel light diverge, as shown in Fig. 23–33. The focal point, F, of a diverging lens is defined as that point from which refracted rays, originating from parallel incident rays, seem to emerge as shown in the figure. And the distance from F to the lens is called the **focal length**, just as for a converging lens.

Optometrists and ophthalmologists, instead of using the focal length, use the reciprocal of the focal length to specify the strength of eyeglass (or contact) lenses. This is called the **power**, P, of a lens:

$$P = \frac{1}{f}. \tag{23-7}$$

The unit for lens power[‡] is the diopter (D), which is an inverse meter: $1\,\text{D} = 1\,\text{m}^{-1}$. For example, a 20-cm-focal-length lens has a power $P = 1/0.20\,\text{m} = 5.0\,\text{D}$. We will mainly use the focal length here, but we will refer again to the power of a lens when we discuss eyeglass lenses in Chapter 25.

[†]We are assuming the lens has an index of refraction greater than that of the surrounding material, such as a glass or plastic lens in air, which is the usual situation.

[‡]Note that lens power has nothing to do with power as the rate of doing work or transforming energy (Section 6–10).

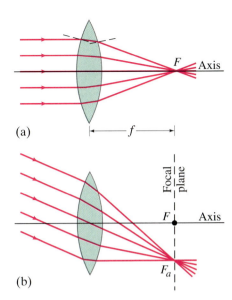

(a)

(b)

FIGURE 23–31 Parallel rays are brought to a focus by a converging thin lens.

Focal length of lens

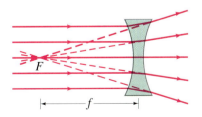

FIGURE 23–32 Image of the Sun burning a hole, almost, on a piece of paper.

Power of lens

FIGURE 23–33 Diverging lens.

The most important parameter of a lens is its focal length f. For a converging lens, f is easily measured by finding the image point for the Sun or other distant objects. Once f is known, the image position can be found for any object. To find the image point by drawing rays would be difficult if we had to determine all the refractive angles. Instead, we can do it very simply by making use of certain facts we already know, such as that a ray parallel to the axis of the lens passes (after refraction) through the focal point. In fact, to find an image point, we need consider only the three rays indicated in Fig. 23–34, which shows an arrow as the object and a converging lens forming an image to the right. These rays, emanating from a single point on the object, are drawn as if the lens were infinitely thin, and we show only a single sharp bend within the lens instead of the refractions at each surface. These three rays are drawn as follows:

➡ **RAY DIAGRAM**

Finding the image position formed by a thin lens

Ray 1 is drawn parallel to the axis; therefore it is refracted by the lens so that it passes along a line through the focal point F, Fig. 23–34a. (See also Fig. 23–31a.)

Ray 2 is drawn on a line passing through the other focal point F' (front side of lens in Fig. 23–34) and emerges from the lens parallel to the axis, Fig. 23–34b.

Ray 3 is directed toward the very center of the lens, where the two surfaces are essentially parallel to each other; this ray therefore emerges from the lens at the same angle as it entered; as we saw in Example 23–8, the ray would be displaced slightly to one side, but since we assume the lens is thin, we draw ray 3 straight through as shown.

FIGURE 23–34 Finding the image by ray tracing for a converging lens. Rays leave each point on the object. Shown are the three most useful rays, leaving the tip of the object, for determining where the image of that point is formed.

Actually, any two of these rays will suffice to locate the image point, which is the point where they intersect. Drawing the third can serve as a check.

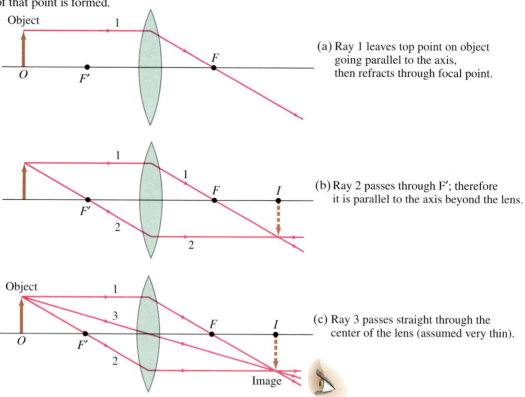

(a) Ray 1 leaves top point on object going parallel to the axis, then refracts through focal point.

(b) Ray 2 passes through F'; therefore it is parallel to the axis beyond the lens.

(c) Ray 3 passes straight through the center of the lens (assumed very thin).

 (a)

 (b)

FIGURE 23–35 (a) A converging lens can form a real image (here of a distant building) on a screen. (b) That real image is also directly visible to the eye. Figure 23–29d shows images seen by the eye made by both diverging and converging lenses.

In this way we can find the image point for one point of the object (the top of the arrow in Fig. 23–34). The image points for all other points on the object can be found similarly to determine the complete image of the object. Because the rays actually pass through the image for the case shown in Fig. 23–34, it is a real image (see page 693). The image could be detected by film, or actually seen on a white surface placed at the position of the image (Fig. 23–30).

The image can also be seen directly by the eye when the eye is placed behind the image, as shown in Fig. 23–34c, so that some of the rays diverging from each point on the image enter the eye.[†] See Fig. 23–35.

By drawing the same three rays we can determine the image position for a diverging lens, as shown in Fig. 23–36. Note that ray 1 is drawn parallel to the axis, but does not pass through the focal point F' behind the lens. Instead it seems to come from the focal point F in front of the lens (dashed line). Ray 2 is directed toward F' and is refracted parallel by the lens. Ray 3 passes directly through the center of the lens. The three refracted rays seem to emerge from a point on the left of the lens. This is the image, I. Because the rays do not pass through the image, it is a **virtual image**. Note that the eye does not distinguish between real and virtual images—both are visible.

➡ **RAY TRACING**

For a diverging lens

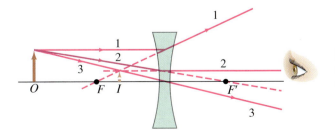

FIGURE 23–36 Finding the image by ray tracing for a diverging lens.

[†]Why, in order to see the image, the rays must be diverging from each point on the image will be discussed in Section 25–2, but is essentially because we see real objects when diverging rays from each point enter the eye as shown in Fig. 23–1.

23-8 The Lens Equation

We now derive an equation that relates the image distance to the object distance and the focal length of the lens. This will make the determination of image position quicker and more accurate than doing ray tracing. Let d_o be the object distance, the distance of the object from the center of the lens, and d_i be the image distance, the distance of the image from the center of the lens; and let h_o and h_i refer to the heights of the object and image. Consider the two rays shown in Fig. 23–37 for a converging lens (assumed to be very thin). The triangles $FI'I$ and FBA (highlighted in yellow in Fig. 23–37) are similar because angle AFB equals angle IFI'; so

$$\frac{h_i}{h_o} = \frac{d_i - f}{f},$$

since length $AB = h_o$. Triangles OAO' and IAI' are similar. Therefore,

$$\frac{h_i}{h_o} = \frac{d_i}{d_o}.$$

We equate the right sides of these two equations, divide by d_i, and rearrange to obtain

LENS EQUATION

$$\frac{1}{d_o} + \frac{1}{d_i} = \frac{1}{f}. \qquad (23–8)$$

This is called the **lens equation**. It relates the image distance d_i to the object distance d_o and the focal length f. It is the most useful equation in geometric optics. (Interestingly, it is exactly the same as the mirror equation, Eq. 23–2). Note that if the object is at infinity, then $1/d_o = 0$, so $d_i = f$. Thus the focal length is the image distance for an object at infinity, as mentioned earlier.

We can derive the lens equation for a diverging lens using Fig. 23–38. Triangles IAI' and OAO' are similar; and triangles IFI' and AFB are similar. Thus (noting that length $AB = h_o$)

$$\frac{h_i}{h_o} = \frac{d_i}{d_o} \qquad \text{and} \qquad \frac{h_i}{h_o} = \frac{f - d_i}{f}.$$

FIGURE 23–37 Deriving the lens equation for a converging lens.

FIGURE 23–38 Deriving the lens equation for a diverging lens.

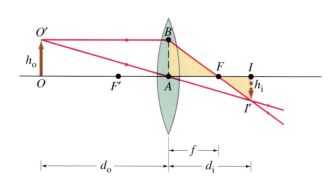

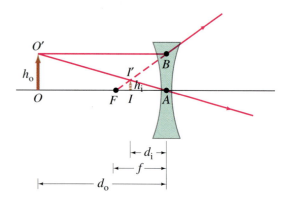

When these are equated and simplified, we obtain

$$\frac{1}{d_o} - \frac{1}{d_i} = -\frac{1}{f}.$$

This equation becomes the same as Eq. 23–8 if we make f and d_i negative. That is, we take f to be *negative for a diverging lens*, and d_i negative when the image is on the same side of the lens as the light comes from. Thus Eq. 23–8 will be valid for both converging and diverging lenses, and for *all* situations, if we use the following **sign conventions**:

➥ **PROBLEM SOLVING**

Sign conventions for lenses

1. The focal length is positive for converging lenses and negative for diverging lenses.

2. The object distance is positive if it is on the side of the lens from which the light is coming (this is usually the case, although when lenses are used in combination, it might not be so); otherwise, it is negative.

3. The image distance is positive if it is on the opposite side of the lens from where the light is coming; if it is on the same side, d_i is negative. Equivalently, the image distance is positive for a real image and negative for a virtual image.

4. The height of the image, h_i, is positive if the image is upright, and negative if the image is inverted relative to the object. (h_o is always taken as positive.)

The **lateral magnification**, m, of a lens is defined as the ratio of the image height to object height, $m = h_i/h_o$. From Figs. 23–37 and 23–38 and the conventions just stated, we have

$$m = \frac{h_i}{h_o} = -\frac{d_i}{d_o}.$$

(23–9)

Lateral magnification of a lens

For an upright image the magnification is positive, and for an inverted image m is negative.

From convention 1 above, it follows that the power (Eq. 23–7) of a converging lens, in diopters, is positive, whereas the power of a diverging lens is negative. A converging lens is sometimes referred to as a **positive lens**, and a diverging lens as a **negative lens**.

23–9 Problem Solving for Lenses

In brief outline, solving problems for thin lenses can be approached as follows:

➥ **PROBLEM SOLVING** **Thin Lenses**

1. As always, read and reread the problem,
2. Draw a ray diagram, precise if possible, but even a rough one can serve as confirmation of analytic results. Draw at least two, and preferably three, of the easy-to-draw rays described in Section 23–7 and shown in Figs. 23–34 and 23–36.
3. For analytic solutions, solve for unknowns in the lens equation (Eq. 23–8) and the magnification (Eq. 23–9). The lens equation (Eq. 23–8) involves reciprocals—avoid the obvious but common error of forgetting to take the reciprocal.
4. Follow the **Sign Conventions** above.
5. Check that your analytic answers are consistent with your ray diagram.

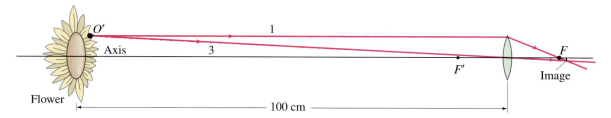

O'

Axis

1

3

Flower

F'

F

Image

|← —————————— 100 cm —————————— →|

FIGURE 23–39
Example 23–11. (Not to scale.)

EXAMPLE 23–11 **Image formed by converging lens.** What is (*a*) the position, and (*b*) the size, of the image of a large 7.6-cm-high flower placed 1.00 m from a +50.0-mm-focal-length camera lens?

SOLUTION Figure 23–39 is a rough ray diagram, showing only rays 1 and 2 for a single point on the flower. We see that the image ought to be a little behind the focal point, *F*, to the right of the lens. (*a*) We find the image position analytically using the lens equation, Eq. 23–8. The camera lens is converging, with $f = +5.00$ cm, and $d_o = 100$ cm, and so Eq. 23–8 gives

$$\frac{1}{d_i} = \frac{1}{f} - \frac{1}{d_o} = \frac{1}{5.00 \text{ cm}} - \frac{1}{100 \text{ cm}} = \frac{20.0 - 1.0}{100 \text{ cm}}.$$

Then

$$d_i = \frac{100 \text{ cm}}{19.0} = 5.26 \text{ cm},$$

or 52.6 mm behind the lens. Notice that the image is 2.6 mm farther from the lens than would be the image for an object at infinity. This is an example of the fact that when focusing a camera lens, the closer the object is to the camera, the farther the lens must be from the film.
(*b*) The magnification is

$$m = -\frac{d_i}{d_o} = -\frac{5.26 \text{ cm}}{100 \text{ cm}} = -0.0526;$$

so

$$h_i = mh_o = (-0.0526)(7.6 \text{ cm}) = -0.40 \text{ cm}.$$

The image is 4.0 mm high and is inverted ($m < 0$), as in Fig. 23–37, and shown in our sketch, Fig. 23–39.

EXAMPLE 23–12 **Object close to converging lens.** An object is placed 10 cm from a 15-cm-focal-length converging lens. Determine the image position and size (*a*) analytically, and (*b*) using a ray diagram.

SOLUTION (*a*) Given $f = 15$ cm and $d_o = 10$ cm, then

$$\frac{1}{d_i} = \frac{1}{15 \text{ cm}} - \frac{1}{10 \text{ cm}} = -\frac{1}{30 \text{ cm}},$$

➡ PROBLEM SOLVING

Don't forget to take the reciprocal

and $d_i = -30$ cm. (Remember to take the reciprocal!) Because d_i is negative, the image must be virtual and on the same side of the lens as the object. The magnification

$$m = -\frac{d_i}{d_o} = -\frac{-30 \text{ cm}}{10 \text{ cm}} = 3.0.$$

The image is three times as large as the object and is upright. (This lens is being used as a simple magnifying glass, which we discuss in more detail in Section 25–3.)

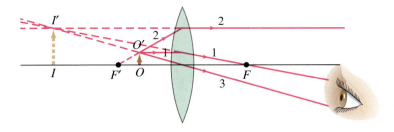

FIGURE 23–40 An object placed within the focal point of a converging lens produces a virtual image. Example 23–12.

(b) The ray diagram is shown in Fig. 23–40, and confirms the result in part (a). For point O' on the top of the object, ray 1 is easy to draw. But ray 2 may take some thought: if we draw it heading toward F', it is going the wrong way—so we have to draw it as if coming from F' (and so dashed), striking the lens, and then going out parallel to the principal axis. We project it backward with a dashed line, as we must do also for ray 1, in order to find where they meet. Ray 3 is easy to draw, through the lens center, and it meets the other two at the image point, I'.

It is a general rule that when an object is placed between a converging lens and its focal point (as in this last Example), the image is virtual.

EXAMPLE 23–13 **Diverging lens.** Where must a small insect be placed if a 25-cm-focal-length diverging lens is to form a virtual image 20 cm in front of the lens?

SOLUTION The ray diagram is basically that of Fig. 23–38 because our lens here is diverging and our image is in front of the lens within the focal distance. (It would be a valuable exercise to draw the ray diagram to scale, precisely, now.) Since $f = -25$ cm and $d_i = -20$ cm, then Eq. 23–8 gives

$$\frac{1}{d_o} = \frac{1}{f} - \frac{1}{d_i} = -\frac{1}{25 \text{ cm}} + \frac{1}{20 \text{ cm}} = \frac{-4 + 5}{100 \text{ cm}} = \frac{1}{100 \text{ cm}}.$$

So the object must be 100 cm in front of the lens. Your diagram should confirm this result.

*23–10 Problem Solving for Combinations of Lenses

We now consider two Examples illustrating how to deal with lenses used in combination. In general, when light passes through more than one lens, we find the image formed by the first lens as if it were alone. This image becomes the *object* for the second lens, and we find the image then formed by this second lens, which is the final image if there are only two lenses. The total magnification will be the product of the separate magnifications of each lens as we shall see.

Multiple lenses:
image formed by first lens
is object for second lens

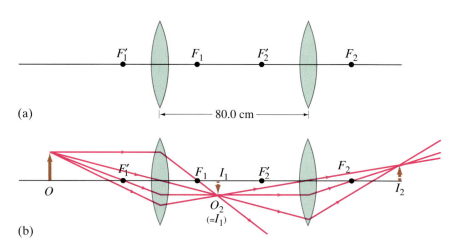

FIGURE 23–41
Example 23–14.

(a)

(b)

EXAMPLE 23–14 **A two-lens system.** Two converging lenses, with focal lengths $f_1 = 20.0$ cm and $f_2 = 25.0$ cm, are placed 80.0 cm apart, as shown in Fig. 23–41a. An object is placed 60.0 cm in front of the first lens as shown. Determine (a) the position, and (b) the magnification, of the final image formed by the combination of the two lenses.

SOLUTION (a) The object is a distance $d_{o1} = +60.0$ cm from the first lens, and this lens forms an image whose position can be calculated using the lens equation:

$$\frac{1}{d_{i1}} = \frac{1}{f_1} - \frac{1}{d_{o1}} = \frac{1}{20.0 \text{ cm}} - \frac{1}{60.0 \text{ cm}} = \frac{3-1}{60.0 \text{ cm}} = \frac{1}{30.0 \text{ cm}}.$$

So the first image is at $d_{i1} = 30.0$ cm behind the first lens. This image becomes the object for the second lens. It is a distance $d_{o2} = 80.0$ cm $- 30.0$ cm $= 50.0$ cm in front of lens 2, as shown in Fig. 23–41b. The image formed by the second lens, again using the lens equation, is at a distance d_{i2} from the second lens:

$$\frac{1}{d_{i2}} = \frac{1}{f_2} - \frac{1}{d_{o2}} = \frac{1}{25.0 \text{ cm}} - \frac{1}{50.0 \text{ cm}} = \frac{4-2}{100.0 \text{ cm}} = \frac{2}{100.0 \text{ cm}}.$$

Careful!
*Note that object distance for second lens is **not** equal to the image distance for first lens*

Hence $d_{i2} = 50.0$ cm behind lens 2. This is the final image—see Fig. 23–41b.
(b) The first lens has a magnification (Eq. 23–9)

$$m_1 = -\frac{d_{i1}}{d_{o1}} = -\frac{30.0 \text{ cm}}{60.0 \text{ cm}} = -0.500.$$

Thus, the image is inverted and is half as high as the object: again Eq. 23–9,

$$h_{i1} = m_1 h_{o1} = -0.500\, h_{o1}.$$

The second lens takes this image as object and changes its height by a factor

$$m_2 = -\frac{d_{i2}}{d_{o2}} = -\frac{50.0 \text{ cm}}{50.0 \text{ cm}} = -1.000.$$

The final image height is (remember h_{o2} is the same as h_{i1}):

$$h_{i2} = m_2 h_{o2} = m_2 h_{i1} = m_2 m_1 h_{o1}.$$

Total magnification is
$m = m_1 m_2$

We see from this equation that the total magnification is the product of m_1 and m_2, which here equals $(-1.000)(-0.500) = +0.500$, or $\frac{1}{2}$ the original height and upright.

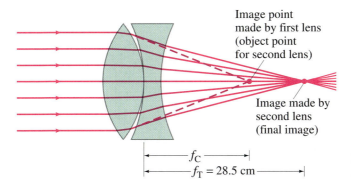

Image point
made by first lens
(object point
for second lens)

Image made by
second lens
(final image)

f_C

$f_T = 28.5$ cm

FIGURE 23–42
Determining the focal length of a
diverging lens. Example 23–15.

EXAMPLE 23–15 **Measuring f for a diverging lens.** To measure the
focal length of a diverging lens, a converging lens is placed in contact
with it, as shown in Fig. 23–42. The Sun's rays are focused by this combination at a point 28.5 cm behind the lenses as shown. If the converging
lens has a focal length f_C of 16.0 cm, what is the focal length f of the diverging lens?

SOLUTION Rays from the Sun are focused 28.5 cm behind the combination, so the focal length of the total combination is $f_T = 28.5$ cm. If the
diverging lens were absent, the converging lens would form the image at
its focal point—that is, at a distance $f_C = 16.0$ cm behind it (dashed lines
in Fig. 23–42). When the diverging lens is placed next to the converging
lens (we assume both lenses are thin and the space between them is negligible), we treat the image formed by the first lens as the *object* for the
second lens. Since this object lies to the right of the diverging lens, this is
a situation where d_o is negative (see the sign conventions, page 707).
Thus, for the diverging lens, the object is virtual and $d_o = -16.0$ cm. The
diverging lens forms the image of this virtual object at a distance
$d_i = 28.5$ cm away (this was given). Thus,

$$\frac{1}{f_D} = \frac{1}{d_o} + \frac{1}{d_i} = \frac{1}{-16.0 \text{ cm}} + \frac{1}{28.5 \text{ cm}} = -0.0274 \text{ cm}^{-1}.$$

We take the reciprocal to find $f_D = -1/(0.0274 \text{ cm}^{-1}) = -36.5$ cm.

Note that the converging lens must be "stronger" than the diverging
lens (that is, it must have a focal length whose magnitude is less than that
of the diverging lens) if this technique is to work.

*23–11 The Lensmaker's Equation

In this optional Section, we will prove that parallel rays are brought to a
focus at a *single* point for a thin lens. At the same time, we will also derive
an equation that relates the focal length of a lens to the radii of curvature
of its two surfaces, which is known as the lens-maker's equation.

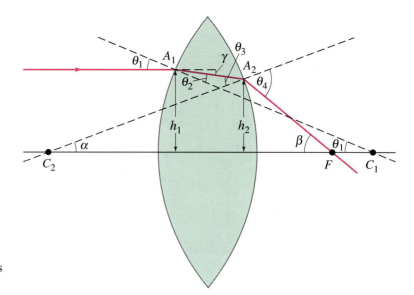

FIGURE 23–43 Diagram of a ray passing through a lens for derivation of the lensmaker's equation.

In Fig. 23–43, a ray parallel to the axis of a lens is refracted at the front surface of the lens at point A_1 and is refracted at the back surface at point A_2. This ray then passes through point F, which we call the focal point for this ray. Point A_1 is a height h_1 above the axis, and point A_2 is height h_2 above the axis. C_1 and C_2 are the centers of curvature of the two lens surfaces; so the length $C_1A_1 = R_1$, the radius of curvature of the front surface, and $C_2A_2 = R_2$ is the radius of the second surface. The thickness of the lens has been grossly exaggerated so the various angles would be clear. But we will assume that the lens is actually very thin and that angles between the rays and the axis are small. In this approximation, $h_1 \approx h_2$, and the sines and tangents of all the angles will be equal to the angles themselves in radians. For example, $\sin \theta_1 \approx \tan \theta_1 \approx \theta_1$ (radians).

To this approximation, then, Snell's law tells us that

$$\theta_1 = n\theta_2$$

$$\theta_4 = n\theta_3$$

where n is the index of refraction of the glass, and we assume that the lens is surrounded by air ($n = 1$). Notice also in Fig. 23–43 that

$$\theta_1 = \sin \theta_1 = \frac{h_1}{R_1}$$

$$\alpha = \frac{h_2}{R_2}$$

$$\beta = \frac{h_2}{f}.$$

The last follows because the distance from F to the lens (assumed very thin) is f. From the diagram, the angle γ was defined as

$$\gamma = \theta_1 - \theta_2.$$

A careful examination of Fig. 23–43 shows also that

$$\alpha = \theta_3 - \gamma.$$

This can be seen by drawing a horizontal line to the left from point A_2, which divides the angle θ_3 into two parts. The upper part equals γ and the lower part equals α. (The opposite angles between an oblique line and two parallel lines are equal.) Thus, $\theta_3 = \gamma + \alpha$. Finally, by drawing a horizontal line to the right from point A_2, we divide θ_4 into two parts. The upper part is α and the lower is β. Thus

$$\theta_4 = \alpha + \beta.$$

We now combine all these equations:

$$\alpha = \theta_3 - \gamma = \frac{\theta_4}{n} - (\theta_1 - \theta_2) = \frac{\alpha}{n} + \frac{\beta}{n} - \theta_1 + \theta_2,$$

or

$$\frac{h_2}{R_2} = \frac{h_2}{nR_2} + \frac{h_2}{nf} - \frac{h_1}{R_1} + \frac{h_1}{nR_1}.$$

Because the lens is thin, $h_1 \approx h_2$ and all h's can be canceled from all the numerators. We then multiply through by n and rearrange to find that

$$\frac{1}{f} = (n - 1)\left(\frac{1}{R_1} + \frac{1}{R_2}\right). \qquad \textbf{(23–10)} \qquad \textit{Lensmaker's equation}$$

This is called the **lensmaker's equation**. It relates the focal length of a lens to the radii of curvature of its two surfaces and its index of refraction. Notice that f does not depend on h_1 or h_2. Thus the position of the point F does not depend on where the ray strikes the lens. Hence, all rays parallel to the axis of a thin lens will pass through the same point F, which we wished to prove.

In our derivation, both surfaces are convex and R_1 and R_2 are considered positive.[†] Equation 23–10 also works for lenses with one or both surfaces concave; but for a concave surface, the radius must be considered *negative*.

Notice in Eq. 23–10 that the equation is symmetrical in R_1 and R_2. Thus, if a lens is turned around so that light impinges on the other surface, the focal length is the same even if the two lens surfaces are different.

[†]Some books use a different convention—for example, R_1 and R_2 are considered positive if their centers of curvature are to the right of the lens, in which case a minus sign appears in their equivalent of Eq. 23–10.

EXAMPLE 23–16 **Calculating _f_ for a converging lens.** A convex meniscus lens (Figs. 23–29a and 23–44) is made from glass with $n = 1.50$. The radius of curvature of the convex surface is 22.4 cm and that of the concave surface is 46.2 cm. (_a_) What is the focal length? (_b_) Where will it focus an object 2.00 m away?

SOLUTION (_a_) $R_1 = 22.4$ cm and $R_2 = -46.2$ cm; the latter is negative because it refers to the concave surface. Then

$$\frac{1}{f} = (1.50 - 1.00)\left(\frac{1}{22.4 \text{ cm}} - \frac{1}{46.2 \text{ cm}}\right)$$

$$= 0.0115 \text{ cm}^{-1}.$$

So

$$f = \frac{1}{0.0115 \text{ cm}^{-1}} = 87 \text{ cm}$$

and the lens is converging. Notice that if we turn the lens around so that $R_1 = -46.2$ cm and $R_2 = +22.4$ cm, we get the same result.
(_b_) From the lens equation, with $f = 0.87$ m and $d_o = 2.00$ m, we have

$$\frac{1}{d_i} = \frac{1}{f} - \frac{1}{d_o} = \frac{1}{0.87 \text{ m}} - \frac{1}{2.00 \text{ m}}$$

$$= 0.65 \text{ m}^{-1},$$

so $d_i = 1/0.65 \text{ m}^{-1} = 1.53$ m.

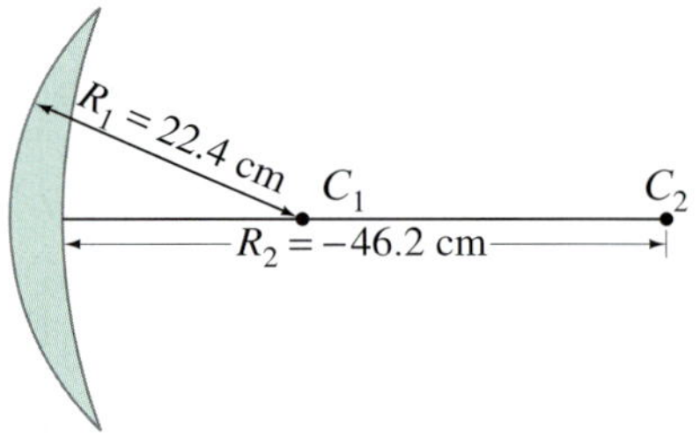

FIGURE 23–44
Example 23–16.

EXAMPLE 23–17 **Calculating _f_ for a diverging lens.** A Lucite planoconcave lens has one flat surface and the other has $R = -18.4$ cm. What is the focal length?

SOLUTION From Table 23–1, n for Lucite is 1.51. A plane surface has infinite radius of curvature; if we call this R_1, then $1/R_1 = 0$. Therefore,

$$\frac{1}{f} = (1.51 - 1.00)\left(-\frac{1}{18.4 \text{ cm}}\right).$$

So $f = (-18.4 \text{ cm})/0.51 = -36$ cm, and the lens is diverging.

S U M M A R Y

Light appears to travel in straight-line paths, called **rays**, at a speed v that depends on the **index of refraction**, n, of the material; that is

$$v = \frac{c}{n},$$

where c is the speed of light in vacuum.

When light reflects from a flat surface, the *angle of reflection equals the angle of incidence*. This **law of reflection** explains why mirrors can form **images**. In a plane mirror, the image is virtual, upright, the same size as the object, and is as far behind the mirror as the object is in front.

A spherical mirror can be concave or convex. A concave spherical mirror focuses parallel rays of light (light from a very distant object) to a point called the **focal point**. The distance of this point from the lens is the **focal length** f of the mirror and

$$f = \frac{r}{2}$$

where r is the radius of curvature of the mirror. Parallel rays falling on a convex mirror reflect from the mirror as if they diverged from a common point behind the mirror. The distance of this point from the mirror is the focal length and is considered negative for a convex mirror. For a given object, the position and size of the image formed by a mirror can be found by ray tracing. Algebraically, the relation between image and object distances, d_i and d_o, and the focal length f, is given by the **mirror equation**:

$$\frac{1}{d_o} + \frac{1}{d_i} = \frac{1}{f}.$$

The ratio of image height to object height, which equals the magnification m, is

$$m = \frac{h_i}{h_o} = -\frac{d_i}{d_o}.$$

If the rays that converge to form an image actually pass through the image, so the image would appear on film or a screen placed there, the image is said to be a **real image**. If the rays do not actually pass through the image, the image is a **virtual image**.

When light passes from one transparent medium into another, the rays bend or refract. The **law of refraction (Snell's law)** states that

$$n_1 \sin \theta_1 = n_2 \sin \theta_2,$$

where n_1 and θ_1 are the index of refraction and angle with the normal to the surface for the incident ray, and n_2 and θ_2 are for the refracted ray.

When light rays reach the boundary of a material where the index of refraction decreases, the rays will be **totally internally reflected** if the incident angle, θ_1, is such that Snell's law would predict $\sin \theta_2 > 1$; this occurs if θ_1 exceeds the critical angle θ_C given by

$$\sin \theta_C = \frac{n_2}{n_1}.$$

A lens uses refraction to produce a real or virtual image. Parallel rays of light are focused to a point, called the **focal point**, by a converging lens. The distance of the focal point from the lens is called the **focal length** f of the lens. After parallel rays pass through a diverging lens, they appear to diverge from a point, its focal point; and the corresponding focal length is considered negative. The **power** P of a lens, which is $P = 1/f$, is given in diopters, which are units of inverse meters (m^{-1}). For a given object, the position and size of the image formed by a lens can be found by ray tracing. Algebraically, the relation between image and object distances, d_i and d_o, and the focal length f, is given by the **lens equation**:

$$\frac{1}{d_o} + \frac{1}{d_i} = \frac{1}{f}.$$

The ratio of image height to object height, which equals the magnification m, is

$$m = \frac{h_i}{h_o} = -\frac{d_i}{d_o}.$$

When using the various equations of geometrical optics, it is important to remember the **sign conventions** for all quantities involved: carefully review them when doing problems.

1. Give arguments to show why the Moon must have a rough surface rather than a polished mirrorlike surface.

2. When you look at the Moon's reflection from a ripply sea, it appears elongated (Fig. 23–45). Explain.

FIGURE 23–45   Question 2.

3. Although a plane mirror reverses left and right, it doesn't reverse up and down. Explain.

4. Archimedes is said to have burned the whole Roman fleet in the harbor of Syracuse by focusing the rays of the Sun with a huge spherical mirror. Is this reasonable?

5. If a concave mirror produces a real image, is the image necessarily inverted?

6. When you use a concave mirror, you cannot see an inverted image of yourself unless you place your head beyond the center of curvature *C*. Yet you can see an inverted image of another object placed between *C* and *F*, as in Fig. 23–14. Explain. [*Hint*: You can see a real image only if your eye is behind the image, so that the image can be formed.]

7. Using the rules for the three rays discussed with reference to Fig. 23–14, draw ray 2 for Fig. 23–17b.

8. What is the focal length of a plane mirror? What is the magnification of a plane mirror?

9. Does the mirror equation, Eq. 23–2, hold for a plane mirror? Explain.

10. How might you determine the speed of light in a solid, rectangular, transparent object?

11. When a wide beam of parallel light enters water at an angle, the beam broadens. Explain.

12. What is the angle of refraction when a light ray meets the boundary between two materials perpendicularly?

13. When you look down into a swimming pool or a lake, are you likely to underestimate or overestimate its depth? Explain. How does the apparent depth vary with the viewing angle? (Use ray diagrams.)

14. Draw a ray diagram to show why a stick looks bent when part of it is under water (Fig. 23–20).

15. Your eye looks into an aquarium and views a fish inside. One ray of light that emerges from the tank from the fish is shown in Fig. 23–46. Also shown is the apparent position of the fish as seen by the eyeball. In the drawing, indicate the approximate position of the actual fish. Briefly justify your answer.

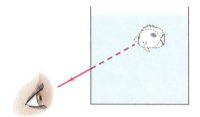

FIGURE 23–46   Question 15.

16. How can you "see" a round drop of water on a table even though the water is transparent and colorless?

17. Can a light ray traveling in air be totally reflected when it strikes a smooth water surface if the incident angle is right?

18. When you look up at an object in air from beneath the surface in a swimming pool, does the object appear to be the same size as when you see it directly in air? Explain.

19. What type of mirror is shown in Fig. 23–47?

FIGURE 23–47
Question 19.

20. Where must the film be placed if a camera lens is to make a sharp image of an object very far away?

21. A photographer moves closer to his subject and then refocuses. Does the camera lens move farther or nearer to the film?

22. Can a diverging lens form a real image under any circumstances? Explain.

23. Show that a real image formed by a thin lens is always inverted, whereas a virtual image is always upright if the object is real.

24. Light rays are said to be "reversible." Is this consistent with the lens equation?

25. Can real images be projected on a screen? Can virtual images? Can either be photographed? Discuss carefully.

26. A thin converging lens is moved closer to a nearby object. Does the real image formed change (*a*) in position, (*b*) in size? If yes, describe how.

27. Compare the mirror equation with the lens equation. Discuss similarities and differences, especially the sign conventions for the quantities involved.

28. A lens is made of a material with an index of refraction $n = 1.30$. In air, it is a converging lens. Will it still be a converging lens if placed in water? Explain, using a ray diagram.

29. A dog stands facing a converging lens with its tail in the air. If the nose and the tail are each focused on a screen in turn, which will have the greater magnification?

30. A cat stands facing a converging lens with its tail in the air. Under what circumstances (if any) would the image of the nose be virtual and the image of the tail be real? Where would the image of the rest of the cat be?

* 31. Why, in Example 23–15, must the converging lens have a shorter focal length than the diverging lens if the latter's focal length is to be determined by combining them?

* 32. An unsymmetrical lens (say, planoconvex) forms an image of a nearby object. Does the image point change if the lens is turned around?

* 33. The thicker a double convex lens is in the center as compared to its edges, the shorter its focal length for a given lens diameter. Explain.

* 34. Does the focal length of a lens depend on the fluid in which it is immersed? What about the focal length of a spherical mirror? Explain.

P R O B L E M S

SECTION 23–2

1. (I) Suppose that you want to take a photograph of yourself as you look at your image in a flat mirror 1.5 m away. For what distance should the camera lens be focused?

2. (I) When you look at yourself in a tall plane mirror, you see the same amount of your body whether you are close to the mirror or far away. (Try it and see.) Use ray diagrams to show why this should be true.

3. (II) Stand up two plane mirrors so they form a right angle as in Fig. 23–48. When you look into this double mirror, you see yourself as others see you, instead of reversed as in a single mirror. Make a ray diagram to show how this occurs.

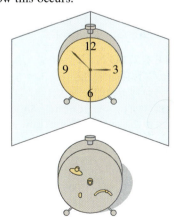

FIGURE 23–48
Problems 3 and 8.

4. (II) A person whose eyes are 1.62 m above the floor stands 2.10 m in front of a vertical plane mirror whose bottom edge is 43 cm above the floor, Fig. 23–49. What is the horizontal distance x to the base of the wall supporting the mirror of the nearest point on the floor that can be seen reflected in the mirror?

1.62 m

43 cm

x

2.10 m

FIGURE 23–49   Problem 4.

5. (II) Two mirrors meet at a 135° angle, Fig. 23–50. If light rays strike one mirror at 40° as shown, at what angle ϕ do they leave the second mirror?

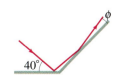

40°

ϕ

FIGURE 23–50   Problem 5.

6. (II) Suppose you are 70 cm from a plane mirror. What area of the mirror is used to reflect the rays entering one eye from a point on the tip of your nose if your pupil diameter is 5.5 mm?

7. (III) Show that if two plane mirrors meet at an angle ϕ, a single ray reflected successively from both mirrors is deflected through an angle of 2ϕ independent of the incident angle. Assume $\phi < 90°$ and that only two reflections, one from each mirror, take place.

8. (III) Suppose a third mirror is placed beneath the two shown in Fig. 23–48, so that all three are perpendicular to each other. (a) Show that for such a "corner reflector," any incident ray will return in its original direction after three reflections. (b) What happens if it makes only two reflections?

SECTION 23–3

9. (I) A solar cooker, really a concave mirror pointed at the Sun, focuses the Sun's rays 17.0 cm in front of the mirror. What is the radius of the spherical surface from which the mirror was made?

10. (I) How far from a concave mirror (radius 27.0 cm) must an object be placed if its image is to be at infinity?

11. (II) Show with ray diagrams that the magnification of a concave mirror is less than 1 if the object is beyond the center of curvature C and is greater than 1 if it is within this point.

12. (II) If you look at yourself in a shiny Christmas tree ball with a diameter of 9.0 cm when your face is 30.0 cm away from it, where is your image? Is it real or virtual? Is it upright or inverted?

13. (II) A mirror at an amusement park shows an upright image of any person who stands 1.3 m in front of it. If the image is three times the person's height, what is the radius of curvature?

14. (II) A dentist wants a small mirror that, when 2.20 cm from a tooth, will produce a $4.5 \times$ upright image. What kind of mirror must be used and what must its radius of curvature be?

15. (II) Some rearview mirrors produce images of cars to your rear that are a bit smaller than they would be if the mirror were flat. Are the mirrors concave or convex? What type and height of image would such a mirror produce of a car that was 1.3 m high and was 15.0 m behind you, assuming the mirror's radius of curvature is 3.2 m?

16. (II) A luminous object 3.0 mm high is placed 20 cm from a convex mirror of radius of curvature 20 cm. (a) Show by ray tracing that the image is virtual, and estimate the image distance. (b) Show that to compute this (negative) image distance from Eq. 23–2, it is sufficient to let the focal length be -10 cm. (c) Compute the image size, using Eq. 23–3.

17. (II) (a) Where should an object be placed in front of a concave mirror so that it produces an image at the same location? (b) Is the image real or virtual? (c) Is the image inverted or erect? (d) What is the magnification of the image?

18. (II) The image of a distant tree is virtual and very small when viewed in a curved mirror. The image appears to be 14.0 cm behind the mirror. What kind of mirror is it, and what is its radius of curvature?

19. (II) Use the mirror equation to show that the magnitude of the magnification of a concave mirror is less than 1 if the object is beyond the center of curvature C ($d_o > r$), and is greater than 1 if the object is within C ($d_o < r$).

20. (II) Show, using a ray diagram, that the magnification m of a convex mirror is $m = -d_i/d_o$, just as for a concave mirror. [Hint: Consider a ray from the top of the object that reflects at the center of the mirror.]

21. (II) Use ray diagrams to show that the mirror equation, Eq. 23–2, is valid for a convex mirror as long as f is considered negative.

22. (II) The magnification of a convex mirror is $+0.80 \times$ for objects 2.2 m from the mirror. What is the focal length of this mirror?

23. (II) (a) A plane mirror can be considered a limiting case of a spherical mirror. Specify what this limit is. (b) Determine an equation that relates the image and object distances in this limit of a plane mirror. (c) Determine the magnification of a plane mirror in this same limit. (d) Are your results in parts (b) and (c) consistent with the discussion of Section 23–2 on plane mirrors?

24. (III) A 4.5-cm tall object is placed 28 cm in front of a spherical mirror. It is desired to produce a virtual image that is erect and 3.5 cm tall. (a) What type of mirror should be used? (b) Where is the image located? (c) What is the focal length of the mirror? (d) What is the radius of curvature of the mirror?

25. (III) A shaving/makeup mirror is designed to magnify your face by a factor of 1.3 when your face is placed 20.0 cm in front of it. (a) What type of mirror is it? (b) Describe the type of image that it makes of your face. (c) Calculate the required radius of curvature for the mirror.

SECTION 23–4

26. (I) What is the speed of light in (a) crown glass, and (b) Lucite?

27. (I) The speed of light in ice is 2.29×10^8 m/s. What is the index of refraction of ice?

28. (II) The speed of light in a certain substance is 85 percent of its value in water. What is the index of refraction of this substance?

29. (II) If $n = 1.00030 \pm 0.000010$ for air, what uncertainty will this introduce in the value for the speed of light in air?

SECTION 23–5

30. (I) A flashlight beam strikes the surface of a pane of glass ($n = 1.50$) at a 63° angle to the normal. What is the angle of refraction?

31. (I) A diver shines a flashlight upward from beneath the water at a 42.5° angle to the vertical. At what angle does the light leave the water?

32. (I) A light beam coming from an underwater spotlight exits the water at an angle of 60°. At what angle of incidence did it hit the air–water interface from below the surface?

33. (I) Rays of the Sun are seen to make a 21.0° angle to the vertical beneath the water. At what angle above the horizon is the Sun?

34. (II) Light is incident on an equilateral crown glass prism at a 45.0° angle to one face, Fig. 23–51. Calculate the angle at which light emerges from the opposite face. Assume that $n = 1.52$.

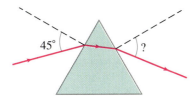

FIGURE 23–51
Problems 34 and 45.

35. (II) In searching the bottom of a pool at night, a watchman shines a narrow beam of light from his flashlight, 1.3 m above the water level, onto the surface of the water at a point 2.7 m from his foot at the edge of the pool (Fig. 23–52). Where does the spot of light hit the bottom of the pool, relative to the edge, if the pool is 2.1 m deep?

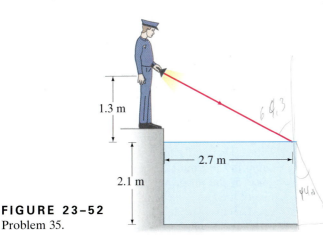

FIGURE 23–52
Problem 35.

36. (II) A beam of light in air strikes a slab of crown glass ($n = 1.52$) and is partially reflected and partially refracted. Find the angle of incidence if the angle of reflection is twice the angle of refraction.

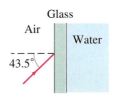

FIGURE 23–53   Problem 37.

37. (II) An aquarium filled with water has flat glass sides whose index of refraction is 1.52. A beam of light from outside the aquarium strikes the glass at a 43.5° angle to the perpendicular (Fig. 23–53). What is the angle of this light ray when it enters (*a*) the glass, and then (*b*) the water? (*c*) What would be the refracted angle if the ray entered the water directly?

38. (II) Prove in general that for a light beam incident on a uniform layer of transparent material, as in Fig. 23–21, the direction of the emerging beam is parallel to the incident beam, independent of the incident angle θ.

39. (III) A light ray is incident on a flat piece of glass as in Fig. 23–21. Show that if the incident angle θ is small, the ray is displaced a distance $d = t\theta(n-1)/n$, where t is the thickness of the glass and θ is in radians.

SECTION 23–6

40. (I) What is the critical angle for the interface between water and Lucite? To be internally reflected, the light must start in which material?

41. (I) The critical angle for a certain liquid-air surface is 44.7°. What is the index of refraction of the liquid?

42. (II) A beam of light is emitted in a pool of water from a depth of 62.0 cm. Where must it strike the air–water interface, relative to the spot directly above it, in order that the light does *not* exit the water?

43. (II) A ray of light enters a light fiber at an angle of 15° with the long axis of the fiber, as in Fig. 23–54. Calculate the distance the light ray travels between successive reflections off the sides of the fiber. Assume that the fiber has an index of refraction of 1.6 and is 10^{-4} m in diameter.

FIGURE 23–54   Problem 43.

44. (II) A beam of light is emitted 8.0 cm beneath the surface of a liquid and strikes the surface 7.0 cm from the point directly above the source. If total internal reflection occurs, what can you say about the index of refraction of the liquid?

45. (III) Suppose a ray strikes the left face of the prism in Fig. 23–51 at 45° as shown, but is totally internally reflected at the opposite side. If the apex angle (at the top) is $\theta = 75°$, what can you say about the index of refraction of the prism?

46. (III) A beam of light enters the end of an optic fiber as shown in Fig. 23–55. Show that we can guarantee total internal reflection at the side surface of the material (at point a), if the index of refraction is greater than about 1.42. In other words, regardless of the angle α, the light beam reflects back into the material at point a.

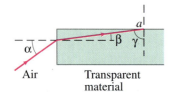

FIGURE 23–55
Problem 46.

Air   Transparent material

47. (III) (a) What is the minimum index of refraction for a glass or plastic prism to be used in binoculars (Fig. 23–25) so that total internal reflection occurs at 45°? (b) Will binoculars work if its prisms (assume $n = 1.50$) are immersed in water? (c) What minimum n is needed if the prisms are immersed in water?

SECTIONS 23–7 TO 23–9

48. (I) A sharp image is located 78.0 mm behind a 65.0-mm-focal-length converging lens. Find the object distance (a) using a ray diagram, (b) by calculation.

49. (I) Sunlight is observed to focus at a point 18.5 cm behind a lens. (a) What kind of lens is it? (b) What is its power in diopters?

50. (I) A certain lens focuses an object 2.25 m away as an image 48.3 cm on the other side of the lens. What type of lens is it and what is its focal length? Is the image real or virtual?

51. (I) (a) What is the power of a 29.5-cm-focal-length lens? (b) What is the focal length of a −6.25-diopter lens? Are these lenses converging or diverging?

52. (II) A stamp collector uses a converging lens with focal length 24 cm to view a stamp 18 cm in front of the lens. (a) Where is the image located? (b) What is the magnification?

53. (II) How large is the image of the Sun on the film used in a camera with (a) a 28-mm-focal-length lens, (b) a 50-mm-focal-length lens, and (c) a 200-mm-focal-length lens? (d) If the 50-mm lens is considered normal for this camera, how would you characterize the other two lenses? The Sun's diameter is 1.4×10^6 km and it is 1.5×10^8 km away.

54. (II) An 80-mm-focal-length lens is used to focus an image on the film of a camera. The maximum distance allowed between the lens and the film plane is 120 mm. (a) How far ahead of the film should the lens be if the object to be photographed is 10.0 m away? (b) 3.0 m away? (c) 1.0 m away? (d) What is the closest object this lens could photograph sharply?

55. (II) A −6.0-diopter lens is held 14.0 cm from an ant 1.0 mm high. What is the position, type, and height of the image?

56. (II) It is desired to magnify reading material by a factor of $2.5 \times$ when a book is placed 8.0 cm behind a lens. (a) Draw a ray diagram and describe the type of image this would be. (b) What type of lens is needed for this? (c) What is the power of the lens in diopters?

57. (II) (a) How far from a 50.0-mm-focal-length lens must an object be placed if its image is to be magnified $2.00 \times$ and be real? (b) What if the image is to be virtual and magnified $2.00 \times$?

58. (II) Repeat Problem 57 for a −50.0-mm-focal-length lens.

59. (II) (a) An object 31.5 cm in front of a certain lens is imaged 8.20 cm in front of that lens (on the same side as the object). What type of lens is this and what is its focal length? Is the image real or virtual? (b) If the image were located, instead, 38.0 cm in front of the lens, what type of lens would it be and what focal length would it have?

60. (II) (a) A 2.20-cm-high insect is 1.20 m from a 135-mm-focal-length lens. Where is the image, how high is it, and what type is it? (b) What if $f = -135$ mm?

61. (II) A bright object and a viewing screen are separated by a distance of 60 cm. At what location(s) between the object and the screen should a lens of focal length 15 cm be placed in order to produce a crisp image on the screen? [Hint: First draw a diagram.]

62. (III) How far apart are an object and an image formed by a 75-cm-focal-length converging lens if the image is $2.75 \times$ larger than the object and is real?

63. (III) (a) Show that the lens equation can be written in the Newtonian form:

$$xx' = f^2,$$

where x is the distance of the object from the focal point on the front side of the lens, and x' is the distance of the image to the focal point on the other side of the lens. Calculate the location of an image if the object is placed 45.0 cm in front of a convex lens with a focal length of 32.0 cm using (b) the standard form of the thin lens formula, and (c) the Newtonian form, derived above.

*64. (II) Two 27.0-cm-focal-length converging lenses are placed 16.5 cm apart. An object is placed 35.0 cm in front of one. Where will the final image formed by the second lens be located? What is the total magnification?

*65. (II) A diverging lens with $f = -31.5$ cm is placed 14.0 cm behind a converging lens with $f = 20.0$ cm. Where will an object at infinity be focused?

*66. (II) A 31.0-cm-focal-length converging lens is 21.0 cm behind a diverging lens. Parallel light strikes the diverging lens. After passing through the converging lens, the light is again parallel. What is the focal length of the diverging lens? [*Hint*: First draw a ray diagram.]

*67. (III) A diverging lens is placed next to a converging lens of focal length f_C, as in Fig. 23–42. If f_T represents the focal length of the combination, show that the focal length of the diverging lens, f_D, is given by

$$\frac{1}{f_D} = \frac{1}{f_T} - \frac{1}{f_C}.$$

*68. (I) A double concave lens has surface radii of 31.2 cm and 23.8 cm. What is the focal length if $n = 1.52$?

*69. (I) Both surfaces of a double convex lens have radii of 31.0 cm. If the focal length is 28.9 cm, what is the index of refraction of the lens material?

*70. (I) Show that if the lens of Example 23–17 is reversed so the light enters the curved face, the focal length is unchanged.

*71. (I) A planoconvex lens (Fig. 23–29) is to have a focal length of 28.5 cm. If made from fused quartz, what must be the radius of curvature of the convex surface?

*72. (I) A glass ($n = 1.50$) planoconcave lens has a focal length of -25.4 cm. What is the radius of the concave surface?

*73. (II) A prescription for a corrective lens calls for $+1.50$ diopters. The lensmaker grinds the lens from a "blank" with $n = 1.56$ and a preformed convex front surface of radius of curvature of 20.0 cm. What should be the radius of curvature of the other surface?

*74. (II) An object is placed 100 cm from a glass lens ($n = 1.56$) with one concave surface of radius 21.0 cm and one convex surface of radius 18.5 cm. Where is the final image? What is the magnification?

*75. (III) A glass lens ($n = 1.50$) in air has a power of $+5.2$ diopters. What would its power be if it were submerged in water?

GENERAL PROBLEMS

76. (II) Two plane mirrors are facing each other at a distance of 2.0 m apart as in Fig. 23–56. You stand 1.5 m away from one of these mirrors and look into it. You will see multiple images of yourself. (*a*) How far away from you are the first three images in the mirror in front of you? (*b*) Which way are these first three images facing, toward you or away from you?

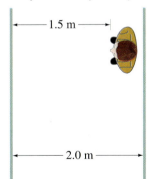

FIGURE 23–56
Problem 76.

77. We wish to determine the depth of a swimming pool filled with water by measuring the width ($x = 5.50$ m) and then noting that the bottom edge of the pool is just visible at an angle of 14.0° above the horizontal as shown in Fig. 23–57. Calculate the depth of the pool.

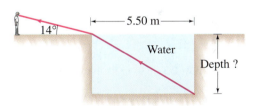

FIGURE 23–57 Problem 77.

78. The critical angle of a certain piece of plastic in air is $\theta_C = 37.3°$. What is the critical angle of the same plastic if it is immersed in water?

79. In a slide or movie projector, the film acts as the object whose image is projected on a screen (Fig. 23–58). If a 100-mm-focal-length lens is to project an image on a screen 7.50 m away, how far from the lens should the slide be? If the slide is 36 mm wide, how wide will the picture be on the screen?

FIGURE 23–58 Problem 79.

FIGURE 23–59
Problem 80.

80. (II) A kaleidoscope makes symmetric patterns with two plane mirrors with a 60° angle between them as shown in Fig. 23–59. Draw the location of the images (some of them images of images) of the object placed between the mirrors.

81. A 35-mm slide (picture size is actually 24 by 36 mm) is to be projected on a screen 1.80 by 2.70 m placed 9.00 m from the projector. What focal-length lens should be used if the image is to cover the screen?

82. Show analytically that the image formed by a converging lens is real and inverted if the object is beyond the focal point ($d_o > f$), and is virtual and upright if the object is within the focal point ($d_o < f$). Describe the image if the object is an image, formed by another lens, so its position is beyond the lens, for which $-d_o > f$, and for which $0 < -d_o < f$.

83. Show analytically that a diverging lens can never form a real image of a real object. Can you describe a situation in which a diverging lens can form a real image?

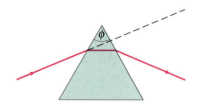

FIGURE 23–60   Problem 84.

84. If the apex angle of a prism is $\phi = 72°$ (see Fig. 23–60), what is the minimum incident angle for a ray if it is to emerge from the opposite side (i.e., not be totally internally reflected), given $n = 1.52$?

* 85. A lighted candle is placed 30 cm in front of a converging lens of focal length $f_1 = 15$ cm, which in turn is 50 cm in front of another converging lens of focal length $f_2 = 10$ cm (see Fig. 23–61). (a) Draw a ray diagram for this problem. Estimate from your diagram, and then calculate (b) the location and the size of the final image.

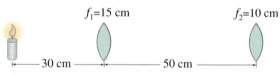

f_1=15 cm f_2=10 cm

|← 30 cm →|← 50 cm →|

FIGURE 23–61   Problem 85.

* 86. (a) Show that if two lenses of focal lengths f_1 and f_2, respectively, are placed next to each other, the focal length of the combination is given by $f_T = f_1 f_2 / (f_1 + f_2)$. (b) Show that the power P of the combination of two lenses is the sum of their separate powers, $P = P_1 + P_2$.

87. A bright object is placed on one side of a converging lens of focal length f, and a white screen for viewing the image is on the opposite side. The distance $d_T = d_i + d_o$ between the object and the screen is kept fixed, but the lens can be moved. Show that (a) if $d_T > 4f$, there will be two positions where the lens can be placed and a sharp image will be produced on the screen, and (b) if $d_T < 4f$, there will be no lens position where a sharp image is formed. (c) Determine the distance between the two lens positions in part (a), and the ratio of the image sizes.

* 88. A converging lens with focal length of 10.0 cm is placed in contact with a diverging lens with a focal length of 20.0 cm. What is the focal length of the combination and is the combination converging or diverging?

89. A 1.65-m-tall person stands 3.25 m from a convex mirror and notices that he looks precisely half as tall as he does in a plane mirror placed at the same distance. What is the radius of curvature of the convex mirror? (Assume that $\sin \theta \approx \theta$.)

90. Each student in a physics lab is assigned to find the location where a bright object may be placed in order that a concave mirror with radius of curvature $r = 40$ cm will produce an image three times the size of the object. Two students complete the assignment at different times using identical equipment, but when they compare notes later, they discover that their answers for the object distance are not the same. Explain why they do not necessarily need to repeat the lab, and justify your response with a calculation.

* 91. A lens whose index of refraction is n is submerged in a material whose index of refraction is n' ($n' \neq 1$). Derive the equivalent of Eqs. 23–8, 23–9, and 23–10 for this lens.

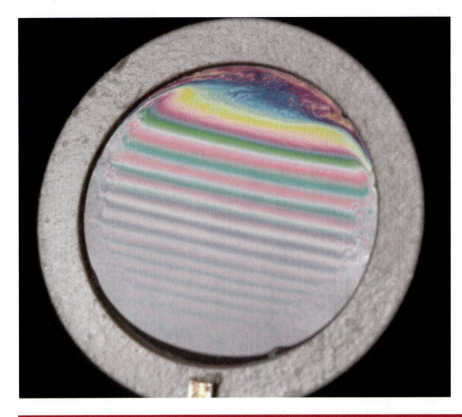

The wave nature of light nicely explains how light reflected from the front and back surfaces of this very thin film of soapy water interferes constructively to produce the bright colors. Which color we see depends on the thickness of the film at that point—the film is very thin at the top and gets thicker going down toward the bottom (because of gravity). Can you use the clues (colors of light) to determine the thickness of the film at any point? You should be able to after reading this chapter.

THE WAVE NATURE OF LIGHT

That light carries energy is obvious to anyone who has focused the Sun's rays with a magnifying glass on a piece of paper and burned a hole in it. But how does light travel, and in what form is this energy carried? In our discussion of waves in Chapter 11, we noted that energy can be carried from place to place in basically two ways: by particles or by waves. In the first case, material bodies or particles can carry energy, such as an avalanche or rushing water. In the second case, water waves and sound waves, for example, can carry energy over long distances even though mass itself does not travel these distances. In view of this, what can we say about the nature of light: does light travel as a stream of particles away from its source, or does it travel in the form of waves that spread outward from the source?

Historically, this question has turned out to be a difficult one. For one thing, light does not reveal itself in any obvious way as being made up of tiny particles nor do we see tiny light waves passing by as we do water waves. The evidence seemed to favor first one side and then the other until about 1830, when most physicists had accepted the wave theory. By the end of the nineteenth century, light was considered to be an *electromagnetic wave* (Chapter 22). In the early twentieth century, light was shown to have a particle nature as well, as we shall discuss in Chapter 27. Nonetheless, the wave theory of light remains valid and has proved very successful. We now investigate the evidence for the wave theory and how it has explained a wide range of phenomena.

Waves Versus Particles; Huygens' Principle and Diffraction

The Dutch scientist Christiaan Huygens (1629–1695), a contemporary of Newton, proposed a wave theory of light that had much merit. Still useful today is a technique he developed for predicting the future position of a wave front when an earlier position is known. This is known as **Huygens' principle** and can be stated as follows: *Every point on a wave front can be considered as a source of tiny wavelets that spread out in the forward direction at the speed of the wave itself. The new wave front is the envelope of all the wavelets*—that is, *the tangent to all of them.*

Huygens' principle

As a simple example of the use of Huygens' principle, consider the wave front *AB* in Fig. 24–1, which is traveling away from a source *S*. We assume the medium is *isotropic*—that is, the speed *v* of the waves is the same in all directions. To find the wave front a short time *t* after it is at *AB*, tiny circles are drawn with radius $r = vt$. The centers of these tiny circles are on the original wave front *AB* and the circles represent Huygens' (imaginary) wavelets. The tangent to all these wavelets, the line *CD*, is the new position of the wave front.

Huygens' principle is particularly useful when waves impinge on an obstacle and the wave fronts are partially interrupted. Huygens' principle predicts that waves bend in behind an obstacle, as shown in Fig. 24–2. This is just what water waves do, as we saw in Chapter 11 (Figs. 11–43 and 11–44). The bending of waves behind obstacles into the "shadow region" is known as **diffraction**. Since diffraction occurs for waves, but not for particles, it can serve as one means for distinguishing the nature of light.

Does light exhibit diffraction? In the mid-seventeenth century, a Jesuit priest, Francesco Grimaldi (1618–1663), had observed that when sunlight entered a darkened room through a tiny hole in a screen, the spot on the opposite wall was larger than would be expected from geometric rays. He also observed that the border of the image was not clear but was surrounded by colored fringes. Grimaldi attributed this to the diffraction of light.

Note that the ray model (Chapter 23) cannot account for diffraction, and it is important to be aware of the limitations of the ray model. Geometric op-

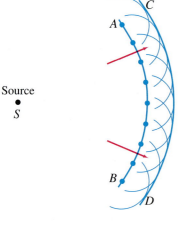

FIGURE 24–1 Huygens' principle used to determine wave front *CD* when *AB* is given.

FIGURE 24–2 Huygens' principle is consistent with diffraction (a) around the edge of an obstacle, (b) through a large hole, (c) through a small hole whose size is on the order of the wavelength of the wave.

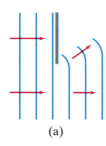

(a)

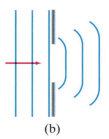

(b)

(c)

tics using rays is so successful in its limited sphere because normal openings and obstacles are much larger than the wavelength of the light, and so relatively little diffraction or bending occurs.

24–2 Huygens' Principle and the Law of Refraction

The laws of reflection and refraction were well known in Newton's time. The law of reflection could not distinguish between the two theories: waves versus particles. For when waves reflect from an obstacle, the angle of incidence equals the angle of reflection (Fig. 11–34). The same is true of particles—think of a tennis ball without spin striking a flat surface.

The law of refraction is another matter. Consider light entering a medium where it is bent toward the normal, as when it travels from air into water. As shown in Fig. 24–3, this effect can be constructed using Huygens' principle if we assume the speed of light is less in the second medium ($v_2 < v_1$). That is, in time t, the point B on wave front AB goes a distance $v_1 t$ to reach point D. Point A, on the other hand, travels a distance $v_2 t$ to reach point C. Huygens' principle is applied to points A and B to obtain the curved wavelets shown at C and D. The wave front is tangent to these two wavelets, so the new wave front is the line CD. Hence the rays, which are perpendicular to the wave fronts, bend toward the normal if $v_2 < v_1$, as drawn.[†] Newton favored a particle theory of light which predicted the opposite result that the speed of light would be greater in the second medium ($v_2 > v_1$). Thus the wave theory predicts that the speed of light in water, for example, is less than in air; and Newton's particle theory predicts the reverse. An experiment to actually measure the speed of light in water was performed in 1850 by the French physicist Jean Foucault, and it confirmed the wave-theory prediction. By then, however, the wave theory was already fully accepted, as we shall see in the next Section.

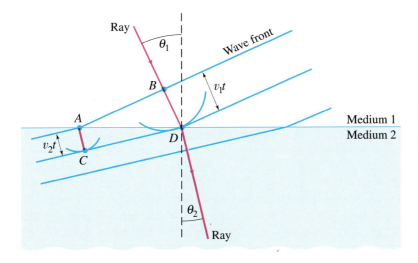

FIGURE 24–3 Refraction explained, using Huygens' principle.

[†]This is basically the same as the discussion around Fig. 11–42.

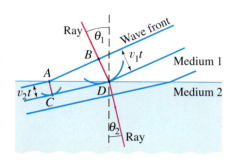

FIGURE 24–3 (Repeated.) Refraction explained, using Huygens' principle.

It is easy to show that Snell's law of refraction follows directly from Huygens' principle, given that the speed of light v in any medium is related to the speed in a vacuum, c, and the index of refraction, n, by Eq. 23–4: that is, $v = c/n$. From the Huygens' construction of Fig. 24–3, angle ADC is equal to θ_2 and angle BAD is equal to θ_1. Then for the two triangles that have the common side AD, we have

$$\sin \theta_1 = \frac{v_1 t}{AD}, \qquad \sin \theta_2 = \frac{v_2 t}{AD}.$$

We divide these two equations and obtain:

$$\frac{\sin \theta_1}{\sin \theta_2} = \frac{v_1}{v_2}.$$

Then, since $v_1 = c/n_1$ and $v_2 = c/n_2$,

$$n_1 \sin \theta_1 = n_2 \sin \theta_2,$$

which is Snell's law of refraction, Eq. 23–5. (The law of reflection can be derived from Huygens' principle in a similar way, and this is given as Problem 1 at the end of the chapter.)

Wavelength depends on n When a light wave travels from one medium to another, its frequency does not change, but its wavelength does. This can be seen from Fig. 24–3, where we assume each of the blue lines representing a wave front corresponds to a crest (peak) of the wave. Then

$$\frac{\lambda_2}{\lambda_1} = \frac{v_2 t}{v_1 t} = \frac{v_2}{v_1} = \frac{n_1}{n_2},$$

where, in the last step, we used Eq. 23–4, $v = c/n$. If medium 1 is a vacuum (or air), so $n_1 = 1$, $v_1 = c$, and we call λ_1 simply λ, then the wavelength in another medium of index of refraction n ($= n_2$) will be

$$\lambda_n = \frac{\lambda}{n}. \tag{24–1}$$

This result is consistent with the frequency f being unchanged since $c = f\lambda$. Combining this with $v = f\lambda_n$ in a medium where $v = c/n$ gives $\lambda_n = v/f = c/nf = f\lambda/nf = \lambda/n$, which checks.

Wave fronts can be used to explain how mirages are produced by refraction of light. Let us explain why, for example, on a hot day motorists sometimes see a mirage of water on the highway ahead of them, with distant vehicles seemingly reflected in it (Fig. 24–4a). On a hot day, there can be a layer of very hot air next to the roadway (made hot by the sun beating on the road). Hot air is less dense than cooler air, so the index of refraction is slightly lower in the hot air. In Fig. 24–4b, we see a diagram of light coming from one point on a distant car (on the right) heading left toward the observer. Wave fronts and two rays are shown. Ray A heads directly at the observer and follows a straight-line path, and represents the normal view of the car. Ray B is a ray initially directed slightly downward. But it does not hit the ground. Instead, ray B is bent slightly as it moves through layers of air of different index of refraction. The wave fronts, shown in blue in Fig. 24–4b, move slightly

(a)

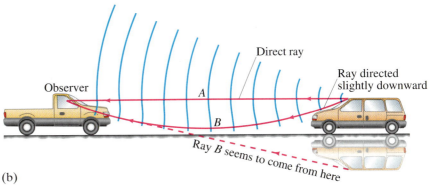

(b)

faster in the layers of air nearer the ground (see Fig. 24–3, and also the soldier analogy in Fig. 11–42). Thus ray B is bent as shown, and seems to the observer to be coming from below (dashed line) as if reflected off the road. Hence the mirage.

FIGURE 24–4 (a) A highway mirage. (b) Drawing (greatly exaggerated) showing wave fronts and rays to explain highway mirages. Note how the sections of the wavefronts near the ground move faster and so are further apart.

24–3 Interference—Young's Double-Slit Experiment

In 1801, the Englishman Thomas Young (1773–1829) obtained convincing evidence for the wave nature of light and was even able to measure the wavelengths for visible light. Figure 24–5a shows a schematic diagram of Young's famous double-slit experiment. Light from a single source (Young used the Sun) falls on a screen containing two closely spaced slits S_1 and S_2. If light consists of tiny particles, we might expect to see two bright lines on a screen placed behind the slits as in (b). But Young observed instead a series of bright lines as in (c). Young was able to explain this result as a **wave-interference** phenomenon. To see this, imagine plane waves of light of a single wavelength—called **monochromatic**, meaning "one color"—falling on the two slits as shown in Fig. 24–6. Because of diffraction, the waves leaving the two small slits spread out as shown. This is equivalent to the interference pattern produced when two rocks are thrown into a lake (Fig. 11–36), or when sound from two loudspeakers interferes (Fig. 12–17).

FIGURE 24–5 (a) Young's double-slit experiment. (b) If light consists of particles, we would expect to see two bright lines on the screen behind the slits. (c) Young observed many lines.

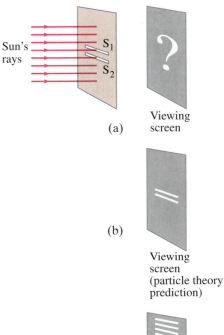

Sun's rays → S_1 S_2

(a)

Viewing screen

(b)

Viewing screen (particle theory prediction)

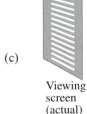

(c)

Viewing screen (actual)

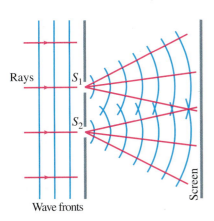

Rays

S_1

S_2

Wave fronts

Screen

FIGURE 24–6 If light is a wave, light passing through one of two slits should interfere with light passing through the other slit.

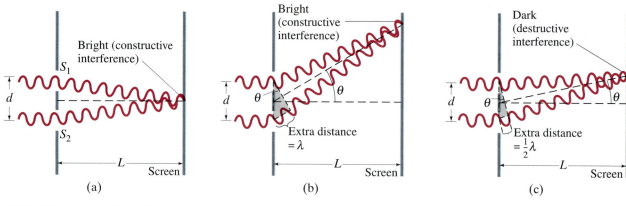

FIGURE 24-7 How the wave theory explains the pattern of lines seen in the double-slit experiment. (a) At the center of the screen the waves from each slit travel the same distance and are in phase. (b) At this angle θ, the lower wave travels an extra distance of one whole wavelength, and the waves are in phase; note from the shaded triangle that the extra distance equals $d \sin \theta$. (c) For this angle θ, the lower wave travels an extra distance equal to one-half wavelength, so the two waves arrive at the screen fully out of phase.

FIGURE 24-8
(a) Constructive interference.
(b) Destructive interference.
(See also Section 11–11.)

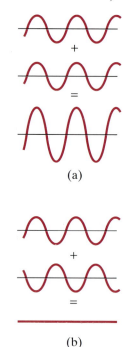

To see how an interference pattern is produced on the screen, we make use of Fig. 24–7. Waves of wavelength λ are shown entering the slits S_1 and S_2, which are a distance d apart. The waves spread out in all directions after passing through the slits, but they are shown only for three different angles θ. In Fig. 24–7a, the waves reaching the center of the screen are shown ($\theta = 0$). The waves from the two slits travel the same distance, so they are in phase: a crest of one wave arrives at the same time as a crest of the other wave. Hence the amplitudes of the two waves add to form a larger amplitude as shown in Fig. 24–8a. This is **constructive interference**, and there is a bright spot at the center of the screen. Constructive interference also occurs when the paths of the two rays differ by one wavelength (or any whole number of wavelengths), as shown in Fig. 24–7b. But if one ray travels an extra distance of one-half wavelength (or $\frac{3}{2}\lambda$, $\frac{5}{2}\lambda$, and so on), the two waves are exactly out of phase when they reach the screen: the crests of one wave arrive at the same time as the troughs of the other wave, and so they add to produce zero amplitude (Fig. 24–8b). This is **destructive interference**, and the screen is dark, Fig. 24–7c. Thus, there will be a series of bright and dark lines (or **fringes**) on the viewing screen.

To determine exactly where the bright lines fall, first note that Fig. 24–7 is somewhat exaggerated; in real situations, the distance d between the slits is very small compared to the distance L to the screen. The rays from each slit for each case will therefore be essentially parallel and θ is the angle they make with the horizontal. From the shaded right triangle in Fig. 24–7b, we can see that the extra distance traveled by the lower ray is $d \sin \theta$. **Constructive interference** will occur, and a bright fringe will appear on the screen, when $d \sin \theta$ equals a whole number of wavelengths:

$$d \sin \theta = m\lambda, \qquad m = 0, 1, 2, \cdots. \qquad \begin{bmatrix} \text{constructive} \\ \text{interference} \end{bmatrix} \quad \textbf{(24–2a)}$$

The value of m is called the **order** of the interference fringe. The first order ($m = 1$), for example, is the first fringe on each side of the central

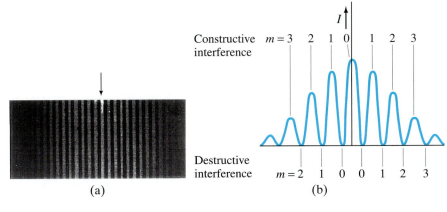

Constructive interference $m = 3$ 2   1   0   1   2   3

Destructive interference $m = 2$ 1   0   0   1   2   3

(a)   (b)

FIGURE 24-9
(a) Interference fringes produced by double-slit experiment and detected by photographic film placed on the viewing screen. The arrow marks the central fringe. (b) Intensity of light in the interference pattern. Also shown are values of m for Eq. 24–2a (constructive interference) and Eq. 24–2b (destructive interference).

fringe (at $\theta = 0$). Destructive interference occurs when the extra distance $d \sin \theta$ is $\frac{1}{2}, \frac{3}{2}$, and so on, wavelengths:

$$d \sin \theta = (m + \tfrac{1}{2})\lambda, \qquad m = 0, 1, 2, \cdots. \qquad \begin{bmatrix} \text{destructive} \\ \text{interference} \end{bmatrix} \quad \textbf{(24–2b)}$$

The intensity of the bright fringes is greatest for the central fringe ($m = 0$) and decreases for higher orders, as shown in Fig. 24–9.

EXAMPLE 24–1   **Line spacing for double-slit interference.**   A screen containing two slits 0.100 mm apart is 1.20 m from the viewing screen. Light of wavelength $\lambda = 500$ nm falls on the slits from a distant source. Approximately how far apart will the bright interference fringes be on the screen?

SOLUTION   Given $d = 0.100$ mm $= 1.00 \times 10^{-4}$ m, $\lambda = 500 \times 10^{-9}$ m, and $L = 1.20$ m, the first-order fringe ($m = 1$) occurs at an angle θ given by

$$\sin \theta_1 = \frac{m\lambda}{d} = \frac{(1)(500 \times 10^{-9} \text{ m})}{1.00 \times 10^{-4} \text{ m}} = 5.00 \times 10^{-3}.$$

This is a very small angle, so we can take $\sin \theta = \theta$, with θ in radians. The first-order fringe will occur a distance x_1 above the center of the screen (see Fig. 24–10) given by $x_1/L = \tan\theta_1 = \theta_1$, so

$$x_1 = L\theta_1 = (1.20 \text{ m})(5.00 \times 10^{-3}) = 6.00 \text{ mm}.$$

The second fringe ($m = 2$) will occur at

$$x_2 = L\theta_2 = l\frac{2\lambda}{d} = 12.0 \text{ mm}$$

above the center, and so on. Thus the fringes are 6.00 mm apart.

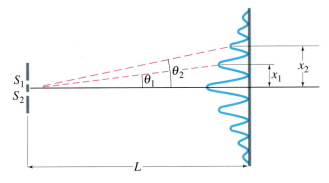

S_1
S_2
θ_1 θ_2
x_1
x_2
L

FIGURE 24-10   For small angles, the interference fringes occur at distance $x = \theta L$ above the center fringe ($m = 0$); θ_1 and x_1 are for the first order fringe ($m = 1$), θ_2 and x_2 are for $m = 2$.

CONCEPTUAL EXAMPLE 24–2 **Changing the wavelength.** (*a*) What happens to the interference pattern shown in Fig. 24–10, Example 24–1, if the incident light (500 nm) is replaced by light of wavelength 700 nm? (*b*) What happens instead if the slits are moved farther apart?

RESPONSE (*a*) When λ increases in Eq. 24–2a but *d* stays the same, then θ increases and the interference pattern spreads out. (*b*) Increasing the slit spacing *d* reduces θ for each order, so the lines are closer together.

Wavelength (or frequency) determines color

From Eqs. 24–2 we can see that, except for the zeroth-order fringe at the center, the position of the fringes depends on wavelength. Consequently, when white light falls on the two slits, as Young found in his experiments, the central fringe is white, but the first- (and higher-) order fringes contain a spectrum of colors like a rainbow; θ was found to be smallest for violet light and largest for red. By measuring the position of these fringes, Young was the first to determine the wavelengths of visible light (using Eqs. 24–2). In doing so, he showed that what distinguishes different colors physically is their wavelength, an idea put forward earlier by Grimaldi in 1665.

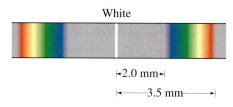

White

|←2.0 mm→|

|←——3.5 mm——→|

FIGURE 24–11 Example 24–3.

EXAMPLE 24–3 **Wavelengths from double-slit interference.** White light passes through two slits 0.50 mm apart and an interference pattern is observed on a screen 2.5 m away. The first-order fringe resembles a rainbow with violet and red light at either end. The violet light falls about 2.0 mm and the red 3.5 mm from the center of the central white fringe (Fig. 24–11). Estimate the wavelengths of the violet and red lights.

SOLUTION We use Eq. 24–2a with $m = 1$ and $\sin \theta = \theta$. Then for violet light, $x = 2.0$ mm, so (see also Fig. 24–10)

$$\lambda = \frac{d\theta}{m} = \frac{d}{m}\frac{x}{L} = \left(\frac{5.0 \times 10^{-4}\,\text{m}}{1}\right)\left(\frac{2.0 \times 10^{-3}\,\text{m}}{2.5\,\text{m}}\right) = 4.0 \times 10^{-7}\,\text{m},$$

or 400 nm. For red light, $x = 3.5$ mm, so

$$\lambda = \frac{d}{m}\frac{x}{L} = \left(\frac{5.0 \times 10^{-4}\,\text{m}}{1}\right)\left(\frac{3.5 \times 10^{-3}\,\text{m}}{2.5\,\text{m}}\right) = 7.0 \times 10^{-7}\,\text{m} = 700\,\text{nm}.$$

Coherent and incoherent sources

The two slits in Fig. 24–7 act as if they were two sources of radiation. They are called **coherent sources** because the waves leaving them bear the same phase relationship to each other at all times (because ultimately the waves come from a single source to the left of the two slits in Fig. 24–7). An interference pattern is observed only when the sources are coherent. If two tiny lightbulbs replaced the two slits, an interference pattern would not be seen. The light emitted by one lightbulb would have a random phase with respect to the second bulb, and the screen would be more or less uniformly illuminated. Two such sources, whose output waves bear no fixed relationship to each other, are called **incoherent sources**.

24–4 The Visible Spectrum and Dispersion

The two most obvious properties of light are readily describable in terms of the wave theory of light: intensity (or brightness) and color. The **intensity** of light is the energy it carries per unit time, and is related to the square of the amplitude of the wave, just as for any wave (see Section 11–9, or Eqs. 22–7 and 22–8). The **color** of the light is related to the wavelength or frequency of the light. Visible light—that to which our eyes are sensitive—falls in the wavelength range of about 400 nm to 750 nm.[†] This is known as the **visible spectrum**, and within it lie the different colors from violet to red, as shown in Fig. 24–12. Light with wavelength shorter than

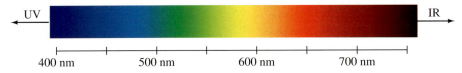

FIGURE 24–12   The spectrum of visible light, showing the range of wavelengths for the various colors.

400 nm is called **ultraviolet** (UV), and light with wavelength greater than 750 nm is called **infrared** (IR).[‡] Although human eyes are not sensitive to UV or IR, some types of photographic film do respond to them.

A prism separates white light into a rainbow of colors, as shown in Fig. 24–13. This happens because the index of refraction of a material depends on the wavelength, as shown for several materials in Fig. 24–14. White light is a mixture of all visible wavelengths, and when incident on a prism, as in Fig. 24–15, the different wavelengths are bent to varying degrees. Because the index of refraction is greater for the shorter wavelengths, violet light is bent the most and red the least as indicated. This spreading of white light into the full spectrum is called **dispersion**.

FIGURE 24–13   White light passing through a prism is broken down into its constituent colors.

Dispersion

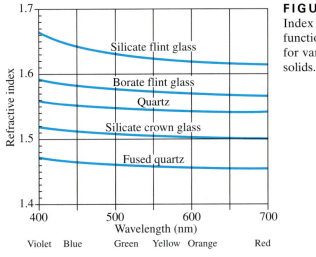

FIGURE 24–14 Index of refraction as a function of wavelength for various transparent solids.

FIGURE 24–15   White light dispersed by a prism into the visible spectrum.

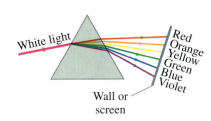

[†]Sometimes the angstrom (Å) unit is used when referring to light: $1 \text{ Å} = 1 \times 10^{-10}$ m. Then visible light falls in the wavelength range of 4000 Å to 7500 Å.

[‡]The complete electromagnetic spectrum is illustrated in Fig. 22–10.

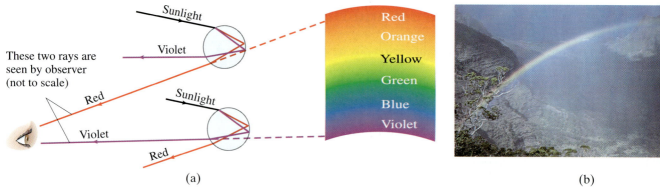

FIGURE 24–16 (a) Ray diagram explaining how a rainbow (b) is formed.

PHYSICS APPLIED

Rainbows

Rainbows are a spectacular example of dispersion—by drops of water. You can see rainbows when you look at falling water droplets with the Sun at your back. Figure 24–16 shows how red and violet rays are bent by spherical water droplets and are reflected off the back surface. Red is bent the least and so reaches the observer's eyes from droplets higher in the sky, as shown in the diagram. Thus the top of the rainbow is red.

Diamonds

Diamonds achieve their brilliance from a combination of dispersion and total internal reflection. Because diamonds have a very high index of refraction of about 2.4, the critical angle for total internal reflection is only 25°. Incident light therefore strikes many of the internal surfaces before it strikes one at less than 25° and emerges. After many such reflections, the light has traveled far enough that the colors have become sufficiently separated to be seen individually and brilliantly by the eye after leaving the crystal.

24–5 Diffraction by a Single Slit or Disk

Young's double-slit experiment put the wave theory of light on a firm footing. But full acceptance came only with studies on diffraction more than a decade later.

We have already discussed diffraction briefly with regard to water waves (Section 11–13) as well as for light (Section 24–1), and we have seen that it refers to the spreading or bending of waves around edges. Let's look in more detail.

A part of the history of the wave theory of light belongs to Augustin Fresnel (1788–1827), who in 1819 presented to the French Academy a wave theory of light that predicted and explained interference and diffraction effects. Almost immediately Siméon Poisson (1781–1840) pointed out a counterintuitive inference: according to Fresnel's wave theory, if light from a point source were to fall on a solid disk, then light diffracted around the edges should interfere constructively at the center of the shadow (Fig. 24–17). That prediction seemed very unlikely. But when the experiment was actually carried out (by Francois Arago), the bright spot was seen at the very center of the shadow! This was strong evidence for the wave theory.

Figure 24–18a is a photograph of the shadow cast by a coin using a (nearly) point source of light (a laser in this case). The bright spot is clearly present at the center. Note also that there are bright and dark fringes beyond the shadow. These resemble the interference fringes of a double slit. Indeed, they are due to interference of waves diffracted around different parts of the disk, and the whole is referred to as a **diffraction pattern**. A diffraction pattern exists around any sharp object illuminated by a point source, as shown in Figs. 24–18b and c. We are not always aware of them because most sources of light in everyday life are not points, so light from different parts of the source washes out the pattern.

To see how a diffraction pattern arises, we will analyze the important case of monochromatic light passing through a narrow slit. We will assume that parallel rays (or plane waves) of light fall on the slit of width D, and that the viewing screen is far away. As we know from studying water waves and from Huygens' principle, the waves passing through the slit spread out in all directions. We will now examine how the waves passing through different parts of the slit interfere with each other.

The (un)expected diffraction spot

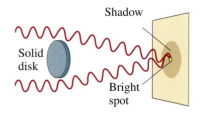

FIGURE 24–17 If light is a wave, a bright spot will appear at the center of the shadow of a solid disk illuminated by a point source of monochromatic light.

FIGURE 24–18 Diffraction pattern of (a) a penny, (b) a razor blade, (c) a single slit, each illuminated by a (nearly) point source of monochromatic light.

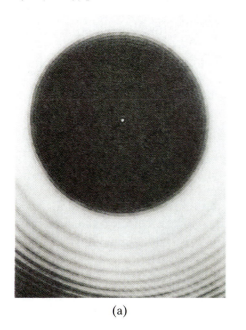

(a)

(b)

(c)

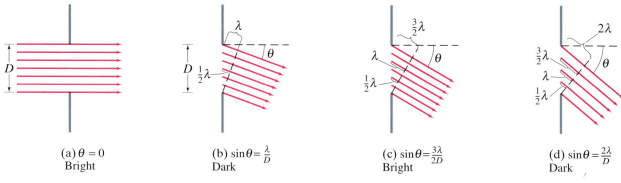

(a) $\theta = 0$
Bright

(b) $\sin\theta = \frac{\lambda}{D}$
Dark

(c) $\sin\theta = \frac{3\lambda}{2D}$
Bright

(d) $\sin\theta = \frac{2\lambda}{D}$
Dark

FIGURE 24–19 Analysis of diffraction pattern formed by light passing through a narrow slit.

Parallel rays of monochromatic light pass through the narrow slit as show in Fig. 24–19a. The light falls on a screen which is assumed to be very far away, so the rays heading for any point are essentially parallel. First we consider rays that pass straight through as in Fig. 24–19a. They are all in phase, so there will be a central bright spot on the screen. In Fig. 24–19b, we consider rays moving at an angle θ such that the ray from the top of the slit travels exactly one wavelength farther than the ray from the bottom edge. The ray passing through the very center of the slit will travel one-half wavelength farther than the ray at the bottom of the slit. These two rays will be exactly out of phase with one another and so will destructively interfere. Similarly, a ray slightly above the bottom one will cancel a ray that is the same distance above the central one. Indeed, each ray passing through the lower half of the slit will cancel with a corresponding ray passing through the upper half. Thus, all the rays destructively interfere in pairs, and so no light will reach the viewing screen at this angle. The angle θ at which this takes place can be seen from the diagram to occur when $\lambda = D \sin \theta$, so

Diffraction equation (angular width of central spot)

$$\sin \theta = \frac{\lambda}{D}.$$ [first minimum]   **(24–3a)**

The light intensity is a maximum at $\theta = 0°$ and decreases to a minimum (intensity = zero) at the angle θ given by Eq. 24–3a.

Now consider a larger angle θ such that the top ray travels $\frac{3}{2}\lambda$ farther than the bottom ray, as in Fig. 24–19c. In this case, the rays from the bottom third of the slit will cancel in pairs with those in the middle third because they will be $\lambda/2$ out of phase. However, light from the top third of the slit will still reach the screen, so there will be a bright spot, but not nearly as bright as the central spot at $\theta = 0°$. For an even larger angle θ such that the top ray travels 2λ farther than the bottom ray, Fig. 24–19d, rays from the bottom quarter of the slit will cancel with those in the quarter just above it because the path lengths differ by $\lambda/2$. And the rays through the quarter of the slit just above center will cancel with those through the top quarter. At this angle there will again be a minimum of zero intensity in the diffraction pattern. A plot of the intensity as a function of angle is shown in Fig. 24–20. This corresponds well with the photo of Fig. 24–18c. Notice that minima (zero intensity) occur at

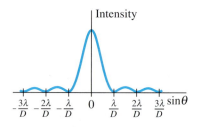

FIGURE 24–20 Intensity in the diffraction pattern of a single slit as a function of $\sin \theta$. Note that the central maximum is not only much higher than the maxima to each side, but it is also twice as wide ($2\lambda/D$ wide) as any of the others (only λ/D wide each).

Single slit diffraction minima

$$D \sin \theta = m\lambda, \qquad m = 1, 2, 3, \cdots, \qquad \text{[minima]} \quad \textbf{(24–3b)}$$

but *not* at $m = 0$ where there is the strongest maximum. Between the minima, smaller intensity maxima occur. [Note that the *minima* for a diffraction pattern, Eq. 24–3b, satisfy a criterion very similar to that for the *maxima* (bright spots) for double-slit interference, Eq. 24–2a.]

EXAMPLE 24–4 **Single-slit diffraction maximum.** Light of wavelength 750 nm passes through a slit 1.0×10^{-3} mm wide. How wide is the central maximum (a) in degrees, and (b) in centimeters, on a screen 20 cm away?

SOLUTION (a) The first minimum occurs at

$$\sin \theta = \frac{\lambda}{D} = \frac{7.5 \times 10^{-7} \text{ m}}{1 \times 10^{-6} \text{ m}} = 0.75.$$

So $\theta = 49°$. This is the angle between the center and the first minimum, Fig. 24–21. The angle subtended by the whole central maximum, between the minima above and below the center, is twice this, or 98°.
(b) The width of the central maximum is $2x$, where $\tan \theta = x/20$ cm. So $2x = 2(20 \text{ cm})(\tan 49°) = 46$ cm. A large width of the screen will be illuminated, but it will not normally be very bright since the amount of light that passes through such a small slit will be small and it is spread over a large area.

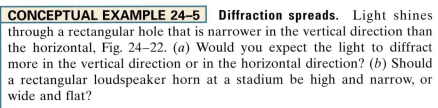

FIGURE 24–21
Example 24–4.

CONCEPTUAL EXAMPLE 24–5 **Diffraction spreads.** Light shines through a rectangular hole that is narrower in the vertical direction than the horizontal, Fig. 24–22. (a) Would you expect the light to diffract more in the vertical direction or in the horizontal direction? (b) Should a rectangular loudspeaker horn at a stadium be high and narrow, or wide and flat?

RESPONSE (a) From Eq. 24–3a we can see that if we make the slit (width D) narrower, the pattern spreads out more. (This is consistent with our earlier study of waves in Chapter 11.) Hence the diffraction through the rectangular hole will be wider vertically, since the aperture in that direction is smaller.
(b) For the loudspeaker, the sound pattern desired is one spread out horizontally, so the horn should be tall and narrow.

FIGURE 24–22
Example 24–5.

24–6 Diffraction Grating

A large number of equally spaced parallel slits is called a **diffraction grating**, although the term "interference grating" might be as appropriate. Gratings can be made by precision machining of very fine parallel lines on a glass plate. The untouched spaces between the lines serve as the slits. Photographic transparencies of an original grating serve as inexpensive gratings. Gratings containing 10,000 lines per centimeter are common today, and are very useful for precise measurements of wavelengths. A diffraction grating containing slits is called a **transmission grating**. **Reflection gratings** are also possible; they can be made by ruling fine lines on a metallic or glass surface from which light is reflected and analyzed. The analysis is basically the same as for a transmission grating, which we now discuss.

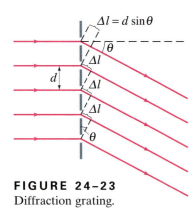

$\Delta l = d \sin\theta$

FIGURE 24–23
Diffraction grating.

*Diffraction grating
maxima (m = order)*

The analysis of a diffraction grating is much like that of Young's double-slit experiment. We assume parallel rays of light are incident on the grating as shown in Fig. 24–23. We also assume that the slits are narrow enough so that diffraction by each of them spreads light over a very wide angle on a distant screen behind the grating, and interference can occur with light from all the other slits. Light rays that pass through each slit without deviation ($\theta = 0°$) interfere constructively to produce a bright line at the center of the screen. Constructive interference also occurs at an angle θ such that rays from adjacent slits travel an extra distance of $\Delta l = m\lambda$, where m is an integer. Thus, if d is the distance between slits, then we see from Fig. 24–23 that $\Delta l = d \sin\theta$, and

$$\sin\theta = \frac{m\lambda}{d}, \qquad m = 0, 1, 2, \cdots \qquad \text{[principal maxima]} \quad \textbf{(24–4)}$$

is the criterion to have a brightness maximum. This is the same equation as for the double-slit situation, and again m is called the order of the pattern.

There is an important difference between a double-slit and a multiple-slit pattern, however. The bright maxima are much *sharper* and *narrower* for a grating. Why this happens can be seen as follows. Suppose that the angle θ

*Why more slits
yield sharper peaks*

is increased just slightly beyond that required for a maximum. In the case of only two slits, the two waves will be only slightly out of phase, so nearly full constructive interference occurs. This means the maxima are wide (see Fig. 24–10). For a grating, the waves from two adjacent slits will also not be significantly out of phase. But waves from one slit and those from a second one a few hundred slits away may be exactly out of phase; all or nearly all the light will cancel in pairs in this way. For example, suppose the angle θ is different from its first-order maximum so that the extra path length for a pair of adjacent slits is not exactly λ but rather 1.0010λ. The wave through one slit and another one 500 slits below will be out of phase by 1.5000λ, or exactly $1\frac{1}{2}$ wavelengths, so the two will cancel. A pair of slits, one below each of these, will also cancel. That is, the light from slit 1 cancels with that from slit 501; light from slit 2 cancels with that from slit 502, and so on. Thus even for a tiny angle[†] corresponding to an extra path length of $\frac{1}{1000}\lambda$, there is much destructive interference, and so the maxima are very narrow. The more lines there are in a grating, the sharper will be the peaks (see Fig. 24–24). Because a grating produces much sharper (and brighter) lines than two slits alone, it is a far more precise device for measuring wavelengths.

Suppose the light striking a diffraction grating is not monochromatic, but rather consists of two or more distinct wavelengths. Then for all orders other than $m = 0$, each wavelength will produce a maximum at a different angle (Fig. 24–25a), just as for a double slit. If white light strikes a grating, the central ($m = 0$) maximum will be a sharp white peak. But for all other orders, there will be a distinct spectrum of colors spread out over a certain angular width, Fig. 24–25b. Because a diffraction grating spreads out light into its component wavelengths, the resulting pattern is called a **spectrum**.

FIGURE 24–24 Intensity as a function of viewing angle θ (or position on screen) for (a) two slits, (b) six slits. For a diffraction grating, the number of slits is very large ($\sim 10^4$) and the peaks are narrower still.

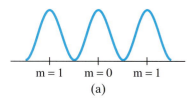

$m = 1 \quad m = 0 \quad m = 1$
(a)

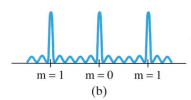

$m = 1 \quad m = 0 \quad m = 1$
(b)

[†]Depending on the total number of slits, there may or may not be complete cancellation for such an angle, so there will be very tiny peaks between the main maxima (see Fig. 24–24b), but they are usually much too small to be seen.

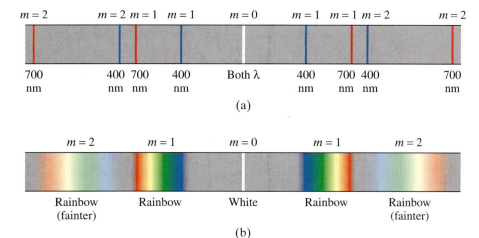

FIGURE 24–25 Spectra produced by a grating: (a) two wavelengths, 400 nm and 700 nm; (b) white light. The second order will normally be dimmer than the first order. (Higher orders are not shown.) If grating spacing is small enough, the second and higher orders will be missing.

EXAMPLE 24–6 **Diffraction grating: lines.** Calculate the first- and second-order angles for light of wavelength 400 nm and 700 nm if the grating contains 10,000 lines/cm.

SOLUTION The grating contains 10^4 lines/cm $= 10^6$ lines/m, which means the separation between slits is $d = (1/10^6)$ m $= 1.0 \times 10^{-6}$ m. In first order ($m = 1$), the angles are

$$\sin \theta_{400} = \frac{m\lambda}{d} = \frac{(1)(4.0 \times 10^{-7} \text{ m})}{1.0 \times 10^{-6} \text{ m}} = 0.400$$

$$\sin \theta_{700} = 0.700$$

so $\theta_{400} = 23.6°$ and $\theta_{700} = 44.4°$. In second order,

$$\sin \theta_{400} = \frac{(2)(4.0 \times 10^{-7} \text{ m})}{1.0 \times 10^{-6} \text{ m}} = 0.800$$

$$\sin \theta_{700} = 1.40$$

so $\theta_{400} = 53.1°$, but the second order does not exist for $\lambda = 700$ nm because $\sin \theta$ cannot exceed 1. No higher orders will appear.

EXAMPLE 24–7 **Spectra overlap.** White light containing wavelengths from 400 nm to 750 nm strikes a grating containing 4000 lines/cm. Show that the blue at $\lambda = 450$ nm of the third-order spectrum overlaps the red at 700 nm of the second order.

SOLUTION The grating spacing is $d = (1/4000)$ cm $= 2.50 \times 10^{-6}$ m. The blue of the third order occurs at an angle θ given by

$$\sin \theta = \frac{m\lambda}{d} = \frac{(3)(4.50 \times 10^{-7} \text{ m})}{(2.50 \times 10^{-6} \text{ m})} = 0.540.$$

Red in second order occurs at

$$\sin \theta = \frac{(2)(7.00 \times 10^{-7} \text{ m})}{(2.50 \times 10^{-6} \text{ m})} = 0.560$$

which is a greater angle; so the second order overlaps into the beginning of the third-order spectrum.

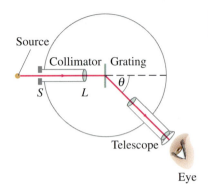

Source
Collimator   Grating
S   L θ
Telescope
Eye

FIGURE 24–26 Spectrometer or spectroscope.

The diffraction grating is the essential component of a spectrometer, a device for precise measurement of wavelengths, and we discuss it next.

24–7 The Spectrometer and Spectroscopy

A **spectrometer** or **spectroscope**, Fig. 24–26, is a device[†] to measure wavelengths accurately using a diffraction grating (discussed in Section 24–6), or a prism, to separate different wavelengths of light. Light from a source passes through a narrow slit S in the collimator. The slit is at the focal point of the lens L, so parallel light falls on the grating. The movable telescope can bring the rays to a focus. Nothing will be seen in the viewing telescope unless it is positioned at an angle θ that corresponds to a diffraction peak (first order is usually used) of a wavelength emitted by the source. The angle θ can be measured[‡] to very high accuracy, so the wavelength of a line can be determined to high accuracy using Eq. 24–4:

$$\lambda = \frac{d}{m} \sin \theta,$$

where m is an integer representing the order, and d is the distance between grating lines. (The line you see in a spectrometer corresponding to each wavelength is actually an image of the slit S. The narrower the slit, the narrower—but dimmer—the line is, and the more precisely we can measure its angular position. If the light contains a continuous range of wavelengths, then a continuous spectrum is seen in the spectroscope.)

In some spectrometers, a reflection grating is used, and sometimes a prism. A prism works because of dispersion (Section 24–4), bending light of different wavelengths into different angles.

An important use of a spectrometer is for the identification of atoms or molecules. When a gas is heated or a large electric current is passed through it, the gas emits a characteristic **line spectrum**. That is, only certain discrete wavelengths of light are emitted, and these are different for different elements and compounds.[§] Figure 24–27 shows the line spectra for a number of elements in the gas state. Line spectra occur only for gases at high temperatures and low pressure. The light from heated solids, such as a lightbulb filament, and even from a dense gaseous object such as the Sun, produces a **continuous spectrum** including a wide range of wavelengths.

Figure 24–27 also shows the Sun's "continuous spectrum," which contains a number of *dark* lines (only the most prominent are shown), called **absorption lines**. Atoms and molecules can absorb light at the same wavelengths at which they emit light. The Sun's absorption lines are due to absorption by atoms and molecules in the cooler outer atmosphere of the Sun, as well as by atoms and molecules in the Earth's atmosphere. A careful analysis of all these thousands of lines reveals that at least two thirds of all elements are present in

Line spectra

[†]If the spectrum of a source is recorded (say, on film) rather than viewed by the eye, the device is called a **spectrometer** or spectrograph, as compared to a **spectroscope**, which is for viewing only; but these terms are often used interchangeably. Devices that can also measure the intensity of light of a given wavelength are called **spectrophotometers**.

[‡]The angle θ for a given wavelength is usually measured on both sides of center because the grating cannot always be aligned precisely; the average of the two values is then taken.

[§]Why atoms and molecules emit line spectra was a great mystery for many years and played a central role in the development of modern quantum theory, as we shall see in Chapter 27.

Atomic hydrogen

Mercury

Sodium

Solar absorption spectrum

FIGURE 24–27 Line spectra for the gases indicated, and spectrum from the Sun showing absorption lines.

the Sun's atmosphere. The presence of elements in the atmosphere of other planets, in interstellar space, and in stars is also determined by spectroscopy.

Spectroscopy is useful for determining the presence of certain types of molecules in laboratory specimens where chemical analysis would be difficult. For example, biological DNA and different types of protein absorb light in particular regions of the spectrum (such as in the UV). The material to be examined, which is often in solution, is placed in a monochromatic light beam whose wavelength is chosen by placement of a prism or diffraction grating. The amount of absorption, as compared to a standard solution without the specimen, can reveal not only the presence of a particular type of molecule, but also its concentration.

Light emission and absorption also occur outside the visible part of the spectrum, such as in the UV and IR regions. Glass absorbs light in these regions, so reflection gratings and mirrors (in place of lenses) are used. Special types of film or detectors are used for detection.

➡ **PHYSICS APPLIED**

Chemical and biochemical analysis

24–8 Interference by Thin Films

Interference of light gives rise to many everyday phenomena such as the bright colors reflected from soap bubbles and from thin oil films on water (Fig. 24–28). In these and other cases, the colors are a result of constructive interference between light reflected from the two surfaces of the thin film.

FIGURE 24–28 Thin film interference patterns seen in (a) soap bubbles, (b) thin films of soapy water, and (c) a thin layer of oil on the water of a street puddle.

(a)

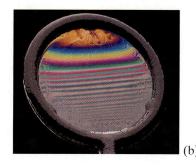

(b)

(c)

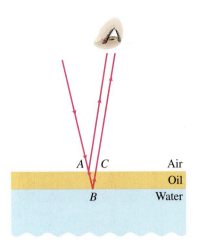

A   C   Air
            Oil
     B   Water

FIGURE 24–29 Light reflected from the upper and lower surfaces of a thin film of oil lying on water. Analysis assumes the light strikes the surface perpendicularly, but is shown here at a slight angle for clarity.

Newton's rings

To see how this happens, consider a smooth surface of water on top of which is a thin uniform layer of another substance, say an oil whose index of refraction is less than that of water (we'll see why we assume this in a moment); see Fig. 24–29. Part of the incident light (say, from the Sun or street lights) is reflected at *A* on the top surface, and part of that transmitted is reflected at *B* on the lower surface. The part reflected at the lower surface must travel the extra distance *ABC*. If the distance *ABC* is equal to one or a whole number of wavelengths, the two waves will interfere constructively and the light will be bright. But if *ABC* equals $\frac{1}{2}\lambda, \frac{3}{2}\lambda$, and so on, the two waves will be exactly out of phase and destructive interference occurs. The wavelength λ is *the wavelength in the film* (see Eq. 24–1).

When white light falls on such a film, the path *ABC* will equal λ (or $m\lambda$, with $m =$ an integer) for only one wavelength at a given viewing angle. This color will be seen as very bright. For light viewed at a slightly different angle, the path *ABC* will be longer or shorter and a different color will undergo constructive interference. Thus, for an extended (nonpoint) source emitting white light, a series of bright colors will be seen next to one another. Variations in thickness of the film will also alter the path length *ABC* and therefore affect the color of light that is most strongly reflected.

When a curved glass surface is placed in contact with a flat glass surface, Fig. 24–30, a series of concentric rings is seen when illuminated from above by monochromatic light. These are called **Newton's rings**[†] and they are due to interference between rays reflected by the top and bottom surfaces of the very thin *air gap* between the two pieces of glass. Because this gap (which is equivalent to a thin film) increases in width from the central contact point out to the edges, the extra path length for the lower ray (equal to *BCD*) varies; where it equals $0, \frac{1}{2}\lambda, \lambda, \frac{3}{2}\lambda, 2\lambda$, and so on, it corresponds to constructive and destructive interference; and this gives rise to the series of bright and dark lines seen in Fig. 24–30b.

The point of contact of the two glass surfaces (*A* in Fig. 24–30a) is dark in Fig. 24–30b. Since the path difference is zero here, we expect the rays reflected from each surface to be in phase and so this central point ought to be bright. But it is dark, which tells us the two rays must be completely out of phase. This can happen only because one of the waves undergoes a change in phase of 180° upon reflection, corresponding to $\frac{1}{2}$ cycle or $\frac{1}{2}\lambda$. Indeed, this and other experiments reveal that *a beam of light reflected by a material whose*

FIGURE 24–30 Newton's rings.

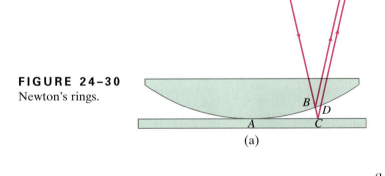

(a)

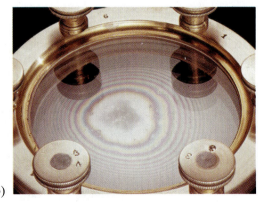

(b)

[†]Although Newton gave an elaborate description of them, they had been first observed and described by his contemporary, Robert Hooke. Newton did not realize their significance in support of a wave theory of light.

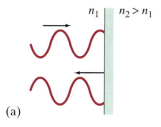
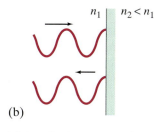

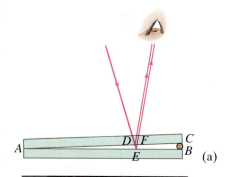

FIGURE 24–31 (a) Reflected ray changes phase by $\frac{1}{2}\lambda$ if $n_2 > n_1$, but does not (b) if $n_2 < n_1$.

FIGURE 24–32 (a) Light rays reflected from the upper and lower surfaces of a thin wedge of air interfere to produce bright and dark bands. (b) Pattern observed when glass plates are optically flat; (c) pattern when plates are not so flat. See Example 24–8.

index of refraction is greater than that of the material in which it is traveling, *changes phase by $\frac{1}{2}$ cycle*. See Fig. 24–31. If the refractive index is less than that of the material in which the light is traveling, no phase change occurs.[†] (This corresponds to the reflection of a wave traveling along a rope when it reaches the end; as we saw in Fig. 11–35, if the end is tied down, the wave changes phase and the pulse flips over, but if the end is free, no phase change occurs.) Thus the ray reflected by the curved surface above the air gap in Fig. 24–30a undergoes no change in phase. That reflected at the lower surface, where the beam in air strikes the glass, undergoes a $\frac{1}{2}\lambda$ phase change. Thus the two rays reflected at the point of contact A of the two glass surfaces (where the air gap approaches zero thickness) will be $\frac{1}{2}\lambda$ out of phase, and a dark spot occurs. Other dark bands will occur when the path difference BCD in Fig. 24–30a is equal to an integral number of wavelengths. Bright bands will occur when the path difference is $\frac{1}{2}\lambda$, $\frac{3}{2}\lambda$, and so on, because the phase change at one surface effectively adds another $\frac{1}{2}\lambda$.

EXAMPLE 24–8 **Thin film of air, wedge-shaped.** A very fine wire 7.35×10^{-3} mm in diameter is placed between two flat glass plates as in Fig. 24–32a. Light whose wavelength in air is 600 nm falls (and is viewed) perpendicular to the plates, and a series of bright and dark bands is seen, Fig. 24–32b. How many light and dark bands will there be in this case? Will the area next to the wire be bright or dark?

SOLUTION The thin film is the wedge of air between the two glass plates. Because of the phase change at the lower surface, there will be a dark band when the path difference is 0, λ, 2λ, 3λ, and so on. Since the light rays are perpendicular to the plates, the extra path length equals $2t$, where t is the thickness of the air gap at any point; so dark bands occur where

$$2t = m\lambda, \qquad m = 0, 1, 2, \cdots.$$

Bright bands occur when $2t = (m + \frac{1}{2})\lambda$, where m is an integer. At the position of the wire, $t = 7.35 \times 10^{-6}$ m. At this point there will be $(2)(7.35 \times 10^{-6}\,\text{m})/(6.00 \times 10^{-7}\,\text{m}) = 24.5$ wavelengths; this is a "half integer," so the area next to the wire will be bright. There will be a total of 25 dark lines along the plates, corresponding to path lengths of 0λ, 1λ, 2λ, 3λ, $\cdots$, 24λ, including the one at the point of contact A ($m = 0$). Between them, there will be 24 bright lines plus the one at the end, or 25. The bright and dark bands will be straight only if the glass plates are extremely flat. If they are not, the pattern is uneven, as in Fig. 24–32c. This is a very precise way of testing a glass surface for flatness. Curved lens surfaces can be tested for precision similarly by placing the lens on a flat glass surface and observing Newton's rings (Fig. 24–30b) for perfect circularity.

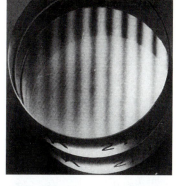

(b)

(c)

[†]Note that in Fig. 24–29, the light reflecting at both interfaces, air–oil and oil–water, underwent a phase change of $\frac{1}{2}\lambda$, since $n_{\text{water}} > n_{\text{oil}} > n_{\text{air}}$. Since the phase changes were equal, they don't affect our analysis.

If the wedge between the two glass plates of Example 24–8 is filled with some transparent substance other than air—say, water—the pattern shifts because the wavelength of the light changes. In a material where the index of refraction is n, the wavelength is $\lambda_n = \lambda/n$ where λ is the wavelength in vacuum (see Eq. 24–1). For instance, if the thin wedge of Example 24–8 were filled with water, $\lambda_n = 600\,\text{nm}/1.33 = 450\,\text{nm}$; and instead of 25 dark lines, there would be 33.

When white light (rather than monochromatic light) is incident on the thin wedge of Figs. 24–30a or 24–32a, a colorful series of fringes is seen. This is because constructive interference occurs in the reflected light at different locations along the wedge for different wavelengths. Such a difference in thickness is part of the reason bright colors appear when light is reflected from a soap bubble or a thin layer of oil on a puddle or lake (Fig. 24–28). Which wavelengths appear brightest also depends on the viewing angle, as we saw earlier.

EXAMPLE 24–9 **Thickness of soap bubble skin.** A soap bubble appears green ($\lambda = 540\,\text{nm}$) at the point on its front surface nearest the viewer. What is its minimum thickness? Assume $n = 1.35$.

⇒ **PROBLEM SOLVING**

A formula is not enough: 0 you must also check for phase changes at surfaces

SOLUTION The light is reflected perpendicularly from the point on a spherical surface nearest the viewer. Therefore the path difference is $2t$, where t is the thickness of the soap film. Light reflected from the first (outer) surface undergoes a $\frac{1}{2}\lambda$ phase change (index of refraction of soap is greater than that of air), whereas that at the second (inner) surface does not. Therefore, green light is bright when the minimum path difference equals $\frac{1}{2}\lambda_n$. Thus, $2t = \lambda/2n$, so

$$t = \frac{\lambda}{4n} = \frac{(540\ \text{nm})}{(4)(1.35)} = 100\ \text{nm}.$$

This is the minimum thickness. The front surface would also appear green if $2t = 3\lambda/2n$, and, in general, if $2t = (2m + 1)\,\lambda/2n$ where m is an integer.

⇒ **PHYSICS APPLIED**

Lens coatings

An important application of thin-film interference is in the coating of glass to make it "nonreflecting," particularly for lenses. A glass surface reflects about 4 percent of the light incident upon it. Good-quality cameras, microscopes, and other optical devices may contain six to ten thin lenses. Reflection from all these surfaces can reduce the light level considerably and multiple reflections produce a background haze that reduces the quality of the image. By reducing reflection, transmission is increased. A very thin coating on the lens surfaces can reduce reflections considerably: the thickness of the film is chosen so that light (at least for one wavelength) reflecting from the front and rear surfaces of the film destructively interferes. The amount of reflection at a boundary depends on the difference in index of refraction between the two materials. Ideally, the coating material should have an index of refraction which is the geometric mean of those for air and glass, so that the amount of reflection at each surface is about equal. Then destructive interference can occur nearly completely for one particular wavelength depending on the thickness of the coating. Nearby wavelengths will at least partially destructively interfere, but it is clear that a single coating cannot eliminate reflections for all wavelengths. Nonetheless, a single coating can reduce total reflection from 4 percent to 1 percent of the incident light. Often the coating is designed to eliminate the center of the reflected spectrum (around 550 nm). The extremes of the spectrum—red and violet—will not be reduced as much. Since a mixture of red and violet produces purple, the light seen reflected from such coated lenses is purple (Fig. 24–33). Lenses containing two or three separate coatings can more effectively reduce a wider range of reflecting wavelengths.

FIGURE 24–33 A coated lens. Note color of light reflected from the front lens surface.

EXAMPLE 24–10 **Nonreflective coating.** What is the thickness of an optical coating of MgF_2 whose index of refraction is $n = 1.38$ and is designed to eliminate reflected light at wavelengths centered at 550 nm when incident normally on glass for which $n = 1.50$?

SOLUTION Figure 24–34 shows an incoming ray and two rays reflected from the front and rear surfaces of the coating on the lens. The rays are drawn not quite perpendicular to the lens so we can see each of them. To eliminate reflection, we want the reflected rays 1 and 2 to be $\frac{1}{2}$ wavelength out of phase with each other so they destructively interfere. Rays 1 and 2 *both* undergo a change of phase by $\frac{1}{2}\lambda$ when they reflect, respectively, from the front and rear surfaces of the coating. Therefore, we want the extra distance traveled by ray 2 ($= 2t$) to be a half integral number of wavelengths. That is, $2t = (m + \frac{1}{2})\lambda_n$, where m is an integer and λ_n is the wavelength inside the MgF_2 coating. The minimum thickness ($m = 0$) is usually chosen because destructive interference will then occur over the widest angle. Then

$$t = \frac{\lambda_n}{4} = \frac{\lambda}{4n} = \frac{(550 \text{ nm})}{(4)(1.38)} = 99.6 \text{ nm}.$$

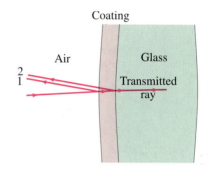

FIGURE 24–34 Incident ray of light is partially reflected at front surface of a lens coating (ray 1) and again partially reflected at the rear surface of the coating (ray 2), with most of the energy passing as the transmitted ray into the glass.

---

*24–9 Michelson Interferometer

Interference by thin films is the basis of the **Michelson interferometer** (Fig. 24–35),[†] invented by the American Albert A. Michelson (Section 22–6). Monochromatic light from a single point on an extended source is shown striking a half-silvered mirror M_S. This **beam splitter** mirror M_S has a thin layer of silver that relects only half the light that hits it, so that half of the beam passes through to a fixed mirror M_2, where it is reflected back. The other half is reflected by M_S up to a mirror M_1 that is movable (by a fine-thread screw), where it is also reflected back. Upon its return, part of beam 1 passes through M_S and reaches the eye; and part of beam 2, on its return, is reflected by M_S into the eye. If the two path lengths are identical, the two coherent beams entering the eye constructively interfere and brightness will be seen. If the movable mirror is moved a distance $\lambda/4$, one beam will travel an extra distance equal to $\lambda/2$ (because it travels back and forth over the distance $\lambda/4$). In this case, the two beams will destructively interfere and darkness will be seen. As M_1 is moved farther, brightness will recur (when the path difference is λ), then darkness, and so on.

FIGURE 24–35 Michelson interferometer.

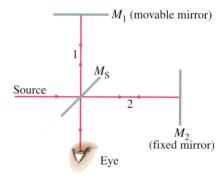

[†]There are other types of interferometer, but Michelson's is the best known.

Very precise length measurements can be made with an interferometer. The motion of mirror M_1 by only $\frac{1}{4}\lambda$ produces a clear difference between brightness and darkness. For $\lambda = 400$ nm, this means a precision of 100 nm or 10^{-4} mm! If mirror M_1 is tilted very slightly, the sequence of bright and dark spots is seen instead as a series of fringes. By counting the number of fringes, or fractions thereof, extremely precise length measurements can be made.

Michelson saw that the interferometer could be used to determine the length of the standard meter in terms of the wavelength of a particular light. In 1960, that standard was chosen to be a particular orange line in the spectrum of krypton-86 (krypton atoms with atomic mass 86). Careful repeated measurements of the old standard meter (the distance between two marks on a platinum–iridium bar kept in Paris) were made to establish 1 meter as being 1,650,763.73 wavelengths of this light, which was *defined* to be the meter. In 1983, the meter was redefined in terms of the speed of light (Section 1–5).

24–10 Polarization

An important and useful property of light is that it can be *polarized*. To see what this means, let us examine waves traveling on a rope. A rope can be set into vibration in a vertical plane as in Fig. 24–36a, or in a horizontal plane as in Fig. 24–36b. In either case, the wave is said to be **plane-polarized**—that is, the oscillations are in a plane.

If we now place an obstacle containing a vertical slit in the path of the wave, Fig. 24–37, a vertically polarized wave passes through, but a horizontal-

FIGURE 24–36 Transverse waves on a rope polarized (a) in a vertical plane and (b) in a horizontal plane.

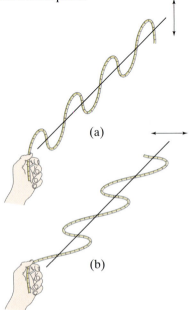
(a)

(b)

FIGURE 24–37 Vertically polarized wave passes through a vertical slit, but a horizontally polarized wave will not. [*Note*: This diagram would apply to the magnetic field **B** in an EM wave—see footnote on next page.]

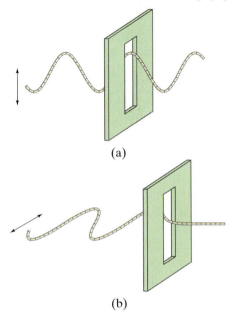
(a)

(b)

ly polarized wave will not. If a horizontal slit were used, the vertically polarized wave would be stopped. If both types of slit were used, both types of wave would be stopped. Note that polarization can exist *only* for *transverse waves*, and not for longitudinal waves such as sound. The latter vibrate only along the direction of motion, and neither orientation of slit would stop them.

Maxwell's theory of light as electromagnetic (EM) waves predicted that light can be polarized since an EM wave is a transverse wave. The direction of polarization in a plane-polarized EM wave is taken as the direction of the electric field vector (the *y* direction in Fig. 22–7).

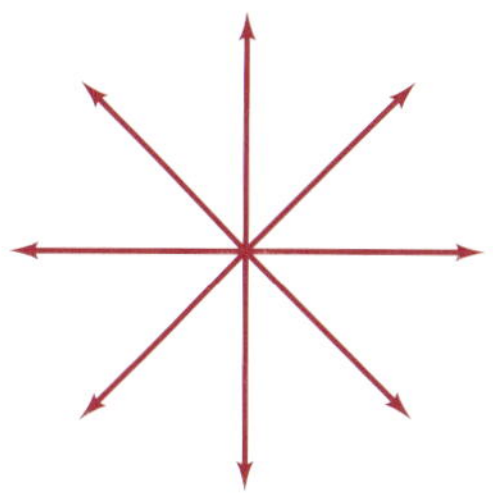

Light is not necessarily polarized. It can be **unpolarized**, which means that the source has vibrations in many planes at once, as shown in Fig. 24–38. An ordinary incandescent lightbulb emits unpolarized light, as does the Sun.

Plane-polarized light can be obtained from unpolarized light using certain crystals such as tourmaline. Or, more commonly today, we can use a **Polaroid sheet**. (Polaroid materials were invented in 1929 by Edwin Land.) A Polaroid sheet consists of complicated long molecules arranged parallel to one another. Such a Polaroid acts like a series of parallel slits to allow one orientation of polarization to pass through nearly undiminished (this direction is called the *axis* of the Polaroid), whereas a perpendicular polarization is absorbed almost completely.[†] If a beam of plane-polarized light strikes a Polaroid whose axis is at an angle θ to the incident polarization direction, the beam will emerge plane-polarized parallel to the Polaroid axis and its amplitude will be reduced by $\cos \theta$, Fig. 24–39. Thus, a Polaroid passes only that component of polarization (the electric field vector, **E**) that is parallel to its axis. Because the intensity of a light beam is proportional to the square of the amplitude (Sections 11–9 and 22–7), we see that the intensity of a plane-polarized beam transmitted by a polarizer is

$$I = I_0 \cos^2 \theta, \qquad \qquad \textbf{(24–5)}$$

where θ is the angle between the polarizer axis and the plane of polarization of the incoming wave, and I_0 is the incoming intensity.[‡]

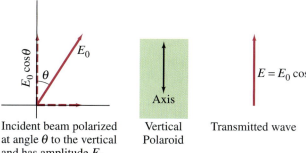

Incident beam polarized at angle θ to the vertical and has amplitude E_0.

Vertical Polaroid

Transmitted wave

FIGURE 24–38 Vibration of the electric field vectors in unpolarized light. The light is traveling into or out of the page.

FIGURE 24–39 Vertical Polaroid transmits only the vertical component of a wave incident upon it.

[†]How this occurs can be explained at the molecular level. An electric field **E** that oscillates parallel to the long molecules can set electrons into motion along the molecules, thus doing work on them and transferring energy. Hence, if **E** is parallel to the molecules, it gets absorbed. An electric field **E** perpendicular to the long molecules does not have this possibility of doing work and transferring its energy, and so passes through freely. When we speak of the *axis* of a Polaroid, we mean the direction for which **E** is passed, so a Polaroid axis is perpendicular to the long molecules. (If we want to think of there being slits between the parallel molecules in the sense of Fig. 24–37, then Fig. 24–37 would apply for the **B** field in the EM wave, not the **E** field.)

[‡]Equation 24–5 is often referred to as **Malus' law**, after Etienne Malus, a contemporary of Fresnel.

A Polaroid can be used as a **polarizer** to produce plane-polarized light from unpolarized light, since only the component of light parallel to the axis is transmitted. A Polaroid can also be used as an **analyzer** to determine (1) if light is polarized and (2) what is the plane of polarization. A Polaroid acting as an analyzer will pass the same amount of light independent of the orientation of its axis if the light is unpolarized; try rotating one lens of a pair of Polaroid sunglasses while looking through it at a lightbulb. If the light is polarized, however, when you rotate the Polaroid the transmitted light will be a maximum when the plane of polarization is parallel to the Polaroid's axis, and a minimum when perpendicular to it. If you do this while looking at the sky, preferably at right angles to the Sun's direction, you will see that skylight is polarized. (Direct sunlight is unpolarized, but don't look directly at the Sun, even through a polarizer, for damage to the eye may occur.)

Unpolarized light consists of light with random directions of polarization (electric field vector). Each of these polarization directions can be resolved into components along two mutually perpendicular directions. Thus, an unpolarized beam can be thought of as two plane-polarized beams of equal magnitude perpendicular to one another. When unpolarized light passes through a polarizer, one of the components is eliminated. So the intensity of the light passing through is reduced by half since half the light is eliminated, $I = \frac{1}{2}I_0$ (Fig. 24–40).

When two Polaroids are *crossed*—that is, their axes are perpendicular to one another—unpolarized light can be entirely stopped. As shown in Fig. 24–41, unpolarized light is made plane-polarized by the first Polaroid (the polarizer). The second Polaroid, the analyzer, then eliminates this component since its axis is perpendicular to the first. You can try this with Polaroid sunglasses (Fig. 24–42). Note that Polaroid sunglasses eliminate 50 percent of unpolarized light because of their polarizing property; they absorb even more because they are colored.

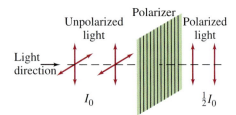

FIGURE 24–40 Unpolarized light has both vertical and horizontal components. After passing through a polarizer, one of these components is eliminated. The intensity of the light is reduced to half.

Crossed Polaroids

FIGURE 24–41 Crossed Polaroids completely eliminate light.

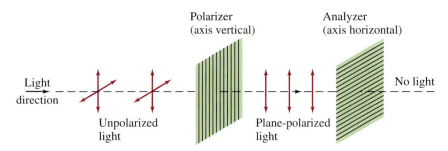

FIGURE 24–42 Crossed Polaroids. When the two polarized sunglass lenses overlap, with axes perpendicular, almost no light passes through.

EXAMPLE 24–11 Two Polaroids at 60°. Unpolarized light passes through two Polaroids; the axis of one is vertical and that of the other is at 60° to the vertical. What is the orientation and intensity of the transmitted light?

SOLUTION The first Polaroid eliminates half the light so the intensity is reduced by half: $I_1 = \frac{1}{2}I_0$. The light reaching the second polarizer is vertically polarized and so is reduced in intensity (Eq. 24–5) to

$$I_2 = I_1(\cos 60°)^2 = \frac{1}{4}I_1.$$

Thus, $I_2 = \frac{1}{8}I_0$. The transmitted light has an intensity one-eighth that of the original and is plane-polarized at a 60° angle to the vertical.

CONCEPTUAL EXAMPLE 24–12 Three Polaroids. We saw in Fig. 24–41 that when unpolarized light falls on two crossed Polaroids (axes at 90°), no light passes through. What happens if a third Polaroid, with axis at 45° to each of the other two, is placed between them?

RESPONSE We start just as in Example 24–11. The first Polaroid changes the unpolarized light to plane-polarized and reduces the intensity from I_0 to $I = \frac{1}{2}I_0$. The second polarizer further reduces the intensity by $(\cos 45°)^2$, Eq. 24–5:

$$I_2 = I_1(\cos 45°)^2 = \frac{1}{2}I_1 = \frac{1}{4}I_0.$$

The light leaving the second polarizer is plane polarized at 45° (Fig. 24–43) relative to the third polarizer, so the latter reduces the intensity to

$$I_3 = I_2(\cos 45°)^2 = \frac{1}{2}I_2$$

or $I_3 = \frac{1}{8}I_0$. Thus $\frac{1}{8}$ of the original intensity gets transmitted. But if we don't put the 45° Polaroid in at all, zero intensity results (Fig. 24–41). The 45° Polaroid must be inserted *between* the other two if transmission is to occur. Placing it before or after the other two results in zero intensity.

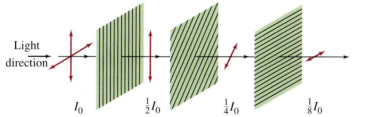

Light direction

I_0 $\frac{1}{2}I_0$ $\frac{1}{4}I_0$ $\frac{1}{8}I_0$

FIGURE 24–43 Example 24–12.

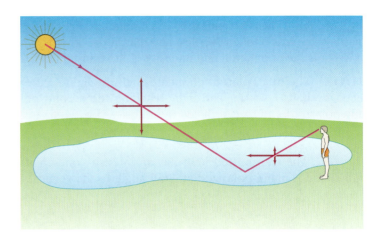

FIGURE 24–44 Light reflected from a nonmetallic surface, such as the smooth surface of water in a lake, is partially polarized parallel to the surface.

FIGURE 24–45 Photographs of a river, (a) allowing all light into the camera lens, and (b) using a polarizer which is adjusted to absorb most of the (polarized) light reflected from the water's surface, allowing the dimmer light from the bottom of the river, and any fish swimming there, to be seen more readily.

(a)

(b)

Another means of producing polarized light from unpolarized light is by reflection. When light strikes a nonmetallic surface at any angle other than perpendicular, the reflected beam is polarized preferentially in the plane parallel to the surface, Fig. 24–44. In other words, the component with polarization in the plane perpendicular to the surface is preferentially transmitted or absorbed. You can check this by rotating Polaroid sunglasses while looking through them at a flat surface of a lake or road. Since most outdoor surfaces are horizontal, Polaroid sunglasses are made with their axes vertical to eliminate the more strongly reflected horizontal component, and thus reduce glare. Fishermen wear Polaroids to eliminate reflected glare from the surface of a lake or stream and thus see beneath the water more clearly (Fig. 24–45).

The amount of polarization in the reflected beam depends on the angle, varying from no polarization at normal incidence to 100 percent polarization at an angle known as the **polarizing angle**, θ_p.[†] This angle is related to the index of refraction of the two materials on either side of the boundary by the equation

$$\tan \theta_p = \frac{n_2}{n_1},$$ **(24–6a)**

where n_1 is the index of refraction of the material in which the beam is traveling, and n_2 is that of the medium beyond the reflecting boundary. If the beam is traveling in air, $n_1 = 1$, and Eq. 24–6a becomes

$$\tan \theta_p = n.$$ **(24–6b)**

The polarizing angle θ_p is also called **Brewster's angle**, and Eqs. 24–6 called *Brewster's law*, after the Scottish physicist David Brewster (1781–1868),

---

[†]Only a fraction of the incident light is reflected at the surface of the transparent medium. Although this reflected light is 100 percent polarized (if $\theta = \theta_p$), the remainder of the light, which is transmitted into the new medium, is only partially polarized.

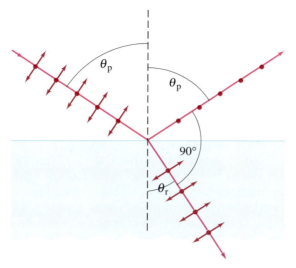

FIGURE 24–46 At θ_p the reflected light is plane-polarized parallel to the surface, and $\theta_p + \theta_r = 90°$, where θ_r is the refraction angle. (The large dots represent vibrations perpendicular to the page.)

who worked it out experimentally in 1812. Equations 24–6 can be derived from the electromagnetic wave theory of light. It is interesting that at Brewster's angle, the reflected and transmitted rays make a 90° angle to each other; that is, $\theta_p + \theta_r = 90°$, Fig. 24–46. This can be seen as follows: we substitute Eq. 24–6a, $n_2 = n_1 \tan \theta_p = n_1 \sin \theta_p/\cos \theta_p$, into Snell's law, $n_1 \sin \theta_p = n_2 \sin \theta_r$, and get $\cos \theta_p = \sin \theta_r$, which can only hold if $\theta_p = 90° - \theta_r$.

EXAMPLE 24–13 **Polarizing angle.** (a) At what incident angle is sunlight reflected fully plane-polarized from a lake? (b) What is the refraction angle?

SOLUTION (a) We use Eq. 24–6b with $n = 1.33$, so $\tan \theta_p = 1.33$ and $\theta_p = 53.1°$.
(b) $\theta_r = 90.0° - \theta_p = 36.9°$.

*24–11 Scattering of Light by the Atmosphere

Sunsets are red, the sky is blue, and skylight is polarized (at least partially). These phenomena can be explained on the basis of the *scattering* of light by the molecules of the atmosphere. In Fig. 24–47 we see unpolarized light from the Sun impinging on a molecule of the Earth's atmosphere. The electric field of the EM wave sets the electric charges within the molecule into motion, and the molecule absorbs some of the incident radiation. But it quickly reemits this light because the charges are oscillating. As discussed in Section 22–3, oscillating electric charges produce EM waves. The electric field of these waves is in the plane that includes the line of oscillation. The intensity is strongest along a line perpendicular to the oscillation, and drops to zero along the line of oscillation (Fig. 22–6). In Fig. 24–47, the motion of the charges is resolved into two components. An observer viewing at right angles to the direction of the sunlight, as shown, will see plane-polarized light because no light is emitted along the line of the other component of the oscillation. (Another way to understand this is to note that when viewing along the line of oscillation, one doesn't see the oscillation, and hence sees no waves made by it.) At other viewing angles, both

FIGURE 24–47 Unpolarized sunlight scattered by molecules of the air. An observer at right angles sees plane-polarized light, because the component of vibration along the line of sight emits no light along that line.

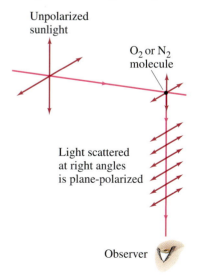

Unpolarized sunlight

O_2 or N_2 molecule

Light scattered at right angles is plane-polarized

Observer

components will be present; one will be stronger, however, so the light appears partially polarized. Thus, the process of scattering explains the polarization of skylight. [It can also explain complete polarization by reflection, Fig. 24–46. At Brewster's angle, as we saw, the angle between the reflected and refracted ray is 90°. If we think of the molecules of the medium oscillating perpendicular to the direction of the refracted ray, we can see that at 90° to this direction (the direction of the reflected ray) there will be only one component of polarization (Fig. 24–46) just as for scattering (Fig. 24–47).]

Scattering of light by the Earth's atmosphere depends on λ. For particles much smaller than the wavelength of light (such as molecules of air), the particles will be less of an obstruction to long wavelengths than to short ones. The scattering decreases, in fact, as $1/\lambda^4$. Red and orange light is thus scattered much less than blue and violet, which is why the sky looks blue. At sunset, on the other hand, the Sun's rays pass through a maximum length of atmosphere. Much of the blue has been taken out by scattering. The light that reaches the surface of the Earth is thus lacking in blue, which is why sunsets appear reddish.

The dependence of scattering on $1/\lambda^4$ is valid only if the scattering objects are much smaller than the wavelength of the light. This is valid for oxygen and nitrogen molecules whose diameters are about 0.2 nm. Clouds, however, contain water droplets or crystals that are much larger than λ; they scatter all frequencies of light nearly uniformly. Hence clouds appear white (or gray, if shadowed).

➡ **PROBLEM SOLVING** Interference

1. Interference effects depend on the simultaneous arrival of two or more waves at the same point in space.

2. Constructive interference occurs when waves arrive in phase with each other: a crest of one wave arrives at the same time as a crest of the other wave. The amplitudes of the waves then add to form a larger amplitude. Constructive interference also occurs when the phase difference is exactly one full wavelength or any integer multiple of a full wavelength: $1\lambda, 2\lambda, 3\lambda, \cdots$.

3. Destructive interference occurs when a crest of one wave arrives at the same time as a trough of the other wave. The amplitudes add, but they are of opposite sign, so the total amplitude is reduced to zero if the two amplitudes are equal. Destructive interference occurs whenever the phase difference is a half-integral number of wavelengths. Thus, the total amplitude will be zero if two identical waves arrive one-half wavelength out of phase, or $(m + \frac{1}{2})\lambda$ out of phase where m is an integer.

4. For thin-film interference, don't forget an extra one-half wavelength phase shift that occurs when light reflects from an optically more dense medium (going from a medium of lesser toward greater index of refraction).

SUMMARY

The wave theory of light is strongly supported by the observations that light exhibits **interference** and **diffraction**. Wave theory also explains the refraction of light and the fact that light travels more slowly in transparent solids and liquids than it does in air. The wavelength of light in a medium with index of refraction n is

$$\lambda_n = \lambda/n,$$

where λ is the wavelength in vacuum; the frequency is not changed.

Young's double-slit experiment clearly demonstrated the interference of light. The observed bright spots of the interference pattern were explained as constructive interference between the beams coming through the two slits, where the beams differ in path length by an integral number of wavelengths. The dark areas in between are due to destructive interference when the path lengths differ by $\frac{1}{2}\lambda$, $\frac{3}{2}\lambda$, and so on. The angles θ at which **constructive interference** occurs are given by

$$\sin\theta = m\frac{\lambda}{d},$$

where λ is the wavelength of the light, d the separation of the slits, and m an integer $(0, 1, 2, \cdots)$. **Destructive interference** occurs at angles θ given by

$$\sin\theta = (m + \tfrac{1}{2})\frac{\lambda}{d}$$

where m is an integer $(0, 1, 2, \cdots)$.

The wavelength of light determines its color. The **visible spectrum** extends from about 400 nm (violet) to about 750 nm (red). Glass prisms break down white light into its constituent colors because the index of refraction varies with wavelength, a phenomenon known as **dispersion**.

The formula $\sin\theta = m\lambda/d$ for constructive interference also holds for a **diffraction grating**, which consists of many parallel slits or lines, separated from each other by a distance d. The peaks of constructive interference are much brighter and sharper for a diffraction grating than for the simple two-slit apparatus. A diffraction grating (or a prism) is used in a **spectroscope** to separate different colors or to observe **line spectra**, since for a given order m, θ depends on λ. Precise determination of wavelength can thus be done with a spectroscope by careful measurement of θ.

Diffraction refers to the fact that light, like other waves, bends around objects it passes, and spreads out after passing through narrow slits. This bending gives rise to a **diffraction pattern** due to interference between rays of light that travel different distances. Light passing through a very narrow slit of width D will produce a pattern with a bright central maximum of half-width θ given by

$$\sin\theta = \lambda/D,$$

flanked by fainter lines to either side.

Light reflected from the front and rear surfaces of a thin film of transparent material can interfere. A phase change of $180°$ $(\frac{1}{2}\lambda)$ occurs when the light reflects at a surface where the index of refraction increases. Such **thin-film interference** has many practical applications, such as lens coatings and Newton's rings.

In **unpolarized light**, the electric field vectors vibrate at all angles. If the electric vector vibrates only in one plane the light is said to be **plane-polarized**. Light can also be partially polarized. When an unpolarized light beam passes through a Polaroid sheet, the emerging beam is plane-polarized. When a light beam is polarized and passes through a Polaroid, the intensity varies as the Polaroid is rotated. Thus a Polaroid can act as a polarizer or as an analyzer.

Light can also be partially or fully **polarized by reflection**. If light traveling in air is reflected from a medium of index of refraction n, the reflected beam will be *completely* plane-polarized if the incident angle θ_p is given by $\tan\theta_p = n$. The fact that light can be polarized shows that it must be a transverse wave.

QUESTIONS

1. Does Huygens' principle apply to sound waves? To water waves?
2. What is the evidence that light is energy?
3. Why is light sometimes described as rays and sometimes as waves?
4. Two rays of light from the same source destructively interfere if their path lengths differ by how much?
5. If Young's double-slit experiment were submerged in water, how would the fringe pattern be changed?
6. Monochromatic red light is incident on a double slit and the interference pattern is viewed on a screen some distance away. Explain how the fringe pattern would change if the red light source is replaced by a blue light source.
7. Why was the observation of the double-slit interference pattern more convincing evidence for the wave theory of light than the observation of diffraction?
8. Compare a double-slit experiment for sound waves to that for light waves. Discuss the similarities and differences.
9. Why doesn't the light from the two headlights of a distant car produce an interference pattern?
10. When white light passes through a flat piece of window glass, it is not broken down into colors as it is by a prism. Explain.
11. For both converging and diverging lenses, discuss how the focal length for red light differs from that for violet light.
12. We can hear sounds around corners, but we cannot see around corners; yet both sound and light are waves. Explain the difference.
13. For diffraction by a single slit, what is the effect of increasing (a) the slit width, and (b) the wavelength?
14. What is the difference in the interference patterns formed by two slits 10^{-4} cm apart and by a diffraction grating containing 10^4 lines/cm?
15. Explain why there are tiny peaks between the main peaks produced by a diffraction grating illuminated with monochromatic light. Why are the peaks so tiny?
16. For a diffraction grating, what is the advantage of (a) many slits, (b) closely spaced slits.
17. White light strikes (a) a diffraction grating, and (b) a prism. A rainbow appears on a wall just below the direction of the horizontal incident beam in each case. What is the color of the top of the rainbow in each case?
18. Why are interference fringes noticeable only for a *thin* film like a soap bubble and not for a thick piece of glass, say?
19. Why are Newton's rings (Fig. 24–30) closer together farther from the center?
20. Some coated lenses appear greenish yellow when seen by reflected light. What wavelengths do you suppose they are designed to eliminate completely?
* 21. Describe how a Michelson interferometer could be used to measure the index of refraction of air.
22. What does polarization tell us about the nature of light?
23. Explain the advantage of polarized sunglasses over normal tinted sunglasses.
24. How can you tell if a pair of sunglasses is polarizing or not?
* 25. What would be the color of the sky if the Earth had no atmosphere?
* 26. If the Earth's atmosphere were 50 times denser than it is, would sunlight still be white, or would it be some other color?

PROBLEMS

SECTION 24-2

1. (II) Derive the law of reflection—namely, that the angle of incidence equals the angle of reflection from a flat surface—using Huygens' principle for waves.

SECTION 24-3

2. (I) Monochromatic light falling on two slits 0.042 mm apart produces the fifth-order fringe at a 7.8° angle. What is the wavelength of the light used?
3. (I) The third-order fringe of 650 nm light is observed at an angle of 15° when the light falls on two narrow slits. How far apart are the slits?
4. (II) Monochromatic light falls on two very narrow slits 0.040 mm apart. Successive fringes on a screen 5.00 m away are 5.5 cm apart near the center of the pattern. What is the wavelength and frequency of the light?
5. (II) A parallel beam of light from a He-Ne laser, with a wavelength 656 nm, falls on two very narrow slits 0.050 mm apart. How far apart are the fringes in the center of the pattern if the screen is 2.6 m away?
6. (II) In a water tank experiment, water waves are generated with their crests 2.5 cm apart and parallel. They pass through two openings 5.0 cm apart in a long wooden board. If the end of the tank is 2.0 m beyond the boards, where would you stand, relative to the "straight-through" direction, so that you received little or no wave action?

7. (II) Light of wavelength 680 nm falls on two slits and produces an interference pattern in which the fourth-order fringe is 48 mm from the central fringe on a screen 1.5 m away. What is the separation of the two slits?

8. (II) If 480-nm and 620-nm light passes through two slits 0.54 mm apart, how far apart are the second-order fringes for these two wavelengths on a screen 1.6 m away?

9. (II) Suppose a thin piece of glass were placed in front of the lower slit in Fig. 24–7 so that the two waves enter the slits 180° out of phase (Fig. 24–48). Describe in detail the interference pattern on the screen.

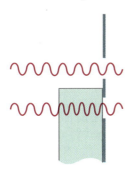

FIGURE 24–48   Problem 9.

10. (II) In a double-slit experiment it is found that blue light of wavelength 460 nm gives a second-order maximum at a certain location on the screen. What wavelength of visible light would have a minimum at the same location?

11. (III) Light of wavelength 400 nm in air falls on two slits 5.00×10^{-2} mm apart. The slits are immersed in water, as is a viewing screen 40.0 cm away. How far apart are the fringes on the screen?

12. (III) A very thin sheet of plastic ($n = 1.60$) covers one slit of a double-slit apparatus illuminated by 540-nm light. The center point on the screen, instead of being a maximum, is dark. What is the (minimum) thickness of the plastic?

SECTION 24–4

13. (I) By what percent, approximately, does the speed of red light (700 nm) exceed that of violet light (400 nm) in silicate flint glass? (See Fig. 24–14).

14. (I) How much less is the speed of light in silicate flint glass (see Fig. 24–14) for blue light of wavelength 450 nm than red light of 700 nm? Express as a percent.

15. (II) A light beam strikes a piece of glass at a 60.00° incident angle. The beam contains two wavelengths, 450.0 nm and 700.0 nm, for which the index of refraction of the glass is 1.4820 and 1.4742, respectively. What is the angle between the two refracted beams?

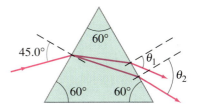

FIGURE 24–49   Problem 16.

16. (II) A parallel beam of light containing two wavelengths, $\lambda_1 = 400$ nm and $\lambda_2 = 650$ nm, enters the silicate flint glass of an equilateral prism as shown in Fig. 24–49. At what angle does each beam leave the prism (give angle with normal to the face)?

* **17.** (III) A double convex lens whose radii of curvature are both 17.5 cm is made of crown glass. Find the distance between the focal points for 400-nm and 700-nm light. [*Hint*: Use the lensmaker's equation, Eq. 23–10, and Fig. 24–14.]

SECTION 24–5

18. (I) If 520-nm light falls on a slit 0.0440 mm wide, what is the angular width of the central diffraction peak?

19. (I) Monochromatic light falls on a slit that is 3.00×10^{-3} mm wide. If the angle between the first dark fringes on either side of the central maximum is 37.0°, what is the wavelength of the light used?

20. (I) Light of wavelength 550 nm falls on a slit that is 3.50×10^{-3} mm wide. How far from the central maximum will the first diffraction maximum fringe be if the screen is 10.0 m away?

21. (II) Monochromatic light of wavelength 633 nm falls on a slit. If the angle between the first bright fringes on either side of the central maximum is 19.5°, what is the slit width?

22. (II) How wide is the central diffraction peak on a screen 2.50 m behind a 0.0348-mm-wide slit illuminated by 589-nm light?

23. (II) When violet light of wavelength 415 nm falls on a single slit, it creates a central diffraction peak that is 9.20 cm wide on a screen that is 2.55 m away. How wide is the slit?

24. (II) If a slit diffracts 550-nm light so that the diffraction maximum is 3.0 cm wide on a screen 1.50 m away, what will be the width of the diffraction maximum for light with a wavelength of 400 nm?

25. (II) (*a*) For a given wavelength λ, what is the maximum slit width for which there will be no diffraction minima? (*b*) What is the maximum slit width so that no visible light exhibits a diffraction minimum?

26. (I) At what angle will 650-nm light produce a second-order maximum when falling on a grating whose slits are 1.15×10^{-3} cm apart?

27. (I) A 3500-line/cm grating produces a third-order fringe at a 22.0° angle. What wavelength of light is being used?

28. (II) The first-order line of 589-nm light falling on a diffraction grating is observed at a 15.5° angle. How far apart are the slits? At what angle will the third order be observed?

29. (II) Light falling normally on a 10,000-line/cm grating is revealed to contain three lines in the first-order spectrum at angles of 31.2°, 36.4°, and 47.5°. What wavelengths are these?

30. (II) How many lines per centimeter does a grating have if the third-order occurs at a 23.0° angle for 630-nm light?

31. (II) A grating has 7000 lines/cm. How many spectral orders can be seen when it is illuminated by white light?

32. (II) What is the highest spectral order that can be seen if a grating with 6000 lines per cm is illuminated with 633-nm laser light? Assume normal incidence.

33. (II) Two and only two full spectral orders can be seen on either side of the central maximum when white light is sent through a diffraction grating. What is the maximum number of lines per cm for the grating?

34. (II) White light containing wavelengths from 400 nm to 750 nm falls on a grating with 7500 lines/cm. How wide is the first-order spectrum on a screen 2.30 m away?

35. (II) The α and δ lines of the atomic hydrogen spectrum have wavelengths of 656 nm and 410 nm. If these fall at normal incidence on a grating with 6600 lines per cm, what will be the angular separation of the two wavelengths in the first-order spectrum?

36. (II) Two first-order spectrum lines are measured by an 8500-line/cm spectroscope at angles, on each side of center, of $+26°38'$, $+41°08'$ and $-26°48'$, $-41°19'$. What are the wavelengths?

37. (III) Suppose the angles measured in Problem 36 were produced when the spectrometer (but not the source) was submerged in water. What then would be the wavelengths?

SECTION 24–8

38. (I) If a soap bubble is 120 nm thick, what color will appear at the center when illuminated normally by white light? Assume that $n = 1.34$.

39. (I) How far apart are the dark fringes in Example 24–8 if the glass plates are each 26.5 cm long?

40. (II) What is the minimum thickness of a soap film ($n = 1.42$) that would appear black if illuminated with 480-nm light? Assume there is air on both sides of the soap film.

41. (II) A lens appears greenish yellow ($\lambda = 570$ nm is strongest) when white light reflects from it. What minimum thickness of coating ($n = 1.25$) do you think is used on such a (glass) lens, and why?

42. (II) A total of 31 bright and 31 dark Newton's rings (not counting the dark spot at the center) are observed when 550-nm light falls normally on a planoconvex lens resting on a flat glass surface (Fig. 24–30). How much thicker is the center than the edges?

43. (II) A fine metal foil separates one end of two pieces of optically flat glass, as in Fig. 24–32. When light of wavelength 670 nm is incident normally, 28 dark lines are observed (with one at each end). How thick is the foil?

44. (II) How thick (minimum) should the air layer be between two flat glass surfaces if the glass is to appear bright when 450-nm light is incident normally? What if the glass is to appear dark?

45. (II) A thin film of alcohol ($n = 1.36$) lies on a flat glass plate ($n = 1.51$). When monochromatic light, whose wavelength can be changed, is incident normally, the reflected light is a minimum for $\lambda = 512$ nm and a maximum for $\lambda = 640$ nm. What is the thickness of the film?

46. (III) When a Newton's ring apparatus (Fig. 24–30) is immersed in a liquid, the diameter of the eighth dark ring decreases from 2.92 cm to 2.48 cm. What is the refractive index of the liquid?

47. (III) Show that the radius r of the mth dark Newton's ring, as viewed from directly above (Fig. 24–30), is given by $r = \sqrt{m\lambda R}$ where R is the radius of curvature of the curved glass surface and λ is the wavelength of light used. Assume that the thickness of the air gap is much less than R at all points and that $r \ll R$.

48. (III) A planoconvex Lucite lens 6.1 cm in diameter is placed on a flat piece of glass as in Fig. 24–30. When 610-nm light is incident normally, 48 bright rings are observed, the last one right at the edge. What is the radius of curvature of the lens surface?

*SECTION 24–9

*49. (II) What is the wavelength of the light entering an interferometer if 644 bright fringes are counted when the movable mirror moves 0.225 mm?

*50. (II) A micrometer is connected to the movable mirror of an interferometer. When the micrometer bears on a thin metal foil, the net number of bright fringes that move, compared to the empty micrometer, is 272. What is the thickness of the foil? The wavelength of light used is 589 nm.

*51. (II) How far must the mirror M_1 in a Michelson interferometer be moved if 850 fringes of 589-nm light are to pass by a reference line?

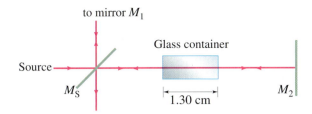

FIGURE 24–50   Problem 52.

* **52.** (III) One of the beams of an interferometer (Fig. 24–50) passes through a small glass container containing a cavity 1.30 cm deep. When a gas is allowed to slowly fill the container, a total of 236 dark fringes are counted to move past a reference line. The light used has a wavelength of 610 nm. Calculate the index of refraction of the gas, assuming that the interferometer is in vacuum.

* **53.** (III) The yellow sodium D lines have wavelengths of 589.0 and 589.6 nm. When they are used to illuminate a Michelson interferometer, it is noted that the interference fringes disappear and reappear periodically as the mirror M_1 is moved. Why does this happen? How far must the mirror move between one disappearance and the next?

SECTION 24–10

54. (I) The axes of two polarizers make a 70° angle to one another. Unpolarized light falls on them. What fraction of the light intensity is transmitted?

55. (I) What is Brewster's angle for an air–glass ($n = 1.52$) surface?

56. (I) What is Brewster's angle for a diamond submerged in water if the light is hitting the diamond while traveling in the water?

57. (II) The critical angle for total internal reflection at a boundary between two materials is 52°. What is Brewster's angle at this boundary?

58. (II) Two Polaroids are aligned so that the light passing through them is a maximum. At what angle should one of them be placed so that the intensity is subsequently reduced by half?

59. (II) At what angle should the axes of two Polaroids be placed so as to reduce the intensity of the incident unpolarized light by an additional factor (after the first Polaroid cuts it in half) of (*a*) 25 percent, (*b*) 10 percent, (*c*) 1 percent?

60. (II) Two polarizers are oriented at 40° to each other and plane-polarized light is incident on them. If only 15 percent of the light gets through both of them, what was the initial polarization direction of the incident light?

61. (II) Two polarizers are oriented at 58.0° to one another. Light polarized at a 29.0° angle to each polarizer passes through both. What reduction in intensity takes place?

62. (II) What would Brewster's angle be for reflections off the surface of water for light coming from beneath the surface? Compare to the angle for total internal reflection, and to Brewster's angle from above the surface.

63. (III) Describe how to rotate the plane of polarization of a plane-polarized beam of light by 90° and produce only a 10 percent loss in intensity using "perfect" polarizers.

GENERAL PROBLEMS

64. Stealth aircraft are designed to not reflect radar, whose wavelength is typically 2 cm, by using an anti-reflecting coating. Ignoring any change in wavelength in the coating, estimate its thickness.

65. A teacher stands well back from an outside doorway 0.88 m wide, and blows a whistle of frequency 750 Hz. Ignoring reflections, estimate at what angle(s) it is *not* possible to hear the whistle clearly on the playground outside the doorway.

66. The wings of a certain beetle have a series of parallel lines across them. When normally incident 460-nm light is reflected from the wing, the wing appears bright when viewed at an angle of 50°. How far apart are the lines?

67. How many lines per centimeter must a grating have if there is to be no second-order spectrum for any visible wavelength?

68. Show that the second- and third-order spectra of white light produced by a diffraction grating always overlap. What wavelengths overlap exactly?

69. Television and radio waves can reflect from nearby mountains or from airplanes, and the reflections can interfere with the direct signal from the station. (*a*) Determine what kind of interference will occur when 75-MHz television signals arrive at a receiver directly from a distant station, and are reflected from an airplane 118 m directly above the receiver. (Assume $\frac{1}{2}\lambda$ change in phase of the signal upon reflection.) (*b*) What kind of interference will occur if the plane is 22 m closer to the receiver?

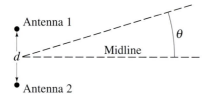

FIGURE 24–51   Problem 70.

70. A radio station operating at 102.1 MHz broadcasts from two identical antennae at the same elevation but separated by a 9.0-m horizontal distance d, Fig. 24–51. A maximum signal is found along the midline, perpendicular to d at its midpoint and extending horizontally in both directions. If the midline is taken as $0°$, at what other angle(s) θ is a maximum signal detected? A minimum signal? Assume all measurements are made much farther than 9 m from the antenna towers.

71. Light of wavelength 590 nm passes through two narrow slits 0.60 mm apart. The screen is 1.70 m away. A second source of unknown wavelength produces its second-order fringe 1.33 mm closer to the central maximum than the 590-nm light. What is the wavelength of the unknown light?

72. At what angle above the horizon is the Sun when light reflecting off a smooth lake is polarized most strongly?

73. Unpolarized light falls on two polarizer sheets whose axes are at right angles. (a) What fraction of the incident light intensity is transmitted? (b) What fraction is transmitted if a third polarizer is placed between the first two so that its axis makes a $60°$ angle with the axis of the first polarizer? (c) What if the third polarizer is in front of the other two?

74. What is the index of refraction of a clear material if a minimum of 150 nm thickness of it, when laid on glass, is needed to reduce reflection to nearly zero when light of 600 nm is incident normally upon it? Do you have a choice for an answer?

75. Monochromatic light of variable wavelength is incident normally on a thin sheet of plastic film in air. The reflected light is a minimum only for $\lambda = 512$ nm and $\lambda = 640$ nm in the visible spectrum. What is the thickness of the film ($n = 1.58$)?

76. If parallel light falls on a single slit of width D at a $30°$ angle to the normal, describe the diffraction pattern.

77. When yellow sodium light, $\lambda = 589$ nm, falls on a diffraction grating, its first-order peak on a screen 60.0 cm away falls 3.32 cm from the central peak. Another source produces a line 3.71 cm from the central peak. What is its wavelength? How many lines/cm are on the grating?

78. Compare the minimum thickness needed for an anti-reflective coating ($n = 1.38$) applied to a glass lens in order to eliminate (a) blue (450 nm), or (b) red (700 nm) reflections for light at normal incidence.

79. Suppose you viewed the light *transmitted* through a thin film on a flat piece of glass. Draw a diagram, similar to Fig. 24–29 or 24–34, and describe the conditions required for maxima and minima; consider all possible values of index of refraction. Discuss the relative size of the minima compared to the maxima and to zero.

80. Light of wavelength λ strikes a screen containing two slits a distance d apart at an angle θ_i to the normal. Determine the angle θ_m at which the mth-order maximum occurs.

81. Four ideal polarizers are placed in succession with their axes vertical, at $30°$ to the vertical, at $60°$ to the vertical, and at $90°$ to the vertical respectively. (a) Calculate what fraction of the incident unpolarized light is transmitted by the four polarizers. (b) Can the transmitted light be *decreased* by removing one of the polarizers? If so, which one? (c) Can the transmitted light intensity be extinguished by removing polarizers? If so which one(s)?

82. Very highly reflective mirrors for a particular wavelength can be made by alternating many layers of *transparent* materials of indices of refraction n_1 and $n_2 (1 < n_1 < n_2)$. What should the minimum thicknesses d_1 and d_2 of Fig. 24–52 be, in terms of the incident wavelength λ to maximize reflection?

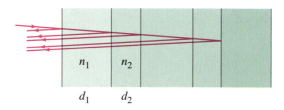

FIGURE 24–52   Problem 82.

83. Light is incident on a diffraction grating with 7500 lines per centimeter and the pattern is viewed on a screen located 2.5 m from the grating. The incident light beam consists of two wavelengths, $\lambda_1 = 4.4 \times 10^{-7}$ m and $\lambda_2 = 6.3 \times 10^{-7}$ m. Calculate the linear distance between the first-order bright fringes of these two wavelengths on the screen.

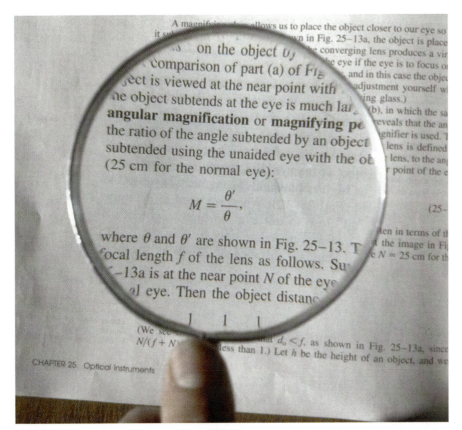

Of the many optical devices we discuss in this chapter, the magnifying glass is the simplest. Here it is magnifying a page that describes how it works according to the ray model.

OPTICAL INSTRUMENTS

I n our discussion of the behavior of light in the two previous chapters, we also described a few instruments such as the spectrometer and the Michelson interferometer. In this chapter, we will discuss some other, more common, instruments, most of which use lenses, such as the camera, telescope, microscope, and the human eye. To describe their operation, we will use ray diagrams. However, we will see that understanding some aspects of their operation will require the wave nature of light.

25–1 The Camera

The basic elements of a **camera** are a lens, a light-tight box, a shutter to let light pass through the lens only briefly, and a sensitized plate or piece of film (Fig. 25–1). When the shutter is opened, light from external objects in the field of view is focused by the lens as an image on the film. The film contains light-sensitive chemicals that undergo change when light strikes them. In the development process, chemical reactions cause the changed areas to turn opaque so that the image is recorded on the film.[†] You can

➡ **PHYSICS APPLIED**

The camera

[†]This is called a *negative*, because the black areas correspond to bright objects and vice versa. The same process occurs during printing to produce a black-and-white "positive" picture from the negative. Color film makes use of three dyes corresponding to the primary colors.

see an image yourself by removing the camera back and viewing through a piece of tissue or wax paper (on which the image can form) placed at the position of the film with the shutter open.

There are three main adjustments on good-quality cameras: shutter speed, f-stop, and focusing, and we now discuss them. Although many cameras today make these adjustments automatically, it is valuable to understand these adjustments to use any camera effectively. For special or top-quality work, a camera that allows manual adjustments is indispensable (see Fig. 25–2).

Shutter speed **Shutter speed.** This refers to how long the shutter is open and the film exposed. It may vary from a second or more ("time exposures") to $\frac{1}{1000}$ s or less. To avoid blurring from camera movement, speeds faster than $\frac{1}{100}$ s are normally used. If the object is moving, faster shutter speeds are needed to "stop" the action. Figure 25–1 shows the common type of shutter just behind the lens. A second type used in SLR (single-lens reflex) cameras (discussed later in this Section) is the "focal-plane" shutter, which is an opaque curtain just in front of the film whose opening can move quickly across the film to expose it.

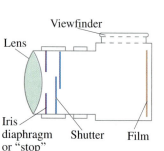

FIGURE 25–1 A simple camera.

***f*-stop.** The amount of light reaching the film must be carefully controlled to avoid **underexposure** (too little light for any but the brightest objects to show up) or **overexposure** (too much light, so that all bright objects look the same, with a consequent lack of contrast and a "washed-out" appearance). To control the exposure, a "stop" or iris diaphragm, whose opening is of variable diameter, is placed behind the lens (Fig. 25–1). The size of the opening is varied to compensate for bright or dark days, the sensitivity of the film[†] used, and for different shutter speeds. The size of the opening is specified by the ***f*-stop**, defined as

f-stop
$$f\text{-stop} = \frac{f}{D},$$

where f is the focal length of the lens and D is the diameter of the opening. For example, when a 50-mm-focal-length lens has an opening $D = 25\,\text{mm}$, we say it is set at $f/2$. When the lens is set at $f/8$, the opening is only $6\frac{1}{4}$ mm $(50/6\frac{1}{4} = 8)$. The faster the shutter speed, or the darker the day, the greater the opening must be to get a proper exposure. This corresponds to a smaller f-stop number. The smaller the f-stop number, the more light passes through the lens to the film. The smallest f-number of a lens (largest opening) is referred to as the *speed* of the lens. It is common to find $f/2.0$ lenses today, and even some as fast as $f/1.0$. Fast lenses are expensive to make and require many elements in order to reduce the defects present in simple thin lenses (Section 25–6). The advantage of a fast lens is that it allows pictures to be taken under poor lighting conditions.

Lenses normally stop down to $f/16$, $f/22$, or $f/32$. Although the lens opening can usually be varied continuously, there are nearly always markings for specific lens openings: the standard f-stop markings are 1.0, 1.4, 2.0, 2.8, 4.0, 5.6, 8, 11, 16, 22, and 32. Notice that each of these stops corresponds to a diameter reduction by a factor of about $\sqrt{2} = 1.4$. Because the amount of light reaching the film is proportional to the *area* of the opening,

FIGURE 25–2 On this camera, the f-stops and the focusing ring are on the camera lens. Shutter speeds are selected on the small wheel on top of the camera body.

[†]Different films have different sensitivities to light, referred to as the "film speed," and specified as an "ASA" number; a "faster" film is more sensitive and needs less light to produce a good image.

(a)

(b)

FIGURE 25–3 Photos with camera focused (a) on nearby object with distant object blurry, and (b) on more distant object with nearby object blurry.

and therefore proportional to the diameter squared, we see that each standard f-stop corresponds to a factor of 2 in light intensity reaching the film.

Focusing.   Focusing is the operation of placing the lens at the correct position relative to the film for the sharpest image. The image distance is a minimum for objects at infinity (the symbol ∞ is used for infinity) and is equal to the focal length. For closer objects, the image distance is greater than the focal length, as can be seen from the lens equation, $1/f = 1/d_o + 1/d_i$. To focus on nearby objects, the lens must therefore be moved away from the film, and this is usually done by turning a ring on the lens.

Focusing

If the lens is focused on a nearby object, a sharp image of it will be formed, but distant objects may be blurry (Fig. 25–3). The rays from a point on the distant object will be out of focus—they will form a circle on the film as shown (exaggerated) in Fig. 25–4. The distant object will thus produce an image consisting of overlapping circles and will be blurred. These circles are called **circles of confusion**. If you want to have near and distant objects sharp at the same time, you can try setting the lens focus at an intermediate position. Neither near nor distant objects will then be perfectly sharp, but the circles of confusion may be small enough that the blurriness is not too noticeable. For a given distance setting, there is a range of distances over which the circles of confusion will be small enough that the images will be reasonably sharp. This is called the **depth of field.** For a particular choice of circle of confusion diameter as upper limit (typically taken to be 0.03 mm for 35-mm cameras), the depth of field varies

Depth of field

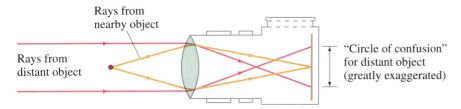

Rays from nearby object

Rays from distant object

"Circle of confusion" for distant object (greatly exaggerated)

FIGURE 25–4 When the lens is positioned to focus on a nearby object, points on a distant object produce circles and are therefore blurred. (The effect is shown greatly exaggerated.)

with the lens opening. If the lens opening is smaller, only rays through the central part of the lens are accepted, and these form smaller circles of confusion for a given object distance. Hence, at smaller lens openings, a greater range of object distances will fit within the circle of confusion criterion, so that the depth of field is greater.

Other factors also affect the sharpness of the image, such as the graininess of the film, diffraction, and lens aberrations relating to the quality of the lens itself. (Lens quality and diffraction effects will be discussed in Sections 25–6 and 25–7.)

Camera lenses are categorized into normal, telephoto, and wide angle, according to focal length and film size. A **normal lens** is one that covers the film with a field of view that corresponds approximately to that of normal vision. A normal lens for 35-mm film has a focal length in the vicinity of 50 mm.[†] A **telephoto lens**, as its name implies, acts like a telescope to magnify images. They have longer focal lengths than a normal lens: as we saw in Chapter 23 (Eq. 23–9), the height of the image for a given object distance is proportional to the image distance, and the image distance will be greater for a lens with longer focal length. For distant objects, the image height is very nearly proportional to the focal length (can you prove this?). Thus a 200-mm telephoto lens for use with a 35-mm camera gives a $4\times$ magnification over the normal 50-mm lens. A **wide-angle lens** has a shorter focal length than normal: a wider field of view is included and objects appear smaller. A **zoom lens** is one whose focal length can be changed so that you seem to zoom up to, or away from, the subject as you change the focal length.

Two types of viewing systems are common in cameras today. In many cameras, you view through a small window just above the lens as in Fig. 25–1. In a **single-lens reflex** camera (SLR), you actually view through the lens with the use of prisms and mirrors (Fig. 25–5). A mirror hangs at a 45° angle behind the lens and flips up out of the way just before the

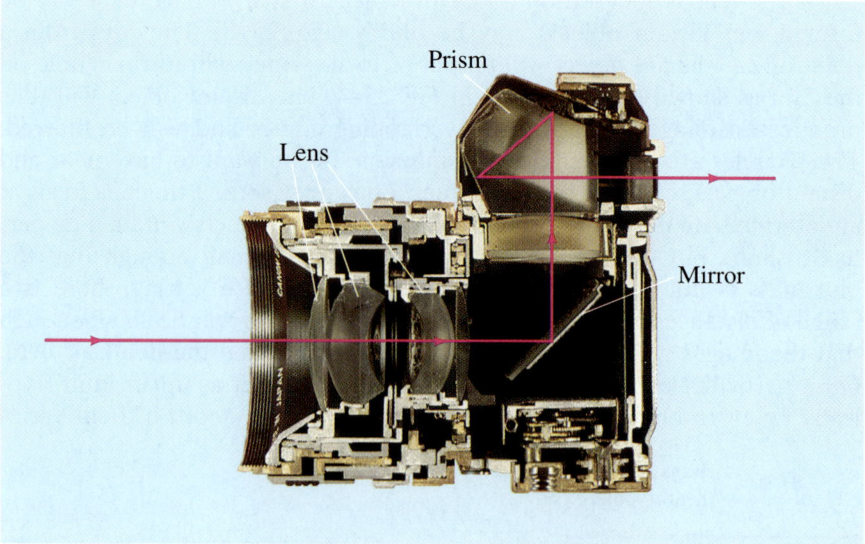

Telephoto and wide angle lenses

FIGURE 25–5 Single-lens reflex (SLR) camera, showing how the image is viewed through the lens with the help of a movable mirror and prism.

[†]Note that a "35-mm camera" uses film that is 35 mm wide; that 35 mm is not to be confused with a focal length.

shutter opens. SLRs have the great advantage that you can see almost exactly what you will get on film.

EXAMPLE 25–1 **Camera focus.** How far must a 50.0-mm-focal-length camera lens be moved from its infinity setting in order to sharply focus an object 3.00 m away?

SOLUTION When focused at infinity, the lens is 50.0 mm from the film. When focused at $d_o = 3.00$ m, the image distance is given by the lens equation,

$$\frac{1}{d_i} = \frac{1}{f} - \frac{1}{d_o} = \frac{1}{50.0 \text{ mm}} - \frac{1}{3000 \text{ mm}}.$$

We solve for d_i and find $d_i = 50.8$ mm, so the lens moves 0.8 mm.

CONCEPTUAL EXAMPLE 25–2 **Shutter speed.** To improve the depth of field, you stop down your camera lens by two f-stops (say, from $f/4$ to $f/8$). What should you do to the shutter speed to maintain the same exposure?

RESPONSE The amount of light admitted by the lens is proportional to the area of the lens opening. Reducing the lens opening by one f-stop reduces the diameter by a factor of $\sqrt{2}$, and the area by a factor of 2. Stopping down by two f-stops reduces the area of the lens opening by a factor of 4. To maintain the same exposure, the shutter must be open 4 times as long. So if the shutter speed had been $\frac{1}{250}$ s, you have to increase it to $\frac{1}{60}$ s.

25–2 The Human Eye; Corrective Lenses

The human eye resembles a camera in its basic structure (Fig. 25–6). The eye is an enclosed volume into which light passes through a lens. A diaphragm, called the **iris** (the colored part of your eye), adjusts automatically to control the amount of light entering the eye. The hole in the iris through which light passes (the **pupil**) is black because no light is reflected from it (it's a hole), and very little light is reflected back out from the interior of the eye. The **retina**, which plays the role of the film in a camera, is on the curved rear surface. It consists of a complex array of nerves and receptors known as *rods* and *cones* which act to change light energy into electrical signals that travel along the nerves. The reconstruction of the image from all these tiny receptors is done mainly in the brain, although some analysis is apparently done in the complex interconnected nerve network at the retina itself. At the center of the retina is a small area called the **fovea**, about 0.25 mm in diameter, where the cones are very closely packed and the sharpest image and best color discrimination are found.

Unlike a camera, the eye contains no shutter. The equivalent operation is carried out by the nervous system, which analyzes the signals to form images at the rate of about 30 per second. This can be compared to motion picture or television cameras, which operate by taking a series of still pictures at a rate of 24 (movies) or 30 (U.S. television) per second. The rapid projection of these on the screen gives the appearance of motion.

➡ **PHYSICS APPLIED**

The eye

Anatomy of the eye

FIGURE 25–6 Diagram of a human eye.

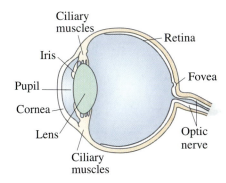

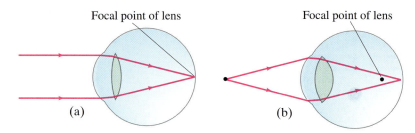

FIGURE 25–7 Accommodation by a normal eye: (a) lens relaxed, focused at infinity; (b) lens thickened, focused on a nearby object.

Focusing the eye

The lens of the eye does little of the bending of the light rays. Most of the refraction is done at the front surface of the **cornea** (index of refraction = 1.376), which also acts as a protective covering. The lens acts as a fine adjustment for focusing at different distances. This is accomplished by the ciliary muscles (Fig. 25–6), which change the curvature of the lens so that its focal length is changed. To focus on a distant object, the muscles are relaxed and the lens is thin, Fig. 25–7a, and parallel rays focus at the focal point (on the retina). To focus on a nearby object, the muscles contract, causing the center of the lens to be thicker, Fig. 25–7b, thus shortening the focal length so that images of nearby objects can be focused on the retina, behind the focal point. This focusing adjustment is called **accommodation**.

The closest distance at which the eye can focus clearly is called the **near point** of the eye. For young adults it is typically 25 cm, although younger children can often focus on objects as close as 10 cm. As people grow older, the ability to accommodate is reduced and the near point increases. A given person's **far point** is the farthest distance at which an object can be seen clearly. For some purposes it is useful to speak of a **normal eye** (a sort of average over the population), which is defined as one having a near point of 25 cm and a far point of infinity. To check your own near point, place this book close to your eye and slowly move it away until the type is sharp.

"Normal eye"

➡ PHYSICS APPLIED

Corrective lenses

The "normal" eye is more of an ideal than a commonplace. A large part of the population have eyes that do not accommodate within the normal range of 25 cm to infinity, or have some other defect. Two common defects are nearsightedness and farsightedness. Both can be corrected to a large extent with lenses—either eyeglasses or contact lenses.

Nearsightedness

Nearsightedness, or *myopia*, refers to an eye that can focus only on nearby objects. The far point is not infinity but some shorter distance, so distant objects are not seen clearly. It is usually caused by an eyeball that is too long, although sometimes it is the curvature of the cornea that is too great. In either case, images of distant objects are focused in front of the retina. A diverging lens, because it causes parallel rays to diverge, allows the rays to be focused at the retina (Fig. 25–8a) and thus corrects this defect.

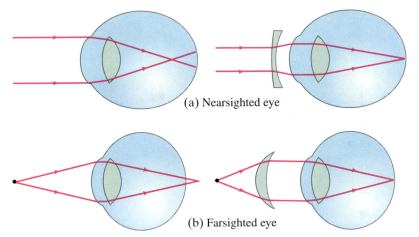

(a) Nearsighted eye

(b) Farsighted eye

FIGURE 25–8 Correcting eye defects with lenses: (a) a nearsighted eye, which cannot focus clearly on distant objects, can be corrected by use of a diverging lens; (b) a farsighted eye, which cannot focus clearly on nearby objects, can be corrected by use of a converging lens.

Farsightedness

Farsightedness, or *hyperopia*, refers to an eye that cannot focus on nearby objects. Although distant objects are usually seen clearly, the near point is somewhat greater than the "normal" 25 cm, which makes reading difficult. This defect is caused by an eyeball that is too short or (less often) by a cornea that is not sufficiently curved. It is corrected by a converging lens, Fig. 25–8b. Similar to hyperopia is *presbyopia*, which refers to the lessening ability of the eye to accommodate as one ages, and the near point moves out. Converging lenses also compensate for this.

Astigmatism

Astigmatism is usually caused by an out-of-round cornea or lens so that point objects are focused as short lines, which blurs the image. It is as if the cornea were spherical with a cylindrical section superimposed. As shown in Fig. 25–9, a cylindrical lens focuses a point into a line parallel to its axis. An astigmatic eye focuses rays in a vertical plane, say, at a shorter distance than it does for rays in a horizontal plane. Astigmatism is corrected with the use of a compensating cylindrical lens. Lenses for eyes that are nearsighted or farsighted as well as astigmatic are ground with superimposed spherical and cylindrical surfaces, so that the radius of curvature of the correcting lens is different in different planes. Astigmatism is tested for by looking with one eye at a pattern like that in Fig. 25–10. Sharply focused lines appear dark, whereas those that are not in focus appear dimmer or gray.

FIGURE 25–9 A cylindrical lens forms a line image of a point object because it is converging in one plane only.

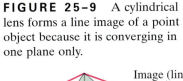
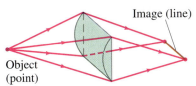

Image (line)

Object (point)

FIGURE 25–10 Test for astigmatism.

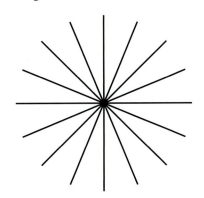

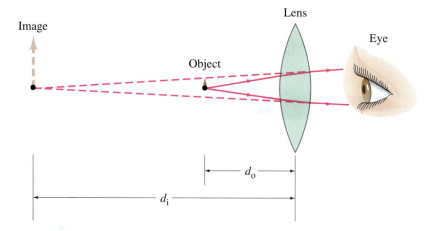

FIGURE 25–11 Lens of reading glasses (Example 25–3).

EXAMPLE 25–3 **Farsighted eye.** A particular farsighted person has a near point of 100 cm. Reading glasses must have what lens power so that this person can read a newspaper at a distance of 25 cm? Assume the lens is very close to the eye.

SOLUTION When the object is placed 25 cm from the lens, we want the image to be 100 cm away on the *same* side of the lens, and so it will be virtual, Fig. 25–11. Thus, $d_o = 25$ cm, $d_i = -100$ cm, and the lens equation gives

$$\frac{1}{f} = \frac{1}{25 \text{ cm}} + \frac{1}{-100 \text{ cm}} = \frac{4 - 1}{100 \text{ cm}} = \frac{1}{33 \text{ cm}}.$$

So $f = 33$ cm $= 0.33$ m. The power P of the lens is $P = 1/f = +3.0$ D. The plus sign indicates that it is a converging lens.

EXAMPLE 25–4 **Nearsighted eye.** A nearsighted eye has near and far points of 12 cm and 17 cm, respectively. What lens power is needed for this person to see distant objects clearly, and what then will be the near point? Assume that the lens is 2.0 cm from the eye (typical for eyeglasses).

SOLUTION (*a*) First we determine the power of the lens needed to focus objects at infinity, when the eye is relaxed. For a distant object ($d_o = \infty$), as shown in Fig. 25–12a, the lens must put the image 17 cm from the eye (its far point), which is 15 cm in front of the lens; hence $d_i = -15$ cm. We use the lens equation to solve for the focal length of the needed lens:

$$\frac{1}{f} = -\frac{1}{15 \text{ cm}} + \frac{1}{\infty} = -\frac{1}{15 \text{ cm}}.$$

So $f = -15$ cm $= -0.15$ m or $P = 1/f = -6.7$ D. The minus sign indicates that it must be a diverging lens.

FIGURE 25–12
Example 25–4.

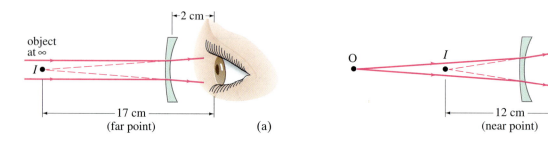

(b) To determine the near point when wearing the glasses, we note that a sharp image will be 12 cm from the eye (its near point, see Fig. 25–12b), which is 10 cm from the lens; so $d_i = -0.10$ m and the lens equation gives

$$\frac{1}{d_o} = \frac{1}{f} - \frac{1}{d_i} = -\frac{1}{0.15\text{ m}} + \frac{1}{0.10\text{ m}} = \frac{1}{0.30\text{ m}}.$$

So $d_o = 30$ cm, which means the near point when the person is wearing glasses is 30 cm in front of the lens.

Contact lenses could be used to correct the eye in Example 25–4. Since contacts are placed directly on the cornea, we would not subtract out the 2.0 cm for the image distances. That is, for distant objects $d_i = -17$ cm, so $P = 1/f = -5.9\ D$ (diopters). The new near point would be 41 cm. Thus we see that a contact lens and an eyeglass lens will require slightly different powers, or focal lengths, for the same eye because of their different placements relative to the eye.

Contact lenses

25–3 The Magnifying Glass

Much of the remainder of this chapter will deal with optical devices that are used to produce magnified images of objects. We first discuss the **simple magnifier**, or **magnifying glass**, which is simply a converging lens, Fig. 25–13.

How large an object appears, and how much detail we can see on it, depends on the size of the image it makes on the retina. This, in turn, depends on the angle subtended by the object at the eye. For example, a penny held 30 cm from the eye looks twice as high as one held 60 cm away because the angle it subtends is twice as great (Fig. 25–14). When we want to examine detail on an object, we bring it up close to our eyes so that it subtends a greater angle. However, our eyes can accommodate only up to a point (the near point), and we will assume a standard distance of 25 cm as the near point in what follows.

FIGURE 25–13 Photo of a magnifying glass and the image it makes.

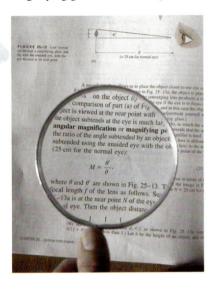

FIGURE 25–14 When the same object is viewed at a shorter distance, the image on the retina is greater, so the object appears larger and more detail can be seen. The angle θ that the object subtends in (a) is greater than in (b).

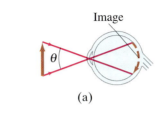

(a)

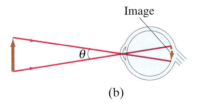

(b)

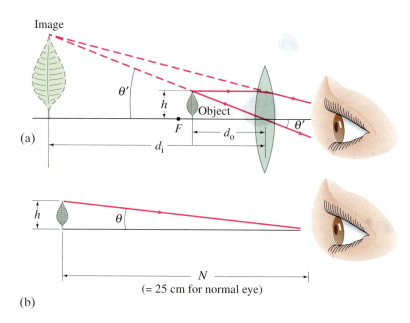

Image

θ'

h Object

F d_o

d_i

(a)

h θ

N

(= 25 cm for normal eye)

(b)

FIGURE 25–15 Leaf viewed (a) through a magnifying glass, and (b) with the unaided eye, with the eye focused at its near point.

➡ **PHYSICS APPLIED**

Magnification of a magnifying glass

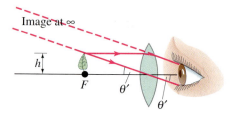

Image at ∞

h

F θ'

θ'

FIGURE 25–16 With the eye relaxed, the object is placed at the focal point, and the image is at infinity. Compare to Fig. 25–15a where the image is at the eye's near point.

Magnification of a simple magnifier

A magnifying glass allows us to place the object closer to our eye so that it subtends a greater angle. As shown in Fig. 25–15a, the object is placed at the focal point or just within it. Then the converging lens produces a virtual image, which must be at least 25 cm from the eye if the eye is to focus on it. If the eye is relaxed, the image will be at infinity, and in this case the object is exactly at the focal point. (You make this slight adjustment yourself when you "focus" on the object by moving the magnifying glass.)

A comparison of part (a) of Fig. 25–15 with part (b), in which the same object is viewed at the near point with the unaided eye, reveals that the angle the object subtends at the eye is much larger when the magnifier is used. The **angular magnification** or **magnifying power**, M, of the lens is defined as the ratio of the angle subtended by an object when using the lens, to the angle subtended using the unaided eye with the object at the near point N of the eye ($N = 25$ cm for the normal eye):

$$M = \frac{\theta'}{\theta}, \qquad (25–1)$$

where θ and θ' are shown in Fig. 25–15. We can write M in terms of the focal length by noting that $\theta = h/N$ (Fig. 25–15b) and $\theta' = h/d_o$ (Fig. 25–15a), where h is the height of the object and we assume the angles are small so θ and θ' equal their sines and tangents. If the eye is relaxed (for least eye strain), the image will be at infinity and the object will be precisely at the focal point; see Fig. 25–16. Then $d_o = f$ and $\theta' = h/f$. Thus

$$M = \frac{\theta'}{\theta} = \frac{h/f}{h/N} = \frac{N}{f}. \qquad \begin{bmatrix} \text{eye focused at } \infty \text{;} \\ N = 25 \text{ cm for normal eye} \end{bmatrix} \quad (25–2a)$$

We see that the shorter the focal length of the lens, the greater the magnification. [However, simple single-lens magnifiers are limited to about 2 or 3× because of distortion due to spherical aberration (Section 25–6).] The magnification of a given lens can be increased a bit by moving the lens and adjusting your eye so it focuses on the image at the eye's near point. In this case, $d_i = -N$ (see Fig. 25–15a) if your eye is very near the magnifier.

Then the object distance d_o is given by

$$\frac{1}{d_o} = \frac{1}{f} - \frac{1}{d_i} = \frac{1}{f} + \frac{1}{N}.$$

(We see from this equation that $d_o < f$, as shown in Fig. 25–15a, since $N/(f + N)$ must be less than 1.) Now $\theta' = h/d_o$ so

$$M = \frac{\theta'}{\theta} = \frac{h/d_o}{h/N} = \frac{N}{d_o} = N\left(\frac{1}{f} + \frac{1}{N}\right)$$

or

$$M = \frac{N}{f} + 1. \qquad \left[\begin{array}{l}\text{eye focused at near point, } N;\\ N = 25 \text{ cm for normal eye}\end{array}\right] \quad \textbf{(25–2b)}$$

We see that the magnification is slightly greater when the eye is focused at its near point, rather than relaxed.

EXAMPLE 25–5 **ESTIMATE** **A jeweler's loupe.** An 8-cm-focal-length converging lens is used as a "jeweler's loupe," which is a magnifying glass. Estimate (a) the magnification when the eye is relaxed, and (b) the magnification if the eye is focused at its near point $N = 25$ cm.

SOLUTION (a) With the relaxed eye focused at infinity, $M = N/f = 25\text{ cm}/8\text{ cm} \approx 3\times$. (b) The magnification when the eye is focused at its near point ($N = 25$ cm), and the lens is near the eye, is:

$$M = 1 + \frac{N}{f} = 1 + \frac{25}{8} \approx 4\times.$$

25–4 Telescopes

A telescope is used to magnify objects that are very far away. In most cases, the object can be considered to be at infinity.

Galileo, although he did not invent it,[†] developed the telescope into a usable and important instrument. He was the first to examine the heavens with the telescope (Fig. 25–17), and he made world-shaking discoveries

➡ **PHYSICS APPLIED**

Telescopes and their magnification

FIGURE 25–17 (a) Objective lens (mounted now in an ivory frame) from the telescope with which Galileo made his world-shaking discoveries, including the moons of Jupiter. (b) Later telescopes made by Galileo.

(a)

(b)

[†]Galileo built his first telescope in 1609 after having heard of such an instrument existing in Holland. The first telescopes magnified only 3 to 4 times, but Galileo soon made a 30-power instrument. The first Dutch telescope seems to date from about 1604, but there is a reference suggesting it may have been copied from an Italian telescope built as early as 1590. Kepler (see Chapter 5) gave a ray description (1611) of the Keplerian telescope, which is named for him because he first described it, although he did not build it.

(the moons of Jupiter, the phases of Venus, sunspots, the structure of the Moon's surface, that the Milky Way is made up of a huge number of individual stars, among others).

Refracting telescope

Several types of **astronomical telescope** exist. The common **refracting** type, sometimes called **Keplerian**, contains two converging lenses located at opposite ends of a long tube, as diagrammed in Fig. 25–18. The lens closest to the object is called the **objective lens** and forms a real image I_1 of the distant object in the plane of its focal point F_o (or near it if the object is not at infinity). Although this image, I_1, is smaller than the original object, it subtends a greater angle and is very close to the second lens, called the **eyepiece**, which acts as a magnifier. That is, the eyepiece magnifies the image produced by the objective to produce a second, greatly magnified image, I_2, which is virtual, and inverted. If the viewing eye is relaxed, the eyepiece is adjusted so the image I_2 is at infinity. Then the real image I_1 is at the focal point F'_e of the eyepiece, and the distance between the lenses is $f_o + f_e$ for an object at infinity.

To find the total magnification of this telescope, we note that the angle an object subtends as viewed by the unaided eye is just the angle θ subtended at the telescope objective. From Fig. 25–18 we can see that $\theta \approx h/f_o$, where h is the height of the image I_1 and we assume θ is small so that $\tan\theta \approx \theta$. Note, too, that the thickest of the three rays drawn in the figure is parallel to the axis before it strikes the eyepiece and therefore passes through the eyepiece focal point F_e. Thus, $\theta' \approx h/f_e$ and the total magnifying power (angular magnification) of this telescope is

Telescope magnification

$$M = \frac{\theta'}{\theta} = -\frac{f_o}{f_e}, \tag{25–3}$$

where we have inserted a minus sign to indicate that the image is inverted. To achieve a large magnification, the objective lens should have a long focal length and the eyepiece a short focal length.

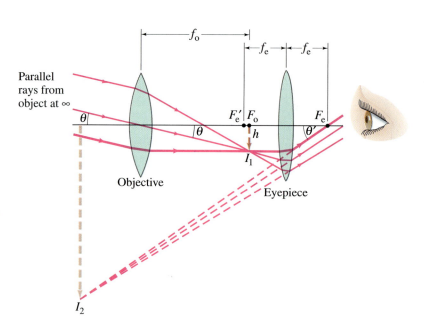

FIGURE 25–18
Astronomical telescope (refracting). Parallel light from one point on a distant object ($d_o = \infty$) is brought to a focus by the objective lens in its focal plane. This image (I_1) is magnified by the eyepiece to form the final image I_2. Only two of the rays shown are standard rays described in Fig. 23–34.

EXAMPLE 25–6 **Telescope magnification.** The largest refracting telescope in the world is located at the Yerkes Observatory in Wisconsin, Fig. 25–19. It is referred to as a "40-inch" telescope, meaning that the diameter of the objective is 40 inches, or 102 cm. The objective has a focal length of 19 m, and the eyepiece has a focal length of 10 cm. (a) Calculate the total magnifying power of this telescope. (b) Estimate the length of the telescope.

SOLUTION (a) From Eq. 25–3 we find,

$$M = -\frac{f_o}{f_e} = -\frac{19\text{ m}}{0.10\text{ m}} = -190\times.$$

(b) For a relaxed eye, the image I_1 is at the focal point of both the eyepiece and the objective lenses. The distance between the two lenses is thus $f_o + f_e \approx 19$ m, which is essentially the length of the telescope.

For an astronomical telescope to produce bright images of distant stars, the objective lens must be large to allow in as much light as possible. Indeed, the diameter of the objective (and hence its "light-gathering power") is the most important parameter for an astronomical telescope, which is why the largest ones are specified by giving the objective diameter (such as the 200-inch Hale telescope on Palomar Mountain). The construction and grinding of large lenses is very difficult. Therefore, the largest telescopes are **reflecting telescopes** that use a curved mirror as the objective, Fig. 25–20, since a mirror has only one surface to be ground and can be supported along its entire surface[†] (a large lens, supported at its edges, would sag under its own weight). Normally, the eyepiece lens or mirror (see Fig. 25–20) is removed so that the real image formed by the objective mirror can be recorded directly on film.

FIGURE 25–19 This large refracting telescope was built in 1897 and is housed at Yerkes Observatory in Wisconsin. The objective lens is 102 cm (40 inches) in diameter, and the telescope tube is about 19 m long.

Reflecting telescopes

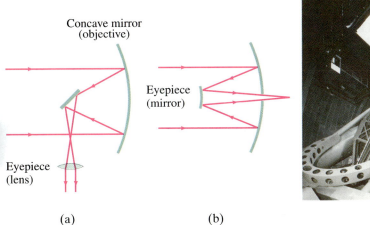

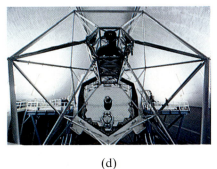

(a)   (b)   (c)   (d)

FIGURE 25–20 A concave mirror can be used as the objective of an astronomical telescope. Either a lens (a) or a mirror (b) can be used as the eyepiece. Arrangement (a) is called the Newtonian focus and (b) the Cassegrainian focus. Other arrangements are also possible. (c) The 200-inch (mirror diameter) Hale telescope on Palomar Mountain in California. (d) The 10-meter Keck telescope on Mauna Kea, Hawaii. The Keck combines thirty-six 1.8 meter six-sided mirrors into the equivalent of a very large (10-m diameter) single reflector.

[†]Another advantage of mirrors is that they exhibit no chromatic aberration (Section 25–6) because the light doesn't pass through them. Also, they can be ground in a parabolic shape to correct for spherical aberration (Section 25–6). The reflecting telescope was first proposed by Newton.

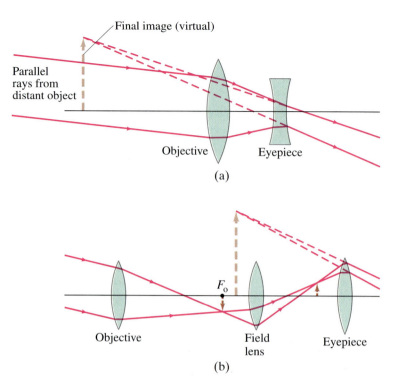

Final image (virtual)

Parallel rays from distant object

Objective

Eyepiece

(a)

F_o

Objective

Field lens

Eyepiece

(b)

FIGURE 25–21 Terrestrial telescopes that produce an upright image: (a) Galilean; (b) spyglass, or field-lens, type.

A **terrestrial telescope** (for use in viewing objects on Earth), unlike its astronomical counterpart, must provide an upright image. Two designs are shown in Fig. 25–21. The **Galilean** type shown in part (a), which Galileo used for his great astronomical discoveries, has a diverging lens as eyepiece which intercepts the converging rays from the objective lens before they reach a focus, and acts to form a virtual upright image. This design is often used in opera glasses. The tube is reasonably short, but the field of view is small. The second type, shown in Fig. 25–21b, is often called a **spyglass** and makes use of a third lens ("field lens") that acts to make the image upright as shown. A spyglass must be quite long. The most practical design today is the **prism binocular** which was shown in Fig. 23–25. The objective and eyepiece are converging lenses. The prisms reflect the rays by total internal reflection and shorten the physical size of the device, and they also act to produce an upright image. One prism reinverts the image in the vertical plane, the other in the horizontal plane.

25–5 Compound Microscope

The compound **microscope**, like the telescope, has both objective and eyepiece (or ocular) lenses, Fig. 25–22. The design is different from that for a telescope because a microscope is used to view objects that are very close, so the object distance is very small. The object is placed just beyond the objective's focal point as shown in Fig. 25–22a. The image I_1 formed by the objective lens is real, quite far from the lens, and much enlarged. This image is magnified by the eyepiece into a very large virtual image, I_2, which is seen by the eye and is inverted.

The overall magnification of a microscope is the product of the magnifications produced by the two lenses. The image I_1 formed by the objec-

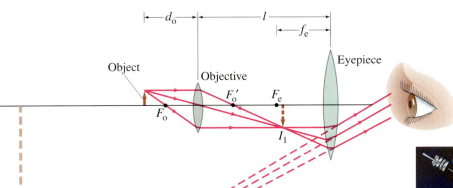

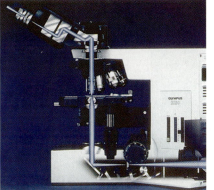

FIGURE 25–22
Compound microscope:
(a) ray diagram, (b) photograph
(illumination comes from the
lower right, then up through
the slide holding the object).

(a)   (b)

tive is a factor m_o greater than the object itself. From Fig. 25–22a and Eq. 23–9 for the lateral magnification of a simple lens, we have

$$m_o = \frac{h_i}{h_o} = \frac{d_i}{d_o} = \frac{l - f_e}{d_o}, \tag{25–4}$$

where d_o and d_i are the object and image distances for the objective lens, l is the distance between the lenses (equal to the length of the barrel), and we ignored the minus sign in Eq. 23–9 which only tells us that the image is inverted. The eyepiece acts like a simple magnifier. If we assume that the eye is relaxed, its angular magnification M_e is (from Eq. 25–2a)

$$M_e = \frac{N}{f_e}, \tag{25–5}$$

where the near point $N = 25$ cm for the normal eye. Since the eyepiece enlarges the image formed by the objective, the overall angular magnification M is the product of the lateral magnification of the objective lens, m_o, times the angular magnification, M_e, of the eyepiece lens (Eqs. 25–4 and 25–5):

$$M = M_e m_o = \left(\frac{N}{f_e}\right)\left(\frac{l - f_e}{d_o}\right) \tag{25–6a}$$

$$\approx \frac{Nl}{f_e f_o}. \tag{25–6b}$$

Magnification

of

microscope

The approximation, Eq. 25–6b, is accurate when f_e and f_o are small compared to l, so $l - f_e \approx l$ and $d_o \approx f_o$ (Fig. 25–22a). This is a good approximation for large magnifications, since these are obtained when f_o and f_e are very small (they are in the denominator of Eq. 25–6b). In order to make lenses of very short focal length, which can be done best for the objective, compound lenses involving several elements must be used to avoid serious aberrations, as discussed in the next Section.

EXAMPLE 25–7 **Microscope.** A compound microscope consists of a 10× eyepiece and a 50× objective 17.0 cm apart. Determine (a) the overall magnification, (b) the focal length of each lens, and (c) the position of the object when the final image is in focus with the eye relaxed. Assume a normal eye, so $N = 25$ cm.

SOLUTION (a) The overall magnification is $10 \times 50 = 500\times$. (b) The eyepiece focal length is (Eq. 25–5) $f_e = N/M_e = 25\,\text{cm}/10 = 2.5$ cm. For the objective lens, it is easier to next find d_o (part (c)) before we find f_o because we can use Eq. 25–4. Solving for d_o, we find

$$d_o = (l - f_e)/m_o = (17.0\,\text{cm} - 2.5\,\text{cm})/50 = 0.29\,\text{cm}.$$

Then, from the lens equation with $d_i = l - f_e = 14.5$ cm (see Fig. 25–22a),

$$\frac{1}{f_o} = \frac{1}{d_o} + \frac{1}{d_i} = \frac{1}{0.29\,\text{cm}} + \frac{1}{14.5\,\text{cm}} = 3.52;$$

so $f_o = 0.28$ cm.
(c) We just calculated $d_o = 0.29$ cm, which is very close to f_o.

Opaque objects are generally illuminated by a source placed above them. If the objects to be viewed are transparent, such as cells or tissue, light is normally passed through the object from a source beneath the microscope stage (see Fig. 25–22b). The illumination system must be carefully designed if maximum sharpness and contrast are to be achieved. Usually, a *condenser* is employed, which is a set of two or three lenses, although inexpensive condensers may be a single lens or curved mirror. The purpose of the condenser is to gather light over a wide angle from the source, and to "condense" it down to a narrow beam that will illuminate the object strongly and uniformly. A number of different designs are employed. The source is often placed in the focal plane of the condenser so that light from each point on the source is parallel when it passes through the object.

Some specialized microscope types are described in Section 25–10.

25–6 Aberrations of Lenses and Mirrors

In Chapter 23, we developed a theory of image formation by a thin lens. We found, for example, that all rays from each point on an object are brought to a single point as the image point. This, and other results, were based on approximations such as that all rays make small angles with one another and we can use $\sin\theta \approx \theta$. Because of these approximations, we expect deviations from the simple theory and these are referred to as **lens aberrations**. There are several types of aberration; we will briefly discuss each of them separately but all may be present at one time.

Consider an object at any point (even at infinity) on the axis of a lens. Rays from this point that pass through the outer regions of the lens are brought to a focus at a different point from those that pass through the center of the lens. This is called **spherical aberration**, and is shown exaggerated in Fig. 25–23. Consequently, the image seen on a piece of film (for

Spherical aberration

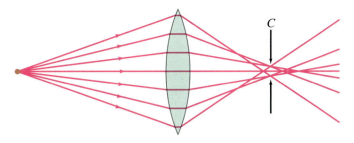

FIGURE 25–23 Spherical aberration (exaggerated). Circle of least confusion is at C.

example) will not be a point but a tiny circular patch of light. If the film is placed at the point C, as indicated, the circle will have its smallest diameter, which is referred to as the **circle of least confusion**. Spherical aberration is present whenever spherical surfaces are used. It can be corrected by using nonspherical lens surfaces, but to grind such lenses is difficult and expensive. It can be minimized with spherical surfaces by choosing the curvatures so that equal amounts of bending occur at each lens surface; a lens can be designed like this for only one particular object distance. Spherical aberration is usually corrected (by which we mean reduced greatly) by the use of several lenses in combination, and by using only the central part of lenses.

For object points off the lens axis, additional aberrations occur. Rays passing through the different parts of the lens cause spreading of the image that is noncircular. We won't go into the details but merely point out that there are two effects: **coma** (because the image of a point is comet-shaped rather than a circle) and **off-axis astigmatism**.[†] Furthermore, the image points for objects off the axis but at the same distance from the lens do not fall on a flat plane but on a curved surface—that is, the focal plane is not flat. (We expect this because the points on a flat plane, such as the film in a camera, are not equidistant from the lens.) This aberration is known as **curvature of field** and is obviously a problem in cameras and other devices where the film is placed in a flat plane. In the eye, however, the retina is curved, which compensates for this effect. Another aberration, known as **distortion**, is a result of variation of magnification at different distances from the lens axis. Thus a straight line object some distance from the axis may form a curved image. A square grid of lines may be distorted to produce "barrel distortion," or "pincushion distortion," Fig. 25–24. The latter is common in extreme wide-angle lenses.

Off-axis aberrations

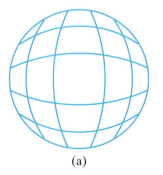
(a)

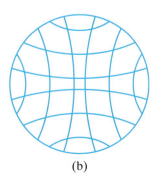
(b)

FIGURE 25–24 Distortion. Lenses may image a square grid of perpendicular lines to produce (a) barrel distortion or (b) pincushion distortion. These distortions can be seen in the photgraphs of Fig. 23-29d.

[†]Although the effect is the same as for astigmatism in the eye (Section 25–2), the cause is different. Off-axis astigmatism is no problem in the eye because objects are clearly seen only at the fovea, which is on the lens axis.

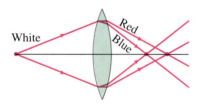

FIGURE 25–25 Chromatic aberration. Different colors are focused at different points.

FIGURE 25–26 Achromatic doublet.

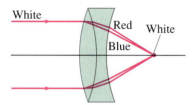

All the above aberrations occur for monochromatic light and hence are referred to as *monochromatic aberrations*. Normal light is not monochromatic, and there will also be **chromatic aberration**. This aberration arises because of dispersion—the variation of index of refraction of transparent materials with wavelength (Section 24–4). For example, blue light is bent more than red light by glass. So if white light is incident on a lens, the different colors are focused at different points, Fig. 25–25, and there will be colored fringes in the image. Chromatic aberration can be eliminated for any two colors (and reduced greatly for all others) by the use of two lenses made of different materials with different indices of refraction and dispersion. Normally one lens is converging and the other diverging, and they are often cemented together (Fig. 25–26). Such a lens combination is called an **achromatic doublet** (or "color-corrected" lens).

It is not possible to fully correct all aberrations. Combining two or more lenses together can reduce them. High-quality lenses used in cameras, microscopes, and other devices are **compound lenses** consisting of many simple lenses (referred to as **elements**). A typical high-quality camera lens may contain six to eight (or more) elements.

For simplicity we will normally indicate lenses in diagrams as if they were simple lenses. But it must be remembered that good-quality lenses are compound.

The human eye is also subject to aberrations, but they are minimal. Spherical aberration, for example, is minimized because (1) the cornea is less curved at the edges than at the center, and (2) the lens is less dense at the edges than at the center. Both effects cause rays at the outer edges to be bent less strongly, and thus help to reduce spherical aberration. Chromatic aberration is partially compensated for because the lens absorbs the shorter wavelengths appreciably and the retina is less sensitive to the blue and violet wavelengths. This is just the region of the spectrum where dispersion—and thus chromatic aberration—is greatest (Fig. 24–14).

Spherical mirrors (Section 23–3) also suffer aberrations including spherical aberration (see Fig. 23–12). Mirrors can be ground in a parabolic shape to correct for spherical aberration, but they are much harder to make and so very expensive. Spherical mirrors do not, however, exhibit chromatic aberration because the light does not pass through them.

25–7 Limits of Resolution; the Rayleigh Criterion

The ability of a lens to produce distinct images of two point objects very close together is called the **resolution** of the lens. The closer the two images can be and still be seen as distinct (rather than overlapping blobs), the higher the resolution. The resolution of a camera lens, for example, is often specified as so many lines per millimeter,[†] and can be determined by photographing a standard set of parallel lines on fine-grain film. The minimum spacing of lines distinguishable on film using the lens gives the resolution.

Two principal factors limit the resolution of a lens. The first is lens aberrations. As we saw, because of spherical and other aberrations, a point object is not a point on the image but a tiny blob. Careful design of compound lenses can reduce aberrations significantly, but they cannot be elim-

[†]This may be specified at the center of the field of view as well as at the edges, where it is usually less because of off-axis aberrations.

inated entirely. The second factor that limits resolution is *diffraction*, which cannot be corrected for because it is a natural result of the wave nature of light. We discuss it now.

In Section 24–5 we saw that because light travels as a wave, light from a point source passing through a slit is spread out into a diffraction pattern (Figs. 24–18 and 24–20). A lens, because it has edges, acts like a slit. When a lens forms the image of a point object, the image of that point is actually a tiny diffraction pattern. Thus, *an image would be blurred even if aberrations were absent.*

Image point is a diffraction pattern (i.e. blurred)

In the analysis that follows we assume that the lens is free of aberrations, so that we can focus our attention on diffraction effects and how much they limit the resolution of a lens. In Fig. 24–20 we saw that the diffraction pattern produced by light passing through a rectangular slit has a central maximum in which most of the light falls. This central peak falls to a minimum on either side of its center at an angle $\theta \approx \sin\theta = \lambda/D$ (this is Eq. 24–3), where D is the width of the slit, λ is the wavelength of light used, and we assume θ is small. There are also low-intensity fringes beyond. For a lens, or any circular hole, the image of a point object will consist of a *circular* central peak (called the *diffraction spot* or *Airy disk*) surrounded by faint circular fringes, as shown in Fig. 25–27a. The central maximum has an angular half width given by

$$\theta = \frac{1.22\lambda}{D},$$

where D is the diameter.

This differs from the formula for a slit (Eq. 24–3) by the factor 1.22. This factor comes from the fact that the width of a circular hole is not uniform (like a rectangular slit) but varies from its diameter D to zero. A careful analysis shows that the "average" width is $D/1.22$. Hence we get the equation above rather than Eq. 24–3. The intensity of light in the diffraction pattern of light from a point source passing through a circular opening is shown in Fig. 25–28. (The image for a non-point source is a superposition of such patterns.) For most purposes, we need consider only the central spot since the concentric rings are much dimmer.

If two point objects are very close, the diffraction patterns of their images will overlap as shown in Fig. 25–27b. As the objects are moved closer,

FIGURE 25–27 Photographs of images (greatly magnified) formed by a lens, showing diffraction pattern of image for: (a) a single point object; (b) two point objects whose images are barely resolved.

FIGURE 25–28 Intensity of light across the diffraction pattern of a circular hole.

(a)

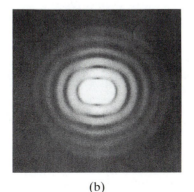

(b)

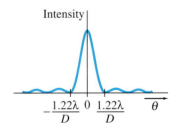

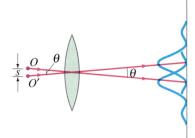

FIGURE 25–29 The *Rayleigh criterion.* Two images are just resolvable when the center of the diffraction peak of one is directly over the first minimum in the diffraction pattern of the other. The two point objects O and O' subtend an angle θ at the lens; only one ray is drawn for each point to indicate the center of the diffraction pattern of its image.

a separation is reached where you can't tell if there are two overlapping images or a single image. The separation at which this happens may be judged differently by different observers. However, a generally accepted criterion is one proposed by Lord Rayleigh (1842–1919). This **Rayleigh criterion** states that *two images are just resolvable when the center of the diffraction disk of one is directly over the first minimum in the diffraction pattern of the other.* This is shown in Fig. 25–29. Since the first minimum is at an angle $\theta = 1.22\lambda/D$ from the central maximum, Fig. 25–29 shows us that two objects can be considered just resolvable if they are separated by this angle θ:

Rayleigh criterion (resolution limit)

$$\theta = \frac{1.22\lambda}{D}. \qquad (25\text{–}7)$$

This is the limit on resolution set by the wave nature of light due to diffraction.

EXAMPLE 25–8 Hubble space telescope. The Hubble Space Telescope (HST) is a reflecting telescope that has been placed in orbit above the Earth's atmosphere, so its resolution is not limited by turbulence in the atmosphere (Fig. 25–30). Its objective diameter is 2.4 m. [The Hubble can also observe radiation in the near ultraviolet (wavelengths as small as 115 nm) and infrared (wavelengths as long as 1 mm) which are ranges of the spectrum blocked by the atmosphere.] For visible light (take $\lambda = 550$ nm), estimate the improvement in resolution the Hubble offers over Earth-bound telescopes, which are limited in resolution by movement of the Earth's atmosphere to about half an arc second (each degree is divided into 60 minutes each containing 60 seconds, so $1° = 3600$ arc seconds).

SOLUTION Earth-bound telescopes are limited to an angular resolution of

$$\theta = \frac{1}{2}\left(\frac{1}{3600}\right)^{\circ}\left(\frac{2\pi\,\text{rad}}{360°}\right) = 2.4 \times 10^{-6}\,\text{rad}.$$

The Hubble, on the other hand, is limited by diffraction (Eq. 25–7) which for $\lambda = 550$ nm is

$$\theta = \frac{1.22(550 \times 10^{-9}\,\text{m})}{2.4\,\text{m}} = 2.8 \times 10^{-7}\,\text{rad},$$

which is almost ten times better.

FIGURE 25–30 Hubble Space Telescope, with Earth in the background. The flat orange panels are solar cells that collect energy from the Sun.

25–8 Resolution of Telescopes and Microscopes

You might think that a microscope or telescope could be designed to produce any desired magnification, depending on the choice of focal lengths and quality of the lenses. But this is not possible, because of diffraction. An increase in magnification above a certain point merely results in magnification of the diffraction patterns. This would be highly misleading since we might think we are seeing details of an object when we are really seeing details of the diffraction pattern. To examine this, we apply the Rayleigh criterion: two objects (or two nearby points on one object) are just resolvable if they are separated by an angle θ (Fig. 25–29) given by Eq. 25–7:

$$\theta = \frac{1.22\lambda}{D}.$$

This is valid for either a microscope or a telescope, where D is the diameter of the objective lens. For a telescope, the resolution is specified by stating θ as given by this equation.[†]

For a microscope, it is more convenient to specify the actual distance, s, between two points that are just barely resolvable, Fig. 25–29. Since objects are normally placed near the focal point of the microscope objective, $\theta = s/f$, or $s = f\theta$. If we combine this with Eq. 25–7, we obtain for the **resolving power (RP)**:

$$RP = s = f\theta = \frac{1.22\lambda f}{D}. \qquad (25\text{–}8)$$

Resolving power

This distance s is called the resolving power of the lens because it is the minimum separation of two object points that can just be resolved, assuming the highest quality lens since this limit is imposed by the wave nature of light.

Equation 25–8 is often written in terms of the *angle of acceptance*, α, of the objective lens as defined in Fig. 25–31. The derivation is long and we only quote the result:

$$RP = s = \frac{1.22\lambda}{2\sin\alpha} = \frac{0.61\lambda}{\sin\alpha}.$$

The resolving power can be increased by placing a drop of oil that encloses the object and the front surface of the objective. This is called an **oil-immersion objective**. In the oil, the wavelength of the light is reduced to λ/n (Eq. 24–1), where n is the oil's index of refraction. Thus the resolving power becomes

$$RP = \frac{0.61\lambda}{n\sin\alpha}. \qquad (25\text{–}9)$$

The oil typically has $n \approx 1.5$, although n may be as great as 1.8. Thus oil immersion increases the resolution by 50 percent or more.

The quantity $(n\sin\alpha)$ is called the **numerical aperture** (NA) of the lens:

$$NA = n\sin\alpha. \qquad (25\text{–}10)$$

It is usually specified on the objective lens housing along with the magnification. The larger the value of the NA, the finer the resolving power.

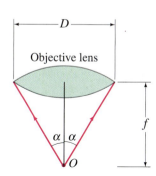

FIGURE 25–31 Objective lens of a microscope, showing the angle of acceptance, α.

[†]Telescopes with large-diameter objectives are usually limited not by diffraction but by other effects such as turbulence in the atmosphere. The resolution of a high-quality microscope, on the other hand, normally *is* limited by diffraction because microscope objectives are complex compound lenses containing many elements of small diameter (since f is small).

FIGURE 25–32 The 300-meter radiotelescope in Arecibo, Puerto Rico, uses radio waves (Fig. 22–10) instead of visible light.

EXAMPLE 25–9 **Telescope resolution (radio wave vs. visible light).** What is the theoretical minimum angular separation of two stars that can just be resolved by: (*a*) the 200-inch telescope on Palomar Mountain (Fig. 25–20c); and (*b*) the Arecibo radiotelescope (Fig. 25–32), whose diameter is 300 m and whose radius of curvature is also 300 m. Assume $\lambda = 550$ nm for the visible-light telescope in part (*a*), and $\lambda = 4$ cm (the shortest wavelength at which the radiotelescope has been operated) in part (*b*).

SOLUTION (*a*) Since $D = 200$ inch $= 5.1$ m, we have from Eq. 25–7 that

$$\theta = \frac{1.22\lambda}{D} = \frac{(1.22)(5.50 \times 10^{-7}\ \text{m})}{(5.1\ \text{m})} = 1.3 \times 10^{-7}\ \text{rad},$$

or 0.75×10^{-5} deg. (Note that this is equivalent to resolving two points less than 1 cm apart from a distance of 100 km!) This is the limit set by diffraction. The resolution is not this good because of aberrations and, more importantly, turbulence in the atmosphere. In fact, large-diameter objectives are not justified by increased resolution, but by their greater light-gathering ability—they allow more light in, so fainter objects can be seen. (*b*) Radiotelescopes are not hindered by atmospheric turbulence, and for radio waves with $\lambda = 0.04$ m the resolution is

$$\theta = \frac{(1.22)(0.04\ \text{m})}{(300\ \text{m})} = 1.6 \times 10^{-4}\ \text{rad}.$$

EXAMPLE 25–10 **Microscope resolution.** Determine the NA and RP of the best oil-immersion microscopes, where the index of refraction of the oil is $n = 1.8$ and $\sin \alpha \approx 0.90$. Assume that $\lambda = 550$ nm.

SOLUTION The NA $= n \sin \alpha = 1.6$. The resolving power is

$$\text{RP} = \frac{0.61\lambda}{\text{NA}} = \frac{(0.61)(5.50 \times 10^{-7}\ \text{m})}{(1.6)} \approx 2 \times 10^{-7}\ \text{m} = 200\ \text{nm}.$$

This is the best resolution that a visible-light microscope can attain.

Diffraction sets an ultimate limit on the detail that can be seen on any object. In Eq. 25–8 we note that the focal length of a lens cannot be made less than (approximately) the radius of the lens, and even that is very difficult—see the lensmaker's equation (Eq. 23–10). In this best case,[†] Eq. 25–8 gives, with $f \approx D/2$,

$$\text{RP} \approx \frac{\lambda}{2}. \tag{25–11}$$

Thus we can say, to within a factor of 2 or so, that

Resolution limited to λ

it is not possible to resolve detail of objects smaller than the wavelength of the radiation being used.

This is an important and useful rule of thumb.

[†]The same result can be obtained from Eq. 25–9 since $\sin \alpha$ can never exceed 1 and typically is 0.6 to 0.9 at most. With oil immersion, Eq. 25–9 gives, *at best*, RP $\approx \lambda/3$, which corresponds to the result of Example 25–10.

Compound lenses are now designed so well that the actual limit on resolution is often set by diffraction—that is, by the wavelength of the light used. To obtain greater detail, one must use radiation of shorter wavelength. The use of UV radiation can increase the resolution by a factor of perhaps 2. Far more important, however, was the discovery in the early twentieth century that electrons have wave properties (Chapter 27) and that their wavelengths can be very small. The wave nature of electrons is utilized in the electron microscope (Section 27–7), which can magnify 100 to 1000 times more than a visible-light microscope because of the much shorter wavelengths. X-rays, too, have very short wavelengths and are often used to study objects in great detail (Section 25–11).

25–9 Resolution of the Human Eye and Useful Magnification

The resolution of the human eye is limited by several factors, all of roughly the same order of magnitude. The resolution is best at the fovea, where the cone spacing is smallest, about $3 \, \mu m \, (= 3000 \, nm)$. The diameter of the pupil varies from about 0.1 cm to about 0.8 cm. So for $\lambda = 550 \, nm$ (where the eye's sensitivity is greatest), the diffraction limit is about $\theta \approx 1.22 \lambda / D \approx 8 \times 10^{-5} \, rad$ to $6 \times 10^{-4} \, rad$. Since the eye is about 2 cm long, this corresponds to a resolving power of $s \approx (8 \times 10^{-5} \, rad)(2 \times 10^{-2} \, m) \approx 2 \, \mu m$ at best, to about $15 \, \mu m$ at worst (pupil small). Spherical and chromatic aberration also limit the resolution to about $10 \, \mu m$. The net result is that the eye can resolve objects whose angular separation is about $5 \times 10^{-4} \, rad$ at best. This corresponds to objects separated by 1 cm at a distance of about 20 m.

The typical near point of a human eye is about 25 cm. At this distance, the eye can just resolve objects that are $(25 \, cm)(5 \times 10^{-4} \, rad) \approx 10^{-4} \, m = \frac{1}{10}$ mm apart. Since the best light microscopes can resolve objects no smaller than about 200 nm (see Example 25–10), the useful magnification [= (resolution by naked eye)/(resolution by microscope)] is limited to about

$$\frac{10^{-4} \, m}{200 \times 10^{-9} \, m} = 500 \times.$$

In practice, magnifications of about 1000× are often used to minimize eyestrain. Any greater magnification would simply make visible the diffraction pattern produced by the microscope objective.

* 25–10 Specialty Microscopes and Contrast

All the resolving power a microscope can attain will be useless if the object to be seen cannot be distinguished from the background. The difference in brightness between the image of an object and the image of the surroundings is called **contrast**. Achieving high contrast is an important problem in microscopy and other forms of imaging. The problem arises in biology, for example, because cells consist largely of water and are almost uniformly transparent to light. We now discuss two special types of microscope that can increase contrast: the interference and phase-contrast microscopes.

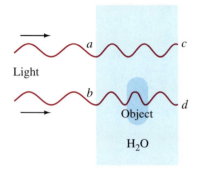

FIGURE 25–33 Object—say, a bacterium—in a water solution.

Interference microscope. The interference microscope makes use of the wave properties of light in a direct way. It is one of the most effective means to increase contrast in a transparent object. To see how it works, let us consider a transparent object—say, a bacterium—in water solution (Fig. 25–33). Light enters uniformly from the left and is coherent (meaning in phase) at all points such as a and b. If the object is as transparent as the water solution, the beam leaving at d will be as bright as that at c. There will be no contrast and the object will not be seen. However, if the object's refractive index is slightly different from that of the surrounding medium, the wavelength within the object will be altered as shown. Hence the waves at points c and d will differ in phase, if not in amplitude. This appears at first to be of no help, since the eye responds only to differences in amplitude or brightness, and does not detect this difference in phase. What the interference microscope does is to change this difference in phase into a difference of amplitude. It does so by superimposing the light that passes through the sample onto a reference beam that does not pass through the object, so that they interfere. One way of doing this is shown in Fig. 25–34. Light from a source is split into two equal beams by a half-silvered mirror, MS_1. One beam passes through the object and the second (comparison beam) passes through an identical system without the object. The two meet again and are superposed by the half-silvered mirror MS_2 before entering the eyepiece and the eye. The path length (and amplitude) of the comparison beam is adjustable. It can be adjusted, for example, so that the background is dark; that is, full destructive interference occurs. Light passing through the object (beam bd in Fig. 25–33) will also interfere with the comparison beam. But because of its different phase, the interference will not be completely destructive. Thus it will appear brighter than the background. Where the object varies in thickness, the phase difference between beams ac and bd in Fig. 25–33 will be different; and this will affect the amount of interference. Hence *variation in the thickness of the object will appear as variations in brightness in the image.* As an example, suppose that the object is a bacterium 1.0 μm thick and has a refractive index of 1.35. Then if light of wavelength $\lambda = 550$ nm (in air) is used, there will be $(1.0\,\mu\text{m})/(550\,\text{nm}/1.35) = 2.46$ wavelengths in the bacterium and $(1.0\mu\text{m})/(550\,\text{nm}/1.33) = 2.42$ wavelengths in the water. Thus the two waves will be out of phase by 0.04 wavelengths, or 14° ($=0.04 \times 360°$).

FIGURE 25–34 Diagram of an interference microscope.

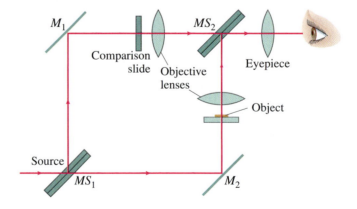

Phase-contrast microscope. The phase-contrast microscope also makes use of interference and differences in phase to produce a high-contrast image. Although it has certain limitations, it is far simpler to construct and operate than an interference microscope. To describe the operation of a phase-contrast microscope in detail, we would have to discuss the diffraction theory of image formation—how the diffraction pattern produced by each point on the object contributes to the final image. This is quite complicated, so we give only a simplified description. Figure 25–35 is a simplified diagram of a phase-contrast microscope. The object to be viewed is illuminated from below as usual. To be specific, we assume that rays from each point on the source are made parallel by a set of condensing lenses (not shown). However, a plate with a ring-shaped hole is placed above the source, so the light can pass only through this annular ring. Light that is not deviated by the object (the beam is shown shaded) is brought into focus by the objective lens in the *source image plane*. If the source is effectively at infinity (because of the condensing lenses), as is assumed here, the source image plane is at the focal point of the lens. Light that strikes the object, on the other hand, is diffracted or scattered by the object. Each point on the object then serves as a source for rays diverging from that point (dashed lines in the figure). These rays are brought to a focus in the *object image plane*, which is behind the source image plane (because the object is so close to the lens). The undeviated light from the source diverges, meanwhile, from its image plane and provides a broad bright background at the object image plane. The object is transparent, however, and the image will not be seen clearly since there will be little contrast. Contrast is achieved by inserting a circular glass *phase plate* at the source image plane. The phase plate has a groove as shown (or a raised portion) in the shape of a ring. This ring is positioned so that all the undeviated rays pass through it. Most of the rays deviated by the object, on the other hand, do not pass through this ring (see Fig. 25–35). Because the rays deviated by the object travel through a different thickness of glass than the undeviated source rays, the two can be out of phase and can interfere destructively at the object image plane. Thus the image of the object will contrast sharply with the background. Actually, because only a small fraction of the light is deviated by the object, the background light will be much stronger, and so the contrast will not be great. To compensate for this, the grooved ring on the phase plate is darkened to absorb a good part of the undeviated light so that its intensity is more nearly equal to that of the deviated light. Then nearly complete destructive interference can occur at particular points, and the contrast will be very high. The chief limitation of the phase-contrast microscope is that images tend to have "halos" around them as a result of diffraction from the phase-plate opening. Because of this artifact, care must be taken in the interpretation of images.

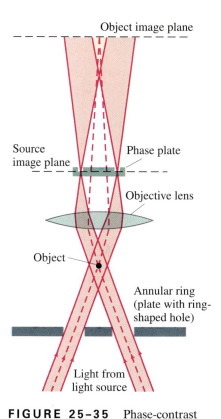

FIGURE 25–35 Phase-contrast microscope. Light beam from the source that is undeviated is shown shaded (pinkish) for clarity. Rays deviated by the object, and which form the image of the object, are shown dashed.

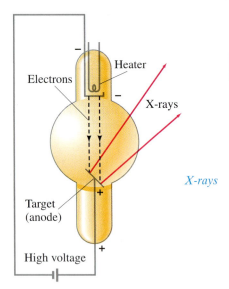

FIGURE 25–36 X-ray tube. Electrons emitted by a heated filament in a vacuum tube are accelerated by high voltage. When they strike the surface of the anode, the "target," X-rays are emitted.

FIGURE 25–37 This X-ray diffraction pattern is one of the first observed by Max von Laue in 1912 when he aimed a beam of X-rays at a zinc sulfide crystal. The diffraction pattern was detected directly on a photographic plate.

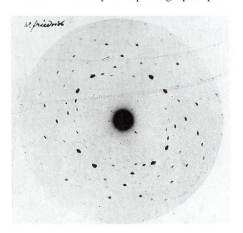

25–11 X-Rays and X-Ray Diffraction

In 1895, W. C. Roentgen (1845–1923) discovered that when electrons were accelerated by a high voltage in a vacuum tube and allowed to strike a glass (or metal) surface inside the tube, fluorescent minerals some distance away would glow, and photographic film would become exposed. Roentgen attributed these effects to a new type of radiation (different from cathode rays). They were given the name **X-rays** after the algebraic symbol x, meaning an unknown quantity. He soon found that X-rays penetrated through some materials better than through others, and within a few weeks he presented the first X-ray photograph (of his wife's hand). The production of X-rays today is usually done in a tube (Fig. 25–36) similar to Roentgen's, using voltages of typically 30 kV to 150 kV.

Investigations into the nature of X-rays indicated they were not charged particles (such as electrons) since they could not be deflected by electric or magnetic fields. It was suggested that they might be a form of invisible light. However, they showed no diffraction or interference effects using ordinary gratings. Of course, if their wavelengths were much smaller than the typical grating spacing of 10^{-6} m ($= 10^3$ nm), no effects would be expected. Around 1912, it was suggested by Max von Laue (1879–1960) that if the atoms in a crystal were arranged in a regular array (see Fig. 13–2a), such a crystal might serve as a diffraction grating for very short wavelengths on the order of the spacing between atoms, estimated to be about 10^{-10} m ($= 10^{-1}$ nm). Experiments soon showed that X-rays scattered from a crystal did indeed show the peaks and valleys of a diffraction pattern (Fig. 25–37). Thus it was shown, in a single blow, that X-rays have a wave nature and that atoms are arranged in a regular way in crystals. Today, X-rays are recognized as electromagnetic radiation with wavelengths in the range of about 10^{-2} nm to 10 nm, the range readily produced in an X-ray tube.

We saw in Sections 25–7 and 25–8 that light of shorter wavelength provides greater resolution when we are examining an object microscopically. Since X-rays have much shorter wavelengths than visible light, in principle, they should offer much greater resolution. However, there seems to be no effective material to use as lenses for the very short wavelengths of X-rays. Instead, the clever but complicated technique of **X-ray diffraction** (or **crystallography**) has proved very effective for examining the microscopic world of atoms and molecules. In a simple crystal such as NaCl, the atoms are arranged in an orderly cubical fashion, Fig. 25–38, with atoms spaced a distance d apart. Suppose that a beam of X-rays is incident on the crystal at an angle ϕ to the surface, and that the two rays shown are reflected from two subsequent planes of atoms as shown. The two rays will constructively interfere if the extra distance ray I travels is a whole number of wavelengths farther than what ray II travels. This extra distance is $2d \sin \phi$. Therefore, constructive interference will occur when

Bragg equation

$$m\lambda = 2d \sin \phi, \qquad m = 1, 2, 3, \cdots, \tag{25–12}$$

where m can be any integer. (Notice that ϕ is *not* the angle with respect to the normal to the surface.) This is called the **Bragg equation** after W. L. Bragg (1890–1971), who derived it and who, together with his father W. H. Bragg (1862–1942), developed the theory and technique of X-ray diffraction by crystals in 1912–13. Thus, if the X-ray wavelength is known and the angle ϕ at which constructive interference occurs is measured, d can be obtained. This is the basis for X-ray crystallography.

Actual X-ray diffraction patterns are quite complicated. First of all, a crystal is a three-dimensional object, and X-rays can be diffracted from different planes at different angles within the crystal, as shown in Fig. 25–39. Although the analysis is complex, a great deal can be learned about any substance that can be put in crystalline form. If the substance is not a single crystal but a mixture of many tiny crystals—as in a metal or a powder—then instead of a series of spots, as in Fig. 25–37, a series of circles is obtained, Fig. 25–40; each circle corresponds to diffraction of a certain order m (Eq. 25–12), from a particular set of parallel planes.

X-ray diffraction has been very useful in determining the structure of biologically important molecules. Often it is possible to make a crystal of such molecules. The analysis is complex, and it is usually necessary to make various guesses of the structure of the molecule. Predictions of the diffraction pattern for each guessed structure can then be compared to that actually obtained. For larger molecules, such as proteins and nucleic acids, an important innovation has been the "heavy-atom technique." Since very large atoms scatter X-rays much more strongly than the ordinary C, N, O, and H atoms of biological molecules, heavy atoms can be used as "markers." The heavy atoms are chemically added to particular spots on the molecule (say, a protein)—hopefully without disturbing its structure significantly. Analysis of the *changes* in the resulting diffraction pattern gives helpful information.

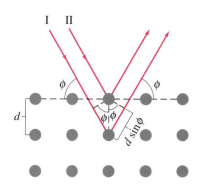

FIGURE 25–38 X-ray diffraction by a crystal.

FIGURE 25–39 There are many possible planes existing within a crystal from which X-rays can be diffracted.

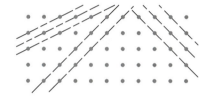

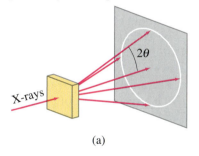

FIGURE 25–40 (a) Diffraction of X-rays from a polycrystalline substance produces a set of circular rings as in (b), which is for polycrystalline sodium acetoacetate.

FIGURE 25–41 X-ray diffraction photo of DNA molecules taken by Rosalind Franklin in the early 1950's. The cross of spots suggested that DNA is a helix.

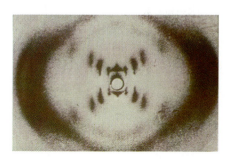

Even when a good crystal cannot be obtained, if the molecule under study has a regularly repeating shape (such as many proteins and DNA have), X-ray diffraction can reveal it. In a sense, each molecule is then like a single crystal and a sample is a collection of such tiny crystals. Indeed, it was with the help of X-ray diffraction that, in 1953, J. D. Watson and F. H. C. Crick worked out the double-helix structure of DNA. See Fig. 25–41, and for models of the double helix, Figs. 16–34a and 16–35.

Around 1960, the first detailed structure of a protein molecule was elucidated with the aid of X-ray diffraction; this was for myoglobin, a relative of the important constituent of blood, hemoglobin. Soon the structure of hemoglobin itself was worked out, and since then the structures of a great many molecules have been determined with the help of X-rays.

For a conventional medical (or dental) X-ray photograph, the X-rays emerging from the tube (Fig. 25–36, Section 25–11) pass through the body and are detected on photographic film or a fluorescent screen, Fig. 25–42a. The rays travel in very nearly straight lines through the body with minimal deviation since at X-ray wavelengths there is little diffraction or refraction. There is absorption (and scattering), however; and the difference in absorption by different structures in the body is what gives rise to the image produced by the transmitted rays. The less the absorption, the greater the transmission and the darker the film. The image is, in a sense, a "shadow" of what the rays have passed through. (The X-ray image is *not* produced by focusing rays with lenses as is the case for the instruments discussed earlier in this chapter.)

Within months of Roentgen's 1895 discovery, X-rays had already become a powerful tool for medical diagnosis, and they have remained so to this day. Although many technical advances have been made over the years, the basic principles for normal X-rays have not changed significantly. However, in the 1970s, a revolutionary new technique called **computerized tomography** (CT) using X-rays was developed.

In conventional X-ray images, the entire thickness of the body is projected onto the film; structures overlap and in many cases are difficult to distinguish. A tomographic image, on the other hand, is an image of a *slice* through the body. (The word **tomography** comes from the Greek: *tomos* = slice, *graph* = picture.) Structures and lesions previously impossible to visualize can now be seen with remarkable clarity. The principle behind CT is shown in Fig. 25–42b: a thin collimated beam of X-rays (to "collimate" means to "make straight") passes through the body to a detector that measures the transmitted intensity. Measurements are made at a large number of points as the source and detector are moved past the body together. The apparatus is then rotated slightly about the body axis and again scanned; this is repeated at (perhaps) 1° intervals for 180°. The intensity of the transmitted beam for the many points of each scan, and for each angle, are sent to a computer that

FIGURE 25–42 (a) Conventional X-ray imaging, which is essentially shadowing. (b) Tomographic imaging: the X-ray source and detector move together across the body, the transmitted intensity being measured at a large number of points. Then the source-detector assembly is rotated slightly (say, 1°) and another scan is made. This process is repeated for perhaps 180°. The computer reconstructs the image of the slice and it is presented on a TV monitor (cathode-ray tube).

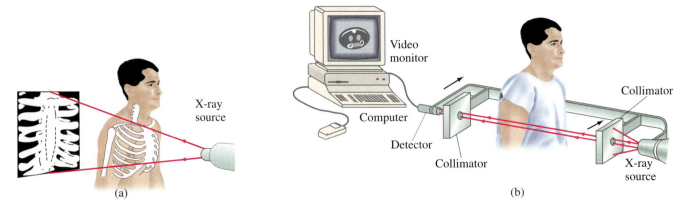

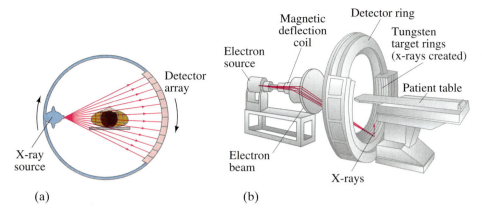

(a)

(b)

FIGURE 25–43 (a) Fan-beam scanner. Rays transmitted through the entire body are measured simultaneously at each angle. The source and detector rotate to take measurements at different angles. In another type of fan-beam scanner, there are detectors around the entire 360° of the circle which remain fixed as the source moves. (b) In another type, a beam of electrons from the source is directed by magnetic fields at tungsten targets surrounding the patient.

reconstructs the image of the slice. Note that the imaged slice is perpendicular to the long axis of the body. For this reason, CT is sometimes called **computerized axial tomography** (CAT), although the abbreviation CAT, as in CAT scan, can also be read as **computer-assisted tomography**.

CAT scans

The use of a single detector as in Fig. 25–42b requires a few minutes for the many scans needed to form a complete image. Much faster scanners use a fan beam, Fig. 25–43a, in which beams passing through the entire cross section of the body are detected simultaneously by many detectors. The source and detectors are then rotated about the patient. At each of the hundreds of angular positions of the apparatus, several hundred detectors can measure the intensity of transmitted rays simultaneously, so an image requires only a few seconds. Even faster, and therefore useful for heart scans, are fixed source machines wherein an electron beam is directed (by magnetic fields) to tungsten targets surrounding the patient, creating the X-rays. See Fig. 25–43b.

But how is the image formed? We can think of the slice to be imaged as being divided into many tiny picture elements (or **pixels**), which could be squares, as in Fig. 25–44. For CT, the width of each pixel is chosen according to the width of the detectors and/or the width of the X-ray beams, and this determines the resolution of the image, which is typically about 2 mm. An X-ray detector measures the intensity of the transmitted beam after it has passed through the body. Subtracting this value from the intensity of the beam at the source, we get the total absorption. Note that only the total absorption (called "a projection") along each beam line can be measured (the absorption by all the pixels in a line). To form an image, we need to determine how much radiation is absorbed at *each* pixel. (How that can be done will be discussed in a moment.) We can then assign a "grayness value" to each pixel according to how much radiation was absorbed. The image, then, is made up of tiny spots (pixels) of varying shades of gray, as is a black-and-white television picture. Often the amount of absorption is color-coded. The colors in the resulting ("false-color") image have nothing to do, however, with the actual color of the object.

Finally, we must discuss how the "grayness" of each pixel can be determined even though all we can measure is the total absorption along each beam line in the slice. It can be done only by using the many beam scans made at a great many different angles. Suppose the image is to be an array of 100×100 elements for a total of 10^4 pixels. If we have 100 detectors and measure the absorption projections at 100 different angles, then we get 10^4 pieces of information. From this information, an image can be reconstructed, but not precisely. If more angles are measured, the reconstruction of the image can be done more accurately.

FIGURE 25–44 Example of an image made up of many small squares called *pixels* (picture elements). This one has rather poor resolution.

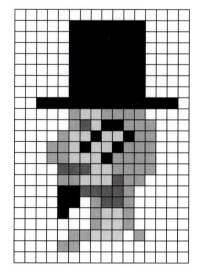

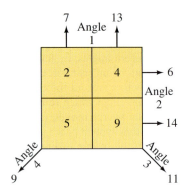

FIGURE 25–45 A simple 2×2 image showing true absorption values and measured projections.

There are a number of mathematical reconstruction techniques, all of which are complicated and require the use of a computer. To suggest how it is done, we consider a very simple case using the so-called "iterative" technique ("to iterate" is from the Latin "to repeat"). Although this technique is less used now than the more direct "Fourier transform" and "back projection" techniques, it is the simplest to explain. Suppose our sample slice is divided into the simple 2×2 pixels as shown in Fig. 25–45. The number in each pixel represents the amount of absorption by the material in that area (say, in tenths of a percent): that is, 4 represents twice as much absorption as 2. But we cannot directly measure these values—they are the unknowns we want to solve for. All we can measure are the projections—the total absorption along each beam line—and these are shown in the diagram as the sum of the absorptions for the pixels along each line at four different angles. These projections (given at the tip of each arrow) are what we can measure, and we now want to work back from them to see how close we can get to the true absorption value for each pixel. We start our analysis with each pixel being assigned a zero value, Fig. 25–46a. In the iterative technique, we use the projections to estimate the absorption value in each square, and repeat for each angle. The angle 1 projections are 7 and 13. We divide each of these equally between their two squares: each square in the left column gets $3\frac{1}{2}$ (half of 7), and each square in the right column gets $6\frac{1}{2}$ (half of 13); see Fig. 25–46b. Next we use the projections at angle 2. We calculate the difference between the measured projections at angle 2 (6 and 14) and the projections based on the previous estimate (top row: $3\frac{1}{2} + 6\frac{1}{2} = 10$; same for bottom row). Then we distribute this difference equally to the squares in that row. For the top row, we have

$$3\frac{1}{2} + \frac{6 - 10}{2} = 1\frac{1}{2} \qquad \text{and} \qquad 6\frac{1}{2} + \frac{6 - 10}{2} = 4\frac{1}{2};$$

and for the bottom row,

$$3\frac{1}{2} + \frac{14 - 10}{2} = 5\frac{1}{2} \qquad \text{and} \qquad 6\frac{1}{2} + \frac{14 - 10}{2} = 8\frac{1}{2}.$$

These values are inserted as shown in Fig. 25–46c. Next, the projection at angle 3 gives

(upper left) $1\frac{1}{2} + \dfrac{11 - 10}{2} = 2$ and   (lower right) $8\frac{1}{2} + \dfrac{11 - 10}{2} = 9;$

and that for angle 4 gives

(lower left) $5\frac{1}{2} + \dfrac{9 - 10}{2} = 5$ and   (upper right) $4\frac{1}{2} + \dfrac{9 - 10}{2} = 4.$

The result, shown in Fig. 25–46d, corresponds exactly to the true values.

FIGURE 25–46 Reconstructing the image using projections in an iterative procedure.

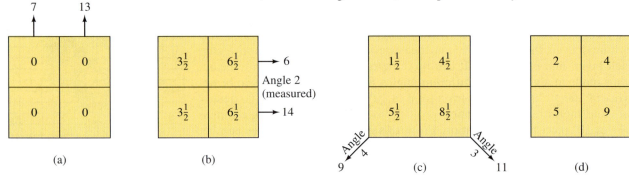

(Note that in real situations, the true values are not known, which is why these computer techniques are required.) To obtain these numbers exactly, we used six pieces of information (two each at angles 1 and 2, one each at angles 3 and 4). For the much larger number of pixels used for actual images, exact values are generally not attained. Many iterations may be needed, and the calculation is considered sufficiently precise when the difference between calculated and measured projections is sufficiently small. The above example illustrates the "convergence" of the process: the first iteration (b to c in Fig. 25–46) changed the values by 2, the last iteration (c to d) by only $\frac{1}{2}$.

Figure 25–47 illustrates what actual CT images look like. It is generally agreed that CT scanning has revolutionized some areas of medicine by providing much less invasive, and/or more accurate, diagnosis.

Computerized tomography can also be applied to ultrasound imaging (Section 12–10) and to emissions from radioisotopes and nuclear magnetic resonance (Sections 31–8 and 31–9).

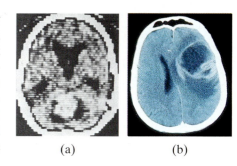

(a)   (b)

FIGURE 25–47 Two CT images, with different resolutions, each showing a cross section of a brain. The left photo is of low resolution; the right photo, of higher resolution, shows a brain tumor (dark area on the right).

SUMMARY

A **camera** lens forms an image on film by allowing light in through a shutter. The lens is focused by moving it relative to the film, and its *f*-stop (or lens opening) must be adjusted for the brightness of the scene and the chosen shutter speed. The *f*-stop is defined as the ratio of the focal length to the diameter of the lens opening.

The human **eye** also adjusts for the available light—by opening and closing the iris. It focuses not by moving the lens, but by adjusting the shape of the lens to vary its focal length. The image is formed on the retina, which contains an array of receptors known as rods and cones. Diverging eyeglass or contact lenses are used to correct the defect of a nearsighted eye, which cannot focus well on distant objects. Converging lenses are used to correct for defects in which the eye cannot focus on close objects.

A **simple magnifier** is a converging lens that forms a virtual image of an object placed at (or within) the focal point. The **angular magnification**, when viewed by a relaxed normal eye, is

$$M = \frac{N}{f},$$

where *f* is the focal length of the lens and *N* is the near point of the eye (25 cm for a "normal" eye).

An **astronomical telescope** consists of an **objective** lens or mirror and an **eyepiece** that magnifies the real image formed by the objective. The **magnification** is equal to the ratio of the objective and eyepiece focal lengths, and the image is inverted:

$$M = -\frac{f_o}{f_e}.$$

A compound **microscope** also uses objective and eyepiece lenses, and the final image is inverted. The total magnification is the product of the magnifications of the two lenses and is approximately

$$M \approx \left(\frac{N}{f_e}\right)\left(\frac{l}{f_o}\right),$$

where *l* is the distance between the lenses, *N* is the near point of the eye, and f_o and f_e are the focal lengths of objective and eyepiece, respectively.

Microscopes, telescopes, and other optical instruments are limited in the formation of sharp images by **lens aberrations**. These include **spherical aberration**, in which rays passing through the edge of a lens are not focused at the same point as those that pass near the center; and **chromatic aberration**, in which different colors are focused at different points. Compound lenses, consisting of several elements, can largely correct for aberrations.

The wave nature of light also limits the sharpness, or **resolution**, of images. Because of diffraction, it is *not possible to discern details smaller than the wavelength* of the radiation being used. This limits the useful magnification of a light microscope to about 1000×.

X-rays are a form of electromagnetic radiation of very short wavelength. They are produced when high-speed electrons, accelerated by high voltage in an evacuated tube, strike a glass or metal target.

QUESTIONS

1. Why is the depth of field greater, and the image sharper, when a camera lens is "stopped down" to a larger f-number? Ignore diffraction.

2. Describe how diffraction affects the statement of Question 1. [*Hint*: See Eq. 24–3.]

3. Why must a camera lens be moved farther from the film to focus on a closer object?

4. Why are bifocals needed mainly by older persons and not generally by younger people?

5. Explain why swimmers with good eyes see distant objects as blurry when they are underwater. Use a diagram and also show why goggles correct this problem.

6. Will a nearsighted person who wears corrective lenses be able to see clearly underwater when wearing glasses? Use a diagram to show why or why not.

7. You can tell whether a person is nearsighted or farsighted by looking at the width of the face through their glasses. If the person's face appears shrunk through the glasses, is the person farsighted or nearsighted?

8. The human eye is much like a camera—yet, when a camera shutter is left open and the camera moved, the image will be blurred; but when you move your head with your eyes open, you still see clearly. Explain.

9. In attempting to discern distant details, people will sometimes squint. Why does this help?

10. Is the image formed on the retina of the human eye upright or inverted? Discuss the implications of this for our perception of objects.

11. Reading glasses use converging lenses. A simple magnifier is also a converging lens. Are reading glasses therefore magnifiers? Discuss the similarities and differences between converging lenses as used for these two different purposes.

12. Inexpensive microscopes for children's use usually produce images that are colored at the edges. Why?

13. Spherical aberration in a thin lens is minimized if rays are bent equally by the two surfaces. If a planoconvex lens is used to form a real image of an object at infinity, which surface should face the object? Use ray diagrams to show why.

14. Which aberrations present in a simple lens are not present (or are greatly reduced) in the human eye?

15. Explain why chromatic aberration occurs for thin lenses but not for mirrors.

16. What are the advantages (give at least two) for the use of large reflecting mirrors in astronomical telescopes?

17. Which color of visible light would give the best resolution in a microscope?

18. Atoms have diameters of about 10^{-8} cm. Can visible light be used to "see" an atom? Why or why not?

PROBLEMS

SECTION 25–1

1. (I) A 55-mm-focal-length lens has f-stops ranging from $f/1.4$ to $f/22$. What is the corresponding range of lens diaphragm diameters?

2. (I) A television camera lens has a 14-cm focal length and a lens diameter of 6.0 cm. What is its f-number?

3. (I) A light meter reports that a camera setting of $\frac{1}{250}$ s at $f/5.6$ will give a correct exposure. But the photographer wishes to use $f/11$ to increase the depth of field. What should the shutter speed be?

4. (I) A properly exposed photograph is taken at $f/16$ and $\frac{1}{60}$ s. What lens opening would be required if the shutter speed were $\frac{1}{1000}$ s?

5. (II) A "pinhole" camera uses a tiny pinhole instead of a lens. Show, using ray diagrams, how reasonably sharp images can be formed using such a pinhole camera. In particular, consider two point objects 2.0 cm apart that are 1.0 m from a 1.0-mm-diameter pinhole. Show that on a piece of film 7.0 cm behind the pinhole, each object produces a tiny, easily resolvable spot.

6. (II) Suppose that a correct exposure is $\frac{1}{250}$ s at $f/11$. Under the same conditions, what exposure time would be needed for a pinhole camera if the pinhole diameter is 1.0 mm and the film is 7.0 cm from the hole?

7. (II) If a 135-mm telephoto lens is designed to cover object distances from 1.2 m to ∞, over what distance must the lens move relative to the plane of the film?

8. (II) A 200-mm-focal-length lens can be adjusted so that it is 200.0 mm to 206.0 mm from the film. For what range of object distances can it be adjusted?

9. (II) A nature photographer wishes to photograph a 22-m tall tree from a distance of 50 m. What focal-length lens should be used if the image is to fill the 24-mm height of the film?

SECTION 25–2

10. (I) A human eyeball is about 2.0 cm long and the pupil has a maximum diameter of about 5.0 mm. What is the "speed" of this lens?

11. (II) A person struggles to read by holding a book at arm's length, a distance of 50 cm away. What power of reading glasses should be prescribed for him, assuming they will be placed 2.0 cm from the eye and he wants to read at the normal near point of 25 cm?

12. (II) Reading glasses of what power are needed for a person whose near point is 120 cm, so that he can read a computer screen at 50 cm? Assume a lens–eye distance of 1.8 cm.

13. (II) Show that if the nearsighted person in Example 25–4 wore contact lenses corrected for the far point ($= \infty$), that the near point would be 41 cm. (Would glasses be better in this case?)

14. (II) A person's left eye is corrected by a -5.0 diopter lens, 2.0 cm from the eye. (a) Is this person near- or farsighted? (b) What is this person's far point without glasses?

15. (II) A person's right eye can see objects clearly only if they are between 20 cm and 70 cm away. (a) What power of contact lens is required so that objects far away are sharp? (b) What will be the near point with the lens in place?

16. (II) About how much longer is the nearsighted eye in Example 25–4 than the 2.0 cm of a normal eye?

17. (II) One lens of a nearsighted person's eyeglasses has a focal length of -25.0 cm and the lens is 1.8 cm from the eye. If the person switches to contact lenses that are placed directly on the eye, what should be the focal length of the corresponding contact lens?

18. (II) What is the focal length of the eye lens system when viewing an object (a) at infinity, and (b) 30 cm from the eye? Assume that the lens–retina distance is 2.0 cm.

SECTION 25–3

19. (I) What is the magnification of a lens used with a relaxed eye if its focal length is 12 cm?

20. (I) What is the focal length of a magnifying glass of 3.0× magnification for a relaxed normal eye?

21. (I) A magnifier is rated at 2.5× for a normal eye focusing on an image at the near point. (a) What is its focal length? (b) What is its focal length if the 2.5× refers to a relaxed eye?

22. (II) Sherlock Holmes is using a 10.0-cm-focal-length lens as his magnifying glass. To obtain maximum magnification, where must the object be placed (assume a normal eye), and what will be the magnification?

23. (II) A 3.10-mm-wide beetle is viewed with a 9.00-cm-focal-length lens. A normal eye views the image at its near point. Calculate (a) the angular magnification, (b) the width of the image, and (c) the object distance from the lens.

24. (II) A small insect is placed 5.35 cm from a $+6.00$-cm-focal-length lens. Calculate (a) the position of the image, and (b) the angular magnification.

25. (II) A magnifying glass with a focal length of 9.5 cm is used to read print placed at a distance of 8.5 cm. Calculate: (a) the position of the image; (b) the linear magnification; and (c) the angular magnification.

26. (II) A magnifying glass is rated at 3.0× for a normal eye that is relaxed. What would be the magnification for a relaxed eye whose near point is (a) 50 cm, and (b) 16 cm? Explain the differences.

27. (II) A nearsighted person has near and far points of 10.0 and 20.0 cm respectively. If she puts on a pair of glasses with power $P = -4.0D$, what are her new near and far points? Neglect the distance between her eye and the lens.

SECTION 25–4

28. (I) What is the magnification of an astronomical telescope whose objective lens has a focal length of 80 cm and whose eyepiece has a focal length of 2.8 cm? What is the overall length of the telescope when adjusted for a relaxed eye?

29. (I) The overall magnification of an astronomical telescope is desired to be 25×. If an objective of 80 cm focal length is used, what must be the focal length of the eyepiece? What is the overall length of the telescope when adjusted for use by the relaxed eye?

30. (I) An 8.0× binocular has 3.0-cm-focal-length eyepieces. What is the focal length of the objective lenses?

31. (II) An astronomical telescope has an objective with focal length 85 cm and a $+40$-D eyepiece. What is the total magnification?

32. (II) An astronomical telescope has its two lenses spaced 76 cm apart. If the objective lens has a focal length of 74.5 cm, what is the magnification of this telescope? Assume a relaxed eye.

33. (II) A Galilean telescope adjusted for a relaxed eye is 33 cm long. If the objective lens has a focal length of 36 cm, what is the magnification?

34. (II) What is the magnifying power of an astronomical telescope using a reflecting mirror whose radius of curvature is 5.0 m and an eyepiece whose focal length is 3.2 cm?

35. (II) The Moon's image appears to be magnified 120× by a reflecting astronomical telescope with an eyepiece having a focal length of 3.5 cm. What are the focal length and radius of curvature of the main mirror?

36. (III) A 180× astronomical telescope is adjusted for a relaxed eye when the two lenses are 1.25 m apart. What is the focal length of each lens?

37. (III) A 6.0× pair of binoculars has an objective focal length of 26 cm. If the binoculars are focused on an object 4.0 m away (from the objective), what is the magnification? (The 6.0× refers to objects at infinity; Eq. 25–3 holds only for objects at infinity and not for nearby ones.)

SECTION 25–5

38. (I) A microscope uses an eyepiece with a focal length of 1.5 cm. Using a normal eye with a final image at infinity, the tube length is 17.5 cm and the focal length of the objective lens is 0.65 cm. What is the magnification of the microscope?

39. (I) A 720× microscope uses a 0.40-cm-focal-length objective lens. If the tube length is 17.5 cm, what is the focal length of the eyepiece? Assume a normal eye and that the final image is at infinity.

40. (II) A microscope has a 12.0× eyepiece and a 62.0× objective 20.0 cm apart. Calculate (a) the total magnification, (b) the focal length of each lens, and (c) where the object must be for a normal relaxed eye to see it in focus.

41. (II) A microscope has a 1.8-cm-focal-length eyepiece and 0.80-cm objective. Calculate (a) the position of the object if the distance between the lenses is 16.0 cm, and (b) the total magnification assuming a relaxed normal eye.

42. (II) Repeat Problem 41 assuming that the final image is located 25 cm from the eyepiece (near point of a normal eye).

43. (II) The eyepiece of a compound microscope has a focal length of 2.70 cm and the objective has $f = 0.740$ cm. If an object is placed 0.790 cm from the objective lens, calculate (a) the distance between the lenses when the microscope is adjusted for a relaxed eye, and (b) the total magnification.

SECTION 25–6

44. (II) An achromatic lens is made of two very thin lenses placed in contact that have focal lengths of $f_1 = -28$ cm and $f_2 = +23$ cm. (a) Is the combination converging or diverging? (b) What is the net focal length?

* **45.** (III) Let's examine spherical aberration in a particular situation. A planoconvex lens of index of refraction 1.50 and a radius of curvature $R = 12.0$ cm is shown in Fig. 25–48. Consider an incoming ray parallel to the principle axis and a height h above it as shown. Determine the distance d, from the flat face of the lens, to where this ray crosses the principle axis if (a) $h = 1.0$ cm, and (b) $h = 6.0$ cm. (c) How far apart are these "focal points"? (d) How large is the "circle of confusion" at the "focal point" for $h = 1.0$ cm?

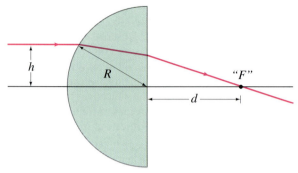

FIGURE 25–48 Problem 45.

SECTIONS 25–7 TO 25–9

46. (I) What is the angular resolution limit set by diffraction for the Mt. Wilson 100-in. (mirror diameter) telescope ($\lambda = 550$ nm)?

47. (I) What is the angle of acceptance α of a microscope oil-immersion objective, and its resolving power, if $n = 1.80$ and the NA = 1.41? Use $\lambda = 500$ nm.

48. (II) A microscope objective is immersed in oil ($n = 1.60$) and accepts light scattered from the object up to 60° on either side of vertical. (a) What is the numerical aperture? (b) What is the approximate resolution of the microscope for 550-nm light?

49. (II) Two stars 20 light-years away are barely resolved by a 12-in. (mirror diameter) amateur telescope. How far apart are the stars? Assume $\lambda = 500$ nm.

50. (II) The numerical aperture of an instrument is 0.95. How far apart are the lines on a certain shell if they are just barely resolved when viewed with green light (540 nm)? Will they be resolved if viewed by violet (400 nm) or red (700 nm) light?

51. (II) A certain sea organism has a pattern of dots on its surface with an average spacing of 0.63 μm. If the specimen is viewed using 480-nm light, what minimum value must the numerical aperture be in order that the dots be resolved?

52. (II) What minimum magnification would be required to see the dots on the organism of Problem 51?

53. (II) (a) How far away can a human eye distinguish two car headlights 2.0 m apart? Consider only diffraction effects and assume an eye diameter of 5.0 mm and a wavelength of 500 nm. (b) What is the minimum angular separation an eye could resolve when viewing two stars, considering only diffraction effects? In reality, it is about 1′ of arc. Why is it not equal to your answer in (b)?

*SECTION 25–10

* **54.** (II) Show that the phase difference (in radians) between the two waves ac and bd in Fig. 25–33 is $\delta = (2\pi/\lambda)(n_2 - n_1)t$, where n_2 and n_1 are the refractive indices of the object and the medium, t the thickness of the object, and λ the wavelength of light used.

*SECTION 25–11

* **55.** (II) X-rays of wavelength 0.135 nm fall on a crystal whose atoms, lying in planes, are spaced 0.280 nm apart. At what angle must the X-rays be directed if the first diffraction maximum is to be observed?

* **56.** (II) X-rays of wavelength 0.0973 nm are directed at an unknown crystal. The second diffraction maximum is recorded at 23.4°. What is the spacing between crystal planes?

* **57.** (II) First-order Bragg diffraction is observed at 23.4° from a crystal with spacing between atoms of 0.24 nm. (a) At what angle will second order be observed? (b) What is the wavelength of the X-rays?

* **58.** (III) If X-ray diffraction peaks corresponding to the first three orders ($m = 1, 2,$ and 3) are measured, can both the X-ray wavelength λ and lattice spacing d be determined? Prove your answer.

*59. (II) (a) Suppose for a conventional X-ray image that the X-ray beam consists of parallel rays. What would be the magnification of the image? (b) Suppose, in- stead, the X-rays come from a point source (as in Fig. 25–42a) that is 15 cm in front of a human body 25 cm thick, and the film is pressed against the per- son's back. Determine and discuss the range of mag- nifications that results.

GENERAL PROBLEMS

60. Sam purchases +3.2 diopter eyeglasses which cor- rect his faulty vision to put his near point at 25 cm. (Assume he wears the lenses 2.0 cm from his eyes.) (a) Is Sam nearsighted or farsighted? (b) Calculate the focal length of Sam's glasses. (c) Calculate Sam's near point without glasses. (d) Pam, who has normal eyes with near point at 25 cm, puts on Sam's glasses. Calculate Pam's near point with Sam's glasses on.

61. As early morning passed toward midday, and the sun- light got more intense, a photographer who was tak- ing repeated shots of the same subject noted that, if she kept her shutter speed constant, she had to change the f-number from f/5.6 to f/22. By how much had the sunlight intensity increased during that time?

62. Show that for objects very far away (assume infini- ty), the magnification of a camera lens is proportion- al to its focal length.

63. For a camera equiped with a 50-mm-focal-length lens, what is the object distance if the image height equals the object height? How far is the object from the film?

64. A woman can see clearly with her right eye only when objects are between 40 cm and 180 cm away. Prescription bifocals should have what powers so that she can see distant objects clearly (upper part) and be able to read a book 25 cm away (lower part)? Assume that the glasses will be 2.0 cm from the eye.

65. A child has a near point of 15 cm. What is the maxi- mum magnification the child can obtain using an 8.0- cm-focal-length magnifier? Compare to that for a normal eye.

66. What is the magnifying power of a +4.0-D lens used as a magnifier? Assume a relaxed normal eye.

67. A physicist lost in the mountains tries to make a tele- scope using the lenses from his reading glasses. They have powers of +2.0 D and +4.5 D, respectively. (a) What maximum magnification telescope is possi- ble? (b) Which lens should be used as the eyepiece?

68. The normal lens on a 35-mm camera has a focal length of 50 mm. Its aperture diameter varies from a maximum of 25 mm (f/2) to a minimum of 3.0 mm (f/16). Determine the resolution limit set by diffrac- tion for f/2 and f/16. Specify as the number of lines per millimeter resolved on the film. Take $\lambda = 500$ nm.

69. A 50-year-old man uses +2.5 diopter lenses to be able to read a newspaper 25 cm away. Ten years later, he finds that he must hold the paper 35 cm away to see clearly with the same lenses. What power lenses does he need now? (Distances are measured from the lens.)

70. Spy planes fly at extremely high altitudes (25 km) to avoid interception. Their cameras are reportedly able to discern features as small as 5 cm. What must be the minimum aperture of the camera lens to af- ford this resolution? (Use $\lambda = 550$ nm.)

71. Suppose that you wish to construct a telescope that can resolve features 10 km across on the Moon, 384,000 km away. You have a 2.2-m-focal-length ob- jective lens whose diameter is 12 cm. What focal- length eyepiece is needed if your eye can resolve objects 0.10 mm apart at a distance of 25 cm? What is the resolution limit set by the size of the objective lens (that is, by diffraction)? Use $\lambda = 500$ nm.

72. Exposure times must be increased for pictures taken at very short distances, because of the increased dis- tance of the lens from the film for a focused image. (a) Show that when the object is so close to the cam- era that the image height equals the object height, the exposure time must be four times longer than when the object is a long distance away (say, ∞), given the same illumination and f-stop. (b) Show that if d_o is at least four or five times the focal length f of the lens, the exposure time is increased negligibly rel- ative to the same object being a great distance away.

73. The objective lens and the eyepiece of a telescope are spaced 85 cm apart. If the eyepiece is 20 diopters, what is the total magnification of the telescope?

*74. A Lucite planoconvex lens (Fig. 23–29a) has one flat surface and the other has $R = 18.4$ cm. It is used to view an object, located 66.0 cm away from the lens, which is a mixture of red and yellow. The index of re- fraction of the glass is 1.5106 for red light and 1.5226 for yellow light. What are the locations of the red and yellow images formed by the lens?

75. A movie star catches a reporter shooting pictures of her at home. She claims the reporter was trespassing and to prove her point, she gives as evidence the film she seized. Her 1.75-m height is 8.25 mm high on the film, and the focal length of the camera lens was 200 mm. How far away from the subject was the re- porter standing?

Albert Einstein (1879–1955), one of the great minds of the twentieth century, creator of the special and general theories of relativity, here shown lecturing.

26 SPECIAL THEORY OF RELATIVITY

FIGURE 26–1 Albert Einstein and his second wife.

Physics at the end of the nineteenth century looked back on a period of great progress. The theories developed over the preceding three centuries had been very successful in explaining a wide range of natural phenomena. Newtonian mechanics beautifully explained the motion of objects on Earth and in the heavens. Furthermore, it formed the basis for successful treatments of fluids, wave motion, and sound. Kinetic theory explained the behavior of gases and other materials. Maxwell's theory of electromagnetism not only brought together and explained electric and magnetic phenomena, but it predicted the existence of electromagnetic (EM) waves that would behave in every way just like light—so light came to be thought of as an electromagnetic wave. Indeed, it seemed that the natural world, as seen through the eyes of physicists, was very well explained. A few puzzles remained, but it was felt that these would soon be explained using already known principles.

But it did not turn out so simply. Instead, these few puzzles were to be solved only by the introduction, in the early part of the twentieth century, of two revolutionary new theories that changed our whole conception of nature: the *theory of relativity* and *quantum theory*.

Physics as it was known at the end of the nineteenth century (what we've covered up to now in this book) is referred to as **classical physics**. The new physics that grew out of the great revolution at the turn of the twentieth century is now called **modern physics**. In this chapter, we present the special theory of relativity, which was first proposed by Albert Einstein (1879–1955; Fig. 26–1) in 1905. In the following chapter, we introduce the equally momentous quantum theory.

Classical vs. modern physics

26–1 Galilean–Newtonian Relativity

Einstein's special theory of relativity deals with how we observe events, particularly how objects and events are observed from different frames of reference.[†] This subject had, of course, already been explored by Galileo and Newton. We first briefly discuss these earlier ideas, before seeing (starting in Section 26–3) how the theory of relativity changed them.

The special theory of relativity deals with events that are observed and measured from so-called **inertial reference frames**, which (as mentioned in Chapter 4) are reference frames in which Newton's first law, the law of inertia, is valid. (Newton's first law states that, if an object experiences no net force, the object either remains at rest or continues in motion with constant velocity in a straight line.) It is easiest to analyze events when they are observed and measured from inertial frames, and the Earth, though not quite an inertial frame (it rotates), is close enough that for most purposes we can consider it an inertial frame. Rotating or otherwise accelerating frames of reference are noninertial frames,[‡] and Einstein dealt with such complicated frames of reference in his general theory of relativity (Chapter 33).

A reference frame that moves with constant velocity with respect to an inertial frame is itself also an inertial frame, since Newton's laws hold in it as well. When we say that we observe or make measurements from a certain reference frame, it means that we are at rest in that reference frame.

Both Galileo and Newton were aware of what we now call the **relativity principle** applied to mechanics: that *the basic laws of physics are the same in all inertial reference frames*. You may have recognized its validity in everyday life. For example, objects move in the same way in a smoothly moving (constant-velocity) train or airplane as they do on Earth. (This assumes no vibrations or rocking—for they would make the reference frame noninertial.) When you walk, drink a cup of soup, play Ping-Pong, or drop a pencil on the floor while traveling in a train, airplane, or ship moving at constant velocity, the bodies move just as they do when you are at rest on Earth. Suppose you are in a car traveling rapidly along at constant velocity. If you release a coin from above your head inside the car, how will it fall? It falls straight downward with respect to the car, and hits the floor

Relativity principle: the laws of physics are the same in all inertial reference frames

---

[†]A reference frame is a set of coordinate axes fixed to some body such as the Earth, a train, the Moon, and so on. See Section 2–1.

[‡]On a rotating platform (say a merry-go-round), for example, an object at rest starts moving outward even though no body exerts a force on it. This is therefore not an inertial frame. See Appendix C, Fig. C–1.

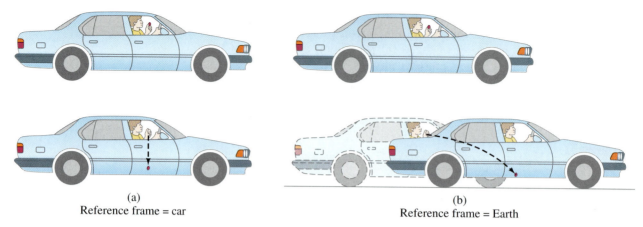

(a)
Reference frame = car

(b)
Reference frame = Earth

FIGURE 26–2 A coin is dropped by a person in a moving car. (a) In the reference frame of the car, the coin falls straight down. (b) In a reference frame fixed on the Earth, the coin follows a curved (parabolic) path. The upper views show the moment of the coin's release, and the lower views are a short time later.

directly below the point of release, Fig. 26–2a. (If you drop the coin out the car's window, this won't happen because the moving air drags the coin backward relative to the car.) This is just how objects fall on the Earth—straight down—and thus our experiment in the moving car is in accord with the relativity principle.

Note in this example, however, that to an observer on the Earth, the coin follows a curved path, Fig. 26–2b. The actual path followed by the coin is different as viewed from different frames of reference. This does not violate the relativity principle because this principle states that the *laws* of physics are the same in all inertial frames. The same law of gravity, and the same laws of motion, apply in both reference frames. And the acceleration of the coin is the same in both reference frames. The difference in Figs. 26–2a and b is that in the Earth's frame of reference, the coin has an initial velocity (equal to that of the car). The laws of physics therefore predict it will follow a parabolic path like any projectile. In the car's reference frame, there is no initial velocity, and the laws of physics predict that the coin will fall straight down. The laws are the same in both reference frames, although the specific paths are different.[†]

Galilean–Newtonian relativity involves certain unprovable assumptions that make sense from everyday experience. It is assumed that the lengths of objects are the same in one reference frame as in another, and that time passes at the same rate in different reference frames. In classical mechanics, then, space and time are considered to be **absolute**: their measurement doesn't change from one reference frame to another. The mass of an object, as well as all forces, are assumed to be unchanged by a change in inertial reference frame.

The position of an object is, of course, different when specified in different reference frames, and so is velocity. For example, a person may walk inside a bus toward the front with a speed of 5 km/h. But if the bus moves 40 km/h with respect to the Earth, the person is then moving with a speed of 45 km/h with respect to the Earth. The acceleration of a body, however, is the same in any inertial reference frame according to classical mechanics. This is because the change in velocity, and the time interval, will be the same.

[†]Galileo, in his great book *Dialogues on the Two Chief Systems of the World*, described a similar experiment and predicted the same results. Galileo's example involved a sailor dropping a knife from the top of the mast of a sailing vessel. If the vessel moves at constant speed, where will the knife hit the deck (ignoring Earth's rotation and air resistance)?

For example, the person in the bus may accelerate from 0 to 5 km/h in 1.0 seconds, so $a = 5$ km/h/s in the reference frame of the bus. With respect to the Earth, the acceleration is $(45 \text{ km/h} - 40 \text{ km/h})/(1.0 \text{ s}) = 5$ km/h/s, which is the same.

Since neither F, m, nor a changes from one inertial frame to another, then Newton's second law, $F = ma$, does not change. Thus Newton's second law satisfies the relativity principle. It is easily shown that the other laws of mechanics also satisfy the relativity principle.

That the laws of mechanics are the same in all inertial reference frames implies that no one inertial frame is special in any sense. We express this important conclusion by saying that **all inertial reference frames are equivalent** for the description of mechanical phenomena. No one inertial reference frame is any better than another. A reference frame fixed to a car or an aircraft traveling at constant velocity is as good as one fixed on the Earth. When you travel smoothly at constant velocity in a car or airplane, it is just as valid to say you are at rest and the Earth is moving as it is to say the reverse. There is no experiment you can do to tell which frame is "really" at rest and which is moving. Thus, there is no way to single out one particular reference frame as being at absolute rest.

All inertial reference frames are equally valid

A complication arose, however, in the last half of the nineteenth century. When Maxwell presented his comprehensive and very successful theory of electromagnetism (Chapter 22), he showed that light can be considered an electromagnetic wave. Maxwell's equations predicted that the velocity of light c would be 3.00×10^8 m/s; and this is just what is measured, within experimental error. The question then arose: in what reference frame does light have precisely the value predicted by Maxwell's theory? For it was assumed that light would have a different speed in different frames of reference. For example, if observers were traveling on a rocket ship at a speed of 1.0×10^8 m/s away from a source of light, we might expect them to measure the speed of the light reaching them to be 3.0×10^8 m/s $- 1.0 \times 10^8$ m/s $= 2.0 \times 10^8$ m/s. But Maxwell's equations have no provision for relative velocity. They predicted the speed of light to be $c = 3.0 \times 10^8$ m/s. This seemed to imply there must be some special reference frame where c would have this value.

We discussed in Chapters 11 and 12 that waves travel on water and along ropes or strings, and sound waves travel in air and other materials. Nineteenth-century physicists viewed the material world in terms of the laws of mechanics, so it was natural for them to assume that light too must travel in some *medium*. They called this transparent medium the **ether** and assumed it permeated all space.[†] It was therefore assumed that the velocity of light given by Maxwell's equations must be with respect to the ether.

The "ether"

However, it appeared that Maxwell's equations did *not* satisfy the relativity principle. They were not the same in all inertial reference frames. They were simplest in the frame where $c = 3.00 \times 10^8$ m/s; that is, in a reference frame at rest in the ether. In any other reference frame, extra terms would have to be added to take into account the relative velocity. Thus, although most of the laws of physics obeyed the relativity principle,

[†]The medium for light waves could not be air, since light travels from the Sun to Earth through nearly empty space. Therefore, another medium was postulated, the ether. The ether was not only transparent, but, because of difficulty in detecting it, was assumed to have zero density.

the laws of electricity and magnetism apparently did not. Instead, they seemed to single out one reference frame that was better than any other—a reference frame that could be considered absolutely at rest.

Scientists soon set out to determine the speed of the Earth relative to this absolute frame, whatever it might be. A number of clever experiments were designed. The most direct were performed by A. A. Michelson and E. W. Morley in the 1880s. The details of their experiment are discussed in the next Section. Briefly, what they did was measure the difference in the speed of light in different directions. They expected to find a difference depending on the orientation of their apparatus with respect to the ether. For just as a boat has different speeds relative to the land when it moves upstream, downstream, or across the stream, so too light would be expected to have different speeds depending on the velocity of the ether past the Earth.

Strange as it may seem, they detected no difference at all. This was a great puzzle. A number of explanations were put forth over a period of years, but they led to contradictions or were otherwise not generally accepted.

Then in 1905, Albert Einstein proposed a radical new theory that reconciled these many problems in a simple way. But at the same time, as we shall soon see, it completely changed our ideas of space and time.

* 26–2 The Michelson–Morley Experiment

The Michelson–Morley experiment was designed to measure the speed of the *ether*—the medium in which light was assumed to travel—with respect to the Earth. The experimenters thus hoped to find an absolute reference frame, one that could be considered to be at rest.

One of the possibilities nineteenth-century scientists considered was that the ether is fixed relative to the Sun, for even Newton had taken the Sun as the center of the universe. If this were the case (there was no guarantee, of course), the Earth's speed of about 3×10^4 m/s in its orbit around the Sun would produce a change of 1 part in 10^4 in the speed of light (3.0×10^8 m/s). Direct measurement of the speed of light to this accuracy was not possible. But A. A. Michelson, later with the help of E. W. Morley, was able to use his interferometer (Section 24–9) to measure the difference in the speed of light in different directions to this accuracy. This famous experiment is based on the principle shown in Fig. 26–3. Part (a) is a diagram of the Michelson interferometer, and it is assumed that the "ether wind" is moving with speed v to the right. (Alternatively, the Earth is assumed to move to the left with respect to the ether at speed v.) The light from the source is split into two beams by the half-silvered mirror M_S. One beam travels to mirror M_1 and the other to mirror M_2. The beams are reflected by M_1 and M_2 and are joined again after passing through M_S. The now superposed beams interfere with each other and the resultant is viewed by the observer's eye as an interference pattern (discussed in Section 24–9).

Whether constructive or destructive interference occurs at the center of the interference pattern depends on the relative phases of the two beams after they have traveled their separate paths. To examine this let us consider an analogy of a boat traveling up and down, and across, a river whose current moves with speed v, as shown in Fig. 26–3b. In still water, the boat can travel with speed c (not the speed of light in this case).

First we consider beam 2 in Fig. 26–3a, which travels parallel to the "ether wind." In its journey from M_S to M_2, we expect the light to travel

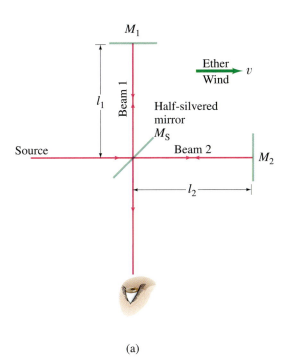

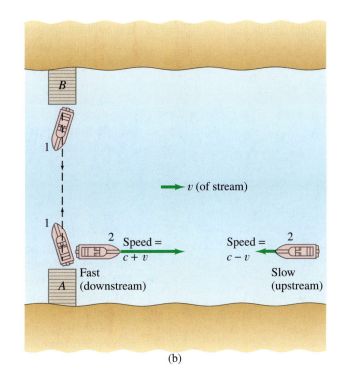

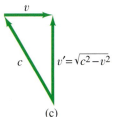

(a)   (b)   (c)

FIGURE 26–3 The Michelson–Morley experiment. (a) Michelson interferometer. (b) Boat analogy: boat 1 goes across the stream and back; boat 2 goes downstream and back upstream. (c) Calculation of the velocity of boat (or light beam) traveling perpendicular to the current (or ether wind).

with speed $c + v$, just as a boat traveling downstream (see Fig. 26–3b) acquires the speed of the river current. Since the beam travels a distance l_2, the time it takes to go from M_S to M_2 is $t = l_2/(c + v)$. To make the return trip from M_2 to M_S, the light must move against the ether wind (like the boat going upstream), so its relative speed is expected to be $c - v$. The time for the return trip is $l_2/(c - v)$. The total time required for beam 2 to travel from M_S to M_2 and back to M_S is

$$t_2 = \frac{l_2}{c + v} + \frac{l_2}{c - v}$$

$$= \frac{2l_2}{c(1 - v^2/c^2)}.$$

The second line was obtained from the first by finding the common denominator and factoring out c^2 in the denominator.

Now let us consider beam 1, which travels crosswise to the ether wind. Here the boat analogy (part b) is especially helpful. The boat is to go from wharf A to wharf B directly across the stream. If it heads directly across, the stream's current will drag it downstream. To reach wharf B, the boat must head at an angle upstream. The precise angle depends on the magnitudes of c and v, but is of no interest to us in itself. Part (c) of Fig. 26–3 shows how to calculate the velocity v' of the boat relative to Earth as it crosses the stream. Since c, v, and v' form a right triangle, we have that $v' = \sqrt{c^2 - v^2}$. The boat has the same velocity when it returns. If we now apply these principles to light beam 1 in Fig. 26–3a, we see that the beam travels with a speed $\sqrt{c^2 - v^2}$ in going from M_S to M_1 and back again. The total distance traveled is $2l_1$, so the time required for beam 1 to make the round trip is $2l_1/\sqrt{c^2 - v^2}$, or

$$t_1 = \frac{2l_1}{c\sqrt{1 - v^2/c^2}}.$$

Notice that the denominator in this equation for t_1 involves a square root, whereas that for t_2 does not.

If $l_1 = l_2 = l$, we see that beam 2 will lag behind beam 1 by an amount

$$\Delta t = t_2 - t_1 = \frac{2l}{c}\left(\frac{1}{1 - v^2/c^2} - \frac{1}{\sqrt{1 - v^2/c^2}}\right).$$

If $v = 0$, then $\Delta t = 0$, and the two beams will return in phase since they were initially in phase. But if $v \neq 0$, then $\Delta t \neq 0$, and the two beams will return out of phase. If this change of phase from the condition $v = 0$ to that for $v = v$ could be measured, then v could be determined. But the Earth cannot be stopped. Furthermore, it is not possible to independently assume $l_1 = l_2$.

Michelson and Morley realized that they could detect the difference in phase (assuming that $v \neq 0$) if they rotated their apparatus by 90°, for then the interference pattern between the two beams should change. In the rotated position, beam 1 would now move parallel to the ether and beam 2 perpendicular to it. Thus the roles could be reversed, and in the rotated position the times (designated by primes) would be

$$t_1' = \frac{2l_1}{c(1 - v^2/c^2)} \quad \text{and} \quad t_2' = \frac{2l_2}{c\sqrt{1 - v^2/c^2}}.$$

The time lag between the two beams in the nonrotated position (unprimed) would be

$$\Delta t = t_2 - t_1 = \frac{2l_2}{c(1 - v^2/c^2)} - \frac{2l_1}{c\sqrt{1 - v^2/c^2}}.$$

In the rotated position, the time difference would be

$$\Delta t' = t_2' - t_1' = \frac{2l_2}{c\sqrt{1 - v^2/c^2}} - \frac{2l_1}{c(1 - v^2/c^2)}.$$

When the rotation is made, the fringes of the interference pattern (Section 24–9) will shift an amount determined by the difference:

$$\Delta t - \Delta t' = \frac{2}{c}(l_1 + l_2)\left(\frac{1}{1 - v^2/c^2} - \frac{1}{\sqrt{1 - v^2/c^2}}\right).$$

This expression can be considerably simplified if we assume that $v/c \ll 1$. For in this case we can use the binomial expansion,[†] so

$$\frac{1}{1 - v^2/c^2} \approx 1 + \frac{v^2}{c^2} \quad \text{and} \quad \frac{1}{\sqrt{1 - v^2/c^2}} \approx 1 + \frac{1}{2}\frac{v^2}{c^2}.$$

[†]The binomial expansion (see Appendix A) states that $(1 \pm x)^n = 1 \pm nx + [n(n - 1)/2]x^2 + \cdots$. In our case we have, therefore, $(1 - x)^{-1} \approx 1 + x$, and $(1 - x)^{-1/2} \approx 1 + \frac{1}{2}x$, where only the first term is kept, since $x = v^2/c^2$ is assumed to be small.

Then

$$\Delta t - \Delta t' \approx \frac{2}{c}(l_1 + l_2)\left(1 + \frac{v^2}{c^2} - 1 - \frac{1}{2}\frac{v^2}{c^2}\right)$$

$$\approx (l_1 + l_2)\frac{v^2}{c^3}.$$

Now we take $v = 3.0 \times 10^4$ m/s, the speed of the Earth in its orbit around the Sun. In Michelson and Morley's experiments, the arms l_1 and l_2 were about 11 m long. The time difference would then be about

$$(22 \text{ m})(3.0 \times 10^4 \text{ m/s})^2/(3.0 \times 10^8 \text{ m/s})^3 \approx 7.0 \times 10^{-16} \text{ s}.$$

For visible light of wavelength $\lambda = 5.5 \times 10^{-7}$ m, say, the frequency would be $f = c/\lambda = (3.0 \times 10^8 \text{ m/s})/(5.5 \times 10^{-7} \text{ m}) = 5.5 \times 10^{14}$ Hz, which means that wave crests pass by a point every $1/(5.5 \times 10^{14} \text{ Hz}) = 1.8 \times 10^{-15}$ s. Thus, with a time difference of 7.0×10^{-16} s, Michelson and Morley should have noted a movement in the interference pattern of $(7.0 \times 10^{-16} \text{ s})/(1.8 \times 10^{-15} \text{ s}) = 0.4$ fringe. They could easily have detected this, since their apparatus was capable of observing a fringe shift as small as 0.01 fringe.

But they found *no significant fringe shift whatever*! They set their apparatus at various orientations. They made observations day and night so that they would be at various orientations with respect to the Sun (due to the Earth's rotation). They tried at different seasons of the year (the Earth at different locations due to its orbit around the Sun). Never did they observe a significant fringe shift. *The null result*

This "null" result was one of the great puzzles of physics at the end of the nineteenth century. To explain it was a difficult challenge. One possibility to explain the null result was to apply an idea put forth independently by G. F. Fitzgerald and H. A. Lorentz (in the 1890s) in which they proposed that any length (including the arm of an interferometer) contracts by a factor $\sqrt{1 - v^2/c^2}$ in the direction of motion through the ether. According to Lorentz, this could be due to the ether affecting the forces between the molecules of a substance, which were assumed to be electrical in nature. This theory was eventually replaced by the far more comprehensive theory proposed by Albert Einstein in 1905—the special theory of relativity.

26–3 Postulates of the Special Theory of Relativity

The problems that existed at the turn of the century with regard to electromagnetic theory and Newtonian mechanics were beautifully resolved by Einstein's introduction of the theory of relativity in 1905. Einstein, however, was apparently not influenced directly by the null result of the Michelson–Morley experiment. What motivated Einstein were certain questions regarding electromagnetic theory and light waves. For example, he asked himself: "What would I see if I rode a light beam?" The answer was that instead of a traveling electromagnetic wave, he would see alternating electric and magnetic fields at rest whose magnitude changed in space, but did not change in time. Such fields, he realized, had never been detected and indeed were not consistent with Maxwell's electromagnetic theory. He argued, therefore, that it was unreasonable to think that the speed of light relative to any observer could be reduced to zero, or in fact reduced at all. This idea became the second postulate of his theory of relativity.

Einstein concluded that the inconsistencies he found in electromagnetic theory were due to the assumption that an absolute space exists. In his famous 1905 paper, he proposed doing away completely with the idea of the ether and the accompanying assumption of an absolute reference frame at rest. This proposal was embodied in two postulates. The first postulate was an extension of the Newtonian relativity principle to include not only the laws of mechanics but also those of the rest of physics, including electricity and magnetism:

First postulate (*the relativity principle*): The laws of physics have the same form in all inertial reference frames.

The second postulate is consistent with the first:

Second postulate (*constancy of the speed of light*): Light propagates through empty space with a definite speed *c* independent of the speed of the source or observer.

These two postulates form the foundation of Einstein's **special theory of relativity**. It is called "special" to distinguish it from his later "general theory of relativity," which deals with noninertial (accelerating) reference frames (discussed in Chapter 33). The special theory, which is what we discuss here, deals only with inertial frames.

The second postulate may seem hard to accept, for it violates commonsense notions. First of all, we have to think of light traveling through empty space. Giving up the ether is not too hard, however, for after all, it had never been detected. But the second postulate also tells us that the speed of light in vacuum is always the same, 3.00×10^8 m/s, no matter what the speed of the observer or the source. Thus, a person traveling toward or away from a source of light will measure the same speed for that light as someone at rest with respect to the source. This conflicts with our everyday notions, for we would expect to have to add in the velocity of the observer. Part of the problem is that in our everyday experience, we do not measure velocities anywhere near as large as the speed of light. Thus we can't expect our everyday experience to be helpful when dealing with such a high velocity. On the other hand, the Michelson–Morley experiment is fully consistent with the second postulate.[†]

Einstein's proposal has a certain beauty. For by doing away with the idea of an absolute reference frame, it was possible to reconcile classical mechanics with Maxwell's electromagnetic theory. The speed of light predicted by Maxwell's equations *is* the speed of light in vacuum in *any* reference frame.

Einstein's theory required giving up commonsense notions of space and time, and in the following sections we will examine some strange but interesting consequences of Einstein's theory. Our arguments for the most part will be simple ones. We will use a technique that Einstein himself did:

[†]The Michelson–Morley experiment can also be considered as evidence for the first postulate, for it was intended to measure the motion of the Earth relative to an absolute reference frame. Its failure to do so implies the absence of any such preferred frame.

we will imagine very simple experimental situations in which little mathematics is needed. In this way, we can see many of the consequences of relativity theory without getting involved in detailed calculations. Einstein called these "gedanken" experiments, which is German for "thought" experiments. Some of the more mathematical details of special relativity are treated in Appendix E.

26–4 Simultaneity

One of the important consequences of the theory of relativity is that we can no longer regard time as an absolute quantity. No one doubts that time flows onward and never turns back. But, as we shall see in this section and the next, the time interval between two events, and even whether two events are simultaneous, depends on the observer's reference frame.

Two events are said to occur simultaneously if they occur at exactly the same time. But how do we know if two events occur precisely at the same time? If they occur at the same point in space—such as two apples falling on your head at the same time—it is easy. But if the two events occur at widely separated places, it is more difficult to know whether the events are simultaneous since we have to take into account the time it takes for the light from them to reach us. Because light travels at finite speed, a person who sees two events must calculate back to find out when they actually occurred. For example, if two events are *observed* to occur at the same time, but one actually took place farther from the observer than the other, then the former must have occurred earlier, and the two events were not simultaneous.

We will now make use of a simple thought experiment. We assume an observer, called O, is located exactly halfway between points A and B where two events occur, Fig. 26–4. The two events may be lightning that strikes the points A and B, as shown, or any other type of events. For brief events like lightning, only short pulses of light will travel outward from A and B and reach O. O "sees" the events when the pulses of light reach point O. If the two pulses reach O at the same time, then the two events had to be simultaneous. This is because the two light pulses travel at the same speed (postulate 2), and since the distance OA equals OB, the time for the light to travel from A to O and B to O must be the same. Observer O can then definitely state that the two events occurred simultaneously. On the other hand, if O sees the light from one event before that from the other, then it is certain the former event occurred first.

A "thought" experiment

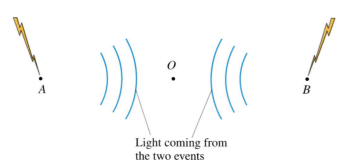

FIGURE 26–4 A moment after lightning strikes points A and B, the pulses of light are traveling toward the observer O, but O "sees" the lightning only when the light reaches O.

Light coming from the two events at A and B

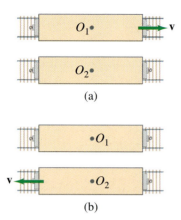

(a)

(b)

FIGURE 26–5 Observers O_1 and O_2, on two different trains (two different reference frames), are moving with relative velocity v. O_2 says that O_1 is moving to the right (a); O_1 says that O_2 is moving to the left (b). Both viewpoints are legitimate—it all depends on your reference frame.

The question we really want to examine is this: if two events are simultaneous to an observer in one reference frame, are they also simultaneous to another observer moving with respect to the first? Let us call the observers O_1 and O_2 and assume they are fixed in reference frames 1 and 2 that move with speed v relative to one another. These two reference frames can be thought of as trains (Fig. 26–5). O_2 says that O_1 is moving to the right with speed v, as in (a); and O_1 says O_2 is moving to the left with speed v, as in (b). Both viewpoints are legitimate according to the relativity principle. (There is, of course, no third point of view which will tell us which one is "really" moving.)

Now suppose two events occur that are observed and measured by both observers. Let us assume again that the two events are the striking of lightning and that the lightning marks both trains where it struck: at A_1 and B_1 on O_1's train, and at A_2 and B_2 on O_2's train. For simplicity, we assume that O_1 happens to be exactly halfway between A_1 and B_1, and that O_2 is halfway between A_2 and B_2. We now put ourselves in one reference frame or the other, from which we make our observations and measurements. Let us put ourselves in O_2's reference frame, so we observe O_1 moving to the right with speed v. Let us also assume that the two events occur *simultaneously* in O_2's frame, and just at the instant when O_1 and O_2 are opposite each other, Fig. 26–6a. A short time later, Fig. 26–6b, the light from A_2 and B_2 reaches O_2 at the same time (we assumed this). Since O_2 knows (or measures) the distances O_2A_2 and O_2B_2 as equal, O_2 knows the two events are simultaneous in the O_2 reference frame.

FIGURE 26–6 Thought experiment on simultaneity. To observer O_2, the reference frame of O_1 is moving to the right. In (a), one lightning bolt strikes the two reference frames at A_1 and A_2, and a second lightning bolt strikes at B_1 and B_2. (b) A moment later, the light from the two events reaches O_2 at the same time, so according to observer O_2, the two bolts of lightning strike simultaneously. But in O_1's reference frame, the light from B_1 has already reached O_1, whereas the light from A_1 has not yet reached O_1. So in O_1's reference frame, the event at B_1 must have preceded the event at A_1. Time is not absolute.

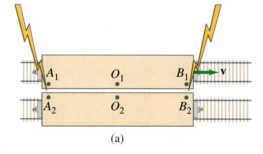

(a)

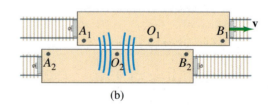

(b)

But what does observer O_1 observe and measure? From our (O_2) reference frame, we can predict what O_1 will observe. We see that O_1 moves to the right during the time the light is traveling to O_1 from A_1 and B_1. As shown in Fig. 26–6b, we can see from our O_2 reference frame that the light from B_1 has already passed O_1, whereas the light from A_1 has not yet reached O_1. Therefore, it is clear that O_1 will observe the light coming from B_1 before he observes the light coming from A_1. Now O_1's frame is as good as O_2's. Light travels at the same speed c for O_1 as for O_2 (the second postulate)[†]; and in the O_1 reference frame, this speed c is of course the same for light traveling from A_1 to O_1 as it is for light traveling from B_1 to O_1. Furthermore the distance $O_1 A_1$ equals $O_1 B_1$. Hence, since O_1 observes the light from B_1 before he observes the light from A_1 (we established this above, looking from the O_2 reference frame, Fig. 26–6b), then observer O_1 can only conclude that the event at B_1 occurred before the event at A_1. The two events are not simultaneous for O_1, even though they are for O_2.

We thus find that *two events* which are simultaneous to one observer are not necessarily simultaneous to a second observer.

Simultaneity is relative

It may be tempting to ask: "Which observer is right, O_1 or O_2?" The answer, according to relativity, is that they are *both* right. There is no "best" reference frame we can choose to determine which observer is right. Both frames are equally good. We can only conclude that simultaneity is not an absolute concept, but is relative. We are not aware of it in everyday life, however, because the effect is noticeable only when the relative speed of the two reference frames is very large (near c), or the distances involved are very large.

Because of the principle of relativity, the argument we gave for the thought experiment of Fig. 26–6 can be done from O_1's reference frame as well. In this case, O_1 will be at rest and will see event B_1 occur before A_1. But O_1 will recognize (by drawing a diagram equivalent to Fig. 26–6—try it and see!) that O_2, who is moving with speed v to the left, will see the two events as simultaneous.

26–5 Time Dilation and the Twin Paradox

The fact that two events simultaneous to one observer may not be simultaneous to a second observer suggests that time itself is not absolute. Could it be that time passes differently in one reference frame than in another? This is, indeed, just what Einstein's theory of relativity predicts, as the following thought experiment shows.

[†]Note that O_1 does not see himself catching up with one light beam and running away from the other (that is O_2's viewpoint of what happens for O_1). O_1 sees both light beams traveling at the same speed, c.

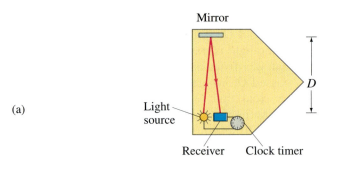

Mirror

(a)

Light source

Receiver   Clock timer

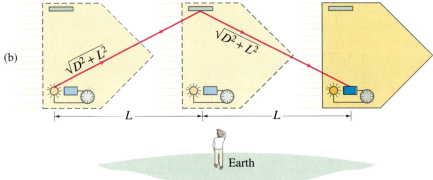

(b)

$\sqrt{D^2 + L^2}$

$\sqrt{D^2 + L^2}$

$\sqrt{D^2 + L^2}$

L   L

Earth

FIGURE 26–7 Time dilation can be shown by a thought experiment: the time it takes for light to travel over and back on a spaceship is longer for the observer on Earth (b) than for the observer on the spaceship (a).

Figure 26–7 shows a spaceship traveling past Earth at high speed. The point of view of an observer on the spaceship is shown in part (a), and that of an observer on Earth in part (b). Both observers have accurate clocks. The person on the spaceship (a) flashes a light and measures the time it takes the light to travel across the spaceship and return after reflecting from a mirror. The light travels a distance $2D$ at speed c, so the time required, which we call Δt_0, is

$$\Delta t_0 = \frac{2D}{c}.$$

This is the time as measured by the observer on the spaceship.

The observer on Earth, Fig. 26–7b, observes the same process. But to this observer, the spaceship is moving. So the light travels the diagonal path shown in going across the spaceship, reflecting off the mirror, and returning to the sender. Although the light travels at the same speed to this observer (the second postulate), it travels a greater distance. Hence the time required, as measured by the observer on Earth, will be *greater* than that measured by the observer on the spaceship. The time interval, Δt, as observed by the observer on Earth can be calculated as follows. In the time Δt, the spaceship travels a distance $2L = v\,\Delta t$ where v is the speed of the spaceship (Fig. 26–7b). Thus, the light travels a total distance on its diagonal path of $2\sqrt{D^2 + L^2}$, and therefore

$$c = \frac{2\sqrt{D^2 + L^2}}{\Delta t} = \frac{2\sqrt{D^2 + v^2(\Delta t)^2/4}}{\Delta t}.$$

We square both sides, and then solve for Δt, to find

$$c^2 = \frac{4D^2}{(\Delta t)^2} + v^2,$$

$$\Delta t = \frac{2D}{c\sqrt{1 - v^2/c^2}}.$$

We combine this with the formula above for Δt_0 ($\Delta t_0 = 2D/c$) and find:

$$\Delta t = \frac{\Delta t_0}{\sqrt{1 - v^2/c^2}}. \qquad\qquad \textbf{(26–1)}$$

Time-dilation formula

Since $\sqrt{1 - v^2/c^2}$ is always less than 1, we see that $\Delta t > \Delta t_0$. That is, the time interval between the two events (the sending of the light, and its reception on the spaceship) is *greater* for the observer on Earth than for the observer on the spaceship. This is a general result of the theory of relativity, and is known as **time dilation**. Stated simply, the time-dilation effect says that

> **clocks moving relative to an observer are measured by that observer to run more slowly (as compared to clocks at rest).**

Time dilation: moving clocks run slowly

However, we should not think that the clocks are somehow at fault. Time is actually measured to pass more slowly in any moving reference frame as compared to your own. This remarkable result is an inevitable outcome of the two postulates of the theory of relativity.

The concept of time dilation may be hard to accept, for it violates our commonsense understanding. We can see from Eq. 26–1 that the time dilation effect is negligible unless v is reasonably close to c. If v is much less than c, then the term v^2/c^2 is much smaller than the 1 in the denominator of Eq. 26–1, and then $\Delta t \approx \Delta t_0$ (see Example 26–2). The speeds we experience in everyday life are much smaller than c, so it is little wonder we don't ordinarily notice time dilation. Experiments have tested the time-dilation effect, and have confirmed Einstein's predictions. In 1971, for example, extremely precise atomic clocks were flown around the world in jet planes. The speed of the planes (10^3 km/h) was much less than c, so the clocks had to be accurate to nanoseconds (10^{-9} s) in order to detect any time dilation. They were this accurate, and they confirmed Eq. 26–1 to within experimental error. Time dilation had been confirmed decades earlier, however, by observation on "elementary particles" (see Chapter 32) which have very small masses (typically 10^{-30} to 10^{-27} kg) and so require little energy to be accelerated to speeds close to c. Many of these elementary particles are not stable and decay after a time into smaller particles. One example is the muon, whose mean lifetime is 2.2 μs when at rest. Careful experiments showed that when a muon is traveling at high speeds, its lifetime is measured to be longer than when it is at rest, just as predicted by the time-dilation formula.

Why we don't usually notice time dilation

EXAMPLE 26–1 **Lifetime of a moving muon.** (*a*) What will be the mean lifetime of a muon as measured in the laboratory if it is traveling at $v = 0.60c = 1.8 \times 10^8$ m/s with respect to the laboratory? Its mean life at rest is 2.2×10^{-6} s. (*b*) How far does a muon travel in the laboratory, on average, before decaying?

SOLUTION (*a*) If an observer were to move along with the muon (the muon would be at rest to this observer), the muon would have a mean life of 2.2×10^{-6} s. To an observer in the lab, the muon lives longer because of time dilation. From Eq. 26–1 with $v = 0.60c$, we have

$$\Delta t = \frac{\Delta t_0}{\sqrt{1 - \dfrac{v^2}{c^2}}} = \frac{2.2 \times 10^{-6} \text{ s}}{\sqrt{1 - \dfrac{0.36c^2}{c^2}}} = \frac{2.2 \times 10^{-6} \text{ s}}{\sqrt{0.64}} = 2.8 \times 10^{-6} \text{ s}.$$

(*b*) At a speed of 1.8×10^8 m/s, classical physics would tell us that with a mean life of $2.2\ \mu$s, an average muon would travel $d = vt = (1.8 \times 10^8 \text{ m/s})(2.2 \times 10^{-6} \text{ s}) = 400$ m. But relativity predicts an average distance of $(1.8 \times 10^8 \text{ m/s})(2.8 \times 10^{-6} \text{ s}) = 500$ m, and it is this longer distance that is measured experimentally.

We need to make a comment about the use of Eq. 26–1 and the meaning of Δt and Δt_0. The equation is true only when Δt_0 represents the time interval between the two events in a reference frame where the two events occur at *the same point in space* (as in Fig. 26–7a where the two events are *Proper* the light flash being sent and being received). This time interval, Δt_0, is *time* called the **proper time**. Then Δt in Eq. 26–1 represents the time interval between the two events as measured in a reference frame moving with speed v with respect to the first. In Example 26–1 above, Δt_0 (and not Δt) was set equal to 2.2×10^{-6} s because it is only in the rest frame of the muon that the two events ("birth" and "decay") occur at the same point in space.

EXAMPLE 26–2 **Time dilation at 100 km/h.** Let's check time dilation for everyday speeds. A car traveling 100 km/h covers a certain distance in 10.00 s according to the driver's watch. What does an observer on Earth measure for the time interval?

SOLUTION The car's speed relative to Earth is 100 km/h = $(1.00 \times 10^5 \text{ m})/(3600 \text{ s}) = 27.8$ m/s. We set $\Delta t_0 = 10.00$ s in the time-dilation formula (the driver is at rest in the reference frame of the car), and then Δt is

$$\Delta t = \frac{\Delta t_0}{\sqrt{1 - \dfrac{v^2}{c^2}}} = \frac{10.00 \text{ s}}{\sqrt{1 - \left(\dfrac{27.8 \text{ m/s}}{3.00 \times 10^8 \text{ m/s}}\right)^2}} = \frac{10.00 \text{ s}}{\sqrt{1 - 8.59 \times 10^{-15}}}.$$

If you put these numbers into a calculator, you will obtain $\Delta t = 10.00$ s, since the denominator differs from 1 by such a tiny amount. Indeed, the time measured by an observer on Earth would be no different from that measured by the driver, even with the best of today's instruments. A computer that could calculate to a large number of decimal places could reveal a difference between Δt and Δt_0. But we can estimate the difference quite easily using the binomial expansion (Appendix A), which says that in a formula of the form $(1 \pm x)^n$, if $x \ll 1$, then to a good approximation,

$$(1 \pm x)^n \approx 1 \pm nx.$$

➡ **PROBLEM SOLVING**

Use of the binomial expansion

In our time-dilation formula, we have the factor $1/\sqrt{1 - v^2/c^2} = (1 - v^2/c^2)^{-1/2}$. Thus (setting $x = v^2/c^2$ and $n = -\frac{1}{2}$ in the binomial expansion):

$$\Delta t = \Delta t_0 \left(1 - \frac{v^2}{c^2}\right)^{-1/2} \approx \Delta t_0 \left(1 + \frac{1}{2}\frac{v^2}{c^2}\right)$$

$$\approx 10.00 \text{ s} \left[1 + \frac{1}{2}\left(\frac{27.8 \text{ m/s}}{3.00 \times 10^8 \text{ m/s}}\right)^2\right] \approx 10.00 \text{ s} + 4 \times 10^{-15} \text{ s}.$$

So the difference between Δt and Δt_0 is predicted to be 4×10^{-15} s, an unmeasurably small amount.

Time dilation has aroused interesting speculation about space travel. According to classical (Newtonian) physics, to reach a star 100 light-years away would not be possible for ordinary mortals (1 light-year is the distance light can travel in 1 year $= 3.0 \times 10^8$ m/s $\times 3.15 \times 10^7$ s $= 9.5 \times 10^{15}$ m). Even if a spaceship could travel at close to the speed of light, it would take over 100 years to reach such a star. But time dilation tells us that the time involved would be less for an astronaut. In a spaceship traveling at $v = 0.999c$, the time for such a trip would be only about $\Delta t_0 = \Delta t\sqrt{1 - v^2/c^2} = (100 \text{ yr})\sqrt{1 - (0.999)^2} = 4.5$ yr. Thus time dilation allows such a trip, but the enormous practical problems of achieving such speeds will not be overcome in the near future.

Notice, in this example, that whereas 100 years would pass on Earth, only 4.5 years would pass for the astronaut on the trip. Is it just the clocks that would slow down for the astronaut? The answer is no. All processes, including life processes, run more slowly for the astronaut according to the Earth observer. But to the astronaut, time would pass in a normal way. The astronaut would experience 4.5 years of normal sleeping, eating, reading, and so on. And people on Earth would experience 100 years of ordinary activity.

Not long after Einstein proposed the special theory of relativity, an apparent paradox was pointed out. According to this **twin paradox**, suppose one of a pair of 20-year-old twins takes off in a spaceship traveling at very high speed to a distant star and back again, while the other twin remains on Earth. According to the Earth twin, the traveling twin will age less. Whereas 20 years might pass for the Earth twin, perhaps only 1 year (depending on the spacecraft's speed) would pass for the traveler. Thus, when the traveler returns, the earthbound twin could expect to be 40 years old whereas the traveling twin would be only 21.

Twin paradox

This is the viewpoint of the twin on the Earth. But what about the traveling twin? If all inertial reference frames are equally good, won't the traveling twin make all the claims the Earth twin does, only in reverse? Can't the astronaut twin claim that since the Earth is moving away at high speed, time passes more slowly on Earth and the twin on Earth will age less? This is the opposite of what the Earth twin predicts. They cannot both be right, for after all the spacecraft returns to Earth and a direct comparison of ages and clocks can be made.

There is, however, not a paradox at all. The consequences of the special theory of relativity—in this case, time dilation—can be applied only by observers in inertial reference frames. The Earth is such a frame (or nearly so), whereas the spacecraft is not. The spacecraft accelerates at the start and end of its trip and, more importantly, when it turns around at the far point of its journey. During these acceleration periods, the spacecraft's predictions based on special relativity are not valid. The twin on Earth is in an inertial frame and can make valid predictions. Thus, there is no paradox. The traveling twin's point of view expressed above is not correct. The predictions of the Earth twin *are* valid, and the prediction that the traveling twin returns having aged less is the proper one.[†]

26–6 Length Contraction

Not only time intervals are different in different reference frames. Space intervals—lengths and distances—are different as well, according to the special theory of relativity, and we illustrate this with a thought experiment.

Observers on Earth watch a spacecraft traveling at speed v from Earth to, say, Neptune, Fig. 26–8a. The distance between the planets, as measured by the Earth observers, is L_0. The time required for the trip, measured from Earth, is $\Delta t = L_0/v$. In Fig. 26–8b we see the point of view of observers on the spacecraft. In this frame of reference, the spaceship is at rest; Earth and Neptune move with speed v. (We assume v is much greater than the relative speed of Neptune and Earth, so the latter can be ignored.) The time between the departure of Earth and arrival of Neptune (as observed from the spacecraft) is the "proper time" (since the two events occur at the same point in space—i.e., on the spacecraft). Therefore the time interval is less for the spacecraft observers than for the Earth observers, because of time dilation. From Eq. 26–1, the time for the trip as viewed by the spacecraft is $\Delta t_0 = \Delta t \sqrt{1 - v^2/c^2}$. Since the spacecraft observers measure the same speed but less time between these two events, they must also measure the

FIGURE 26–8 (a) A spaceship traveling at very high speed from Earth to Neptune, as seen from Earth's frame of reference. (b) As viewed by an observer on the spaceship, Earth and Neptune are moving at the very high velocity v: Earth leaves the spaceship, and a time Δt_0 later planet Neptune arrives at the spaceship. [Note in (b) that each planet does not look shortened because at high speeds we see the trailing edge (as in Fig. 26–10), and the net effect is to leave its appearance as a circle.]

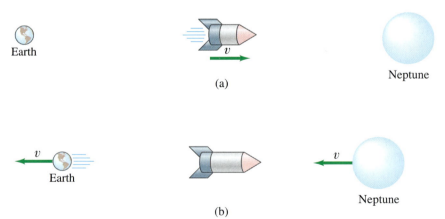

[†]Einstein's general theory of relativity, which deals with accelerating reference frames, confirms this result.

distance as less. If we let L be the distance between the planets as viewed by the spacecraft observers, then $L = v\,\Delta t_0$. We have already seen that $\Delta t_0 = \Delta t\sqrt{1 - v^2/c^2}$ and $\Delta t = L_0/v$, so we have $L = v\,\Delta t_0 = v\,\Delta t\sqrt{1 - v^2/c^2} = L_0\sqrt{1 - v^2/c^2}$. That is,

$$L = L_0\sqrt{1 - v^2/c^2}.$$ (26–2) *Length-contraction formula*

This is a general result of the special theory of relativity and applies to lengths of objects as well as to distance. The result can be stated most simply in words as:

the length of an object is measured to be shorter when it is moving relative to the observer than when it is at rest.

Length contraction: moving objects are shorter (in the direction of motion)

This is called **length contraction**. The length L_0 in Eq. 26–2 is called the **proper length**. It is the length of the object—or distance between two points whose *positions are measured at the same time*—as measured by observers at rest with respect to it. Equation 26–2 gives the length L that will be measured by observers when the object travels past them at speed v. It is important to note, however, that length contraction occurs *only along the direction of motion*. For example, the moving spaceship in Fig. 26–8a is shortened in length, but its height is the same as when it is at rest.

Length contraction, like time dilation, is not noticeable in everyday life because the factor $\sqrt{1 - v^2/c^2}$ in Eq. 26–2 differs from 1.00 significantly only when v is very large.

EXAMPLE 26–3   Painting's contraction. A rectangular painting measures 1.00 m tall and 1.50 m wide. It is hung on the side wall of a spaceship which is moving past the Earth at a speed of $0.90c$. See Fig. 26–9a. (*a*) What are the dimensions of the picture according to the captain of the spaceship? (*b*) What are the dimensions as seen by an observer on the Earth?

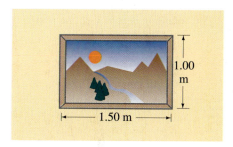

(a)

SOLUTION   (*a*) The painting (as well as everything else in the spaceship) looks perfectly normal to everyone on the spaceship, so the captain sees a 1.00 m by 1.50 m painting.
(*b*) Only the dimension in the direction of motion is shortened, so the height is unchanged at 1.00 m, Fig. 26–9b. The length, however, is contracted to

$$L = L_0\sqrt{1 - \frac{v^2}{c^2}}$$

$$= (1.50\text{ m})\sqrt{1 - (0.90)^2} = 0.65\text{ m}.$$

So the picture has dimensions 1.00 m $\times$ 0.65 m.

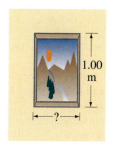

(b)

FIGURE 26–9   Example 26–3.

Equation 26–2 tells us what the length of an object will be *measured* to be when traveling at speed v. The *appearance* of the object is another matter. Suppose, for example, you are traveling to the left past a small building at speed $v = 0.85c$. This is equivalent to the building moving past

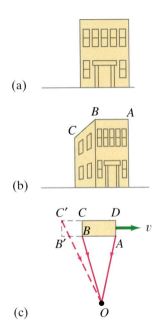

(a)

(b)

(c)

FIGURE 26–10 Building seen (a) at rest, and (b) moving at high speed. (c) Diagram explains why the side of the building is seen (see the text).

you to the right at speed v. The building will look narrower (and the same height), but you will also be able to see the side of the building even if you are directly in front of it. This is shown in Fig. 26–10b—part (a) shows the building at rest. The fact that you see the side is not really a relativistic effect, but is due to the finite speed of light. To see how this occurs, we look at Fig. 26–10c which is a top view of the building, looking down. At the instant shown, the observer O is directly in front of the building. Light from points A and B reach O at the same time. If the building were at rest, light from point C could never reach O. But the building is moving at very high speed and does "get out of the way" so that light from C can reach O. Indeed, at the instant shown, light from point C when it was at an earlier location (C' on the diagram) can reach O because the building has moved. In order to reach the observer at the same time as light from A and B, light from C had to leave at an earlier time since it must travel a greater distance. Thus it is light from C' that reaches the observer at the same time as light from A and B. This, then, is how an observer might see both the front and side of an object at the same time even when directly in front of it.[†] It can be shown, by the same reasoning, that spherical objects will actually still have a circular outline even at high speeds. That is why the planets in Fig. 26–8b are drawn round rather than contracted.

26–7 Four-Dimensional Space–Time

Let us imagine a person is on a train moving at a very high speed, say $0.65c$, Fig. 26–11. This person begins a meal at 7:00 and finishes at 7:15, according to a clock on the train. The two events, beginning and ending the meal, take place at the same point on the train. So the proper time between these two events is 15 min. To observers on Earth, the meal will take longer—20 min according to Eq. 26–1. Let us assume that the meal was served on a 20-cm-diameter plate. To observers on the Earth, the plate

FIGURE 26–11 According to an accurate clock on a fast-moving train, a person (a) begins dinner at 7:00 and (b) finishes at 7:15. At the beginning of the meal, observers on Earth set their watches to correspond with the clock on the train. These observers measure the eating time as 20 minutes.

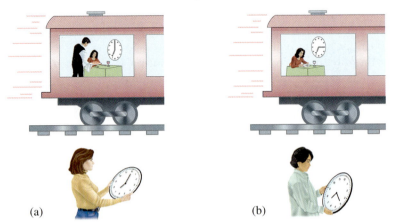

(a)

(b)

[†]It would be an error to think that the building in Fig. 26–10b would look rotated. This is not correct since in that case side A would look shorter than side B. In fact, if the observer is directly in front, these sides appear equal in height. Thus the building looks contracted in its front face, but we also see the side, as described above. Also, though not shown in Fig. 26–10b, the walls of the building would appear curved, because of differing distances from the observer's eye of the various points from top to bottom along a vertical wall.

is only 15 cm wide (length contraction). Thus, to observers on the Earth, the meal looks smaller but lasts longer.

In a sense these two effects, time dilation and length contraction, balance each other. When viewed from the Earth, what the meal seems to lose in size it gains in length of time it lasts. Space, or length, is exchanged for time.

Considerations like this led to the idea of **four-dimensional space–time**: space takes up three dimensions and time is a fourth dimension. Space and time are intimately connected. Just as when we squeeze a balloon we make one dimension larger and another smaller, so when we examine objects and events from different reference frames, a certain amount of space is exchanged for time, or vice versa.

Although the idea of four dimensions may seem strange, it refers to the idea that any object or event is specified by four quantities—three to describe where in space, and one to describe when in time. The really unusual aspect of four-dimensional space–time is that space and time can intermix: a little of one can be exchanged for a little of the other when the reference frame is changed.

It is difficult for most of us to understand the idea of four-dimensional space–time. Somehow we feel, just as physicists did before the advent of relativity, that space and time are completely separate entities. Yet we have found in our thought experiments that they are not completely separate. Our difficulty in accepting this is reminiscent of the situation in the seventeenth century at the time of Galileo and Newton. Before Galileo, the vertical direction, that in which objects fall, was considered to be distinctly different from the two horizontal dimensions. Galileo showed that the vertical dimension differs only in that it happens to be the direction in which gravity acts. Otherwise, all three dimensions are equivalent, a viewpoint we all accept today. Now we are asked to accept one more dimension, time, which we had previously thought of as being somehow different. This is not to say that there is no distinction between space and time. What relativity has shown is that space and time determinations are not independent of one another.

26–8 Momentum and Mass

The three basic mechanical quantities are length, time intervals, and mass. The first two have been shown to be relative—their value depends on the reference frame from which they are measured. We might ask if mass, too, is a relative quantity.

Analysis of collision processes between two particles shows that if we want to preserve conservation of momentum as a principle also in relativity, we must redefine momentum as

$$p = \frac{m_0 v}{\sqrt{1 - \dfrac{v^2}{c^2}}}.$$

(26–3) *Relativistic momentum*

For speeds much less than the speed of light, Eq. 26–3 gives the classical momentum, $p = m_0 v$. We have written m_0 rather than m because Eq. 26–3 suggests a relativistic interpretation of mass. Namely, that *the mass of an object is measured to increase as its speed increases* according to the formula

Mass increase formula

$$m = \frac{m_0}{\sqrt{1 - v^2/c^2}}. \tag{26-4}$$

In this **mass-increase** formula, m_0 is the **rest mass** of the object—the mass it has as measured in a reference frame in which it is at rest; and m is the mass it will be measured to have in a reference frame in which it moves at speed v.

Relativistic momentum and mass increase have been tested countless times on tiny elementary particles (such as muons), and they have been found to increase in accord with Eqs. 26–3 and 26–4.

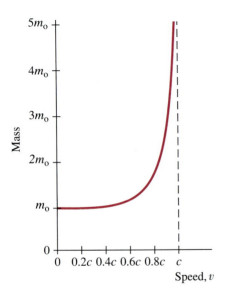

FIGURE 26–12 Mass of a particle (rest mass m_0) as a function of speed v (given as a fraction of c, the speed of light).

EXAMPLE 26–4 **Mass of moving electron.** Calculate the mass of an electron when it has a speed of (*a*) 4.00×10^7 m/s in the CRT of a television set, and (*b*) $0.98c$ in an accelerator used for cancer therapy.

SOLUTION The rest mass of an electron is $m_0 = 9.11 \times 10^{-31}$ kg. (*a*) At $v = 4.00 \times 10^7$ m/s, the electron's mass will be

$$m = \frac{m_0}{\sqrt{1 - \dfrac{v^2}{c^2}}} = \frac{9.11 \times 10^{-31} \text{ kg}}{\sqrt{1 - \dfrac{(4.00 \times 10^7 \text{ m/s})^2}{(3.00 \times 10^8 \text{ m/s})^2}}} = 9.19 \times 10^{-31} \text{ kg}.$$

Even at such a high speed ($v \approx 0.1c$), the electron's mass is only about 1 percent higher than its rest mass. But in (*b*), we have

$$m = \frac{m_0}{\sqrt{1 - \dfrac{v^2}{c^2}}} = \frac{m_0}{\sqrt{1 - \dfrac{(0.98c)^2}{c^2}}} = \frac{m_0}{\sqrt{1 - (0.98)^2}} = 5.0 m_0.$$

An electron traveling at 98 percent the speed of light has a mass five times its rest mass!

Figure 26–12 is a graph of mass versus speed for any particle.

26–9 The Ultimate Speed

A basic result of the special theory of relativity is that the speed of an object cannot equal or exceed the speed of light. That the speed of light is a natural upper speed limit in the universe can be seen from any one of Eqs. 26–1 through 26–4. It is perhaps easiest to see it from Eq. 26–4, the mass-increase formula, $m = m_0/\sqrt{1 - v^2/c^2}$. As an object is accelerated to greater and greater speeds, its mass becomes larger and larger. Indeed, if v were to equal c, the denominator in this equation would be zero and the mass m would become infinite. To accelerate an object up to $v = c$ would thus require infinite energy, and so is not possible. Similarly, Eqs. 26–1 and 26–2 tell us that length would disappear and time become infinite as v approaches c.

26–10 $E = mc^2$; Mass and Energy

When a steady net force is applied to an object of rest mass m_0, the object increases in speed. Since the force is acting over a distance, work is done on the object and its kinetic energy increases. As the speed of the object approaches c, the speed cannot increase indefinitely because it cannot exceed c. On the other hand, the mass of the object increases with increasing speed. That is, the work done on an object not only increases its speed but also contributes to increasing its *mass*. Since the work done on an object increases its energy, this new twist from the theory of relativity leads to the idea that mass is a form of energy, a crucial part of Einstein's theory.

To find the mathematical relationship between mass and energy, Einstein assumed that the work–energy theorem (Chapter 6) is still valid in relativity. That is, the net work done on a particle is equal to its change in kinetic energy (KE). Using this theorem, Einstein showed that at high speeds the formula $\text{KE} = \frac{1}{2}mv^2$ is not correct. You might think that using Eq. 26–4 for m would give $\text{KE} = \frac{1}{2}m_0 v^2 / \sqrt{1 - v^2/c^2}$, but this formula, too, is wrong. Instead, Einstein showed that the kinetic energy of a particle is given by

$$\text{KE} = mc^2 - m_0 c^2, \qquad \text{(26–5)}$$

Relativistic kinetic energy

where m is the mass of the particle traveling at speed v and m_0 is its rest mass.

But what does the second term in Eq. 26–5—the $m_0 c^2$—mean? Consistent with the idea that mass is a form of energy, Einstein called $m_0 c^2$ the **rest energy** of the object. We can rearrange Eq. 26–5 to get $mc^2 = m_0 c^2 + \text{KE}$. We call mc^2 the *total energy* E of the particle (assuming no potential energy), and we see that the total energy equals the rest energy plus the kinetic energy:

$$E = mc^2, \qquad \text{(26–6a)}$$

or

$$E = m_0 c^2 + \text{KE}. \qquad \text{(26–6b)}$$

$E = mc^2$, mass related to energy

Here we have Einstein's famous formula $E = mc^2$.

For a particle at rest in a given reference frame, its total energy is $E_0 = m_0 c^2$, which we have called its rest energy. This formula mathematically relates the concepts of energy and mass. But if this idea is to have any meaning from a practical point of view, then mass ought to be convertible to energy and vice versa. That is, if mass is just one form of energy, then it should be convertible to other forms of energy just as other types of energy are interconvertible. Einstein suggested that this might be possible, and indeed changes of mass to other forms of energy, and vice versa, have been experimentally confirmed countless times. The interconversion of mass and energy is most easily detected in nuclear and elementary particle physics. For example, the neutral pion (π^0) of rest mass $2.4 \times 10^{-28}\,\text{kg}$ is observed to decay into pure electromagnetic radiation (photons). The π^0 completely disappears in the process. The amount of electromagnetic energy produced is found to be exactly equal to that predicted by Einstein's formula, $E = m_0 c^2$. The reverse process is also commonly observed in the laboratory: electromagnetic radiation under certain conditions can be converted into material particles such as electrons. On a larger scale, the energy produced in nuclear power plants is a result of the loss in mass of the uranium fuel as it undergoes the process called fission

Mass and energy interchangeable

(Chapter 31). Even the radiant energy we receive from the Sun is an example of $E = mc^2$; the Sun's mass is continually decreasing as it radiates electromagnetic energy outward.

The relation $E = mc^2$ is now believed to apply to all processes, although the changes are often too small to measure. That is, when the energy of a system changes by an amount ΔE, the mass of the system changes by an amount Δm given by

$$\Delta E = (\Delta m)(c^2).$$

In a chemical reaction where heat is gained or lost, the masses of the reactants and the products will be different. Even when water is heated on a stove, the mass of the water increases very slightly. This example is also easy to understand from the point of view of kinetic theory (Chapter 13), because as heat is added, the temperature and therefore the average speed of the molecules increases; and Eq. 26–4 tells us that the mass also increases.

EXAMPLE 26–5 **Pion's KE.** A π^0 meson ($m_0 = 2.4 \times 10^{-28}$ kg) travels at a speed $v = 0.80c = 2.4 \times 10^8$ m/s. What is its kinetic energy? Compare to a classical calculation.

SOLUTION The mass of the π^0 moving with a speed of $v = 0.80c$ is

$$m = \frac{m_0}{\sqrt{1 - v^2/c^2}} = \frac{2.4 \times 10^{-28} \text{ kg}}{\sqrt{1 - (0.80)^2}} = 4.0 \times 10^{-28} \text{ kg}.$$

➡ **PROBLEM SOLVING**

Relativistic KE

Thus its KE is

$$\text{KE} = (m - m_0)c^2 = (4.0 \times 10^{-28} \text{ kg} - 2.4 \times 10^{-28} \text{ kg})(3.0 \times 10^8 \text{ m/s})^2$$
$$= 1.4 \times 10^{-11} \text{ J}.$$

Notice that the units of mc^2 are kg·m²/s², which is the joule. A classical calculation would give $\text{KE} = \frac{1}{2}m_0 v^2 = \frac{1}{2}(2.4 \times 10^{-28} \text{ kg})(2.4 \times 10^8 \text{ m/s})^2 = 6.9 \times 10^{-12}$ J, about half as much, but this is not a correct result.

EXAMPLE 26–6 **Energy from pion mass.** How much energy would be released if the π^0 meson in the last example is transformed completely into electromagnetic radiation?

SOLUTION The rest energy of the π^0 is

Rest energy

$$E_0 = m_0 c^2 = (2.40 \times 10^{-28} \text{ kg})(3.00 \times 10^8 \text{ m/s})^2 = 2.16 \times 10^{-11} \text{ J}.$$

This is how much energy would be released if the π^0 decayed at rest. We saw in Chapter 17, Section 17–4, that the energies of atomic particles are often expressed in terms of the electron volt (eV) unit:

$$1 \text{ eV} = 1.60 \times 10^{-19} \text{ J}, \quad \text{and} \quad 1 \text{ MeV} = 10^6 \text{ eV} = 1.60 \times 10^{-13} \text{ J}.$$

Thus the rest mass of the π^0 is equivalent to

$$\frac{2.16 \times 10^{-11} \text{ J}}{1.60 \times 10^{-13} \text{ J/MeV}} = 135 \text{ MeV}$$

of energy. If the π^0 had $\text{KE} = 1.4 \times 10^{-11}$ J, the total energy released would be $(2.16 + 1.4) \times 10^{-11} \text{ J} = 3.6 \times 10^{-11}$ J, or 230 MeV.

EXAMPLE 26–7 **Energy from nuclear decay.** The energy required or released in nuclear reactions and decays comes from a change in mass between the initial and final particles. In one type of radioactive decay (Chapter 30), an atom of uranium ($m = 232.03714$ u) decays to an atom of thorium ($m = 228.02873$ u) plus an atom of helium ($m = 4.00260$ u) where the masses given are in atomic mass units (1 u $= 1.6605 \times 10^{-27}$ kg). Calculate the energy released in this decay.

SOLUTION The initial mass is 232.03714 u, and after the decay it is 228.02873 u $+ 4.00260$ u $= 232.03133$ u, so there is a decrease in mass of 0.00581 u. This mass, which equals $(0.00581 \text{ u})(1.66 \times 10^{-27} \text{ kg}) = 9.64 \times 10^{-30}$ kg, is changed into energy. By $E = mc^2$, we have

Energy released in nuclear process

$$E = (9.64 \times 10^{-30} \text{ kg})(3.0 \times 10^8 \text{ m/s})^2 = 8.68 \times 10^{-13} \text{ J}.$$

Since 1 MeV $= 1.60 \times 10^{-13}$ J, the energy released is 5.4 MeV.

Equation 26–5 for the kinetic energy can be written in terms of the speed v of the object with the help of Eq. 26–4:

$$\text{KE} = m_0 c^2 \left(\frac{1}{\sqrt{1 - v^2/c^2}} - 1 \right). \tag{26–7}$$

At low speeds, $v \ll c$, we can expand the square root in Eq. 26–7 using the binomial expansion (see Appendix A or Example 26–2). Then we get

$$\text{KE} \approx m_0 c^2 \left(1 + \frac{1}{2} \frac{v^2}{c^2} + \cdots - 1 \right)$$

$$\approx \tfrac{1}{2} m_0 v^2,$$

where the dots in the first expression represent very small terms in the expansion which we have neglected since we assumed that $v \ll c$. Thus at low speeds, the relativistic form for kinetic energy reduces to the classical form, $\text{KE} = \tfrac{1}{2} m_0 v^2$. This is, of course, what we would like. It makes relativity a more valuable theory in that it can predict accurate results at low speed as well as at high. Indeed, the other equations of special relativity also reduce to their classical equivalents at ordinary speeds: length contraction, time dilation, and mass increase all disappear for $v \ll c$ since $\sqrt{1 - v^2/c^2} \approx 1$.

A useful relation between the total energy E of a particle and its momentum p can also be derived. The relativistic momentum of a particle of mass m and speed v is given by Eq. 26–3:

$$p = mv = \frac{m_0 v}{\sqrt{1 - v^2/c^2}}.$$

Relativistic momentum

Then, since $E = mc^2$, we can write (in the first line we insert "$v^2 - v^2$" which is zero, but will help us):

$$E^2 = m^2 c^4 = m^2 c^2 (c^2 + v^2 - v^2)$$

$$= m^2 c^2 v^2 + m^2 c^2 (c^2 - v^2)$$

$$= p^2 c^2 + \frac{m_0^2 c^4 (1 - v^2/c^2)}{1 - v^2/c^2},$$

or

$$E^2 = p^2 c^2 + m_0^2 c^4, \tag{26–8}$$

Energy–momentum relation

where we have assumed there is no potential energy. Thus, the total energy can be written in terms of the momentum p, or in terms of the kinetic energy (Eq. 26–6).

Units:
eV/c for p
eV/c² for m

In the tiny world of atoms and nuclei, it is common to quote energies in eV (electron volts) or multiples such as MeV (10^6 eV). Momentum (see Eq. 26–8) can be quoted in units of eV/c (or MeV/c). And mass can be quoted (from $E = mc^2$) in units of eV/c² (or MeV/c²).

26–11 Relativistic Addition of Velocities

Consider a rocket ship that travels away from the Earth with speed v, and assume that this rocket has fired off a second rocket that travels at speed u' with respect to the first (Fig. 26–13). We might expect that the speed u of rocket 2 with respect to Earth is $u = v + u'$, which in the case shown in the figure is $u = 0.60c + 0.60c = 1.20c$. But, as discussed in Section 26–9, no object can travel faster than the speed of light in any reference frame. Indeed, Einstein showed that since length and time are different in different reference frames, the old addition-of-velocities formula is no longer valid. Instead, the correct formula is

*Relativistic addition of velocities formula (**u** and **v** along same line)*

$$u = \frac{v + u'}{1 + vu'/c^2} \tag{26–9}$$

for motion along a straight line. We derive this formula in Appendix E. If u' is in the opposite direction from v, then u' must have a minus sign and $u = (v - u')/(1 - vu'/c^2)$.

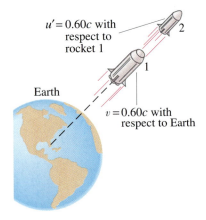

$u' = 0.60c$ with respect to rocket 1

Earth

$v = 0.60c$ with respect to Earth

FIGURE 26–13 Rocket 2 is fired from rocket 1 with speed $u' = 0.60c$. What is the speed of rocket 2 with respect to the Earth?

EXAMPLE 26–8 Relative velocity, relativistically. Calculate the speed of rocket 2 in Fig. 26–13 with respect to Earth.

SOLUTION Rocket 2 moves with speed $u' = 0.60c$ with respect to rocket 1. Rocket 1 has speed $v = 0.60c$ with respect to Earth. The speed of rocket 2 with respect to Earth is therefore

$$u = \frac{0.60c + 0.60c}{1 + \frac{(0.60c)(0.60c)}{c^2}} = \frac{1.20c}{1.36} = 0.88c.$$

Notice that Eq. 26–9 reduces to the classical form for velocities small compared to the speed of light since $1 + vu'/c^2 \approx 1$ for v and $u' \ll c$. Thus, $u \approx v + u'$.

Let us test our formula in one more case, that of the speed of light. Suppose that rocket 1 in Fig. 26–13 sends out a beam of light so that

$u' = c$. Equation 26–9 tells us that the speed of this light with respect to Earth is

$$u = \frac{0.60c + c}{1 + \frac{(0.60c)(c)}{c^2}} = \frac{1.60c}{1.60} = c,$$

which is fully consistent with the second postulate of relativity.

26–12 The Impact of Special Relativity

A great many experiments have been performed to test the predictions of the special theory of relativity. Within experimental error, no contradictions have been found. Scientists have therefore accepted relativity as an accurate description of nature.

At speeds much less than the speed of light, the relativistic formulas reduce to the old classical ones, as we have discussed. We would, of course, hope—or rather, insist—that this be true since Newtonian mechanics works so well for objects moving with speeds $v \ll c$. This insistence that a more general theory (such as relativity) give the same results as a more restricted theory (such as classical mechanics which works for $v \ll c$) is called the **correspondence principle**. The two theories must correspond where their realms of validity overlap. Relativity thus does not contradict classical mechanics. Rather, it is a more general theory, of which classical mechanics is now considered to be a limiting case.

Correspondence principle

The importance of relativity is not simply that it gives more accurate results, especially at very high speeds. Much more than that, it has changed the way we view the world. The concepts of space and time are now seen to be relative, and intertwined with one another, whereas before they were considered absolute and separate. Even our concepts of matter and energy have changed: either can be converted to the other. The impact of relativity extends far beyond physics. It has influenced the other sciences, and even the world of art and literature; it has, indeed, entered the general culture.

From a practical point of view, we do not have much opportunity in our daily lives to use the mathematics of relativity. For example, the factor $\sqrt{1 - v^2/c^2}$, which appears in many relativistic formulas, has a value of 0.995 when $v = 0.10c$. Thus, for speeds even as high as $0.10c = 3.0 \times 10^7$ m/s, the factor $\sqrt{1 - v^2/c^2}$ in relativistic formulas gives a numerical correction of less than 1 percent. For speeds less than $0.10c$, or unless mass and energy are interchanged, we thus don't usually need to use the more complicated relativistic formulas, and can use the simpler classical formulas.

The special theory of relativity we have studied in this chapter deals with inertial (nonaccelerating) reference frames. In Chapter 33 we will discuss briefly the more complicated "general theory of relativity" which can deal with noninertial reference frames.

SUMMARY

An **inertial reference frame** is one in which Newton's law of inertia holds. Inertial reference frames can move at constant velocity relative to one another; accelerating reference frames are noninertial.

The **special theory of relativity** is based on two principles: the **relativity principle**, which states that the laws of physics are the same in all inertial reference frames, and the principle of the **constancy of the speed of light**, which states that the speed of light in empty space has the same value in all inertial reference frames.

One consequence of relativity theory is that two events that are simultaneous in one reference frame may not be simultaneous in another. Other effects are **time dilation**: moving clocks are measured to run slowly; **length contraction**: the length of a moving object is measured to be shorter (in its direction of motion) than when it is at rest; *mass increase*: the mass of a body is measured to increase with speed. Quantitatively,

$$L = L_0\sqrt{1 - v^2/c^2}$$

$$\Delta t = \frac{\Delta t_0}{\sqrt{1 - v^2/c^2}}$$

$$m = \frac{m_0}{\sqrt{1 - v^2/c^2}}$$

where L, Δt, and m are the length, time interval, and mass of objects (or events) that are observed as they move by at speed v; L_0, Δt_0, and m_0 are the **proper length**, **proper time**, and **rest mass**—that is, the same quantities as measured in the rest frame of the objects or events. Velocity addition also must be done in a special way. All these effects are significant only at high speeds, close to the speed of light, which itself is the ultimate speed in the universe.

The theory of relativity has changed our notions of space and time, and of mass and energy. Space and time are seen to be intimately connected, with time being the fourth dimension in addition to the three dimensions of space. Mass and energy are interconvertible. The equation

$$E = mc^2$$

tells how much energy E is needed to create a mass m, or vice versa. Said another way, $E = mc^2$ is the amount of energy an object has because of its mass m. The law of conservation of energy must include mass as a form of energy. The kinetic energy of an object moving at speed v is given by

$$\text{KE} = mc^2 - m_0c^2 = \left(\frac{1}{\sqrt{1 - v^2/c^2}} - 1\right)m_0c^2,$$

where m_0 is the rest mass of the object. The momentum p of an object is related to its total energy E (assuming no potential energy) by

$$E^2 = p^2c^2 + m_0^2c^4.$$

QUESTIONS

1. You are in a windowless car in an exceptionally smooth train. Is there any physical experiment you can do in the train car to determine whether you are moving?

2. You might have had the experience of being at a red light when, out of the corner of your eye, you see the car beside you creep forward. Instinctively you stomp on the brake pedal, thinking that you are rolling backward. What does this say about absolute and relative motion?

3. A worker stands on top of a moving railroad car, and throws a heavy ball straight up (from his point of view). Ignoring air resistance, will the ball land on the car or behind it?

4. Does the Earth really go around the Sun? Or is it also valid to say that the Sun goes around the Earth? Discuss in view of the first principle of relativity (that there is no best reference frame).

5. If you were on a spaceship traveling at $0.5c$ away from a star, at what speed would the starlight pass you?

6. Will two events that occur at the same place and same time for one observer be simultaneous to a second observer moving with respect to the first?

7. Analyze the thought experiment of Section 26–4 from O_1's point of view. (Make a diagram analogous to Fig. 26–6.)

8. The time-dilation effect is sometimes expressed as "moving clocks run slowly." Actually, this effect has nothing to do with motion affecting the functioning of clocks. What then does it deal with?

9. Does time dilation mean that time actually passes more slowly in moving reference frames or that it only *seems* to pass more slowly?

10. A young-looking woman astronaut has just arrived home from a long trip. She rushes up to an old gray-haired man and in the ensuing conversation refers to him as her son. How might this be possible?

11. If you were traveling away from Earth at speed 0.5c, would you notice a change in your heartbeat? Would your mass, height, or waistline change? What would observers on Earth using telescopes say about you?

12. Discuss how our everyday lives would be different if the speed of light were only 25 m/s.

13. Do mass increase, time dilation, and length contraction occur at ordinary speeds, say 90 km/h?

14. Suppose the speed of light were infinite. What would happen to the relativistic predictions of length contraction, time dilation, and mass increase?

15. Explain how the length-contraction and time-dilation formulas might be used to indicate that c is the limiting speed in the universe.

16. Consider an object of mass m to which is applied a constant force for an indefinite period of time. Discuss how its velocity and mass change with time.

17. A white-hot iron bar is cooled to room temperature. Does its mass change?

18. Does the equation $E = mc^2$ conflict with the conservation of energy principle? Explain.

19. Does $E = mc^2$ apply to particles that travel at the speed of light? Does it apply only to them?

20. An electron is limited to travel at speeds less than c. Does this put an upper limit on the momentum of an electron? If so, what is this upper limit?

21. If mass is a form of energy, does this mean that a spring has more mass when compressed than when relaxed?

22. It is not correct to say that "matter can neither be created nor destroyed." What must we say instead?

23. Is our intuitive notion that velocities simply add, as we did in Section 3–8, completely wrong?

PROBLEMS

SECTIONS 26–5 AND 26–6

1. (I) Lengths and time intervals (as well as mass) depend on the factor

$$\sqrt{1 - v^2/c^2}$$

according to the theory of relativity (Eqs. 26–1, 26–2, 26–4). Evaluate this correction factor for speeds of: (a) $v = 20{,}000$ m/s (typical speed of a satellite); (b) $v = 0.0100c$; (c) $v = 0.100c$; (d) $v = 0.900c$; (e) $v = 0.990c$; (f) $v = 0.999c$.

2. (I) A spaceship passes you at a speed of 0.850c. You measure its length to be 48.2 m. How long would it be when at rest?

3. (I) A beam of a certain type of elementary particle travels at a speed of 2.70×10^8 m/s. At this speed, the average lifetime is measured to be 4.76×10^{-6} s. What is the particle's lifetime at rest?

4. (I) If you were to travel to a star 100 light-years from Earth at a speed of 2.60×10^8 m/s, what would you measure this distance to be?

5. (II) You are sitting in your car when a very fast sports car passes you at a speed of 0.37c. A person in that car says his car is 6.00 m long and yours is 6.21 m long. What do you measure for these two lengths?

6. (II) What is the speed of a beam of pions if their average lifetime is measured to be 4.10×10^{-8} s? At rest, their lifetime is 2.60×10^{-8} s.

7. (II) Suppose you decide to travel to a star 90 light-years away. How fast would you have to travel so the distance would be only 25 light-years?

8. (II) At what speed do the relativistic formulas for length and time intervals differ from classical values by 1.00 percent? (This is a reasonable way to estimate when to do relativistic calculations rather than classical.)

9. (II) Suppose a news report stated that starship *Enterprise* had just returned from a 5-year voyage while traveling at 0.89c. (a) If the report meant 5.0 years of *Earth time*, how much time elapsed on the ship? (b) If the report meant 5.0 years of *ship time*, how much time passed on Earth?

10. (II) A certain star is 75.0 light-years away. How long would it take a spacecraft traveling 0.950c to reach that star from Earth, as measured by observers: (a) on Earth, (b) on the spacecraft? (c) What is the distance traveled according to observers on the spacecraft? (d) What will the spacecraft occupants compute their speed to be from the results of (b) and (c)?

11. (II) A friend of yours travels by you in her fast sports vehicle at a speed of $0.580c$. You measure it to be 5.80 m long and 1.20 m high. (a) What will be its length and height at rest? (b) How many seconds would you say elapsed on your friend's watch when 20.0 s passed on yours? (c) How fast did you appear to be traveling to your friend? (d) How many seconds would she say elapsed on your watch when she saw 20.0 s pass on hers?

12. (III) How fast must a pion be moving, on average, to travel 10.0 m before it decays? The average lifetime, at rest, is 2.60×10^{-8} s.

SECTION 26-8

13. (I) What is the mass of a proton traveling at $v = 0.90c$?

14. (I) At what speed will an object's mass be twice its rest mass?

15. (II) At what speed v will the mass of an object be 10 percent greater than its rest mass?

16. (II) Escape velocity from the Earth is 40,000 km/h. What would be the percent increase in mass of a 7.2×10^5-kg spacecraft traveling at that speed?

17. (II) (a) What is the speed of an electron whose mass is 10,000 times its rest mass? Such speeds are reached in the Stanford Linear Accelerator, SLAC. (b) If the electrons travel in the lab through a tube 3.0 km long (as at SLAC), how long is this tube in the electron's reference frame?

SECTION 26-10

18. (I) What is the kinetic energy of an electron whose mass is 3.0 times its rest mass?

19. (I) A certain chemical reaction requires 4.82×10^4 J of energy input for it to go. What is the increase in rest mass of the products over the reactants?

20. (I) When a uranium nucleus at rest breaks apart in the process known as fission in a nuclear reactor, the resulting fragments have a total kinetic energy of about 200 MeV. How much mass was lost in the process?

21. (I) Calculate the rest energy of an electron in joules and in MeV (1 MeV = 1.60×10^{-13} J).

22. (I) Calculate the rest mass of a proton in MeV/c^2.

23. (I) The total annual energy consumption in the United States is about 8×10^{19} J. How much mass would have to be converted to energy to fuel this need?

24. (II) How much energy can be obtained from conversion of 1.0 gram of mass? How much mass could this energy raise to a height of 100 m?

25. (II) Show that when the kinetic energy of a particle equals its rest energy, the speed of the particle is about $0.866c$.

26. (II) (a) How much work is required to accelerate a proton from rest up to a speed of $0.998c$? (b) What would be the momentum of this proton?

27. (II) (a) By how much does the mass of the Earth increase each year as a result solely of the sunlight reaching it? (b) How much mass does the Sun lose per year? (Radiation from the Sun reaches the Earth at a rate of about 1400 W/m^2 of area perpendicular to the energy flow.)

28. (II) Calculate the kinetic energy and momentum of a proton traveling 2.50×10^8 m/s.

29. (II) What is the momentum of a 750-MeV proton (that is, one with KE = 750 MeV)?

30. (II) What is the speed of a proton accelerated by a potential difference of 75 MV?

31. (II) What is the speed of an electron whose KE is 1.00 MeV?

32. (II) What is the speed and apparent rest mass of an electron when it hits a television screen after being accelerated by the 25,000 V of the picture tube?

33. (II) Two identical particles of rest mass m_0 approach each other at equal and opposite speeds, v. The collision is completely inelastic and results in a single particle at rest due to momentum conservation. What is the rest mass of the new particle? How much energy was lost in the collision? How much kinetic energy is lost in this collision?

34. (II) Calculate the mass of a proton ($m_0 = 1.67 \times 10^{-27}$ kg) whose kinetic energy is half its total energy. How fast is it traveling?

35. (II) What is the speed and momentum of an electron ($m_0 = 9.11 \times 10^{-31}$ kg) whose kinetic energy equals its rest energy?

36. (II) Suppose a spacecraft of rest mass 37,000 kg is accelerated to $0.21c$. (a) How much kinetic energy would it have? (b) If you used the classical formula for KE, by what percentage would you be in error?

37. (II) Calculate the kinetic energy and momentum of a proton ($m_0 = 1.67 \times 10^{-27}$ kg) traveling 9.8×10^7 m/s. By what percentages would your calculations have been in error if you had used classical formulas?

38. (II) The americium nucleus, $^{241}_{95}$Am, decays to a neptunium nucleus, $^{237}_{93}$Np, by emitting an alpha particle of mass 4.00260 u and kinetic energy 5.5 MeV. Estimate the mass of the neptunium nucleus, ignoring its recoil, given that the americium mass is 241.05682 u.

39. (II) An electron ($m_0 = 9.11 \times 10^{-31}$ kg) is accelerated from rest to speed v by a conservative force. In this process, its potential energy decreases by 7.60×10^{-14} J. Determine the electron's speed, v.

40. (II) Make a graph of the kinetic energy versus momentum for (a) a particle of nonzero rest mass, and (b) a particle with zero rest mass.

41. (II) What magnetic field intensity is needed to keep 900-GeV protons revolving in a circle of radius 1.0 km (at, say, the Fermilab synchrotron)? Use the relativistic mass. The proton's rest mass is $0.938\ \text{GeV}/c^2$. (1 GeV $= 10^9$ eV.)

42. (II) A negative muon traveling at 33 percent the speed of light collides head on with a positive muon traveling at 50 percent the speed of light. The two muons (each of mass $105.7\ \text{MeV}/c^2$) annihilate, and produce electromagnetic energy of what total amount?

43. (II) Show that the energy of a particle of charge e revolving in a circle of radius r in a magnetic field B is given by $E\ (\text{in eV}) = Brc$ in the relativistic limit $(v \approx c)$.

44. (III) Show that the kinetic energy (KE) of a particle of rest mass m_0 is related to its momentum p by the equation $p = \sqrt{(\text{KE})^2 + 2(\text{KE})(m_0 c^2)}/c$.

SECTION 26–11

45. (I) A person on a rocket traveling at $0.50c$ (with respect to the Earth) observes a meteor come from behind and pass her at a speed she measures as $0.50c$. How fast is the meteor moving with respect to the Earth?

46. (II) Two spaceships leave the Earth in opposite directions, each with a speed of $0.50c$ with respect to the Earth. (a) What is the velocity of spaceship 1 relative to spaceship 2? (b) What is the velocity of spaceship 2 relative to spaceship 1?

47. (II) An observer on Earth sees an alien vessel approach at a speed of $0.60c$. The *Enterprise* comes to the rescue (Fig. 26–14), overtaking the aliens while moving directly toward Earth at a speed of $0.90c$ relative to Earth. What is the relative speed of one vessel as seen by the other?

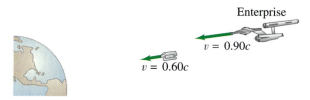

FIGURE 26–14 Problem 47.

48. (II) A spaceship leaves Earth traveling $0.65c$. A second spaceship leaves the first at a speed of $0.91c$ with respect to the first. Calculate the speed of the second ship with respect to Earth if it is fired (a) in the same direction the first spaceship is already moving, (b) directly backward toward Earth.

GENERAL PROBLEMS

49. As a rule of thumb, anything traveling faster than about $0.1c$ is called *relativistic*—i.e., for which the correction using special relativity is a significant effect. Is the electron in a hydrogen atom (radius 0.5×10^{-10} m) relativistic? (Treat the electron as though it were in a circular orbit around the proton.)

50. An atomic clock is taken to the North Pole, while another stays at the Equator. How far will they be out of synchronization after a year has elapsed?

51. The nearest star to Earth is Proxima Centauri, 4.3 light-years away. (a) At what constant velocity must a spacecraft travel from Earth if it is to reach the star in 4.0 years, as measured by travelers on the spacecraft? (b) How long does the trip take according to Earth observers?

52. Derive a formula showing how the density of an object changes with speed v relative to an observer.

53. An airplane travels 1500 km/h around the world, returning to the same place, in a circle of radius essentially equal to that of the Earth. Estimate the difference in time to make the trip as seen by Earth and airplane observers. [*Hint:* Use the binomial expansion, Appendix A.]

54. How many grams of matter would have to be totally destroyed to run a 100-W lightbulb for 1 year?

55. What minimum amount of electromagnetic energy is needed to produce an electron and a positron together? A positron is a particle with the same rest mass as an electron, but has the opposite charge. (Note that electric charge is conserved in this process. See Section 27–4.)

56. A 1.68-kg mass oscillates on the end of a spring whose spring constant is $k = 48.7$ N/m. If this system is in a spaceship moving past Earth at $0.900c$, what is its period of oscillation according to (a) observers on the ship, and (b) observers on Earth?

57. An electron ($m_0 = 9.11 \times 10^{-31}$ kg) enters a uniform magnetic field $B = 1.8$ T, and moves perpendicular to the field lines with a speed $v = 0.92c$. What is the radius of curvature of its path?

58. A free neutron can decay into a proton, an electron, and a neutrino. The neutrino's rest mass is zero, and the other masses can be found in the table inside the front cover. Determine the total kinetic energy shared among the three particles when a neutron decays at rest.

59. The Sun radiates energy at a rate of about 4×10^{26} W. (a) At what rate is the Sun's mass decreasing? (b) How long does it take for the Sun to lose a mass equal to that of Earth? (c) Estimate how long the Sun could last if it radiated constantly at this rate.

60. An unknown particle is measured to have a negative charge and a speed of 2.24×10^8 m/s. Its momentum is determined to be 3.07×10^{-22} kg·m/s. Identify the particle by finding its rest mass.

61. How much energy would be required to break a helium nucleus into its constituents, two protons and two neutrons? The rest masses of a proton (including an electron), a neutron, and helium are, respectively, 1.00783 u, 1.00867 u, and 4.00260 u. (This is called the *total binding energy* of the $^4_2$He nucleus.)

62. What is the percentage increase in the mass of a car traveling 110 km/h as compared to at rest?

63. Two protons, each having a speed of $0.935c$ in the laboratory, are moving toward each other. Determine (a) the momentum of each proton in the laboratory, (b) the total momentum of the two protons in the laboratory, and (c) the momentum of one proton as seen by the other proton.

64. A pi meson of rest mass m_π decays at rest into a muon (rest mass m_μ) and a neutrino of zero rest mass. Show that the kinetic energy of the muon is $\text{KE}_\mu = (m_\pi - m_\mu)^2 c^2 / 2m_\pi$.

65. A farm boy studying physics believes that he can fit a 15.0-m-long pole into a 12.0-m-long barn if he runs fast enough (carrying the pole). Can he do it? Explain in detail. How does this fit with the idea that when he is running the barn looks even shorter than 12.0 m?

66. Show analytically that a particle with momentum p and energy E has a speed given by

$$v = \frac{pc^2}{E} = \frac{pc}{\sqrt{m_0^2 c^2 + p^2}}.$$

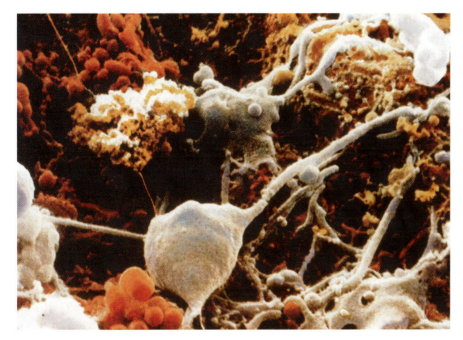

Electron microscopes produce images based on wave properties of electrons. Since the wavelength of electrons can be much smaller than that of visible light, much greater resolution and magnification can be obtained. The scanning electron microscope can produce images with a three-dimensional quality, as for these brain cells. The large grey objects are neurons. Magnification is about 2000×.

CHAPTER
27

EARLY QUANTUM THEORY AND MODELS OF THE ATOM

The second aspect of the revolution that shook the world of physics in the early part of the twentieth century (the first half was Einstein's theory of relativity) was the quantum theory. Unlike the special theory of relativity, the revolution of quantum theory required almost three decades to unfold, and many scientists contributed to its development. It began in 1900 with Planck's quantum hypothesis, and culminated in the mid-1920s with the theory of quantum mechanics of Schrödinger and Heisenberg which has been so effective in explaining the structure of matter. The discovery of the electron in the 1890s, with which we begin this chapter, might be said to mark the beginning of modern physics, and is a sort of precursor to the quantum theory.

27–1 Discovery and Properties of the Electron

Toward the end of the nineteenth century, studies were being done on the discharge of electricity through rarefied gases. One apparatus, diagrammed

823

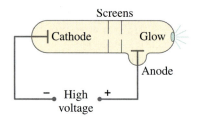

FIGURE 27–1 Discharge tube. In some models, one of the screens is the anode (positive plate).

in Fig. 27–1, was a glass tube fitted with electrodes and evacuated so only a small amount of gas remained inside. The negative electrode is called the *cathode*, and the positive one the *anode* (Section 17–10). When a very high voltage was applied to the electrodes, a dark space seemed to extend outward from the cathode toward the opposite end of the tube; and that far end of the tube would glow. If one or more screens containing a small hole were inserted as shown, the glow was restricted to a tiny spot on the end of the tube. It seemed as though something being emitted by the cathode traveled to the opposite end of the tube. These "somethings" were given the name **cathode rays**.

There was much discussion at the time about what these rays might be. Some scientists thought they might resemble light. But the observation that the bright spot at the end of the tube could be deflected to one side by an electric or magnetic field suggested that cathode rays could be charged particles; and the direction of the deflection was consistent with a negative charge. Furthermore, if the tube contained certain types of rarefied gas, the path of the cathode rays was made visible by a slight glow.

e/m measured

Estimates of the charge e of the (assumed) cathode-ray particles, as well as of their charge-to-mass ratio, e/m, had been made by 1897. But in that year, J. J. Thomson (1856–1940) was able to measure e/m directly, using the apparatus shown in Fig. 27–2. Cathode rays are accelerated by a high voltage and then pass between a pair of parallel plates built into the tube. The voltage applied to the plates produces an electric field, and a pair of coils produces a magnetic field. When only the electric field is present, say with the upper plate positive, the cathode rays are deflected upward as in path a in the figure. If only a magnetic field exists, say inward in the figure, the rays are deflected downward along path c. These observations are just what is expected for a negatively charged particle. The force on the rays due to the magnetic field B is (Eq. 20–4)

$$F = evB,$$

where e is the charge and v is the velocity of the cathode rays. In the absence

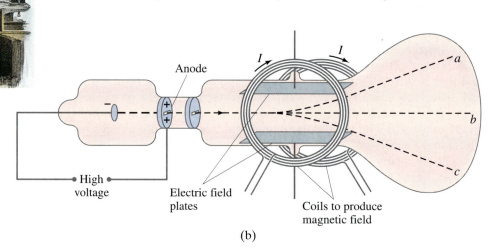

FIGURE 27–2 (a) J. J. Thomson and his cathode ray tube. (b) Cathode rays deflected by electric and magnetic fields.

of an electric field, the rays are bent into a curved path, so from $F = ma$, we have

$$evB = \frac{mv^2}{r},$$

and thus

$$\frac{e}{m} = \frac{v}{Br}.$$

The radius of curvature r can be measured, and so can B. The velocity v is found by applying an electric field in addition to the magnetic field. The electric field E is adjusted so that the cathode rays are undeflected and follow path b in Fig. 27–2. In this situation, the force due to the electric field, $F = eE$, is just balanced by the force due to the magnetic field, $F = evB$. Thus we have $eE = evB$ and

$$v = \frac{E}{B}.$$

Combining this with the above equation, we have

$$\frac{e}{m} = \frac{E}{B^2 r}. \tag{27–1}$$

The quantities on the right side can all be measured, so that although e and m could not be determined separately, the ratio e/m could be determined. The accepted value today is $e/m = 1.76 \times 10^{11}$ C/kg. Cathode rays soon came to be recognized as beams of particles, which we now call **electrons**.

It is worth noting that the "discovery" of the electron, like many others in science, is not quite so obvious as discovering gold or oil. Should the discovery of the electron be credited to the person who first saw a glow in the tube? Or to the person who first called them cathode rays? Perhaps neither one, for they had no conception of the electron as we know it today. In fact, the credit for the discovery is generally given to Thomson, but not because he was the first to see the glow in the tube. Rather it is because he believed that this phenomenon was due to tiny negatively charged particles and made careful measurements on them; furthermore he argued that these particles were constituents of atoms, and not ions or atoms themselves as many thought, and he developed an electron theory of matter. His view is close to what we accept today, and this is why Thomson is credited with the "discovery." Note, however, that neither he nor anyone else ever actually saw an electron itself. We discussed this, briefly, for it illustrates that discovery in science is not always a clear-cut matter. In fact, some philosophers of science think the word "discovery" is not always appropriate, such as in this case.

"Discovery" of the electron

Thomson believed that an electron was not an atom, but rather a constituent, or part, of an atom. Convincing evidence for this came soon with the determination of the charge and the mass of the cathode ray particles. Thomson's student, J. S. Townsend, made the first direct (but rough) measurements of e in 1897. But it was the more refined **oil-drop experiment** of Robert A. Millikan (1868–1953) that yielded a precise value for the charge on the electron and showed that electric charge comes in discrete amounts.

Millikan oil-drop experiment to determine e

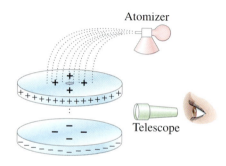

FIGURE 27–3 Millikan's oil-drop experiment.

Atomizer

Telescope

In this experiment, tiny droplets of mineral oil carrying an electric charge were allowed to fall under gravity between two parallel plates, Fig. 27–3. The electric field E between the plates was adjusted until the oil drop was suspended in midair. The downward pull of gravity, mg, was then just balanced by the upward force due to the electric field. Thus, $qE = mg$, so the charge $q = mg/E$. The mass of the droplet was determined by measuring its terminal velocity in the absence of the electric field (see Problem 4). Sometimes the drop was charged negatively and sometimes positively, suggesting that the drop had acquired or lost electrons (by friction, leaving the atomizer). Millikan's painstaking observations and analysis presented convincing evidence that any electric charge was an integral multiple of a smallest charge, e, that was ascribed to the electron, and that the value of e was 1.6×10^{-19} C. (Today's value of e, as mentioned in Chapter 16, is $e = 1.602 \times 10^{-19}$ C.)

This value for e, combined with the measurement of e/m (see above), gives the mass of the electron to be $(1.6 \times 10^{-19}$ C$)/(1.76 \times 10^{11}$ C/kg$) = 9.1 \times 10^{-31}$ kg. This mass is less than a thousandth the mass of the smallest atom, and thus confirmed the idea that the electron is only a part of an atom. The accepted value today for the mass of the electron is $m_e = 9.11 \times 10^{-31}$ kg. The experimental results that any charge seems to be an integral multiple of e means that electric charge is *quantized* (exists only in discrete amounts), as we discussed in Chapter 16.

27–2 Planck's Quantum Hypothesis

One of the observations that was unexplained at the end of the nineteenth century was the spectrum of light emitted by hot objects. We saw in Section 14–9 that all objects emit radiation whose total intensity is proportional to the fourth power of the Kelvin temperature (T^4). At normal temperatures, we are not aware of this electromagnetic radiation because of its low intensity. At higher temperatures, there is sufficient infrared radiation that we can feel heat if we are close to the object. At still higher temperatures (on the order of 1000 K), objects actually glow, such as a red-hot electric stove burner or the element in a toaster. At temperatures above 2000 K, objects glow with a yellow or whitish color, such as white-hot iron and the filament of a lightbulb. As the temperature increases, the electromagnetic radiation emitted by bodies is strongest at higher and higher frequencies.

Blackbody radiation

The spectrum of light emitted by a hot dense object is shown in Fig. 27–4 for an idealized **blackbody**. A blackbody is a body that would absorb all the radiation falling on it (and so would appear black under reflection when illuminated from outside). The radiation such a blackbody would emit when hot and luminous, called **blackbody radiation** (though not necessarily black in color), is the easiest to deal with, and the radiation approximates that of many real objects. As can be seen, the spectrum contains a continuous range of frequencies. Such a continuous spectrum is emitted by any heated solid or liquid, and even by dense gases. The 6000-K curve in Fig. 27–4, corresponding to the temperature of the surface of the Sun, peaks in the visible part of the spectrum. For lower temperatures, the total

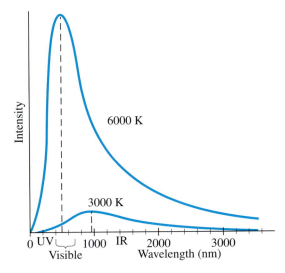

FIGURE 27-4 Spectrum of frequencies emitted by a blackbody at two different temperatures.

radiation drops considerably and the peak occurs at longer wavelengths. Hence the blue end of the visible spectrum (and the UV) is relatively weaker. (This is why objects glow with a red color at around 1000 K.) It is found that the wavelength at the peak of the spectrum, λ_P, is related to the Kelvin temperature T by

$$\lambda_P T = 2.90 \times 10^{-3}\ \text{m·K}. \qquad (27\text{-}2)$$

This is known as **Wien's law**.

EXAMPLE 27–1 **The Sun's surface temperature.** Estimate the temperature of the surface of our Sun, given that the Sun emits light whose peak is in the visible spectrum at around 500 nm.

SOLUTION Wien's law gives

$$T = \frac{2.90 \times 10^{-3}\ \text{m·K}}{\lambda_P} = \frac{2.90 \times 10^{-3}\ \text{m·K}}{500 \times 10^{-9}\ \text{m}} \approx 6000\ \text{K}.$$

EXAMPLE 27–2 **Star color.** Suppose a star has a surface temperature of 32,500 K. What color would this star appear?

SOLUTION From Wien's law we have,

$$\lambda_P = \frac{2.90 \times 10^{-3}\ \text{m·K}}{T} = \frac{2.90 \times 10^{-3}\ \text{m·K}}{3.25 \times 10^{4}\ \text{K}} = 89.2\ \text{nm}.$$

The peak is in the UV range of the spectrum. In the visible region, the curve will be descending (see Fig. 27–4), so the shortest visible wavelengths will be strongest. Hence the star will appear bluish (or blue-white).

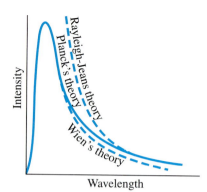

FIGURE 27–5 Comparison of the Wien and the Rayleigh–Jeans theories to that of Planck, which closely follows experiment.

A major problem facing scientists in the 1890s was to explain black-body radiation. Maxwell's electromagnetic theory had predicted that oscillating electric charges produce electromagnetic waves, and the radiation emitted by a hot object could be due to the oscillations of electric charges in the molecules of the material. Although this would explain where the radiation came from, it did not correctly predict the observed spectrum of emitted light. Two important theoretical curves based on classical ideas were those proposed by W. Wien (in 1896) and by Lord Rayleigh (in 1900). The latter was slightly modified later by J. Jeans and since then has been known as the Rayleigh–Jeans theory. As experimental data came in, it became clear that neither Wien's nor the Rayleigh–Jeans formulations were in accord with experiment. Wien's was accurate at short wavelengths but deviated from experiment at longer wavelengths, whereas the reverse was true for the Rayleigh–Jeans theory (see Fig. 27–5).

The break came in late 1900 when Max Planck (1858–1947; Fig. 27–6) proposed an empirical formula that nicely fit the data. He then sought a theoretical basis for the formula and within two months found that he could obtain the formula by making a new and radical (though not so recognized at the time) assumption: that the energy distributed among the oscillating electric charges of the molecules is not continuous, but instead consists of a finite number of very small discrete amounts, each related to the frequency of oscillation by

$$E_{\min} = hf.$$

Here h is a constant, now called **Planck's constant**, whose value was estimated by Planck by fitting his formula for the blackbody radiation curve to experiment. The value accepted today is

$$h = 6.626 \times 10^{-34} \, \text{J·s}.$$

Planck's assumption suggests that the energy of any molecular vibration could be only some whole number multiple of hf:

Planck's quantum hypothesis

$$E = nhf, \qquad n = 1, 2, 3, \cdots. \tag{27–3}$$

This idea is often called **Planck's quantum hypothesis** ("quantum" means "fixed amount"), although little attention was brought to this point at the time. In fact, it appears that Planck considered it more as a mathematical device to get the "right answer" rather than as a discovery comparable to those of Newton. Planck himself continued to seek a classical explanation for the introduction of h. The recognition that this was an important and radical innovation did not come until later, after about 1905 when others, particularly Einstein, entered the field.

FIGURE 27–6 Max Planck.

The quantum hypothesis, Eq. 27–3, states that the energy of an oscillator can be $E = hf$, or $2hf$, or $3hf$, and so on, but there cannot be vibrations whose energy lies between these values. That is, energy would not be a continuous quantity as had been believed for centuries; rather it is **quantized**—it exists only in discrete amounts. The smallest amount of energy possible (hf) is called the **quantum of energy**. Another way of expressing the quantum hypothesis is that not just any amplitude of vibration is possible. The possible values for the amplitude are related to the frequency f.

A simple analogy may help. A stringed instrument such as a violin or guitar can be played over a continuous range of frequencies by moving

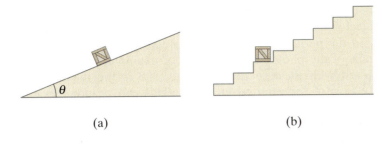

(a)   (b)

your finger along the string. A flute or piano, on the other hand, is "quantized" in the sense that only certain frequencies (notes) can be played. Or compare a ramp, on which a box can be placed at any height, to a flight of stairs on which the box can have only certain discrete amounts of potential energy, as shown in Fig. 27–7.

27–3 Photon Theory of Light and the Photoelectric Effect

In 1905, the same year that he introduced the special theory of relativity, Einstein made a bold extension of the quantum idea by proposing a new theory of light. Planck's work had suggested that the vibrational energy of molecules in a radiating object is quantized with energy $E = nhf$, where n is an integer. Einstein argued that therefore, when light is emitted by a molecular oscillator, its energy of nhf must decrease by an amount hf (or by $2hf$, etc.) to another integer times hf, namely $(n - 1)hf$. Then to conserve energy, the light ought to be emitted in packets or quanta, each with an energy

$$E = hf. \qquad (27\text{–}4)$$

Again h is Planck's constant. Since all light ultimately comes from a radiating source, this suggests that perhaps *light is transmitted as tiny particles*, or **photons**, as they are now called, rather than as waves. This, too, was a radical departure from classical ideas. Einstein proposed a test of the quantum theory of light: quantitative measurements on the photoelectric effect.

The **photoelectric effect** is the phenomenon that when light shines on a metal surface, electrons are emitted from the surface. (The photoelectric effect occurs in other materials, but is most easily observed with metals.) It can be observed using the apparatus shown in Fig. 27–8. A metal plate P and a smaller electrode C are placed inside an evacuated glass tube, called a **photocell**. The two electrodes are connected to an ammeter and a source of emf, as shown. When the photocell is in the dark, the ammeter reads zero. But when light of sufficiently high frequency is shone on the plate, the ammeter indicates a current flowing in the circuit. To explain completion of the circuit, we can imagine electrons, ejected by the impinging radiation, flowing across the tube from the plate to the "collector" C as shown in the diagram.

That electrons should be emitted when light shines on a metal is consistent with the electromagnetic (EM) wave theory of light: the electric

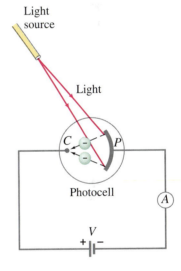

FIGURE 27–8   The photoelectric effect.

Photoelectric effect

field of an EM wave could exert a force on electrons in the metal and eject some of them. Einstein pointed out, however, that the wave theory and the photon theory of light give very different predictions on the details of the photoelectric effect. For example, one thing that can be measured with the apparatus of Fig. 27–8 is the maximum kinetic energy (KE_{max}) of the emitted electrons. This can be done by using a variable voltage source and reversing the terminals so that electrode C is negative and P is positive. The electrons emitted from P will be repelled by the negative electrode, but if this reverse voltage is small enough, the fastest electrons will still reach C and there will be a current in the circuit. If the reversed voltage is increased, a point is reached where the current reaches zero—no electrons have sufficient kinetic energy to reach C. This is called the *stopping potential*, or *stopping voltage*, V_0, and from its measurement, KE_{max} can be determined using conservation of energy (loss of KE = gain in PE):

$$\text{KE}_{max} = eV_0.$$

Now let us examine the details of the photoelectric effect from the point of view of the wave theory versus Einstein's particle theory. First the wave theory, assuming monochromatic light. The two important properties of a light wave are its intensity and its frequency (or wavelength). When these two quantities are varied, the wave theory makes the following predictions:

Wave

theory

predictions

1. If the light intensity is increased, the number of electrons ejected and their maximum KE should be increased because the higher intensity means a greater electric field amplitude, and the greater electric field should eject electrons with higher speed.

2. The frequency of the light should not affect the KE of the ejected electrons. Only the intensity should affect KE_{max}.

The photon theory makes completely different predictions. First we note that in a monochromatic beam, all photons have the same energy ($=hf$). Increasing the intensity of the light beam means increasing the number of photons in the beam, but does not affect the energy of each photon as long as the frequency is not changed. According to Einstein's theory, an electron is ejected from the metal by a collision with a single photon. In the process, all the photon energy is transferred to the electron and the photon ceases to exist. Since electrons are held in the metal by attractive forces, some minimum energy W_0 (called the **work function**, which is on the order of a few electron volts for most metals) is required just to get an electron out through the surface. If the frequency f of the incoming light is so low that hf is less than W_0, then the photons will not have enough energy to eject any electrons at all. If $hf > W_0$, then electrons will be ejected and energy will be conserved in the process. That is, the input energy (of the photon), hf, will equal the outgoing KE of the electron plus the energy required to get it out of the metal, W:

$$hf = \text{KE} + W. \tag{27–5a}$$

For the least tightly held electrons, W is the work function W_0, and KE in this equation becomes KE_{max}:

$$hf = \text{KE}_{max} + W_0. \qquad \textbf{(27–5b)}$$

Many electrons will require more energy than the bare minimum (W_0) to get out of the metal, and thus the KE of such electrons will be less than the maximum.

From these considerations, the photon theory makes the following predictions:

1. An increase in intensity of the light beam means more photons are incident, so more electrons will be ejected; but since the energy of each photon is not changed, the maximum KE of electrons is not changed.

2. If the frequency of the light is increased, the maximum KE of the electrons increases linearly, according to Eq. 27–5b. That is,

$$\text{KE}_{max} = hf - W_0.$$

This relationship is plotted in Fig. 27–9.

3. If the frequency f is less than the "cutoff" frequency f_0, where $hf_0 = W_0$, no electrons will be ejected at all, no matter how great the intensity.

These predictions of the photon theory are clearly very different from the predictions of the wave theory. In 1913–1914, careful experiments were carried out by R. A. Millikan. The results were fully in agreement with Einstein's photon theory.

Photon

theory

predictions

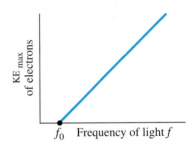

FIGURE 27–9 Photoelectric effect: maximum kinetic energy of ejected electrons increases linearly with frequency of incident light. No electrons are emitted if $f < f_0$.

EXAMPLE 27–3 **Photon energy.** Calculate the energy of a photon of blue light, $\lambda = 450\ \text{nm}$.

SOLUTION Since $f = c/\lambda$, we have

$$E = hf = \frac{hc}{\lambda} = \frac{(6.63 \times 10^{-34}\ \text{J·s})(3.0 \times 10^8\ \text{m/s})}{(4.5 \times 10^{-7}\ \text{m})} = 4.4 \times 10^{-19}\ \text{J},$$

or $(4.4 \times 10^{-19}\ \text{J})/(1.6 \times 10^{-19}\ \text{J/eV}) = 2.7\ \text{eV}$.

EXAMPLE 27–4 **ESTIMATE** **Photons from a lightbulb.** Estimate how many visible light photons a 100-W lightbulb emits per second.

SOLUTION Let's assume an average wavelength in the middle of the visible spectrum, $\lambda \approx 500\ \text{nm}$. The energy emitted in one second ($= 100\ \text{J}$) is $E = nhf$ when n is the number of photons emitted per second and $f = c/\lambda$. Hence

$$n = \frac{E}{hf} = \frac{E\lambda}{hc} = \frac{(100\ \text{J})(500 \times 10^{-9}\ \text{m})}{(6.63 \times 10^{-34}\ \text{J·s})(3.0 \times 10^8\ \text{m/s})} = 2.5 \times 10^{20}.$$

This is an overestimate since much of the 100 J of electric energy input is transformed into heat rather than light. If the efficiency is between 1 percent and 10 percent, then the number of photons emitted is on the order of 10^{19}, still an enormous number.

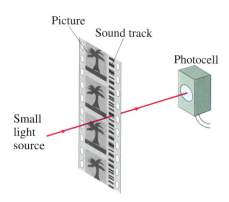

Picture

Sound track

Photocell

Small
light
source

FIGURE 27–10 Optical sound track on movie film. In the projector, light from a small source (different from that for the picture) passes through the sound track on the moving film. The light and dark areas on the sound track vary the intensity of the transmitted light which reaches the photocell, whose current output is then a replica of the original sound. This output is amplified and sent to the loudspeakers. High-quality projectors can show movies containing several parallel sound tracks to go to different speakers around the theater.

➡ **PHYSICS APPLIED**

Alarms,
door openers,
smoke detectors,
light meters,
film soundtrack

➡ **PHYSICS APPLIED**

Photosynthesis

EXAMPLE 27–5 **Photoelectron speed and energy.** What is the maximum kinetic energy and speed of an electron ejected from a sodium surface whose work function is $W_0 = 2.28\ \text{eV}$ when illuminated by light of wavelength: (a) 410 nm; (b) 550 nm?

SOLUTION (a) For $\lambda = 410\ \text{nm}$,

$$hf = \frac{hc}{\lambda} = 4.85 \times 10^{-19}\ \text{J} \qquad \text{or} \qquad 3.03\ \text{eV}.$$

From Eq. 27–5b, $\text{KE}_{\text{max}} = 3.03\ \text{eV} - 2.28\ \text{eV} = 0.75\ \text{eV}$, or $1.2 \times 10^{-19}\ \text{J}$. Since $\text{KE} = \frac{1}{2}mv^2$ where $m = 9.1 \times 10^{-31}\ \text{kg}$,

$$v = \sqrt{2\,\text{KE}/m} = 5.1 \times 10^5\ \text{m/s}.$$

Notice that we used the nonrelativistic equation for KE. If v had turned out to be more than about $0.1c$, our calculation would have been inaccurate by more than a percent or so, and we would probably prefer to redo it using the relativistic form (Eq. 26–7).
(b) For $\lambda = 550\ \text{nm}$, $hf = 3.60 \times 10^{-19}\ \text{J} = 2.25\ \text{eV}$. Since this photon energy is less than the work function, no electrons are ejected.

The photoelectric effect, besides playing an important historical role in confirming the photon theory of light, also has many practical applications. Burglar alarms and automatic door openers often make use of the photocell circuit of Fig. 27–8. When a person interrupts the beam of light, the sudden drop in current in the circuit activates a switch—often a solenoid—which operates a bell or opens the door. UV or IR light is sometimes used in burglar alarms because of its invisibility. Many smoke detectors use the photoelectric effect to detect tiny amounts of smoke that interrupt the flow of light and so alter the electric current. Photographic light meters use this circuit as well. Photocells are used in many other devices, such as absorption spectrophotometers, to measure light intensity. One type of film sound track is a variably shaded narrow section at the side of the film. Light passing through the film is thus "modulated," and the output electrical signal of the photocell detector follows the frequencies on the sound track. See Fig. 27–10. For many applications today, the vacuum-tube photocell of Fig. 27–8 has been replaced by a semiconductor device known as a **photodiode**. In these semiconductors, the absorption of a photon liberates a bound electron, which changes the conductivity of the material, so the current through a photodiode is altered.

The photon theory of light is useful in biology and medicine. Here is an Example.

EXAMPLE 27–6 **Photosynthesis.** In **photosynthesis**, which is the process by which pigments such as chlorophyll in plants capture the energy of sunlight to change CO_2 to useful carbohydrate, about nine photons are needed to transform one molecule of CO_2 to carbohydrate and O_2. Assuming light of wavelength $\lambda = 670\ \text{nm}$ (chlorophyll absorbs most strongly in the range 650 nm to 700 nm), how efficient is the photosynthetic process? The reverse chemical reaction has a heat of combustion of 4.9 eV/molecule of CO_2.

SOLUTION The energy of nine photons, each of energy $hf = hc/\lambda$, is $(9)(6.6 \times 10^{-34}\ \text{J·s})(3.0 \times 10^8\ \text{m/s})/(6.7 \times 10^{-7}\ \text{m}) = 2.7 \times 10^{-18}\ \text{J}$, or 17 eV. Thus the process is $(4.9\ \text{eV}/17\ \text{eV}) = 29$ percent efficient.

27–4 Photon Interactions; Compton Effect and Pair Production

A number of other experiments were carried out in the early twentieth century that also supported the photon theory. One of these was the **Compton effect** (1923), named after its discoverer, A. H. Compton (1892–1962). Compton scattered short-wavelength light (actually X-rays) from various materials. He found the scattered light had a slightly lower frequency than did the incident light, indicating a loss of energy. This finding, he showed, could be explained on the basis of the photon theory, as incident photons colliding with electrons of the material, Fig. 27–11. He applied the laws of conservation of energy and momentum to such collisions and found that the predicted energies of scattered photons was in accord with experimental results.

To analyze the Compton effect (and other photon interactions), we must recognize that the photon is truly a relativistic particle—it travels at the speed of light. Thus we must use relativistic formulas for dealing with its mass, energy, and momentum. The mass m of any particle is given by $m = m_0/\sqrt{1 - v^2/c^2}$. Since $v = c$ for a photon, the denominator is zero. So the rest mass, m_0, of a photon must also be zero, or its energy $E = mc^2$ would be infinite. Of course, a photon is never at rest. The momentum of a photon, from Eq. 26–8 with $m_0 = 0$, is $E^2 = p^2c^2$, or

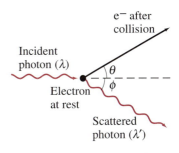

FIGURE 27–11 The Compton effect. A single photon of wavelength λ strikes an electron in some material, knocking it out of its atom. The scattered photon has less energy (since some is given to the electron) and hence has a longer wavelength λ'. Experiments found scattered X-rays of just the wavelengths predicted by conservation of energy and momentum using the photon model.

$$p = \frac{E}{c}.$$

Since $E = hf$, the momentum of a photon is related to its wavelength by

$$p = \frac{hf}{c} = \frac{h}{\lambda}. \tag{27–6}$$

Using Eq. 27–6 for momentum of a photon, Compton applied the laws of conservation of momentum and energy to the collision of Fig. 27–11 and derived the following equation for the wavelength of the scattered photons:

$$\lambda' = \lambda + \frac{h}{m_0 c}(1 - \cos \phi), \tag{27–7}$$

where m_0 is the rest mass of the electron. (The quantity h/m_0c, which has the dimensions of length, is called the **Compton wavelength** of the electron.) We see that the wavelength of scattered photons depends on the angle ϕ at which they are detected. Compton's measurements of 1923 were consistent with this formula. The wave theory of light predicts no such shift: an incoming EM wave of frequency f should set electrons into oscillation at frequency f; and such oscillating electrons would reemit EM waves of this same frequency f (Section 22–3), which would not change with angle (ϕ). Hence the Compton effect adds to the firm experimental foundation for the photon theory of light.

EXAMPLE 27–7 **X-ray scattering.** X-rays of wavelength 0.140 nm are scattered from a block of carbon. What will be the wavelengths of X-rays scattered at (a) 0°, (b) 90°, (c) 180°?

SOLUTION (a) For $\phi = 0°$, $\cos\phi = 1$, and Eq. 27–7 gives $\lambda' = \lambda = 0.140$ nm. This makes sense since for $\phi = 0°$, there really isn't any collision as the photon goes straight through without interacting.
(b) For $\phi = 90°$, $\cos\phi = 0$, so

$$\lambda' = \lambda + \frac{h}{m_0 c} = 0.140 \text{ nm} + \frac{6.63 \times 10^{-34} \text{ J·s}}{(9.11 \times 10^{-31} \text{ kg})(3.00 \times 10^8 \text{ m/s})}$$

$$= 0.140 \text{ nm} + 2.4 \times 10^{-12} \text{ m} = 0.142 \text{ nm};$$

that is, the wavelength is longer by one Compton wavelength ($= 0.0024$ nm for an electron).
(c) For $\phi = 180°$, which means the photon is scattered backward, returning in the direction from which it came (a direct "head-on" collision), $\cos\phi = -1$, so

$$\lambda' = \lambda + 2\frac{h}{m_0 c} = 0.140 \text{ nm} + 2(0.0024 \text{ nm}) = 0.145 \text{ nm}.$$

When a photon passes through matter, it interacts with the atoms and electrons. There are four important types of interactions that a photon can undergo:

Photon

interactions

1. The photon can be scattered off an electron (or a nucleus) and in the process lose some energy; this is the *Compton effect* (Fig. 27–11). But notice that the photon is not slowed down. It still travels with speed c, but its frequency will be lower.
2. The *photoelectric effect*: a photon may knock an electron out of an atom and in the process itself disappear.
3. The photon may knock an atomic electron to a higher energy state in the atom if its energy is not sufficient to knock the electron out altogether. In this process the photon also disappears, and all its energy is given to the atom. Such an atom is then said to be in an *excited state*, and we shall discuss this more later.
4. *Pair production*: A photon can actually create matter, such as the production of an electron and a positron, Fig. 27–12. (A positron has the same mass as an electron, but the opposite charge, $+e$.)

Pair production

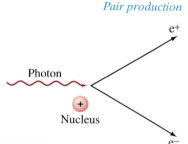

FIGURE 27–12 Pair production: a photon disappears and produces an electron and a positron.

This last process is called **pair production**, and the photon disappears in the process of creating the electron-positron pair. This is an example of rest mass being created from pure energy, and it occurs in accord with Einstein's equation $E = mc^2$. Notice that a photon cannot create an electron alone since electric charge would not then be conserved. The inverse of pair production also occurs: if an electron collides with a positron, the two **annihilate** each other and their energy, including their mass, appears as electromagnetic energy of photons. Because of this process, positrons usually do not last long in nature.

EXAMPLE 27–8 **Pair production.** What is (a) the minimum energy of a photon that can produce an electron–positron pair? (b) What is this photon's wavelength?

SOLUTION (a) Because $E = mc^2$, and the mass created is equal to two electron masses, the photon must have energy

$$E = 2(9.11 \times 10^{-31}\,\text{kg})(3.0 \times 10^8\,\text{m/s})^2 = 1.64 \times 10^{-13}\,\text{J}$$
$$= 1.02\,\text{MeV}.$$

A photon with less energy cannot undergo pair production.

(b) Since $E = hf = hc/\lambda$, the wavelength of a 1.02-MeV photon is

$$\lambda = \frac{hc}{E} = \frac{(6.6 \times 10^{-34}\,\text{J}\cdot\text{s})(3.0 \times 10^8\,\text{m/s})}{(1.64 \times 10^{-13}\,\text{J})} = 1.2 \times 10^{-12}\,\text{m},$$

which is 0.0012 nm. Such photons are in the gamma-ray (or very short X-ray) region of the electromagnetic spectrum.

Pair production cannot occur in empty space, for energy and momentum could not simultaneously be conserved. In Example 27–8, for instance, energy is conserved, but the electron–positron pair would be created with no momentum to carry away the initial momentum of the photon. Indeed, it can be shown that at any energy, an additional massive object, such as an atomic nucleus, must take part in the interaction to carry off some of the momentum.

The Compton effect has been used to diagnose bone disease such as osteoporosis. Gamma rays, which are photons of even shorter wavelength than X-rays, coming from a radioactive source are scattered off bone material. The total intensity of the scattered radiation is proportional to the density of electrons which is in turn proportional to the bone density. Changes in the density of bone material can indicate the onset of osteoporosis.

27–5 Wave–Particle Duality; the Principle of Complementarity

The photoelectric effect, the Compton effect, and other experiments (see, for example, Section 28–9 on X-rays) have placed the particle theory of light on a firm experimental basis. But what about the classic experiments of Young and others (Chapter 24) on interference and diffraction which showed that the wave theory of light also rests on a firm experimental basis?

We seem to be in a dilemma. Some experiments indicate that light behaves like a wave; others indicate that it behaves like a stream of particles. These two theories seem to be incompatible, but both have been shown to have validity. Physicists have finally come to the conclusion that this duality of light must be accepted as a fact of life. It is referred to as the **wave–particle duality**. Apparently, light is a more complex phenomenon than just a simple wave or a simple beam of particles.

To clarify the situation, the great Danish physicist Niels Bohr (1885–1962, Fig. 27–13) proposed his famous **principle of complementarity**. It states that to understand any given experiment, we must use either the wave or the photon theory, but not both. Yet we must be aware of both the wave and particle aspects of light if we are to have a full understanding of light. Therefore these two aspects of light complement one another.

It is not possible to "visualize" this duality. We cannot picture a combination of wave and particle. Instead, we must recognize that the two aspects of light are different "faces" that light shows to experimenters.

FIGURE 27–13 Niels Bohr, walking with Enrico Fermi along the Appian Way outside Rome.

Wave–particle duality

Principle of complementarity

Part of the difficulty stems from how we think. Visual pictures (or models) in our minds are based on what we see in the everyday world. We apply the concepts of waves and particles to light because in the macroscopic world we see that energy is transferred from place to place by these two methods. We cannot see directly whether light is a wave or particle—so we do indirect experiments. To explain the experiments, we apply the models of waves or of particles to the nature of light. But these are abstractions of the human mind. When we try to conceive of what light really "is," we insist on a visual picture. Yet there is no reason why light should conform to these models (or visual images) taken from the macroscopic world. The "true" nature of light—if that means anything—is not possible to visualize. The best we can do is recognize that our knowledge is limited to the indirect experiments, and that in terms of everyday language and images, light reveals both wave and particle properties.

It is worth noting that Einstein's equation $E = hf$ itself links the particle and wave properties of a light beam. In this equation, E refers to the energy of a particle; and on the other side of the equation, we have the frequency f of the corresponding wave.

27–6 | Wave Nature of Matter

In 1923, Louis de Broglie (1892–1987; Fig. 27–14) extended the idea of the wave–particle duality. He sensed deeply the symmetry in nature and argued that if light sometimes behaves like a wave and sometimes like a particle, then perhaps those things in nature thought to be particles—such as electrons and other material objects—might also have wave properties. De Broglie proposed that the wavelength of a material particle would be related to its momentum in the same way as for a photon,[†] Eq. 27–6, $p = h/\lambda$. That is, for a particle of mass m traveling with speed v, the wavelength λ is given by

de Broglie wavelength

$$\lambda = \frac{h}{mv}.$$ (27–8)

This is sometimes called the **de Broglie wavelength** of a particle.

EXAMPLE 27–9   Wavelength of a ball. Calculate the de Broglie wavelength of a 0.20-kg ball moving with a speed of 15 m/s.

SOLUTION $\lambda = \dfrac{h}{mv} = \dfrac{(6.6 \times 10^{-34} \text{ J·s})}{(0.20 \text{ kg})(15 \text{ m/s})} = 2.2 \times 10^{-34}$ m.

This is an incredibly small wavelength. Even if the speed were extremely small, say 10^{-4} m/s, the wavelength would be about 10^{-29} m. Indeed, the wavelength of any ordinary object is much too small to be measured and detected. The problem is that the properties of waves, such as interference and diffraction, are significant only when the size of objects or slits is not much larger than the wavelength. And there are no known objects or slits to diffract waves only 10^{-30} m long, so the wave properties of ordinary objects go undetected.

But tiny elementary particles, such as electrons, are another matter. Since the mass m appears in the denominator in Eq. 27–8, a very small mass should have a much larger wavelength.

FIGURE 27–14
Louis de Broglie.

[†]De Broglie chose this formula (rather than, say, $E = hf$, which is not consistent with $p = h/\lambda$ for a particle with non-zero rest mass), because it allowed him to explain, or give a reason for, Bohr's model of the atom. This will be discussed in Section 27–11.

EXAMPLE 27–10 **Wavelength of an electron.** Determine the wavelength of an electron that has been accelerated through a potential difference of 100 V.

SOLUTION We assume that the speed of the electron will be much less than c, so we use nonrelativistic mechanics. (If this assumption were to come out wrong, we would have to recalculate using relativistic formulas—see Section 26–10.) The gain in KE will equal the loss in PE, so $\frac{1}{2}mv^2 = eV$ and

$$v = \sqrt{2eV/m} = \sqrt{(2)(1.6 \times 10^{-19}\,\text{C})(100\,\text{V})/(9.1 \times 10^{-31}\,\text{kg})}$$

$$= 5.9 \times 10^6\,\text{m/s}.$$

Then

$$\lambda = \frac{h}{mv} = \frac{(6.6 \times 10^{-34}\,\text{J·s})}{(9.1 \times 10^{-31}\,\text{kg})(5.9 \times 10^6\,\text{m/s})} = 1.2 \times 10^{-10}\,\text{m},$$

or 0.12 nm.

From this Example, we see that electrons can have wavelengths on the order of 10^{-10} m. Although small, this wavelength can be detected: the spacing of atoms in a crystal is on the order of 10^{-10} m and the orderly array of atoms in a crystal could be used as a type of diffraction grating, as was done earlier for X-rays (see Section 25–11). C. J. Davisson and L. H. Germer performed the crucial experiment; they scattered electrons from the surface of a metal crystal and, in early 1927, observed that the electrons were scattered into a pattern of regular peaks. When they interpreted these peaks as a diffraction pattern, the wavelength of the diffracted electron wave was found to be just that predicted by de Broglie, Eq. 27–8. In the same year, G. P. Thomson (son of J. J. Thomson, who is credited with the discovery of the particle nature of electrons as we saw in Section 27–1), using a different experimental arrangement, also detected diffraction of electrons. (See Fig. 27–15.) Later experiments showed that protons, neutrons, and other particles also have wave properties.

Thus the wave–particle duality applies to material objects as well as to light. The principle of complementarity applies to matter as well. That is, we must be aware of both the particle and wave aspects in order to have an understanding of matter, including electrons. But again we must recognize that a visual picture of a "wave–particle" is not possible.

We might ask ourselves: "What is an electron?" The early experiments of J. J. Thomson (see Section 27–1) indicated a glow in a tube that moved when a magnetic field was applied. The results of these and other experiments were best interpreted as being caused by tiny negatively charged particles which we now call electrons. No one, however, has actually seen an electron directly. The drawings we sometimes make of electrons as tiny spheres with a negative charge on them are merely convenient pictures (now recognized to be inaccurate). Again we must rely on experimental results, some of which are best interpreted using the particle model and others using the wave model. These models are mere pictures that we use to extrapolate from the macroscopic world to the tiny microscopic world of the atom. And there is no reason to expect that these models somehow reflect the reality of an electron. We thus use a wave or a particle model (whichever works best in a situation) so that we can talk about what is happening. But we shouldn't be led to believe that an electron *is* a wave or

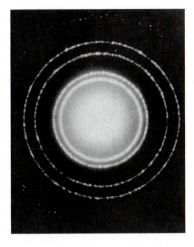

FIGURE 27–15 Diffraction pattern of electrons scattered from aluminum foil, as recorded on film.

Wave-particle duality and the complementarity principle apply also to matter

What is an electron?

a particle. Instead, we could say that an electron is the set of its properties that we can measure. Bertrand Russell said it well when he wrote that an electron is "a logical construction".

* 27–7 # Electron Microscopes

➡ **PHYSICS APPLIED**

Electron microscope

The idea that electrons have wave properties led to the development of the **electron microscope**, which can produce images of much greater magnification than a light microscope. Figures 27–16 and 27–17 are diagrams of two types, the **transmission electron microscope**, which produces a two-dimensional image, and the **scanning electron microscope**, which produces images with a three-dimensional quality. In each design, the objective and eyepiece lenses are actually magnetic fields that exert forces on the electrons to bring them to a focus. The fields are produced by carefully designed current-carrying coils of wire. Photographs using each type are shown in Fig. 27–18.

As discussed in Sections 25–7 and 25–8, the best resolution of details on an object is on the order of the wavelength of the radiation used to view it. Electrons accelerated by voltages on the order of 10^5 V have wavelengths on the order of 0.004 nm. The maximum resolution obtainable would be on this order, but in practice, aberrations in the magnetic lenses limit the resolution in transmission electron microscopes to at best about 0.2 to 0.5 nm. This is still 10^3 times finer than that attainable

FIGURE 27–17 Scanning electron microscope. Scanning coils move an electron beam back and forth across the specimen. Secondary electrons produced when the beam strikes the specimen are collected and modulate the intensity of the beam in the CRT to produce a picture.

FIGURE 27–16 Transmission electron microscope. The magnetic-field coils are designed to be "magnetic lenses," which bend the electron paths and bring them to a focus, as shown.

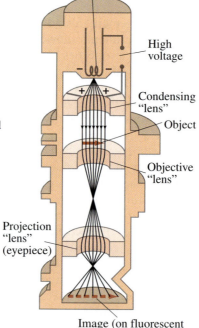

Hot filament (source of electrons)
High voltage
Condensing "lens"
Object
Objective "lens"
Projection "lens" (eyepiece)
Image (on fluorescent screen or film)

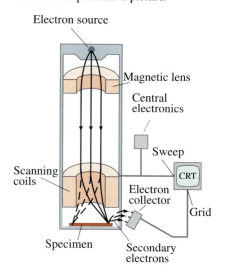

Electron source
Magnetic lens
Central electronics
Sweep
CRT
Scanning coils
Electron collector
Grid
Specimen
Secondary electrons

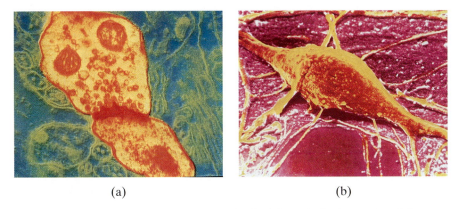

(a)   (b)

FIGURE 27–18 Electron micrographs (in false color) of neurons of the human cerebral cortex: (a) transmission electron micrograph of synapse (junction) between two neurons ($\approx 40{,}000\times$); (b) scanning electron micrograph of a single neuron ($\approx 4000\times$).

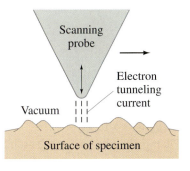

FIGURE 27–19 Probe tip of scanning tunneling electron microscope moves up and down to maintain constant tunneling current, producing an image of the surface.

with a light microscope, and corresponds to a useful magnification of about a million. Such magnifications are difficult to attain, and more common magnifications are 10^4 to 10^5. The maximum resolution attainable with a scanning electron microscope is somewhat less, about 5 to 10 nm at best.

The **scanning tunneling electron microscope** (STM), developed in the 1980s, contains a tiny probe, whose tip may be only one (or a few) atoms wide, that is moved across the specimen to be examined in a series of linear passes, like those made by the electron beam in a TV tube (CRT, Section 17–10). The tip, as it scans, remains very close to the surface of the specimen, about 1 nm above it, Fig. 27–19. A small voltage applied between the probe and the surface causes electrons to leave the surface and pass through the vacuum to the probe, by a process known as *tunneling* (discussed in Section 30–12). This "tunneling" current is very sensitive to the gap width, so that a feedback mechanism can be used to raise and lower the probe to maintain a constant electron current. The probe's vertical motion, following the surface of the specimen, is then plotted as a function of position, producing a three-dimensional image of the surface. (See, for example, Fig. 27–20.) Surface features as fine as the size of an atom can be resolved: a resolution better than 0.1 nm laterally and 10^{-2} to 10^{-3} nm vertically. This kind of resolution was not available previously and has given a great impetus to the study of the surface structure of materials. The "topographic" image of a surface actually represents the distribution of electron charge.

The new **atomic force microscope** (AFM) is in many ways similar to an STM, but can be used on a wider range of sample materials. Instead of detecting an electric current, the AFM measures the force between a cantilevered tip and the sample, a force which depends strongly on the tip-sample separation at each point. The tip is moved as for the STM.

➡ **PHYSICS APPLIED**

STM and AFM

FIGURE 27–20 Image of cellular DNA, magnified about 2 million times, taken with a scanning tunneling microscope. Three turns of the DNA double helix can be seen in this false-color image.

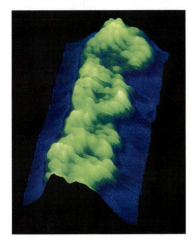

|← ≈10^{-10}m →|

FIGURE 27–21
Plum-pudding model of the atom.

FIGURE 27–22
Ernest Rutherford (left), with
J. J. Thomson.

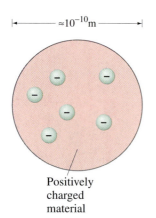

FIGURE 27–23
(a) Experimental setup for
Rutherford's experiment: α
particles emitted by radium
strike metallic foil and some
rebound backward; (b) backward
rebound of α particles explained
as repulsion from heavy
positively charged nucleus.

Early Models of the Atom

The idea that matter is made up of atoms was accepted by most scientists
by 1900. With the discovery of the electron in the 1890s, scientists began to
think of the atom itself as having a structure and electrons as part of that
structure. We now trace, in the remainder of this chapter and the next, the
development of our modern understanding of the atom, and of the quantum theory with which it is intertwined.[†]

A typical model of the atom in the 1890s visualized the atom as a homogeneous sphere of positive charge inside of which there were tiny negatively charged electrons, a little like plums in a pudding, Fig. 27–21.
J. J. Thomson, soon after his discovery of the electron in 1897, argued that
the electrons in this model should be moving.

Around 1911, Ernest Rutherford (1871–1937; Fig. 27–22) and his col-
leagues performed experiments whose results contradicted Thomson's
model of the atom. In these experiments a beam of positively charged
"alpha (α) particles" was directed at a thin sheet of metal foil such as gold,
Fig. 27–23a. (These newly discovered α particles were emitted by certain
radioactive materials and were soon shown to be doubly ionized helium
atoms—that is, having a charge of $+2e$; more on alpha particles in Chapter
30.) It was expected from Thomson's model that the alpha particles would
not be deflected significantly since electrons are so much lighter than alpha
particles, and the alpha particles should not have encountered any massive
concentration of positive charge in that model to strongly repel them. The
experimental results completely contradicted these predictions. It was
found that most of the alpha particles passed through the foil unaffected, as
if the foil were mostly empty space. And of those deflected, a few were deflected at very large angles—some even nearly back in the direction from

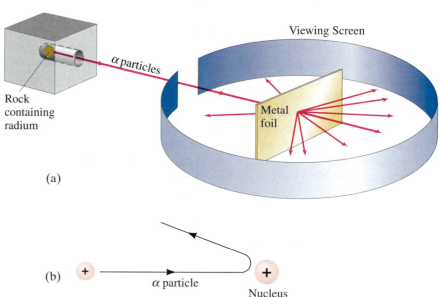

(a)

(b)

[†]Some readers may say: "Tell us the facts as we know them today, and don't bother us with
the historical background and its outmoded theories." Such an approach would ignore the
creative aspect of science and thus give a false impression of how science develops. More-
over, it is not really possible to understand today's view of the atom without insight into the
concepts that led to it.

which they had come. This could happen, Rutherford reasoned, only if the positively charged alpha particles were being repelled by a massive positive charge concentrated in a very small region of space (see Fig. 27–23b). He theorized that the atom must consist of a tiny but massive positively charged nucleus, containing over 99.9 percent of the mass of the atom, surrounded by electrons some distance away. The electrons would be moving in orbits about the nucleus—much as the planets move around the Sun—because if they were at rest, they would fall into the nucleus due to electrical attraction, Fig. 27–24. Rutherford's experiments suggested that the nucleus must have a radius of about 10^{-15} to 10^{-14} m. From kinetic theory, and especially Einstein's analysis of Brownian movement (see Section 13–1), the radius of atoms was estimated to be about 10^{-10} m. Thus the electrons would seem to be at a distance from the nucleus of about 10,000 to 100,000 times the radius of the nucleus itself (if the nucleus were the size of a baseball, the atom would have the diameter of a big city several kilometers across). So an atom would be mostly empty space.

Rutherford's "planetary" model of the atom (also called the "nuclear model of the atom") was a major step toward how we view the atom today. It was not, however, a complete model and presented some major problems, as we shall see.

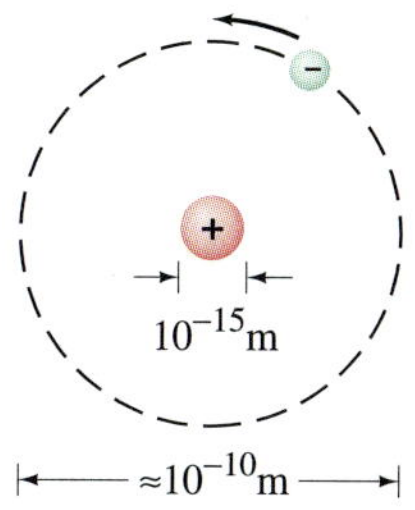

Rutherford's planetary model

FIGURE 27–24 Rutherford's model of the atom, in which electrons orbit a tiny positive nucleus (not to scale). The atom is visualized as mostly empty space.

27–9 Atomic Spectra: Key to the Structure of the Atom

Earlier in this chapter we saw that heated solids (as well as liquids and dense gases) emit light with a continuous spectrum of wavelengths. This radiation is assumed to be due to oscillations of atoms and molecules, which are largely governed by the interaction of each atom or molecule with its neighbors.

Rarefied gases can also be excited to emit light. This is done by intense heating, or more commonly by applying a high voltage to a "discharge tube" containing the gas at low pressure, Fig. 27–25. The radiation from excited gases had been observed early in the nineteenth century, and it was found that the spectrum was not continuous, but *discrete*. Since excited gases emit light of only certain wavelengths, when

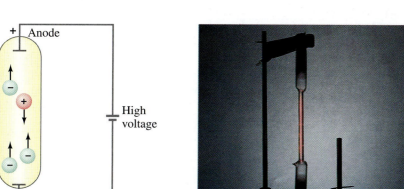

(a)

(b)

FIGURE 27–25 Gas-discharge tube: (a) is a diagram; (b) photo of actual discharge tube for hydrogen.

(a)

(b)

(c)

FIGURE 27–26 Emission spectra of the gases (a) atomic hydrogen, (b) helium, and (c) the *solar absorption* spectrum.

this light is analyzed through the slit of a spectroscope or spectrometer, a **line spectrum** is seen rather than a continuous spectrum. The line spectra emitted by a number of elements in the visible region are shown here in Fig. 27–26, and also in Chapter 24, Fig. 24–27. The **emission spectrum** is characteristic of the material and can serve as a type of "fingerprint" for identification of the gas.

We also saw (Chapter 24) that if a continuous spectrum passes through a gas, dark lines are observed in the emerging spectrum, at wavelengths corresponding to lines normally emitted by the gas. This is called an **absorption spectrum** (Fig. 27–26c), and it became clear that gases can absorb light at the same frequencies at which they emit. Using film sensitive to ultraviolet and to infrared light, it was found that gases emit and absorb discrete frequencies in these regions as well as in the visible.

For our purposes here, the importance of the line spectra is that they are emitted (or absorbed) by gases with low density. In such thin gases, the atoms are far apart on the average and hence the light emitted or absorbed is assumed to be by *individual atoms* rather than through interactions between atoms, as in a solid, liquid, or dense gas. Thus the line spectra serve as a key to the structure of the atom: any theory of atomic structure must be able to explain why atoms emit light only of discrete wavelengths, and it should be able to predict what these wavelengths are.

Hydrogen is the simplest atom—it has only one electron orbiting its nucleus. It also has the simplest spectrum. The spectrum of most atoms shows little apparent regularity. But the spacing between lines in the hydrogen spectrum decreases in a regular way (Fig. 27–27). Indeed, in 1885, J. J. Balmer (1825–1898) showed that the four visible lines in the hydrogen spectrum (with measured wavelengths 656 nm, 486 nm, 434 nm, and 410 nm) fit the following formula

$$\frac{1}{\lambda} = R\left(\frac{1}{2^2} - \frac{1}{n^2}\right), \qquad n = 3, 4, \cdots, \qquad \textbf{(27–9)}$$

where n takes on the values 3, 4, 5, 6 for the four lines, and R, called the **Rydberg constant**, has the value $R = 1.097 \times 10^7\,\text{m}^{-1}$. Later it was found

FIGURE 27–27 Balmer series of lines for hydrogen.

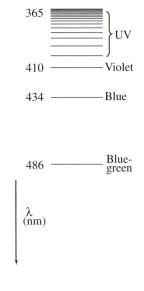

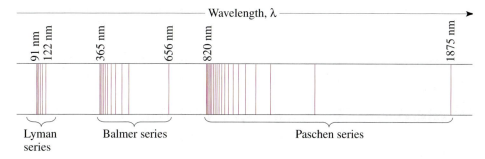

FIGURE 27-28 Line spectrum of atomic hydrogen. Each series fits the formula $\frac{1}{\lambda} = R\left(\frac{1}{n'^2} - \frac{1}{n^2}\right)$ where $n' = 1$ for the Lyman series, $n' = 2$ for the Balmer series, $n' = 3$ for the Paschen series, and so on; n can take on all integer values from $n = n' + 1$ up to infinity. The only lines in the visible region of the electromagnetic spectrum are part of the Balmer series.

that this **Balmer series** of lines extended into the UV region, ending at $\lambda = 365$ nm, as shown in Fig. 27–27. Balmer's formula, Eq. 27–9, also worked for these lines with higher integer values of n. The lines near 365 nm became too close together to distinguish, but the limit of the series at 365 nm corresponds to $n = \infty$ (so $1/n^2 = 0$ in the formula).

Later experiments on hydrogen showed that there were similar series of lines in the UV and IR regions, and each series had a pattern just like the Balmer series, but at different wavelengths, Fig. 27–28. Each of these series was found to fit a formula with the same form as Eq. 27–9 but with the $1/2^2$ replaced by $1/1^2$, $1/3^2$, $1/4^2$, and so on. For example, the so-called **Lyman series** contains lines with wavelengths from 91 nm to 122 nm (in the UV region) and fits the formula

$$\frac{1}{\lambda} = R\left(\frac{1}{1^2} - \frac{1}{n^2}\right), \qquad n = 2, 3, \cdots.$$

And the wavelengths of the **Paschen series** (in the IR region) fit

$$\frac{1}{\lambda} = R\left(\frac{1}{3^2} - \frac{1}{n^2}\right), \qquad n = 4, 5, \cdots.$$

The Rutherford model, as it stood, was unable to explain why atoms emit line spectra. It had other difficulties as well. According to the Rutherford model, electrons orbit the nucleus, and since their paths are curved the electrons are accelerating. Hence they should give off light like any other accelerating electric charge (Chapter 22). Then, since energy is conserved, the electron's own energy must decrease to compensate. Hence electrons would be expected to spiral into the nucleus. As they spiraled inward, their frequency would increase in a short time and so too would the frequency of the light emitted. Thus the two main difficulties of the Rutherford model are these: (1) it predicts that light of a continuous range of frequencies will be emitted, whereas experiment shows line spectra; (2) it predicts that atoms are unstable—electrons should quickly spiral into the nucleus—but we know that atoms in general are stable, since the matter around us is stable.

Clearly Rutherford's model was not sufficient. Some sort of modification was needed, and it was Niels Bohr who provided it by adding an essential idea—the quantum hypothesis.

Bohr had studied in Rutherford's laboratory for several months in 1912 and was convinced that Rutherford's planetary model of the atom had validity. But in order to make it work, he felt that the newly developing quantum theory would somehow have to be incorporated in it. The work of Planck and Einstein had shown that in heated solids, the energy of oscillating electric charges must change discontinuously—from one discrete energy state to another, with the emission of a quantum of light. Perhaps, Bohr argued, the electrons in an atom also cannot lose energy continuously, but must do so in quantum "jumps." In working out his theory during the next year, Bohr postulated that electrons move about the nucleus in circular orbits, but that only certain orbits are allowed. He further postulated that an electron in each orbit would have a definite energy and would move in the orbit *without radiating energy* (even though this violated classical ideas since accelerating electric charges are supposed to emit EM waves; see Chapter 22). He thus called the possible orbits **stationary states**. Light is emitted, he hypothesized, only when an electron jumps from one stationary state to another of lower energy. When such a jump occurs, a single photon of light is emitted whose energy, since energy is conserved, is given by

Stationary state

$$hf = E_u - E_l, \qquad \textbf{(27–10)}$$

where E_u refers to the energy of the upper state and E_l the energy of the lower state. See Fig. 27–29.

Bohr set out to determine what energies these orbits would have, since the spectrum of light emitted could then be predicted from Eq. 27–10. In the Balmer formula he had the key he was looking for. Bohr quickly found that his theory would be in accord with the Balmer formula if he assumed that the electron's angular momentum L is quantized and equal to an integer n times $h/2\pi$. As we saw in Chapter 8, angular momentum is given by $L = I\omega$, where I is the moment of inertia and ω is the angular velocity. For a single particle of mass m moving in a circle of radius r with speed v, $I = mr^2$ and $\omega = v/r$; hence, $L = I\omega = (mr^2)(v/r) = mvr$. Bohr's **quantum condition** is

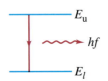

E_u

hf

E_l

FIGURE 27–29 An atom emits a photon (energy $= hf$) when its energy changes from E_u to a lower energy E_l.

Angular momentum

$$L = mvr_n = n\frac{h}{2\pi}, \qquad n = 1, 2, 3, \cdots, \qquad \textbf{(27–11)}$$

where n is an integer and r_n is the radius of the nth possible orbit. The allowed orbits are numbered $1, 2, 3, \cdots$, according to the value of n, which is called the **quantum number** of the orbit.

Quantum number, n

Equation 27–11 did not have a firm theoretical foundation. Bohr had searched for some "quantum condition," and such tries as $E = hf$ (where E represents the energy of the electron in an orbit) did not give results in accord with experiment. Bohr's reason for using Eq. 27–11 was simply that it worked; and we now look at how.

An electron in a circular orbit of radius r_n (Fig. 27–30) would have a centripetal acceleration v^2/r_n produced by the electrical force of attraction between the negative electron and the positive nucleus. This force is given by Coulomb's law,

$$F = k\frac{(Ze)(e)}{r^2},$$

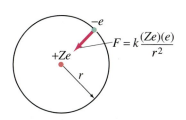

FIGURE 27–30 Electric force (Coulomb's law) keeps the negative electron in orbit around the positively charged nucleus.

where $k = 1/4\pi\epsilon_0 = 9.00 \times 10^9\,\text{N·m}^2/\text{C}^2$. The charge on the electron is $q_1 = -e$, and that on the nucleus is $q_2 = +Ze$, where Z is the number of positive charges[†] (i.e., protons). For the hydrogen atom, $Z = +1$.

In Newton's second law, $F = ma$, we substitute $a = v^2/r_n$ and Coulomb's law for F, and obtain

$$F = ma$$

$$k\frac{Ze^2}{r_n^2} = \frac{mv^2}{r_n}.$$

We solve this for r_n, and then substitute for v from Eq. 27–11 (which says $v = nh/2\pi mr_n$):

$$r_n = \frac{kZe^2}{mv^2} = \frac{kZe^2 4\pi^2 mr_n^2}{n^2h^2}.$$

We solve for r_n (it appears on both sides, so we cancel one of them) and find

$$r_n = \frac{n^2h^2}{4\pi^2 mkZe^2} = \frac{n^2}{Z}r_1 \qquad\qquad \textbf{(27–12)}$$

where

$$r_1 = \frac{h^2}{4\pi^2 mke^2}. \qquad\qquad \textbf{(27–13a)}$$

Equation 27–12 gives the radii of all possible orbits. The smallest orbit is for $n = 1$, and for hydrogen ($Z = 1$) has the value

$$r_1 = \frac{(1)^2(6.626 \times 10^{-34}\,\text{J·s})^2}{4(3.14)^2(9.11 \times 10^{-31}\,\text{kg})(9.00 \times 10^9\,\text{N·m}^2/\text{C}^2)(1)(1.602 \times 10^{-19}\,\text{C})^2}$$

$$r_1 = 0.529 \times 10^{-10}\,\text{m}. \qquad\qquad \textbf{(27–13b)}$$

The radius of the smallest orbit in hydrogen, r_1, is sometimes called the **Bohr radius**. From Eq. 27–12, we see that the radii of the larger orbits[‡] increase as n^2, so

Bohr radius

$$r_2 = 4r_1 = 2.12 \times 10^{-10}\,\text{m},$$

$$r_3 = 9r_1 = 4.76 \times 10^{-10}\,\text{m},$$

$$\vdots$$

$$r_n = n^2 r_1.$$

[†]We include Z in our derivation so that we can treat other single-electron ("hydrogenlike") atoms such as the ions He⁺ ($Z = 2$) and Li²⁺ ($Z = 3$). Helium in the neutral state has two electrons; if one electron is missing, the remaining He⁺ ion consists of one electron revolving around a nucleus of charge $+2e$. Similarly, doubly ionized lithium, Li²⁺, also has a single electron, and in this case $Z = 3$.

[‡]Be careful not to believe that these well-defined orbits actually exist. The Bohr model is only a model, not reality. The idea of electron orbits was rejected a few years later, and today electrons are better thought of as forming "clouds," as discussed in Chapter 28.

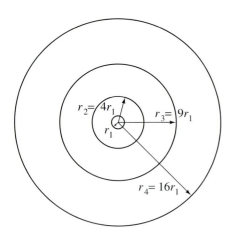

FIGURE 27–31 Possible orbits in the Bohr model of hydrogen; $r_1 = 0.529 \times 10^{-10}$ m.

The first four orbits are shown in Fig. 27–31. Notice that, according to Bohr's model, an electron can exist only in the orbits given by Eq. 27–12. There are no allowable orbits in between.

In each of its possible orbits, the electron would have a definite energy, as the following calculation shows. The total energy equals the sum of the kinetic and potential energies. The potential energy of the electron is given by $PE = qV = -eV$, where V is the potential due to a point charge $+Ze$ as given by Eq. 17–3: $V = kQ/r = kZe/r$. So

$$PE = -eV = -k\frac{Ze^2}{r}.$$

The total energy E_n for an electron in the nth orbit of radius r_n is the sum of the kinetic and potential energies:

$$E_n = \frac{1}{2}mv^2 - \frac{kZe^2}{r_n}.$$

When we substitute v from Eq. 27–11 and r_n from Eq. 27–12 into this equation, we obtain

Energy levels
$$E_n = -\frac{2\pi^2 Z^2 e^4 m k^2}{h^2}\frac{1}{n^2} \qquad n = 1, 2, 3, \cdots \qquad \textbf{(27–14)}$$

or

$$E_n = \frac{Z^2}{n^2}E_1 \qquad n = 1, 2, 3, \cdots$$

where E_1 is the lowest energy level ($n = 1$) for hydrogen ($Z = 1$). The value of E_1 is

Ground state of hydrogen
$$E_1 = -\frac{2\pi^2 e^4 m k^2}{h^2} = -2.17 \times 10^{-18} \text{ J} = -13.6 \text{ eV},$$

where we have converted joules to electron volts, as is customary in atomic physics. Since n^2 appears in the denominator of Eq. 27–14, the energies of the larger orbits in hydrogen ($Z = 1$) are given by

$$E_n = \frac{-13.6 \text{ eV}}{n^2}.$$

For example,

Excited states (first two)
$$E_2 = \frac{-13.6 \text{ eV}}{4} = -3.40 \text{ eV},$$

$$E_3 = \frac{-13.6 \text{ eV}}{9} = -1.51 \text{ eV}.$$

We see that not only are the orbit radii quantized, but from Eq. 27–14 so is the energy. The quantum number n that labels the orbit radii also labels the energy levels. The lowest **energy level** or **energy state** has energy E_1, and is called the **ground state**. The higher states, E_2, E_3, and so on, are called **excited states**.

Notice that although the energy for the larger orbits has a smaller numerical value, all the energies are less than zero. Thus, -3.4 eV is a greater energy than -13.6 eV. Hence the orbit closest to the nucleus (r_1) has the lowest energy. The reason the energies have negative values has to do with the way we defined the zero for potential energy. For two point charges, PE $= kq_1q_2/r$ corresponds to zero PE when the two charges are infinitely far apart. Thus, an electron that can just barely be free from the atom by reaching $r = \infty$ (or, at least, far from the nucleus) with zero KE will have $E = 0$, corresponding to $n = \infty$ in Eq. 27–14. If an electron is free and has kinetic energy, then $E > 0$. To remove an electron that is part of an atom requires an energy input (otherwise atoms would not be stable). Since $E \geq 0$ for a free electron, then an electron bound to an atom must have $E < 0$. That is, energy must be added to bring its energy up, from a negative value, to at least zero in order to free it.

The minimum energy required to remove an electron from the ground state of an atom is called the **binding energy** or **ionization energy**. The ionization energy for hydrogen has been measured to be 13.6 eV, and this corresponds precisely to removing an electron from the lowest state, $E_1 = -13.6$ eV, up to $E = 0$ where it can be free.

Binding energy
Ionization energy

It is useful to show the various possible energy values as horizontal lines on an energy-level diagram.[†] This is shown for hydrogen in Fig. 27–32. The electron in a hydrogen atom can be in any one of these levels according to Bohr theory. But it could never be in between, say at -9.0 eV. At room temperature, nearly all H atoms will be in the ground state ($n = 1$). At higher temperatures, or during an electric discharge when there are many collisions between free electrons and atoms, many atoms can be in excited states ($n > 1$). Once in an excited state, an atom's electron can jump down to a lower state, and give off a photon in the process. This is, according to the Bohr model, the origin of the emission spectra of excited gases. The vertical arrows in Fig. 27–32 represent the

Line spectra
emission
explained

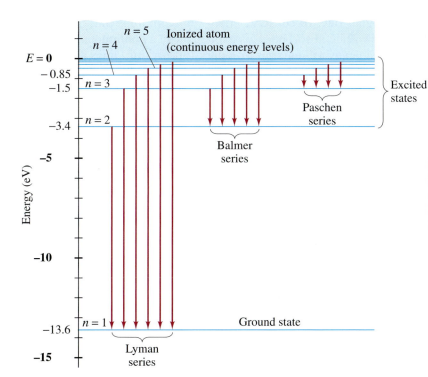

FIGURE 27–32 Energy-level diagram for the hydrogen atom, showing origin of spectral lines for the Lyman, Balmer, and Paschen series.

[†]Note that above $E = 0$, an electron is free and can have any energy (E is not quantized). Thus there is a continuum of energy states above $E = 0$, as indicated in the energy-level diagram of Fig. 27–32.

transitions or jumps that correspond to the various observed spectral lines. For example, an electron jumping from the level $n = 3$ to $n = 2$ would give rise to the 656-nm line in the Balmer series, and the jump from $n = 4$ to $n = 2$ would give rise to the 486-nm line (see Fig. 27–27). We can predict wavelengths of the spectral lines emitted by combining Eq. 27–10 with Eq. 27–14. Since $hf = hc/\lambda$, we have from Eq. 27–10

$$\frac{1}{\lambda} = \frac{hf}{hc} = \frac{1}{hc}(E_n - E_{n'}),$$

and then using Eq. 27–14,

$$\frac{1}{\lambda} = \frac{2\pi^2 Z^2 e^4 m k^2}{h^3 c}\left(\frac{1}{n'^2} - \frac{1}{n^2}\right), \tag{27-15}$$

where n refers to the upper state and n' to the lower state. This theoretical formula has the same form as the experimental Balmer formula, Eq. 27–9, with $n' = 2$. Thus we see that the Balmer series of lines corresponds to transitions or "jumps" that bring the electron down to the second energy level. Similarly, $n' = 1$ corresponds to the Lyman series and $n' = 3$ to the Paschen series (see Fig. 27–32). When the constant $(2\pi^2 Z^2 e^4 m k^2/h^3 c)$ in Eq. 27–15 is evaluated with $Z = 1$, it is found to have the measured value of the Rydberg constant, $R = 1.0974 \times 10^7\,\text{m}^{-1}$ in Eq. 27–9, in accord with experiment (see Problem 52).

The great success of Bohr's theory is that it explains why atoms emit line spectra and accurately predicts, for hydrogen, the wavelengths of emitted light. The Bohr theory also explains absorption spectra: photons of just the right wavelength can knock an electron from one energy level to a higher one. To conserve energy, only photons that have just the right energy will be absorbed. This explains why a continuous spectrum of light entering a gas will emerge with dark (absorption) lines at frequencies that correspond to emission lines.

Absorption lines explained

The Bohr theory also ensures the stability of atoms. It establishes stability by fiat: the ground state is the lowest state for an electron and there is no lower energy level to which it can go and emit more energy. Finally, as we saw above, the Bohr theory accurately predicts the ionization energy of 13.6 eV for hydrogen. However, the Bohr theory was not so successful for other atoms, and has been superseded as we shall discuss in the next chapter.

EXAMPLE 27–11 **Wavelength of a Lyman line.** Use Fig. 27–32 to determine the wavelength of the first Lyman line, the transition from $n = 2$ to $n = 1$. In what region of the electromagnetic spectrum does this lie?

SOLUTION In this case, $hf = E_2 - E_1 = 13.6\,\text{eV} - 3.4\,\text{eV} = 10.2\,\text{eV} = 1.63 \times 10^{-18}\,\text{J}$. Since $\lambda = c/f$, we have

$$\lambda = \frac{c}{f} = \frac{hc}{E_2 - E_1} = \frac{(6.63 \times 10^{-34}\,\text{J·s})(3.00 \times 10^8\,\text{m/s})}{1.63 \times 10^{-18}\,\text{J}}$$

$$= 1.22 \times 10^{-7}\,\text{m},$$

or 122 nm, which is in the UV region. See Fig. 27–28.

EXAMPLE 27–12 **Wavelength of a Balmer line.** Determine the wavelength of light emitted when a hydrogen atom makes a transition from the $n = 6$ to the $n = 2$ energy level according to the Bohr model.

SOLUTION We can use Eq. 27–15 or its equivalent, Eq. 27–9, with $R = 1.097 \times 10^7 \, \text{m}^{-1}$. Thus

$$\frac{1}{\lambda} = (1.097 \times 10^7 \, \text{m}^{-1})\left(\frac{1}{4} - \frac{1}{36}\right) = 2.44 \times 10^6 \, \text{m}^{-1}.$$

So $\lambda = 1/(2.44 \times 10^6 \, \text{m}^{-1}) = 4.10 \times 10^{-7} \, \text{m}$ or 410 nm. This is the fourth line in the Balmer series, Fig. 27–27, and is violet in color.

EXAMPLE 27–13 **Absorption wavelength.** Use Figure 27–32 to determine the maximum wavelength that hydrogen in its ground state can absorb. What would be the next smaller wavelength that would work?

SOLUTION Maximum λ corresponds to minimal energy, and thus the jump from the ground state to the first excited state (Fig. 27–32) for which the energy is $13.6 \, \text{eV} - 3.4 \, \text{eV} = 10.2 \, \text{eV}$; the required wavelength, as we saw in Example 27–11, is 122 nm. The next possibility is to jump from the ground state to the second excited state, which requires $13.6 \, \text{eV} - 1.5 \, \text{eV} = 12.1 \, \text{eV}$ and corresponds to a wavelength

$$\lambda = \frac{c}{f} = \frac{hc}{E_3 - E_1}$$

$$= \frac{(6.63 \times 10^{-34} \, \text{J·s})(3.00 \times 10^8 \, \text{m/s})}{(12.1 \, \text{eV})(1.60 \times 10^{-19} \, \text{J/eV})} = 103 \, \text{nm}.$$

EXAMPLE 27–14 **Ionization energy.** Use the Bohr model to determine the ionization energy of the He^+ ion, which has a single electron. Also calculate the minimum wavelength a photon must have to cause ionization.

SOLUTION We want to determine the minimum energy required to lift the electron from its ground state and to barely reach the free state at $E = 0$. The ground state energy of He^+ is given by Eq. 27–14 with $n = 1$ and $Z = 2$. Since all the symbols in Eq. 27–14 are the same as for the calculation for hydrogen, except that Z is 2 instead of 1, we see that E_1 will be $Z^2 = 2^2 = 4$ times the E_1 for hydrogen. That is,

$$E_1 = 4(-13.6 \, \text{eV}) = -54.4 \, \text{eV}.$$

Thus, to ionize the He^+ ion should require 54.4 eV, and this value agrees with experiment. The minimum wavelength photon that can cause ionization will have energy $hf = 54.4 \, \text{eV}$ and wavelength $\lambda = c/f = hc/hf = (6.63 \times 10^{-34} \, \text{J·s})(3.00 \times 10^8 \, \text{m/s})/(54.4 \, \text{eV})(1.60 \times 10^{-19} \, \text{J/eV}) = 22.8 \, \text{nm}$. If the atom absorbed a photon of greater energy (wavelength shorter than 22.8 nm), the atom could still be ionized and the freed electron would have kinetic energy of its own.

In this last Example, we saw that E_1 for the He$^+$ ion is four times more negative than that for hydrogen. Indeed, the energy-level diagram for He$^+$ looks just like that for hydrogen, Fig. 27–32, except that the numerical values for each energy level are four times larger. It is important to note, however, that we are talking here about the He$^+$ *ion*. Normal (neutral) helium has two electrons and its energy level diagram is entirely different.

| **CONCEPTUAL EXAMPLE 27–15** | **Hydrogen at 20° C.** Estimate the average kinetic energy of hydrogen atoms (or molecules) at room temperature, and use the result to explain why nearly all H atoms are in the ground state at room temperature, and hence emit no light.

RESPONSE According to kinetic theory (Chapter 13), the average KE of atoms or molecules in a gas is given by Eq. 13–8:

$$\overline{KE} = \tfrac{3}{2}kT,$$

where $k = 1.38 \times 10^{-23}$ J/K is Boltzmann's constant, and T is the kelvin (absolute) temperature. Room temperature is about $T = 300$ K, so

$$\overline{KE} = \tfrac{3}{2}(1.38 \times 10^{-23} \text{ J/K})(300 \text{ K}) = 6.2 \times 10^{-21} \text{ J},$$

or, in electron volts:

$$\overline{KE} = \frac{6.2 \times 10^{-21} \text{ J}}{1.6 \times 10^{-19} \text{ J/eV}} = 0.04 \text{ eV}.$$

The average KE is thus very small compared to the energy between the ground state and the next higher energy state (13.6 eV − 3.4 eV = 10.2 eV). Any atoms in excited states emit light and eventually fall to the ground state. Once in the ground state, collisions with other atoms can transfer energy of only 0.04 eV on the average. A small fraction of atoms can have much more energy (see Section 13–12 on the distribution of molecular speeds), but even a KE that is 10 times the average is not nearly enough to excite atoms above the ground state. Thus, at room temperature, nearly all atoms are in the ground state. Atoms can be excited to upper states by very high temperatures, or by passing a current of high energy electrons through the gas, as in a discharge tube (Fig. 27–25).

We should note that Bohr made some radical assumptions that were at variance with classical ideas. He assumed that electrons in fixed orbits do not radiate light even though they are accelerating (moving in a circle), and he assumed that angular momentum is quantized. Furthermore, he was not able to say how an electron moved when it made a transition from one energy level to another. On the other hand, there is no real reason to expect that in the tiny world of the atom electrons would behave as ordinary-sized objects do. Nonetheless, he felt that where quantum theory overlaps with the macro-scopic world, it should predict classical results. This is the **correspondence** *Correspondence* **principle**, already mentioned in regard to relativity (Section 26–12). This *principle* principle does work for Bohr's theory of the hydrogen atom. The orbit sizes and energies are quite different for $n = 1$ and $n = 2$, say. But orbits with $n = 100{,}000{,}000$ and $100{,}000{,}001$ would be very close in size and energy (see Fig. 27–32). Indeed, jumps between such large orbits, which would approach macroscopic sizes, would be imperceptible. Such orbits would thus appear to be continuously spaced, which is what we expect in the everyday world.

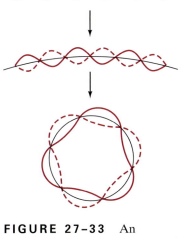

27–11 de Broglie's Hypothesis Applied to Atoms

Bohr's theory was largely of an *ad hoc* nature. Assumptions were made so that theory would agree with experiment. But Bohr could give no reason why the orbits were quantized, nor why there should be a stable ground state. Finally, ten years later, a reason was proposed by de Broglie.

We saw in Section 27–6 that in 1923, Louis de Broglie proposed that material particles, such as electrons, have a wave nature; and that this hypothesis was confirmed by experiment several years later.

One of de Broglie's original arguments in favor of the wave nature of electrons was that it provided an explanation for Bohr's theory of the hydrogen atom. According to de Broglie, a particle of mass m moving with speed v would have a wavelength (Eq. 27–8) of

$$\lambda = \frac{h}{mv}.$$

FIGURE 27–33 An ordinary standing wave compared to a circular standing wave.

Each electron orbit in an atom, he proposed, is actually a standing wave. As we saw in Chapter 11, when a violin or guitar string is plucked, a vast number of wavelengths are excited. But only certain ones—those that have nodes at the ends—are sustained. These are the *resonant* modes of the string. Waves with other wavelengths interfere with themselves upon reflection and their amplitudes quickly drop to zero. With electrons moving in circles, according to Bohr's theory, de Broglie argued that the electron wave was a *circular* standing wave that closes on itself, Fig. 27–33. If the wavelength of a wave does not close on itself, as in Fig. 27–34, destructive interference takes place as the wave travels around the loop, and the wave quickly dies out. Thus, the only waves that persist are those for which the circumference of the circular orbit contains a whole number of wavelengths, Fig. 27–35. The circumference of a Bohr orbit of radius r_n is $2\pi r_n$, so we have

FIGURE 27–34 When a wave does not close (and hence interferes destructively with itself), it rapidly dies out.

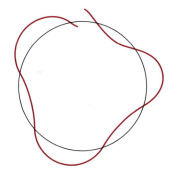

$$2\pi r_n = n\lambda, \qquad n = 1, 2, 3, \cdots.$$

When we substitute $\lambda = h/mv$, we get

$$2\pi r_n = \frac{nh}{mv},$$

or

$$mvr_n = \frac{nh}{2\pi}.$$

This is just the *quantum condition* proposed by Bohr on an *ad hoc* basis, Eq. 27–11. And it is from this equation that the discrete orbits and energy

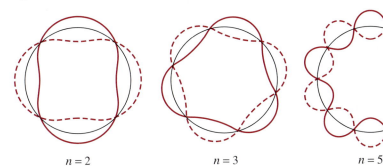

$n = 2$ $\qquad$ $n = 3$ $\qquad$ $n = 5$

FIGURE 27–35 Standing circular waves for two, three, and five wavelengths on the circumference; n, the number of wavelengths, is also the quantum number.

levels were derived. Thus we have an explanation for the quantized orbits and energy states in the Bohr model: they are due to the wave nature of the electron, and that only resonant "standing" waves can persist. This implies that the *wave–particle duality* is at the root of atomic structure.

It should be noted in viewing the circular electron waves of Fig. 27–35 that the electron is not to be thought of as following the oscillating wave pattern. In the Bohr model of hydrogen, the electron, considered as a particle, moves in a circle. The circular wave, on the other hand, represents the *amplitude* of the electron "matter wave," and in Fig. 27–35 the wave amplitude is shown superimposed on the circular path of the particle orbit for convenience.

Bohr's theory worked well for hydrogen and for one-electron ions. It did not prove as successful for multielectron atoms. We will discuss this and other problems with the Bohr theory in the next chapter, and we will see how a new and radical theory, quantum mechanics, finally solved the problem of atomic structure and gave us a very different view of the atom: the idea of electrons in well-defined orbits was replaced with the idea of electron "clouds." And this new theory of quantum mechanics has given us a wholly different view of the basic mechanisms underlying physical processes.

■ SUMMARY

Quantum theory has its origins in **Planck's quantum hypothesis** that molecular oscillations are quantized: their energy E can only be integer (n) multiples of hf, where h is Planck's constant and f is the natural frequency of oscillation: $E = nhf$. This hypothesis explained the spectrum of radiation emitted by (black) bodies at high temperature.

Einstein proposed that for some experiments, light could be pictured as being emitted and absorbed as quanta (particles), which we now call **photons,** each with energy

$$E = hf.$$

He proposed the photoelectric effect as a test for the photon theory of light. In the **photoelectric effect**, the photon theory says that each incident photon can strike an electron in a material and eject it if it has sufficient energy. The maximum energy of ejected electrons is then linearly related to the frequency of the incident light. The photon theory is also supported by the Compton effect and the observation of electron–positron **pair production**.

The wave–particle duality refers to the idea that light and matter (such as electrons) have both wave and particle properties. The wavelength of a material object is given by

$$\lambda = \frac{h}{mv},$$

where mv is the momentum of the object. The **principle of complementarity** states that we must be aware of both the particle and wave properties of light and of matter for a complete understanding of them.

Early models of the atom include the plum-pudding model and Rutherford's planetary (or nuclear) model. Rutherford's model, which was created to explain the backscattering of alpha particles from thin metal foils, assumes that an atom consists of a tiny but massive positively charged nucleus surrounded (at a relatively great distance) by electrons.

To explain the line spectra emitted by atoms, as well as the stability of atoms, **Bohr theory** postulated (1) that electrons bound in an atom can only occupy orbits for which the angular momentum is

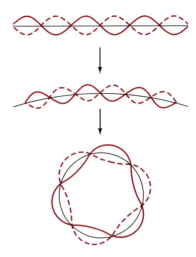

quantized, which results in discrete values for the radius and energy; (2) that an electron in such a **stationary state** emits no radiation; (3) that, if an electron jumps to a lower state, it emits a photon whose energy equals the difference in energy between the two states; (4) that the angular momentum L of atomic electrons is quantized by the rule

$$L = \frac{nh}{2\pi},$$

where n is an integer called a **quantum number**. The $n = 1$ state in hydrogen is the **ground state**, which has an energy $E_1 = -13.6\ \text{eV}$; higher values of n correspond to **excited states** and their energies are

$$E_n = -\frac{13.6\ \text{eV}}{n^2}.$$

Atoms are excited to these higher states by collisions with other atoms or electrons or by absorption of a photon of just the right frequency.

de Broglie's hypothesis that electrons (and other matter) have a wavelength $\lambda = h/mv$ gave an explanation for Bohr's quantized orbits by bringing in the wave–particle duality: the orbits correspond to circular standing waves in which the circumference of the orbit equals a whole number of wavelengths.

QUESTIONS

1. What can be said about the relative temperature of whitish-yellow, reddish, and bluish stars?

2. If energy is radiated by all objects, why can't we see them in the dark? (See also Section 14–9.)

3. Does a lightbulb at a temperature of 2500 K produce as white a light as the Sun at 6000 K? Explain.

4. An ideal blackbody can be approximated by a small hole in an otherwise enclosed cavity. Explain. [*Hint*: The pupil of your eye is an approximate case.]

5. Why do jewelers often examine diamonds in daylight rather than with indoor light?

6. Darkrooms for developing black and white film are sometimes lit by a red bulb. Why red? Would such a bulb work in a darkroom for developing color photographs?

7. If the threshold wavelength in the photoelectric effect increases when the emitting metal is changed to a different metal, what can you say about the work functions of the two metals?

8. Explain why the existence of a cutoff frequency in the photoelectric effect more strongly favors a particle theory rather than a wave theory of light.

9. UV light causes sunburn, whereas visible light does not. Explain.

10. If an X-ray photon is scattered by an electron, does its wavelength change? If so, does it increase or decrease?

11. In both the photoelectric effect and in the Compton effect we have a photon colliding with an electron causing the electron to fly off. What then, is the difference between the two processes?

12. Explain how the photoelectric circuit of Fig. 27–8 could be used in (*a*) a burglar alarm, (*b*) a smoke detector, (*c*) a photographic light meter, and (*d*) a spectrophotometer (see Section 24–7).

13. Why do we say that light has wave properties? Why do we say that light has particle properties?

14. Why do we say that electrons have wave properties? Why do we say that electrons have particle properties?

15. What is the difference between a photon and an electron? Be specific: make a list.

16. If an electron and a proton travel at the same speed, which has the shorter wavelength?

17. In Rutherford's planetary model of the atom, what keeps the electrons from flying off into space?

18. Which of the following can emit a line spectrum: (*a*) gases, (*b*) liquids, (*c*) solids? Which can emit a continuous spectrum?

19. Why doesn't the O_2 gas in the air around us give off light?

20. How can you tell if there is oxygen on the Sun?

21. When a wide spectrum of light passes through hydrogen gas at room temperature, absorption lines are observed that correspond only to the Lyman series. Why don't we observe the other series?

22. Explain how the closely spaced energy levels for hydrogen near the top of Fig. 27–32 correspond to the closely spaced spectral lines at the top of Fig. 27–27.

23. Discuss the differences between Rutherford's and Bohr's theory of the atom.

24. In a helium atom, which contains two electrons, do you think that on the average the electrons are closer to the nucleus or farther away than in a hydrogen atom? Why?

25. How can the spectrum of hydrogen contain so many lines when hydrogen contains only one electron?

26. The Lyman series is brighter than the Balmer series, because this series of transitions ends up in the most common state for hydrogen, the ground state. Why then was the Balmer series discovered first?

27. The shortest wavelength light in the Lyman *absorption* spectrum is 91 nm. Could light of 89 nm (slightly higher energy) also be absorbed by hydrogen? What would be the final state of the electron? Why isn't the light from the sun absorbed at this frequency?

28. Use conservation of momentum to explain why photons emitted by hydrogen atoms have slightly less energy than that predicted by Eq. 27–10.

PROBLEMS

SECTION 27–1

1. (I) What is the value of e/m for a particle that moves in a circle of radius 7.0 mm in a 0.86-T magnetic field if a crossed 300-V/m electric field will make the path straight?

2. (II) What is the velocity of a beam of electrons that go undeflected when passing through crossed electric and magnetic fields of magnitude 1.38×10^4 V/m and 2.90×10^{-3} T, respectively? What is the radius of the electron orbit if the electric field is turned off?

3. (II) An oil drop whose mass is determined to be 2.8×10^{-15} kg is held at rest between two large plates separated by 1.0 cm when the potential difference between them is 340 V. How many excess electrons does this drop have?

4. (III) In Millikan's oil-drop experiment, the mass of the oil drop is obtained by observing the terminal speed v_T of the freely falling drop in the absence of an electric field. Under these circumstances, the "effective" weight equals the viscous force given by Stokes's law, $F = 6\pi\eta r v_T$, where η is the viscosity of air and r the radius of the drop. Also, the actual weight, $mg = \frac{4}{3}\pi r^3 \rho g$, must be corrected for the buoyant force of the air. This is done by replacing ρ with $\rho - \rho_A$, where ρ is the density of the oil and ρ_A the density of air. With these preliminaries, show that the charge on the drop is given by

$$q = 18\pi \frac{d}{V} \sqrt{\frac{\eta^3 v_T^3}{2(\rho - \rho_A)g}},$$

where d is the separation of the plates (Fig. 27–3) and V is the voltage across them that just keeps the drop stationary. All of the quantities on the right side of this equation are known or can be measured. The terminal velocity v_T is determined by measuring the time it takes the drop to fall a measured distance, which is observed through a small telescope.

SECTION 27–2

5. (I) How hot is a metal being welded if it radiates most strongly at 410 nm?

6. (I) (*a*) At what temperature will the peak of a blackbody spectrum be at 15.0 nm? (*b*) What is the wavelength at the peak of a blackbody spectrum if the body is at a temperature of 2000 K?

7. (I) Estimate the peak wavelength for radiation from (*a*) ice at 0° C, (*b*) a floodlamp at 3000 K, (*c*) helium at 4 K, assuming blackbody emission. In what region of the EM spectrum is each?

8. (I) An HCl molecule vibrates with a natural frequency of 8.1×10^{13} Hz. What is the difference in energy (in joules and electron volts) between possible values of the oscillation energy?

9. (II) A child's swing has a natural frequency of 0.90 Hz. (*a*) What is the separation between possible energy values (in joules)? (*b*) If the swing reaches a vertical height of 45 cm above its lowest point and has a mass of 20 kg (including the child), what is the value of the quantum number n? (*c*) What is the fractional change in energy between levels whose quantum numbers are n (as just calculated) and $n + 1$? Would quantization be measurable in this case?

10. (II) The steps of a flight of stairs are 20.0 cm high (vertically). If a 58.0-kg person stands with both feet on the same step, what is the gravitational potential energy of this person, relative to the ground, on (*a*) the first step, (*b*) the second step, (*c*) the third step, (*d*) the nth step? (*e*) What is the change in energy as the person descends from step 6 to step 2?

11. (II) Estimate the peak wavelength of light issuing from the pupil of the human eye (which approximates a blackbody) assuming normal body temperature.

SECTION 27–3

12. (I) What is the energy range in eV, of photons in the visible spectrum, of wavelength 400 nm to 700 nm?

13. (I) What is the energy of photons (in eV) emitted by a 102.1-MHz FM radio station?

14. (I) A typical gamma ray emitted from a nucleus during radioactive decay may have an energy of 200 keV. What is its wavelength? Would we expect significant diffraction of this type of light when it passes through an everyday opening, like a door?

15. (I) About 0.1 eV is required to break a "hydrogen bond" in a protein molecule. What are the minimum frequency and maximum wavelength of a photon that can accomplish this?

16. (I) What minimum frequency of light is needed to eject electrons from a metal whose work function is 4.3×10^{-19} J?

17. (II) What is the longest wavelength of light that will emit electrons from a metal whose work function is 3.10 eV?

18. (II) Barium has a work function of 2.48 eV. What is the maximum kinetic energy of electrons if the metal is illuminated by light of wavelength 390 nm? What is their speed?

19. (II) The work functions for sodium, cesium, copper and iron are 2.3, 2.1, 4.7 and 4.5 eV respectively. Which of these metals will not emit electrons when visible light shines on it?

20. (II) In a photoelectric effect experiment it is observed that no current flows unless the wavelength is less than 570 nm. (a) What is the work function of this material? (b) What is the stopping voltage required if light of wavelength 400 nm is used?

21. (II) What is the maximum kinetic energy of electrons ejected from barium ($W_0 = 2.48$ eV) when illuminated by white light, $\lambda = 400$ to 700 nm?

22. (II) When UV light of wavelength 225 nm falls on a metal surface, the maximum kinetic energy of emitted electrons is 1.40 eV. What is the work function of the metal?

23. (II) The threshold wavelength for emission of electrons from a given surface is 320 nm. What will be the maximum kinetic energy of ejected electrons when the wavelength is changed to (a) 250 nm, (b) 350 nm?

24. (II) A certain type of film is sensitive only to light whose wavelength is less than 660 nm. What is the energy (eV and kcal/mol) needed for the chemical reaction to occur which causes the film to change?

25. (II) When 230-nm light falls on a metal, the current through a photoelectric circuit (Fig. 27–8) is brought to zero at a reverse voltage of 1.64 V. What is the work function of the metal?

SECTION 27–4

26. (I) How much total kinetic energy will an electron–positron pair have if produced by a 2.54-MeV photon?

27. (II) What is the momentum and the effective mass of a 0.35-nm X-ray photon?

28. (II) What is the longest wavelength photon that could produce a proton–antiproton pair? (Each has a mass of 1.67×10^{-27} kg.)

29. (II) What is the minimum photon energy needed to produce a $\mu^+ - \mu^-$ pair? The mass of each μ (muon) is 207 times the mass of the electron. What is the wavelength of such a photon?

30. (II) A gamma-ray photon produces an electron–positron pair, each with a kinetic energy of 345 keV. What was the energy and wavelength of the photon?

31. (II) The quantity h/m_0c, which has the dimensions of length, is called the *Compton wavelength*. Determine the Compton wavelength for (a) an electron, (b) a proton. (c) Show that if a photon has wavelength equal to the Compton wavelength of a particle, the photon's energy is equal to the rest energy of the particle.

32. (II) X-rays of wavelength $\lambda = 0.120$ nm are scattered from a carbon block. What is the Compton wavelength shift for photons detected at angles (relative to the incident beam) of (a) 45°, (b) 90°, (c) 180°?

33. (III) For each of the scattering angles in the previous Problem, determine (a) the fractional energy loss of the photon, and (b) the energy given to the scattered electron.

34. (III) Derive Eq. 27–7 using conservation of energy and momentum.

35. (III) In the Compton effect, a 0.100-nm photon strikes a free electron in a head-on collision and knocks it into the forward direction. The rebounding photon recoils directly backward. Use conservation of (relativistic) energy and momentum to determine (a) the kinetic energy of the electron, and (b) the wavelength of the recoiling photon. Assume the electron's kinetic energy is given by the nonrelativistic formula. (Note: use Eq. 27–6, but not Eq. 27–7.)

SECTION 27–6

36. (I) Calculate the wavelength of a 0.21-kg ball traveling at 0.10 m/s.

37. (I) What is the wavelength of a neutron ($m_0 = 1.67 \times 10^{-27}$ kg) traveling 5.5×10^4 m/s?

38. (I) Through how many volts of potential difference must an electron be accelerated to achieve a wavelength of 0.23 nm?

39. (II) Calculate the de Broglie wavelength of an electron in your TV picture tube if it is accelerated by 20,000 V. Is it relativistic? How does its wavelength compare to the size of the "neck" of the tube, typically 5 cm? Do we have to worry about diffraction problems blurring our picture on the screen?

40. (II) What is the wavelength of an electron of energy (a) 10 eV, (b) 100 eV, (c) 1.0 keV?

41. (II) Show that if an electron and a proton have the same nonrelativistic kinetic energy, the proton has the shorter wavelength.

42. (II) Calculate the ratio of the KE of an electron to that of a proton if their wavelengths are equal. Assume that the speeds are non-relativistic.

43. (II) What is the wavelength of an O_2 molecule in the air at room temperature? [*Hint*: See Chapter 13.]

44. (III) A Cadillac with a mass of 2000 kg approaches a freeway underpass that is 10 m across. At what speed must the car be moving, in order for it to have a wavelength such that it might somehow "diffract" after passing through this single "slit"? How do these conditions compare to normal freeway speeds of 30 m/s?

45. (III) Show that for a particle of rest mass m_0, if $\lambda = h/mv$ then it cannot be true that $E = hf$ where E is (a) kinetic energy, or (b) KE plus rest mass energy, and $v = f\lambda$ is the speed of the particle.

*SECTION 27–7

* **46.** (II) What voltage is needed to produce electron wavelengths of 0.10 nm? (Assume that the electrons are nonrelativistic.)

* **47.** (III) Electrons are accelerated by 2250 V in an electron microscope. To achieve a resolution of 5.0 nm, what numerical aperture (Section 25–8) is required?

SECTION 27–10

48. (I) For the three hydrogen transitions indicated below, with n being the initial state and n' being the final state, is the transition an absorption or an emission? Which is higher, the initial state energy or the final state energy of the atom? Finally, which of these transitions involves the largest energy photon? (a) $n = 1, n' = 3$ (b) $n = 6, n' = 2$ (c) $n = 4, n' = 5$.

49. (I) How much energy is needed to ionize a hydrogen atom in the $n = 2$ state?

50. (I) The third longest wavelength in the Paschen series in hydrogen (Fig. 27–32) corresponds to what transition?

51. (I) Calculate the ionization energy of doubly ionized lithium, Li^{2+}, which has $Z = 3$.

52. (I) Evaluate the Rydberg constant R using Bohr theory (compare Eqs. 27–9 and 27–15) and show that its value is $R = 1.0974 \times 10^7 \, m^{-1}$.

53. (II) What wavelength photon would be required to ionize a hydrogen atom in the ground state and give the ejected electron a kinetic energy of 10.0 eV?

54. (II) (a) Determine the wavelength of the second Balmer line ($n = 4$ to $n = 2$ transition) using Fig. 27–32. Determine likewise (b) the wavelength of the second Lyman line and (c) the wavelength of the third Balmer line.

55. (II) In the Sun, an ionized helium (He^+) atom makes a transition from the $n = 6$ state to the $n = 2$ state, emitting a photon. Can that photon be absorbed by hydrogen atoms present in the Sun? If so, between what energy states will the hydrogen atom jump?

56. (II) What is the longest wavelength light capable of ionizing a hydrogen atom in the ground state?

57. (II) Construct the energy-level diagram for the He^+ ion (see Fig. 27–32).

58. (II) Construct the energy-level diagram for doubly ionized lithium, Li^{2+}.

59. (II) What is the potential energy and the kinetic energy of an electron in the ground state of the hydrogen atom?

60. (II) An excited hydrogen atom could, in principle, have a radius of 1.00 mm. What would be the value of n for a Bohr orbit of this size? What would its energy be?

61. (II) Is the use of nonrelativistic formulas justified in the Bohr atom? To check, calculate the electron's velocity, v, in terms of c for the ground state of hydrogen, and then calculate $\sqrt{1 - v^2/c^2}$.

62. (II) Suppose an electron was bound to a proton, as in the hydrogen atom, by the gravitational force rather than by the electric force. What would be the radius, and energy, of the first Bohr orbit?

63. (III) Show that the magnitude of the PE of an electron in any Bohr orbit of a hydrogen atom is twice the magnitude of its KE in that orbit.

64. (III) *Correspondence principle*: Show that for large values of n, the difference in radius Δr between two adjacent orbits (with quantum numbers n and $n - 1$) is given by

$$\Delta r = r_n - r_{n-1} \approx \frac{2r_n}{n},$$

so $\Delta r/r_n \to 0$ as $n \to \infty$, in accordance with the correspondence principle. [Note that we can check the correspondence principle by either considering large values of $n(n \to \infty)$ or by letting $h \to 0$. Are these equivalent?]

SECTION 27–11

65. (III) Suppose a particle of mass m is confined to a one-dimensional box of width L. According to quantum theory, the particle's wave (with $\lambda = h/mv$) is a standing wave with nodes at the edges of the box. (a) Show the possible modes of vibration on a diagram. (b) Show that the kinetic energy of the particle has quantized energies given by KE $= n^2h^2/8mL^2$, where n is an integer.

(c) Calculate the ground-state energy ($n = 1$) for an electron confined to a box of width 0.50×10^{-10} m. (d) What is the ground-state energy, and speed, of a baseball ($m = 140$ g) in a box 0.50 m wide? (e) An electron confined to a box has a ground-state energy of 20 eV. What is the width of the box?

GENERAL PROBLEMS

66. The Big Bang theory states that the beginning of the Universe was accompanied by a huge burst of photons. Those photons are still present today and make up the so called Cosmic Microwave Background Radiation. The Universe radiates like a blackbody with a temperature of about 2.7 K. Calculate the peak wavelength of this radiation.

67. At low temperatures, nearly all the atoms in hydrogen gas will be in the ground state. What minimum frequency photon is needed if the photoelectric effect is to be observed?

68. A beam of 85-eV electrons is scattered from a crystal, as in X-ray diffraction, and a first-order peak is observed at $\theta = 38°$. What is the spacing between planes in the diffracting crystal? (See Section 25–11.)

69. Show that the energy E (in electron volts) of a photon whose wavelength is λ (meters) is given by E (in eV) $= 1.24 \times 10^{-6}/\lambda$.

70. A microwave oven produces electromagnetic radiation at $\lambda = 12.2$ cm and produces a power of 760 W. Calculate the number of microwave photons produced by the microwave oven each second.

71. Sunlight reaching the Earth has an intensity of about 1000 W/m^2. Estimate how many photons per square meter per second this represents. Take the average wavelength to be 550 nm.

72. A beam of red laser light ($\lambda = 633$ nm) hits a black wall and is fully absorbed. If this exerts a total force $F = 5.5$ nN on the wall, how many photons per second are hitting the wall?

73. If a 100-W light bulb emits 3.0 percent of the input energy as visible light (average wavelength 550 nm) uniformly in all directions, estimate how many photons per second of visible light will strike the pupil (4.0 mm diameter) of the eye of an observer 1.0 km away.

74. By what potential difference must (a) a proton ($m_0 = 1.67 \times 10^{-27}$ kg), and (b) an electron ($m_0 = 9.11 \times 10^{-31}$ kg), be accelerated to have a wavelength $\lambda = 5.0 \times 10^{-12}$ m?

75. In certain of Rutherford's experiments (Fig. 27–23), the α particles (mass $= 6.64 \times 10^{-27}$ kg) had a kinetic energy of 4.8 MeV. How close could they get to a gold nucleus (charge $= +79e$)? Ignore the recoil motion of the nucleus.

76. By what fraction does the mass of an H atom decrease when it makes an $n = 4$ to $n = 1$ transition?

77. For what maximum kinetic energy is a collision between an electron and a hydrogen atom in its ground state definitely elastic?

78. Using Bohr theory, derive an equation for the angular velocity ω, and frequency f, of an electron in a hydrogen atom. Determine (a) for the ground state and (b) for the first excited state ($n = 2$).

79. Calculate the ratio of the gravitational to electric force for the electron in a hydrogen atom. Can the gravitational force be safely ignored?

80. Electrons accelerated by a potential difference of 12.3 V pass through a gas of hydrogen atoms at room temperature. What wavelengths of light will be emitted?

81. Atoms can be formed in which a muon (mass $= 207$ times the mass of an electron) replaces one of the electrons in an atom. Calculate, using Bohr theory, the energy of the photon emitted when a muon makes a transition from $n = 2$ to $n = 1$ in a muonic $^{208}_{82}$Pb atom (lead whose nucleus has a mass 208 times the proton mass and charge $+82e$).

82. In an X-ray tube (see Fig. 25–36 and discussion in Section 25–11), the high voltage between filament and target is V. After being accelerated through this voltage, an electron strikes the target where it is decelerated (by positively charged nuclei) and in the process one or more X-ray photons are emitted. (*a*) Show that the photon of shortest wavelength will have

$$\lambda_0 = \frac{hc}{eV}.$$

(*b*) What is the shortest wavelength of X-ray emitted when accelerated electrons strike the face of a 27-kV television picture tube?

83. Show that the wavelength of a particle of mass m_0 with kinetic energy KE is given by the relativistic formula $\lambda = hc/\sqrt{(\text{KE})^2 + 2m_0c^2(\text{KE})}$.

84. What is the kinetic energy and wavelength of a "thermal" neutron (one that is in equilibrium at room temperature—see Chapter 13)?

85. What is the theoretical limit of resolution for an electron microscope whose electrons are accelerated through 60 kV? (Relativistic formulas should be used.)

86. The intensity of the Sun's light in the vicinity of the Earth is about 1000 W/m². Imagine a spacecraft with a mirrored square sail of dimension 1.0 km. Estimate how much thrust (in newtons) this craft will experience due to collisions with the Sun's photons. [*Hint:* Assume the photons bounce off the sail with no change in magnitude of their momentum.]

Neon tubes are thin glass tubes (moldable into various shapes) filled with neon (or other) gas that glow with a particular color when high voltage current passes through them. The atoms of the gas are excited to upper energy levels, and when their electrons jump down to lower energy levels they emit light (photons) whose wavelengths (color) are characteristic of the type of gas. A tube filled with neon produces a red-orange color, helium produces pink, and a mixture of Ne, Ar, and Hg produces blue, for example. See Fig. 24–27 or 27–26. The color of the glass, if not clear, also affects the color seen.

QUANTUM MECHANICS OF ATOMS

Bohr's model of the atom gave us a first (though rough) picture of what an atom is like. It proposed an explanation for why there should be emission and absorption of light by atoms at discrete wavelengths, as well as for the stability of atoms. The wavelengths of the line spectra and the ionization energy for hydrogen (and one-electron ions) that is predicted are in excellent agreement with experiment. But the Bohr theory had important limitations. It was not able to predict the line spectra for more complex atoms—not even for the neutral helium atom, which has only two electrons. Nor could it explain why emission lines, when viewed with great precision, consist of two or more very closely spaced lines (referred to as *fine structure*). The Bohr theory also didn't explain why some spectral lines were brighter than others. And it couldn't explain the bonding of atoms in molecules or in solids and liquids.

From a theoretical point of view, too, the Bohr theory was not satisfactory. For it was a strange mixture of classical and quantum ideas. And the wave–particle duality was still not really resolved.

FIGURE 28–1 Erwin Schrödinger with Lise Meitner (see Chapter 31).

FIGURE 28–2 Werner Heisenberg (center) on Lake Como with Wolfgang Pauli (right) and Enrico Fermi (left).

We mention these limitations of the Bohr theory not to disparage it—for it was a landmark in the history of science. Rather, we mention them to show why, in the early 1920s, it became increasingly evident that a new, more comprehensive theory was needed. It was not long in coming. Less than two years after de Broglie gave us his matter–wave hypothesis, Erwin Schrödinger (1887–1961; Fig. 28–1) and Werner Heisenberg (1901–1976; Fig. 28–2) independently developed a new comprehensive theory. Their separate approaches were quite different but were soon shown to be fully compatible.

28–1 Quantum Mechanics—A New Theory

The new theory, called **quantum mechanics**, unifies the wave–particle duality into a single consistent theory. As a theory, quantum mechanics has been extremely successful. It has successfully dealt with the spectra emitted by complex atoms, even the fine details. It explains the relative brightness of spectral lines and how atoms form molecules. It is also a much more general theory that covers all quantum phenomena from blackbody radiation to atoms and molecules. It has explained a wide range of natural phenomena and from its predictions many new practical devices have become possible. Indeed, it has been so successful that it is accepted today by nearly all physicists as the fundamental theory underlying physical processes.

Quantum mechanics deals mainly with the microscopic world of atoms and light. But in our everyday macroscopic world, we do perceive light and we accept that ordinary objects are made up of atoms. This new theory must therefore also account for the verified results of classical physics. That is, when it is applied to macroscopic phenomena, quantum mechanics must be able to produce the old classical laws. This, the **correspondence principle** (already mentioned in Section 27–10), is met fully by quantum mechanics. This doesn't mean we throw away classical theories such as Newton's laws. In the everyday world, the latter are far easier to apply and they give an accurate description. But when we deal with high speeds, close to the speed of light, we must use the theory of relativity; and when we deal with the tiny world of the atom, we use quantum mechanics.

Although we won't go into the detailed mathematics of quantum mechanics, we will discuss the main ideas and how they involve the wave and particle properties of matter to explain atomic structure and other applications.

The Wave Function and Its Interpretation; the Double-Slit Experiment

The important properties of any wave are its wavelength, frequency, and amplitude. For an electromagnetic wave, the wavelength determines whether the light is visible or not, and if so, what color it is. We also have seen that the wavelength (or frequency) is a measure of the energy of the corresponding photon ($E = hf$). The amplitude of an electromagnetic wave at any point is the strength of the electric (or magnetic) field at that point, and is related to the intensity of the wave (the brightness of the light).

For material particles such as electrons, quantum mechanics relates the wavelength to momentum according to de Broglie's formula, $\lambda = h/mv$. But what about the amplitude of a matter wave? In quantum mechanics the amplitude of, say, an electron wave is called the **wave function** and is given the symbol Ψ (the Greek letter psi). Thus Ψ represents the amplitude, as a function of time and position, of a new kind of field which we might call a "matter" field or a matter wave.

To calculate the wave function Ψ in a given situation (say, for an electron in an atom) is one of the basic tasks of quantum mechanics. Indeed, the development of an equation to do so was Schrödinger's great contribution. The *Schrödinger wave equation*, as it is called, is a differential equation (so we will not deal with it here) and is considered to be the basic equation for the description of nonrelativistic material particles. Since our treatment of quantum mechanics is necessarily brief (whole books are needed for the subject), the actual form of the Schrödinger equation, and how it is solved, will not concern us. What will be useful are the solutions obtained and the meaning of the wave function, Ψ, itself. One way to interpret Ψ is simply as the amplitude, at any point in space and time, of a "matter wave," so it plays the role that E (the electric field) does for an electromagnetic wave. Another interpretation is possible, however, based on the wave–particle duality. To understand this, we make an analogy with light.

We saw in Chapter 11 that the intensity I of any wave is proportional to the square of the amplitude. This holds true for light waves as well, as we saw in Chapter 22; that is,

$$I \propto E^2,$$

where E is the electric field strength. From the *particle* point of view, the intensity of a light beam (of given frequency) is proportional to the number of photons, N, that pass through a given area per unit time. The more photons there are, the greater the intensity. Thus

$$I \propto E^2 \propto N.$$

This proportion can be turned around so that we have

$$N \propto E^2.$$

That is, the number of photons (striking a page of this book, say) is proportional to the square of the electric-field strength.

If the light beam is very weak, only a few photons will be involved. Indeed, it is possible to "build up" a photograph in a camera using very weak

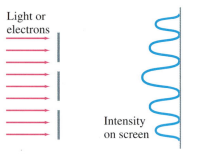

FIGURE 28–3 Parallel beam, of light or electrons, falls on two slits whose sizes are comparable to the wavelength. An interference pattern is observed.

Light or electrons

Intensity on screen

Probability $\propto \Psi^2$

FIGURE 28–4 Young's double-slit experiment done with electrons—note that the pattern is not evident with only a few electrons (top photo), but with more and more electrons (second and third photos), the familiar double-slit interference pattern (Chapter 24) is seen.

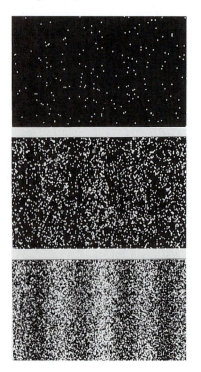

light so the effect of individual photons can be seen. If we are dealing with only one photon, the relationship above ($N \propto E^2$) can be interpreted in a slightly different way. At any point the square of the electric field strength, E^2, is a measure of the *probability* that a photon will be at that location. At points where E^2 is large, there is a high probability the photon will be there; where E^2 is small, the probability is low.

We can interpret matter waves in the same way, as was first suggested by Max Born (1882–1970) in 1927. The wave function Ψ may vary in magnitude from point to point in space and time. If Ψ describes a collection of many electrons, then Ψ^2 at any point will be proportional to the number of electrons expected to be found at that point. When dealing with small numbers of electrons we can't make very exact predictions, so Ψ^2 takes on the character of a probability. If Ψ, which depends on time and position, represents a single electron (say, in an atom), then Ψ^2 is interpreted as follows: Ψ^2 *at a certain point in space and time represents the probability of finding the electron at the given position and time.*

To understand this better, we take as a thought experiment the familiar double-slit experiment, and consider it both for light and for electrons.

Consider two slits whose size and separation are on the order of the wavelength of whatever we direct at them, either light or electrons, Fig. 28–3. We know very well what would happen in this case for light, since this is just Young's double-slit experiment (Section 24–3): an interference pattern would be seen on the screen behind. If light were replaced by electrons with wavelength comparable to the slit size, they too would produce an interference pattern (recall Fig. 27–15). In the case of light, the pattern would be visible to the eye or could be recorded on film. For electrons, a fluorescent screen could be used (it glows where an electron strikes).

Now, if we reduced the flow of electrons (or photons) so that they passed through the slits one at a time, we would see a flash each time one struck the screen. At first, the flashes would seem random. Indeed, there is no way to predict just where any one electron would hit the screen. If we let the experiment run for a long time, however, and kept track of where each electron hit the screen, we would soon see a pattern emerging—namely the interference pattern predicted by the wave theory; see Fig. 28–4. Thus, although we could not predict where a given electron would strike the screen, we could predict probabilities. (The same can be said for photons.) The probability, as mentioned before, is proportional to Ψ^2. Where Ψ^2 is zero, we would get a minimum in the interference pattern. And where Ψ^2 is a maximum, we would get a peak in the interference pattern.

Since the interference pattern would occur even when electrons (or photons) passed through the slits one at a time, it is clear that the interference pattern would not arise from the interaction of one electron with another. It is as if the electron passed through both slits at the same time, interfering with itself. This is possible because, remember, an electron is not precisely a particle. It is as much a wave as it is a particle, and a wave could certainly travel through both slits at once. But what would happen if we covered one of the slits so we knew that the electron passed through the other one, and after a time we covered the second slit so the electron had to have passed through the first? The result would be that no interference pattern would be seen. We would see, instead, two bright areas (or diffraction patterns) on the screen behind the slits. This confirms our idea that if both slits were open, each electron would pass through both, as if it were a wave, forming an interference pattern. Yet each electron would make a tiny spot on the screen as if it were a particle.

The main point of this discussion is this: if we treat electrons (and other matter) as if they were waves, then Ψ represents the wave amplitude. If we treat them as particles, then we must treat them on a *probabilistic* basis. The square of the wave function, Ψ^2, gives the probability of finding a given electron at a given point. We cannot predict—or even follow—the path of a single electron precisely through space and time.

28–3 The Heisenberg Uncertainty Principle

Whenever a measurement is made, some uncertainty or error is always involved. For example, you cannot make an absolutely exact measurement of the length of a table. Even with a measuring stick that has markings 1 mm apart, there will be an inaccuracy of about $\frac{1}{2}$ mm or so. More precise instruments will produce more precise measurements. But there is always some uncertainty involved in a measurement no matter how good the measuring device. We expect that by using more precise instruments, the uncertainty in a measurement can be made indefinitely small.

But according to quantum mechanics, there is actually a limit to the accuracy of certain measurements. This limit is not a restriction on how well instruments can be made; rather, it is inherent in nature. It is the result of two factors: the wave–particle duality, and the unavoidable interaction between the thing observed and the observing instrument. Let us look at this in more detail.

To make a measurement of an object without somehow disturbing it, at least a little, is not possible. Consider trying to locate a Ping-pong ball in a completely dark room, You grope about trying to find its position; and just when you touch it with your finger, it bounces away. Whenever we measure the position of an object, whether it's a Ping-pong ball or an electron, we always touch it with something else that gives us the information about its position. To locate a lost Ping-pong ball in a dark room, you could probe about with your hand or a stick; or you could shine a light and detect the light reflecting off the ball. When you search with your hand or a stick, you find the ball's position when you touch it. But when you touch the ball you unavoidably bump it, and give it some momentum. Thus you won't know its *future* position. The same would be true, but to a much lesser extent, if you observe the Ping-pong ball using light. In order to "see" the ball, at least one photon must scatter from it, and the reflected photon must enter your eye or some other detector. When a photon strikes an ordinary-sized object, it does not appreciably alter the motion or position of the object. But when a photon strikes a very tiny object like an electron, it can transfer momentum to the object and thus greatly change the object's motion and position in an unpredictable way. The mere act of measuring the position of an object at one time makes our knowledge of its future position imprecise.

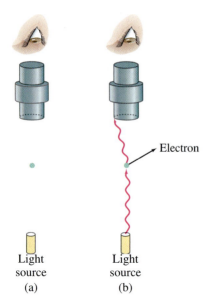

FIGURE 28–5 Thought experiment for observing an electron with a powerful light microscope. At least one photon must scatter from the electron (transferring some momentum to it) and enter the microscope.

Now let us see where the wave–particle duality comes in. Imagine a thought experiment in which we are trying to measure the position of an object, say an electron, with photons, Fig. 28–5. (The arguments would be similar if we were using, instead, an electron microscope.) As we saw in Chapter 25, objects can be seen to an accuracy at best of about the wavelength of the radiation used. If we want an accurate position measurement, we must use a short wavelength. But a short wavelength corresponds to high frequency and high energy (since $E = hf$); and the more energy the photons have, the more momentum they can give the object when they strike it. If photons of longer wavelength, and correspondingly lower energy are used, the object's motion when struck by the photons will not be affected as much. But the longer wavelength means lower resolution, so the object's position will be less accurately known. Thus the act of observing produces a significant uncertainty in either the *position* or the *momentum* of the electron. This is the essence of the *uncertainty principle* first enunciated by Heisenberg in 1927.

Quantitatively, we can make an approximate calculation of the magnitude of this effect. If we use light of wavelength λ, the position can be measured at best to an accuracy of about λ. That is, the uncertainty in the position measurement, Δx, is approximately

$$\Delta x \approx \lambda.$$

Suppose that the object can be detected by a single photon. The photon has a momentum $p = h/\lambda$. When the photon strikes our object, it will give some or all of this momentum to the object, Fig. 28–5. Therefore, the final momentum of our object will be uncertain in the amount

$$\Delta p \approx \frac{h}{\lambda}$$

since we can't tell beforehand how much momentum will be transferred. The product of these uncertainties is

$$(\Delta x)(\Delta p) \approx h.$$

Of course, the uncertainties could be worse than this, depending on the apparatus and the number of photons needed for detection. In Heisenberg's more careful calculation, he found that at the very best

UNCERTAINTY PRINCIPLE
(Δx and Δp)

$$(\Delta x)(\Delta p) \gtrsim \frac{h}{2\pi}. \qquad \textbf{(28–1)}$$

This is a mathematical statement of the **Heisenberg uncertainty principle**, or, as it is sometimes called, the **indeterminancy principle**. It tells us that we cannot measure both the position and momentum of an object precisely at the same time. The more accurately we try to measure the position, so that Δx is small, the greater will be the uncertainty in momentum, Δp. If we try to measure the momentum very precisely, then the uncertainty in the position becomes large. The uncertainty principle does not forbid single exact measurements, however. For example, in principle we could measure the position of an object exactly. But then its momentum would be completely unknown. Thus, although we might know the position of the object exactly at one instant, we could have no idea at all where it would be a moment later.

Another useful form of the uncertainty principle relates energy and time, and we examine this as follows. The object to be detected has an uncertainty in position $\Delta x \approx \lambda$. Now the photon used to detect it travels with speed c, and it takes a time $\Delta t \approx \Delta x/c \approx \lambda/c$ to pass through the distance of uncertainty. Hence, the measured time when our object is at a given position is uncertain by about

$$\Delta t \approx \frac{\lambda}{c}.$$

Since the photon can transfer some or all of its energy ($= hf = hc/\lambda$) to our object, the uncertainty in energy of our object as a result is

$$\Delta E \approx \frac{hc}{\lambda}.$$

The product of these two uncertainties is

$$(\Delta E)(\Delta t) \approx h.$$

Heisenberg's more careful calculation gives

$$(\Delta E)(\Delta t) \gtrsim \frac{h}{2\pi}. \qquad (28\text{--}2)$$

UNCERTAINTY PRINCIPLE (ΔE and Δt)

This form of the uncertainty principle tells us that the energy of an object can be uncertain, or may even be nonconserved, by an amount ΔE for a time $\Delta t \approx h/(2\pi \Delta E)$.

The quantity ($h/2\pi$) appears so often in quantum mechanics that for convenience it is given the symbol $\hbar$ ("h-bar"). That is

$$\hbar = \frac{h}{2\pi} = \frac{6.626 \times 10^{-34}\ \text{J}\cdot\text{s}}{2\pi} = 1.055 \times 10^{-34}\ \text{J}\cdot\text{s}.$$

By using this notation, Eqs. 28–1 and 28–2 for the uncertainty principle can be written

$$(\Delta x)(\Delta p) \gtrsim \hbar \qquad \text{and} \qquad (\Delta E)(\Delta t) \gtrsim \hbar.$$

We have been discussing the position and velocity of an electron as if it were a particle. But it isn't a particle. Indeed, we have the uncertainty principle because an electron—and matter in general—has wave as well as particle properties. What the uncertainty principle really tells us is that if we insist on thinking of the electron as a particle, then there are certain limitations on this simplified view—namely, that the position and velocity cannot both be known precisely at the same time; and that the energy can be uncertain (or nonconserved) in the amount ΔE for a time $\Delta t \approx \hbar/\Delta E$.

Because Planck's constant, h, is so small, the uncertainties expressed in the uncertainty principle are usually negligible on the macroscopic level. But at the level of the atom, the uncertainties are significant. Because we consider ordinary objects to be made up of atoms containing nuclei and electrons, the uncertainty principle is relevant to our understanding of all of nature. The uncertainty principle expresses, perhaps most clearly, the probabilistic nature of quantum mechanics. It thus is often used as a basis for philosophic discussion.

SECTION 28–3   The Heisenberg Uncertainty Principle   **865**

EXAMPLE 28–1 **Position uncertainty of electron.** An electron moves in a straight line with a constant speed $v = 1.10 \times 10^6$ m/s which has been measured to a precision of 0.10 percent. What is the maximum precision with which its position could be simultaneously measured?

SOLUTION The momentum of the electron is $p = mv = (9.11 \times 10^{-31}$ kg$)\cdot(1.10 \times 10^6$ m/s$) = 1.00 \times 10^{-24}$ kg·m/s. The uncertainty in the momentum is 0.10 percent of this, or

$$\Delta p = 1.0 \times 10^{-27} \text{ kg·m/s}.$$

From the uncertainty principle, the best simultaneous position measurement will have an uncertainty of

$$\Delta x = \frac{\hbar}{\Delta p} = \frac{1.06 \times 10^{-34} \text{ J·s}}{(1.0 \times 10^{-27} \text{ kg·m/s})} = 1.1 \times 10^{-7} \text{ m},$$

or 110 nm. This is about 1000 times the diameter of an atom.

EXAMPLE 28–2 **Position uncertainty of a baseball.** What is the uncertainty in position, imposed by the uncertainty principle, on a 150-g baseball thrown at (93 ± 2) mph $= (42 \pm 1)$ m/s? (Should the umpire be concerned? Can he use Heisenberg as an excuse?)

SOLUTION The uncertainty in the momentum is

$$\Delta p = m\,\Delta v = (0.150 \text{ kg})(1 \text{ m/s}) = 0.15 \text{ kg·m/s}.$$

Hence the uncertainty in a position measurement could be as small as

$$\Delta x = \frac{\hbar}{\Delta p} = \frac{1.06 \times 10^{-34} \text{ J·s}}{(0.15 \text{ kg·m/s})} = 7 \times 10^{-34} \text{ m},$$

which is a distance incredibly smaller than any we could imagine observing or measuring. Indeed, the uncertainty principle sets no relevant limit on measurement for macroscopic objects. (The ump is on his own for excuses.)

EXAMPLE 28–3 **ESTIMATE** **J/ψ lifetime calculated.** The J/ψ meson, discovered in 1974, was measured to have an average mass of 3100 MeV/c^2 (note the use of energy units since $E_0 = m_0 c^2$) and an intrinsic width of 63 keV/c^2. By this we mean that the masses of different J/ψ mesons were actually measured to be slightly different from one another. This mass "width" is related to the very short lifetime of the J/ψ before it decays into other particles. Estimate its lifetime using the uncertainty principle.

SOLUTION The uncertainty of 63 keV/c^2 in the J/ψ's mass is an uncertainty in its rest energy, which in joules is

$$\Delta E = (63 \times 10^3 \text{ eV})(1.60 \times 10^{-19} \text{ J/eV}) = 1.01 \times 10^{-14} \text{ J}.$$

Then we expect its lifetime $\tau\,(=\Delta t$ here) to be

$$\tau \approx \frac{\hbar}{\Delta E} = \frac{1.06 \times 10^{-34} \text{ J·s}}{1.01 \times 10^{-14} \text{ J}} \approx 1.01 \times 10^{-20} \text{ s}.$$

Lifetimes this short are difficult to measure directly, and the assignment of very short lifetimes depends on this use of the uncertainty principle. (See Chapter 32.)

**Philosophic Implications;
Probability versus Determinism**

The classical Newtonian view of the world is a deterministic one (see Section 5–9). One of its basic ideas is that once the position and velocity of an object are known at a particular time, its future position can be predicted if the forces on it are known. For example, if a stone is thrown a number of times with the same initial velocity and angle, and the forces on it remain the same, the path of the projectile will always be the same. If the forces are known (gravity and air resistance, if any), the stone's path can be precisely predicted. This mechanistic view implies that the future unfolding of the universe, assumed to be made up of particulate bodies, is completely determined.

This classical deterministic view of the physical world has been radically altered by quantum mechanics. As we saw in the analysis of the double-slit experiment (Section 28–2), electrons all prepared in the same way will not all end up in the same place. According to quantum mechanics, certain probabilities exist that an electron will arrive at different points. This is very different from the classical view, in which the path of a particle is precisely predictable from the initial position and velocity and the forces exerted on it. According to quantum mechanics, the position and velocity of an object cannot even be known accurately at the same time. This is expressed in the uncertainty principle, and arises because basic entities, such as electrons, are not considered simply as particles: they have wave properties as well. Quantum mechanics allows us to calculate only the probability[†] that, say, an electron (when thought of as a particle) will be observed at various places. Quantum mechanics says there is some inherent unpredictability in nature.

Since matter is considered to be made up of atoms, even ordinary-sized objects are expected to be governed by probability, rather than by strict determinism. For example, there is a finite (but very small) probability that when you throw a stone, its path will suddenly curve upward instead of following the downward-curved parabola of normal projectile motion. Quantum mechanics predicts with very high probability that ordinary objects will behave just as the classical laws of physics predict. But these predictions are considered probabilities, not certainties. The reason that macroscopic objects behave in accordance with classical laws with very high probability is due to the large number of molecules involved: when large numbers of objects are present in a statistical situation, deviations from the average (or most probable) are negligible. It is the average configuration of vast numbers of molecules that follows the so-called fixed laws of classical physics with such high probability, and gives rise to an apparent "determinism." Deviations from classical laws are observed when small numbers of molecules are dealt with. We can say, then, that although there are no precise deterministic laws in quantum mechanics, there are statistical laws based on probability.

It is important to note that there is a difference between the probability imposed by quantum mechanics and that used in the nineteenth century to understand thermodynamics and the behavior of gases in terms of molecules (Chapters 13 and 15). In thermodynamics, probability is used because there

---

[†]Note that these probabilities can be calculated precisely, just like exact predictions of probabilities at dice or playing cards, but unlike predictions of probabilities at sporting events or for natural or man-made disasters, which are only estimates.

are far too many molecules to be kept track of. But the molecules were still assumed to move and interact in a deterministic way according to Newton's laws. Probability in quantum mechanics is quite different. It is seen as *inherent* in nature, and not as a limitation on our abilities to calculate.

Although a few physicists have not given up the deterministic view of nature and have been reluctant to accept quantum mechanics as a complete theory—one was Einstein—nonetheless, the vast majority of physicists do accept quantum mechanics and the probabilistic view of nature. This view, which as presented here is the generally accepted one, is called the **Copenhagen interpretation** of quantum mechanics in honor of Niels Bohr's home, since it was largely developed there through discussions between Bohr and other prominent physicists.

Because electrons are not simply particles, they cannot be thought of as following particular paths in space and time. This suggests that a description of matter in space and time may not be completely correct. This deep and far-reaching conclusion has been a lively topic of discussion among philosophers. Perhaps the most important and influential philosopher of quantum mechanics was Bohr. He argued that a space–time description of actual atoms and electrons is not possible. Yet a description of experiments on atoms or electrons must be given in terms of space and time and other concepts familiar to ordinary experience, such as waves and particles. We must not let our *descriptions* of experiments lead us into believing that atoms or electrons themselves actually exist in space and time as particles. This distinction between our interpretations of experiments and what is "really" happening in nature is crucial.

28–5 Quantum-Mechanical View of Atoms

At the beginning of this chapter, we discussed the limitations of the Bohr theory of atomic structure. Now we examine the quantum-mechanical theory of atoms, which is far more complete than the old Bohr theory. Although the Bohr model has been discarded as an accurate description of nature, nonetheless, quantum mechanics reaffirms certain aspects of the older theory, such as that electrons in an atom exist only in discrete states of definite energy, and that a photon of light is emitted (or absorbed) when an electron makes a transition from one state to another. But quantum mechanics is a much deeper theory, and has provided us with a very different view of the atom. According to quantum mechanics, electrons do not exist in well-defined circular orbits as in the Bohr theory. Rather, the electron (because of its wave nature) can be thought of as spread out in space as a "cloud." The size and shape of the electron cloud can be calculated for a given state of an atom. For the ground state in the hydrogen atom, the electron cloud is spherically symmetric, as shown in Fig. 28–6. The electron cloud roughly indicates the "size" of an atom; but just as a cloud may not have a distinct border, atoms do not have a precise boundary or a well-defined size. Not all electron clouds have a spherical shape, as we shall see later in this chapter.

The electron cloud can be interpreted using either the particle or the wave viewpoint. Remember that by a particle we mean something that is localized in space—it has a definite position at any given instant. By contrast, a wave is spread out in space. The electron cloud, spread out in space

FIGURE 28–6
Electron cloud or "probability distribution" for the ground state of the hydrogen atom. The circle represents the Bohr radius.

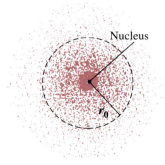

Nucleus

r_0

as in Fig. 28–6, is a result of the wave nature of electrons. Electron clouds can also be interpreted as **probability distributions** for a particle. If you were to make 500 different measurements of the position of an electron (considering it as a particle), the majority of the results would show the electron at points where the probability is high (darker areas with more dots in Fig. 28–6). Only occasionally would the electron be found where the probability is low. We cannot predict the path an electron will follow. As we saw in Section 28–3, after one measurement of its position we cannot predict exactly where it will be at a later time. We can only calculate the probability that it will be found at different points. This is clearly different from classical Newtonian physics. Indeed, as Bohr later pointed out, it is meaningless even to ask how an electron gets from one state to another when the atom emits a photon of light.

Probability distributions

28–6 Quantum Mechanics of the Hydrogen Atom; Quantum Numbers

We now look more closely at what quantum mechanics tells us about the hydrogen atom. Much of what we say here also applies to more complex atoms, which are discussed in the next Section.

Quantum mechanics predicts basically the same energy levels (Fig. 27–32) for the hydrogen atom as does the Bohr theory. That is,

$$E_n = -\frac{13.6 \text{ eV}}{n^2}, \qquad n = 1, 2, 3, \cdots,$$

where n is an integer. In the simple Bohr theory, there was only one quantum number, n. In quantum mechanics, it turns out that four different quantum numbers are needed to specify each state in the atom.

Quantum numbers

The *quantum number*, n, from the Bohr theory is found also in quantum mechanics and is called the **principal quantum number**. It can have any integer value from 1 to ∞. The total energy of a state in the hydrogen atom depends on n, as we saw above.

The **orbital quantum number**, l, is related to the magnitude of the angular momentum of the electron; l can take on integer values from 0 to $(n - 1)$. For the ground state ($n = 1$), l can only be zero.[†] But for $n = 3$, l can be 0, 1, or 2. The actual magnitude of the angular momentum L is related to the quantum number l by

$$L = \sqrt{l(l + 1)}\,\hbar \tag{28–3}$$

(where again $\hbar = h/2\pi$). The value of l has almost no effect on the total energy in the hydrogen atom; only n does to any appreciable extent.[‡] But in atoms with two or more electrons, the energy does depend on l as well as n, as we shall see in the next Section.

The **magnetic quantum number**, m_l, is related to the direction of the electron's angular momentum, and it can take on integer values ranging from $-l$ to $+l$. For example, if $l = 2$, then m_l can be $-2, -1, 0, +1,$ or $+2$. Since angular momentum is a vector, it is not surprising that both its mag-

[†] Contrast this with the Bohr theory, which assigned $l = 1$ to the ground state (Eq. 27–11).
[‡] See discussion of *fine structure* on the next page.

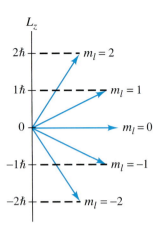

FIGURE 28–7 Quantization of angular momentum direction for $l = 2$.

FIGURE 28–8 When a magnetic field is applied, an $n = 3$, $l = 2$ energy level is split into five separate levels, corresponding to the five values of m_l (2, 1, 0, −1, −2). An $n = 2$, $l = 1$ level is split into three levels ($m_l = 1, 0, −1$). Transitions can occur between levels (not all transitions are shown), with photons of several slightly different frequencies being given off (the Zeeman effect).

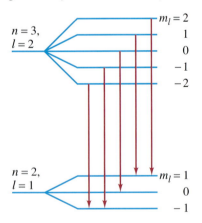

nitude and its direction would be quantized. For $l = 2$, the five different directions allowed can be represented by the diagram of Fig. 28–7. This limitation on the direction of **L** is often called **space quantization**. In quantum mechanics, the direction of the angular momentum is usually specified by giving its component along the z axis (this choice is arbitrary). Then L_z is related to m_l by the equation

$$L_z = m_l \hbar.$$

The name for m_l derives not from theory (which relates it to L_z), but from experiment. It was found that when a gas-discharge tube was placed in a magnetic field, the spectral lines were split into several very closely spaced lines. This splitting, known as the **Zeeman effect**, implies that the energy levels must be split (Fig. 28–8), and thus that the energy of a state depends not only on n but also on m_l when a magnetic field is applied—hence the name "magnetic quantum number." (Why the energy should depend on the direction of **L** can be seen from a semiclassical view of a moving electron as an electric current that interacts with the magnetic field.)

Finally, there is the **spin quantum number**, m_s, which for an electron can have only two values, $m_s = +\frac{1}{2}$ and $m_s = -\frac{1}{2}$. The existence of this quantum number did not come out of Schrödinger's original theory, as did n, l, and m_l. Instead, a subsequent modification by P. A. M. Dirac (1902–1984) explained its presence as a relativistic effect. The first hint that m_s was needed, however, came from experiment. A careful study of the spectral lines of hydrogen showed that each actually consisted of two (or more) very closely spaced lines even in the absence of an external magnetic field. It was at first hypothesized that this tiny splitting of energy levels, called **fine structure**, was due to angular momentum associated with a spinning of the electron. That is, the electron might spin on its axis as well as orbit the nucleus, just as the Earth spins on its axis as it orbits the Sun. The interaction between the tiny current of the spinning electron could then interact with the magnetic field due to the orbiting charge and cause the small observed splitting of energy levels. (So the energy depends slightly on m_l and m_s.) Today we consider this picture of a spinning electron as not legitimate. We cannot even view an electron as a localized object, much less a spinning one. What is important is that the electron can have two different states due to some intrinsic property that behaves as an angular momentum, and we still call this property "spin." The two possible values of m_s ($+\frac{1}{2}$ and $-\frac{1}{2}$) are often said to be "spin up" and "spin down," referring to the two possible directions of the spin angular momentum.

The possible values of the four quantum numbers for an electron in the hydrogen atom are summarized in Table 28–1.

TABLE 28–1 Quantum Numbers for an Electron

Name	Symbol	Possible Values
Principal	n	$1, 2, 3, \cdots, \infty$.
Orbital	l	For a given n: l can be $0, 1, 2, \cdots, n - 1$.
Magnetic	m_l	For given n and l: m_l can be $l, l - 1, \cdots, 0, \cdots, -l$.
Spin	m_s	For each set of n, l, and m_l: m_s can be $+\frac{1}{2}$ or $-\frac{1}{2}$.

CONCEPTUAL EXAMPLE 28–4 **Possible states for n = 3.** How many different states are possible for an electron whose principal quantum number is $n = 3$?

RESPONSE For $n = 3$, l can have the values $l = 2, 1, 0$. For $l = 2$, m_l can be $2, 1, 0, -1, -2$, which is five different possibilities. For each of these, m_s can be either up or down $(+\frac{1}{2}$ or $-\frac{1}{2})$; so for $l = 2$, there are $2 \times 5 = 10$ states. For $l = 1$, m_l can be $1, 0, -1$, and since m_s can be $+\frac{1}{2}$ or $-\frac{1}{2}$ for each of these, we have six more possible states. Finally, for $l = 0$, m_l can only be 0, and there are only two states corresponding to $m_s = +\frac{1}{2}$ and $-\frac{1}{2}$. The total number of states is $10 + 6 + 2 = 18$, as detailed in the following table:

n	l	m_l	m_s	n	l	m_l	m_s
3	2	2	$\frac{1}{2}$	3	2	-2	$-\frac{1}{2}$
3	2	2	$-\frac{1}{2}$	3	1	1	$\frac{1}{2}$
3	2	1	$\frac{1}{2}$	3	1	1	$-\frac{1}{2}$
3	2	1	$-\frac{1}{2}$	3	1	0	$\frac{1}{2}$
3	2	0	$\frac{1}{2}$	3	1	0	$-\frac{1}{2}$
3	2	0	$-\frac{1}{2}$	3	1	-1	$\frac{1}{2}$
3	2	-1	$\frac{1}{2}$	3	1	-1	$-\frac{1}{2}$
3	2	-1	$-\frac{1}{2}$	3	0	0	$\frac{1}{2}$
3	2	-2	$\frac{1}{2}$	3	0	0	$-\frac{1}{2}$

EXAMPLE 28–5 **E and L for n = 3.** Determine (a) the energy and (b) the orbital angular momentum for each of the states in Example 28–4.

SOLUTION (a) The energy of a state depends only on n, except for the very small corrections mentioned above, which we will ignore. Since $n = 3$ for all these states, they all have the same energy,

$$E_3 = -\frac{13.6 \text{ eV}}{(3)^2} = -1.51 \text{ eV}.$$

(b) For $l = 0$,

$$L = \sqrt{l(l + 1)}\hbar = 0.$$

For $l = 1$,

$$L = \sqrt{1(1 + 1)}\hbar = \sqrt{2}\hbar$$
$$= 1.49 \times 10^{-34} \text{ J·s}.$$

Atomic angular momenta are generally given as a multiple of $\hbar$ (e.g., $\sqrt{2}\hbar$ in this case), rather than in SI units. But note that for $l = 1$, L is on the order of 10^{-34} J·s. This means that macroscopic angular momenta will have such extremely high quantum numbers that the quantization of angular momentum will not be detectable: L will appear continuous, in accordance with the correspondence principle. Finally, for $l = 2$,

$$L = \sqrt{2(2 + 1)}\hbar = \sqrt{6}\hbar.$$

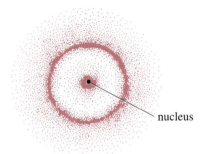

(a) $n = 2$, $l = 0$, $m_l = 0$

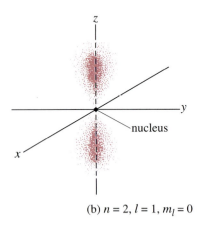

(b) $n = 2$, $l = 1$, $m_l = 0$

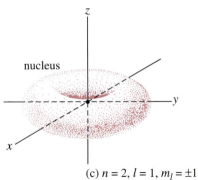

(c) $n = 2$, $l = 1$, $m_l = \pm 1$

FIGURE 28–9 Electron cloud, or probability distribution, for $n = 2$ states in hydrogen.

Another prediction of quantum mechanics is that when a photon is emitted or absorbed, transitions can occur only between states with values of l that differ by one unit:

$$\Delta l = \pm 1.$$

According to this **selection rule**, an electron in an $l = 2$ state can jump only to a state with $l = 1$ or $l = 3$. It cannot jump to a state with $l = 2$ or $l = 0$. A transition such as $l = 2$ to $l = 0$ is called a **forbidden transition**. Actually, such a transition is not absolutely forbidden and can occur, but only with very low probability compared to **allowed transitions**—those that satisfy the selection rule $\Delta l = \pm 1$. Since the orbital angular momentum of an H atom must change by one unit when it emits a photon, conservation of angular momentum tells us that the photon must carry off angular momentum. Indeed, experimental evidence of many sorts shows that the photon can be assigned a spin of 1.

Although l and m_l do not significantly affect the energy levels in hydrogen, they do affect the electron probability distribution in space. For $n = 1$, l and m_l can only be zero and the electron distribution is as shown in Fig. 28–6. For $n = 2$, l can be 0 or 1. The distribution for $n = 2$, $l = 0$ is shown in Fig. 28–9a, and it is seen to differ from that for the ground state, although it is still spherically symmetric. For $n = 2$, $l = 1$, the distributions are not spherically symmetric as shown in Figs. 28–9b (for $m_l = 0$) and 28–9c (for $m_l = +1$ or -1).

Although the spatial distributions of the electron can be calculated for the various states, it is difficult to measure them experimentally. Most of the experimental information about the atom has come from a careful examination of the emission spectra under various conditions.

28–7 Complex Atoms; the Exclusion Principle

We have discussed the hydrogen atom in detail because it is the simplest to deal with. Now we briefly discuss more complex atoms, those that contain more than one electron, and whose energy levels can be determined experimentally from an analysis of the emission spectra. The energy levels are *not* the same as in the H atom, since the electrons interact with each other as well as with the nucleus. For atoms with more than one electron, the energy levels depend on both n and l.

The number of electrons in a neutral atom is called its **atomic number**, Z; Z is also the number of positive charges (protons) in the nucleus, and determines what kind of atom it is. That is, Z determines most of the properties that distinguish one atom from another.

Although modifications of the Bohr theory had been attempted in order to deal with complex atoms, the development of quantum mechanics in the years after 1925 proved far more successful. The mathematics becomes very difficult, however, since in multielectron atoms, each electron is not only attracted to the nucleus but is repelled by the other electrons.

The simplest approach has been to treat each electron in an atom as occupying a particular state characterized by the quantum numbers n, l, m_l, and m_s. But to understand the possible arrangements of electrons in an atom, a new principle was needed. It was introduced by Wolfgang Pauli (1900–1958; Fig. 28–2) and is called the **Pauli exclusion principle**. It states:

No two electrons in an atom can occupy the same quantum state.

Pauli exclusion principle

Thus, no two electrons in an atom can have exactly the same set of the quantum numbers $n, l, m_l,$ and m_s. The Pauli exclusion principle[†] forms the basis not only for understanding complex atoms, but also for understanding molecules and bonding, and other phenomena as well.

Let us now look at the structure of some of the simpler atoms when they are in the ground state. After hydrogen, the next simplest atom is *helium* with two electrons. Both electrons can have $n = 1$, since one can have spin up ($m_s = +\frac{1}{2}$) and the other spin down ($m_s = -\frac{1}{2}$), thus satisfying the exclusion principle. Of course, since $n = 1$, l and m_l must be zero (Table 28–1). Thus the two electrons have the quantum numbers indicated in the table in the margin.

Lithium has three electrons, two of which can have $n = 1$. But the third cannot have $n = 1$ without violating the exclusion principle. Hence the third electron must have $n = 2$. Since it happens that the $n = 2, l = 0$ level has a lower energy than $n = 2, l = 1$, the electrons in the ground state have the quantum numbers indicated in the table in the margin. Of course, the quantum numbers of the third electron could also be, say, $(3, 1, -1, \frac{1}{2})$. But the atom in this case would be in an excited state since it would have greater energy. It would not be long before it jumped to the ground state with the emission of a photon. At room temperature, unless extra energy is supplied (as in a discharge tube), the vast majority of atoms are in the ground state.

We can continue in this way to describe the quantum numbers of each electron in the ground state of larger and larger atoms. That for sodium, with its eleven electrons, is shown in the table in the margin.

Figure 28–10 shows a simple energy level diagram where occupied states are shown as up or down arrows ($m_s = +\frac{1}{2}$ or $-\frac{1}{2}$), and possible empty states are shown as a small circle.

The ground-state configuration for all atoms is given in the **periodic table**, which is displayed inside the back cover of this book, and discussed in the next Section.

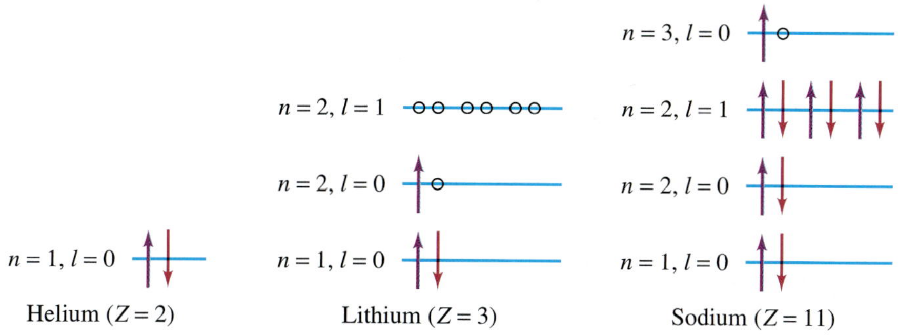

Helium, $Z = 2$

n	l	m_l	m_s
1	0	0	$\frac{1}{2}$
1	0	0	$-\frac{1}{2}$

Lithium, $Z = 3$

n	l	m_l	m_s
1	0	0	$\frac{1}{2}$
1	0	0	$-\frac{1}{2}$
2	0	0	$\frac{1}{2}$

Sodium, $Z = 11$

n	l	m_l	m_s
1	0	0	$\frac{1}{2}$
1	0	0	$-\frac{1}{2}$
2	0	0	$\frac{1}{2}$
2	0	0	$-\frac{1}{2}$
2	1	1	$\frac{1}{2}$
2	1	1	$-\frac{1}{2}$
2	1	0	$\frac{1}{2}$
2	1	0	$-\frac{1}{2}$
2	1	-1	$\frac{1}{2}$
2	1	-1	$-\frac{1}{2}$
3	0	0	$\frac{1}{2}$

$n = 3, l = 0$

$n = 2, l = 1$ $n = 2, l = 1$

$n = 2, l = 0$ $n = 2, l = 0$

$n = 1, l = 0$ $n = 1, l = 0$ $n = 1, l = 0$

Helium ($Z = 2$)   Lithium ($Z = 3$)   Sodium ($Z = 11$)

FIGURE 28–10 Energy level diagram showing occupied states (arrows) and unoccupied states (○) for He, Li, and Na. Note that we have shown the $n = 2, l = 1$ level of Li even though it is empty.

[†]The exclusion principle applies to identical particles whose spin quantum number is a half-integer ($\frac{1}{2}, \frac{3}{2}$, and so on), including electrons, protons, and neutrons; such particles are called **fermions** (after E. Fermi who derived a statistical theory describing them). The exclusion principle does not apply to particles with integer spin (0, 1, 2, and so on), such as the photon and π meson, all of which are referred to as **bosons** (after S. N. Bose, who derived a statistical theory for them).

TABLE 28–2 Values of *l*

Value of *l*	Letter Symbol	Maximum Number of Electrons in Subshell
0	*s*	2
1	*p*	6
2	*d*	10
3	*f*	14
4	*g*	18
5	*h*	22
⋮	⋮	⋮

Shells and subshells

TABLE 28–3 Electron Configuration of Some Elements

Z (No. of Electrons)	Element[†]	Ground State Configuration (outer electrons)
1	H	$1s^1$
2	He	$1s^2$
3	Li	$2s^1$
4	Be	$2s^2$
5	B	$2s^2 2p^1$
6	C	$2s^2 2p^2$
7	N	$2s^2 2p^3$
8	O	$2s^2 2p^4$
9	F	$2s^2 2p^5$
10	Ne	$2s^2 2p^6$
11	Na	$3s^1$
12	Mg	$3s^2$
13	Al	$3s^2 3p^1$
14	Si	$3s^2 3p^2$
15	P	$3s^2 3p^3$
16	S	$3s^2 3p^4$
17	Cl	$3s^2 3p^5$
18	Ar	$3s^2 3p^6$
19	K	$4s^1$
20	Ca	$4s^2$
21	Sc	$3d^1 4s^2$
22	Ti	$3d^2 4s^2$
23	V	$3d^3 4s^2$
24	Cr	$3d^5 4s^1$
25	Mn	$3d^5 4s^2$
26	Fe	$3d^6 4s^2$

[†]Names of elements can be found in Appendix F.

28–8 The Periodic Table of Elements

A century ago, Dmitri Mendeleev (1834–1907) arranged the then known elements into what we now call the **periodic table** of the elements. The atoms were arranged according to increasing mass, but also so that elements with similar chemical properties would fall in the same column. Today's version is shown inside the back cover. Each square contains the atomic number Z, the symbol for the element, and the atomic mass (in atomic mass units). Finally, in the lower left corner the configuration of the ground state of the atom is given. This requires some explanation. Electrons with the same value of n are referred to as being in the same **shell**. Electrons with $n = 1$ are in one shell (the K shell), those with $n = 2$ are in a second shell (the L shell), those with $n = 3$ are in the third (M shell), and so on. Electrons with the same values of n and l are referred to as being in the same **subshell**. Letters are often used to specify the value of l as shown in Table 28–2. That is, $l = 0$ is the s subshell; $l = 1$ is the p subshell; $l = 2$ is the d subshell; beginning with $l = 3$, the letters follow the alphabet, f, g, h, i, and so on. (The first letters s, p, d, and f were originally abbreviations of "sharp," "principal," "diffuse," and "fundamental," experimental terms referring to the spectra.)

The Pauli exclusion principle limits the number of electrons possible in each shell and subshell. For any value of l, there are $2l + 1$ different m_l values (m_l can be any integer from 1 to l, from -1 to $-l$, or zero), and two different m_s values. There can be, therefore, at most $2(2l + 1)$ electrons in any l subshell. For example, for $l = 2$, five m_l values are possible 2, 1, 0, -1, -2), and for each of these, m_s can be $+\frac{1}{2}$ or $-\frac{1}{2}$, for a total of $2(5) = 10$ states. Table 28–2 lists the maximum number of electrons that can occupy each subshell.

Since the energy levels depend almost entirely on the values of n and l, it is customary to specify the electron configuration simply by giving the n value and the appropriate letter for l, with the number of electrons in each subshell given as a superscript. The ground-state configuration of sodium, for example, is written as $1s^2 2s^2 2p^6 3s^1$. This is simplified in the periodic table by specifying the configuration only of the outermost electrons and any other nonfilled subshells (see Table 28–3 here, and the periodic table inside the back cover).

CONCEPTUAL EXAMPLE 28–6 **Electron configurations.** Which of the following electron configurations are possible, and which forbidden? (*a*) $1s^2 2s^2 2p^6 3s^3$; (*b*) $1s^2 2s^2 2p^6 3s^2 3p^5 4s^2$; (*c*) $1s^2 2s^2 2p^6 2d^1$.

RESPONSE (*a*) This is not allowed, because there are too many electrons in the s subshell in the M ($n = 3$) shell. The s subshell has $m_l = 0$, with two slots only: for "spin up" and "spin down" electrons. (*b*) This is allowed, but it is an excited state. One of the electrons in the $3p$ subshell has jumped up to the $4s$ subshell. Since there are 19 electrons, the element is potassium. (*c*) This is not allowed, because there is no $d(l = 2)$ subshell in the $n = 2$ shell (Table 28–1). The outermost electron will have to be (at least) in the $n = 3$ shell.

The grouping of atoms in the periodic table is according to increasing mass. There is also a regularity according to chemical properties. And although this is treated in chemistry textbooks, we discuss it here briefly because it is a result of quantum mechanics. See the periodic table on the inside back cover.

All the noble gases (in the last column of the periodic table) have completely filled shells or subshells. That is, their outermost subshell is completely full, and the electron distribution is spherically symmetric. With such full spherical symmetry, other electrons are not attracted nor are electrons readily lost (ionization energy is high). This is why the noble gases are nonreactive (more on this when we discuss molecules and bonding in Chapter 29). Column seven contains the **halogens**, which lack one electron from a filled shell. Because of the shapes of the orbits (see Section 29–1), an additional electron can be accepted from another atom, and hence these elements are quite reactive. They have a valence of -1, meaning that when an extra electron is acquired, the resulting ion has a net charge of $-1e$. At the left of the periodic table, column I contains the **alkali metals**, all of which have a single outer s electron. This electron spends most of its time outside the inner closed shells and subshells which shield it from most of the nuclear charge. Indeed, it is relatively far from the nucleus and is attracted to it by a net charge of only about $+1e$, because of the shielding effect of the other electrons. Hence this outer electron is easily removed and can spend much of its time around another atom, forming a molecule. This is why the alkali metals are highly reactive and have a valence of $+1$. The other columns of the table can be treated similarly.

The presence of the transition elements in the center of the table, as well as the lanthanides (rare earths) and actinides below, is a result of incomplete inner shells. For the lowest Z elements, the subshells are filled in a simple order: first $1s$, then $2s$, followed by $2p$, $3s$, and $3p$. You might expect that $3d$ ($n = 3, l = 2$) would be filled next, but it isn't. Instead, the $4s$ level actually has a slightly lower energy than the $3d$ (due to electrons interacting with each other), so it fills first (K and Ca). Only then does the $3d$ shell start to fill up, beginning with Sc. (The $4s$ and $3d$ levels are close, so some elements have only one $4s$ electron, such as Cr.) Most of the chemical properties of these **transition elements** are governed by the relatively loosely held $4s$ electrons and hence they usually have valences of $+1$ or $+2$. See also Table 28–3. A similar effect is responsible for the rare earths, which are shown at the bottom of the periodic table for convenience. All have very similar chemical properties, which are determined by their two outer $6s$ or $7s$ electrons, whereas the different numbers of electrons in the unfilled inner shells have little effect.

28–9 X-Ray Spectra and Atomic Number

The line spectra of atoms in the visible, UV, and IR regions of the EM spectrum are mainly due to transitions between states of the outer electrons. Much of the charge of the nucleus is shielded from these electrons by the negative charge on the inner electrons. But the innermost electrons in the $n = 1$ shell "see" the full charge of the nucleus. Since the energy of a level is proportional to Z^2 (see Eq. 27–14), for an atom with $Z = 50$, we would expect wavelengths about $50^2 = 2500$ times shorter than those found in the Lyman series of hydrogen (around 100 nm), or 10^{-2} to 10^{-1} nm. Such short wavelengths lie in the X-ray region of the spectrum.

X-rays are produced when electrons accelerated by a high voltage strike the metal target inside the X-ray tube (Section 25–11). If we look at the spectrum of wavelengths emitted by an X-ray tube, we see that the spectrum consists of two parts: a continuous spectrum with a cutoff at some λ_0

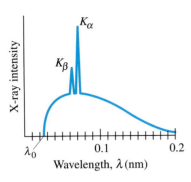

FIGURE 28–11 Spectrum of X-rays emitted from a molybdenum target in an X-ray tube operated at 50 kV.

FIGURE 28–12 Plot of $\sqrt{1/\lambda}$ vs. Z for K_α X-ray lines.

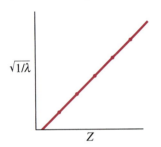

which depends only on the voltage across the tube, and a series of peaks superimposed. A typical example is shown in Fig. 28–11. The smooth curve and the cutoff wavelength λ_0 move to the left as the voltage across the tube increases. The peaks (labeled K_α and K_β in Fig. 28–11), however, remain at the same wavelength when the voltage is changed, although they are located at different wavelengths when different target materials are used. This observation suggests that the peaks are characteristic of the material used. Indeed, we can explain them by imagining that the electrons accelerated by the high voltage of the tube can reach sufficient energies that when they collide with the atoms of the target, they can knock out one of the very tightly held inner electrons. Then we explain these **characteristic X-rays** (the peaks in Fig. 28–11) as photons emitted when an electron in an upper state drops down to fill the vacated lower state.

The K lines result from transitions *into* the K shell ($n = 1$). The K_α line is a transition that originates from the $n = 2$ (L) shell, the K_β line from the $n = 3$ (M) shell. An L line is due to a transition into the L shell, and so on.

Measurement of the characteristic X-ray spectra has allowed a determination of the inner energy levels of atoms. It has also allowed the determination of Z values for many atoms, since (as we have seen) the wavelength of the shortest X-rays emitted will be inversely proportional to Z^2. Actually, for an electron jumping from, say, the $n = 2$ to the $n = 1$ level, the wavelength is inversely proportional to $(Z - 1)^2$ because the nucleus is shielded by the one electron that still remains in the 1s level. In 1914, H. G. J. Moseley (1887–1915) found that a plot of $\sqrt{1/\lambda}$ versus Z produced a straight line, Fig. 28–12. The Z values of a number of elements were determined by fitting them to such a **Moseley plot**. The work of Moseley put the concept of atomic number on a firm experimental basis.

EXAMPLE 28–7 **X-ray wavelength.** Estimate the wavelength for an $n = 2$ to $n = 1$ transition in molybdenum ($Z = 42$). What is the energy of such a photon?

SOLUTION We use the Bohr formula, Eq. 27–15, with Z^2 replaced by $(Z - 1)^2 = (41)^2$. Or, more simply, we can use the result of Example 27–11 for the $n = 2$ to $n = 1$ transition in hydrogen ($Z = 1$). Since

$$\lambda \propto \frac{1}{(Z - 1)^2},$$

we will have

$$\lambda = (1.22 \times 10^{-7}\,\text{m})/(41)^2 = 0.073\,\text{nm}.$$

This is close to the measured value (Fig. 28–11) of 0.071 nm. Each of these photons would have energy (in eV) of:

$$E = hf = \frac{hc}{\lambda} = \frac{(6.63 \times 10^{-34}\,\text{J·s})(3.00 \times 10^8\,\text{m/s})}{(7.3 \times 10^{-11}\,\text{m})(1.60 \times 10^{-19}\,\text{J/eV})} = 17\,\text{keV}.$$

EXAMPLE 28–8 **Determining atomic number.** High-energy photons are used to bombard an unknown material. The strongest peak is found for X-rays with an energy of 66 keV. Guess what the material is.

SOLUTION The strongest X-rays are generally for the K_α line (see Fig. 28–11) which occurs when photons knock out K shell electrons (the innermost orbit) and their place is taken by electrons from the L shell. We use the Bohr model, and assume the electrons "see" a nuclear charge of $Z - 1$ (screened by one electron). The hydrogen transition $n = 2$ to $n = 1$ would yield about 10.2 eV (see Fig. 27–32 or Example 27–11). Then since energy E is proportional to Z^2 (Eq. 27–14), or rather $(Z - 1)^2$ as we've just discussed, we can write

$$\frac{(Z - 1)^2}{1^2} = \frac{66 \times 10^3 \, \text{eV}}{10.2 \text{eV}} = 6.5 \times 10^3,$$

so $Z - 1 = \sqrt{6500} = 81$, and $Z = 82$, which makes it lead.

Now we briefly analyze the continuous part of an X-ray spectrum (Fig. 28–11) based on the photon theory of light. When electrons strike the target, they collide with atoms of the material and give up most of their energy as heat (about 99 percent, so X-ray tubes must be cooled). However, electrons can also give up energy by emitting a photon of light. An electron can be decelerated by interaction with atoms of the target (Fig. 28–13). But an accelerating charge can emit radiation (Chapter 22), and in this case it is called **bremsstrahlung** (German for "braking radiation"). Because energy is conserved, the energy of the emitted photon, hf, must equal the loss of kinetic energy of the electron, $\Delta \text{KE} = \text{KE} - \text{KE}'$, so

$$hf = \Delta \text{KE}.$$

An electron may lose all or a part of its energy in such a collision. The continuous X-ray spectrum (Fig. 28–11) is explained as being due to such bremsstrahlung collisions in which varying amounts of energy are lost by the electrons. The shortest-wavelength X-ray (the highest frequency) must be due to an electron that gives up *all* its kinetic energy to produce one photon in a single collision. Since the initial kinetic energy of an electron is equal to the energy given it by the accelerating voltage, V, then $\text{KE} = eV$. In a single collision in which the electron is brought to rest, we have

$$hf_0 = eV$$

or

$$\lambda_0 = \frac{hc}{eV}, \tag{28–4}$$

where $\lambda_0 = c/f_0$ is the cutoff wavelength, Fig. 28–11. This prediction for λ_0 corresponds precisely with that observed experimentally. This result is further evidence that X-rays are a form of electromagnetic radiation (light)[†] and that the photon theory of light is valid.

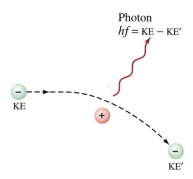

Photon
$hf = \text{KE} - \text{KE}'$

KE

KE'

FIGURE 28–13
Bremsstrahlung photon produced by an electron decelerated by interaction with a target atom.

EXAMPLE 28–9 Cutoff wavelength. What is the shortest-wavelength X-ray photon emitted in an X-ray tube subjected to 50 kV?

SOLUTION From Eq. 28–4,

$$\lambda_0 = \frac{(6.6 \times 10^{-34} \, \text{J·s})(3.0 \times 10^8 \, \text{m/s})}{(1.6 \times 10^{-19} \, \text{C})(5.0 \times 10^4 \, \text{V})} = 2.5 \times 10^{-11} \, \text{m},$$

or 0.025 nm. This agrees well with experiment, Fig. 28–11.

[†]If X-rays were not photons but rather neutral particles with rest mass m_0, Eq. 28–4 would not hold.

X-rays have many applications, some of which we discussed in Chapter 25, including imaging techniques.

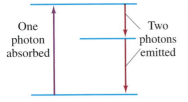

One photon absorbed

Two photons emitted

FIGURE 28–14 Fluorescence.

FIGURE 28–15 When UV light illuminates these various "fluorescent" rocks, they fluoresce in the visible region of the spectrum.

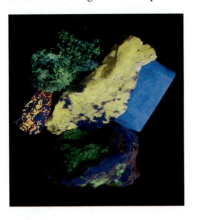

* 28–10 Fluorescence and Phosphorescence

When an atom is excited from one energy state to a higher one by the absorption of a photon, it may return to the lower level in a series of two (or more) jumps if there is an energy level in between (Fig. 28–14). The photons emitted will consequently have lower energy and frequency than the absorbed photon. When the absorbed photon is in the UV and the emitted photons are in the visible region of the spectrum, this phenomenon is called **fluorescence** (Fig. 28–15).

The wavelength for which fluorescence will occur depends on the energy levels of the particular atoms. Because the frequencies are different for different substances, and because many substances fluoresce readily, fluorescence is a powerful tool for identification of compounds. It is also used for assaying—determining how much of a substance is present—and for following substances along a natural pathway as in plants and animals. For detection of a given compound, the stimulating light must be monochromatic, and solvents or other materials present must not fluoresce in the same region of the spectrum. Sometimes the observation of fluorescent light being emitted is sufficient to detect a compound. In other cases, spectrometers are used to measure the wavelengths and intensities of the emitted light.

Fluorescent lightbulbs work in a two-step process. The applied voltage accelerates electrons that strike atoms of the gas in the tube and cause them to be excited. When the excited atoms jump down to their normal levels, they emit UV photons which strike a fluorescent coating on the inside of the tube. The light we see is a result of this material fluorescing in response to the UV light striking it.

Materials such as those used for luminous watch dials are said to be **phosphorescent**. When an atom is raised to a normal excited state, it drops back down within about 10^{-8} s. In phosphorescent substances, atoms can be excited by photon absorption to energy levels, said to be **metastable**, which are states that last much longer—even a few seconds or longer. In a collection of such atoms, many of the atoms will descend to the lower state fairly soon, but many will remain in the excited state for over an hour. Hence light will be emitted even after long periods. When you put your watch dial close to a bright lamp, it excites many atoms to metastable states, and you can see the glow a long time after.

* 28–11 Lasers

A laser is a device that can produce a very narrow intense beam of monochromatic coherent light. (By *coherent*, we mean that across any cross section of the beam, all parts have the same phase.) The emitted beam is a nearly perfect plane wave. An ordinary light source, on the other hand, emits light in all directions (so the intensity decreases rapidly with distance), and the emitted light is incoherent (the different parts of the beam are not in phase with each other). The excited atoms that emit the light in an ordinary lightbulb act independently, so each photon emitted can be considered as a short wave train, typically 30 cm long and lasting 10^{-8} s. These wave trains bear no phase relation to one another. Just the opposite is true of lasers.

The action of a laser is based on quantum theory. We have seen that a photon can be absorbed by an atom if (and only if) its energy hf corresponds to the energy difference between an occupied energy level of the atom and an available excited state, Fig. 28–16a. This is, in a sense, a resonant situation. If the atom is already in the excited state, it may of course jump spontaneously (i.e. no apparent stimulus) to the lower state with the emission of a photon. However, if a photon with this same energy strikes the excited atom, it can stimulate the atom to make the transition sooner to the lower state, Fig. 28–16b. This phenomenon is called **stimulated emission**, and it can be seen that not only do we still have the original photon, but also a second one of the same frequency as a result of the atom's transition. And these two photons are exactly *in phase*, and they are moving in the same direction. This is how coherent light is produced in a laser. Hence the name "laser," which is an acronym for **l**ight **a**mplification by **s**timulated **e**mission of **r**adiation.

Normally, most atoms are in the lower state, so incident photons will mostly be absorbed. In order to obtain the coherent light from stimulated emission, two conditions must be satisfied. First, the atoms must be excited to the higher state. That is, an **inverted population** is needed, one in which more atoms are in the upper state than in the lower one (Fig. 28–17), so that *emission* of photons will dominate over absorption. And second, the higher state must be a **metastable state**—a state in which the electrons remain longer than usual[†] so that the transition to the lower state occurs by stimulated emission rather than spontaneously. How these conditions are achieved for different lasers will be discussed shortly. For now, we assume that the atoms have been excited to an upper state. Figure 28–18 is a schematic diagram of a laser: the "lasing" material is placed in a long narrow tube at the ends of which are two mirrors, one of which is partially transparent (perhaps 1 or 2 percent). Some of the excited atoms drop down fairly soon after being excited. One of these is the atom shown on the far left in Fig. 28–18. If the emitted photon strikes another atom in the excited state, it stimulates this atom to emit a photon of the *same* frequency, moving in the *same* direction, and *in phase* with it. These two photons then move on to strike other atoms

Stimulated emission

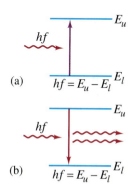

(a) $hf = E_u - E_l$

(b) $hf = E_u - E_l$

FIGURE 28–16 (a) Absorption of a photon. (b) Stimulated emission.

FIGURE 28–17 Two energy levels for a collection of atoms. Each dot represents the energy state of one atom. (a) A normal situation; (b) an inverted population.

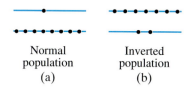

Normal population (a)

Inverted population (b)

FIGURE 28–18 Laser diagram, showing excited atoms stimulated to emit light.

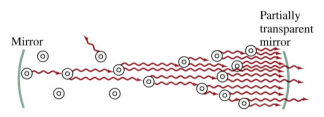

Mirror

Partially transparent mirror

[†]An excited atom may land in such a state and can jump to a lower state only by a so-called forbidden transition (discussed in Section 28–6), which is why its lifetime is longer than normal.

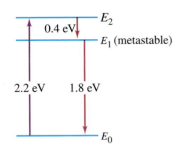

FIGURE 28–19 Energy levels of chromium in a ruby crystal. Photons of energy 2.2 eV "pump" atoms from E_0 to E_2, which then decay to metastable state E_1. Lasing action occurs by stimulated emission of photons in transition from E_1 to E_0.

Ruby laser

He–Ne laser

causing more stimulated emission. As the process continues, the number of photons multiplies. When the photons strike the end mirrors, most are reflected back, and as they move in the opposite direction, they continue to stimulate more atoms to emit photons. As the photons move back and forth between the mirrors, a small percentage passes through the partially transparent mirror at one end. These photons make up the narrow coherent external laser beam.

Inside the tube, some spontaneously emitted photons will be emitted at an angle to the axis, and these will merely go out the side of the tube and not affect the narrowness of the main beam. In a well-designed laser, the spreading of the beam is limited only by diffraction, so the angular spread is $\approx \lambda/D$ (see Eq. 24–3) where D is the diameter of the end mirror. The diffraction spreading can be incredibly small. The light energy, instead of spreading out in space as it does for an ordinary light source, is directed in a pencil-thin beam.

The excitation of the atoms in a laser can be done in several ways to produce the necessary inverted population. In a ruby laser, the lasing material is a ruby rod consisting of Al_2O_3 with a small percentage of aluminum (Al) atoms replaced by chromium (Cr) atoms. The Cr atoms are the ones involved in lasing. The atoms are excited by strong flashes of light of wavelength 550 nm, which corresponds to a photon energy of 2.2 eV. As shown in Fig. 28–19, the atoms are excited from state E_0 to state E_2. This process is called **optical pumping**. The atoms quickly decay either back to E_0 or to the intermediate state E_1, which is metastable with a lifetime of about 3×10^{-3} s (compared to 10^{-8} s for ordinary levels). With strong pumping action, more atoms can be forced into the E_1 state than are in the E_0 state. Thus we have the inverted population needed for lasing. As soon as a few atoms in the E_1 state jump down to E_0, they emit photons that produce stimulated emission of the other atoms and the lasing action begins. A ruby laser thus emits a beam whose photons have energy 1.8 eV and a wavelength of 694.3 nm (or "ruby-red" light).

In a helium–neon (He–Ne) laser, the lasing material is a gas, a mixture of about 15 percent He and 85 percent Ne. The atoms are excited by applying a high voltage to the tube so that an electric discharge takes place within the gas. In the process, some of the He atoms are raised to the metastable state E_1 shown in Fig. 28–20, which corresponds to a jump of 20.61 eV, almost exactly equal to an excited state in neon, 20.66 eV. The

FIGURE 28–20 Energy levels for He and Ne. He is excited in the electric discharge to the E_1 state. This energy is transferred to the E_3' level of the Ne by collision. E_3' is metastable and decays to E_2' by stimulated emission.

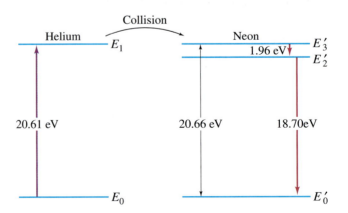

He atoms do not quickly return to the ground state by spontaneous emission, but instead often give their excess energy to a Ne atom when they collide—see Fig. 28–20. In such a collision, the He drops to the ground state and the Ne atom is excited to the state E_3' (the prime refers to neon states). The slight difference in energy (0.05 eV) is supplied by the kinetic energy of the moving molecules. In this manner, the E_3' state in Ne—which is metastable—becomes more populated than the E_2' level. This inverted population between E_3' and E_2' is what is needed for lasing.

Other lasers

Other types of laser include: chemical lasers, in which the energy input comes from the chemical reaction of highly reactive gases; dye lasers, whose frequency is tunable; CO_2 gas lasers, capable of high power output in the infrared; rare-earth solid-state lasers such as the high-power Nd:Yag laser; and the *pn* junction laser in which the transitions occur between the bottom of the conduction band and the upper part of the valence band (Section 29–6).

The excitation of the atoms in a laser can be done continuously or in pulses. In a **pulsed laser**, the atoms are excited by periodic inputs of energy. The multiplication of photons continues until all the atoms have been stimulated to jump down to the lower state, and the process is repeated with each input pulse. In a **continuous laser**, the energy input is continuous so that as atoms are stimulated to jump down to the lower level, they are soon excited back up to the upper level so that the output is a continuous laser beam. Any laser, of course, is not a source of energy. Energy must be put in, and the laser converts a part of this input energy into an intense narrow beam output.

The unique feature of light from a laser is, as mentioned before, that it is a coherent narrow beam of a single frequency (or several distinct frequencies). Because of this feature, the laser has found many applications. Lasers are a useful surgical tool. The narrow intense beam can be used to destroy tissue in a localized area, or to break up gallstones and kidney stones. Because of the heat produced, a laser beam can be used to "weld" broken tissue, such as a detached retina. For some types of internal surgery, the laser beam can be carried by an optical fiber (Section 23–6) to the surgical point, sometimes as an additional fiber-optic path on an endoscope, Fig. 28–21 (also discussed in Section 23–6). An example is the removal of plaque clogging human arteries. Tiny organelles within a living cell have been destroyed using lasers by researchers studying how the absence of that organelle affects the behavior of the cell. Laser beams have been used to destroy cancerous and precancerous cells; at the same time, the heat seals off capillaries and lymph vessels, thus "cauterizing" the wound in the process to prevent spread of the disease. The intense heat produced in a small area by a laser beam is also used for welding and machining metals and for drilling tiny holes in hard materials. The beam of a laser is narrow in itself (typically, a few mm). But because the beam is coherent, monochromatic, and essentially parallel and narrow, lenses can be used to focus the light into incredibly small areas without the usual aberration problems. The limiting factor thus becomes diffraction, and the energy crossing unit area per unit time can be very large. The precise straightness of a laser beam is also useful to surveyors for lining up equipment precisely, especially in inaccessible places.

➡ **PHYSICS APPLIED**

Medical and other uses of lasers

FIGURE 28–21 Laser being used in eye surgery.

FIGURE 28–22 Reading a CD. The fine beam of a laser, focused even more finely with lenses, is directed at the undersurface of a rotating compact disk. The beam is reflected back from the areas between pits but reflects much less from pits. The reflected light is detected as shown, reflected by a half-reflecting mirror MS. The strong and weak reflections correspond to the 0s and 1s of the binary code representing the audio or video signal.

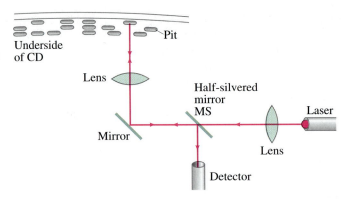

→ **PHYSICS APPLIED**

CD players and bar codes

In everyday life, lasers are used as bar-code readers (at store checkout stands) and in compact disc (CD) players. The laser beam reflects off the stripes and spaces of a bar code, and off the tiny pits of a CD as shown in Fig. 28–22. The recorded information on a CD is a series of pits and spaces representing 0s and 1s (or "off" and "on") of a digitized code that must be decoded electronically before being sent to the audio or video system. A bar-code reader is similar.

* 28–12 Holography

→ **PHYSICS APPLIED**

Holography

One of the most interesting applications of laser light is the production of three-dimensional images called **holograms** (see Fig. 28–23). In an ordinary photograph, the film simply records the intensity of light reaching it at each point. When the photograph or transparency is viewed, light reflecting from it or passing through it gives us a two-dimensional picture. In holography, the images are formed by interference, without lenses. When a laser hologram is made on film, a broadened laser beam is split into two parts by a half-silvered mirror, Fig. 28–23. One part goes directly to the film; the rest passes to the object to be photographed, from which it is reflected to the film. Light from every point on the object reaches each point on the film, and the interference of the two beams allows the film to record both the intensity and relative phase of the light at each point. It is crucial that the light be coherent—that is, in phase at all points—which is why a laser is used. After the film is developed, it is placed again in a laser beam and a three-dimensional image of the object is created. You can walk around such an

FIGURE 28–23 Making a hologram. Light reflected from various points on the object interferes (at the film) with light from the direct beam.

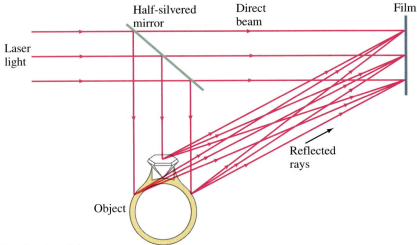

image and see it from different sides as if it were the original object. Yet, if you try to touch it with your hand, there will be nothing material there.

So-called **volume** or **white-light holograms** do not require a laser to see the image, but can be viewed with ordinary white light (preferably a nearby point source, such as the Sun or a clear bulb with a small bright filament). Such holograms must be made, however, with a laser. They are made not on thin film, but on a *thick* emulsion. The interference pattern, instead of being two-dimensional as for an ordinary hologram, is actually three-dimensional (hence the name "volume hologram"). The interference pattern in the film emulsion can be thought of as an array of bands or ribbons (consisting of the silver grains from the development process) where constructive interference occurred. This array, and the reconstruction of the image, can be compared to Bragg scattering of X-rays from the atoms in a crystal (see Section 25–11). White light can reconstruct the image because the Bragg condition ($m\lambda = 2d\sin\theta$) selects out the appropriate single wavelength. If the hologram is originally produced by lasers emitting the three additive primary colors (red, green, and blue), the three-dimensional image can be seen in full color when viewed with white light.

White-light holograms

SUMMARY

In 1925, Schrödinger and Heisenberg separately worked out a new theory, **quantum mechanics**, which is now considered to be the basic theory at the atomic level. It is a statistical theory rather than a deterministic one.

An important aspect of quantum mechanics is the **Heisenberg uncertainty principle**. It results from the wave–particle duality and the unavoidable interaction between the observed object and the observer. One form of the uncertainty principle states that both the position x and momentum p of an object cannot be measured precisely at the same time. The products of the uncertainties, $(\Delta x)(\Delta p)$, can be no less than $\hbar \, (= h/2\pi)$:

$$(\Delta p)(\Delta x) \gtrsim \hbar.$$

Another form states that the energy can be uncertain or nonconserved, by an amount ΔE for a time Δt where

$$(\Delta E)(\Delta t) \gtrsim \hbar.$$

In the quantum mechanical view of the atom, the electrons do not have well-defined orbits, but instead exist as a "cloud." Electron clouds can be interpreted as an electron wave spread out in space, or as a **probability distribution** for electrons considered as particles.

According to quantum mechanics, the state of an electron in an atom is specified by four **quantum numbers**: $n, l, m_l,$ and m_s. The principal quantum number, n, can take on any integer value $(1, 2, 3, \cdots)$ and corresponds to the quantum number of the old Bohr theory; l can take on values from 0 up to $n - 1$; m_l can take on integer values from $-l$ to $+l$; and m_s can be $+\frac{1}{2}$ or $-\frac{1}{2}$. The energy levels in the hydrogen atom depend on n, whereas in other atoms they depend on n and l. When an external magnetic field is applied, the spectral lines are split (the **Zeeman effect**), indicating that the energy depends also on m_l in this case. Even in the absence of a magnetic field, precise measurements of spectral lines show a tiny splitting of the lines called **fine structure**, whose explanation is that the energy depends very slightly on m_l and m_s.

The arrangement of electrons in multielectron atoms is governed by the **Pauli exclusion principle**, which states that no two electrons can occupy the same quantum state—that is, they cannot have the same set of quantum numbers $n, l, m_l,$ and m_s. Electrons, as a result, are grouped into **shells** (according to the value of n) and **subshells** (according to l).

Electron configurations are specified using the numerical values of n, but using letters for l: s, p, d, f, etc., for $l = 0, 1, 2, 3$, and so on, plus a superscript for the number of electrons in that subshell. Thus, the ground state of hydrogen is $1s^1$, whereas that for oxygen is $1s^2 2s^2 2p^4$. The **periodic table** arranges the elements in horizontal rows according to increasing atomic number (number of electrons in the neutral atom); each vertical column contains elements with similar chemical properties.

X-rays, which are a form of electromagnetic radiation of very short wavelength, are produced when high-speed electrons strike a target. The spectrum of X-rays so produced consists of two parts, a continuous spectrum produced when the electrons are decelerated by atoms of the target, and peaks representing photons emitted by atoms of the target after being excited by collision with the high-speed electrons.

QUESTIONS

1. Compare a matter wave Ψ to (a) a wave on a string, (b) an EM wave. Discuss similarities and differences.

2. Explain why Bohr's theory of the atom is not compatible with quantum mechanics, particularly the uncertainty principle.

3. Explain why it is that the more massive an object is, the easier it becomes to predict its future position.

4. In view of the uncertainty principle, why does a baseball seem to have a well-defined position and speed whereas an electron does not?

5. Would it ever be possible to balance a very sharp needle precisely on its point? Explain.

6. A cold thermometer is placed in a hot bowl of soup. Will the temperature reading of the thermometer be the same as the temperature of the hot soup before the measurement was made?

7. When you check the pressure in a tire, doesn't some air inevitably escape? Is is possible to avoid this escape of air altogether? What is the relation to the uncertainty principle?

8. It has been said that the ground-state energy in the hydrogen atom can be precisely known but the excited states have some uncertainty in their values (an "energy width"). Is this consistent with the uncertainty principle in its energy form? Explain.

9. Which model of the hydrogen atom, the Bohr model or the quantum-mechanical model, predicts that the electron spends more time near the nucleus?

10. The size of atoms varies by only a factor of three or so from largest to smallest, yet the number of electrons varies from one to over 100. Why?

11. Excited hydrogen and excited helium atoms both radiate light as they jump down to the $n = 1$, $l = 0$, $m_l = 0$ state. Yet the two elements have very different emission spectra. Why?

12. The 589-nm yellow line in sodium is actually two very closely spaced lines. This splitting is due to an "internal" Zeeman effect. Can you explain this? [*Hint*: Put yourself in the reference frame of the electron.]

13. Which of the following electron configurations are forbidden? (a) $1s^2 2s^2 2p^4 3s^2 4p^2$; (b) $1s^2 2s^2 2p^8 2s^1$, (c) $1s^2 2s^2 2p^6 3s^2 3p^5 4s^2 4d^5 4f^1$.

14. Give the complete electron configuration for a uranium atom (careful scrutiny across the periodic table on the inside back cover will provide useful hints).

15. In what column of the periodic table would you expect to find the atom with each of the following configurations? (a) $1s^2 2s^2 2p^6 3s^2$; (b) $1s^2 2s^2 2p^6 3s^2 3p^6$; (c) $1s^2 2s^2 2p^6 3s^2 3p^6 4s^1$; (d) $1s^2 2s^2 2p^5$.

16. The ionization energy for neon ($Z = 10$) is 21.6 eV and that for sodium ($Z = 11$) is 5.1 eV. Explain the large difference.

17. Why do chlorine and iodine exhibit similar properties?

18. Explain why potassium and sodium exhibit similar properties.

19. Why are the chemical properties of the rare earths so similar?

20. Why do we not expect perfect agreement between measured values of X-ray line wavelengths and those calculated using Bohr theory, as in Example 28–7?

21. How would you figure out which lines in an X-ray spectrum correspond to K_α, K_β, L, etc., transitions?

22. Why do we expect electron transitions deep within an atom to produce shorter wavelengths than transitions by outer electrons?

* 23. Compare spontaneous emission to stimulated emission.

* 24. How does laser light differ from ordinary light? How is it the same?

* 25. Explain how a 0.0005-W laser beam, photographed at a distance, can seem much stronger than a 1000-W street lamp.

PROBLEMS

SECTION 28–2

1. (II) The neutrons in a parallel beam, each having kinetic energy $\frac{1}{40}$ eV, are directed through two slits 0.50 mm apart. How far apart will the interference peaks be on a screen 1.0 m away?

2. (II) Bullets of mass 3.0 g are fired in parallel paths with speeds of 200 m/s through a hole 3.0 mm in diameter. How far from the hole must you be to detect a 1.0-cm diameter spread in the beam?

3. (I) A proton is traveling with a speed of $(6.560 \pm 0.012) \times 10^5$ m/s. With what maximum accuracy can its position be ascertained?

4. (I) If an electron's position can be measured to an accuracy of 2.0×10^{-8} m, how accurately can its velocity be known?

5. (I) An electron remains in an excited state of an atom for typically 10^{-8} s. What is the minimum uncertainty in the energy of the state (in eV)?

6. (I) The Z^0 boson, discovered in 1985, is the mediator of the weak nuclear force, and it typically decays very quickly. Its average rest energy is 91.19 GeV, but its short lifetime shows up as an intrinsic width of 2.5 GeV. What is the lifetime of this particle?

7. (II) What is the uncertainty in the mass of a muon $(m = 105.7 \text{ MeV}/c^2)$, specified in eV/c^2, given its lifetime of 2.20 μs?

8. (II) A free neutron $(m = 1.67 \times 10^{-27}$ kg) has a mean life of 900 s. What is the uncertainty in its mass (in kg)?

9. (II) An electron and a 140-g baseball are each traveling 150 m/s measured to an accuracy of 0.055 percent. Calculate and compare the uncertainty in position of each.

10. (II) Estimate the lowest possible energy of a neutron contained in a typical nucleus of radius 1.0×10^{-15} m. [*Hint*: A particle can have an energy at least as large as its uncertainty.]

11. (II) Use the uncertainty principle to show that if an electron were present in the nucleus $(r \approx 10^{-15}$ m), its kinetic energy (use relativity) would be hundreds of MeV. (Since such electron energies are not observed, we conclude that electrons are not present in the nucleus.) [*Hint*: See hint for Problem 10.]

12. (III) How accurately can the position of a 3.00-keV electron be measured assuming its energy is known to 1.00 percent?

13. (III) The uncertainty principle can be stated in terms of angular quantities as follows:

$$\Delta L \Delta \phi \gtrsim \hbar.$$

Here, L stands for angular momentum along a given axis, and ϕ for the angular position measured in a plane perpendicular to that axis. (*a*) Make a plausibility argument for this relation. (*b*) Electrons in atoms have well-defined quantized values of angular momentum, with no uncertainty. What does this say about the uncertainty in angular position and the concept of electron orbits?

14. (I) For $n = 6$, what values can l have?

15. (I) For $n = 5$, $l = 4$, what are the possible values of m_l and m_s?

16. (I) How many electrons can be in the $n = 6$, $l = 3$ subshell?

17. (I) List the quantum numbers for each electron in the ground state of nitrogen ($Z = 7$).

18. (I) List the quantum numbers for each electron in the ground state of magnesium ($Z = 12$).

19. (I) How many different states are possible for an electron whose principal quantum number is $n = 4$? Write down the quantum numbers for each state.

20. (I) A hydrogen atom is known to have $l = 4$. What are the possible values for n, m_l, and m_s?

21. (I) If a hydrogen atom has $m_l = -3$, what are the possible values of n, l, and m_s?

22. (I) Calculate the magnitude of the angular momentum of an electron in the $n = 4$, $l = 3$ state of hydrogen.

23. (II) Show that there can be 18 electrons in a "g" subshell.

24. (II) What is the full electron configuration in the ground state for elements with Z equal to (*a*) 27, (*b*) 36, (*c*) 38? [*Hint*: See the periodic table inside the back cover.]

25. (II) What is the full electron configuration for (*a*) selenium (Se), (*b*) gold (Au), (*c*) uranium (U)? [*Hint*: See the periodic table inside the back cover.]

26. (II) For each of the following atomic transitions, state whether the transition is *allowed* or *forbidden*, and if forbidden, what rule is being violated: (*a*) $4p \rightarrow 3p$; (*b*) $2p \rightarrow 1s$; (*c*) $3d \rightarrow 2d$; (*d*) $4d \rightarrow 3s$; (*e*) $4s \rightarrow 2p$.

27. (II) A hydrogen atom is in the $6h$ state. Determine (*a*) the principal quantum number, (*b*) the energy of the state, (*c*) the orbital angular momentum and its quantum number l, and (*d*) the possible values for the magnetic quantum number.

28. (II) Estimate the binding energy of the third electron in lithium using the Bohr theory. [*Hint*: This electron has $n = 2$ and "sees" a net charge of approximately $+1e$.] The measured value is 5.36 eV.

29. (II) The ionization (binding) energy of the outermost electron in boron is 8.26 eV. (*a*) Use the Bohr model to estimate the "effective charge," Z_{eff}, seen by this electron. (*b*) Estimate the average orbital radius.

30. (II) Show that the total angular momentum is zero for a filled subshell.

31. (II) Show that the maximum number of electrons allowed in any subshell is equal to $2(2l + 1)$ where l is the angular momentum quantum number of the subshell.

32. (I) What are the shortest-wavelength X-rays emitted by electrons striking the face of a 30-kV TV picture tube? What are the longest wavelengths?

33. (I) If the shortest-wavelength bremsstrahlung X-rays emitted from an X-ray tube have $\lambda = 0.030$ nm, what is the voltage across the tube?

34. (I) Show that the cutoff wavelength λ_0 is given by

$$\lambda_0 = 1240 \text{ nm}/V,$$

where V is the X-ray tube voltage in volts.

35. (II) Use the result of Example 28–7 to estimate the X-ray wavelength emitted when a Co ($Z = 27$) atom jumps from $n = 2$ to $n = 1$.

36. (II) Estimate the wavelength for an $n = 2$ to $n = 1$ transition in chromium ($Z = 24$).

37. (II) Use the Bohr theory to estimate the wavelength for an $n = 3$ to $n = 1$ transition in molybdenum. The measured value is 0.063 nm. Why do we not expect perfect agreement?

38. (II) A mixture of iron and an unknown material is bombarded with electrons. The wavelength of the K_α lines are 194 pm for iron and 229 pm for the unknown. What is the unknown material?

*SECTION 28–11

*39. (II) A laser used to weld detached retinas puts out 30-ms-long pulses of 640-nm light which average 0.60-W output during a pulse. How much energy can be deposited per pulse and how many photons does each pulse contain?

*40. (II) Estimate the angular spread of a laser beam due to diffraction if the beam emerges through a 3.0-mm-diameter mirror. Assume that $\lambda = 694$ nm. What would be the diameter of this beam if it struck a satellite 300 km above the Earth?

GENERAL PROBLEMS

41. Use the uncertainty principle to estimate the position uncertainty for the electron in the ground state of the hydrogen atom. [*Hint:* Determine the momentum using the Bohr model of Section 27–10 and assume the momentum can be anywhere between this value and zero.] How does this result compare to the Bohr radius?

42. An electron in the $n = 2$ state of hydrogen remains there on the average about 10^{-8} s before jumping to the $n = 1$ state. (*a*) Estimate the uncertainty in the energy of the $n = 2$ state. (*b*) What fraction of the transition energy is this? (*c*) What is the wavelength, and width (in nm), of this line in the spectrum of hydrogen?

43. What are the largest and smallest possible values for the angular momentum L of an electron in the $n = 5$ shell?

44. Estimate (*a*) the quantum number l for the orbital angular momentum of the Earth about the Sun, and (*b*) the number of possible orientations for the plane of Earth's orbit.

45. A 12-g bullet leaves a rifle at a speed of 180 m/s. (*a*) What is the wavelength of this bullet? (*b*) If the position of the bullet is known to an accuracy of 0.60 cm (radius of the barrel), what is the minimum uncertainty in its momentum? (*c*) If the accuracy of the bullet were determined only by the uncertainty principle (an unreasonable assumption), by how much might the bullet miss a pinpoint target 200 m away?

46. Using the Bohr formula for the radius of an electron orbit, estimate the average distance from the nucleus for an electron in the innermost ($n = 1$) orbit of a lead atom ($Z = 82$). Approximately how much energy would be required to remove this innermost electron?

47. An X-ray tube operates at 100 kV with a current of 25 mA and nearly all the electron energy goes into heat. If the specific heat capacity of the 0.085-kg plate is 0.11 kcal/kg·C°, what will be the temperature rise per minute if no cooling water is used?

48. Use the Bohr theory (especially Eq. 27–15) to show that the Moseley plot (Fig. 28–12) can be written

$$\sqrt{\frac{1}{\lambda}} = a(Z - b),$$

where $b \approx 1$, and evaluate a.

49. In the so-called *vector model* of the atom, space quantization of angular momentum (Fig. 28–7) is illustrated as shown in Fig. 28–24. The angular momentum vector of magnitude $L = \sqrt{l(l + 1)}\,\hbar$ is thought of as precessing around the z axis (like a spinning top or gyroscope) in such a way that the z component of angular momentum, $L_z = m_l\hbar$, also stays constant. Calculate the possible values for the angle θ between **L** and the z axis (*a*) for $l = 1$, (*b*) $l = 2$, and (*c*) $l = 3$. (*d*) Determine the minimum value of θ for $l = 100$ and $l = 10^6$. Is this consistent with the correspondence principle?

FIGURE 28–24
The vector model for orbital angular momentum. The orbital angular momentum vector **L** is imagined to precess about the z axis; L and L_z remain constant, but L_x and L_y continually change. Problem 49.

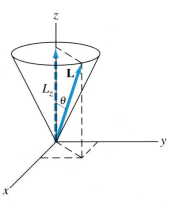

50. (II) Show that the diffractive spread of a laser beam, $\approx \lambda/D$ as described in Section 28–11, is precisely what you might expect from the uncertainty principle. [*Hint:* Since the beam's width is constrained by the dimension of the aperture D, the component of the light's momentum perpendicular to the laser axis is uncertain.]

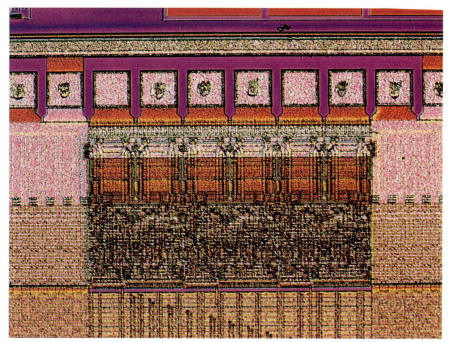

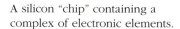
A silicon "chip" containing a complex of electronic elements.

CHAPTER 29

MOLECULES AND SOLIDS

Since its development in the 1920s, quantum mechanics has had a profound influence on our lives, both intellectually and technologically. Even the way we view the world has changed, as we saw in Chapter 28. In the present chapter, we will discuss how quantum mechanics has given us an understanding of the structure of molecules and matter in bulk, as well as a number of important applications including semiconductor devices and applications to biology. Our discussions will necessarily be qualitative for the most part.

* 29–1 Bonding in Molecules

One of the great successes of quantum mechanics was to give scientists, at last, an understanding of the nature of chemical bonds. Since it is based in physics, and because this understanding is so important in many fields, we discuss it here.

By a molecule, we mean a group of two or more atoms that are strongly held together so as to function as a single unit. When atoms make such an attachment, we say that a chemical **bond** has been formed. There are two main types of strong chemical bond: covalent and ionic. Many bonds are actually intermediate between these two types.

Covalent Bonds. To understand how covalent bonds are formed, we take the simplest case, the bond that holds two hydrogen atoms together to form the hydrogen molecule, H_2. The mechanism is basically the same for other covalent bonds. As two H atoms approach each other, the electron clouds begin to overlap, and the electrons from each atom can "orbit" both nuclei. (This is sometimes called "sharing" electrons.) If both electrons are in the ground state ($n = 1$) of their respective atoms, there are two possibilities: their spins can be parallel (both up or both down), in which case the total spin is $S = \frac{1}{2} + \frac{1}{2} = 1$; or their spins can be opposite ($m_s = +\frac{1}{2}$ for one, $m_s = -\frac{1}{2}$ for the other), so that the total spin $S = 0$. We shall now see that a bond is formed only for the $S = 0$ state, when the spins are opposite. First we consider the $S = 1$ state, for which the spins are the same. The two electrons cannot both be in the lowest energy state and be attached to the same atom, for then they would have identical quantum numbers in violation of the exclusion principle. The exclusion principle tells us that since no two electrons can occupy the same quantum state, if two electrons have the same quantum numbers, they must be different in some other way—namely, by being in different places in space (for example, attached to different atoms). When the two atoms approach, the electrons will stay away from each other as shown by the probability distribution of Fig. 29–1. The positively charged nuclei then repel each other, and no bond is formed.

For the $S = 0$ state, on the other hand, the spins are opposite and the two electrons are consequently in different quantum states. Hence they can come close together spatially. In this case, the probability distribution looks like Fig. 29–2. As can be seen, the electrons spend much of their time between the two positively charged nuclei, and the latter are attracted to the negatively charged electron cloud between them. This attraction, which holds the two atoms together to form a molecule, constitutes a *covalent bond*.

The probability distributions of Figs. 29–1 and 29–2 can perhaps be better understood on the basis of waves. What the exclusion principle requires is that when the spins are the same, there is destructive interference of the electron wave functions in the region between the two atoms. But when the spins are opposite, constructive interference occurs in the region between the two atoms, resulting in a large amount of negative charge there. Thus a covalent bond can be said to be the result of constructive interference of the electron wave functions in the space between the two atoms, and of the electrostatic attraction of the two positive nuclei for the negative charge concentration between them.

Why a bond is formed can also be understood from the energy point of view. When the two H atoms approach close to one another, if the spins of their electrons are opposite, the electrons can occupy the same space, as discussed above. This means that each electron can now move about in the space of two atoms instead of in the volume of only one. Because each electron now occupies more space, it is less well localized. From the uncertainty principle, with Δx increased, the momentum, and hence the energy, can be less. Another way of understanding this is from the point of view of de Broglie waves (Section 27–11). Because each electron has a larger "orbit," its wavelength λ can be longer, so its momentum $p = h/\lambda$ (Eq. 27–8) can be less. With less momentum, each electron has less energy when the two atoms combine than when they are separate. That is, the molecule has less energy than the two separate atoms, and so is more stable. An energy input is required to break the H_2 molecule into two separate H atoms, so the H_2 molecule is a stable entity. This is what we mean by a *bond*. The energy required to break a bond is called the **bond energy**, the **binding energy**, or the **dissociation energy**.

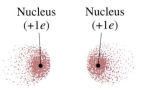

Nucleus
(+1e)   Nucleus
               (+1e)

FIGURE 29–1 Electron probability distribution (electron cloud) for two H atoms when their spins are the same ($S = 1$).

FIGURE 29–2 Electron probability distribution (cloud) around two H atoms when their spins are opposite ($S = 0$). In this case, a bond is formed because the positive nuclei are attracted to the concentration of negative charge between them. This is a hydrogen molecule, H_2.

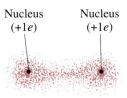

Nucleus
(+1e)   Nucleus
               (+1e)

Bond energy

Ionic Bonds. An ionic bond is, in a sense, a special case of the covalent bond. Instead of the electrons being shared equally, they are shared unequally. For example, in sodium chloride (NaCl), the outer electron of the sodium spends nearly all its time around the chlorine (Fig. 29–3). The chlorine atom acquires a net negative charge as a result of the extra electron, whereas the sodium atom is left with a net positive charge. The electrostatic attraction between these two charged atoms holds them together. The resulting bond is called an *ionic bond* because it is created by the attraction between the two ions (Na^+ and Cl^-). But to understand the ionic bond, we must understand why the extra electron from the sodium spends so much of its time around the chlorine. After all, the chlorine is neutral; why should it attract another electron?

The answer lies in the probability distributions of the two neutral atoms. Sodium contains 11 electrons, 10 of which are in spherically symmetric closed shells (Fig. 29–4). The last electron spends most of its time beyond these closed shells. Because the closed shells have a total charge of $-10e$ and the nucleus has charge $+11e$, the outermost electron in sodium "feels" a net attraction due to $+1e$. It is not held very strongly. On the other hand, 12 of chlorine's 17 electrons form closed shells, or subshells (corresponding to $1s^2 2s^2 2p^6 3s^2$). These 12 form a spherically symmetric shield around the nucleus. The other five electrons are in $3p$ states whose probability distributions are not spherically symmetric and have a form similar to those for the $2p$ states in hydrogen shown in Fig. 28–9b and c. Four of these $3p$ electrons can have "doughnut-shaped" distributions symmetric about the z axis, as shown in Fig. 29–5. The fifth can have a "barbell-shaped" distribution (as for $m_l = 0$ in Fig. 28–9b), which in Fig. 29–5 is shown only in faint outline because it is half empty. That is, the exclusion principle allows one more electron to be in this state (it will have spin opposite to that of the electron already there). If an extra electron—say from a Na atom—happens to be in the vicinity, it can be in this state, say at point x in Fig. 29–5. It could experience an attraction due to as much as $+5e$ because the $+17e$ of the nucleus is partly shielded at this point by the 12 inner electrons. Thus, the outer electron of a Na atom will be more strongly attracted by the $+5e$ of the chlorine atom than by the $+1e$ of its own atom. This, combined with the strong attraction between the two ions when the extra electron stays with the Cl^-, produces the charge distribution of Fig. 29–3, and hence the ionic bond.

Ionic bond

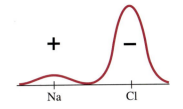

FIGURE 29–3 Probability distribution for last electron of Na in NaCl.

FIGURE 29–4 In a neutral sodium atom, the 10 inner electrons shield the nucleus, so the single outer electron is attracted by a net charge of $+1e$.

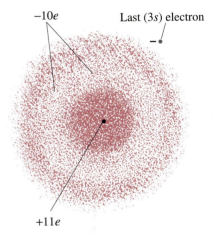

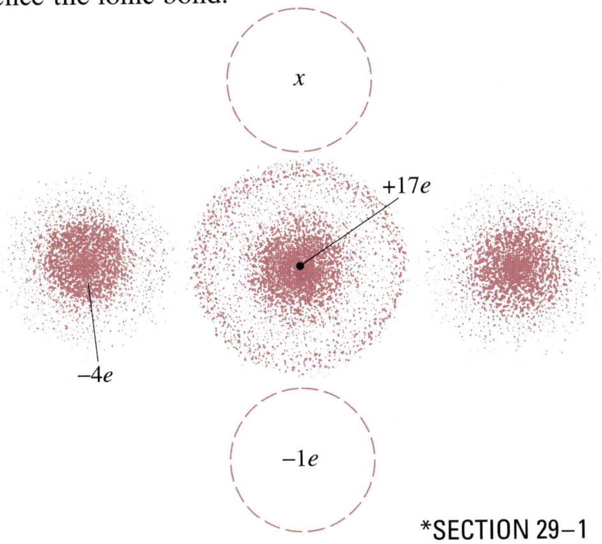

$-10e$ Last ($3s$) electron

$+11e$

FIGURE 29–5 Neutral chlorine atom. The $+17e$ of the nucleus is shielded by the 12 electrons in the inner shells and subshells. Four of the five $3p$ electrons are shown in doughnut-shaped clouds, and the fifth is in the (dashed-line) cloud concentrated about the z axis (vertical). An extra electron at x will be attracted by a net charge that can be as much as $+5e$.

x

$+17e$

$-4e$

$-1e$

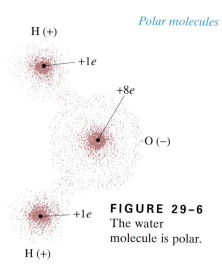

Polar molecules

H (+)

+1e

+8e

O (−)

+1e

FIGURE 29–6
The water molecule is polar.

H (+)

Partial Ionic Character of Covalent Bonds. A pure covalent bond in which the electrons are shared equally occurs mainly in symmetrical molecules such as H_2, O_2, and Cl_2. When the atoms involved are different from each other, it is usual to find that the shared electrons are more likely to be in the vicinity of one atom than the other. The extreme case is an ionic bond; in intermediate cases the covalent bond is said to have a *partial ionic character*. The molecules themselves are **polar**—that is, one part (or parts) of the molecule has a net positive charge and other parts a net negative charge. An example is the water molecule, H_2O (Fig. 29–6). The shared electrons are more likely to be found around the oxygen atom than around the two hydrogens. The reason is similar to that discussed above in connection with ionic bonds. Oxygen has eight electrons ($1s^2 2s^2 2p^4$), of which four form a spherically symmetric core and the other four could have, for example, a doughnut-shaped distribution. The barbell-shaped distribution on the z axis (like that shown dashed in Fig. 29–5) could be empty, so electrons from hydrogen atoms can be attracted by a net charge of $+4e$. They are also attracted by the H nuclei, so they partly orbit the H atoms as well as the O atom. The net effect is that there is a net positive charge on each H atom (less than $+1e$), because the electrons spend only part of their time there. And, there is a net negative charge on the O atom.

FIGURE 29–7 Potential energy as a function of separation for two point charges of (a) like sign and (b) opposite sign.

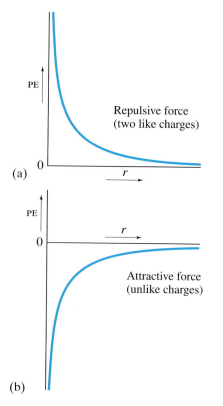

PE

Repulsive force
(two like charges)

0

r

(a)

PE

r

0

Attractive force
(unlike charges)

(b)

* 29–2 Potential-Energy Diagrams for Molecules

It is useful to analyze the interaction between two objects—say, between two atoms or molecules—with the use of a potential-energy diagram, a plot of the potential energy versus the separation distance.

For the simple case of two point charges, q_1 and q_2, the PE is given by (see Chapter 17):

$$PE = \frac{1}{4\pi\epsilon_0} \frac{q_1 q_2}{r},$$

where r is the distance between the charges, and the constant $(1/4\pi\epsilon_0)$ is equal to $9.0 \times 10^9 \, \text{N·m}^2/\text{C}^2$. If the two charges have the same sign, the PE is positive for all values of r, and a graph of PE versus r in this case is shown in Fig. 29–7a. The force is repulsive (the charges have the same sign) and the curve rises as r decreases; this makes sense since work is done to bring the charges together, thereby increasing their potential energy. If, on the other hand, the two charges are of opposite sign, the PE is negative because the product $q_1 q_2$ is negative. The force is attractive in this case and the graph of PE versus r looks like Fig. 29–7b. The PE becomes more *negative* as r decreases.

Now let us look at the potential-energy diagram for the formation of a covalent bond, such as for the hydrogen molecule. The potential energy of one H atom in the presence of the other is plotted in Fig. 29–8. Starting at large r, the PE decreases as the atoms approach, because the electrons concen-

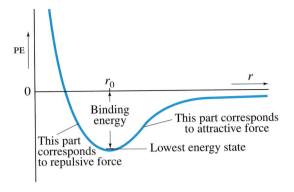

FIGURE 29–8 Potential-energy diagram for H_2 molecule; r is the separation of the two H atoms. The binding energy (the energy difference between PE = 0 and the lowest energy state near the bottom of the well) is 4.5 eV, and $r_0 = 0.074$ nm.

trate between the two nuclei (Fig. 29–2), so attraction occurs. However, at very short distances, the electrons would be "squeezed out"—there is no room for them between the two nuclei. Without the electrons between them, each nucleus would feel a repulsive force due to the other, so the curve rises as r decreases further. There is an optimum separation of the atoms, r_0 in Fig. 29–8, at which the energy is lowest. This is the point of greatest stability for the hydrogen molecule, and r_0 is the average separation of atoms in the H_2 molecule. The depth of this "well" is the *binding energy*,[†] as shown. This is how much energy must be put into the system to separate the two atoms to infinity, where the PE = 0.

For many bonds, the potential-energy curve has the shape shown in Fig. 29–9. There is still an optimum distance r_0 at which the molecule is stable. But when the atoms approach from a large distance, the force is initially repulsive rather than attractive. The atoms thus do not interact spontaneously. Instead, some additional energy must be injected into the system to get it over the "hump" (or barrier) in the potential-energy diagram. This required energy is called the **activation energy**.

The curve of Fig. 29–9 is much more common than that of Fig. 29–8. The activation energy often reflects a need to break other bonds, before

Binding energy

Activation energy

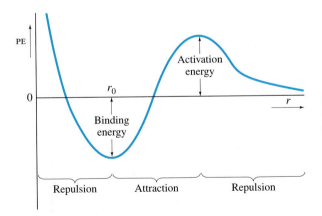

FIGURE 29–9 Potential-energy diagram for a bond requiring an activation energy.

[†]The binding energy corresponds not quite to the bottom of the PE curve, but to the lowest energy state, slightly above it, as shown in Fig. 29–8.

the one under discussion can be made. For example, to make water from O_2 and H_2, the H_2 and O_2 molecules must first be broken into H and O atoms by an input of energy; this is what the activation energy represents. Then the H and O atoms can combine to form H_2O with the release of a great deal more energy that was put in initially. The initial activation energy can be provided by applying an electric spark to a mixture of H_2 and O_2, breaking a few of these molecules into H and O atoms. The resulting explosive release of energy when these atoms combine to form H_2O quickly provides the activation energy needed for further reactions, so additional H_2 and O_2 molecules are broken up and recombined to form H_2O.

The potential-energy diagrams for ionic bonds can have similar shapes. In NaCl, for example, the Na^+ and Cl^- ions attract each other at distances a bit larger than some r_0, but at shorter distances the overlapping of inner electron shells gives rise to repulsion. The two atoms thus are most stable at some intermediate separation, r_0, and there often is an activation energy.

Sometimes the potential energy of a bond looks like that of Fig. 29–10. In this case, the energy of the bonded molecule, at a separation r_0, is greater than when there is no bond ($r = \infty$). That is, an energy *input* is required to make the bond (hence the binding energy is negative), and there is energy release when the bond is broken. Such a bond is stable only because there is the barrier of the activation energy. This type of bond is important in living cells, for it is in such bonds that energy can be stored efficiently in certain molecules, particularly ATP (adenosine triphosphate). The bond that connects the last phosphate group (designated Ⓟ in Fig. 29–10) to the rest of the molecule (ADP, meaning adenosine diphosphate, since it contains only two phosphates) is of the form shown in Fig. 29–10. Energy is actually stored in this bond. When the bond is broken (ATP → ADP + Ⓟ), energy is released and this energy can be used to make other chemical reactions "go."

In living cells, many chemical reactions have activation energies that are often on the order of several eV. Such energy barriers are not easy to overcome in the cell. This is where enzymes come in. They act as catalysts, which means that they act to lower the activation energy so that reactions can occur that otherwise would not. Enzymes act by distorting the bonding electrons so that the initial bonds are easily broken.

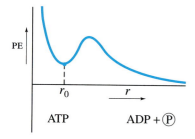

FIGURE 29–10 Potential-energy diagram for the formation of ATP from ADP and phosphate (Ⓟ).

* 29–3 Weak (van der Waals) Bonds

Once a bond between two atoms or ions is made, energy must normally be supplied to break the bond and separate the atoms. As mentioned in Section 29–1, this energy is called the *bond energy* or *binding energy*. The binding energy for covalent and ionic bonds is typically 2 to 5 eV. These bonds, which hold atoms together to form molecules, are often called **strong bonds** to distinguish them from so-called "weak bonds." The term **weak bond**, as we use it here, refers to an attachment between molecules due to simple electrostatic attraction—such as *between* polar molecules (and not *within* a polar molecule, which is a strong bond). The strength of the attachment is much less than for the strong bonds. Binding energies are typically in the range 0.04 to 0.3 eV—hence their name "weak bonds."

Weak bonds are generally the result of attraction between dipoles. For example, Fig. 29–11 shows two molecules, which have permanent dipole

FIGURE 29–11 The C^+—O^- and H^+—N^- dipoles attract each other. (These dipoles may be part of, for example, cytosine and guanine molecules. See Fig. 29–12.)

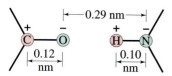

moments, attracting one another. Besides such **dipole–dipole bonds**, there can also be **dipole–induced dipole bonds**, in which a polar molecule with a permanent dipole moment can induce a dipole moment in an otherwise electrically balanced (nonpolar) molecule, just as a single charge can induce a separation of charge in a nearby object (see Fig. 16–7). There can even be an attraction between two nonpolar molecules. Even though a molecule may not have a permanent dipole moment on the average, we can think of its electrons as moving about so that at any instant there may be a separation of charge. Such transient dipoles can induce a dipole moment in a nearby molecule, creating a weak attraction. All these weak bonds are referred to as **van der Waals bonds**, and the forces involved **van der Waals forces**. The potential energy has the general shape shown in Fig, 29–8, with the attractive van der Waals PE varying as $1/r^6$.

When one of the atoms in a dipole–dipole bond is hydrogen, as in Fig. 29–11, it is called a **hydrogen bond**. A hydrogen bond is generally the strongest of the weak bonds. This is because the hydrogen atom is the smallest atom and thus can be approached more closely. Hydrogen bonds also have a partial "covalent" character. That is, electrons between the two dipoles may be shared to a small extent.

Weak bonds are important in liquids and solids when strong bonds are absent (see Section 29–5). They are also very important for understanding the activities of cells, such as the double helix shape of DNA (Fig. 29–12), and DNA replication (see Section 16–10). The average kinetic energy of molecules in a cell is around $\frac{3}{2}kT \approx 0.04\,\text{eV}$, about the magnitude of weak bonds. This means that a weak bond can readily be broken just by a molecular collision. Hence weak bonds are not very permanent—they are, instead, brief attachments. But because of this, they play an important role in the cell. On the other hand, strong bonds—those that hold molecules together—are almost never broken simply by molecular collision. Thus they are relatively permanent. They can be broken by chemical action (the making of even stronger bonds), and this usually happens in the cell with the aid of an enzyme, which is a protein molecule.

FIGURE 29–12 (a) Section of a DNA double helix. The red dots represent hydrogen bonds between the two chains. (b) "Close-up" view of helix: cytosine (C) and guanine (G) molecules on separate chains of a DNA double helix are held together by the hydrogen bonds (red dots) involving an H^+ on one molecule attracted to an N^- or C^+—O^- on the other molecule. See also Section 16–10 and Figs. 16–34 and 16–35.

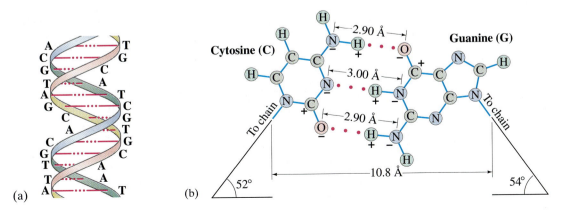

EXAMPLE 29–1 **Nucleotide energy.** Calculate the interaction energy between the C=O dipole of thymine and the H—N dipole of adenine, assuming that the two dipoles are lined up as shown in Fig. 29–11. Dipole moment measurements (see Table 17–2) give $q_H = -q_N = 0.19e = 3.0 \times 10^{-20}$ C, and $q_C = -q_O = 0.41e = 6.6 \times 10^{-20}$ C.

SOLUTION The interaction energy, for which we use the symbol U, will be equal to the potential energy of one dipole in the presence of the other, since this will be equal to the work needed to pull them infinitely far apart. U will consist of four terms:

$$U = U_{CH} + U_{CN} + U_{OH} + U_{ON},$$

where U_{CH} means the potential energy of C in the presence of H, and similarly for the other terms. We do not have terms corresponding to C and O, or N and H, because the two dipoles are assumed to be stable entities. Since the potential energy $U = qV$, where V is the electric potential (Eq. 17–3), then for two point charges $U_{12} = kq_1q_2/r$, where r is the distance between them. So

$$U = k\frac{q_C q_H}{r_{CH}} + k\frac{q_C q_N}{r_{CN}} + k\frac{q_O q_H}{r_{OH}} + k\frac{q_O q_N}{r_{ON}}.$$

Using the distances shown in Fig. 29–11, we get:

$$U = (9.0 \times 10^9 \, \text{N}\cdot\text{m}^2/\text{C}^2)\left(\frac{(6.6)(3.0)}{0.31} + \frac{(6.6)(-3.0)}{0.41} + \frac{(-6.6)(3.0)}{0.19} + \frac{(-6.6)(-3.0)}{0.29}\right)\frac{(10^{-20} \, \text{C})^2}{(10^{-19} \, \text{m})}$$

$$= -1.83 \times 10^{-20} \, \text{J} = -0.11 \, \text{eV}.$$

The PE is negative, meaning 0.11 eV of work (or energy input) is required to separate the molecules. That is, the binding energy of this "weak" or hydrogen bond is 0.11 eV. This is only an estimate, of course, since other charges in the vicinity would have an influence too.

Weak bonds, especially hydrogen bonds, are crucial to the process of protein synthesis. Proteins serve as structural parts of the cell and as enzymes to catalyze chemical reactions needed for the growth and survival of the organism. A protein molecule consists of one or more chains of small molecules known as *amino acids*. Each gene of the DNA contains the information needed to produce a particular type of protein molecule. The four types of bases (A, T, C, G) of the DNA are arranged according to a code, the "genetic code," which is translated into the amino acids that form the protein molecule, as discussed in Fig. 29–13. This process of protein synthesis is often presented as if it occurred in clockwork fashion—as if each molecule knew its role and went to its assigned place. But this is not the case. The forces of attraction between the electric charges of the molecules are rather weak and become significant only when the molecules can come close together and several weak bonds can be made. Indeed, if the shapes are not just right, there is almost no electrostatic attraction, which is why there are few mistakes. The fact that weak bonds are weak is very important. If they were strong, collisions with other molecules

FIGURE 29–13 Model of protein synthesis: messenger-RNA (m-RNA), a molecule much like DNA except that uracil (U) replaces T, is synthesized (on the right), one base (A, C, G, U) after the other, in just the order they appear on the DNA; the bases are selected via weak bonds and "shapes," much as in DNA replication (Section 16–10). The completed m-RNA (for one gene) then is buffeted about in the cell until it is attracted, and becomes attached to, a "ribosome" by electrostatic attraction. Transfer RNA (t-RNA) molecules, each carrying a particular amino acid and a corresponding "anticodon" of three bases, "line up" via electrostatic attraction at certain positions on the ribosome. Only the proper anticodon will line up with an m-RNA codon, as shown. The t-RNA molecules act as "translators" of the genetic code; the aligned amino acids (on the other end of the t-RNA) are attached together by enzymes (at which point the t-RNA, held tenuously by weak bonds, drifts away and the ribosome can move to the next codon on the m-RNA), thus making a protein chain molecule.

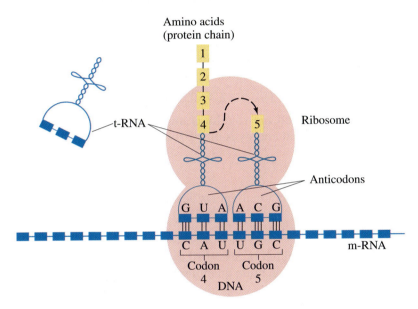

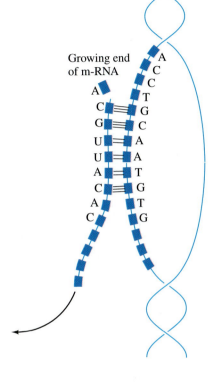

would not allow a t-RNA molecule to be released from the ribosome, or the m-RNA to be released from the DNA. If they were not temporary encounters, metabolism would grind to a halt.

* 29–4 Molecular Spectra

When atoms combine to form molecules, the probability distributions of the outer electrons overlap and this interaction alters the energy levels. Nonetheless, molecules can undergo transitions between electron energy levels just as atoms do. For example, the H_2 molecule can absorb a photon of just the right frequency to excite one of its $1s$ electrons to a $2p$ state. The excited electron can then return to the ground state, emitting a photon. The energy of photons emitted by molecules is of the same order of magnitude as for atoms, typically 1 to 10 eV.

Additional energy levels become possible for molecules (but not for atoms) because the molecule as a whole can rotate, and the atoms of the molecule can vibrate relative to each other. The energy levels for both rotational and vibrational levels are quantized, and are generally spaced much more closely (10^{-3} to 10^{-1} eV) than the electronic levels. Each atomic energy level thus becomes a set of closely spaced levels corresponding to the vibrational and rotational motions, Fig. 29–14. Transitions from one level to another appear as many very closely spaced lines. In fact, the lines are not always distinguishable, and these spectra are called **band spectra**. Each type of molecule has its own characteristic spectrum, which can be used for identification and for determination of structure.

Molecular Rotation.   Let us now look in more detail at rotational and vibrational states in molecules. We begin with rotation, considering here only diatomic molecules, although the analysis can be extended to polyatomic molecules. When a diatomic molecule rotates about its center of mass, as shown in Fig. 29–15, its kinetic energy of rotation (see Section 8–7) is

$$E_{rot} = \frac{1}{2}I\omega^2 = \frac{(I\omega)^2}{2I},$$

where ($I\omega$) is the angular momentum (Section 8–8). Quantum mechanics predicts quantization of angular momentum just as in atoms (see Eq. 28–3):

$$I\omega = \sqrt{L(L+1)}\hbar, \qquad L = 0, 1, 2, \cdots,$$

where L is an integer called the **rotational angular momentum quantum number**. Thus the rotational energy is quantized:

$$E_{rot} = \frac{(I\omega)^2}{2I} = L(L+1)\frac{\hbar^2}{2I}, \qquad L = 0, 1, 2, \cdots. \qquad \textbf{(29–1)}$$

Transitions between rotational energy levels are subject to the *selection rule*:

$$\Delta L = \pm 1.$$

The energy of a photon emitted or absorbed for a transition between rotational states with angular-momentum quantum number L and $L - 1$ will be

$$\Delta E_{rot} = E_L - E_{L-1} = \frac{\hbar^2}{2I}L(L+1) - \frac{\hbar^2}{2I}(L-1)(L)$$

$$= \frac{\hbar^2}{I}L. \qquad \textbf{(29–2)}$$

We see that the transition energy increases directly with L. Figure 29–16 shows some of the allowed rotational energy levels and transitions. Measured

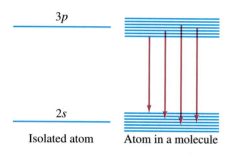

FIGURE 29–14   The individual energy levels of an isolated atom become bands of closely spaced levels in molecules, as well as in solids and liquids.

FIGURE 29–15   Diatomic molecule rotating about a vertical axis.

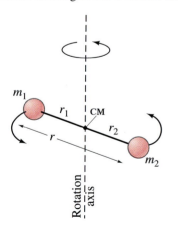

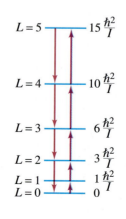

FIGURE 29–16   Rotational energy levels and allowed transitions (emission and absorption) for a diatomic molecule. Upward-pointing arrows represent absorption of a photon, and downward arrows represent emission.

absorption lines fall in the microwave or far-infrared regions of the spectrum, and their frequencies are generally 2, 3, 4, ⋯ times higher than the lowest one, as predicted by Eq. 29–2.

EXAMPLE 29–2 **Rotational transition.** A rotational transition $L = 1$ to $L = 0$ for the molecule CO has a measured absorption wavelength $\lambda_1 = 2.60$ mm (microwave region). Use this to calculate (a) the moment of inertia of the CO molecule, and (b) the CO bond length, r. (c) Calculate the wavelengths of the next three rotational transitions, and the energies of the photon emitted for each of these four transitions

SOLUTION (a) From Eq. 29–2, we can write

$$\frac{\hbar^2}{I}L = \Delta E = hf = \frac{hc}{\lambda_1}.$$

With $L = 1$ (the upper state) in this case, we solve for I:

$$I = \frac{\hbar^2 L}{hc}\lambda_1 = \frac{h\lambda_1}{4\pi^2 c} = \frac{(6.63 \times 10^{-34}\,\text{J·s})(2.60 \times 10^{-3}\,\text{m})}{4\pi^2(3.00 \times 10^8\,\text{m/s})}$$

$$= 1.46 \times 10^{-46}\,\text{kg·m}^2.$$

(b) We write the moment of inertia of the rotating molecule, which rotates about its center of mass (CM), in terms of the CO separation r. In Fig. 29–15, let m_1 be the mass of the C, which is $m_1 = 12$ u and let m_2 be the mass of the O ($m_2 = 16$ u). The distance of the CM from the C atom, which is r_1 in Fig. 29–16, is given by the CM formula, Eq. 7–9:

$$r_1 = \frac{0 + m_2 r}{m_1 + m_2} = \frac{16}{12 + 16}r = 0.57r.$$

The O atom is a distance $r_2 = r - r_1 = 0.43r$ from the CM. The moment of inertia of the CO molecule about its CM is then (see Example 8–11)

$$I = m_1 r_1^2 + m_2 r_2^2$$

$$= [(12\,\text{u})(0.57r)^2 + (16\,\text{u})(0.43r)^2][1.66 \times 10^{-27}\,\text{kg/u}]$$

$$= (1.14 \times 10^{-26}\,\text{kg})r^2.$$

We solve for r and use the result of part (a) for I:

$$r = \sqrt{\frac{1.46 \times 10^{-46}\,\text{kg·m}^2}{1.14 \times 10^{-26}\,\text{kg}}} = 1.13 \times 10^{-10}\,\text{m},$$

or 0.113 nm.

(c) From Eq. 29–2, $\Delta E \propto L$. Hence $\lambda = c/f = hc/\Delta E$ is proportional to $1/L$. Thus, for $L = 2$ to $L = 1$ transitions, $\lambda_2 = \frac{1}{2}\lambda_1 = 1.30$ mm. For $L = 3$ to $L = 2$, $\lambda_3 = \frac{1}{3}\lambda_1 = 0.87$ mm. And for $L = 4$ to $L = 3$, $\lambda_4 = 0.65$ mm. All are close to measured values. The energies of the photons, $hf = hc/\lambda$, are respectively 4.8×10^{-4} eV, 9.5×10^{-4} eV, 1.4×10^{-3} eV, and 1.9×10^{-3} eV.

The spacing of rotational energy levels is thus of the order 10^{-3} eV for the CO molecule. For the H_2 molecule, the wavelengths are on the order of 10^{-4} m, which is in the far-infrared region of the spectrum, and the transition energies are on the order of 10^{-2} eV (see Problem 10).

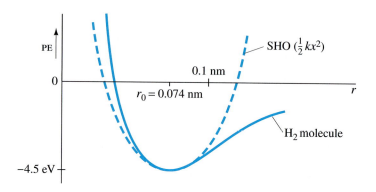

FIGURE 29–17 Potential energy for the H_2 molecule and for a simple harmonic oscillator (PE $= \frac{1}{2}kx^2$, with $|x| = |r - r_0|$).

Vibrational levels **Molecular Vibration.** The potential energy of the two atoms in a typical diatomic molecule has the shape shown in Fig. 29–8 or 29–9, and Fig. 29–17 again shows the PE for the H_2 molecule. We note that the PE, at least in the vicinity of the equilibrium separation r_0, closely resembles the potential energy of a harmonic oscillator, PE $= \frac{1}{2}kx^2$, which is shown superimposed in dashed lines. Thus, for small displacements from r_0, each atom experiences a restoring force proportional to the displacement, and the molecule vibrates as a simple harmonic oscillator (SHO)—see Chapter 11. According to quantum mechanics, the possible energy levels are quantized according to

$$E_{\text{vib}} = (\nu + \tfrac{1}{2})hf, \qquad \nu = 0, 1, 2, \cdots, \qquad \textbf{(29–3)}$$

where f is the classical frequency (see Chapter 11—f depends on the mass of the atoms and on the bond strength or "stiffness") and ν is an integer called the **vibrational quantum number**. The lowest energy state ($\nu = 0$) is not zero (as for rotation), but has $E = \frac{1}{2}hf$. This is called the **zero-point energy**. Higher states have energy $\frac{3}{2}hf, \frac{5}{2}hf$, and so on, as shown in Fig. 29–18. Transitions are subject to the *selection rule*:

$$\Delta\nu = \pm 1,$$

so allowed transitions occur only between adjacent states and all give off

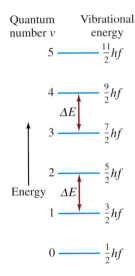

Quantum
number v

Vibrational
energy

5 ———— $\frac{11}{2}hf$

4 ———— $\frac{9}{2}hf$

ΔE

3 ——— $\frac{7}{2}hf$

2 ——— $\frac{5}{2}hf$

Energy $\quad \Delta E$

1 ——— $\frac{3}{2}hf$

0 ———— $\frac{1}{2}hf$

FIGURE 29–18 Allowed vibrational energies for a diatomic molecule, where f is the fundamental frequency of vibration (see Chapter 11). The energy levels are equally spaced. Transitions are allowed only between adjacent levels ($\Delta v = \pm 1$).

photons of energy

$$\Delta E_{\text{vib}} = hf. \tag{29–4}$$

This is very close to experimental values for small v, but for higher energies, the PE curve (Fig. 29–17) begins to deviate from a perfect SHO curve, and this then affects the wavelengths and frequencies of the transitions. Typical transition energies are on the order of 10^{-1} eV, about 10 times larger than for rotational transitions, with wavelengths in the infrared region of the spectrum ($\approx 10^{-5}$ m).

EXAMPLE 29–3 **Vibrational energy levels in hydrogen.** The hydrogen molecule emits infrared radiation of wavelength around 2300 nm. (*a*) What is the separation in energy between different vibrational levels? (*b*) What is the lowest vibrational energy state?

SOLUTION

(*a*) $\Delta E_{\text{vib}} = hf = \dfrac{hc}{\lambda}$

$$= \frac{(6.63 \times 10^{-34} \text{ J·s})(3.00 \times 10^8 \text{ m/s})}{(2300 \times 10^{-9} \text{ m})(1.60 \times 10^{-19} \text{ J/eV})} = 0.54 \text{ eV}.$$

(*b*) The lowest vibrational energy has $v = 0$ in Eq. 29–3: $E = \frac{1}{2}hf = 0.27$ eV.

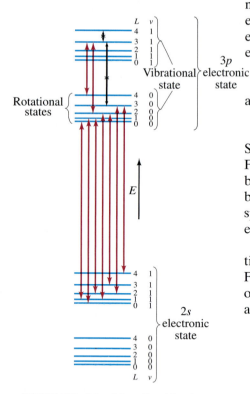

FIGURE 29–19 Combined electronic, vibrational, and rotational energy levels. Transitions marked with an × are not allowed by the selection rules.

When energy is imparted to a molecule, both the rotational and vibrational modes can be excited. Because rotational energies are an order of magnitude or so smaller than vibrational energies, which in turn are smaller than the electronic energy levels, we can represent the grouping of levels as shown in Fig. 29–19. Transitions between energy levels, with emission of a photon, are subject to the *selection rules*:

$$\Delta \nu = \pm 1$$

and

$$\Delta L = \pm 1.$$

Some allowed and forbidden (marked ×) transitions are indicated in Fig. 29–19. Not all transitions and levels are shown, and the separation between vibrational levels, and (even more) between rotational levels, has been exaggerated. But we can clearly see the origin of the very closely spaced lines that give rise to the band spectra, as mentioned with reference to Fig. 29–14 earlier in this Section.

The spectra are quite complicated, so we consider briefly only transitions within the same electronic level, such as those at the top of Fig. 29–19. A transition from a state with quantum numbers ν and L, to one with quantum numbers $\nu + 1$ and $L \pm 1$ (see the selection rules above), will absorb[†] a photon of energy:

$$\Delta E = \Delta E_{\text{vib}} + \Delta E_{\text{rot}}$$

$$= hf + (L + 1)\frac{\hbar^2}{I} \qquad [L \to L + 1], \qquad L = 0, 1, 2, \cdots$$

$$\qquad\qquad\qquad\qquad\qquad\qquad\qquad\qquad\qquad\qquad\qquad\qquad \textbf{(29–5)}$$

$$= hf - L\frac{\hbar^2}{I} \qquad [L \to L - 1], \qquad L = 1, 2, 3, \cdots,$$

where we have used Eqs. 29–2 and 29–4. Note that for $L \to L - 1$ transitions, L cannot be zero since there is then no state with $L = -1$. Equations 29–5 predict an absorption spectrum like that shown schematically in Fig. 29–20, with transitions $L \to L - 1$ on the left and $L \to L + 1$ on the

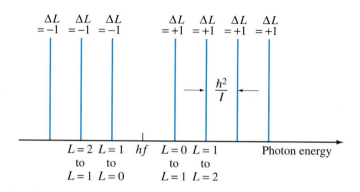

FIGURE 29–20 Expected spectrum for transitions between combined rotational and vibrational states.

[†]This is for absorption; for emission of a photon, the transition would be $\nu \to \nu - 1, L \to L \pm 1$.

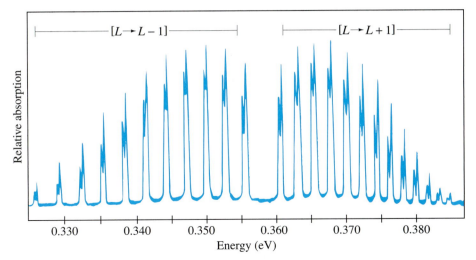

right. Figure 29–21 shows the molecular absorption spectrum of HCl, which follows this pattern very well. (Each line in that spectrum is split into two because Cl consists of two isotopes of different mass; hence there are two kinds of HCl molecule with different moments of inertia I.)

* 29–5 Bonding in Solids

Quantum mechanics has been a great tool for understanding the structure of solids. This active field of research today is called **solid-state physics**, or **condensed-matter physics** so as to include liquids as well. The rest of this chapter is devoted to this subject, and we begin with a brief look at the structure of solids and the bonds that hold them together.

Although some solid materials are *amorphous* in structure (such as glass), in that the atoms and molecules show no long-range order, we will be interested here in the large class of *crystalline* substances whose atoms, ions, or molecules are generally accepted to form an orderly array in a geometric arrangement known as a **lattice**. Figure 29–22 shows three of the possible arrangements of atoms in a crystal: simple cubic, face-centered cubic, and body-centered cubic. The NaCl crystal lattice is face-centered

FIGURE 29–22 Arrangement of atoms in (a) a simple cubic crystal, (b) face-centered cubic crystal (note the atom at the center of each face), and (c) body-centered cubic crystal. Each diagram shows the relationship of the bonds. Each of these "cells" is repeated in three dimensions to the edges of the macroscopic crystal.

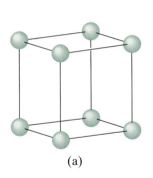

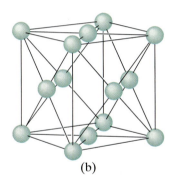

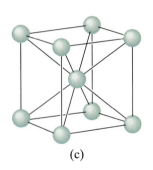

(a)　　　　　　　(b)　　　　　　　(c)

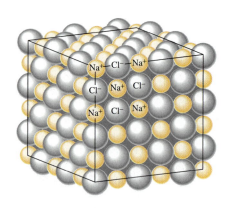

FIGURE 29–23 Diagram of NaCl crystal.

cubic, with one Na^+ ion or one Cl^- ion at each lattice point (see Fig. 29–23).

The molecules of a solid are held together in a number of ways. The most common are by *covalent* bonding (as between the carbon atoms of the diamond crystal) or *ionic* bonding (as in a NaCl crystal). Often the bonds are partially covalent and partially ionic. Our discussion of these bonds earlier in this chapter for molecules applies equally well here to solids.

Let us look for a moment at the NaCl crystal of Fig. 29–23. Each Na^+ ion feels an attractive Coulomb potential due to each of the six "nearest neighbor" Cl^- ions surrounding it. Note that one Na^+ does not "belong" exclusively to one Cl^-, so we must not think of ionic solids as consisting of individual molecules. Each Na^+ also feels a repulsive Coulomb potential due to other Na^+ ions, although this is weaker since the Na^+ ions are farther away.

A different type of bond occurs in metals. Metal atoms have relatively loosely held outer electrons. Present-day **metallic bond** theories propose that in a metallic solid, these outer electrons roam rather freely among all the metal atoms which, without their outer electrons, act like positive ions. The electrostatic attraction between the metal ions and this negative electron "gas" is at least in part responsible for holding the solid together. The binding energy of metal bonds is typically 1 to 3 eV, somewhat weaker than ionic or covalent bonds (5 to 10 eV in solids). The "free electrons" are responsible for the high electrical and thermal conductivity of metals. This theory also nicely accounts for the shininess of smooth metal surfaces: the free electrons can vibrate at any frequency, so when light of a range of frequencies falls on a metal, the electrons can vibrate in response and reemit light of those same frequencies. Hence, the reflected light will consist largely of the same frequencies as the incident light. Compare this to nonmetallic materials that have a distinct color—the atomic electrons exist only in certain energy states, and when white light falls on them, the atoms absorb at certain frequencies, and reflect other frequencies which make up the color we see.

Here is a brief comparison of important strong bonds:

- ionic: an electron is stolen from one atom by another
- covalent: electrons are shared by atoms within a single molecule
- metallic: electrons are shared by all atoms in the metal

The atoms or molecules of some materials, such as the noble gases, can form only **weak bonds** with each other. As we saw in Section 29–3, weak bonds have very low binding energies and would not be expected to hold atoms together as a liquid or solid at room temperature. Indeed, the noble gases condense only at very low temperatures, where the atomic (thermal) kinetic energy is small and the weak attraction can then hold the atoms together.

*29–6 Band Theory of Solids

We saw in Section 29–1 that when two hydrogen atoms approach each other, the wave functions overlap, and the two $1s$ states (one for each atom) divide into two states of different energy. (As we saw, only one of these states, $S = 0$, has low enough energy to give a bound H_2 molecule.)

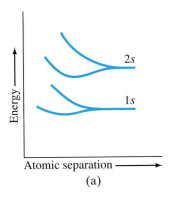

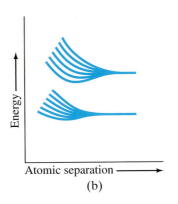

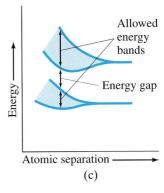

(a)

(b)

(c)

FIGURE 29-24 The splitting of 1s and 2s atomic energy levels as (a) two atoms approach each other (the atomic separation decreases, moving toward the left); (b) the same for six atoms, and (c) for many atoms when they come together to form a solid.

Figure 29–24a shows this situation for 1s and 2s states for two atoms. If six atoms come together, as in Fig. 29–24b, each of the states splits into six levels. If a large number of atoms come together to form a solid, then each of the original atomic levels becomes a **band** as shown in Fig. 29–24c. The energy levels are so close together in each band that they seem essentially continuous. This is why the spectrum of heated solids (Section 27–2) appears continuous.

The crucial aspect of a good **conductor** is that the highest energy band containing electrons is only partially filled. Consider sodium, for example, whose energy bands are shown in Fig. 29–25. The 1s, 2s, and 2p bands are full (just as in a Na atom) and don't concern us. The 3s band, however, is only half full. To see why, recall that the exclusion principle stipulates that in an atom, only two electrons can be in the 3s state, one with spin up and one with spin down. These two states have slightly different energy. For a solid consisting of N atoms, the 3s band will contain 2N possible energy states. Now a sodium atom has a single 3s electron, so in a sample of sodium metal containing N atoms, there are N electrons in the 3s band, and N unoccupied states. When a potential difference is applied across the metal, electrons can respond by accelerating and increasing their energy, since there are plenty of unoccupied states of slightly higher energy available. Hence, a current flows readily and sodium is a good conductor. The characteristic of all good conductors is that the highest energy band is only partially filled, or two bands overlap so that unoccupied states are available. An example of the latter is magnesium, which has two 3s electrons, so its 3s band is filled. But the unfilled 3p band overlaps the 3s band in energy, so there are lots of available states for the electrons to move into. Thus magnesium, too, is a good conductor.

In a material that is a good **insulator**, on the other hand, the highest band containing electrons, called the **valence band**, is completely filled. The next higher energy band, called the **conduction band**, is separated from the valence band by a "forbidden" **energy gap**, (or **band gap**), E_g, of typically 5 to 10 eV. So at room temperature (300 K), where thermal energies (that is, average kinetic energy—see Chapter 13) are on the order of $\frac{3}{2}kT \approx 0.04$ eV, almost no electrons can acquire the 5 eV needed to reach the conduction band. When a potential difference is applied across the material, no available states are accessible to the electrons, and no current flows. Hence, the material is a good insulator.

Conductors

FIGURE 29-25 Energy bands for sodium.

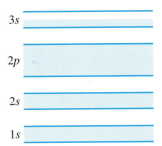

Insulators

Valence and conduction bands

Energy gap

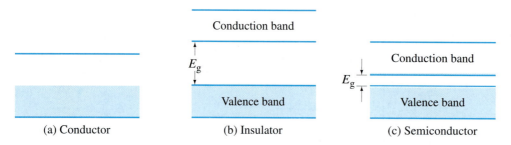

FIGURE 29–26 Energy bands for (a) a conductor, (b) an insulator, and (c) a semiconductor. Shading represents occupied states. Pale shading in (c) represents electrons that can pass from the valence band to the conduction band due to thermal agitation at room temperature (exaggerated).

Semiconductors (pure)

Figure 29–26 compares the relevant energy bands (a) for conductors, (b) for insulators, and also (c) for the important class of materials known as **semiconductors**. The bands for a pure (or **intrinsic**) semiconductor, such as silicon or germanium, are like those for an insulator, except that the unfilled conduction band is separated from the filled valence band by a much smaller energy gap, E_g, typically on the order of 1 eV. At room temperature, a few electrons can acquire enough thermal energy to reach the conduction band, and so a very small current may flow when a voltage is applied. At higher temperatures, more electrons have enough energy to jump the gap. This effect can often more than offset the effects of more frequent collisions due to increased disorder at higher temperature, so that the resistivity of semiconductors can *decrease* with increasing temperature (see Table 18–1). But this is not the whole story of semiconductor conduction. When a potential difference is applied to a semiconductor, the few electrons in the conduction band move toward the positive electrode. Electrons in the valence band try to do the same thing, and a few can because there are a small number of unoccupied states which were left empty by the electrons reaching the conduction band. Such unfilled electron states are called **holes**. Each electron in the valence band that fills a hole in this way as it moves toward the positive electrode leaves behind a hole, so that the holes migrate toward the negative electrode. As the electrons tend to accumulate at one side of the material, the holes tend to accumulate on the opposite side. We will look at this phenomenon in more detail in the next Section.

Holes (in semiconductor)

EXAMPLE 29–4 **Calculating the energy gap.** It is found that the conductivity of a certain semiconductor increases when light of wavelength 345 nm or shorter strikes it, suggesting that electrons are being promoted from the valence band to the conduction band. What is the energy gap, E_g, for this semiconductor?

SOLUTION The longest wavelength, or lowest energy, photon to cause an increase in conductivity has $\lambda = 345$ nm, and it can transfer to an electron an energy

$$E_g = hf = \frac{hc}{\lambda} = \frac{(6.63 \times 10^{-34}\,\text{J·s})(3.00 \times 10^8\,\text{m/s})}{(345 \times 10^{-9}\,\text{m})(1.60 \times 10^{-19}\,\text{J/eV})} = 3.6\,\text{eV}.$$

* 29–7 Semiconductors and Doping

The most commonly used semiconductors in modern electronics are silicon (Si) and germanium (Ge). An atom of silicon or germanium has four outer electrons that act to hold the atoms in the regular lattice structure of the crystal, shown schematically in Fig. 29–27a. Germanium and silicon acquire useful properties for use in electronics only when a tiny amount of impurity is introduced into the crystal structure (perhaps 1 part in 10^6 or 10^7). This is called **doping** the semiconductor. Two kinds of doped semiconductor can be made, depending on the type of impurity used. If the impurity is an element whose atoms have five outer electrons, such as arsenic, we have the situation shown in Fig. 29–27b. Only four of arsenic's electrons fit into the bonding structure. The fifth does not fit in and can move relatively freely, somewhat like the electrons in a conductor. Because of this small number of extra electrons, a doped semiconductor becomes slightly conducting. The density of conduction electrons in an intrinsic (pure) semiconductor is about 1 per 10^{10} atoms. With an impurity concentration of 1 in 10^6 or 10^7 when doped, the conductivity will be much higher and it can be controlled with great precision. An arsenic-doped silicon crystal is called an **n-type semiconductor** because negative charges (electrons) carry the electric current.

Doped semiconductors

n-type

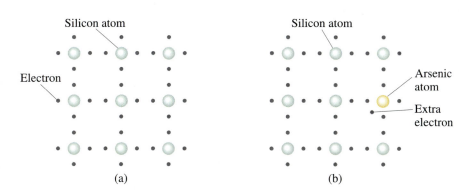

Silicon atom

Electron

(a)

Silicon atom

Arsenic atom

Extra electron

(b)

FIGURE 29–27   Two-dimensional representation of a silicon crystal. (a) Four (outer) electrons surround each silicon atom. (b) Silicon crystal doped with a few arsenic atoms: the extra electron doesn't fit into the crystal lattice and so is free to move about. This is an *n*-type semiconductor.

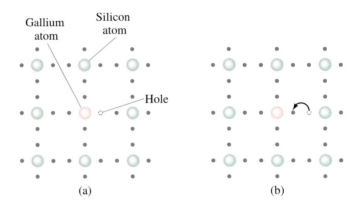

FIGURE 29–28 A *p*-type semiconductor, gallium-doped silicon. (a) Gallium has only three outer electrons, so there is an empty spot, or *hole*, in the structure. (b) Electrons from silicon atoms can jump into the hole and fill it. As a result, the hole moves to a new location (to the right in this figure), to where the electron used to be.

Gallium atom

Silicon atom

Hole

(a)   (b)

p-type

In a ***p*-type semiconductor**, a small amount of an element with three outer electrons—such as gallium—is added to the semiconductor. As shown in Fig. 29–28a, there is a "hole" in the lattice structure near a gallium atom since it has only three outer electrons. Electrons from nearby silicon atoms can jump into this hole and fill it. But this leaves a hole where that electron had previously been, Fig. 29–28b. The vast majority of atoms are silicon, so holes are almost always next to a silicon atom. Since silicon atoms require four outer electrons to be neutral, this means that there is a net positive charge at the hole. Whenever an electron moves to fill a hole, the positive hole is then at the previous position of that electron. Another electron can then fill this hole, and the hole thus moves to a new location; and so on. This type of semiconductor is called *p-type* because it is the *positive holes* that seem to carry the electric current. Note, however, that both *p*-type and *n*-type semiconductors have *no net charge* on them.

Holes are positive

According to the band theory (Section 29–6), in a doped semiconductor, the impurity provides additional energy states between the bands, as shown in Fig. 29–29. In an *n*-type semiconductor, the impurity energy level lies just below the conduction band. Electrons in this energy level need only about 0.05 eV (in Si; even less in Ge) of energy to reach the conduction band; this is on the order of $\frac{3}{2}kT$ ($= \overline{\text{KE}} = 0.04$ eV) at 300 K, so transitions occur readily at room temperature. Since this energy level supplies electrons to the conduction band, it is called a **donor** level. In *p*-type semiconductors, the impurity energy level is just above the valence band (Fig. 29–29). It is called an **acceptor** level because electrons from the valence band can easily jump into it. Positive holes are left behind in the valence band, and as other electrons move into these holes, the holes move about as discussed earlier.

FIGURE 29–29 Impurity energy levels in doped semiconductors.

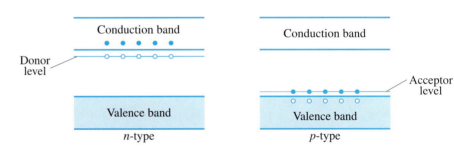

Conduction band

Donor level

Valence band

n-type

Conduction band

Acceptor level

Valence band

p-type

Semiconductor Diodes

Semiconductor diodes and transistors are essential components of modern electronic devices. The miniaturization achieved today allows many thousands of diodes, transistors, resistors, and so on, to be placed on a single *chip* only a millimeter on a side. We now discuss, briefly and qualitatively, the operation of diodes and transistors.

When an *n*-type semiconductor is joined to a *p*-type, a ***p-n* junction diode** is formed. Separately, the two semiconductors are electrically neutral. When joined, a few electrons near the junction diffuse from the *n*-type into the *p*-type semiconductor, where they fill a few of the holes. The *n*-type is left with a positive charge, and the *p*-type acquires a net negative charge. Thus a potential difference is established, with the *n* side positive relative to the *p* side, and this prevents further diffusion of electrons.

If a battery is connected to a diode with the positive terminal to the *p* side and the negative terminal to the *n* side as in Fig. 29–30a, the externally applied voltage opposes the internal potential difference and the diode is said to be **forward biased**. If the voltage is great enough (about 0.3 V for Ge, 0.6 V for Si at room temperature), a current will flow. The positive holes in the *p*-type semiconductor are repelled by the positive terminal of the battery and the electrons in the *n*-type are repelled by the negative terminal of the battery. The holes and electrons meet at the junction, and the electrons cross over and fill the holes. A current is flowing. Meanwhile, the positive terminal of the battery is continually pulling electrons off the *p* end, forming new holes, and electrons are being supplied by the negative terminal at the *n* end. Consequently, a large current flows through the diode.

When the diode is **reverse biased**, as in Fig. 29–30b, the holes in the *p* end are attracted to the battery's negative terminal and the electrons in the *n* end are attracted to the positive terminal. The current carriers do not meet near the junction and, ideally, no current flows.

A graph of current versus voltage for a typical diode is shown in Fig. 29–31. As can be seen, a real diode does allow a small amount of reverse current to flow. For most practical purposes, this is negligible.[†]

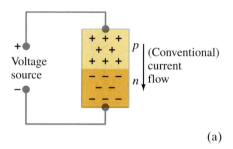

(a)

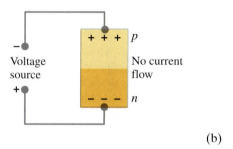

(b)

FIGURE 29–30 Schematic diagram showing how a semiconductor diode operates. Current flows when the voltage is connected in forward bias, as in (a), but not when connected in reverse bias, as in (b).

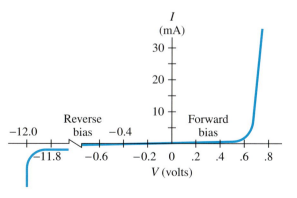

FIGURE 29–31 Current through a diode as a function of applied voltage.

[†]Reverse current in Ge at room temperature is typically a few μA; for Si, a few pA. The reverse current increases rapidly with temperature, however, and may render a diode ineffective above 200°C.

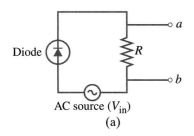

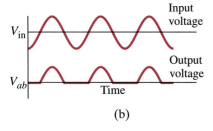

FIGURE 29–32 (a) A simple (half-wave) rectifier circuit using a semiconductor diode. (b) AC source input voltage, and output voltage across R, as functions of time.

Rectifier

EXAMPLE 29–6 **A diode.** The diode whose current–voltage characteristics are shown in Fig. 29–31 is connected in series with a 4.0-V battery and a resistor. If a current of 10 mA is to pass through the diode, what resistance must the resistor have?

SOLUTION In Fig. 29–31, we see that the voltage drop across the diode is about 0.7 V when the current is 10 mA. Therefore, the voltage drop across the resistor is 4.0 V − 0.7 V = 3.3 V, so $R = V/I = (3.3\ \text{V})/(1.0 \times 10^{-2}\ \text{A}) = 330\ \Omega$.

If the voltage across a diode connected in reverse bias is increased greatly, a point is reached where breakdown occurs. The electric field across the junction becomes so large that ionization of atoms results. The electrons thus pulled off their atoms contribute to a larger and larger current as breakdown continues. The voltage remains constant over a wide range of currents. This is shown on the far left in Fig. 29–31. This property of diodes can be used to accurately regulate a voltage supply. A diode designed for this purpose is called a **zener diode**. When placed across the output of an unregulated power supply, a zener diode can maintain the voltage at its own breakdown voltage as long as the supply voltage is always above this point. Zener diodes can be obtained corresponding to voltages of a few volts to hundreds of volts.

Since a *p-n* junction diode allows current to flow only in one direction (as long as the voltage is not too high), it can serve as a **rectifier**—to change ac into dc. A simple rectifier circuit is shown in Fig. 29–32a where the arrow inside the symbol for a diode indicates the direction in which a diode conducts conventional (+) current. The ac source applies a voltage across the diode alternately positive and negative. Only during half of each cycle will a current pass through the diode; so only then is there a current through the resistor R. Hence, a graph of the voltage V_{ab} across R as a function of time looks like Fig. 29–32b). This **half-wave rectification** is not exactly dc, but it is unidirectional. More useful is a **full-wave rectifier** circuit, which uses two diodes (or sometimes four) as shown in Fig. 29–33a. At any given instant, either one diode or the other will conduct current to the right. Therefore, the output across the load resistor R will be as shown in Fig. 29–33b. Actually this is the voltage if the capacitor C were not in the circuit. The capacitor tends to store charge, and thus helps to smooth out the current as shown in Fig. 29–33c.

FIGURE 29–33 (a) Full-wave-rectifier circuit (including a transformer so the magnitude of the voltage can be changed). (b) Output voltage in the absence of capacitor C. (c) Output voltage with the capacitor in the circuit.

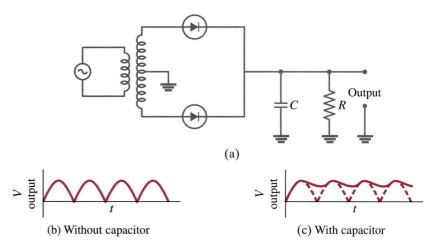

(a)

(b) Without capacitor

(c) With capacitor

Rectifier circuits are important because most line voltage is ac, and most electronic devices require a dc voltage for their operation. Hence, diodes are found in nearly all electronic devices including radio and TV sets, calculators, and computers.

Another useful device is a **light-emitting diode** (LED). When a *p-n* junction is forward biased, a current begins to flow. Electrons cross from the *n* region into the *p* region and combine with holes, and a photon can be emitted with an energy approximately equal to the band gap, E_g (see Figs. 29–26c and 29–29). Often the energy, and hence the wavelength, is in the red region of the visible spectrum, producing the familiar LED displays on VCRs, CD players, car instrument panels, clocks, and so on. Infrared (i.e., nonvisible) LEDs are used in remote controls for TV, VCRs, and stereos.

LED

Photodiodes and **solar cells** are *p-n* junctions used in the reverse way. Photons are absorbed, creating electron–hole pairs if the photon energy is greater than the band gap energy, E_g. The created electrons and holes produce a current that, when connected to an external circuit, becomes a source of emf and power. Applications were discussed in Section 15–12, and you may find solar cells as a source of emf and power in a handheld calculator.

Solar cells

A diode is called a **nonlinear device** because the current is not proportional to the voltage; that is, a graph of current versus voltage (Fig. 29–31) is not a straight line as it is for a resistor (which ideally *is* linear). Transistors are also *nonlinear* devices.

---

*29–9 Transistors and Integrated Circuits

A simple **junction transistor** consists of a crystal of one type of doped semiconductor sandwiched between two crystals of the opposite type. Both *pnp* and *npn* transistors are made, and they are shown schematically in Fig. 29–34a. The three semiconductors are given the names *collector*, *base*, and *emitter*. The symbols for *npn* and *pnp* transistors are shown in Fig. 29–34b. The arrow is always placed on the emitter and indicates the direction of (conventional) current flow in normal operation.

Transistors

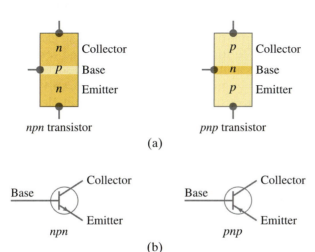

FIGURE 29–34
(a) Schematic diagram of *npn* and *pnp* transistors. (b) Symbols for *npn* and *pnp* transistors.

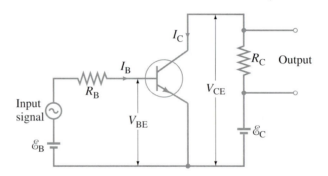

FIGURE 29–35 An *npn* transistor used as an amplifier.

The operation of a transistor can be analyzed qualitatively—very briefly—as follows. Consider an *npn* transistor connected as shown in Fig. 29–35. A voltage V_{CE} is maintained between the collector and emitter by the battery $\mathscr{E}_C$. The voltage applied to the base is called the *base bias voltage*, V_{BE}. If V_{BE} is positive, conduction electrons in the emitter are attracted into the base. Since the base region is very thin (perhaps 1 μm), most of these electrons flow right across into the collector, which is maintained at a positive voltage. A large current, I_C, flows between collector and emitter and a much smaller current, I_B, through the base. A small variation in the base voltage due to an input signal causes a large change in the collector current and therefore a large change in the voltage drop across the output resistor R_C. Hence a transistor can *amplify* a small signal into a larger one. In fact, transistors are the basic elements in modern electronic **amplifiers** of all sorts.

Amplifiers

A *pnp* transistor operates in the same fashion, except that holes move instead of electrons. The collector voltage is negative, and so is the base voltage in normal operation.

In Fig. 29–35 two batteries were shown: $\mathscr{E}_C$ supplied the collector voltage and $\mathscr{E}_B$ the base bias voltage. In practice, only one source is often used, and the base bias voltage can be obtained using a resistance voltage divider as in Fig. 29–36. Transistors can be connected in many other ways as well, and many new and innovative uses have been found for them.

FIGURE 29–36 Typical transistor circuit involving an *npn* transistor. $\mathscr{E}_C = +12 \text{ V}$ and $\mathscr{E}_B$ is determined by the resistors.

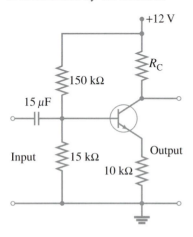

Transistors were a great advance in miniaturization of electronic circuits. Although individual transistors are very small compared to the once used vacuum tubes, they are huge compared to **integrated circuits** or **chips** (see photo at start of this chapter). Tiny amounts of impurities can be placed at particular locations within a single silicon crystal. These can be arranged to form diodes, transistors, and resistors (undoped semiconductors). Capacitors and inductors can also be formed, although they are often connected separately. A tiny chip, only 1 mm on a side, may contain thousands of transistors and other circuit elements. Integrated circuits are now the heart of computers, television, calculators, cameras, and the electronic instruments that control aircraft, space vehicles, and automobiles. The "miniaturization" produced by integrated circuits not only allows extremely complicated circuits to be placed in a small space, but also has allowed a great increase in the speed of operation of, say, computers, because the distances the electronic signals travel are so tiny.

SUMMARY

Quantum mechanics explains the bonding together of atoms to form **molecules**. In a **covalent bond**, the electron clouds of two or more atoms overlap because of constructive interference between the electron waves. The positive nuclei are attracted to this concentration of negative charge between them, forming the bond. An **ionic bond** is an extreme case of a covalent bond in which one or more electrons from one atom spend much more time around the other atom than around their own. The atoms then act as oppositely charged ions that attract each other, forming the bond. These **strong bonds** hold molecules together, and also hold atoms and molecules together in solids. Also important are **weak bonds** (or **van der Waals bonds**), which are generally dipole attractions between molecules.

When atoms combine to form molecules, the energy levels of the outer electrons are altered because they now interact with each other. Additional energy levels also become possible because the atoms can vibrate with respect to each other, and the molecule as a whole can rotate. The energy levels for both vibrational and rotational motion are quantized, and are very close together (typically, 10^{-1} eV to 10^{-3} eV apart). Each atomic energy level thus becomes a set of closely spaced levels corresponding to the vibrational and rotational motions. Transitions from one level to another appear as many very closely spaced lines. The resulting spectra are called **band spectra**. The quantized rotational energy levels are given by

$$E_{\text{rot}} = L(L+1)\frac{\hbar^2}{2I}, \qquad L = 0, 1, 2, \cdots,$$

where I is the moment of inertia of the molecule. The energy levels for vibrational motion are given by

$$E_{\text{vib}} = (\nu + \tfrac{1}{2})hf, \qquad \nu = 0, 1, 2, \cdots,$$

where f is the classical natural frequency of vibration for the molecule. Transitions between energy levels are subject to the selection rules $\Delta L = \pm 1$ and $\Delta \nu = \pm 1$.

Some **solids** are bound together by covalent and ionic bonds, just as molecules are. In metals, the electrostatic force between free electrons and positive ions helps form the **metallic bond**.

In a crystalline solid, the possible energy states for electrons are arranged in **bands**. Within each band the levels are very close together, but between the bands there may be forbidden **energy gaps**. Good conductors are characterized by the highest occupied band (the **conduction band**) being only partially full, so there are many accessible states available to electrons to move about and accelerate when a voltage is applied. In a good insulator, the highest occupied energy band (the **valence band**) is completely full, and there is a large energy gap (5 to 10 eV) to the next highest band, the *conduction band*. At room temperature, molecular kinetic energy (thermal energy) available due to collisions is only about 0.04 eV, so almost no electrons can jump from the valence to the conduction band. In a **semiconductor**, the gap between valence and conduction bands is much smaller, on the order of 1 eV, so a few electrons can make the transition from the essentially full valence band to the nearly empty conduction band.

In a **doped** semiconductor, a small percentage of impurity atoms with five or three valence electrons replace a few of the normal silicon atoms with their four valence electrons. A five-electron impurity produces an **n-type** semiconductor with negative electrons as carriers of current. A three-electron impurity produces a **p-type** semiconductor in which positive **holes** carry the current. The energy level of impurity atoms lies slightly below the conduction band in an *n*-type semiconductor, and acts as a **donor** from which electrons readily pass into the conduction band. The energy level of impurity atoms in a *p*-type semiconductor lies slightly above the valence band and acts as an **acceptor** level, since electrons from the valence band easily reach it, leaving holes behind to act as charge carriers.

A semiconductor **diode** consists of a **p-n-junction** and allows current to flow in one direction only; it can be used as a **rectifier** to change ac to dc. Common **transistors** consist of three semiconductor sections, either as **pnp** or **npn**. Transistors can amplify electrical signals and find many other uses. An integrated circuit consists of a tiny semiconductor crystal or "chip" on which many transistors, diodes, resistors, and other circuit elements have been constructed using careful placement of impurities.

QUESTIONS

*1. What type of bond would you expect for (*a*) the N_2 molecule, (*b*) the HCl molecule, (*c*) Fe atoms in a solid?

*2. Describe how the molecule $CaCl_2$ could be formed.

*3. Does the H_2 molecule have a permanent dipole moment? Does O_2? Does H_2O? Explain.

*4. Although the molecule H_3 is not stable, the ion H_3^+ is. Explain, using the Pauli exclusion principle.

*5. The energy of a molecule can be divided into four categories. What are they?

*6. If conduction electrons are free to roam about in a metal, why don't they leave the metal entirely?

*7. A silicon semiconductor is doped with phosphorus. Will these atoms be donors or acceptors? What type of semiconductor will this be?

*8. Explain why the resistivity of metals increases with temperature whereas the resistivity of semiconductors decreases with increasing temperature.

*9. Can a diode be used to amplify a signal?

*10. Figure 29–37 shows a "bridge-type" full-wave rectifier. Explain how the current is rectified and how current flows during each half cycle.

*11. Compare the resistance of a *p-n* junction diode connected in forward bias to its resistance when connected in reverse bias.

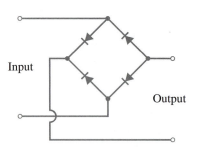

FIGURE 29–37   Question 10.

*12. Explain how a transistor could be used as a switch.

*13. If $\mathscr{E}_C$ were reversed in Fig. 29–35, how would the amplification be altered?

*14. Describe how a *pnp* transistor can operate as an amplifier.

*15. In a transistor, the base–emitter junction and the base–collector junction are essentially diodes. Are these junctions reverse-biased or forward-biased in the application shown in Fig. 29–35?

*16. What purpose does the capacitor in Fig. 29–36 serve?

17. Do diodes and transistors obey Ohm's law?

PROBLEMS

*SECTIONS 29–1 TO 29–3

*1. (I) Estimate the binding energy of a KCl molecule by calculating the electrostatic potential energy when the K^+ and Cl^- ions are at their stable separation of 0.28 nm. Assume each has a charge of magnitude $1.0e$.

*2. (II) The measured binding energy of KCl is 4.43 eV. From the result of Problem 1, estimate the contribution to the binding energy of the repelling electron clouds at the equilibrium distance $r_0 = 0.28$ nm.

*3. (II) Estimate the binding energy of the H_2 molecule, assuming the two H nuclei are 0.074 nm apart and the two electrons spend 33 percent of their time midway between them.

*4. (II) Binding energies are often measured experimentally in kcal per mole, and then the binding energy in eV per molecule is calculated from that result. What is the conversion factor in going from kcal per mole to eV per molecule? What is the binding energy of KCl $(= 4.43\text{ eV})$ in kcal per mole?

*5. (II) Estimate the total binding energy for the three hydrogen bonds between cytosine and guanine shown in Fig. 29–12b. Assume each of the bonds is between the H—N dipole $(q_H = -q_N = 0.19e)$ and an O or N atom of charge $-1.0e$.

*6. (III) Apply reasoning similar to that in the text for the $S = 0$ and $S = 1$ states in the formation of the H_2 molecule to show why the molecule He_2 is *not* formed. Show also why the He_2^+ molecular ion *is* formed (with a binding energy of 3.1 eV at $r_0 = 0.11$ nm).

*SECTION 29–4

*7. (I) Show that the quantity $\hbar^2/I$ has units of energy.

*8. (II) The so-called "characteristic rotational energy," $\hbar^2/2I$, for N_2 is 2.48×10^{-4} eV. Calculate the N_2 bond length.

*9. (II) (a) Calculate the characteristic rotational energy, $\hbar^2/2I$, for the O_2 molecule whose bond length is 0.121 nm. (b) What are the energy and wavelength of photons emitted in a $L = 2$ to $L = 1$ transition?

*10. (II) The equilibrium separation of H atoms in the H_2 molecule is 0.074 nm (Fig. 29–8). Calculate the energies and wavelengths of photons for the rotational transitions (a) $L = 1$ to $L = 0$, (b) $L = 2$ to $L = 1$, and (c) $L = 3$ to $L = 2$.

*11. (II) Calculate the bond length for the NaCl molecule given that three successive wavelengths for rotational transitions are 23.1 mm, 11.6 mm, and 7.71 mm.

*12. (II) In the absorption spectrum of the HCl molecule (Fig. 29–21), the lines are separated by 2.6×10^{-3} eV. Determine (a) the moment of inertia of the HCl molecule, and (b) the bond length. Assume that Cl has mass of 35 u.

*13. (II) Derive Eqs. 29–5.

*14. (II) (a) Show that the moment of inertia of a diatomic molecule can be written as $I = \mu r_0^2$ where r_0 is the distance between atoms and $\mu = m_1 m_2/(m_1 + m_2)$ is called the reduced mass. Determine I for the following molecules: (b) H_2 ($r_0 = 0.074$ nm); (c) O_2 ($r_0 = 0.121$ nm); (d) NaCl ($r_0 = 0.24$ nm); (e) CO ($r_0 = 0.113$ nm).

*15. (III) (a) Use the curve of Fig. 29–17 to estimate the stiffness constant k for the H_2 molecule. (Recall that $PE = \frac{1}{2}kx^2$.) (b) Then estimate the fundamental wavelength for vibrational transitions using the classical formula (Chapter 11).

*SECTION 29–5

*16. (II) The spacing between "nearest neighbor" Na and Cl ions in a NaCl crystal is 0.24 nm. What is the spacing between two nearest neighbor Na ions?

*17. (II) Common salt, NaCl, has a density of 2.165 g/cm³. The molecular weight of NaCl is 58.44. Estimate the distance between nearest neighbor Na and Cl ions. [Hint: Each ion can be considered to have one "cube" or "cell" of side s (our unknown) extending out from it.]

*18. (II) Repeat the previous problem for KCl whose density is 1.99 g/cm³.

*SECTION 29–6

*19. (I) Explain on the basis of energy bands why the sodium chloride crystal is a good insulator. [Hint: Consider the shells of Na⁺ and Cl⁻ ions.]

*20. (I) A semiconductor, bombarded with light of slowly increased frequency, begins to conduct when the wavelength of light is 640 nm; estimate the size of the energy gap E_g.

*21. (I) Calculate the longest-wavelength photon that can cause an electron in silicon ($E_g = 1.1$ eV) to jump from the valence band to the conduction band.

*22. (II) The energy gap between valence and conduction bands in germanium is 0.72 eV. What range of wavelengths can a photon have to excite an electron from the top of the valence band into the conduction band?

*23. (II) A TV remote control emits IR light. If the detector on the TV set is *not* to react to visible light, could it make use of silicon as a "window" with its energy gap $E_g = 1.14$ eV? What is the shortest-wavelength light that can strike silicon without causing electrons to jump from the valence band to the conduction band?

*24. (II) We saw that there are $2N$ possible electron states in the $3s$ band of Na, where N is the total number of atoms. How many possible electron states are there in the (a) $2s$ band, (b) $2p$ band, and (c) $3p$ band? (d) State a general formula for the total number of possible states in any given electron band.

*25. (II) The energy gap E_g in germanium is 0.72 eV. When used as a photon detector, roughly how many electrons can be made to jump from the valence to the conduction band by the passage of a 660-keV photon that loses all its energy in this fashion?

*SECTION 29–7

*26. (II) Suppose that a silicon semiconductor is doped with phosphorus so that one silicon atom in 10^6 is replaced by a phosphorus atom. Assuming that the "extra" electron in every phosphorus atom is donated to the conduction band, by what factor is the density of conduction electrons increased? The density of silicon is 2330 kg/m³, and the density of conduction electrons in pure silicon is about 10^{16} m⁻³ at room temperature.

*SECTION 29–8

*27. (I) At what wavelength will an LED radiate if made from a material with an energy gap $E_g = 1.4$ eV?

*28. (I) If an LED emits light of wavelength $\lambda = 650$ nm, what is the energy gap (in eV) between valence and conduction bands?

*29. (II) A silicon diode, whose current–voltage characteristics are given in Fig. 29–31, is connected in series with a battery and a 660-Ω resistor. What battery voltage is needed to produce a 12-mA current?

*30. (II) Suppose that the diode of Fig. 29–31 is connected in series to a 100-Ω resistor and a 2.0-V battery. What current flows in the circuit? [Hint: Draw a line on Fig. 29–31 representing the current in the resistor as a function of the voltage across the diode; the intersection of this line with the characteristic curve will give the answer.]

31. (II) Sketch the resistance as a function of current, for $V > 0$, for the diode shown in Fig. 29–31.

* 32. (II) An ac voltage of 120 V rms is to be rectified. Estimate very roughly the average current in the output resistor R (18 kΩ) for (a) a half-wave rectifier (Fig. 29–32), and (b) a full-wave rectifier (Fig. 29–33) without capacitor.

* 33. (III) A silicon diode passes significant current only if the forward-bias voltage exceeds about 0.6 V. Make a rough estimate of the average current in the output resistor R of (a) a half-wave rectifier (Fig. 29–32), and (b) a full-wave rectifier (Fig. 29–33) without a capacitor. Assume that $R = 150\,\Omega$ in each case and that the ac voltage is 12.0 V rms in each case.

* 34. (III) A 120-V rms 60-Hz voltage is to be rectified with a full-wave rectifier as in Fig. 29–33, where $R = 18\,k\Omega$, and $C = 25\,\mu F$. (a) Make a rough estimate of the average current. (b) What happens if $C = 0.10\,\mu F$? [Hint: See Section 19–7.]

*SECTION 29–9

* 35. (II) Suppose that the *current gain* of the transistor in Fig. 29–35 is $\beta = I_C/I_B = 80$. If $R_C = 3.3\,k\Omega$, calculate the output voltage for a time-varying input current of $2.0\,\mu A$.

* 36. (II) If the current gain of the transistor amplifier in Fig. 29–35 is $\beta = I_C/I_B = 100$, what value must R_C have if a 1.0-μA base current is to produce an output voltage of 0.40 V?

* 37. (II) A transistor, whose current gain $\beta = I_C/I_B = 90$, is connected as in Fig. 29–35 with $R_B = 2.2\,k\Omega$ and $R_C = 6.2\,k\Omega$. Calculate (a) the voltage gain, and (b) the power amplification.

* 38. (II) An amplifier has a *voltage gain* of 70 and a 14-kΩ load (output) resistance. What is the peak output current through the load resistor if the input voltage is an ac signal with a peak of 0.080 V?

GENERAL PROBLEMS

* 39. Estimate the binding energy of the H_2 molecule by calculating the difference in KE of the electrons between when they are in separate atoms and when they are in the molecule, using the uncertainty principle. Take Δx for the electrons in the separated atoms to be the radius of the first Bohr orbit, 0.053 nm, and for the molecule take Δx to be the separation of the nuclei, 0.074 nm.

* 40. The average translational kinetic energy of an atom or molecule is about $\overline{KE} = \frac{3}{2}kT$ (see Chapter 13), where $k = 1.38 \times 10^{-23}\,J/K$ is Boltzmann's constant. At what temperature T will $\overline{KE}$ be on the order of the bond energy (and hence the bond able to be broken by thermal motion) for (a) a covalent bond (say H_2) of binding energy 4.5 eV, and (b) a "weak" hydrogen bond of binding energy 0.15 eV?

* 41. In the ionic salt KF, the separation distance between ions is about 0.27 nm. (a) Estimate the electrostatic potential energy between the ions assuming them to be point charges (magnitude $1e$). (b) It is known that F releases 4.07 eV of energy when it "grabs" an electron, and 4.34 eV is required to ionize K. Find the binding energy of KF relative to free K and F atoms, neglecting the energy of repulsion.

* 42. Most of the Sun's radiation has wavelengths shorter than 1000 nm. For a solar cell to absorb all this, what energy gap ought the material have?

* 43. For an arsenic donor atom in a doped silicon semiconductor, assume that the "extra" electron moves in a Bohr orbit about the arsenic ion. For this electron in the ground state, take into account the dielectric constant $K = 12$ of the Si lattice (which represents the weakening of the Coulomb force due to all the other atoms or ions in the lattice), and estimate (a) the binding energy, and (b) the orbit radius for this extra electron. [Hint: Substitute $\epsilon = K\epsilon_0$ in Coulomb's law; see Section 17–8.]

* 44. A full-wave rectifier (Fig. 29–33) uses two diodes to rectify a 95-V rms 60 Hz ac voltage. If $R = 10\,k\Omega$ and $C = 30\,\mu F$, what will be the approximate percent variation in the output voltage? The variation in output voltage (Fig. 29–33c) is called *ripple voltage*. [Hint: See Section 19–7 and assume the discharge of the capacitor is approximately linear.]

* 45. A zener diode voltage regulator is shown in Fig. 29–38. Suppose that $R = 1.80\,k\Omega$ and that the diode breakdown voltage is 130 V; the diode is rated at a maximum current of 100 mA. (a) If $R_{load} = 15.0\,k\Omega$, over what range of supply voltages will the circuit maintain the voltage at 130 V? (b) If the supply voltage is 200 V, over what range of load resistance will the voltage be regulated?

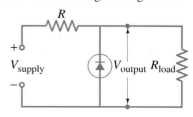

FIGURE 29–38 Problem 45.

*46. Do we need to consider quantum effects for everyday rotating objects? Estimate the differences between rotational energy levels for a spinning baton compared to the energy of the baton. Assume the baton consists of a 30-cm bar with a mass of 200 g and two end masses (each points), each of mass 300 g, and that it rotates at 2.0 rev/s.

*47. A strip of silicon 1.5 cm wide and 1.0 mm thick is immersed in a magnetic field of strength 1.5 T perpendicular to the strip (Fig. 29–39). When a current of 0.20 mA is run through the strip, there is a resulting Hall effect voltage of 18 mV across the strip (Section 20–11). How many electrons per silicon atom are in the conduction band? The density of silicon is 2330 kg/m^3.

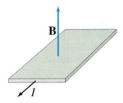

FIGURE 29–39 Problem 47.

*48. When solid argon melts at −189°C, its latent heat of fusion goes directly into breaking the bonds between the atoms. Solid argon is a weakly bound cubic lattice, with each atom connected to six neighbors, each bond having a binding energy of 3.9×10^{-3} eV. What is the latent heat of fusion for argon, in J/kg? [*Hint*: Show that in a simple cubic lattice (Fig. 29–40), there are *three* times as many bonds as there are atoms, when the number of atoms is large.]

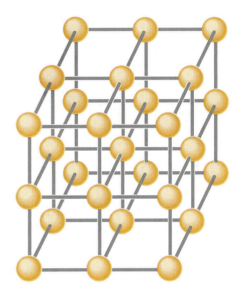

FIGURE 29–40 Problem 48.

A geologist can search for radioactive elements (perhaps uranium) using a portable Geiger counter or scintillation counter.

30 NUCLEAR PHYSICS AND RADIOACTIVITY

In the early part of the twentieth century, Rutherford's experiments led to the idea that at the center of an atom there is a tiny but massive nucleus. At the same time that the quantum theory was being developed and scientists were attempting to understand the structure of the atom and its electrons, investigations into the nucleus itself had also begun. In this chapter and the next, we take a brief look at *nuclear physics*.

30–1 Structure and Properties of the Nucleus

An important question to physicists in the early part of this century was whether the nucleus had a structure, and what that structure might be. It turns out that the nucleus is a complicated entity and is not fully understood even today. However, by the early 1930s, a model of the nucleus had been developed that is still useful. According to this model, a nucleus is considered as an aggregate of two types of particles: protons and neutrons. (Of course, we must remember that these "particles" also have wave properties, but for ease of visualization and language, we often refer to them simply as "particles.") A **proton** is the nucleus of the simplest atom, hydro-

Proton

gen. It has a positive charge ($= +e = +1.60 \times 10^{-19}$ C) and a mass

$$m_p = 1.6726 \times 10^{-27} \text{ kg.}$$

Neutron

The **neutron**, whose existence was ascertained only in 1932 by the Englishman James Chadwick (1891–1974), is electrically neutral ($q = 0$), as its name implies. Its mass, which is almost identical to that of the proton, is

$$m_n = 1.6749 \times 10^{-27} \text{ kg.}$$

These two constituents of a nucleus, neutrons and protons, are referred to collectively as **nucleons**.

Nucleons

Although the hydrogen nucleus consists of a single proton alone, the nuclei of all other elements consist of both neutrons and protons. The different types of nuclei are often referred to as **nuclides**. The number of protons in a nucleus (or nuclide) is called the **atomic number** and is designated by the symbol Z. The total number of nucleons, neutrons plus protons, is designated by the symbol A and is called the **atomic mass number**. This name is used since the mass of a nucleus is very closely A times the mass of one nucleon. A nuclide with 7 protons and 8 neutrons thus has $Z = 7$ and $A = 15$. The **neutron number** N is $N = A - Z$.

Z and A

To specify a given nuclide, we need give only A and Z. A special symbol is commonly used which takes the form

$$_Z^A X,$$

where X is the chemical symbol for the element (see Appendix F, and the periodic table inside the back cover), A is the atomic mass number, and Z is the atomic number. For example, $_7^{15}N$ means a nitrogen nucleus containing 7 protons and 8 neutrons for a total of 15 nucleons. In a neutral atom, the number of electrons orbiting the nucleus is equal to the atomic number Z (since the charge on an electron has the same magnitude but opposite sign to that of a proton). The main properties of an atom, and how it interacts with other atoms, are largely determined by the number of electrons. Hence Z determines what kind of atom it is: carbon, oxygen, gold, or whatever. It is redundant to specify both the symbol of a nucleus and its atomic number Z as described above. If the nucleus is nitrogen, for example, we know immediately that $Z = 7$. The subscript Z is thus sometimes dropped and $_7^{15}N$ is then written simply ^{15}N; in words we say "nitrogen fifteen."

For a particular type of atom (say, carbon), nuclei are found to contain different numbers of neutrons, although they all have the same number of protons. For example, carbon nuclei always have 6 protons, but they may have 5, 6, 7, 8, 9, or 10 neutrons. Nuclei that contain the same number of protons but different numbers of neutrons are called **isotopes**. Thus, $_6^{11}C$, $_6^{12}C$, $_6^{13}C$, $_6^{14}C$, $_6^{15}C$, and $_6^{16}C$ are all isotopes of carbon. Of course, the isotopes of a given element are not all equally common. For example, 98.9 percent of naturally occurring carbon (on Earth) is the isotope $_6^{12}C$ and about 1.1 percent is $_6^{13}C$. These percentages are referred to as the **natural abundances**.[†] Many isotopes that do not occur naturally can be produced in the laboratory by means of nuclear reactions (more on this later). Indeed, all elements beyond uranium ($Z > 92$) do not occur naturally and are only produced artificially, as are many nuclides with $Z \le 92$.

Isotopes

The approximate size of nuclei was determined originally by Rutherford from the scattering of charged particles. Of course, we cannot speak about a definite size for nuclei because of the wave–particle duality: their spatial extent must remain somewhat fuzzy. Nonetheless a rough "size" can be measured by scattering high-speed electrons off nuclei. It is found

---

[†]The mass value for each element as given in the periodic table (inside back cover) is an average weighted according to the natural abundances of its isotopes.

that nuclei have a roughly spherical shape with a radius that increases with A according to the approximate formula

Nuclear radii

$$r \approx (1.2 \times 10^{-15}\,\text{m})(A^{\frac{1}{3}}).$$ (30–1)

Since the volume of a sphere is $V = \frac{4}{3}\pi r^3$, we see that the volume of a nucleus is proportional to the number of nucleons, $V \propto A$. This is what we would expect if nucleons were like impenetrable billiard balls: if you double the number of balls, you double the total volume. Hence, all nuclei have nearly the same density.

EXAMPLE 30–1 ESTIMATE **Nuclear sizes.** Estimate the diameter of the following nuclei: (a) ^1_1H, (b) $^{40}_{20}\text{Ca}$, (c) $^{208}_{82}\text{Pb}$, (d) $^{235}_{92}\text{U}$.

SOLUTION (a) For hydrogen, $A = 1$, Eq. 30–1 gives

$$d = \text{diameter} = 2r \approx 2.4 \times 10^{-15}\,\text{m}$$

since $A^{\frac{1}{3}} = 1^{\frac{1}{3}} = 1$.

(b) For calcium

$$d = 2r \approx (2.4 \times 10^{-15}\,\text{m})(40)^{\frac{1}{3}} = 8.2 \times 10^{-15}\,\text{m}.$$

(c) For lead

$$d \approx (2.4 \times 10^{-15}\,\text{m})(208)^{\frac{1}{3}} = 14 \times 10^{-15}\,\text{m}.$$

(d) For uranium

$$d \approx (2.4 \times 10^{-15}\,\text{m})(235)^{\frac{1}{3}} = 15 \times 10^{-15}\,\text{m}.$$

The masses of nuclei can be determined, in one method, by measuring the radius of curvature of fast-moving nuclei in a magnetic field using a mass spectrometer, as discussed in Section 20–12. Indeed, as mentioned there, the existence of different isotopes of the same element was discovered using this device. Nuclear masses are specified in **unified atomic mass units** (u). On this scale, a neutral $^{12}_{6}\text{C}$ atom is given the precise value 12.000000 u. A neutron then has a measured mass of 1.008665 u, a proton 1.007276 u, and a neutral hydrogen atom, ^1_1H (proton plus electron) 1.007825 u. The masses of many nuclides are given in Appendix F. It should be noted that the masses in this table, as is customary, are for the *neutral atom*, and not for a bare nucleus.

Masses are for neutral atom

Masses are often specified using the electron-volt energy unit. This can be done because mass and energy are related, and the precise relationship is given by Einstein's equation $E = mc^2$ (Chapter 26). Since the mass of a neutral ^1_1H atom is $1.67353 \times 10^{-27}\,\text{kg}$, or 1.007825 u then $1.0000\,\text{u} = (1.00000/1.007825)\cdot(1.67353 \times 10^{-27}\,\text{kg}) = 1.66054 \times 10^{-27}\,\text{kg}$; this is equivalent to an energy $E = mc^2 = (1.66054 \times 10^{-27}\,\text{kg}) \cdot (2.9979 \times 10^8\,\text{m/s})^2/(1.6022 \times 10^{-19}\,\text{J/eV}) = 931.5\,\text{MeV}$. Thus

Atomic mass unit

$$1\,\text{u} = 1.6605 \times 10^{-27}\,\text{kg}$$
$$= 931.5\,\text{MeV}/c^2.$$

The rest masses of some of the basic particles are given in Table 30–1.

Just as an electron has an intrinsic spin and angular momentum, so too do nuclei and their constituents, the proton and neutron. Both the proton and the neutron are spin-$\frac{1}{2}$ particles. A nucleus, made up of protons and neutrons, has a **nuclear spin**, I, that can be either integer or half integer, depending on whether it is made up of an even or an odd number of nucleons. The *nuclear angular momentum* of a nucleus is given, as might be expected (see Section 28–6), by $\sqrt{I(I+1)}\hbar$.

TABLE 30-1
Rest Masses in Kilograms, Unified Atomic Mass Units, and MeV/c^2

Object	Mass		
	kg	u	MeV/c^2
Electron	9.1094×10^{-31}	0.00054858	0.51100
Proton	1.67262×10^{-27}	1.007276	938.27
$^1_1$H atom	1.67353×10^{-27}	1.007825	938.78
Neutron	1.67493×10^{-27}	1.008665	939.57

30-2 Binding Energy and Nuclear Forces

The total mass of a stable nucleus is always less than the sum of the masses of its constituent protons and neutrons, as the following Example shows.

EXAMPLE 30-2 $^4_2$**He mass compared to its constituents.** Compare the mass of a $^4_2$He nucleus to that of its constituent nucleons.

SOLUTION The mass of a neutral $^4_2$He atom, from Appendix F, is 4.002602 u. The mass of two neutrons and two protons (including the two electrons) is

$$2m_n = 2.017330 \text{ u}$$
$$2m(^1_1\text{H}) = \underline{2.015650 \text{ u}}$$
$$4.032980 \text{ u}.$$

(We almost always deal with masses of neutral atoms—that is, nuclei with Z electrons—since this is how masses are measured. We must therefore be sure to balance out the electrons when we compare masses, which is why we used the mass of $^1_1$H in this Example rather than that of the proton alone.)

➡ **PROBLEM SOLVING**

Keep track of electron masses

Thus the mass of $^4_2$He is measured to be $4.032980 \text{ u} - 4.002602 \text{ u} = 0.030378 \text{ u}$ less than the masses of its constituents. How can this be? If four nucleons come together to form $^4_2$He, where can this lost mass have gone?

It has, in fact, gone into energy of another kind (such as radiation, or kinetic energy, for example). The mass (or energy) difference in the case of $^4_2$He, given in energy units, is $(0.030378 \text{ u})(931.5 \text{ MeV/u}) = 28.30 \text{ MeV}$. This difference is referred to as the **total binding energy** of the nucleus. The total binding energy represents the amount of energy that must be put into a nucleus in order to break it apart into its constituent protons and neutrons. If the mass of, say, a $^4_2$He nucleus were exactly equal to the mass of two neutrons plus two protons, the nucleus could fall apart without any input of energy. To be stable, the mass of a nucleus *must* be less than that of its constituent nucleons, so that energy input is needed to break it apart. Note that the binding energy is not something a nucleus has—it is energy it "lacks" relative to the total mass of its separate constituents.

[Nuclear binding energy can be compared to the binding energy of electrons in an atom. We saw in Chapter 27 that the binding energy of the one electron in the hydrogen atom, for example, is 13.6 eV. The mass of a $^1_1$H atom is less than that of a single proton plus a single electron by

Binding energy

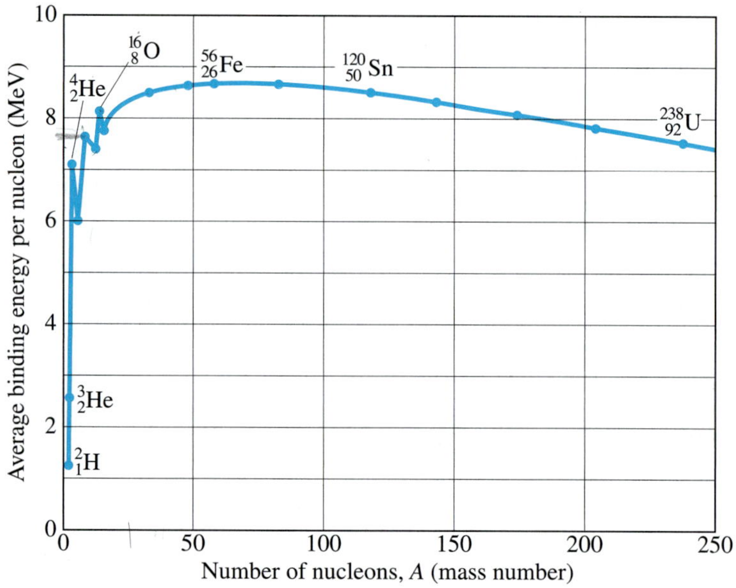

FIGURE 30–1 Average binding energy per nucleon as a function of mass number A for stable nuclei.

13.6 eV. Compared to the total mass of the atom (938 MeV), this is incredibly small (1 part in 10^8), and for practical purposes the mass difference can be ignored. The binding energies of nuclei are on the order of 10^6 times greater than the binding energies of electrons in atoms.]

Binding energy per nucleon   The **average binding energy per nucleon** is defined as the total binding energy of a nucleus divided by A, the total number of nucleons. For $^4_2$He, it is 28.3 MeV/4 = 7.1 MeV. Figure 30–1 shows the average binding energy per nucleon as a function of A for stable nuclei. The curve rises as A increases and reaches a plateau at about 8.7 MeV per nucleon above about $A \approx 40$. Beyond about $A \approx 80$, the curve decreases slowly, indicating that larger nuclei are held together a little less tightly than those in the middle of the periodic table. (We will see later that these figures allow the release of nuclear energy in the processes of fission and fusion.)

EXAMPLE 30–3 **Binding energy for iron.** Calculate the total binding energy and the average binding energy per nucleon for $^{56}_{26}$Fe, the most common stable isotope of iron.

SOLUTION $^{56}_{26}$Fe has 26 protons and 30 neutrons whose separate masses are

$$(26)(1.007825 \text{ u}) = \quad 26.2035 \text{ u (includes electrons)}$$
$$(30)(1.008665 \text{ u}) = \quad \underline{30.2600 \text{ u}}$$
$$\text{Total} = \quad 56.4635 \text{ u.}$$
$$\text{Subtract mass of } ^{56}_{26}\text{Fe:} \quad -55.9349 \text{ u (Appendix F)}$$
$$\Delta m = \quad 0.5286 \text{ u.}$$

The total binding energy is thus

$$(0.5286 \text{ u})(931.5 \text{ MeV/u}) = 492.4 \text{ MeV}$$

and the average binding energy per nucleon is

$$\frac{492.4 \text{ MeV}}{56 \text{ nucleons}} = 8.8 \text{ MeV.}$$

EXAMPLE 30–4 **Binding energy of last neutron.** What is the binding energy of the last neutron in $^{13}_{6}\text{C}$?

SOLUTION We compare the mass of $^{13}_{6}\text{C}$ to that of the atom with one less neutron, $^{12}_{6}\text{C}$, plus a free neutron (Appendix F):

$$\text{Mass } ^{12}_{6}\text{C} = \quad 12.000000 \text{ u}$$

$$\text{Mass } ^{1}_{0}\text{n} = \quad \underline{1.008665 \text{ u}}$$

$$13.008665 \text{ u}.$$

$$\text{Subtract mass of } ^{13}_{6}\text{C}: \quad \underline{-13.003355 \text{ u}}$$

$$\Delta m = \quad 0.005310 \text{ u}$$

which in energy is $(931.5 \text{ MeV/u})(0.005310 \text{ u}) = 4.95 \text{ MeV}$. That is, it would require 4.95 MeV input of energy to remove one neutron from $^{13}_{6}\text{C}$.

We can analyze nuclei not only from the point of view of energy, but also from the point of view of the forces that hold them together. We would not expect a collection of protons and neutrons to come together spontaneously, since protons are all positively charged and thus exert repulsive forces on each other. Indeed, the question arises as to how a nucleus stays together at all in view of the fact that the electric force between protons would tend to break it apart. Since stable nuclei *do* stay together, it is clear that another force must be acting. Because this new force is stronger than the electric force (which, in turn, is much stronger than gravity at the nuclear level) it is called the **strong nuclear force**. The strong nuclear force is an attractive force that acts between all nucleons—protons and neutrons alike. Thus protons attract each other via the nuclear force at the same time they repel each other via the electric force. Neutrons, since they are electrically neutral, only attract other neutrons or protons via the nuclear force.

Strong nuclear force

The nuclear force turns out to be far more complicated than the gravitational and electromagnetic forces. A precise mathematical description is not yet possible. Nonetheless, a great deal of work has been done to try to understand the nuclear force. One important aspect of the strong nuclear force is that it is a **short-range** force: it acts only over a very short distance. It is very strong between two nucleons if they are less than about 10^{-15} m apart, but it is essentially zero if they are separated by a distance greater than this. Compare this to electric and gravitational forces, which can act over great distances and are therefore called **long-range** forces. The strong nuclear force has some strange quirks. For example, if a nuclide contains too many or too few neutrons relative to the number of protons, the binding of the nucleons is reduced; nuclides that are too unbalanced in this regard are unstable. As shown in Fig. 30–2, stable nuclei tend to have the same number of protons as neutrons ($N = Z$) up to about $A \approx 30$ or 40. Beyond this, stable nuclei contain more neutrons than protons. This makes sense since, as Z increases, the electrical repulsion increases, so a greater number of neutrons—which exert only the attractive nuclear force—are required to maintain stability. For very large Z, no number of neutrons can overcome the greatly increased electric repulsion. Indeed, there are no completely stable nuclides above $Z = 82$.

What we mean by a stable nucleus is one that stays together indefinitely. What then is an unstable nucleus? It is one that comes apart; and

Long- and short-range forces

FIGURE 30–2 Number of neutrons versus number of protons for stable nuclides, which are represented by dots.

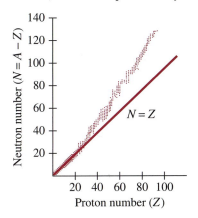

this results in radioactive decay. Before we discuss the important subject of radioactivity (the next Section), we note that there is a second type of nuclear force that is much weaker than the strong nuclear force. It is *Weak force* called the **weak nuclear force**, and we are aware of its existence only because it shows itself in certain types of radioactive decay. These two nuclear forces, the strong and the weak, together with the gravitational and electromagnetic forces, comprise the four known types of force in nature (more on this in Chapter 32).

30–3 Radioactivity

Discovery of radioactivity Nuclear physics had its beginnings in 1896. In that year, Henri Becquerel (1852–1908) made an important discovery: in his studies of phosphorescence, he found that a certain mineral (which happened to contain uranium) would darken a photographic plate even when the plate was wrapped to exclude light. It was clear that the mineral emitted some new kind of radiation that, unlike X-rays, occurred without any external stimulus. This new phenomenon eventually came to be called **radioactivity**.

Soon after Becquerel's discovery, Marie Curie (1867–1934) and her husband, Pierre Curie (1859–1906), isolated two previously unknown elements that were very highly radioactive (Fig. 30–3). These were named polonium and radium. Other radioactive elements were soon discovered as well. The radioactivity was found in every case to be unaffected by the strongest physical and chemical treatments, including strong heating or cooling and the action of strong chemical reagents. It was clear that the source of radioactivity must be deep within the atom, that it must emanate from the nucleus. And it became apparent that radioactivity is the result of the *disintegration* or *decay* of an unstable nucleus. Certain isotopes are not stable under the action of the nuclear force, and they decay with the emission of some type of radiation or "rays."

Many unstable isotopes occur in nature, and such radioactivity is called "natural radioactivity." Other unstable isotopes can be produced in the laboratory by nuclear reactions (Section 31–1); these are said to be produced "artificially" and to have "artificial radioactivity."

FIGURE 30–3 Marie and Pierre Curie in their laboratory (about 1906) where radium was discovered.

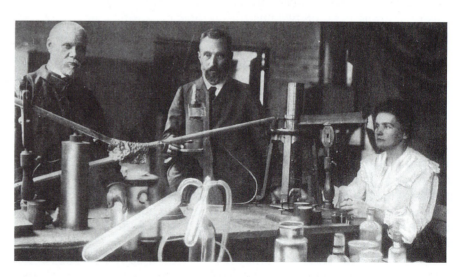

Rutherford and others began studying the nature of the rays emitted in radioactivity about 1898. They found that the rays could be classified into three distinct types according to their penetrating power. One type of radiation could barely penetrate a piece of paper. The second type could pass through as much as 3 mm of aluminum. The third was extremely penetrating: it could pass through several centimeters of lead and still be detected on the other side. They named these three types of radiation alpha (α), beta (β), and gamma (γ), respectively, after the first three letters of the Greek alphabet.

Each type of ray was found to have a different charge and hence is bent differently in a magnetic field, Fig. 30–4; α rays are positively charged, β rays are negatively charged, and γ rays are neutral. It was soon found that all three types of radiation consisted of familiar kinds of particles. Gamma rays are very high-energy photons whose energy is even higher than that of X-rays. Beta rays are electrons, identical to those that orbit the nucleus (but they are created within the nucleus itself). Alpha rays (or α particles) are simply the nuclei of helium atoms, $^4_2$He; that is, an α ray consists of two protons and two neutrons bound together.

We now discuss each of these three types of radioactivity, or decay, in more detail.

α, β, γ radiation

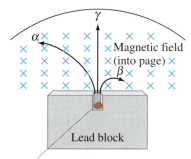

Radioactive
sample (radium)

FIGURE 30–4 Alpha and beta rays are bent in opposite directions by a magnetic field, whereas gamma rays are not bent at all.

30–4 Alpha Decay

When a nucleus emits an α particle ($^4_2$He), it is clear that the remaining nucleus will be different from the original: for it has lost two protons and two neutrons. Radium 226 ($^{226}_{88}$Ra), for example, is an α emitter. It decays to a nucleus with $Z = 88 - 2 = 86$ and $A = 226 - 4 = 222$. The nucleus with $Z = 86$ is radon (Rn)—see Appendix F or the periodic table. Thus the radium decays to radon with the emission of an α particle. This is written

$$^{226}_{88}\text{Ra} \rightarrow\ ^{222}_{86}\text{Rn} + ^4_2\text{He}.$$

FIGURE 30–5 Radioactive decay of radium to radon with emission of an alpha particle.

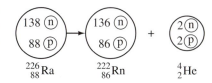

α decay

See Fig. 30–5.

It is clear that when α decay occurs, a new element is formed. The **daughter** nucleus ($^{222}_{86}$Rn in this case) is different from the **parent** nucleus ($^{226}_{88}$Ra in this case). This changing of one element into another is called **transmutation**.

Alpha decay can be written

$$^A_Z N \rightarrow\ ^{A-4}_{Z-2} N' + ^4_2\text{He} \qquad\qquad [\alpha\ \text{decay}]$$

where N is the parent, N' the daughter, and Z and A are the atomic number and atomic mass number, respectively, of the parent.

Alpha decay occurs because the strong nuclear force is unable to hold very large nuclei together. Because the nuclear force is a short-range force, it acts only between neighboring nucleons. But the electric force can act all the way across a large nucleus. For very large nuclei, the large Z means the repulsive electric force becomes very large (Coulomb's law); and it acts between all protons. The strong nuclear force, since it acts only between neighboring nucleons, is overpowered and is unable to hold the nucleus together.

We can express the instability in terms of energy (or mass): the mass of the parent nucleus is greater than the mass of the daughter nucleus plus the mass of the α particle. The mass difference appears as kinetic energy, which is carried away by the α particle and the recoiling daughter nucleus.

Daughter nucleus
Parent nucleus
Transmutation

Why the strong nuclear force cannot hold a nucleus together

The total energy released is called the **disintegration energy**, Q, or the **Q-value** of the decay. Q is defined as

Q-value

$$Q = (M_P - M_D - m_\alpha)c^2, \tag{30-2}$$

where M_P, M_D, and m_α are the masses of the parent, daughter, and α particle, respectively. If the parent had *less* mass than the daughter plus the α particle (so $Q < 0$), the decay could not occur, for the conservation of energy law would be violated.

EXAMPLE 30–5 **Uranium decay energy release.** Calculate the disintegration energy when $^{232}_{92}U$ (mass = 232.037131 u) decays to $^{228}_{90}Th$ (228.028716 u) with the emission of an α particle. (As always, masses are for neutral atoms.)

SOLUTION Since the mass of the ^4_2He is 4.002602 u (Appendix F), the total mass in the final state is

$$228.028716\text{ u} + 4.002602\text{ u} = 232.031318\text{ u}.$$

The mass lost when the $^{232}_{92}U$ decays is

$$232.037131\text{ u} - 232.031318\text{ u} = 0.005813\text{ u}.$$

Since 1 u = 931.5 MeV, the energy Q released is

$$Q = (0.005813\text{ u})(931.5\text{ MeV/u}) \approx 5.4\text{ MeV},$$

and this energy appears as KE of the α particle and the daughter nucleus. (Using conservation of momentum, it can be shown that the α particle in this decay has a KE of about 5.3 MeV. Thus, the daughter nucleus—which recoils in the opposite direction from the emitted α particle—has about 0.1 MeV of kinetic energy. See Problem 34.)

Why α particles?

Why, you may wonder, do nuclei emit this combination of four nucleons called an α particle? Why not just four nucleons, or even one? The answer is that the α particle is very strongly bound, so that its mass is significantly less than that of four separate nucleons. As we saw in Example 30–2, two protons and two neutrons separately have a total mass of about 4.032980 u. The total mass of a $^{228}_{90}Th$ nucleus plus four separate nucleons is 232.061696 u, which is greater than the mass of the parent nucleus. Such a decay could not occur because it would violate the conservation of energy. Similarly, it is almost always true that the emission of a single nucleon is energetically not possible. Or, to put it another way, for any nuclide for which it could be possible (mass of parent > mass of daughter + mass of nucleon), the decay happens so fast after formation of the parent that we don't see the parent in nature.

One widespread application of nuclear physics is present in nearly every home in the form of the ordinary **smoke detector**. There are two technologies on which these devices are based. One involves an infrared light source and detector to measure the attenuation of the light if smoke particles are present. This type does not work nearly as well as the more common approach, which is based upon ionization of air by radioactive decay. This type of detector contains about 0.2 mg of the americium isotope $^{241}_{95}Am$, in the form of AmO_2, which is radioactive. The radiation ionizes the nitrogen and oxygen molecules in the air space between two oppositely charged plates. The resulting conductivity allows a small steady current. If smoke enters, the radiation is absorbed by the smoke particles rather than by the air molecules,

PHYSICS APPLIED

Smoke detector

thus reducing the current. The current drop is detected by the device's electronics and sets off the alarm. The radiation dose that escapes from an intact americium smoke detector is much less than the natural radioactive background, and so can be considered harmless. There is no question that smoke detectors save lives and reduce property damage.

30–5 Beta Decay

Transmutation of elements also occurs when a nucleus decays by β decay—that is, with the emission of an electron or β^- particle. The nucleus $^{14}_{6}C$, for example, emits an electron when it decays:

$$^{14}_{6}C \rightarrow {}^{14}_{7}N + e^- + \text{a neutrino,}$$

where e$^-$ is the symbol for the electron. (The symbol $_{-1}^{0}e$ is sometimes used for the electron whose charge corresponds to $Z = -1$ and, since it is not a nucleon and has very small mass, has $A = 0$.) The particle known as the neutrino, with rest mass $m_0 = 0$ and charge $q = 0$, was not initially detected and was only later hypothesized to exist, as we shall discuss later in this section. No nucleons are lost when an electron is emitted, and the total number of nucleons, A, is the same in the daughter nucleus as in the parent. But because an electron has been emitted, the charge on the daughter nucleus is different from that on the parent. The parent nucleus had $Z = +6$. In the decay, the nucleus loses a charge of -1; so, from charge conservation, the nucleus remaining behind must have an extra positive charge for a total of 7. So the daughter nucleus has $Z = 7$, which is nitrogen.

It must be carefully noted that the electron emitted in β decay is *not* an orbital electron. Instead, the electron is created *within the nucleus itself.* What happens is that one of the neutrons changes to a proton and in the process (to conserve charge) throws off an electron. Indeed, free neutrons actually do decay in this fashion: n $\rightarrow$ p + e$^-$ (plus a neutrino). Because of their origin in the nucleus, the electrons emitted in β decay are often referred to as "β particles," rather than as electrons, to remind us of their origin. They are, nonetheless, indistinguishable from orbital electrons.

Emitted e$^-$ comes from nucleus

EXAMPLE 30–6 Energy release in $^{14}_{6}C$ decay. How much energy is released when $^{14}_{6}C$ decays to $^{14}_{7}N$ by β emission? Use Appendix F.

SOLUTION The masses given in Appendix F are those of the neutral atom, and we have to keep track of the electrons involved. Assume the parent nucleus has six orbiting electrons so it is neutral, and its mass is 14.003242 u. The daughter, in this decay $^{14}_{7}N$, is not neutral since it has the same six electrons circling it but the nucleus has a charge of $+7e$. However, the mass of this daughter with its six electrons, plus the mass of the emitted electron (which makes a total of seven electrons), is just the mass of a neutral nitrogen atom. That is, the total mass in the final state is

(mass of $^{14}_{7}N$ nucleus + 6 electrons) + (mass of 1 electron),

and this is equal to

mass of neutral $^{14}_{7}N$ (includes 7 electrons),

which, from Appendix F is a mass of 14.003074 u. (Note that the neutrino doesn't contribute to either the mass or charge balance since it has $m_0 = 0$ and $q = 0$.) Hence the mass after decay is 14.003074 u, whereas before decay, it was 14.003242 u. So the mass difference is 0.000168 u, which corresponds to 0.156 MeV or 156 keV.

FIGURE 30–6 Enrico Fermi. Fermi contributed significantly to both theoretical and experimental physics, a feat almost unique in this century.

According to this Example, we would expect the emitted electron to have a kinetic energy of 156 keV. (The daughter nucleus, because its mass is very much larger than that of the electron, recoils with very low velocity and hence gets very little of the kinetic energy.) Indeed, very careful measurements indicate that a few emitted β particles do have kinetic energy close to this calculated value. But the vast majority of emitted electrons have somewhat less energy. In fact, the energy of the emitted electron can be anywhere from zero up to the maximum value as calculated above. This range of electron KE was found for any β decay. It was as if the law of conservation of energy was being violated, and indeed Bohr actually considered this possibility. Careful experiments indicated that linear momentum and angular momentum also did not seem to be conserved. Physicists were troubled at the prospect of having to give up these laws, which had worked so well in all previous situations. In 1930, Wolfgang Pauli proposed an alternate solution: perhaps a new particle that was very difficult to detect was emitted during β decay in addition to the electron. This hypothesized particle could be carrying off the energy, momentum, and angular momentum required to maintain the conservation laws. This new particle was named the **neutrino**—meaning "little neutral one"—by the great Italian physicist Enrico Fermi (1901–1954; Fig. 30–6), who in 1934 worked out a detailed theory of β decay. (It was Fermi who, in this theory, postulated the existence of the fourth force in nature, which we call the weak nuclear force.) The neutrino has zero charge and seems to have zero rest mass, although we cannot yet rule out the possibility that it might have a very tiny rest mass. If its rest mass is zero, it is much like a photon in that it is neutral and travels at the speed of light. But the neutrino is far more difficult to detect. In 1956, complex experiments produced further evidence for the existence of the neutrino; but by then, most physicists had already accepted its existence.

The symbol for the neutrino is the Greek letter nu (ν). The correct way of writing the decay of $^{14}_{6}C$ is then

β⁻ decay

$$^{14}_{6}C \rightarrow {}^{14}_{7}N + e^- + \bar{\nu}.$$

The bar ($^-$) over the neutrino symbol is to indicate that it is an "antineutrino." (Why this is called an antineutrino rather than simply a neutrino need not concern us now; it is discussed in Chapter 32.)

Many isotopes decay by electron emission. They are always isotopes that have too many neutrons compared to the number of protons. That is, they are isotopes that lie above the stable isotopes plotted in Fig. 30–2. But what about unstable isotopes that have too few neutrons compared to their number of protons—those that fall below the stable isotopes of Fig. 30–2? These, it turns out, decay by emitting a **positron** instead of an electron. A positron (sometimes called an e^+ or β^+ particle) has the same mass as the electron, but it has a positive charge of $+1e$. Because it is so like an electron, except for its charge, the positron is called the **antiparticle**[†] to the electron. An example of a β^+ decay is that of $^{19}_{10}Ne$:

Positron (β⁺) decay

$$^{19}_{10}Ne \rightarrow {}^{19}_{9}F + e^+ + \nu,$$

where e^+ (or $^{0}_{1}e$) stands for a positron. (Note that the ν emitted here is a neutrino, whereas that emitted in β^- decay is called an antineutrino. Thus

[†]Discussed in Chapter 32. Briefly, an antiparticle has the same mass as its corresponding particle, but opposite charge.

an antielectron (= positron) is emitted with a neutrino, whereas an anti-neutrino is emitted with an electron; this is discussed in Chapter 32.)

We can write β^- and β^+ decay, in general, as follows:

$$^A_Z N \rightarrow {}^A_{Z+1} N' + e^- + \bar{\nu} \qquad [\beta^- \text{ decay}]$$

$$^A_Z N \rightarrow {}^A_{Z-1} N' + e^+ + \nu, \qquad [\beta^+ \text{ decay}]$$

where N is the parent nucleus and N' the daughter.

Besides β^- and β^+ emission, there is a third related process. This is **electron capture** (abbreviated EC in Appendix F) and occurs when a nucleus absorbs one of its orbiting electrons. An example is $^7_4$Be, which as a result becomes $^7_3$Li. The process is written

$$^7_4 \text{Be} + e^- \rightarrow {}^7_3 \text{Li} + \nu,$$

Electron capture

or, in general,

$$^A_Z \text{N} + e^- \rightarrow {}^A_{Z-1} \text{N}' + \nu \qquad [\text{electron capture}]$$

Usually it is an electron in the innermost (K) shell that is captured, in which case it is called "K-capture." The electron disappears in the process and a proton in the nucleus becomes a neutron; a neutrino is emitted as a result. This process is inferred experimentally by detection of emitted X-rays (due to electrons jumping down to fill the empty state) of just the proper energy.

K-capture

In β decay, it is the weak nuclear force that plays the crucial role. The neutrino is unique in that it interacts with matter only via the weak force, which is why it is so hard to detect.

30–6 Gamma Decay

Gamma rays are photons having very high energy. The decay of a nucleus by emission of a γ ray is much like emission of photons by excited atoms. Like an atom, a nucleus itself can be in an excited state. When it jumps down to a lower energy state, or to the ground state, it emits a photon. The possible energy levels of a nucleus are much farther apart than those of an atom: on the order of keV or MeV, as compared to a few eV for electrons in an atom. Hence, the emitted photons have energies that can range from a few keV to several MeV. For a given decay, the γ ray always has the same energy. Since a γ ray carries no charge, there is no change in the element as a result of a γ decay.

How does a nucleus get into an excited state? It may occur because of a violent collision with another particle. More commonly, the nucleus remaining after a previous radioactive decay may be in an excited state. A typical example is shown in the energy-level diagram of Fig. 30–7. $^{12}_5$B can decay by β decay directly to the ground state of $^{12}_6$C; or it can go by β decay to an excited state of $^{12}_6$C, which then decays by emission of a 4.4 MeV γ ray to the ground state.

We can write γ decay as

$$^A_Z N^* \rightarrow {}^A_Z N + \gamma. \qquad [\gamma \text{ decay}]$$

In some cases, a nucleus may remain in an excited state for some time before it emits a γ ray. The nucleus is then said to be in a **metastable state** and is called an **isomer**.

Isomer

FIGURE 30–7 Energy-level diagram showing how $^{12}_5$B can decay to the ground state of $^{12}_6$C by β decay (total energy released = 13.4 MeV), or can instead β-decay to an excited state of $^{12}_6$C (indicated by *), which subsequently decays to its ground state by emitting a 4.4-MeV γ ray.

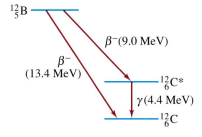

Internal conversion

An excited nucleus can sometimes return to the ground state by another process known as **internal conversion** with no γ ray emitted. In this process, the excited nucleus interacts with one of the orbital electrons and ejects this electron from the atom with the same KE (minus the binding energy of the electron) that an emitted γ ray would have had.

What, you may wonder, is the difference between a γ ray and an X-ray? They both are electromagnetic radiation (photons) and, though γ rays usually have higher energy than X-rays, their range of energies overlap to some extent. The difference is not intrinsic. We use the term X-ray if the photon is produced by an electron–atom interaction, and γ ray if the photon is produced in a nuclear process.

TABLE 30–2 The Three Types of Radioactive Decay

α *decay:*
$$^A_Z N \rightarrow {}^{A-4}_{Z-2} N' + {}^4_2 He$$

β *decay:*
$$^A_Z N \rightarrow {}^A_{Z+1} N' + e^- + \bar{\nu}$$

$$^A_Z N \rightarrow {}^A_{Z-1} N' + e^+ + \nu$$

$$^A_Z N + e^- \rightarrow {}^A_{Z-1} N' + \nu \text{ [EC]}^\dagger$$

γ *decay:*
$$^A_Z N^* \rightarrow {}^A_Z N + \gamma$$

*Indicates the excited state of a nucleus.
†Electron capture.

30–7 Conservation of Nucleon Number and Other Conservation Laws

In all three types of radioactive decay, the classical conservation laws hold. Energy, linear momentum, angular momentum, and electric charge are all conserved. These quantities are the same before the decay as after. But a new conservation law is also revealed, the **law of conservation of nucleon number**. According to this law, the total number of nucleons (A) remains constant in any process, although one type can change into the other type (protons into neutrons or vice versa). This law holds in all three types of decay. Table 30–2 gives a summary of α, β, and γ decay.

30–8 Half-Life and Rate of Decay

A macroscopic sample of any radioactive isotope consists of a vast number of radioactive nuclei. These nuclei do not all decay at one time. Rather, they decay one by one over a period of time. This is a random process: we can't predict exactly when a given nucleus will decay. But we can determine, on a probabilistic basis, approximately how many nuclei in a sample will decay over a given time period, by assuming that each nucleus has the same probability of decaying in each second that it exists.

The number of decays ΔN that occur in a very short time interval Δt is then proportional to Δt and to the total number N of radioactive nuclei present:

$$\Delta N = -\lambda N \Delta t. \tag{30–3}$$

In this equation, λ is a constant of proportionality called the **decay constant**, which is different for different isotopes. The greater λ is, the greater the rate of decay and the more radioactive that isotope is said to be. The number of decays that occur in the short time interval Δt is designated ΔN because each decay that occurs corresponds to a decrease by one in the number N of nuclei present. That is, radioactive decay is a "one-shot" process, Fig. 30–8. Once a particular parent nucleus decays into its daughter, it cannot do it again. The minus sign in Eq. 30–3 is needed to indicate that N is decreasing.

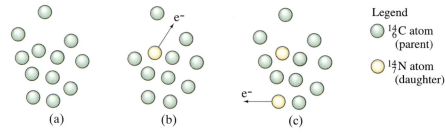

Equation 30–3 can be solved for N (using calculus) and the result is

$$N = N_0 e^{-\lambda t}, \qquad \text{(30–4)}$$

Radioactive decay law

where N_0 is the number of nuclei present at time $t = 0$, and N the number remaining after a time t. The symbol e is the natural exponential (encountered earlier in Sections 19–7 and 21–11) whose value is $e = 2.718 \cdots$. Thus the number of parent nuclei in a sample decreases exponentially in time, as shown in Fig. 30–9a for $^{14}_{6}C$ decay. Equation 30–4 is called the **radioactive decay law**.

The number of decays per second, $\Delta N/\Delta t$, is called the **activity** of the sample. Since $\Delta N/\Delta t$ is proportional to N (see Eq. 30–3), it, too, decreases exponentially in time with the same time constant (Fig. 30–9b). The activity at time t is given by

Activity

$$\frac{\Delta N}{\Delta t} = \left(\frac{\Delta N}{\Delta t}\right)_0 e^{-\lambda t}, \qquad \text{(30–5)}$$

where $(\Delta N/\Delta t)_0$ is the activity at $t = 0$.

The rate of decay of any isotope is often specified by giving its "half-life" rather than the decay constant. The **half-life** of an isotope is defined as the time it takes for half the original amount of isotope in a given sample to decay. For example, the half-life of $^{14}_{6}C$ is 5730 years. If at some time a piece of petrified wood contains, say, 1.00×10^{22} nuclei of $^{14}_{6}C$, then 5730 years later it will contain only 0.50×10^{22} of these nuclei. After another 5730 years it will contain 0.25×10^{22} nuclei, and so on. This is shown in Fig. 30–9a. Since the rate of decay $\Delta N/\Delta t$ is proportional to N, it, too, decreases by a factor of 2 every half-life (Fig. 30–9b).

Half-life

FIGURE 30–9 (a) The number N of parent nuclei in a given sample of $^{14}_{6}C$ decreases exponentially. (b) The number of decays per second also decreases exponentially. The half-life of $^{14}_{6}C$ is 5730 yr, which means that the number of parent nuclei, N, and the rate of decay, $\Delta N/\Delta t$, decrease by half every 5730 yr.

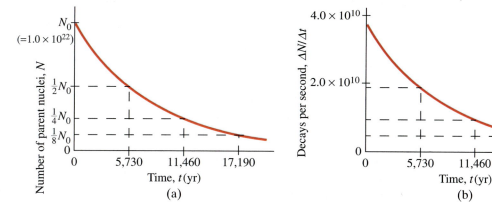

The half-lives of known radioactive isotopes vary from about 10^{-22} s to 10^{28} s (about 10^{21} yr). The half-lives of many isotopes are given in Appendix F. It should be clear that the half-life (which we designate $T_{\frac{1}{2}}$) bears an inverse relationship to the decay constant. The longer the half-life of an isotope, the more slowly it decays, and hence λ is smaller. Conversely, very active isotopes (large λ) have very short half-lives. The precise relationship between half-life and decay constant is

$$T_{\frac{1}{2}} = \frac{0.693}{\lambda}. \tag{30–6}$$

We can derive this starting with Eq. 30–4. At $t = 0$, $N = N_0 e^0 = N_0$. At $t = T_{\frac{1}{2}}$, $N = N_0/2$ by definition of $T_{\frac{1}{2}}$ (half the parent nuclei remain). Then, from Eq. 30–4 evaluated at time $T_{\frac{1}{2}}$,

$$\frac{N_0}{2} = N_0 e^{-\lambda T_{\frac{1}{2}}} \qquad \text{or} \qquad \frac{1}{2} = e^{-\lambda T_{\frac{1}{2}}},$$

so

$$e^{\lambda T_{\frac{1}{2}}} = 2.$$

We take natural logs of both sides (remember that "ln" and "e" are inverse operations, meaning $\ln(e^x) = x$) and find

$$\ln\left(e^{\lambda T_{\frac{1}{2}}}\right) = \ln 2,$$

so

$$\lambda T_{\frac{1}{2}} = 0.693 \qquad \text{and} \qquad T_{\frac{1}{2}} = \frac{0.693}{\lambda},$$

which is Eq. 30–6. In the next Section we will see how to make calculations involving $T_{\frac{1}{2}}$.

30–9 Calculations Involving Decay Rates and Half-Life

Let us now consider Examples of what we can determine about a sample of radioactive material if we know the half-life.

EXAMPLE 30–7 **Sample activity.** The isotope $^{14}_{6}\text{C}$ has a half-life of 5730 yr. If at some time a sample contains 1.00×10^{22} carbon-14 nuclei, what is the activity of the sample?

SOLUTION First we calculate the decay constant λ from Eq. 30–6, and obtain

$$\lambda = \frac{0.693}{T_{\frac{1}{2}}} = \frac{0.693}{(5730 \text{ yr})(3.156 \times 10^7 \text{s/yr})} = 3.83 \times 10^{-12} \text{ s}^{-1},$$

since the number of seconds in a year is $(60)(60)(24)(365\frac{1}{4}) = 3.156 \times 10^7$ s. From Eq. 30–3, the activity or rate of decay (we ignore the minus sign) is

$$\frac{\Delta N}{\Delta t} = \lambda N = (3.83 \times 10^{-12} \text{ s}^{-1})(1.00 \times 10^{22})$$

$$= 3.83 \times 10^{10} \text{ decays/s.}$$

(The unit "decays/s" is often written simply as s^{-1} since "decays" is not a unit but refers only to the number.) Note that the graph of Fig. 30–9b starts at this value, corresponding to the original value of $N = 1.0 \times 10^{22}$ nuclei in Fig. 30–9a.

EXAMPLE 30–8 **A sample of radioactive $^{13}_7N$.** A laboratory has $1.49\,\mu g$ of pure $^{13}_7N$, which has a half-life of 10.0 min (600 s). (a) How many nuclei are present initially? (b) What is the activity initially? (c) What is the activity after 1.00 h? (d) After approximately how long will the activity drop to less than one per second?

SOLUTION (a) Since the atomic mass is 13.0, then 13.0 g will contain 6.02×10^{23} nuclei (Avogadro's number). Since we have only 1.49×10^{-6} g, the number of nuclei, N_0, that we have initially is given by the ratio

$$\frac{N_0}{1.49 \times 10^{-6}\,g} = \frac{6.02 \times 10^{23}}{13.0\,g},$$

so $N_0 = 6.90 \times 10^{16}$ nuclei.
(b) From Eq. 30–6, $\lambda = (0.693)/(600\,s) = 1.16 \times 10^{-3}\,s^{-1}$. Then, at $t = 0$ (Eq. 30–3),

$$\left(\frac{\Delta N}{\Delta t}\right)_0 = \lambda N_0 = (1.16 \times 10^{-3}\,s^{-1})(6.90 \times 10^{16}) = 8.00 \times 10^{13}\ \text{decays/s}.$$

(c) Since the half-life is 10.0 min, the decay rate decreases by half every 10.0 min. We can make the following table of activity after given periods of time:

Time (min)	Activity (decays/s)
0	8.00×10^{13}
10	4.00×10^{13}
20	2.00×10^{13}
30	1.00×10^{13}
40	0.500×10^{13}
50	0.250×10^{13}
60	0.125×10^{13}

Thus, after 1.0 h, the activity is 1.25×10^{12} decays/s. A simpler way of seeing this is to note that 60 min is 6 half-lives, so the activity will decrease to $(\frac{1}{2})(\frac{1}{2})(\frac{1}{2})(\frac{1}{2})(\frac{1}{2})(\frac{1}{2}) = (\frac{1}{2})^6 = \frac{1}{64}$ of its original value, or $(8.00 \times 10^{13})/(64) = 1.25 \times 10^{12}$ per second. A third way to arrive at this answer is to use Eq. 30–5 (setting $t = 60.0$ min $= 3600$ s):

$$\frac{\Delta N}{\Delta t} = \left(\frac{\Delta N}{\Delta t}\right)_0 e^{-\lambda t}$$

$$= (8.00 \times 10^{13}\,s^{-1})e^{-(1.16 \times 10^{-3}\,s^{-1})(3600\,s)} = 1.23 \times 10^{12}\,s^{-1}.$$

(The slight discrepancy in results arises because we kept only three significant figures.)

(*d*) After each hour, the activity drops by a factor of 64, so our table is extended to read as follows:

Time (h)	Activity (s^{-1})
0	8.00×10^{13}
1	1.25×10^{12}
2	1.95×10^{10}
3	3.05×10^{8}
4	4.77×10^{6}
5	7.45×10^{4}
6	1.16×10^{3}
7	$1.8 \ \times 10^{1} = 18$
8	< 1

So the activity drops to less than one per second within 8 h. We can obtain a more precise result using the properties of the exponential and its inverse, the natural logarithm (ln). We want to determine the time t when $\Delta N/\Delta t = 1.00\,\text{s}^{-1}$. From Eq. 30–5, we have

$$e^{-\lambda t} = \frac{(\Delta N/\Delta t)}{(\Delta N/\Delta t)_0} = \frac{1.00\,\text{s}^{-1}}{8.00 \times 10^{13}\,\text{s}^{-1}} = 1.25 \times 10^{-14}.$$

We take the natural log (ln) of both sides (remember $\ln e^{-\lambda t} = -\lambda t$) and divide by λ to find

$$t = -\frac{\ln(1.25 \times 10^{-14})}{\lambda} = 2.76 \times 10^{4}\,\text{s} = 7.66\,\text{h}.$$

30–10 Decay Series

It is often the case that one radioactive isotope decays to another isotope that is also radioactive. Sometimes this daughter decays to yet a third isotope which also is radioactive. Such successive decays are said to form a **decay series**. An important example is illustrated in Fig. 30–10. As can be seen, $^{238}_{92}\text{U}$ decays by α emission to $^{234}_{90}\text{Th}$, which in turn decays by β decay to $^{234}_{91}\text{Pa}$.. The series continues as shown, with several possible branches near the bottom. For example, $^{218}_{84}\text{Po}$ can decay either by α decay to $^{214}_{82}\text{Pb}$ or by β decay to $^{218}_{85}\text{At}$. The series ends at the stable lead isotope $^{206}_{82}\text{Pb}$. Other radioactive series also exist.

Because of such decay series, certain radioactive elements are found in nature that otherwise would not be. For when the solar system acquired its present form about 5 billion years ago, it is believed that nearly all nuclides were formed (by the fusion process, Sections 31–3 and 33–2). Many isotopes with short half-lives decayed quickly and no longer exist in nature today. But long-lived isotopes, such as $^{238}_{92}\text{U}$ with a half-life of $4.5 \times 10^{9}\,\text{yr}$, still do exist in nature today. Indeed, about half of the original $^{238}_{92}\text{U}$ still remains (assuming that the origin of the solar system was about $5 \times 10^{9}\,\text{yr}$ ago). We might expect, however, that radium ($^{226}_{88}\text{Ra}$), with a half-life of 1600 yr, would long since have disappeared from the Earth. Indeed, the original $^{226}_{88}\text{Ra}$ nuclei must by now have all decayed. However, because $^{238}_{92}\text{U}$ decays (in several steps) to

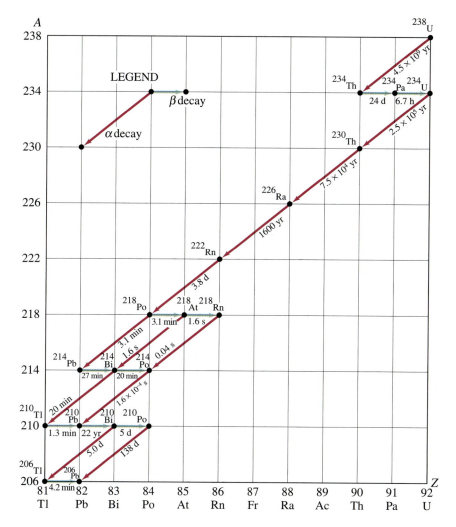

FIGURE 30–10 Decay series beginning with $^{238}_{92}$U. Nuclei in the series are specified by a dot representing A and Z values. Half-lives are given in seconds (s), minutes (min), hours (h), days (d), or years (yr). Note that a horizontal arrow represents β decay (A does not change), whereas a diagonal line represents α decay (A changes by 4, Z changes by 2).

$^{226}_{88}$Ra, the supply of $^{226}_{88}$Ra is continually replenished, which is why it is still found on Earth today. The same can be said for many other radioactive nuclides.

CONCEPTUAL EXAMPLE 30–9 | Decay chain. The decay chain starting with ^{234}U in Fig. 30–10 has nuclides with half-lives of 250,000 yr, 75,000 yr, 1600 yr, and a little under 4 days, respectively. Each decay in the chain has an alpha particle of a characteristic energy, and so we can monitor the radioactive decay rate of each nuclide. Given a sample that was pure ^{234}U a million years ago, which alpha decay would you expect to have the highest activity rate in the sample?

RESPONSE The first instinct is to say that the process with the shortest half-life would show the highest activity. Surprisingly, however, the activity rates in this sample are all the same! The reason is that in each case the decay of the parent acts as a bottleneck to the decay of the daughter. Compared to the 1600-yr half-life of ^{226}Ra, for example, its daughter ^{222}Rn decays almost immediately, but it cannot decay until it is made. (This is like an automobile assembly line: If worker A takes 20 minutes to do a task and then worker B takes only 1 minute to do the next task, worker B still does only one car every 20 minutes).

30–11 Radioactive Dating

Radioactive decay has many interesting applications. One is the technique of *radioactive dating* by which the age of ancient materials can be determined.

➥ PHYSICS APPLIED

Carbon–14 dating

The age of any object made from once-living matter, such as wood, can be determined using the natural radioactivity of $^{14}_{6}C$. All living plants absorb carbon dioxide (CO_2) from the air and use it to synthesize organic molecules. The vast majority of these carbon atoms are $^{12}_{6}C$, but a small fraction, about 1.3×10^{-12}, is the radioactive isotope $^{14}_{6}C$. The ratio of $^{14}_{6}C$ to $^{12}_{6}C$ in the atmosphere has remained roughly constant over many thousands of years, in spite of the fact that $^{14}_{6}C$ decays with a half-life of about 5730 yr. This is because neutrons in the cosmic radiation that impinges on the Earth from outer-space collide with atoms of the atmosphere. In particular, collisions with nitrogen nuclei produce the following nuclear transformation: $n + {}^{14}_{7}N \rightarrow {}^{14}_{6}C + p$. That is, a neutron strikes and is absorbed by a $^{14}_{7}N$ nucleus, and a proton is knocked out in the process. The remaining nucleus is $^{14}_{6}C$. This continual production of $^{14}_{6}C$ in the atmosphere roughly balances the loss of $^{14}_{6}C$ by radioactive decay. As long as a plant or tree is alive, it continually uses the carbon from carbon dioxide in the air to build new tissue and to replace old. Animals eat plants, so they too are continually receiving a fresh supply of carbon for their tissues. Organisms cannot distinguish[†] $^{14}_{6}C$ from $^{12}_{6}C$, and since the ratio of $^{14}_{6}C$ to $^{12}_{6}C$ in the atmosphere remains nearly constant, the ratio of the two isotopes within the living organism remains nearly constant as well. But when an organism dies, carbon dioxide is no longer absorbed and utilized. Because the $^{14}_{6}C$ decays radioactively, the ratio of $^{14}_{6}C$ to $^{12}_{6}C$ in a dead organism decreases in time. Since the half-life of $^{14}_{6}C$ is about 5730 yr, the $^{14}_{6}C/^{12}_{6}C$ ratio decreases by half every 5730 yr. If, for example, the $^{14}_{6}C/^{12}_{6}C$ ratio of an ancient wooden tool is half of what it is in living trees, then the object must have been made from a tree that was felled about 5700 years ago. Actually, corrections must be made for the fact that the $^{14}_{6}C/^{12}_{6}C$ ratio in the atmosphere has not remained precisely constant over time. The determination of what this ratio has been over the centuries has required using techniques such as comparing the expected ratio to the actual ratio for objects whose age is known, such as very old trees whose annual rings can be counted.

➥ PHYSICS APPLIED

Archeological dating

EXAMPLE 30–10 **An ancient animal.** An animal bone fragment found in an archeological site has a carbon mass of 200 g. It registers an activity of 15 decays/s. What is the age of the bone?

SOLUTION When the animal was alive, the ratio of $^{14}_{6}C$ to $^{12}_{6}C$ in the 200-g piece of bone was 1.3×10^{-12}. The number of $^{14}_{6}C$ nuclei at that time was

$$N_0 = \left(\frac{6.02 \times 10^{23}\ \text{atoms}}{12\ \text{g}} \right)(200\ \text{g})(1.3 \times 10^{-12}) = 1.3 \times 10^{13}.$$

From Eq. 30–3 we can see that

$$\left(\frac{\Delta N}{\Delta t} \right)_0 = \lambda N_0,$$

[†]Organisms operate almost exclusively via chemical reactions—which involve only the outer orbital electrons of the atom; extra neutrons in the nucleus have essentially no effect.

where $\lambda = 3.83 \times 10^{-12} \, \text{s}^{-1}$ (Example 30–7). So the original activity was

$$\left(\frac{\Delta N}{\Delta t}\right)_0 = (3.83 \times 10^{-12} \, \text{s}^{-1})(1.3 \times 10^{13}) = 50 \, \text{s}^{-1}.$$

Combining Eq. 30–3 with Eq. 30–4, and ignoring the minus sign, gives

$$\frac{\Delta N}{\Delta t} = \lambda N = \lambda N_0 e^{-\lambda t},$$

and we rewrite this as

$$e^{\lambda t} = \frac{\lambda N_0}{\Delta N / \Delta t}.$$

Now we take the natural log (ln) of both sides to get

$$t = \frac{1}{\lambda} \ln \frac{\lambda N_0}{\Delta N / \Delta t}$$

$$= \frac{1}{3.83 \times 10^{-12} \, \text{s}^{-1}} \ln \left[\frac{(3.83 \times 10^{-12} \, \text{s}^{-1})(1.3 \times 10^{13})}{15 \, \text{s}^{-1}} \right]$$

$$= 3.13 \times 10^{11} \, \text{s} = 9900 \, \text{yr},$$

which is the time elapsed since the death of the animal.

Carbon dating is useful only for determining the age of objects less than about 60,000 years old. The amount of $^{14}_{6}\text{C}$ remaining in older objects is usually too small to measure accurately, although new techniques are allowing detection of even smaller amounts of $^{14}_{6}\text{C}$, pushing the time frame further back. On the other hand, radioactive isotopes with longer half-lives can be used in certain circumstances to obtain the age of older objects. For example, the decay of $^{238}_{92}\text{U}$, because of its long half-life of 4.5×10^9 years, is useful in determining the ages of rocks on a geologic time scale. When molten material solidified into rock as the temperature dropped, different compounds solidified according to the melting points, and thus different compounds separated to some extent. Uranium present in a material became fixed in position and the daughter nuclei that result from the decay of uranium were also fixed in that position. Thus, by measuring the amount of $^{238}_{92}\text{U}$ remaining in the material relative to the amount of daughter nuclei, the time when the rock solidified can be determined.

➡ **PHYSICS APPLIED**

Geological dating

Radioactive dating methods using $^{238}_{92}\text{U}$ and other isotopes have shown the age of the oldest Earth rocks to be about 4×10^9 yr. The age of rocks in which the oldest fossilized organisms are embedded indicates that life appeared at least 3 billion years ago. The earliest fossilized remains of mammals are found in rocks 200 million years old, and the first humanlike creatures seem to have appeared about 2 million years ago. Radioactive dating has been indispensable for the reconstruction of Earth's history and the evolution of its biological organisms.

➡ **PHYSICS APPLIED**

Oldest Earth rocks and earliest life

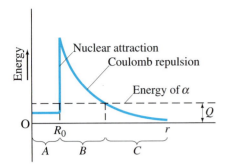

FIGURE 30–11 Potential energy for alpha particle and nucleus, showing the Coulomb barrier through which the α particle must tunnel to escape. The Q-value of the reaction is also shown.

We have seen that radioactive decay occurs only when the mass of the parent nucleus is greater than the sum of the masses of the daughter nucleus and all particles emitted. For example, $^{238}_{92}U$ can decay to $^{234}_{90}Th$ because the mass of $^{238}_{92}U$ is greater than the mass of the $^{234}_{90}Th$ plus the mass of the α particle. Since systems tend to go in the direction that reduces their internal or potential energy (a ball rolls downhill, a positive charge moves toward a negative charge), you may wonder why an unstable nucleus doesn't fall apart immediately. In other words, why do $^{238}_{92}U$ nuclei ($T_{\frac{1}{2}} = 4.5 \times 10^9 \, \text{yr}$) and other isotopes have such long half-lives? Why don't parent nuclei all decay at once?

The answer has to do with quantum theory and the nature of the forces involved. One way to view the situation is with the aid of a potential-energy diagram, as in Fig. 30–11. Let us consider the particular case of the decay $^{238}_{92}U \rightarrow ^{234}_{90}Th + ^4_2He$. The blue line represents the potential energy, including rest mass, where we imagine the α particle as a separate entity within the $^{238}_{92}U$ nucleus. The region labeled A in Fig. 30–11 represents the PE of the α particle when it is held within the uranium nucleus by the nuclear force (R_0 is the nuclear radius). Region C represents the PE when the α particle is free of the nucleus. The downward-curving PE (proportional to $1/r$) represents the electrical (Coulomb's law) repulsion between the positively charged α and the $^{234}_{90}Th$ nucleus. In order to get to region C, the α particle has to get by the barrier shown. Since the PE just beyond $r = R_0$ (region B) is greater than the energy of the alpha particle (dashed line), the α particle could not escape the nucleus if it were governed by classical physics. It could escape only if there were an input of energy equal to the height of the barrier. Nuclei decay spontaneously, however, without any input of energy. How, then, does the α particle get from region A to region C? *Tunneling* It actually passes through the barrier in a process known as **tunneling**. Classically, this could not happen, because an α particle in region B (within the barrier) would be violating the conservation-of-energy principle.[†] The uncertainty principle, however, tells us that energy conservation can be violated by an amount ΔE for a length of time Δt given by

$$(\Delta E)(\Delta t) \approx \frac{h}{2\pi}.$$

We saw in Section 28–3 that this is a result of the wave–particle duality. Thus quantum mechanics allows conservation of energy to be violated for brief periods that may be long enough for an α particle to "tunnel" through the barrier. ΔE would represent the energy difference between the average barrier height and the particle's energy, and Δt the time to pass through the barrier. The higher and wider the barrier, the less time the α particle has to escape and the less likely it is to do so. It is therefore the height and width of this barrier that controls the rate of decay and half-life of an isotope.

[†]Since $\text{KE} = \frac{1}{2}mv^2 \geq 0$, then $E = \text{KE} + \text{PE} \geq \text{PE}$. But in Fig. 30–11, the total E (dashed line) is less than the PE in region B, so a particle reaching region B would violate conservation of energy.

30–13 Detection of Radiation

Individual particles such as electrons, protons, α particles, neutrons, and γ rays are not detected directly by our senses. Consequently, a variety of instruments have been developed to detect them.

One of the most common is the **Geiger counter**. As shown in Fig. 30–12, it consists of a cylindrical metal tube filled with a certain type of gas. A long wire runs down the center and is kept at a high positive voltage ($\approx 10^3$ V) with respect to the outer cylinder. The voltage is just slightly less than that required to ionize the gas atoms. When a charged particle enters through the thin "window" at one end of the tube, it ionizes a few atoms of the gas. The freed electrons are attracted toward the positive wire and as they are accelerated they strike and ionize additional atoms. An "avalanche" of electrons is quickly produced, and when it reaches the wire anode, it produces a voltage pulse. The pulse, after being amplified, can be sent to an electronic counter, which counts how many particles have been detected. Or the pulses can be sent to a loudspeaker and each detection of a particle is heard as a "click."

A **scintillation counter** makes use of a solid, liquid, or gas known as a **scintillator** or **phosphor**. The atoms of a scintillator are easily excited when struck by an incoming particle and emit visible light when they return to their ground states. Typical scintillators are crystals of NaI and certain plastics. One face of a solid scintillator is cemented to a photomultiplier tube, and the whole is wrapped with opaque material to keep it light-tight or is placed within a light-tight container. The **photomultiplier** (PM) **tube** converts the energy of the scintillator-emitted photon(s) into an electric signal. A PM tube is a vacuum tube containing several electrodes (typically 8 to 14), called *dynodes*, which are maintained at successively higher voltages as shown in Fig. 30–13. At its top surface is a photoelectric surface, called the *photocathode*, whose work function (Section 27–3) is low enough that an electron is easily released when struck by a photon from the scintillator. Such an electron is accelerated toward the first dynode. When it strikes the first dynode, the electron has acquired sufficient kinetic energy so that it can eject two to five more electrons. These, in turn, are accelerated to the second dynode, and a multiplication process begins. The number of electrons striking the last dynode may be 10^6 or more. Thus the passage of a particle through the scintillator results in an electric signal at the output of the PM tube that can be sent to an electronic counter just as for a Geiger tube. Because a scintillator crystal is a solid and therefore much more dense than the gas of a Geiger counter, it is a more efficient detector—especially for γ rays, which interact less with matter than do β rays.

In tracer work and other biological experiments (Section 31–7), **liquid scintillators** are often used. Radioactive samples taken at different times or from different parts of an organism are placed directly in small bottles containing the liquid scintillator. This is particularly convenient for detection of β rays from $^3_1$H and $^{14}_6$C, which have very low energies and have difficulty passing through the outer covering of a crystal scintillator or Geiger tube. A PM tube is still used to produce the electric signal.

A **semiconductor detector** consists of a reverse-biased *p-n* junction diode (Section 29–8). A particle passing through the junction can excite electrons into the conduction band, leaving holes in the valence band. The freed charges produce a short electrical pulse that can be counted just as for Geiger and scintillation counters.

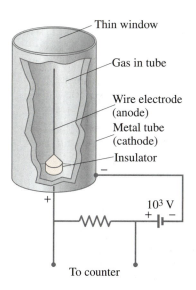

FIGURE 30–12 Diagram of a Geiger counter.

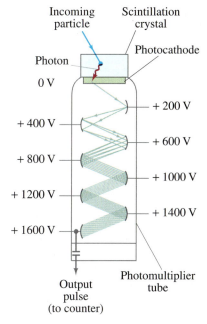

FIGURE 30–13 Scintillation counter with a photomultiplier tube.

The three devices discussed so far are used for counting the number of particles (or decays of a radioactive isotope). Other devices have been developed that allow the track of a charged particle to be *seen*. The simplest is the **photographic emulsion** (which, being small and simple and therefore portable, is now used particularly for cosmic-ray studies from balloons). A particle passing through a layer of photographic emulsion ionizes the atoms along its path. This results in a chemical change at these points, and when the emulsion is developed, the particle's path is revealed.

In a **cloud chamber**, a gas is cooled to a temperature slightly below its usual condensation point. (It is said to be "supercooled.") The gas molecules begin to condense on any ionized molecules present. Thus the ions produced when a charged particle passes through serve as centers on which tiny droplets form (Fig. 30–14). Light scatters more from these droplets than from the gas background, so a photo of the cloud chamber at the right moment shows the track of the particle. An important instrument in the early days of nuclear physics, it is little used today.

The **bubble chamber**, invented in 1952 by D. A. Glaser (1926–   ), makes use of a superheated liquid. The liquid is kept close to its normal boiling point and the bubbles characteristic of boiling form around ions produced by the passage of a charged particle (Fig. 30–15). A photograph of the interior of the chamber thus reveals the paths of particles that recently passed through. Because the bubble chamber uses a liquid—often liquid hydrogen—the density of atoms is much greater than in a cloud chamber. Hence it is a much more efficient device for observing the tracks of charged particles and their interactions with the nuclei of the liquid. Usually, a magnetic field is applied across the chamber and the momentum of the moving particles can be determined from the radius of curvature of their paths.

A **wire chamber** consists of a set of closely spaced fine parallel wires. Alternate planes of wires are grounded and the ones in between are kept at very high voltage. When a charged particle passes through, the ions produced in the gas between the wires become an avalanche, and a large current results which produces a visible spark. Thus the path of the particle is made visible and can be photographed. Alternatively, the sparks produce electric pulses in the wires and the positions of the sparks can be determined by the time it takes the pulses to reach detectors placed at the ends of the wires. Then the particle's path is reconstructed electronically with computers which can then "draw" a picture of the tracks.

FIGURE 30–14 In a cloud or bubble chamber, droplets or bubbles are formed around ions produced by the passage of a charged particle.

Path of particle

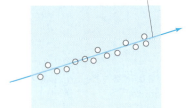

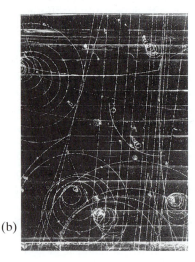

FIGURE 30–15 (a) Bubble chamber; (b) particle tracks in a bubble chamber.

(a)

(b)

SUMMARY

Nuclear physics is the study of atomic nuclei. Nuclei contain **protons** and **neutrons**, which are collectively known as **nucleons**. The total number of nucleons, A, is the **atomic mass number**. The number of protons, Z, is the **atomic number**. The number of neutrons equals $A - Z$. **Isotopes** are nuclei with the same Z, but with different numbers of neutrons. For an element X, an isotope of given Z and A is represented by

$$\frac{A}{Z}X.$$

The nuclear radius is proportional to $A^{\frac{1}{3}}$, indicating that all nuclei have about the same density. Nuclear masses are specified in **unified atomic mass units** (u), where the mass of $^{12}_{6}C$ (including its 6 electrons) is defined as exactly 12.000000 u, or in terms of their energy equivalent (because $E = mc^2$), where

$$1\,u = 931.5\,MeV/c^2 = 1.66 \times 10^{-27}\,kg.$$

The mass of a stable nucleus is less than the sum of the masses of its constituent nucleons. The difference in mass (times c^2) is the **total binding energy**. It represents the energy needed to break the nucleus into its constituent nucleons. The **binding energy per nucleon** averages about 8 MeV per nucleon, and is lowest for low mass and high mass nuclei.

Unstable nuclei undergo **radioactive decay**; they change into other nuclei with the emission of an α, β, or γ particle. An α particle is a $^4_2$He nucleus; a β particle is an electron or positron; and a γ ray is a high-energy photon. In β decay, a **neutrino** is also emitted. The transformation of the parent into the daughter nucleus is called **transmutation** of the elements. Radioactive decay occurs spontaneously only when the rest mass of the products is less than the mass of the parent nucleus. The loss in mass appears as kinetic energy of the products.

Nuclei are held together by the **strong nuclear force**. The **weak nuclear force** makes itself apparent in β decay. These two forces, plus the gravitational and electromagnetic forces, are the four known types of force. Electric charge, linear and angular momentum, mass–energy, and **nucleon number** are **conserved** in all decays.

Radioactive decay is a statistical process. For a given type of radioactive nucleus, the number of nuclei that decay (ΔN) in a time Δt is proportional to the number N of parent nuclei present:

$$\Delta N = -\lambda N\,\Delta t.$$

The proportionality constant, λ, is called the **decay constant** and is characteristic of the given nucleus. The number N of nuclei remaining after a time t decreases exponentially

$$N \propto e^{-\lambda t}$$

as does the **activity**, $\Delta N/\Delta t$. The **half-life**, $T_{\frac{1}{2}}$, is the time required for half the nuclei of a radioactive sample to decay. It is related to the decay constant by $T_{\frac{1}{2}} = 0.693/\lambda$.

QUESTIONS

1. What do different isotopes of a given element have in common? How are they different?

2. What are the elements represented by the X in the following: (a) $^{232}_{92}X$; (b) $^{18}_{7}X$; (c) $^{1}_{1}X$; (d) $^{82}_{38}X$; (e) $^{247}_{97}X$?

3. How many protons and how many neutrons do each of the isotopes in Question 2 have?

4. Why are the atomic masses of many elements (see the periodic table) not close to whole numbers?

5. How do we know there is such a thing as the strong nuclear force?

6. What are the similarities and the differences between the strong nuclear force and the electric force?

7. What is the experimental evidence in favor of radioactivity being a nuclear process?

8. The isotope $^{64}_{29}Cu$ is unusual in that it can decay by γ, β^-, and β^+ emission. What is the resulting nuclide for each case?

9. A $^{238}_{92}U$ nucleus decays to a nucleus containing how many neutrons?

10. Describe, in as many ways as possible, the difference between α, β, and γ rays.

11. What element is formed by the radioactive decay of (a) $^{24}_{11}Na$ (β^-); (b) $^{22}_{11}Na$ (β^+); (c) $^{210}_{84}Po$ (α)?

12. What element is formed by the decay of (a) $^{32}_{15}P$ (β^-); (b) $^{35}_{16}S$ (β^-); (c) $^{211}_{83}Bi$ (α)?

13. Fill in the missing particle or nucleus:
    (a) $^{45}_{20}Ca \rightarrow ? + e^- + \bar{\nu}$
    (b) $^{58}_{29}Cu \rightarrow ? + \gamma$
    (c) $^{46}_{24}Cr \rightarrow ^{46}_{23}V + ?$
    (d) $^{234}_{94}Pu \rightarrow ? + \alpha$
    (e) $^{239}_{93}Np \rightarrow ^{239}_{94}Pu + ?$

14. Immediately after a $^{238}_{92}$U nucleus decays to $^{234}_{90}$Th $+$ $^{4}_{2}$He, the daughter thorium nucleus still has 92 electrons circling it. Since thorium normally holds only 90 electrons, what do you suppose happens to the two extra ones?

15. When a nucleus undergoes beta decay, what happens to the energy levels of the atomic electrons? What is likely to happen to these electrons following the decay?

16. The alpha particles from a given alpha-emitting nuclide are monoenergetic; that is, they all have the same kinetic energy. But the beta particles from a beta-emitting nuclide have a spectrum of energies. Explain the difference between these two cases.

17. Do isotopes that undergo electron capture generally lie above or below the line of stability in Fig. 30–2?

18. Can hydrogen or deuterium emit an α particle?

19. Why are many artificially produced radioactive isotopes rare in nature?

20. An isotope has a half-life of one month. After two months, will a given sample of this isotope have completely decayed? If not, how much remains?

21. Explain the absence of β^+ emitters in the radioactive decay series of Fig. 30–10.

22. Can $^{14}_{6}$C dating be used to measure the age of stone walls and tablets of ancient civilizations?

23. What assumptions are made in carbon-dating? What do you think could affect these assumptions?

* **24.** Describe how the potential energy curve for an α particle in an α-emitting nucleus differs from that for a stable nucleus.

PROBLEMS

SECTION 30–1

1. (I) What is the rest mass of an α particle in MeV/c^2?

2. (I) A pi meson has a mass of $139\,\text{MeV}/c^2$. What is this in atomic mass units?

3. (I) What is the approximate radius of an alpha particle ($^{4}_{2}$He)?

4. (I) By what percentage is the radius of the isotope $^{14}_{6}$C greater than that of its sister $^{12}_{6}$C?

5. (II) (a) What is the fraction of the hydrogen atom's mass that is in the nucleus? (b) What is the fraction of the hydrogen atom's volume that is occupied by the nucleus? (c) What is the density of nuclear matter? Compare this with water.

6. (II) (a) What is the approximate radius of a $^{64}_{29}$Cu nucleus? (b) Approximately what is the value of A for a nucleus whose radius is 3.7×10^{-15} m?

7. (II) How much energy must an α particle have to just "touch" the surface of a $^{238}_{92}$U nucleus?

8. (II) If an alpha particle were released from rest near the surface of a $^{243}_{95}$Am nucleus, what would its kinetic energy be when far away?

9. (II) What stable nucleus has approximately half the radius of a uranium nucleus? [*Hint*: Find A and use Appendix F to get Z.]

SECTION 30–2

10. (I) Estimate the total binding energy for $^{40}_{20}$Ca, using Fig. 30–1.

11. (I) Use Fig. 30–1 to estimate the total binding energy of (a) $^{238}_{92}$U, and (b) $^{84}_{36}$Kr.

12. (II) Use Appendix F to calculate the binding energy of $^{2}_{1}$H (deuterium).

13. (II) Calculate the binding energy per nucleon for a $^{14}_{7}$N nucleus.

14. (II) Calculate the total binding energy and the binding energy per nucleon for $^{6}_{3}$Li. Use Appendix F.

15. (II) Calculate the binding energy of the last neutron in a $^{12}_{6}$C nucleus. [*Hint*: Compare the mass of $^{12}_{6}$C with that of $^{11}_{6}$C $+ \, ^{1}_{0}$n; use Appendix F.]

16. (II) Compare the binding energy of a neutron in $^{23}_{11}$Na to that in $^{24}_{11}$Na.

17. (II) How much energy is required to remove (a) a proton, (b) a neutron, from $^{16}_{8}$O? Explain the difference in your answers.

18. (II) (a) Show that the nucleus $^{8}_{4}$Be (mass $= 8.005305$ u) is unstable to decay into two α particles. (b) Is $^{12}_{6}$C stable against decay into three α particles? Show why or why not.

SECTIONS 30–3 TO 30–7

19. (I) $^{60}_{27}$Co in an excited state emits a 1.33-MeV γ ray as it jumps to the ground state. What is the mass (in u) of the excited cobalt atom?

20. (I) How much energy is released when tritium, $^{3}_{1}$H, decays by β^- emission?

21. (I) What is the maximum KE of an electron emitted in the β decay of a free neutron?

22. (I) Show that the decay $^{11}_{6}$C $\rightarrow \,^{10}_{5}$B $+$ p is not possible because energy would not be conserved.

23. (II) $^{22}_{11}$Na is radioactive. Is it a β^- or β^+ emitter? Write down the decay reaction, and estimate the maximum KE of the emitted β.

24. (II) Give the result of a calculation that shows whether or not the following decays are possible: (a) $^{236}_{92}$U $\rightarrow \,^{235}_{92}$U $+$ n; (b) $^{16}_{8}$O $\rightarrow \,^{15}_{8}$O $+$ n; (c) $^{23}_{11}$Na $\rightarrow \,^{22}_{11}$Na $+$ n.

25. (II) A $^{232}_{92}$U nucleus emits an α particle with KE $=$ 5.32 MeV. What is the final nucleus and what is the approximate atomic mass (in u) of the final atom?

26. (II) When $^{23}_{10}$Ne (mass = 22.9945 u) decays to $^{23}_{11}$Na (mass = 22.9898 u), what is the maximum kinetic energy of the emitted electron? What is its minimum energy? What is the energy of the neutrino in each case?

27. (II) The nuclide $^{32}_{15}$P decays by emitting an electron whose maximum kinetic energy can be 1.71 MeV. (a) What is the daughter nucleus? (b) What is its atomic mass (in u)?

28. (II) The isotope $^{218}_{84}$Po can decay by either α or β^- emission. What is the energy release in each case? The mass of $^{218}_{84}$Po is 218.008965 u.

29. (II) How much energy is released in electron capture by beryllium: $^7_4$Be + $_{-1}^0$e → $^7_3$Li + ν?

30. (II) What is the energy of the α particle emitted in the decay $^{210}_{84}$Po → $^{206}_{82}$Pb + α?

31. (III) The α particle emitted when $^{238}_{92}$U decays has KE = 4.20 MeV. Calculate the recoil KE of the daughter nucleus and the Q-value of the decay.

32. (III) Show that when a nucleus decays by β^+ decay, the total energy released is equal to $\sim$

$$(M_P - M_D - 2m_e)c^2,$$

where M_P and M_D are the masses of the parent and daughter atoms (neutral), and m_e is the mass of an electron or positron.

33. (III) Use the result of Problem 32 to determine the maximum kinetic energy of β^+ particles released when $^{11}_6$C decays to $^{11}_5$B. What is the maximum energy the neutrino can have? What is its minimum energy?

34. (III) In α decay of, say, a $^{226}_{88}$Ra nucleus, show that the nucleus carries away a fraction $1/(1 + A_D/4)$ of the total energy available, where A_D is the mass number of the daughter nucleus. [Hint: Use conservation of momentum as well as conservation of energy.] Approximately what percentage of the energy available is thus carried off by the α particle in the case cited?

SECTIONS 30–8 TO 30–11

35. (I) A radioactive material produces 1280 decays per minute at one time, and 6 h later produces 320 decays per minute. What is its half-life?

36. (I) Look up the half life of $^{238}_{92}$U in Appendix F, and then determine its decay constant.

37. (I) The decay constant of a given nucleus is 5.4×10^{-3} s^{-1}. What is its half-life?

38. (I) What is the activity of a sample of $^{14}_6$C that contains 4.1×10^{20} nuclei?

39. (I) What fraction of a sample of $^{68}_{32}$Ge, whose half-life is about 9 months, will remain after 3.0 yr?

40. (I) How many nuclei of $^{238}_{92}$U remain in a rock if the activity registers 275 decays per second?

41. (II) What fraction of a sample is left after (a) exactly 4 half-lives, (b) exactly $4\frac{1}{2}$ half-lives?

42. (II) In a series of decays, the nuclide $^{235}_{92}$U becomes $^{207}_{82}$Pb. How many α and β^- particles are emitted in this series?

43. (II) The iodine isotope $^{131}_{53}$I is used in hospitals for diagnosis of thyroid function. If 532 μg are ingested by a patient, determine the activity (a) immediately, (b) 1.0 h later when the thyroid is being tested, and (c) 6 months later. Use Appendix F.

44. (II) $^{124}_{55}$Cs has a half-life of 30.8 s. (a) If we have 7.8 μg initially, how many nuclei are present? (b) How many are present 2.0 min later? (c) What is the activity at this time? (d) After how much time will the activity drop to less than about 1 per second?

45. (II) Calculate the activity of a pure 4.7-μg sample of $^{32}_{15}$P ($T_{\frac{1}{2}} = 1.23 \times 10^6$ s).

46. (II) The activity of a sample of $^{35}_{16}$S ($T_{\frac{1}{2}} = 7.56 \times 10^6$ s) is 3.55×10^5 decays per second. What is the mass of sample present?

47. (II) A sample of $^{233}_{92}$U ($T_{\frac{1}{2}} = 1.59 \times 10^5$ yr) contains 6.50×10^{19} nuclei. (a) What is the decay constant? (b) Approximately how many disintegrations will occur per minute?

48. (II) The activity of a sample drops by a factor of 10 in 9.6 minutes. What is its half-life?

49. (II) A 135-g sample of pure carbon contains 1.3 parts in 10^{12} (atoms) of $^{14}_6$C. How many disintegrations occur per second?

50. (II) A radioactive nuclide produces 2880 decays per minute at one time, and 1.6 h later produces 820 decays per minute. What is the half-life of the nuclide?

51. (II) A sample of $^{40}_{19}$K is decaying at a rate of 8.70×10^2 decays/s. What is the mass of the sample?

52. (II) The rubidium isotope $^{87}_{37}$Rb, a β emitter with a half-life of 4.75×10^{10} yr, is used to determine the age of rocks and fossils. Rocks containing fossils of early animals contain a ratio of $^{87}_{38}$Sr to $^{87}_{37}$Rb of 0.0160. Assuming that there was no $^{87}_{38}$Sr present when the rocks were formed, calculate the age of these fossils. [Hint: Use Eq. 30–3.]

53. (II) Use Fig. 30–10 and calculate the relative decay rates for α decay of $^{218}_{84}$Po and $^{214}_{84}$Po.

54. (II) $^7_4$Be decays with a half-life of about 53 d. It is produced in the upper atmosphere, and filters down onto the Earth's surface. If a plant leaf is detected to have 350 decays/s of $^7_4$Be, how long do we have to wait for the decay rate to drop to 10/s? Estimate the initial mass of $^7_4$Be on the leaf.

55. (II) Which radioactive isotope of lead is being produced in a reaction where the measured activity of a sample drops to 1.050 percent of its original activity in 4.00 h?

56. (III) An ancient club is found that contains 170 g of carbon and has an activity of 5.0 decays per second. Determine its age assuming that in living trees the ratio of ^{14}C/^{12}C atoms is about 1.3×10^{-12}.

57. (III) At $t = 0$, a pure sample of radioactive nuclei contains N_0 nuclei whose decay constant is λ. Determine a formula for the number of daughter nuclei, N_D, as a function of time; assume that $N_D = 0$ at $t = 0$.

58. Show that the radius of the largest nuclei (say, $^{238}_{92}U$) is only about 6 times greater than that of the smallest ($^{1}_{1}H$).

59. (*a*) Determine the density of nuclear matter in kg/m^3, and show that it is essentially the same for all nuclei. (*b*) What would be the radius of the Earth if it had its actual mass but had the density of nuclei? (*c*) What would be the radius of a $^{238}_{92}U$ nucleus if it had the density of the Earth?

60. Using the uncertainty principle, use the size of the nucleus to estimate the kinetic energy of a nucleon in, say, iron. Ignore relativistic corrections. [*Hint*: A particle can have an energy at least as large as its uncertainty.]

61. How much recoil energy does a $^{40}_{19}K$ nucleus get when it emits a 1.46 MeV gamma ray?

62. An old wooden tool is found to contain only 10 percent of $^{14}_{6}C$ that a sample of fresh wood would. How old is the tool?

63. The $^{3}_{1}H$ isotope of hydrogen, which is called *tritium* (because it contains three nucleons), has a half-life of 12.33 yr. It can be used to measure the age of objects up to about 100 yr. It is produced in the upper atmosphere by cosmic rays and brought to Earth by rain. As an application, determine approximately the age of a bottle of wine whose $^{3}_{1}H$ radiation is about $\frac{1}{10}$ that present in new wine.

64. A neutron star consists of neutrons at approximately nuclear density. Estimate, for a 10-km-diameter neutron star, (*a*) its mass number, (*b*) its mass (kg), and (*c*) the acceleration of gravity at its surface.

65. Recent elementary particle theories (Section 32–11) suggest that the proton may be unstable, with a half-life $\geq 10^{32}$ yr. How long would you expect to wait for one proton in your body to decay (consider that your body is all water)?

66. When water is placed near an intense neutron source, the neutrons can be slowed down by collisions with the water molecules and eventually captured by a hydrogen nucleus to form the stable isotope called deuterium, $^{2}_{1}H$, giving off a gamma ray. What is the energy of the gamma ray?

67. How long must you wait (in half-lives) for a radioactive sample to drop to 1.00 percent of its original activity?

68. If the potassium isotope $^{40}_{19}K$ gives 50 decays/s in a liter of milk, estimate how much $^{40}_{19}K$ and regular $^{39}_{19}K$ are in a liter of milk. Use Appendix F.

69. Decay series, such as that shown in Fig. 30–10, can be classified into four families, depending on whether the mass numbers have the form $4n$, $4n + 1$, $4n + 2$, $4n + 3$, where n is an integer. Justify this statement and show that for a nuclide in any family, all its daughters will be in the same family.

70. Strontium-90 is produced as a nuclear fission product of uranium in both reactors and atomic bombs. Look at its location in the periodic table to see what other elements it might be similar to chemically, and tell why you think it might be dangerous to ingest. It has too many neutrons, and it decays with a half-life of about 29 yr. How long will we have to wait for the amount of $^{90}_{38}Sr$ on the Earth's surface to reach 1 percent of its current level, assuming no new material is scattered about? Write down the decay reaction, including the daughter nucleus. Is the daughter radioactive? If so, write down its decay. Finish the decay scheme until you reach a stable nucleus.

71. The nuclide $^{191}_{76}Os$ decays with β^- energy of 0.14 MeV accompanied by γ rays of energy 0.042 MeV and 0.129 MeV. (*a*) What is the daughter nucleus? (*b*) Draw an energy-level diagram showing the ground states of the parent and daughter and excited states of the daughter. To which of the daughter states does β^- decay of $^{191}_{76}Os$ occur?

72. Use the uncertainty principle to argue why electrons are unlikely to be found in the nucleus. Use relativity. [See hint for Problem 60.]

73. Estimate the total binding energy for copper and then estimate the energy, in joules, needed to break a 3-g copper penny into its constituent nucleons.

74. Instead of giving atomic masses for nuclides as in Appendix F, some tables give the *mass excess*, Δ, defined as $\Delta = M - A$, where A is the atomic number and M is the mass in u. Determine the mass excess, in u and in MeV/c^2, for: (*a*) $^{4}_{2}He$; (*b*) $^{12}_{6}C$; (*c*) $^{107}_{47}Ag$; (*d*) $^{235}_{92}U$. (*e*) From a glance at Appendix F, can you make a generalization about the sign of Δ as a function of Z or A?

75. (*a*) A 100-gram sample of natural carbon contains the usual fraction of $^{14}_{6}C$. How long will it take on average before there is only one $^{14}_{6}C$ nucleus left? (*b*) How does the answer in (*a*) change if the sample is 200 grams? What does this tell you about the limits of carbon dating?

76. If the mass of the proton were just a little closer to the mass of the neutron, the following reaction would be possible even at low collision energies:

$$e^- + p \rightarrow n + \nu.$$

Why would this situation be catastrophic? By what percentage would the proton's mass have to be increased to make this reaction possible?

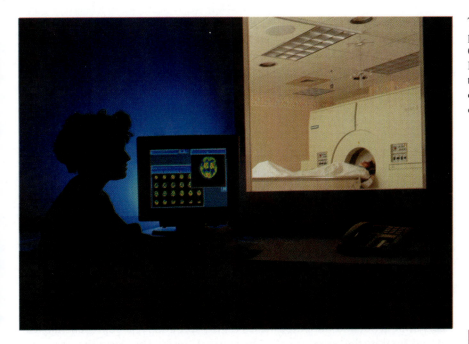

Technician is looking at a positron emission tomography (PET) image of a patient's brain. PET is one of several powerful types of medical imaging based on physics used by doctors to diagnose illnesses.

NUCLEAR ENERGY; EFFECTS AND USES OF RADIATION

31

We continue our study of nuclear physics in this chapter. We begin with a discussion of nuclear reactions, after which we examine the important large energy-releasing processes of fission and fusion. The remainder of the chapter deals with the effects of nuclear radiation when it passes through matter, particularly biological matter, and how radiation is used medically for therapy and diagnosis, including recently developed imaging techniques.

31–1 Nuclear Reactions and the Transmutation of Elements

When a nucleus undergoes α or β decay, the daughter nucleus is that of a different element from the parent. The transformation of one element into another, called *transmutation,* also occurs by means of nuclear reactions. A **nuclear reaction** is said to occur when a given nucleus is struck by another nucleus, or by a simpler particle such as a γ ray or neutron, so that an interaction takes place. Ernest Rutherford was the first to report seeing a nuclear reaction. In 1919 he observed that some of the α particles passing through nitrogen gas were absorbed and protons emitted. He concluded that nitrogen nuclei had been transformed into oxygen nuclei via the reaction

$$^4_2\text{He} + {}^{14}_7\text{N} \rightarrow {}^{17}_8\text{O} + {}^1_1\text{H},$$

where ^4_2He is an α particle, and ^1_1H is a proton.

Since then, a great many nuclear reactions have been observed. Indeed, many of the radioactive isotopes used in the laboratory are made by means of nuclear reactions. Nuclear reactions can be made to occur in the laboratory, but they also occur regularly in nature. In Chapter 30 we saw an example of this: $^{14}_6\text{C}$ is continually being made in the atmosphere via the reaction $\text{n} + {}^{14}_7\text{N} \rightarrow {}^{14}_6\text{C} + \text{p}$.

Nuclear reactions are sometimes written in a shortened form: for example, the reaction

$$\text{n} + {}^{14}_7\text{N} \rightarrow {}^{14}_6\text{C} + \text{p}$$

is written

$$^{14}_7\text{N}\,(\text{n, p})^{14}_6\text{C}.$$

The symbols outside the parentheses on the left and right represent the initial and final nuclei, respectively. The symbols inside the parentheses represent the bombarding particle (first) and the emitted small particle (second).

In any nuclear reaction, both electric charge and nucleon number are conserved. These conservation laws are often useful, as the following Example shows.

Deuterium

CONCEPTUAL EXAMPLE 31–1 **Deuterium reaction.** A neutron is observed to strike an $^{16}_8\text{O}$ nucleus and a deuteron is given off. (A **deuteron**, or **deuterium**, is the isotope of hydrogen containing one proton and one neutron, ^2_1H; it is sometimes given the symbol d or D.) What is the nucleus that results?

RESPONSE We have the reaction $\text{n} + {}^{16}_8\text{O} \rightarrow ? + {}^2_1\text{H}$. The total number of nucleons initially is $1 + 16 = 17$, and the total charge is $0 + 8 = 8$. The same totals apply to the right side of the reaction. Hence the product nucleus must have $Z = 7$ and $A = 15$. From the periodic table, we find that it is nitrogen that has $Z = 7$, so the nucleus produced is $^{15}_7\text{N}$. The reaction can be written $^{16}_8\text{O}\,(\text{n, d})^{15}_7\text{N}$, where d represents deuterium, ^2_1H.

Energy (as well as momentum) is conserved in nuclear reactions, and we can use this to determine whether a given reaction can occur or not. For example, if the total mass of the products is less than the total mass of the initial particles, this decrease in mass (recall $E = mc^2$) appears as kinetic energy of the outgoing particles. But if the total mass of the products is greater than the total mass of the initial reactants, the reaction requires energy. The reaction will then not occur unless the bombarding particle has sufficient kinetic energy. Consider a nuclear reaction of the general form

$$\text{a} + \text{X} \rightarrow \text{Y} + \text{b},$$

where a is a projectile particle (or small nucleus) that strikes nucleus X, producing nucleus Y and particle b (typically, p, n, α, γ). We define the **reaction energy**, or **Q-value**, in terms of the masses involved, as

$$Q = (M_a + M_X - M_b - M_Y)c^2. \quad \textbf{(31-1a)} \qquad \textit{Q-value}$$

Since energy is conserved, Q is equal to the change in kinetic energy (final minus initial):

$$Q = \text{KE}_b + \text{KE}_Y - \text{KE}_a - \text{KE}_X. \quad \textbf{(31-1b)}$$

Normally, $\text{KE}_X = 0$ since X is the target nucleus at rest (or nearly so) struck by an incoming particle a. For $Q > 0$, the reaction is said to be *exothermic* or *exoergic*; energy is released in the reaction, so the total KE is greater after the reaction than before. If Q is negative ($Q < 0$), the reaction is said to be *endothermic* or *endoergic*. In this case the final total KE is less than the initial KE, and an energy input is required to make the reaction happen; the energy input comes from the kinetic energy of the initial colliding particles (a and X).

EXAMPLE 31-2 **A slow-neutron reaction.** The nuclear reaction

$$n + {}^{10}_{5}\text{B} \rightarrow {}^{7}_{3}\text{Li} + {}^{4}_{2}\text{He}$$

is observed to occur even when very slow-moving neutrons ($M_n = 1.0087\,\text{u}$) strike a boron atom at rest. For a particular reaction in which $\text{KE}_n \approx 0$, the helium ($M_{\text{He}} = 4.0026\,\text{u}$) is observed to have a speed of $9.30 \times 10^6\,\text{m/s}$. Determine (*a*) the KE of the lithium ($M_{\text{Li}} = 7.0160\,\text{u}$), and (*b*) the Q-value of the reaction.

SOLUTION (*a*) Since the neutron and boron are both essentially at rest, the total momentum before the reaction is zero, and afterward is also zero. Therefore,

$$M_{\text{Li}}v_{\text{Li}} = M_{\text{He}}v_{\text{He}}.$$

We solve this for v_{Li} and substitute it into the equation for kinetic energy. We can use classical KE with little error, rather than relativistic formulas, because $v_{\text{He}} = 9.30 \times 10^6\,\text{m/s}$ is not close to the speed of light c, and v_{Li} will be even less since $M_{\text{Li}} > M_{\text{He}}$. Thus we can write:

$$\text{KE}_{\text{Li}} = \frac{1}{2}M_{\text{Li}}v_{\text{Li}}^2 = \frac{1}{2}M_{\text{Li}}\left(\frac{M_{\text{He}}v_{\text{He}}}{M_{\text{Li}}}\right)^2 = \frac{M_{\text{He}}^2 v_{\text{He}}^2}{2M_{\text{Li}}}.$$

We put in numbers, changing the mass in u to kg and recalling that $1.60 \times 10^{-13}\,\text{J} = 1\,\text{MeV}$:

$$\text{KE}_{\text{Li}} = \frac{(4.0026\,\text{u})^2(1.66 \times 10^{-27}\,\text{kg/u})^2(9.30 \times 10^6\,\text{m/s})^2}{2(7.0160\,\text{u})(1.66 \times 10^{-27}\,\text{kg/u})}$$

$$= 1.64 \times 10^{-13}\,\text{J} = 1.02\,\text{MeV}.$$

(*b*) We are given the data $\text{KE}_a = \text{KE}_X = 0$ in Eq. 31-1b, so $Q = \text{KE}_{\text{Li}} + \text{KE}_{\text{He}}$, where

$$\text{KE}_{\text{He}} = \tfrac{1}{2}M_{\text{He}}v_{\text{He}}^2 = \tfrac{1}{2}(4.0026\,\text{u})(1.66 \times 10^{-27}\,\text{kg/u})(9.30 \times 10^6\,\text{m/s})^2$$

$$= 2.84 \times 10^{-13}\,\text{J} = 1.78\,\text{MeV}.$$

Hence, $Q = 1.02\,\text{MeV} + 1.78\,\text{MeV} = 2.80\,\text{MeV}.$

> **EXAMPLE 31–3** **Will the reaction "go"?** Can the reaction $^{13}_{6}C\,(p, n)\,^{13}_{7}N$ occur when $^{13}_{6}C$ is bombarded by 2.0-MeV protons?
>
> **SOLUTION** We use Eq. 31–1a, looking up the masses of the nuclei in Appendix F. The total masses before and after the reaction are:
>
> | Before | After |
> |---|---|
> | $M(^{13}_{6}C) = 13.003355$ | $M(^{13}_{7}N) = 13.005738$ |
> | $M(^{1}_{1}H) = \underline{1.007825}$ | $M(n) = \underline{1.008665}$ |
> | 14.011180 | 14.014403 |
>
> (We must use the mass of the $^{1}_{1}H$ atom rather than that of the bare proton because the masses of $^{13}_{6}C$ and $^{13}_{7}N$ include the electrons, and we must include an equal number of electrons on each side of the equation since none are created or destroyed.) The products have an excess mass of $(14.014403 - 14.011180)\,u = 0.003223\,u \times 931.5\,MeV/u = 3.00\,MeV$. Thus $Q = -3.00\,MeV$, and the reaction is endoergic. This reaction requires energy, and the 2.0 MeV protons do not have enough to make it go.

The proton in this last Example would have to have somewhat more than 3.00 MeV of KE to make this reaction go; 3.00 MeV would be enough to conserve energy, but a proton of this energy would produce the $^{13}_{7}N$ and n with no KE and hence no momentum. Since an incident 3.0-MeV proton has momentum, conservation of momentum would be violated. A more complicated calculation shows that to conserve both energy and momentum, the minimum proton energy, called the **threshold energy**, is 3.23 MeV in this case (see Problem 15).

The artificial transmutation of elements took a great leap forward in the 1930s when Enrico Fermi realized that neutrons would be the most effective projectiles for causing nuclear reactions and in particular for producing new elements. Because neutrons have no net electric charge, they are not repelled by positively charged nuclei as are protons or alpha particles. Hence the probability of a neutron reaching the nucleus and causing a reaction is much greater than for charged projectiles,[†] particularly at low energies. Between 1934 and 1936, Fermi and his co-workers in Rome produced many previously unknown isotopes by bombarding different elements with neutrons. Fermi realized that if the heaviest known element, uranium, were bombarded with neutrons, it might be possible to produce new elements whose atomic numbers were greater than that of uranium. After several years of hard work, it was suspected that two new elements had been produced, neptunium ($Z = 93$) and plutonium ($Z = 94$). The full confirmation that such "transuranic" elements could be produced came several years later at the University of California, Berkeley. The reactions are shown in Fig. 31–1.

It was soon shown that what Fermi actually had observed when he bombarded uranium was an even stranger process—one that was destined to play an extraordinary role in the world at large.

FIGURE 31–1 Neptunium and plutonium are produced in this series of reactions, after bombardment of $^{238}_{92}U$ by neutrons.

(a) $n + {}^{238}_{92}U \rightarrow {}^{239}_{92}U$

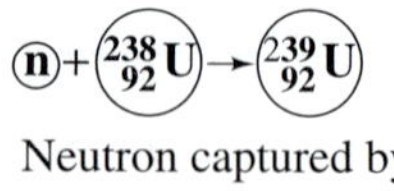

Neutron captured by $^{238}_{92}U$.

(b) $^{239}_{92}U \rightarrow {}^{239}_{93}Np + e^- + \bar{\nu}$

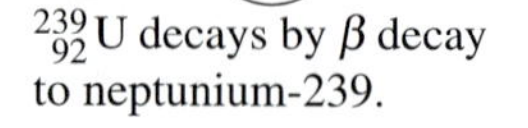

$^{239}_{92}U$ decays by β decay to neptunium-239.

(c) $^{239}_{93}Np \rightarrow {}^{239}_{94}Pu + e^- + \bar{\nu}$

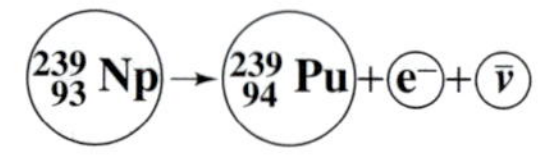

$^{239}_{93}Np$ itself decays by β decay to produce plutonium-239.

[†]That is, positively charged particles. Electrons rarely cause nuclear reactions because they do not interact via the strong nuclear force.

31–2 Nuclear Fission; Nuclear Reactors

In 1938, the German scientists Otto Hahn and Fritz Strassmann made an amazing discovery. Following up on Fermi's work, they found that uranium bombarded by neutrons sometimes produced smaller nuclei that were roughly half the size of the original uranium nucleus. Lise Meitner and Otto Frisch, two refugees from Nazi Germany working in Scandinavia, quickly realized what had happened: the uranium nucleus, after absorbing a neutron, actually had split into two roughly equal pieces. This was startling, for until then the known nuclear reactions involved knocking out only a tiny fragment (for example, n, p, or α) from a nucleus.

This new phenomenon was named **nuclear fission** because of its resemblance to biological fission (cell division). It occurs much more readily for $^{235}_{92}U$ than for the more common $^{238}_{92}U$. The process can be visualized by imaging the uranium nucleus to be like a liquid drop. According to this **liquid-drop model**, the neutron absorbed by the $^{235}_{92}U$ nucleus gives the nucleus extra internal energy (like heating a drop of water). This intermediate state, or **compound nucleus**, is $^{236}_{92}U$ (because of the absorbed neutron). The extra energy of this nucleus—it is in an excited state—appears as increased motion of the individual nucleons, which causes the nucleus to take on abnormal elongated shapes, Figure 31–2. When the nucleus elongates (in this model) into the shape shown in Fig. 31–2c, the attraction of the two ends via the short-range nuclear force is greatly weakened by the increased separation distance, and the electric repulsive force becomes dominant. So the nucleus splits in two. The two resulting nuclei, N_1 and N_2, are called **fission fragments**, and in the process a number of neutrons (typically two or three) are also given off. The reaction can be written

$$n + {}^{235}_{92}U \rightarrow {}^{236}_{92}U \rightarrow N_1 + N_2 + \text{neutrons.} \qquad \textbf{(31–2a)}$$

The compound nucleus, $^{236}_{92}U$, exists for less than 10^{-12} s, so the process occurs very quickly. The two fission fragments more often split the original uranium mass as about 40%–60% rather than precisely half and half. A typical fission reaction is

$$n + {}^{235}_{92}U \rightarrow {}^{141}_{56}Ba + {}^{92}_{36}Kr + 3n, \qquad \textbf{(31–2b)}$$

although many others also occur.

A tremendous amount of energy is released in a fission reaction because the mass of $^{235}_{92}U$ is considerably greater than the total mass of the fission fragments plus neutrons. This can be seen from the binding-energy-per-nucleon curve of Fig. 30–1; the binding energy per nucleon for uranium is about 7.6 MeV/nucleon, but for fission fragments that have intermediate mass (in the center portion of the graph, $A \approx 100$), the average binding energy per nucleon is about 8.5 MeV/nucleon. Since the fission fragments are more tightly bound, they have less mass. The difference in mass (or energy) between the original uranium nucleus and the fission fragments is about $8.5 - 7.6 = 0.9$ MeV per nucleon. Since there are 236 nucleons involved in each fission, the total energy released per fission is

$$(0.9 \text{ MeV/nucleon})(236 \text{ nucleons}) \approx 200 \text{ MeV.}$$

This is an enormous amount of energy for one single nuclear event. At a practical level, the energy from one fission is, of course, tiny. But if many

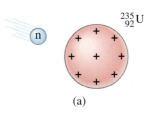

(a)

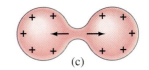

(b) ^{236}U (excited)

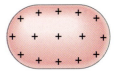

(c)

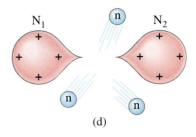

(d)

FIGURE 31–2 Fission of a $^{235}_{92}U$ nucleus after capture of a neutron, according to the liquid drop model.

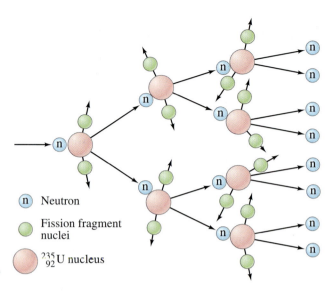

FIGURE 31–3 Chain reaction.

FIGURE 31–4 Color painting of the first nuclear reactor, built by Fermi under the grandstand of Stagg Field at the University of Chicago. (There are no photographs of the original reactor because of military secrecy.) Natural uranium was used with graphite as moderator. On December 2, 1942, Fermi slowly withdrew the cadmium control rods and the reactor went critical. This first self-sustaining chain reaction was announced to Washington, by telephone, by Arthur Compton who witnessed the event and reported: "The Italian navigator has just landed in the new world."

Chain reaction

Moderator

such fissions could occur in a short time, an enormous amount of energy at the macroscopic level would be available. A number of physicists, including Fermi, recognized that the neutrons released in each fission (Eqs. 31–2) could be used to create a **chain reaction**. That is, one neutron initially causes one fission of a uranium nucleus; the two or three neutrons released can go on to cause additional fissions, so the process multiplies as shown schematically in Fig. 31–3. If a **self-sustaining chain reaction** was actually possible in practice, the enormous energy available in fission could be released on a larger scale. Fermi and his co-workers (at the University of Chicago) showed it was possible by constructing the first **nuclear reactor** in 1942 (Fig. 31–4).

Several problems have to be overcome to make any nuclear reactor function. First, the probability that a $^{235}_{92}U$ nucleus will absorb a neutron is large only for slow neutrons, but the neutrons emitted during a fission, and which are needed to sustain a chain reaction, are moving very fast. A substance known as a **moderator** must be used to slow down the neutrons. The most effective moderator will consist of atoms whose mass is as close as possible to that of the neutrons. (To see why this is true, recall from Chapter 7 that a billiard ball striking an equal mass ball at rest can itself be stopped in one collision; but a billiard ball striking a heavy object bounces off with nearly the same speed it had.) The best moderator would thus contain ^1_1H atoms. Unfortunately, ^1_1H tends to absorb neutrons. But the isotope of hydrogen called *deuterium*, ^2_1H, does not absorb many neutrons and is thus an almost ideal moderator. Either ^1_1H or ^2_1H can be used in the form of water. In the latter case, it is **heavy water**, in which the hydrogen atoms have been replaced by deuterium. Another common moderator is *graphite,* which consists of $^{12}_6C$ atoms.

A second problem is that the neutrons produced in one fission may be absorbed and produce other nuclear reactions with other nuclei in the reactor, rather than produce further fissions. In a "light-water" reactor, the ^1_1H nuclei absorb neutrons, as does $^{238}_{92}U$ to form $^{239}_{92}U$ in the reaction $n + ^{238}_{92}U \rightarrow ^{239}_{92}U + \gamma$. Naturally occurring uranium[†] contains 99.3 percent $^{238}_{92}U$ and only 0.7 percent fissionable $^{235}_{92}U$. To increase the probability of fission of $^{235}_{92}U$ nuclei, natural uranium can be **enriched** to increase the percentage of $^{235}_{92}U$ using processes such as diffusion or centrifugation. (Enrichment is not usually necessary for reactors using heavy water as moderator since heavy water doesn't absorb neutrons.)

The third problem is that some neutrons will escape through the surface of the reactor core before they cause further fissions (Fig. 31–5). Thus the mass of fuel must be sufficiently large for a self-sustaining chain reaction to take place. The minimum mass of uranium needed is called the **critical mass**. The value of the critical mass depends on the moderator, the fuel ($^{239}_{94}Pu$ may be used instead of $^{235}_{92}U$), and how much the fuel is enriched, if at all. Typical values are on the order of a few kilograms (that is, not grams nor thousands of kilograms).

To have a self-sustaining chain reaction, it is clear that on the average at least one neutron produced in each fission must go on to produce another fission. The average number of neutrons per each fission that do go on to produce further fissions is called the **multiplication factor**, f. For a self-sustaining chain reaction, we must have $f \geq 1$. If $f < 1$, the reactor is "subcritical." If $f > 1$, it is "supercritical." Reactors are equipped with movable **control rods** (usually of cadmium or boron), whose function is to absorb neutrons and maintain the reactor at just barely "critical," $f = 1$. The release of neutrons and subsequent fissions caused by them occurs so quickly that manipulation of the control rods to maintain $f = 1$ would not be possible if it weren't for the small percentage ($\sim 1\%$) of so-called **delayed neutrons**. They come from the decay of neutron-rich fission fragments (or their daughters) having lifetimes on the order of seconds—sufficient to allow enough reaction time to operate the control rods and maintain $f = 1$.

Nuclear reactors have been built for use in research and to produce electric power. Fission produces many neutrons and a "research reactor" is basically an intense source of neutrons. These neutrons can be used as projectiles in nuclear reactions to produce nuclides not found in nature, including isotopes used as tracers and for medical therapy. A "power reactor" is used to produce electric power. The energy released in the fission

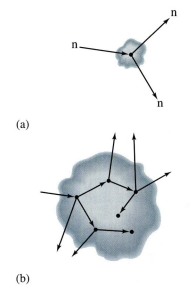

(a)

(b)

FIGURE 31–5 If the amount of uranium exceeds the critical mass, as in (b), a sustained chain reaction is possible. If the mass is less than critical, as in (a), most neutrons escape before additional fissions occur, and the chain reaction is not sustained.

Critical mass

Control rods

Delayed neutrons

Types of nuclear reactor

[†]$^{238}_{92}U$ will fission, but only with fast neutrons ($^{238}_{92}U$ is more stable than $^{235}_{92}U$). The probability of absorbing a fast neutron and producing a fission is too low to produce a self-sustaining chain reaction.

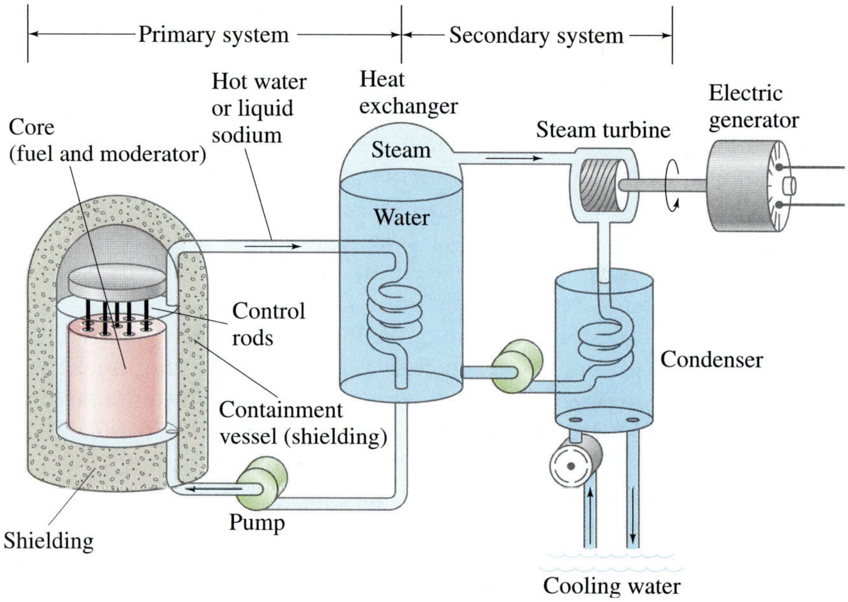

FIGURE 31–6 A nuclear reactor. The heat generated by the fission process in the fuel rods is carried off by hot water or liquid sodium and is used to boil water to steam in the heat exchanger. The steam drives a turbine to generate electricity and is then cooled in the condenser.

Primary system — Secondary system

Core (fuel and moderator)
Hot water or liquid sodium
Heat exchanger
Electric generator
Steam turbine
Steam
Water
Control rods
Containment vessel (shielding)
Condenser
Shielding
Pump
Cooling water

FIGURE 31–7 Devastation around Chernobyl in Russia, after the nuclear power plant disaster in 1986.

process appears as heat, which is used to boil water and produce steam to drive a turbine connected to an electric generator (Fig. 31–6). The **core** of a nuclear reactor consists of the fuel and a moderator (water in most U.S. commercial reactors). The fuel is usually uranium enriched so that it contains 2 to 4 percent $^{235}_{92}$U. Water or other liquid (such as liquid sodium) is allowed to flow through the core. The thermal energy it absorbs is used to produce steam in the heat exchanger, so the fissionable fuel acts as the heat input for a heat engine (see Sections 15–5 and 15–12).

Many problems are associated with nuclear power plants. Besides the usual thermal pollution associated with any heat engine (Section 15–12), there is the problem of disposal of the radioactive fission fragments produced in the reactor, plus radioactive nuclides produced by neutrons interacting with the structural parts of the reactor. Fission fragments, like their uranium or plutonium parents, have about 50 percent more neutrons than protons. Nuclei with atomic number in the typical range for fission fragments ($Z \approx 30$ to 60) are stable only if they have more nearly equal numbers of protons and neutrons (see Fig. 30–2). Hence the highly neutron-rich fission fragments are very unstable and decay radioactively. The accidental release of highly radioactive fission fragments into the atmosphere poses a serious threat to human health (Section 31–4), as does possible leakage of the radioactive wastes when they are disposed of. The accidents at Three Mile Island (1979) and at Chernobyl (1986) have illustrated some of the dangers and have shown that nuclear plants must be constructed, maintained, and operated with great care and precision (Fig. 31–7). Finally, the lifetime of nuclear power plants is limited to 30-some years, due to buildup of radioactivity and the fact that the structural materials themselves are weakened by the intense conditions inside. Decommissioning of a power plant could take a number of forms, but the cost of any method of decommissioning a large plant will be very great.

So-called **breeder reactors** were proposed as a solution to the problem of limited supplies of fissionable uranium. A breeder reactor is one in which some of the neutrons produced in the fission of $^{235}_{92}$U are absorbed by $^{238}_{92}$U, and $^{239}_{94}$Pu is produced via the set of reactions shown in Fig. 31–1. $^{239}_{94}$Pu is fissionable with slow neutrons, so after separation it can be used as a fuel in a nuclear reactor. Thus a breeder reactor "breeds" new fuel[†] ($^{239}_{94}$Pu) from otherwise useless $^{238}_{92}$U. Since natural uranium is 99.3 percent $^{238}_{92}$U, this means that the supply of fissionable fuel could be increased by more than a factor of 100. But breeder reactors not only have the same problems as other reactors, but in addition present other serious problems. Not only is plutonium considered by many to be a serious health hazard in itself (radioactive with a half-life of 24,000 years), but plutonium produced in a reactor can readily be used in a bomb. Thus the use of a breeder reactor, even more than a conventional uranium reactor, presents the danger of nuclear proliferation, and the possibility of theft of fuel by terrorists who could produce a bomb.

It is clear that nuclear power presents many risks. Other large-scale energy-conversion methods, such as conventional coal-burning steam plants, also present health and environmental hazards, some of which were discussed in Chapter 15, including air pollution, oil spills, and the release of CO_2 gas which may be trapping heat as in a greenhouse and raising the Earth's temperature. The solution to the world's needs for energy is not only technological, but economic and political as well. A major factor surely is to "conserve"—to not waste energy and use as little as possible.

EXAMPLE 31–4 **ESTIMATE** **Uranium fuel amount.** Estimate the minimum amount of $^{235}_{92}$U that needs to undergo fission in order to run a 1000-MW power reactor per year of continuous operation. Assume an efficiency (Chapter 15) of about 33%.

SOLUTION For 1000 MW output, the total power generation is 3000 MW, of which 2000 MW is dumped as "waste" heat. Thus the total energy release in 1 yr (3×10^7 s) from fission is about

$$(3 \times 10^9 \text{ J/s})(3 \times 10^7 \text{ s}) \approx 10^{17} \text{ J}.$$

If each fission releases 200 MeV of energy, the number of fissions required is

$$\frac{(10^{17} \text{ J})}{(2 \times 10^8 \text{ eV/fission})(1.6 \times 10^{-19} \text{ J/eV})} \approx 3 \times 10^{27} \text{ fissions}.$$

The mass of a single uranium atom is about $(235 \text{ u})(1.66 \times 10^{-27} \text{ kg/u}) \approx 4 \times 10^{-25}$ kg, so the total mass needed is $(4 \times 10^{-25} \text{ kg})(3 \times 10^{27} \text{ fissions}) \approx 1000$ kg, or about a ton. Since $^{235}_{92}$U is only a fraction of normal uranium, and even when enriched it is never more than 10 percent of the total, the yearly requirement for uranium is on the order of tens of tons. This is orders of magnitude less than coal, both in mass and volume (see Problem 26 in Chapter 15).

The first use of fission, however, was not to produce electric power. Instead, it was first used as a fission bomb (the "atomic bomb"). In early 1940, with Europe already at war, Hitler banned the sale of uranium from

Atom bomb

[†]A breeder reactor does *not* produce more fuel than it uses.

FIGURE 31–8 J. Robert Oppenheimer, on the left, with General Leslie Groves, who was the administrative head of Los Alamos during the war. The photograph was taken at the Trinity site in the New Mexico desert, where the first atomic bomb was exploded.

FIGURE 31–9 Photo taken a month after the bomb was dropped on Nagasaki. The shacks were constructed afterwards from debris in the ruins.

the Czech mines he had recently taken over. Research into the fission process suddenly was enshrouded in secrecy. Physicists in the United States were alarmed. A group of them approached Einstein—a man whose name was a household word—to send a letter to President Roosevelt about the possibilities of using nuclear fission for a bomb far more powerful than any previously known, and inform him that Germany might already have begun development of such a bomb. Roosevelt responded by authorizing the program known as the Manhattan Project, to see if a bomb could be built. Work began in earnest after Fermi's demonstration in 1942 that a sustained chain reaction was possible. A new secret laboratory was developed on an isolated mesa in New Mexico known as Los Alamos. Under the direction of J. Robert Oppenheimer (1904–1967; Fig. 31–8), it became the home of famous scientists from all over Europe and the United States.

To build a bomb that was subcritical during transport but that could be made supercritical (to produce a chain reaction) at just the right moment, two pieces of uranium were used, each less than the critical mass but together greater than the critical mass. The two masses would be kept separate until the moment of detonation arrived. Then a kind of gun would force the two pieces together very quickly, a chain reaction of explosive proportions would occur, and a tremendous amount of energy would be released very suddenly. The first fission bomb was tested in the New Mexico desert in July 1945. It was successful. In early August, a fission bomb using uranium was dropped on Hiroshima and a second, using plutonium, was dropped on Nagasaki (Fig. 31–9). World War II ended shortly thereafter, but the dropping of these bombs aroused controversy.

Scientists were later criticized for working on the bomb, and many regretted having done so. Critical judgment, however, should perhaps be placed in the context of the times. In the early 1940s, Hitler's armies were overrunning Europe, and it was believed that German scientists were trying to build a bomb (they never did). Japan had entered the war and the world seemed in danger of being overwhelmed by oppressive powerful regimes. The scientists at Los Alamos were aware they were developing a dangerous weapon. But they had to make a choice. Stopping Hitler was uppermost in their minds.

Besides its great destructive power, a fission bomb produces many highly radioactive fission fragments, as does a nuclear reactor. When a fission bomb explodes, these radioactive isotopes are released into the atmosphere and are known as *radioactive fallout*.

Testing of nuclear bombs in the atmosphere after World War II was a cause of concern, for the movement of air masses spread the fallout all over the globe. Radioactive fallout eventually settles to the Earth, particularly in rainfall, and is absorbed by plants and grasses and enters the food chain. This is a far more serious problem than the same radioactivity on the exterior of our bodies, since α and β particles are largely absorbed by clothing and the outer (dead) layer of skin. But once inside our bodies via food, the isotopes are in direct contact with living cells. One particularly dangerous radioactive isotope is $^{90}_{38}Sr$, which is chemically much like calcium and becomes concentrated in bone, where it causes bone cancer and destruction of bone marrow. The 1963 treaty signed by over 100 nations that bans nuclear weapons testing in the atmosphere was motivated because of the hazards of fallout.

31–3 Fusion

The mass of every stable nucleus is less than the sum of the masses of its constituent protons and neutrons. For example, the mass of the helium isotope ^4_2He is less than the mass of two protons plus the mass of two neutrons, as we saw in Example 30–2. Thus, if two protons and two neutrons were to come together to form a helium nucleus there would be a loss of mass. This mass loss is manifested in the release of a large amount of energy. The process of building up nuclei by bringing together individual protons and neutrons, or building larger nuclei by combining small nuclei, is called **nuclear fusion**. A glance at Fig. 31–10 (same as Fig. 30–1) shows why small nuclei can combine to form larger ones with the release of energy: it is because the binding energy per nucleon is smaller for light nuclei than it is for those of increasing mass (up to about $A \approx 60$). It is believed that many of the elements in the universe were originally formed through the process of fusion (see Chapter 33), and that today, fusion is continually taking place within the stars, including our Sun, producing the prodigious amounts of radiant energy they emit.

EXAMPLE 31–5 Fusion energy release. One of the simplest fusion reactions involves the production of deuterium, ^2_1H, from a neutron and a proton: $^1_1\text{H} + \text{n} \rightarrow {}^2_1\text{H} + \gamma$. How much energy is released in this reaction?

SOLUTION From Appendix F, the initial rest mass is $1.007825\,\text{u} + 1.008665\,\text{u} = 2.016490\,\text{u}$, and after the reaction the mass is that of the ^2_1H, namely $2.014102\,\text{u}$. The mass difference is $0.002388\,\text{u}$, so the energy released is $(0.002388\,\text{u})(931.5\,\text{MeV/u}) = 2.22\,\text{MeV}$, and it is carried off by the ^2_1H nucleus and the γ ray.

FIGURE 31–10 Average binding energy per nucleon as a function of mass number A for stable nuclei. Same as Fig. 30–1.

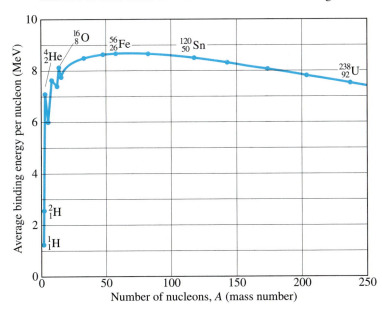

The energy output of our Sun is believed to be due principally to the following sequence of fusion reactions:

Fusion reactions

in the Sun

(proton–proton cycle)

$$^1_1\text{H} + ^1_1\text{H} \rightarrow ^2_1\text{H} + e^+ + \nu \qquad (0.42\,\text{MeV}) \quad \textbf{(31–3a)}$$

$$^1_1\text{H} + ^2_1\text{H} \rightarrow ^3_2\text{He} + \gamma \qquad (5.49\,\text{MeV}) \quad \textbf{(31–3b)}$$

$$^3_2\text{He} + ^3_2\text{He} \rightarrow ^4_2\text{He} + ^1_1\text{H} + ^1_1\text{H} \qquad (12.86\,\text{MeV}) \quad \textbf{(31–3c)}$$

where the energy released (Q-value) for each reaction is given in parentheses. The net effect of this sequence, which is called the **proton–proton cycle**, is for four protons to combine to form one ^4_2He nucleus plus two positrons, two neutrinos, and two gamma rays:

$$4\,^1_1\text{H} \rightarrow ^4_2\text{He} + 2e^+ + 2\nu + 2\gamma. \qquad \textbf{(31–4)}$$

Note that it takes two of each of the first two reactions (Eqs. 31–3a and b) to produce the two ^3_2He for the third reaction, so the total energy release for the net reaction, Eq. 31–4, is $(2 \times 0.42\,\text{MeV} + 2 \times 5.49\,\text{MeV} + 12.86\,\text{MeV}) = 24.7\,\text{MeV}$. However, each of the two e^+ (formed in reaction Eq. 31–3a) quickly annihilates with an electron to produce $2m_e c^2 = 1.02\,\text{MeV}$, so the total energy released is $(24.7\,\text{MeV} + 2 \times 1.02\,\text{MeV}) = 26.7\,\text{MeV}$. The first reaction, the formation of deuterium from two protons (Eq. 31–3a), has a very low probability, and the infrequency of that reaction serves to limit the rate at which the Sun produces energy.

In stars hotter than the Sun, it is more likely that the energy output comes principally from the **carbon cycle**, which comprises the following sequence of reactions:

Carbon

cycle

(some stars)

$$^{12}_6\text{C} + ^1_1\text{H} \rightarrow ^{13}_7\text{N} + \gamma$$

$$^{13}_7\text{N} \rightarrow ^{13}_6\text{C} + e^+ + \nu$$

$$^{13}_6\text{C} + ^1_1\text{H} \rightarrow ^{14}_7\text{N} + \gamma$$

$$^{14}_7\text{N} + ^1_1\text{H} \rightarrow ^{15}_8\text{O} + \gamma$$

$$^{15}_8\text{O} \rightarrow ^{15}_7\text{N} + e^+ + \nu$$

$$^{15}_7\text{N} + ^1_1\text{H} \rightarrow ^{12}_6\text{C} + ^4_2\text{He}.$$

(See Problem 34.) The theory of the proton–proton cycle and of the carbon cycle as the source of energy for the Sun and stars was first worked out by Hans Bethe (1906–   ) in 1939.

CONCEPTUAL EXAMPLE 31–6 **Stellar fusion.** What is the heaviest element likely to be produced in fusion processes in stars?

RESPONSE Fusion is possible as long as the final product has more binding energy (less mass) than the reactants, for then there is net release of energy. Since the binding energy curve in Fig. 31–10 (or Fig. 30–1) peaks near $A \approx 56$, which corresponds to iron, it would not be energetically favorable to produce elements heavier than that. Nevertheless, in the center of massive stars or in supernova explosions, there is enough energy available to drive endothermic reactions that produce heavier elements, as well.

The possibility of utilizing the energy released in fusion to make a power reactor is very attractive. The fusion reactions most likely to succeed in a reactor involve the isotopes of hydrogen, ^2_1H (deuterium) and ^3_1H (tritium), and are as follows, with the energy released given in parentheses:

Fusion reactor

$$^2_1\text{H} + {}^2_1\text{H} \rightarrow {}^3_1\text{H} + {}^1_1\text{H} \qquad\qquad (4.03\,\text{MeV}) \quad \textbf{(31–5a)}$$

Fusion reactions

$$^2_1\text{H} + {}^2_1\text{H} \rightarrow {}^3_2\text{He} + \text{n} \qquad\qquad (3.27\,\text{MeV}) \quad \textbf{(31–5b)}$$

for possible

$$^2_1\text{H} + {}^3_1\text{H} \rightarrow {}^4_2\text{He} + \text{n}. \qquad\qquad (17.59\,\text{MeV}) \quad \textbf{(31–5c)}$$

reactor

Comparing these energy yields with that for the fission of $^{235}_{92}\text{U}$, we can see that the energy released in fusion reactions can be greater for a given mass of fuel than in fission. Furthermore, as fuel, a fusion reactor could use deuterium, which is very plentiful in the water of the oceans (the natural abundance of ^2_1H is 0.015 percent, or about 1 g of deuterium per 60 L of water). The simple proton–proton reaction of Eq. 31–3a, which could use a much more plentiful source of fuel, ^1_1H, has such a small probability of occurring that it cannot be considered a possibility on Earth.

Although a successful fusion reactor has not yet been achieved, considerable progress has been made in overcoming the inherent difficulties. The problems are associated with the fact that all nuclei have a positive charge and thus repel each other. However, if they can be brought close enough together so that the short-range attractive nuclear force can come into play, the latter can pull the nuclei together and fusion will occur. Thus, in order for the nuclei to get close enough together, they must have rather large kinetic energy to overcome the electric repulsion. High kinetic energies are easily attainable with particle accelerators such as the cyclotron, but the number of particles involved is too small. To produce realistic amounts of energy, we must deal with matter in bulk, for which high KE means higher temperatures. Indeed, very high temperatures are required for fusion to occur, and fusion devices are often referred to as **thermonuclear devices**. The Sun and other stars are very hot, many millions of degrees, so the nuclei are moving fast enough for fusion to take place, and the energy released keeps the temperature high so that further fusion reactions can occur. The Sun and the stars represent huge self-sustaining thermonuclear reactors, but on Earth the high temperatures and densities required are not easily attained in a controlled manner.

It was realized after World War II that the temperature produced within a fission (or "atomic") bomb was close to $10^8\,\text{K}$. This suggested that a fission bomb could be used to ignite a fusion bomb (popularly known as a thermonuclear or hydrogen bomb) to release the vast energy of fusion. The uncontrollable release of fusion energy in an H-bomb was relatively easy to obtain. But to realize usable energy from fusion at a slow and controlled rate turned out to be difficult. The temperature required for a usable fusion reactor has been estimated to be in the range $T = 2$ to $4 \times 10^8\,\text{K}$.

EXAMPLE 31–7 **ESTIMATE** **Temperature needed for d–t fusion.** Estimate the temperature required for deuterium–tritium fusion (d–t) to occur.

SOLUTION We assume the nuclei approach head-on, each with kinetic energy KE, and that their "surfaces" must just "touch"—that is, the distance between them is equal to the sum of their radii. From Eq. 30–1, $r_d \approx 1.5$ fm and $r_t \approx 1.7$ fm. The electrostatic potential energy (Chapter 17) of the 2 particles at this distance must equal the total KE of the 2 particles when far apart:

$$2 \text{ KE} \approx k\frac{e^2}{(r_d + r_t)}$$

$$\approx \left(9.0 \times 10^9 \frac{\text{N·m}^2}{\text{C}^2}\right) \frac{(1.6 \times 10^{-19} \text{ C})^2}{(3.2 \times 10^{-15} \text{ m})(1.6 \times 10^{-19} \text{ J/eV})}$$

$$\approx 0.45 \text{ MeV}.$$

Thus, KE ≈ 0.22 MeV, and if we ask that the average KE be this high, then from Eq. 13–8, $\frac{3}{2}kT = \overline{\text{KE}}$, we have

$$T = \frac{2 \, \overline{\text{KE}}}{3 \, k} = \frac{2(0.22 \text{ MeV})(1.6 \times 10^{-13} \text{ J/MeV})}{3(1.38 \times 10^{-23} \text{ J/K})} \approx 2 \times 10^9 \text{ K}.$$

The required temperature is actually about an order of magnitude less, partly because it is not necessary that the *average* KE be 0.22 MeV—a small percentage with this much energy would be sufficient.

It is not only a high temperature that is required for a fusion reactor. There must also be a high density of nuclei to ensure a sufficiently high collision rate. A real difficulty with controlled fusion is to contain nuclei long enough and at a high enough density for sufficient reactions to occur that a usable amount of energy is obtained. At the temperatures needed for fusion, the atoms are ionized, and the resulting collection of nuclei and electrons is referred to as a **plasma**. Ordinary materials vaporize at a few *Plasma* thousand degrees at best, and hence cannot be used to contain a high-temperature plasma. Two major containment techniques are being investigated at present: *magnetic confinement* and *inertial confinement*.

In **magnetic confinement**, magnetic fields are used to try to contain *Magnetic* the hot plasma. One possibility is the "magnetic bottle" shown in *confinement* Fig. 31–11. The paths of the charged particles in the plasma are bent by the magnetic field, and where the lines are close together, the force on the particles is such that they are reflected back toward the center by this "magnetic mirror." Unfortunately, magnetic bottles develop "leaks" and the charged particles slip out before sufficient fusion takes place. A more complex device called the **tokamak**, first developed in the USSR, shows considerable promise. A tokamak (Fig. 31–12) is toroid-shaped and utilizes a combination of two magnetic fields: one directed along the axis of the toroid produced by current-carrying conductors ("toroidal" field); and a second produced by a current that passes through the plasma

FIGURE 31–11 "Magnetic bottle" used to confine a plasma.

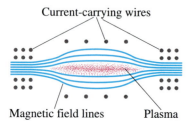

Current-carrying wires

Magnetic field lines   Plasma

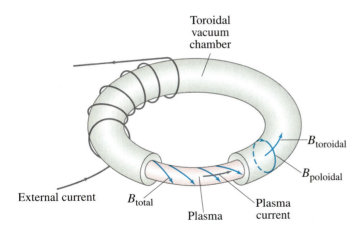

FIGURE 31–12 Tokamak configuration showing total **B** field due to external current plus current in the plasma.

("poloidal" field), which not only helps confinement, but heats the plasma as well. The combination of these two fields produces a helical field as shown in Fig. 31–12.

In 1957, J. D. Lawson showed that the product of ion density n and confinement time τ must satisfy, at a minimum, approximately

$$n\tau \gtrsim 3 \times 10^{20} \text{ s/m}^3.$$

Lawson criterion

This **Lawson criterion** must be reached to produce **ignition**, by which we mean a self-sustaining thermonuclear "burn" that continues after all external heating is turned off. To reach **break-even**, the point at which the energy output due to fusion is equal to the energy input to heat the plasma, requires an $n\tau$ about an order of magnitude less. The break-even point was very closely approached at the Tokamak Fusion Test Reactor (TFTR) at Princeton. The very high temperature needed for ignition (4×10^8 K) was exceeded, and the Lawson criterion closely approached—although not both of these at the same time. Nonetheless, the great progress made in recent years, after more than 40 years of research, gives hope that a working fusion reactor might be a reality by the early part of the next century.

Ignition

Break-even

The second method for containing the fuel for fusion is **inertial confinement**: a small pellet of deuterium and tritium is struck simultaneously from several directions by very intense laser beams. The intense influx of energy in such a laser fusion device heats and ionizes the pellet into a plasma. The outer layers evaporate, but the collisions they make with ions in the core of the pellet drive the latter inward. This implosion raises the density to about 10^3 times normal and further heats the core to temperatures at which fusion occurs. The confinement time is very short, on the order of

(a)

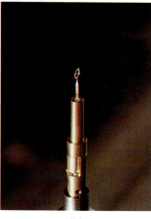

(b)

FIGURE 31–13 (a) Target chamber (5 m in diameter) of NOVA laser at Lawrence Livermore Laboratory, into which 10 laser beams converge on a target. (b) A 1-mm-diameter DT (deuterium–tritium) target, on its support, at the center of the target chamber.

10^{-11} to 10^{-9} s, during which time the ions don't move appreciably because of their own inertia. Very soon thereafter fusion takes place and the pellet explodes. The NOVA laser used for fusion research at the Lawrence Livermore Laboratory (Fig. 31–13) can deposit 10^5 J into the pellet over a 10^{-9} s pulse. This is a power input of 10^{14} W, more than the total electric-generator capacity of the United States! Of course, it is sustained only over an extremely short interval of time. A future reactor might implode 100 such pellets per second, thus requiring an input of 10^5 J $\times$ $100\,\text{s}^{-1} = 10^7$ W on the average. Inertial confinement depends on high particle density, n, over very short time intervals, τ. Recently, the Lawson criterion has been achieved, but at a temperature less than that needed for ignition.

Besides the problems of confinement, the building of a practical fusion reactor will require the development of materials used in its construction that can withstand high temperatures and high levels of radiation.

31–4 Passage of Radiation Through Matter; Radiation Damage

When we speak of *radiation,* we include α, β, γ, and X-rays, as well as protons, neutrons, and other particles such as pions (see Chapter 32). Because charged particles can ionize the atoms or molecules of any material they pass through, they are referred to as **ionizing radiation**. And because radiation produces ionization, it can cause considerable damage to materials, particularly to biological tissue.

Charged particles, such as α and β rays and protons, cause ionization because of the electric force. That is, when they pass through a material, they can attract or repel electrons strongly enough to remove them from the atoms of the material. Since the α and β rays emitted by radioactive substances have energies on the order of 1 MeV (10^4 to 10^7 eV), whereas ionization of atoms and molecules requires on the order of 10 eV, it is clear that a single α or β particle can cause thousands of ionizations.

Neutral particles also give rise to ionization when they pass through materials. For example, X-ray and γ-ray photons can ionize atoms by knocking out electrons by means of the photoelectric and Compton effects (Chapter 27). Furthermore, if a γ ray has sufficient energy (greater than 1.02 MeV), it can undergo pair production: an electron and a positron are produced (Section 27–4). The charged particles produced in all three of these processes can themselves go on to produce further ionization. Neutrons, on the other hand, interact with matter mainly by collisions with nuclei, with which they interact strongly. Often the nucleus is broken apart by such a collision, altering the molecule of which it was a part. And the fragments produced can in turn cause ionization.

Radiation passing through matter can do considerable damage. Metals and other structural materials become brittle and their strength can be weakened if the radiation is very intense, as in nuclear reactor power plants and for space vehicles that must pass through areas of intense cosmic radiation.

Biological damage

The radiation damage produced in biological organisms is due primarily to ionization produced in cells. Several related processes can occur. Ions or radicals are produced that are highly reactive and take part in chemical reactions that interfere with the normal operation of the cell. All forms of radiation can ionize atoms by knocking out electrons. If these are bonding electrons, the molecule may break apart, or its structure may be altered so that it does not perform its normal function or performs a harmful function. In the case of proteins, the loss of one molecule is not serious if there are other copies of that particular one in the cell, and additional ones can be made from its gene. However, large doses of radiation may damage so many molecules that new copies cannot be made quickly enough, and the cell dies. Damage to the DNA is more serious, since a cell may have only one copy. Each alteration in the DNA affects a gene and can alter the molecule it codes for, so that needed proteins or other materials may not be made at all. Again the cell may die. The death of a single cell is not normally a problem, since the body can replace it with a new one. (There are exceptions, such as neurons, which are *not* replaceable, so their loss is serious.) But if many cells die, the organism may not be able to recover. On the other hand, a cell may survive but be defective. It may go on dividing and produce many more defective cells, to the detriment of the whole organism. Thus radiation can cause cancer—the rapid production of defective cells.

Radiation damage to biological organisms is often separated into categories according to its location in the body: "somatic" and "genetic." *Somatic damage* refers to any part of the body except the reproductive organs. Somatic damage affects that particular organism, causing cancer and, at high doses, radiation sickness (characterized by nausea, fatigue, loss of body hair, and other symptoms) or even death. *Genetic damage* refers to damage to reproductive cells and so affects an individual's offspring. Damage to the genes results in mutations, the majority of which are harmful; and mutations, if they occur in the reproductive organs, are transmitted to future generations. Radiation, including that from diagnostic use of X-rays, is commonplace, and in the latter case, the possible damage done by the radiation must be balanced against the medical benefits and prolongation of life as a result of their diagnostic use.

31–5 Measurement of Radiation—Dosimetry

Although the passage of ionizing radiation through the human body can cause considerable damage, radiation can also be used to treat certain diseases, particularly cancer, often by using very narrow beams directed at the cancerous tumor to destroy it (Section 31–6). It is therefore important to be able to quantify the amount, or **dose**, of radiation. This is the subject of **dosimetry**.

The strength of a source can be specified at a given time by stating the **source activity**, or how many disintegrations occur per second. The traditional unit is the **curie** (Ci), defined as

Source activity

$$1 \text{ Ci} = 3.70 \times 10^{10} \text{ disintegrations per second.}$$

(This figure comes from the original definition as the activity of exactly 1 gram of radium.) Although the curie is still in common use, the proper SI unit for source activity is the **becquerel** (Bq), defined as

$$1 \text{ Bq} = 1 \text{ disintegration/s.}$$

Commercial suppliers of **radionuclides** (radioactive nuclides) specify the activity at a given time. Since the activity decreases over time, particularly for short-lived isotopes, it is important to take this into account.

The source activity ($\Delta N/\Delta t$) is related to the half-life, $T_{\frac{1}{2}}$, by (see Section 30–8):

$$\frac{\Delta N}{\Delta t} = \lambda N = \frac{0.693}{T_{\frac{1}{2}}} N.$$

EXAMPLE 31–8 **Radioactivity taken up by cells.** In a certain experiment, $0.016 \, \mu\text{Ci}$ of $^{32}_{15}\text{P}$ is injected into a medium containing a culture of bacteria. After 1 h, the cells are washed and a detector that is 70 percent efficient (counts 70 percent of emitted β rays) records 720 counts per minute from all the cells. What percentage of the original $^{32}_{15}\text{P}$ was taken up by the cells?

SOLUTION The total number of disintegrations per second originally was $(0.016 \times 10^{-6})(3.7 \times 10^{10}) = 590$. The counter could be expected to count 70 percent of this, or 410 per second. Since it counted $720/60 = 12$ per second, then $12/410 = 0.029$ or 2.9 percent was incorporated into the cells.

Absorbed dose

Another type of measurement is the exposure or **absorbed dose**—that is, the effect the radiation has on the absorbing material. The earliest unit of dosage was the **roentgen** (R), which was defined in terms of the amount of ionization produced by the radiation (1.6×10^{12} ion pairs per gram of dry air at standard conditions). Today, 1 R is defined as the amount of X or γ radiation that deposits 0.878×10^{-2} J of energy per kilogram of air. The roentgen was largely superseded by another unit of absorbed dose applic-

The rad

able to any type of radiation, the **rad**: *1 rad is that amount of radiation which deposits energy at a rate of 1.00×10^{-2} J/kg in any absorbing material.* (This is quite close to the roentgen for X- and γ rays.) The proper SI unit for absorbed dose is the **gray** (Gy):

$$1 \text{ Gy} = 1 \text{ J/kg} = 100 \text{ rad}, \tag{31–6}$$

and is slowly coming into use. The absorbed dose depends not only on the strength of a given radiation beam (number of particles per second) and the energy per particle, but also on the type of material that is absorbing the radiation. Since bone, for example, is denser than flesh and absorbs more of the radiation normally used, the same beam passing through a human body deposits a greater dose (in rads or grays) in bone than in flesh.

The gray and the rad are physical units of dose—the energy deposited per unit mass of material. They are, however, not the most meaningful units for measuring the biological damage produced by radiation. This is because equal doses of different types of radiation cause differing amounts of damage. For example, 1 rad of α radiation does 10 to 20 times the amount of damage as 1 rad of β or γ rays. This difference arises largely because α rays (and other heavy particles such as protons and neutrons) move much more slowly than β and γ rays of equal energy due to their greater mass. Hence, ionizing collisions occur closer together, so more irreparable damage can be done. The **relative biological effectiveness** (RBE) or **quality factor** (QF) of a given type of radiation is defined as the number of rads of X or γ radiation that produces the same biological damage as 1 rad of the given radiation. Table 31–1 gives the QF for several types of radiation. The numbers are approximate since they depend somewhat on the energy of the particles and on the type of damage that is used as the criterion.

RBE or QF

The **effective dose** can be given as the product of the dose in rads and the QF, and this unit is known as the **rem** (which stands for *rad equivalent man*):

Effective dose

$$\text{effective dose (in rem)} = \text{dose (in rad)} \times \text{QF.} \qquad \textbf{(31–7a)}$$

The rem

This unit is being replaced by the SI unit for "effective dose," the **sievert** (Sv):

$$\text{effective dose (Sv)} = \text{dose (Gy)} \times \text{QF.} \qquad \textbf{(31–7b)}$$

By this definition, 1 rem of any type of radiation does approximately the same amount of biological damage. For example, 50 rem of fast neutrons does the same damage as 50 rem of γ rays. But note that 50 rem of fast neutrons is only 5 rads, whereas 50 rem of γ rays is 50 rads.

We are constantly exposed to low-level radiation from natural sources: cosmic rays, natural radioactivity in rocks and soil, and naturally occurring radioactive isotopes in our food, such as $^{40}_{19}\text{K}$. Radon, $^{222}_{86}\text{Rn}$, is of considerable concern today. It is the product of radium decay and is an intermediate in the decay series from uranium (see Fig. 30–10). Most intermediates remain in the rocks where formed, but radon is a gas that can escape from rock (and from building material like concrete) to enter the atmosphere we breathe. Although radon is inert chemically (it is a noble gas), it is not inert physically—it decays by alpha emission, and its products, also radioactive, are *not* chemically inert and can attach to the interior of the lung.

Natural radioactivity

Radon

The natural radioactive background averages about 0.36 rem (360 mrem) per year per person, although there are large variations. From medical X-rays, the average person receives about 40 mrem per year. The U.S. government specifies the recommended upper limit of allowed radiation for an individual in the general populace at about 0.5 rem (500 mrem) per year, exclusive of natural sources. It is not known if low doses of radiation increase the chances of cancer or genetic defects, so the attitude today is to play safe and keep the radiation dose as low as possible.

People who work around radiation—in hospitals, in power plants, in research—often are subjected to much higher doses than 0.5 rem/yr. The upper limit for such occupational exposures has been set somewhat higher, on the order of 5 rem/yr whole-body dose (presumably because such people know what they are getting into). To protect them from excess exposure, those people who work around radiation generally carry some type of dosimeter, typically a **radiation film badge**, which is a piece of film wrapped in light-tight material. The passage of ionizing radiation through the film (see Section 30–13) changes it so that the film is darkened upon development, and so indicates the received dose.

Large doses of radiation can cause reddening of the skin, drop in white-blood-cell count, and a large number of unpleasant symptoms such as nausea, fatigue, and loss of body hair. Such effects are sometimes referred to as **radiation sickness**. Large doses can also be fatal, although the

Radiation sickness

TABLE 31–1
Quality Factor (QF) of Different Kinds of Radiation

Type	QF
X and γ rays	≈ 1
β (electrons)	≈ 1
Fast protons	1
Slow neutrons	≈ 3
Fast neutrons	Up to 10
α particles and heavy ions	Up to 20

time span of the dose is important. A short dose of 1000 rem is nearly always fatal. A 400-rem dose in a short period of time is fatal in 50 percent of the cases. However, the body possesses remarkable repair processes, so that a 400-rem dose spread over several weeks is not usually fatal. It will, nonetheless, cause considerable damage to the body.

CONCEPTUAL EXAMPLE 31–9 | **Limiting the dose.** A worker in an environment with a radioactive source is warned that she is accumulating a dose too quickly and will have to lower her exposure by a factor of ten to continue working for the rest of the year. If the worker is able to work farther away from the source, how much farther away is necessary?

RESPONSE If the energy is radiated uniformly in all directions, then the intensity (dose/area) should fall off proportionately to the distance squared, just as it does for sound and light. Thus, moving four times farther away lowers the exposure by a factor of sixteen, enough to do the trick.

EXAMPLE 31–10 **Whole-body dose.** What whole-body dose is received by a 70-kg laboratory worker exposed to a 40 mCi $^{60}_{27}$Co source, assuming the person's body has cross-sectional area 1.5 m^2 and is, on average, 4.0 m from the source for 4.0 h per day? $^{60}_{27}$Co emits γ rays of energy 1.33 MeV and 1.17 MeV in quick succession. Approximately 50 percent of the γ rays interact in the body and deposit all their energy. (The rest pass through.)

SOLUTION The total γ-ray energy per decay is (1.33 + 1.17) MeV = 2.50 MeV, so the total energy emitted by the source is

$$(0.040\ \text{Ci})(3.7 \times 10^{10}\ \text{decays/Ci·s})(2.50\ \text{MeV}) = 3.7 \times 10^{9}\ \text{MeV/s}.$$

The proportion of this intercepted by the body is its 1.5 m^2 area divided by the area of a sphere of radius 4.0 m (Fig. 31–14)

$$\frac{1.5\ \text{m}^2}{4\pi r^2} = \frac{1.5\ \text{m}^2}{4\pi(4.0\ \text{m})^2} = 7.5 \times 10^{-3}.$$

So the rate energy is deposited in the body (remembering that only $\frac{1}{2}$ of the γ rays interact in the body) is

$$E = (\tfrac{1}{2})(7.5 \times 10^{-3})(3.7 \times 10^{9}\ \text{MeV/s})(1.6 \times 10^{-13}\ \text{J/MeV})$$
$$= 2.2 \times 10^{-6}\ \text{J/s}.$$

Since 1 Gy = 1 J/kg, the whole-body dose rate for this 70-kg person is $(2.2 \times 10^{-6}\ \text{J/s})/(70\ \text{kg}) = 3.1 \times 10^{-8}\ \text{Gy/s}$. In the space of 4.0 h, this amounts to a dose of $(4.0\ \text{h})(3600\ \text{s/h})(3.1 \times 10^{-8}\ \text{Gy/s}) = 4.5 \times 10^{-4}\ \text{Gy}$. Since QF $\approx$ 1 for gammas, the effective dose (Eq. 31–7) is 450 μSv or (see Eq. 31–6):

$$(100\ \text{rad/Gy})(4.5 \times 10^{-4}\ \text{Gy})(1) = 45\ \text{mrem}.$$

This 45-mrem effective dose is nearly 10 percent of the normal allowed dose for a whole year (500 mrem/yr), or 1 percent of the yearly allowance for radiation workers. This worker should not receive such a dose every day and should seek ways to reduce it (shield the source, vary the work, work farther away, etc.).

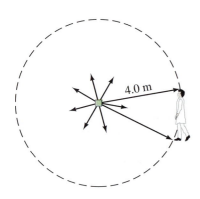

FIGURE 31–14 Radiation spreads out in all directions. A person 4.0 m away intercepts only a fraction: his cross-sectional area divided by the area of a sphere of radius 4.0 m.

* 31–6 Radiation Therapy

The applications of radioactivity and radiation to human beings and other organisms is a vast field that has filled many books. In the medical field there are two basic aspects: (1) **radiation therapy**—the treatment of disease (mainly cancer)—which we discuss in this Section; and (2) the *diagnosis* of disease, which we discuss in the following Sections of this chapter.

Radiation can cause cancer. It can also be used to treat it. Rapidly growing cancer cells are especially susceptible to destruction by radiation. Nonetheless, large doses are needed to kill the cancer cells, and some of the surrounding normal cells are inevitably killed as well. It is for this reason that cancer patients receiving radiation therapy often suffer side effects characteristic of radiation sickness. To minimize the destruction of normal cells, a narrow beam of γ or X-rays is often used when the cancerous tumor is well localized. The beam is directed at the tumor, and the source (or body) is rotated so that the beam passes through various parts of the body to keep the dose at any one place as low as possible—except at the tumor and its immediate surroundings, where the beam passes at all times (Fig. 31–15). The radiation may be from a radioactive source such as $^{60}_{27}\text{Co}$, or it may be from an X-ray machine that produces photons in the range 200 keV to 5 MeV. Protons, neutrons, electrons, and pions, which are produced in particle accelerators (Section 32–2), are also being used in cancer therapy.

In some cases, a tiny radioactive source may be inserted directly inside a tumor, which will eventually kill the majority of the cells. A similar technique is used to treat cancer of the thyroid with the radioactive isotope $^{131}_{53}\text{I}$. The thyroid gland tends to concentrate any iodine present in the bloodstream; so when $^{131}_{53}\text{I}$ is injected into the blood, it becomes concentrated in the thyroid, particularly in any area where abnormal growth is taking place. The intense radioactivity emitted can then destroy the defective cells.

Although radiation can increase the lifespan of many patients, it is not always completely effective. It may not be possible to kill all the diseased cancer cells, so a recurrence of the disease is possible. Many cases, especially when the cancerous cells are not well localized in one area, are difficult to treat at all without damaging the rest of the organism.

Another application of radiation is for sterilizing bandages, surgical equipment, and even packaged foods, since bacteria and viruses can be killed or deactivated by large doses of radiation.

* 31–7 Tracers and Imaging in Research and Medicine

Radioactive isotopes are commonly used in biological and medical research as **tracers**. A given compound is artificially synthesized using a radioactive isotope such as $^{14}_{6}\text{C}$ or $^{3}_{1}\text{H}$. Such "tagged" molecules can then be traced as they move through an organism or as they undergo chemical reaction. The presence of these tagged molecules (or parts of them, if they undergo chemical change) can be detected by a Geiger or scintillation counter, which detects emitted radiation (see Section 30–13). The details of how food molecules are digested, and to what parts of the body they are diverted, can be traced in this way. Radioactive tracers have been used to determine how amino acids and other essential compounds are synthe-

➡ **PHYSICS APPLIED**

Radiation therapy

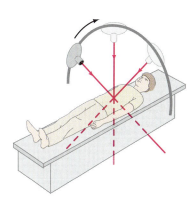

FIGURE 31–15 Radiation source rotates so that the beam always passes through the diseased tissue, but minimizing the dose in the rest of the body.

➡ **PHYSICS APPLIED**

Tracers in medicine and biology

FIGURE 31–16
(a) Autoradiograph of a mature leaf of the squash plant *Cucurbita melopepo* exposed for 30 s to $^{14}CO_2$. The photosynthetic (green) tissue has become radioactive; the nonphotosynthetic tissue of the veins is free of ^{14}C and therefore does not blacken the X-ray sheet. This technique is very useful in following patterns of nutrient transport in plants. (b) An autoradiograph of a fiber of chromosomal DNA isolated from the higher plant *Arabidopsis thaliana*. The dashed arrays of silver grains show the Y-shaped growing point of replicating DNA.

(a)

(b)

sized by organisms. The permeability of cell walls to various molecules and ions can be determined using radioactive isotopes: the tagged molecule or ion is injected into the extracellular fluid and the radioactivity present inside and outside the cells is measured as a function of time.

In many cases, the cells or chemicals involved are removed from the organism and placed in the Geiger or scintillation counter. To follow a process in time, different sets of organisms are treated for different periods of time before the cells or chemicals are extracted. (See Example 31–8.)

In a technique known as **autoradiography**, the position of the radioactive isotopes is detected on film. For example, the distribution of carbohydrates produced in the leaves of plants from absorbed CO_2 can be observed by keeping the plant in an atmosphere where the carbon atom in the CO_2 is $^{14}_{6}C$. After a time, a leaf is placed firmly on a photographic plate and the emitted radiation darkens the film most strongly where the isotope is most strongly concentrated (Fig. 31–16a). Autoradiography using labeled nucleotides (components of DNA) has revealed much about the details of DNA replication (Fig. 31–16b).

For medical diagnosis, the radionuclide commonly used today is $^{99m}_{43}Tc$, a long-lived excited state of technetium-99 (the "m" in the symbol stands for "metastable" state). It is formed when $^{99}_{42}Mo$ decays. The great usefulness of $^{99m}_{43}Tc$ derives from its convenient half-life of 6 h (short, but not too short) and the fact that it can combine with a large variety of compounds. The compound to be labeled with the radionuclide is so chosen because it concentrates in the organ or region of the anatomy to be studied. Detectors outside the body then record, or image, the distribution of the radioactively labeled compound. The detection can be done by a single detector (Fig. 31–17) which is moved across the body, measuring the intensity of radioactivity at a large number of points. The image represents the relative intensity of radioactivity at each point. The relative radioactivity is a diagnostic tool. For example, high or low radioactivity may represent overactivity or underactivity of an organ or part of an organ, or in another case may represent a lesion or tumor. More complex *gamma cameras* make use of many detectors which simultaneously record the radioactivity at many points. The measured intensities can be displayed on a CRT (TV monitor or oscilloscope screen), and allow "dynamic" studies (that is, images that change in time) to be performed.

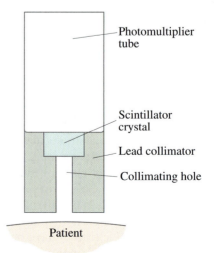

Photomultiplier tube

Scintillator crystal

Lead collimator

Collimating hole

Patient

FIGURE 31–17 Collimated gamma-ray detector for scanning (moving) over patient. The collimator is necessary to select γ rays that come in a straight line from the patient. Without the collimator, γ rays from all parts of the body could strike the scintillator, producing a very poor image.

31–8 Emission Tomography

The images formed using the standard techniques of nuclear medicine, as briefly discussed in the previous section, are produced from radioactive tracer sources within the *volume* of the body. It is also possible to image the radioactive emissions in a single plane or slice through the body using the computed tomography techniques discussed in Section 25–12. A basic gamma camera is moved around the patient to measure the radioactive intensity from the tracer at many points and angles; the data are processed in much the same way as for X-ray CT scans (Section 25–12). This technique is referred to as **single photon emission tomography** (SPET).[†]

Another important technique is **positron emission tomography** (PET), which makes use of positron emitters such as $^{11}_{6}C$, $^{13}_{7}N$, $^{15}_{8}O$, and $^{18}_{9}F$. These isotopes are incorporated into molecules that, when inhaled or injected, accumulate in the organ or region of the body to be studied. When such a nuclide β decays, the emitted positron travels at most a few millimeters before it collides with a normal electron. In this collision, the positron and electron are annihilated, producing two γ rays ($e^+ + e^- \rightarrow 2\gamma$), each having energy of 510 keV. The two γ rays fly off in opposite directions ($180° \pm 0.25°$) since they must have almost exactly equal and opposite momenta to conserve momentum (the momenta of the e^+ and e^- are essentially zero compared to the momenta of the γ rays). Because the photons travel along the same line in opposite directions, their detection in coincidence by rings of detectors surrounding the patient (Fig. 31–18) readily establishes the line along which the emission took place. If the difference in time of arrival of the two photons could be determined accurately, the actual position of the emitting nuclide along that line could be calculated. Present-day electronics can measure times to at best ± 300 ps, so at the γ ray's speed ($c = 3 \times 10^8$ m/s), the actual position could be determined to an accuracy on the order of about $vt \approx (3 \times 10^8 \text{ m/s}) \cdot (300 \times 10^{-2} \text{ s}) \approx 10$ cm, which is not very useful. Although there may be future potential for time-of-flight measurements to determine position, today computed tomography techniques are used instead, similar to those for X-ray CT, which can reconstruct PET images with a resolution on the order of 3–5 mm. One big advantage of PET is that no collimators are needed (as for detection of a single photon—see Fig. 31–17). Thus, fewer photons are "wasted" and lower doses can be administered to the patient with PET.

➡ **PHYSICS APPLIED**

Medical imaging:
PET and SPET

SPET

PET

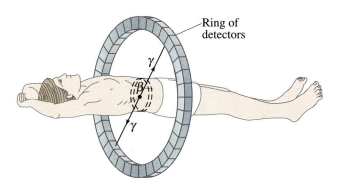

Ring of detectors

FIGURE 31–18 Positron emission tomography (PET) system showing a ring of detectors to detect the two annihilation γ rays ($e^+ + e^- \rightarrow 2\gamma$) emitted at 180° to each other. See also photo at start of this chapter.

[†]Also known as SPECT, "single photon emission computed tomography."

Both PET and SPET systems can give images related to biochemistry, metabolism, and function. This is to be compared to X-ray CT scans (Section 25–12), whose images reflect shape and structure—that is, the anatomy of the imaged region.

*31–9 Nuclear Magnetic Resonance (NMR) and Magnetic Resonance Imaging (MRI)

NMR **Nuclear magnetic resonance** (NMR) is a phenomenon that soon after its discovery in 1946 became a powerful research tool in a variety of fields from physics to chemistry and biochemistry. In recent years, it has been developed into an important and widely used medical imaging technique. We first briefly discuss the phenomenon, and then we will look at its applications.

We saw in Chapter 20 (Section 20–9) that a loop of wire carrying an electric current has a torque on it when placed in a magnetic field. The loop is said to have a magnetic moment, and the torque on the loop tends to align it so that the direction of its magnetic moment (taken to be perpendicular to the plane of the coil) is parallel to the field, **B**. (The coil's plane is perpendicular to **B**.) When the coil is so aligned, it has its lowest potential energy. Then in Chapter 28 (Section 28–6), we saw that atomic electrons in their orbital motions act as tiny current loops, and furthermore that they have an intrinsic angular momentum called *spin*, which also acts like the magnetic moment of a current loop. When atoms are placed in a magnetic field, atomic energy levels split into several closely spaced levels (see Fig. 28–8).

Nuclei, too, exhibit these magnetic properties. Many nuclides have magnetic moments, but we will examine only the simplest, the hydrogen (H) nucleus, since it is the one most used even for medical imaging. The $^1_1$H nucleus consists of a single proton. Its spin angular momentum (and its magnetic moment), like that of the electron, can take on only two values when placed in a magnetic field: we call these "spin up" (parallel to the field) and "spin down" (antiparallel to the field), as suggested in Fig. 31–19. When a magnetic field is present, the energy of the nucleus splits into two levels as shown in Fig. 31–20, with the spin up (parallel to field) having the lower energy. (This is like the Zeeman effect for atomic levels—see Fig. 28–8.) The difference in energy, ΔE, between these two levels is proportional to the total magnetic field, B_T, at the nucleus:

$$\Delta E = kB_T,$$

where k is a proportionality constant that is different for different nuclides.

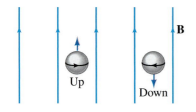

FIGURE 31–19 Schematic picture of a proton represented in a magnetic field **B** (pointing upward) with its two possible states of spin, up and down.

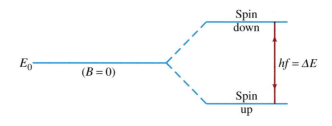

FIGURE 31–20 Energy E_0 in the absence of a magnetic field splits into two levels in the presence of a magnetic field.

In a standard nuclear magnetic resonance (NMR) setup, the sample to be examined is placed in a static magnetic field. A radiofrequency (RF) pulse of electromagnetic radiation (that is, photons) is applied to the sample. If the frequency, f, of this pulse corresponds precisely to the energy difference between the two energy levels (Fig. 31–20), so that

$$hf = \Delta E = kB_{\mathrm{T}}, \tag{31–8}$$

then the photons of the RF beam will be absorbed, exciting many of the nuclei from the lower state to the upper state. This is a resonance phenomenon since there is significant absorption only if f is very near $f = kB_{\mathrm{T}}/h$. Hence the name "nuclear magnetic resonance." For free $_1^1\mathrm{H}$ nuclei, the frequency is 42.58 MHz for a field $B_{\mathrm{T}} = 1.0\,\mathrm{T}$. If the H atoms are bound in a molecule, on the other hand, the total magnetic field B_{T} at the H nuclei will be the sum of the external applied field plus the magnetic field due to electrons and nuclei of neighboring atoms. Since f is proportional to B_{T}, the value of f will be slightly different for the bound H atoms than for free atoms in the same external field. This change in frequency, which can be measured, is called the "chemical shift." A great deal has been learned about the structure of molecules and bonds using such NMR measurements.

For producing medically useful NMR images—now commonly called MRI, or **magnetic resonance imaging**—the element most used is hydrogen since it is the commonest element in the human body and gives the strongest NMR signals. The experimental apparatus is shown in Fig. 31–21. The large coils set up the static magnetic field, and the RF coils produce the RF pulse of electromagnetic waves (photons) that cause the nuclei to jump from the lower state to the upper one (Fig. 31–20). These same coils (or another coil) can detect the absorption of energy or the emitted radiation (also of frequency $f = \Delta E/h$, Eq. 31–8) when the nuclei jump back down to the lower state.

The formation of a two-dimensional or three-dimensional image can be done using techniques similar to those for computed tomography (Section 25–12). The simplest thing to measure for creating an image is the intensity of absorbed and/or reemitted radiation from many different points of the

➡ **PHYSICS APPLIED**

NMR imaging (MRI)

FIGURE 31–21 Typical NMR imaging setup: (a) diagram; (b) photograph.

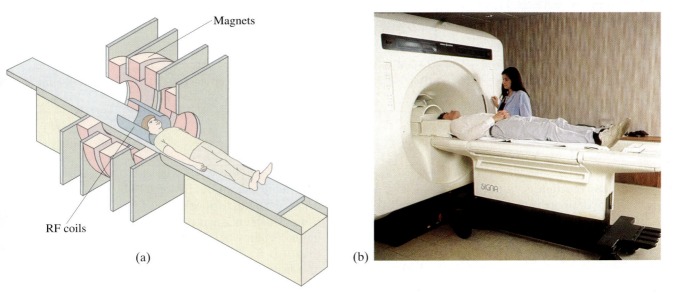

Magnets

RF coils

(a)   (b)

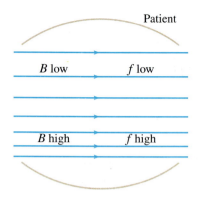

FIGURE 31–22 A static field that is stronger at the bottom than at the top. The frequency of absorbed or emitted radiation is proportional to B in NMR.

FIGURE 31–23 False-color NMR image (MRI) of a vertical section through the head showing structures in the normal brain.

body, and this would be a measure of the density of H atoms at each point. But how do we determine from what part of the body a given photon comes? One technique is to give the static magnetic field a gradient; that is, instead of applying a uniform magnetic field, B_T, the field is made to vary with position across the width of the sample (or patient). Since the frequency absorbed by the H nuclei is proportional to B_T (Eq. 31–8), only one plane within the body will have the proper value of B_T to absorb photons of a particular frequency f. By varying f, absorption by different planes can be measured. Alternately, if the field gradient is applied *after* the RF pulse, the frequency of the emitted photons will be a measure of where they were emitted. See Fig. 31–22. If a magnetic field gradient in one direction is applied during excitation (absorption of photons) and photons of a single frequency are transmitted, only H nuclei in one thin slice will be excited. By applying a gradient in a different direction, perpendicular to the first, during reemission, the frequency f of the reemitted radiation will represent depth in that slice. Other ways of varying the magnetic field throughout the volume of the body can be used in order to correlate NMR frequency with position.

A reconstructed image based on the density of H atoms (that is, the intensity of absorbed or emitted radiation) is not very interesting. More useful are images based on the rate at which the nuclei decay back to the ground state, and such images can produce resolution of 1 mm or better. This NMR technique (sometimes called **spin-echo**) is producing images of great diagnostic value, both in the delineation of structure (anatomy) and in the study of metabolic processes. An NMR image is shown in Fig. 31–23.

Table 31–2 lists the recently developed techniques we have discussed for imaging the interior of the body, along with the optimum resolution attainable today. Of course, resolution is only one factor that must be considered; it must be remembered that the different imaging techniques provide different types of information, useful for different types of diagnosis.

TABLE 31–2 Medical Imaging Techniques

Technique	Where Discussed In This Book	Resolution
Conventional X-ray	Section 25–12	$\frac{1}{2}$ mm
CT scan, X-ray	Section 25–12	$\frac{1}{2}$ mm
Nuclear medicine (tracers)	Section 31–7	1 cm
SPET (single photon emission)	Section 31–8	1 cm
PET (positron emission)	Section 31–8	3–5 mm
NMR	Section 31–9	$\frac{1}{2}$–1 mm
Ultrasound	Section 12–10	2 mm

SUMMARY

A **nuclear reaction** occurs when two nuclei collide and two or more other nuclei (or particles) are produced. In this process, as in radioactivity, **transmutation** (change) of elements occurs.

In **fission**, a heavy nucleus such as uranium splits into two intermediate-sized nuclei after being struck by a neutron. $^{235}_{92}U$ is fissionable by slow neutrons, whereas some fissionable nuclei require fast neutrons. Much energy is released in fission because the binding energy per nucleon is lower for heavy nuclei than it is for intermediate-sized nuclei, so the mass of a heavy nucleus is greater than the total mass of its fission products. The fission process releases neutrons, so that a **chain reaction** is possible. The **critical mass** is the minimum mass of fuel needed to sustain a chain reaction. In a **nuclear reactor** or nuclear bomb, a moderator is needed to slow down the released neutrons.

The **fusion** process, in which small nuclei combine to form larger ones, also releases energy. The energy from our Sun is believed to originate in the fusion reactions known as the **proton–proton cycle** in which four protons fuse to form a ^4_2He nucleus producing over 25 MeV. A working fusion reactor for power generation has not yet proved possible because of the difficulty in containing the fuel (e.g., deuterium) long enough at the high temperature required. Nonetheless, great progress has been made in confining the collection of charged ions known as a **plasma**. The two main methods are **magnetic confinement**, using a magnetic field in a device such as the toroidal-shaped **tokamak**, and **inertial confinement**, in which intense laser beams compress a fuel pellet of deuterium and tritium.

Radiation can cause damage to materials, including biological tissue. Quantifying amounts of radiation is the subject of **dosimetry**. The **curie** (Ci) and the **becquerel** (Bq) are units that measure the **source activity** or rate of decay of a sample: 1 Ci = 3.70×10^{10} disintegrations per second, whereas 1 Bq = 1 disintegration/s. The **absorbed dose**, often specified in **rads**, measures the amount of energy deposited per unit mass of absorbing material: 1 rad is the amount of radiation that deposits energy at the rate of 10^{-2} J/kg of material. The SI unit of absorbed dose is the **gray**: 1 Gy = 1 J/kg = 100 rad. The **effective dose** is often specified by the **rem** = rad × QF, where QF is the "quality factor" of a given type of radiation; 1 rem of any type of radiation does approximately the same amount of biological damage. The average dose received per person per year in the United States is about 0.36 rem. The SI unit for effective dose is the **sievert**: 1 Sv = 10^2 rem.

QUESTIONS

1. Fill in the missing particles or nuclei: (a) $^{137}_{56}\text{Ba}(\text{n}, \gamma)$?; (b) $^{137}_{56}\text{Ba}(\text{n}, ?)^{137}_{55}\text{Cs}$; (c) $^2_1\text{H}(\text{d}, ?)^4_2\text{He}$; (d) $^{197}_{79}\text{Au}(\alpha, \text{d})$? where d stands for deuterium.

2. The isotope $^{32}_{15}\text{P}$ is produced by an (n, p) reaction. What must be the target nucleus?

3. When $^{22}_{11}\text{Na}$ is bombarded by deuterons (^2_1H), an α particle is emitted. What is the resulting nuclide?

4. Why are neutrons such good projectiles for producing nuclear reactions?

5. A proton strikes a $^{20}_{10}\text{Ne}$ nucleus, and an α particle is observed to emerge. What is the residual nucleus? Write down the reaction equation.

6. Are fission fragments β^+ or β^- emitters?

7. If $^{235}_{92}\text{U}$ released only 1.5 neutrons per fission on the average, would a chain reaction be possible? If so, what would be different?

8. $^{235}_{92}\text{U}$ releases an average of 2.5 neutrons per fission compared to 2.7 for $^{239}_{94}\text{Pu}$. Pure samples of which of these two nuclei do you think would have the smaller critical mass?

9. The energy from nuclear fission appears in the form of thermal energy—but the thermal energy of what?

10. Why can't uranium be enriched by chemical means?

11. How can a neutron, with practically no KE, excite a nucleus to the extent shown in Fig. 31–2?

12. Why would a porous block of uranium be more likely to explode if kept under water rather than in air?

13. A reactor that uses highly enriched uranium can use ordinary water (instead of heavy water) as a moderator and still have a self-sustaining chain reaction. Explain.

14. Why must the fission process release neutrons if it is to be useful?

15. Discuss the relative merits and disadvantages, including pollution and safety, of power generation by fossil fuels, nuclear fission, and nuclear fusion.

16. What is the reason for the "secondary system" in a nuclear reactor, Fig. 31–6? That is, why is the water heated by the fuel in a nuclear reactor not used directly to drive the turbines?

17. Discuss how the course of history might have been changed if, during World War II, scientists had refused to work on developing a nuclear bomb. Do you think it would have been possible to delay the building of a bomb indefinitely?

18. Research in molecular biology is moving toward the ability to perform genetic manipulations on human beings. The moral implications of future discoveries along these lines has led to a warning that this may be the molecular biologists' "Hiroshima." Discuss.

19. Does $E = mc^2$ apply in (a) fission, (b) fusion, (c) nuclear reactions?

20. Light energy emitted by the Sun and stars comes from the fusion process. What conditions in the interior of stars makes this possible?

21. How do stars, and our Sun, maintain confinement of the plasma for fusion?

22. What is the basic difference between fission and fusion?

23. People who work around metals that emit alpha particles are trained that there is little danger from proximity or even touching the material, but that they must take extreme precautions against ingesting it. Hence,

there are strong rules against eating and drinking while working, and against machining the metal. Why?

24. Why is the recommended maximum radiation dose higher for women beyond the child-bearing age than for younger women?

* **25.** How might radioactive tracers be used to find a leak in a pipe?

▮ P R O B L E M S

S E C T I O N 31–1

1. (I) Natural aluminum is all $^{27}_{13}\text{Al}$. If it absorbs a neutron, what does it become? Does it decay by β^+ or β^-? What will be the product nucleus?

2. (I) Determine whether the reaction $^2_1\text{H}(\text{d}, \text{n})^3_2\text{He}$ requires a threshold energy. (d stands for deuterium, ^2_1H.)

3. (I) Is the reaction $^{238}_{92}\text{U}\,(\text{n}, \gamma)^{239}_{92}\text{U}$ possible with slow neutrons? Explain.

4. (II) Does the reaction $^7_3\text{Li}(\text{p}, \alpha)^4_2\text{He}$ require energy or does it release energy? How much energy?

5. (II) The reaction $^{18}_8\text{O}(\text{p}, \text{n})^{18}_9\text{F}$ requires an input of energy equal to 2.453 MeV. What is the mass of $^{18}_9\text{F}$?

6. (II) Calculate the energy released (or energy input required) for the reaction $^9_4\text{Be}(\alpha, \text{n})^{12}_6\text{C}$.

7. (II) (*a*) Can the reaction $^{24}_{12}\text{Mg}(\text{n}, \text{d})^{23}_{11}\text{Na}$ occur if the bombarding particles have 10.00 MeV of KE? (*b*) If so, how much energy is released?

8. (II) (*a*) Can the reaction $^7_3\text{Li}(\text{p}, \alpha)^4_2\text{He}$ occur if the incident proton has KE = 2500 keV? (*b*) If so, what is the total kinetic energy of the products?

9. (II) In the reaction $^{14}_7\text{N}(\alpha, \text{p})^{17}_8\text{O}$, the incident α particles have 7.68 MeV of kinetic energy. (*a*) Can this reaction occur? (*b*) If so, what is the total kinetic energy of the products? The mass of $^{17}_8\text{O}$ is 16.999131 u.

10. (II) Calculate the Q-value for the "capture" reaction $^{16}_8\text{O}(\alpha, \gamma)^{20}_{10}\text{Ne}$.

11. (II) Calculate the total KE of the products of the reaction $^{13}_6\text{C}(\text{d}, \text{n})^{14}_7\text{N}$ if the incoming deuteron (d) has KE = 36.3 MeV.

12. (II) An example of a "stripping" nuclear reaction is $^6_3\text{Li}(\text{d}, \text{p})\text{X}$. (*a*) What is X, the resulting nucleus? (*b*) Why is it called a "stripping" reaction? (*c*) What is the Q-value of this reaction? Is the reaction endothermic or exothermic?

13. (II) An example of a "pick-up" nuclear reaction is $^{12}_6\text{C}(^3_2\text{He}, \alpha)\text{X}$. (*a*) Why is it called a "pickup" reaction? (*b*) What is the resulting nucleus? (*c*) What is the Q-value of this reaction? Is the reaction endothermic or exothermic?

14. (II) (*a*) Complete the following nuclear reaction, $?(\text{p}, \gamma)^{32}_{16}\text{S}$. (*b*) What is the Q-value?

15. (III) Use conservation of energy and momentum to show that a bombarding proton must have an energy of 3.23 MeV in order to make the reaction $^{13}_6\text{C}(\text{p}, \text{n})^{13}_7\text{N}$ occur. (See Example 31–3.)

16. (III) Show, using the laws of conservation of energy and momentum, that for a nuclear reaction requiring energy, the minimum kinetic energy of the bombarding particle (the *threshold energy*) is equal to $[Qm_{pr}/(m_{pr} - m_b)]$, where $-Q$ is the energy required (difference in total mass between products and reactants), m_b is the rest mass of the bombarding particle, and m_{pr} the total rest mass of the products. Assume the target nucleus is at rest before an interaction takes place, and that all particles are nonrelativistic.

S E C T I O N 31–2

17. (I) Calculate the energy released in the fission reaction $\text{n} + {}^{235}_{92}\text{U} \rightarrow {}^{88}_{38}\text{Sr} + {}^{136}_{54}\text{Xe} + 12\text{n}$. Use Appendix F; assume the initial KE of the neutron is very small.

18. (I) What is the energy released in the fission reaction of Eq. 31–2b? (The masses of $^{141}_{56}\text{Ba}$ and $^{92}_{36}\text{Kr}$ are 140.91440 u and 91.92630 u, respectively.)

19. (I) How many fissions take place per second in a 200-MW reactor? Assume 200 MeV is released per fission.

20. (II) Suppose that the average power consumption, day and night, in a typical house is 300 W. What initial mass of $^{235}_{92}\text{U}$ would have to undergo fission to supply the electrical needs of such a house for a year? (Assume 200 MeV is released per fission.)

21. (II) What initial mass of $^{235}_{92}\text{U}$ is required to operate a 500-MW reactor for 1 yr? Assume 40% efficiency.

22. (II) If a 1.0-MeV neutron emitted in a fission reaction loses one-half of its KE in each collision with moderator nuclei, how many collisions must it make to reach thermal energy $(\tfrac{3}{2}kT = 0.040\text{ eV})$?

23. (III) Suppose that the neutron multiplication factor is 1.0004. If the average time between successive fissions in a chain of reactions is 1.0 ms, by what factor will the reaction rate increase in 1.0 s?

S E C T I O N 31–3

24. (I) What is the average kinetic energy of protons at the center of a star where the temperature is 10^7 K?

25. (II) Show that the energy released in the fusion reaction $^2_1\text{H} + {}^3_1\text{H} \rightarrow {}^4_2\text{He} + \text{n}$ is 17.59 MeV.

26. (II) Show that the energy released when two deuterium nuclei fuse to form $^3_2$He with the release of a neutron is 3.27 MeV.

27. (II) Verify the energy output stated for each of the reactions of Eq. 31–3. [*Hint*: Be careful with electrons.]

28. (II) Calculate the energy release per gram of fuel for the reactions of Eqs. 31–5a, b, and c. Compare to the energy release per gram of uranium in fission.

29. (II) If a typical house requires 300 W of electric power on average, how much deuterium fuel would have to be used in a year to supply these electrical needs? Assume the reaction of Eq. 31–5b.

30. (II) Show that the energies carried off by the $^4_2$He nucleus and the neutron for the reaction of Eq. 31–5c are about 3.5 MeV and 14 MeV, respectively. Are these fixed values, independent of the plasma temperature?

31. (II) Some stars, in a later stage of evolution, may begin to fuse two $^{12}_6$C nuclei into one $^{24}_{12}$Mg nucleus. (*a*) How much energy would be released in such a reaction? (*b*) What KE must two carbon nuclei each have when far apart, if they can then approach each other to within 6.0 fm, center-to-center? (*c*) Approximately what temperature would this require?

32. (III) Suppose a fusion reactor ran on "d–d" reactions, Eqs. 31–5a and b. Estimate how much water, for fuel, would be needed per hour to run a 1000-MW reactor, assuming 30 percent efficiency.

33. (III) How much energy (J) is contained in 1.00 kg of water if its natural deuterium is used in the fusion reaction of Eq. 31–5a? Compare to the energy obtained from the burning of 1.0 kg of gasoline, about 5×10^7 J.

34. (III) The energy output of massive stars is believed to be due to the *carbon cycle* (see text). (*a*) Show that no carbon is consumed in this cycle and that the net effect is the same as for the proton–proton cycle. (*b*) What is the total energy release? (*c*) Determine the energy output for each reaction and decay. (*d*) Why does the carbon cycle require a higher temperature ($\approx 2 \times 10^7$ K) than the proton–proton cycle ($\approx 1.5 \times 10^7$ K)?

35. (III) The deuterium–tritium pellet in a laser fusion device contains equal numbers of $^2_1$H and $^3_1$H atoms raised to a density of 200×10^3 kg/m^3 by the laser pulses. Estimate (*a*) the density of particles in this compressed state, and (*b*) the length of time τ the pellet must be confined to meet the Lawson criterion for ignition.

SECTION 31–5

36. (I) A dose of 500 rem of γ rays in a short period would be lethal to about half the people subjected to it. How many rads is this?

37. (I) Fifty rads of α-particle radiation is equivalent to how many rads of X-rays in terms of biological damage?

38. (I) How many rads of slow neutrons will do as much biological damage as 50 rads of fast neutrons?

39. (I) How much energy is deposited in the body of a 70-kg adult exposed to a 50-rad dose?

40. (II) A 0.018-μCi sample of $^{32}_{15}$P is injected into an animal for tracer studies. If a Geiger counter intercepts 20 percent of the emitted β particles, what will be the counting rate, assumed 90% efficient.

41. (II) A 1.0-mCi source of $^{32}_{15}$P (in NaHPO$_4$), a β^- emitter, is implanted in an organ where it is to administer 5000 rad. The half-life of $^{32}_{15}$P is 14.3 days and 1 mCi delivers about 1 rad/min. Approximately how long should the source remain implanted?

42. (II) About 35 eV is required to produce one ion pair in air. Show that this is consistent with the two definitions of the roentgen given in the text.

43. (II) $^{57}_{27}$Co emits 122-keV γ rays. If a 70-kg person swallowed 2.0 μCi of $^{57}_{27}$Co, what would be the dose rate (rad/day) averaged over the whole body? Assume that 50 percent of the γ-ray energy is deposited in the body. [*Hint*: Determine the rate of energy deposited in the body and use the definition of the rad.]

44. (II) What is the mass of a 1.00-μCi $^{14}_6$C source?

45. (II) Huge amounts of radioactive $^{131}_{53}$I were released in the accident at Chernobyl in 1986. Chemically, iodine goes to the human thyroid. (Doctors can use it for diagnosis and treatment of thyroid problems.) In a normal thyroid, $^{131}_{53}$I absorption can cause damage to the thyroid. (*a*) Write down the decay scheme for $^{131}_{53}$I. (*b*) Its half-life is 8.0 d; how long would it take for ingested $^{131}_{53}$I to become 10 percent of the initial value? (*c*) Absorbing 1 mCi of $^{131}_{53}$I can be harmful; what mass of iodine is this?

46. (II) Assume a liter of milk typically has an activity of 2000 pCi due to $^{40}_{19}$K. If a person drinks 2 glasses (0.5 L) per day, estimate the total dose (rem) received in a year. As a crude model, assume the milk stays in the stomach 12 hr and is then released. Assume also that very roughly 10 percent of the 1.5 MeV released per decay is absorbed by the body. Compare your result to the normal allowed dose per year. Make your estimate for (*a*) a 50-kg adult, and (*b*) a 5-kg baby.

47. (II) Radon gas, $^{222}_{86}$Rn, is considered a serious health hazard (see discussion in text). It decays by α-emission. (*a*) What is the daughter nucleus? (*b*) Is the daughter nucleus stable or radioactive? If the latter, how does it decay, and what is its lifetime? (*c*) Is the daughter nucleus also a noble gas, or is it chemically reacting? (*d*) Suppose 1.0 ng of $^{222}_{86}$Rn seeps into a basement. What will be its activity? If the basement is then sealed, what will be the activity 1 month later? [*Hint*: See the periodic table as well as Section 30–10 and Fig. 30–10.]

*SECTION 31–9

48. (II) Calculate the wavelength of photons needed to produce NMR transitions in free protons in a 1.000-T field. In what region of the spectrum does it lie?

49. J. Chadwick discovered the neutron by bombarding $^9_4$Be with the popular projectile of the day, alpha particles. (*a*) If one of the reaction products was the then unknown neutron, what was the other product? (*b*) What is the Q of this reaction?

50. Fusion temperatures are often given in keV. Determine the conversion factor from kelvins to keV using, as is usual in this field, $\text{KE} = kT$ without the factor $\frac{3}{2}$.

51. One means of enriching uranium is by diffusion of the gas UF_6. Calculate the ratio of the speeds of molecules of this gas containing $^{235}_{92}$U and $^{238}_{92}$U, on which this process depends.

52. (*a*) What mass of $^{235}_{92}$U was actually fissioned in the first atomic bomb, whose energy was the equivalent of about 20 kilotons of TNT (1 kiloton of TNT releases 5×10^{12} J)? (*b*) What was the actual mass transformed to energy?

53. In a certain town the average yearly background radiation consists of 25 mrad of X-rays and γ rays plus 3.0 mrad of particles having a QF of 10. How many rem will a person receive per year on the average?

54. Deuterium makes up 0.015 percent of natural hydrogen. Make a rough estimate of the total deuterium in the Earth's oceans and estimate the total energy released if all of it were used in fusion reactors.

55. A shielded γ-ray source yields a dose rate of 0.050 rad/h at a distance of 1.0 m for an average-sized person. If workers are allowed a maximum dose rate of 5.0 rem/yr, how close to the source may they operate, assuming a 40-h work week? Assume that the intensity of radiation falls off as the square of the distance. (It actually falls off more rapidly than $1/r^2$ because of absorption in the air, so the answer above will give a better-than-permissible value.)

56. Radon gas, $^{222}_{86}$Rn, is formed by α decay. (*a*) Write the decay equation. (*b*) Ignoring the KE of the daughter nucleus (it's so massive), estimate the KE of the α particle produced. (*c*) Estimate the momentum of the alpha and of the daughter nucleus. (*d*) Estimate the KE of the daughter, and show that your approximation in (*b*) was valid.

57. Consider a system of nuclear power plants that produce 4000 MW. (*a*) What total mass of $^{235}_{92}$U fuel would be required to operate these plants for 1 yr, assuming that 200 MeV is released per fission? (*b*) Typically 6 percent of the $^{235}_{92}$U nuclei that fission produce $^{90}_{38}$Sr, a β^- emitter with a half-life of 29 yr. What is the total radioactivity of the $^{90}_{38}$Sr, in curies, produced in 1 yr? (Neglect the fact that some of it decays during the 1-yr period.)

58. In the net reaction, Eq. 31–4, for the proton–proton cycle in the Sun, the neutrinos escape from the Sun with energy of about 0.5 MeV. The remaining energy, 26.2 MeV, is available within the Sun. Use this value to calculate the "heat of combustion" per kilogram of hydrogen fuel and compare it to the heat of combustion of coal, about 3×10^7 J/kg.

59. Energy reaches the Earth from the Sun at a rate of about 1400 W/m². Calculate (*a*) the total energy output of the Sun, and (*b*) the number of protons consumed per second in the reaction of Eq. 31–4, assuming that this is the source of all the Sun's energy. (*c*) Assuming that the Sun's mass of 2.0×10^{30} kg was originally all protons and that all could be involved in nuclear reactions in the Sun's core, how long would you expect the Sun to "glow" at its present rate?

60. Estimate how many solar neutrinos pass through a 100 m² room in a year. [*Hint:* See Problems 58 and 59.]

61. An average adult body contains about 0.10 μCi of $^{40}_{19}$K, which comes from food. (*a*) How many decays occur per second? (*b*) The potassium decays produce beta particles with energies of around 1.4 MeV. Calculate the dose per year in rem for a 50-kg adult. Is this a significant fraction of the 0.36 rem/year background rate?

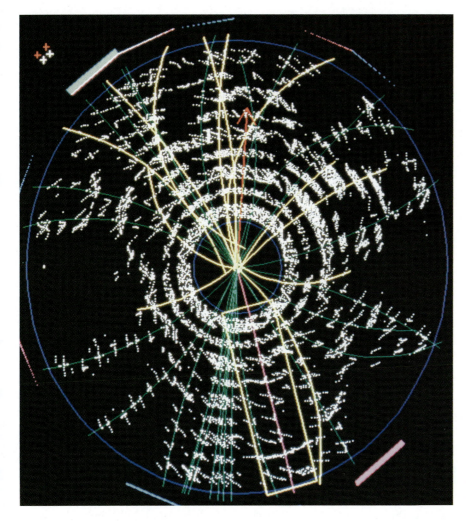

This computer-generated reconstruction of a proton-antiproton collision at Fermilab (Fig. 32–4) occurred at a combined energy of nearly 2 TeV. It is one of the events that provide evidence for the top quark, and that led to the announcement of its discovery in 1995. Because the detector is in a magnetic field, the particle tracks are curved; the radius of curvature is a measure of each particle's momentum (Chapter 20). The top quark (t) has too brief a lifetime ($\approx 10^{-23}$ s) to be detected itself, so experimenters look for its possible decay products. This photo is believed to show the following interaction and subsequent decays:

$$p + \bar{p} \longrightarrow t + \bar{t}$$

The tracks visible in the photo include the jets (each a group of particles moving in roughly the same direction), and a muon (μ^-) whose track is the pink one enclosed by a yellow rectangle to make it stand out. After reading this chapter, try to give the name for each symbol above and comment on whether all conservation laws hold.

ELEMENTARY PARTICLES 32

I n the final two chapters of this book we discuss two of the most excit-ing areas of contemporary physics: (1) elementary particles, and (2) cosmology and astrophysics. These are subjects at the forefront of knowledge—one treats the smallest objects in the universe, the other the largest (and oldest) aspects of the universe. Both of these chapters could be considered optional. Nonetheless, the reader who wants an understand-ing of the great beauties of present-day science—and/or wants to be a good citizen—will want to read these chapters, even if there is not time to cover them in a physics course.

In this penultimate chapter we discuss *elementary particle* physics, which represents the human endeavor to understand the basic building blocks of all matter. In the years after World War II, it was found that if the incoming particle in a nuclear reaction has sufficient energy, new types of particles can be produced. The earliest experiments used **cosmic rays**—particles that impinge on the Earth from space. But in order to produce high-energy particles in the laboratory, various types of particle accelerators have been constructed. Most commonly they accelerate protons or electrons, although heavy ions can also be accelerated, depending on the design. These high-energy accelerators have been used to probe the nucleus more deeply, to produce and study new particles, and to give us information about the basic forces and constituents of nature. Because the projectile particles are at high-energy, this field is sometimes called **high-energy physics**.

32–1 High-Energy Particles

Particles accelerated to high energy are projectiles which can probe the interior of nuclei and nucleons that they strike. An important factor is that faster-moving projectiles can reveal more detail. The wavelength of projectile particles is given by de Broglie's wavelength formula (Eq. 27–8),

de Broglie wavelength

$$\lambda = \frac{h}{mv},$$

showing that the greater the momentum of the bombarding particle, the shorter its wavelength. As discussed in Chapter 25 on optical instruments, resolution of details in images is limited by the wavelength: the shorter the wavelength, the finer the detail that can be obtained. This is one reason why particle accelerators of higher and higher energy have been built in recent years.

> **EXAMPLE 32–1** **High resolution with electrons.** What is the wavelength, and hence the expected resolution, for a beam of 1.3-GeV electrons?
>
> **SOLUTION** 1.3 GeV = 1300 MeV, which is about 2500 times the mass of the electron ($0.51 \text{ MeV}/c^2$). We are clearly dealing with relativistic speeds here, and it is easily shown that the speed of the electron is nearly $c = 3.0 \times 10^8$ m/s. [From Eq. 26–5, KE $= mc^2 - m_0c^2 \approx mc^2$; and from Eq. 26–8, $E^2 = p^2c^2 + m_0^2c^4 \approx p^2c^2$ since m_0c^2 is small compared to pc. Hence, $p^2c^2 = m^2v^2c^2 \approx m^2c^4$, so $v \approx c$.] Therefore
>
> $$\lambda = \frac{h}{mv} \approx \frac{h}{mc} = \frac{hc}{mc^2}$$
>
> where $mc^2 = 1.3$ GeV. Hence
>
> $$\lambda = \frac{(6.6 \times 10^{-34}\,\text{J}\cdot\text{s})(3.0 \times 10^8\,\text{m/s})}{(1.3 \times 10^9\,\text{eV})(1.6 \times 10^{-19}\,\text{J/eV})} = 0.96 \times 10^{-15}\,\text{m},$$
>
> or 0.96 fm. The maximum possible resolution of this beam of electrons is far greater than for a light beam in a light microscope ($\lambda \approx 500$ nm). Indeed, this resolution of about 1 fm is on the order of the size of nuclei (see Eq. 30–1).

Another major reason for building high-energy accelerators is that new particles of greater mass can be produced at higher energies, as we will discuss shortly. Now we look briefly at several types of particle accelerator.

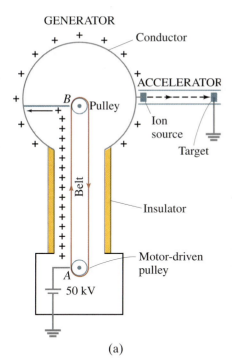

FIGURE 32–1 Van de Graaff generator and accelerator: (a) diagram; (b) in use at Brookhaven National Laboratory.

32–2 Particle Accelerators

Van de Graaff accelerator. The key component of a Van de Graaff accelerator (Fig. 32–1) is a Van de Graaff generator, invented in 1931. A large, hollow, spherical conductor is supported by an insulating column above the base. The sphere is charged to a high potential by a nonconducting moving belt in the following way. A high voltage of typically 50,000 V is applied to a pointed conductor A, which "sprays" positive charge onto the moving belt. (Actually, electrons are pulled off the belt onto the electrode A.) The belt carries the positive charge into the interior of the sphere, where it is "wiped off" the belt at B and races to the outer surface of the spherical conductor. (Remember that charge collects on the outer surface of any conductor, since the charges repel each other and try to get as far from each other as possible.) As more and more charge is brought upward, the sphere becomes more highly charged and reaches greater voltage. The process requires energy, of course, since the upward-moving charged belt is repelled by the charged sphere. The energy is supplied by the motor driving the belt. Very high potential differences can be obtained in this way, often giving the accelerated particles as much as 30 MeV of KE when two or more generators are used in tandem. Connected to the Van de Graaff generator is an evacuated tube which serves as the particle accelerator. A source of H or He ions (p or α) is located inside the tube, and the large positive voltage repels them so they are accelerated toward the grounded target at the far end of the tube (Fig. 32–1).

Cyclotron. The cyclotron was developed in 1930 by E. O. Lawrence (1901–1958; Fig. 32–2) at the University of California, Berkeley. It uses a magnetic field to maintain charged ions—usually protons—in nearly circular paths (Chapter 20). The protons move within two D-shaped cavities, as

Van de Graaff generator and accelerator

FIGURE 32–2 Ernest O. Lawrence holding the first cyclotron, around 1930.

Cyclotron

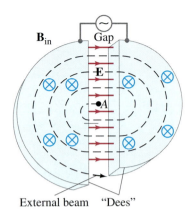

B_{in} Gap

E

•A

External beam   "Dees"

FIGURE 32–3 Diagram of a cyclotron. The magnetic field, applied by a large electromagnet, points into the page. A is the ion source. The field lines shown are for the electric field in the gap.

shown in Fig. 32–3. Each time they pass into the gap between the "dees," a voltage is applied that accelerates them via the electric force. This increases their speed and also increases the radius of curvature of their path. After many revolutions, the protons acquire high kinetic energy and reach the outer edge of the cyclotron. They then either strike a target placed inside the cyclotron or leave the cyclotron with the help of a carefully placed "bending magnet" and are directed to an external target. Acceleration only occurs when the protons are in the gap *between* the dees. (The magnetic field merely keeps the protons revolving so they can be repeatedly accelerated when they reach the gap.) The voltage applied to the dees to produce the acceleration must be alternating. When the protons are moving to the right across the gap in Fig. 32–3, the right dee must be electrically negative and the left one positive. A half-cycle later, the protons are moving to the left, so the left dee must be negative in order to accelerate them. The frequency, f, of the applied voltage must be equal to that of the circulating protons, and this can be determined as follows. When ions of charge q (for a proton, $q = +e$) are circulating within the dees, the net force F on each is simply that due to the magnetic field B, so $F = qvB$, where v is the speed of the ion at a given moment (Eq. 20–4). Since the ions move in circles, the acceleration is centripetal and equals v^2/r, where r is the radius of the ion's path at a given moment. We use Newton's second law, $F = ma$, and find that

$$F = ma$$

$$qvB = \frac{mv^2}{r}$$

or

$$v = \frac{qBr}{m}.$$

The time required for a complete revolution is the period T and is equal to

$$T = \frac{\text{distance}}{\text{speed}} = \frac{2\pi r}{qBr/m} = \frac{2\pi m}{qB}.$$

Hence the frequency of revolution f is

Cyclotron frequency

$$f = \frac{1}{T} = \frac{qB}{2\pi m}. \qquad (32\text{–}1)$$

This is known as the **cyclotron frequency**.

EXAMPLE 32–2 **Cyclotron.** A small cyclotron of maximum radius $R = 0.25\,\text{m}$ accelerates protons in a 1.7-T magnetic field. Calculate (*a*) what frequency is needed for the applied alternating voltage, and (*b*) the kinetic energy of protons when they leave the cyclotron.

SOLUTION (*a*) From Eq. 32–1,

$$f = \frac{qB}{2\pi m}$$

$$= \frac{(1.6 \times 10^{-19}\,\text{C})(1.7\,\text{T})}{(6.28)(1.67 \times 10^{-27}\,\text{kg})} = 2.6 \times 10^7\,\text{Hz} = 26\,\text{MHz},$$

which is in the radio wave region of the EM spectrum (Fig. 22–10).

(b) The protons leave the cyclotron at $r = R = 0.25$ m. Then, since $v = qBr/m$,

$$\text{KE} = \frac{1}{2}mv^2 = \frac{1}{2}m\frac{q^2B^2R^2}{m^2} = \frac{q^2B^2R^2}{2m}$$

$$= \frac{(1.6 \times 10^{-19}\,\text{C})^2(1.7\,\text{T})^2(0.25\,\text{m})^2}{(2)(1.67 \times 10^{-27}\,\text{kg})} = 1.4 \times 10^{-12}\,\text{J} = 8.7\,\text{MeV}.$$

Note that the magnitude of the voltage applied to the dees does not affect the final energy. But the higher this voltage, the fewer revolutions are required to bring the protons to full energy.

An interesting aspect of the cyclotron is that the frequency of the applied voltage, as given by Eq. 32–1, does not depend on the radius r. That is, the frequency does not have to be changed as the ions start from the source and are accelerated to paths of larger and larger radii. Unfortunately, this is only true at nonrelativistic energies. For at higher speeds, the mass of the ions will increase according to Einstein's formula, $m = m_0/\sqrt{1 - v^2/c^2}$, where m_0 is the rest mass. As can be seen from Eq. 32–1, as the mass increases, the frequency of the applied voltage must be reduced. To achieve large energies, complex electronics is needed to decrease the frequency as a packet of protons increases in speed and reaches larger orbits. Such a modified cyclotron is called a **synchrocyclotron**.

Synchrocyclotron

Synchrotron. Another way to deal with the relativistic increase in mass with speed is to increase the magnetic field B as the particles speed up. Such devices are called *synchrotrons*, and today they can be enormous. The Fermi National Accelerator Laboratory (Fermilab) at Batavia, Illinois, has a radius of 1.0 km, and that at CERN (European Center for Nuclear Research) in Geneva, Switzerland, is 1.1 km in radius. These machines can accelerate protons to energies of 500 GeV. The new *Tevatron* at Fermilab uses superconducting magnets to accelerate protons to about 1000 GeV = 1 TeV (hence its name; 1 TeV = 10^{12} eV). These large synchrotrons do not use enormous magnets 1 km in radius. Instead, a narrow ring of magnets is used (see Fig. 32–4) with each magnet placed at the same radius from the center of the circle. The magnets are interrupted by gaps where high voltage accelerates the particles. Thus, once the particles are injected, they must move in a

Synchrotron

FIGURE 32–4 (a) Aerial view of Fermilab at Batavia, Illinois; the accelerator is a circular ring 1.0 km in radius. (b) The interior of the tunnel of the main accelerator at Fermilab. The upper (rectangular-shaped) ring of magnets is for the 500-GeV accelerator. Below it is the ring of superconducting magnets for the Tevatron.

(a)

(b)

circle of constant radius. This is accomplished by giving them considerable energy initially in a much smaller accelerator, and then slowly increasing the magnetic field as they speed up in the large synchrotron.

One problem of any accelerator is that accelerating electric charges radiate electromagnetic energy (see Chapter 22). Since ions or electrons are accelerated in an accelerator, we can expect considerable energy to be lost by radiation. The effect increases with speed and is especially important in circular machines, where centripetal acceleration is present, particularly in synchrotrons, and hence is called **synchrotron radiation**. Synchrotron radiation can actually be useful, however. Intense beams of photons are sometimes needed, and they are usually obtained from an electron synchrotron.

Linac **Linear accelerators.** A Van de Graaff accelerator is essentially a linear accelerator since the ions move in a linear path. But the name *linear accelerator* is usually reserved for a more complex arrangement in which particles are accelerated many, many times along a straight-line path. Figure 32–5 is a diagram of a simple "linac." The ions pass through a series of tubular conductors. The voltage applied to the tubes must be alternating so that when positive ions (say) reach a gap, the tube in front of them is negative and the one they just left is positive. This assures that they are accelerated at each gap. As the ions increase in speed, they cover more distance in the same amount of time. Consequently, the tubes must be longer the farther they are from the source. Linear accelerators are of particular importance for accelerating electrons. Because of their small mass, electrons reach high speeds very quickly; an electron linac such as the one shown in Fig. 32–5a would have tubes nearly equal in length, since the electrons would be traveling close to $c = 3.0 \times 10^8$ m/s for almost the entire distance. The amount of energy radiated away by electrons in a linear machine is much less than for a circular machine (see mention of synchrotron radiation above). The largest electron linear accelerator is that at Stanford (Stanford Linear Accelerator Center, or SLAC), Fig. 32–5b. It is over 3 km (2 miles) long and can accelerate electrons to 50 GeV (recall 1 GeV = 10^9 eV). Many hospitals have 10-MeV electron linacs that produce photons to irradiate tumors.

Colliders **Colliding beams.** The typical way in which high-energy-physics experiments are carried out is to allow the beam of particles from an accelerator to strike a stationary target. An alternative method, which increases the energy of a collision, is to use **colliding beams**—that is, the target particles as well as the projectile particles are moving. The two

FIGURE 32–5 (a) Diagram of a simple linear accelerator. (b) Photo of the Stanford Linear Accelerator (SLAC) in California.

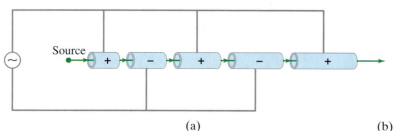

(a)

(b)

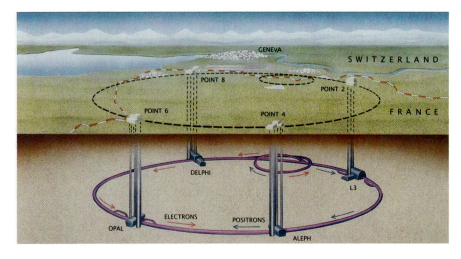

FIGURE 32–6 The 8.5 km diameter LEP collider, about 100 m below ground, straddles the French–Swiss border near Geneva.

beams of particles are made to collide head-on. One way to accomplish this with only one accelerator is through the use of **storage rings**. The accelerator accelerates one type of particle (say electrons or protons) to a maximum energy and then magnets are used to steer these particles into a circular storage ring where the particles can continue to circulate for many hours. It then accelerates a second type of particle (say positrons or antiprotons), or another group of the first type (such as protons). The two beams are directed so they intersect and collide head-on with each other. Colliding-beam experiments are in use at a number of facilities around the world, including CERN, Fermilab, and the Stanford Linear Collider (SLC), and they have played an important role in recent advances in elementary particle physics. For example, in the recent experiments that provided strong evidence for the top quark (see chapter opening photo and Section 32–9), the Fermilab Tevatron accelerated protons and antiprotons each to 900 GeV, so that the combined energy of head-on collisions was 1.8 TeV. The largest collider accelerator today is the truly enormous Large Electron-Positron (LEP) Collider at CERN, which has a circumference of 26.7 km (Fig. 32–6). Inaugurated in 1989, it produces oppositely revolving beams of e^+ and e^-, each of energy 93 GeV, for a total interaction energy of 186 GeV.

32–3 Beginnings of Elementary Particle Physics—the Yukawa Particle

By the mid-1930s, it was recognized that all atoms can be considered to be made up of neutrons, protons, and electrons. The basic constituents of the universe were no longer considered to be atoms but rather the proton, neutron, and electron. Besides these three *elementary particles*, several others were also known: the positron (a positive electron), the neutrino, and the γ particle (or photon), for a total of six elementary particles.

In the decades that followed, hundreds of other subnuclear particles were discovered. The properties and interactions of these particles, and which ones should be considered as fundamental or "elementary," became the substance of research in **elementary particle physics**. Today, many of these particles are considered to be made up of two or three more fundamental entities called *quarks*. But some of the early-discovered particles are also considered elementary, as we shall see.

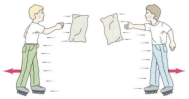

(a) Repulsive force (children throwing pillows)

(b) Attractive force (children grabbing pillows from each other's hands)

FIGURE 32–7 Forces equivalent to particle exchange. (a) Repulsive force (children throwing pillows at each other). (b) Attractive force (children grabbing pillows from each other's hands).

Feynman diagram

QED

Particles that mediate or "carry" forces

Elementary particle physics, as it exists today, can be said to have begun in 1935 when the Japanese physicist Hideki Yukawa (1907–1981) predicted the existence of a new particle that would in some way mediate the strong nuclear force. To understand Yukawa's idea, we first look at the electromagnetic force. When we first discussed electricity, we saw that the electric force acts over a distance, without contact. To better perceive how a force can act over a distance, we saw that Faraday introduced the idea of a *field*. The force that one charged particle exerts on a second can be said to be due to the electric field set up by the first. Similarly, the magnetic field can be said to carry the magnetic force. Later (Chapter 22), we saw that electromagnetic fields can travel through space as waves. Finally, in Chapter 27, we saw that electromagnetic radiation (light) can be considered as either a wave or as a collection of particles called *photons*. Because of this wave–particle duality, it is possible to imagine that the electromagnetic force between charged particles is due (1) to the EM field set up by one charged particle and felt by the other, or (2) to an exchange of photons or γ particles between them. It is (2) that we want to concentrate on here, and a crude analogy for how an exchange of particles could give rise to a force is suggested in Fig. 32–7. In part (a), two children start throwing heavy pillows at each other; each catch results in the child being pushed backward by the impulse. This is the equivalent of a repulsive force. On the other hand, if the two children exchange pillows by grabbing them out of the other person's hand, they will be pulled toward each other, as when an attractive force acts.

For the electromagnetic force between two charged particles, it is photons that are exchanged between the two particles that give rise to the force. A simple diagram describing this photon exchange is shown in Fig. 32–8. Such a diagram, called a **Feynman diagram** (after its inventor, the American physicist, Richard Feynman (1918–1988)), is based on the theory called **quantum electrodynamics** (QED). Figure 32–8 represents the simplest case, in which a single photon is exchanged. One of the charged particles emits the photon and recoils somewhat as a result; and the second particle absorbs the photon. In any collision or *interaction*, energy and momentum are transferred from one particle to the other, carried by the photon. Because the photon is absorbed by the second particle very shortly after it is emitted by the first, it is not observable, and is referred to as a *virtual* photon, in contrast to one that is free and can be detected by instruments. The photon is said to *mediate*, or *carry*, the electromagnetic force.

Now to Yukawa's prediction. By analogy with photon exchange that mediates the electromagnetic force, Yukawa argued that there ought to be a particle that mediates the strong nuclear force—the force that holds nucleons together in the nucleus. Just as the photon is called the quantum of the electromagnetic field or force, so the Yukawa particle would represent the quantum of the strong nuclear force.

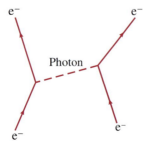

FIGURE 32–8 Feynman diagram showing a photon acting as the carrier of the electromagnetic force between two electrons. This is sort of an *x* vs. *t* graph, with *t* increasing upward. Starting at the bottom, two electrons approach each other (the distance between them decreases in time). As they get close, momentum and energy get transferred from one to the other, carried by a photon (or, perhaps, by more than one), and the two electrons bounce apart.

Yukawa predicted that this new particle would have a mass intermediate between that of the electron and the proton. Hence it was called a **meson**, meaning "in the middle," and Fig. 32–9 is a Feynman diagram of meson exchange. We can make a rough estimate of the mass of the meson as follows. Suppose the proton on the left in Fig. 32–9 is at rest. For it to emit a meson would require energy (to make the mass) that would have to come from nowhere; such a process would violate conservation of energy. But the uncertainty principle allows nonconservation of energy by an amount ΔE if it occurs only for a time Δt given by $(\Delta E)(\Delta t) \approx h/2\pi$. We set ΔE equal to the energy needed to create the mass m of the meson: $\Delta E = mc^2$. Now conservation of energy is violated only as long as the meson exists, which is the time Δt required for the meson to pass from one nucleon to the other. If we assume the meson travels at relativistic speed, close to the speed of light c, then Δt need be at most about $\Delta t = d/c$, where d is the maximum distance that can separate the interacting nucleons. Thus we have

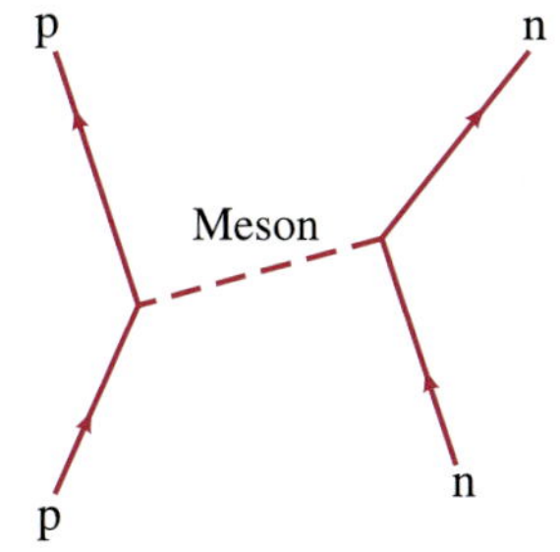

FIGURE 32–9 Meson exchange when a proton and neutron interact via the strong nuclear force.

$$\Delta E \, \Delta t \approx \frac{h}{2\pi}$$

$$mc^2\left(\frac{d}{c}\right) \approx \frac{h}{2\pi}$$

or

$$mc^2 \approx \frac{hc}{2\pi d}. \tag{32–2}$$

The range of the strong nuclear force (the maximum distance away it can be felt), is small—not much more than the size of a nucleon or small nucleus (see Eq. 30–1); so let us take $d \approx 1.5 \times 10^{-15}$ m, and then from Eq. 32–2,

$$mc^2 \approx \frac{hc}{2\pi d} = \frac{(6.6 \times 10^{-34}\,\text{J·s})(3.0 \times 10^{8}\,\text{m/s})}{(6.28)(1.5 \times 10^{-15}\,\text{m})}$$

$$\approx 2.2 \times 10^{-11}\,\text{J} = 130\,\text{MeV}.$$

The mass of the predicted meson is thus very roughly $130\,\text{MeV}/c^2$ or about 250 times the electron mass of $0.51\,\text{MeV}/c^2$. [Note, incidentally, that since the electromagnetic force has infinite range ($d = \infty$), Eq. 32–2 tells us that the exchanged particle for the electromagnetic force, the photon, will have zero rest mass.]

Just as photons can be observed as free particles, as well as acting in an exchange, so it was expected that mesons might be observed directly. Such a meson was searched for in the cosmic radiation that enters the Earth's atmosphere from the Sun and other sources in the universe. In 1937 a new particle was discovered whose mass was $106\,\text{MeV}/c^2$ (207 times the electron mass). This is close to the mass predicted. But it turned out that this new particle, called the **muon**, did not interact strongly with matter. It could hardly mediate the strong nuclear force if it didn't interact via the strong nuclear force. Thus the muon, which can have either a positive or a negative charge and seems to be nothing more than a very massive electron, is not the Yukawa particle.

Muon

The particle predicted by Yukawa was finally discovered in cosmic rays by C. F. Powell and G. Occhialini in 1947. It is called the "π" or pi meson, or simply the **pion**. It comes in three charge states: $+$, $-$, or 0. The π^+ and π^- have mass of $139.6\,\text{MeV}/c^2$ and the π^0 a mass of $135.0\,\text{MeV}/c^2$. All three

Pion

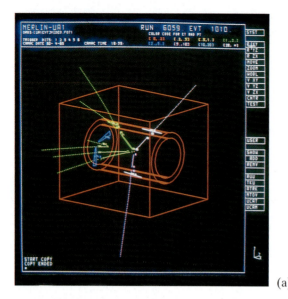
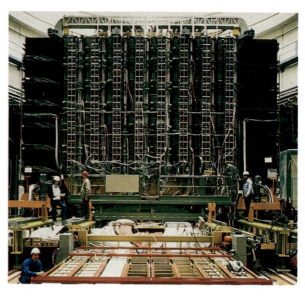

(a)

(b)

FIGURE 32–10
(a) Computer reconstruction of a Z-particle decay into an electron and a positron ($Z^0 \rightarrow e^+e^-$) whose tracks are shown in white, which took place in the UA1 detector at CERN. (b) Photo of the UA1 detector at CERN as it was being built.

interact strongly with matter. Reactions observed in the laboratory, using a particle accelerator, include

$$p + p \rightarrow p + p + \pi^0,$$
$$p + p \rightarrow p + n + \pi^+. \tag{32–3}$$

The incident proton from the accelerator must have sufficient energy to produce the additional mass of the pion. A number of other mesons were discovered in subsequent years which were also considered to mediate the strong nuclear force. The recent theory of quantum chromodynamics, however, which involves quarks, has replaced mesons with *gluons* as the basic carriers of the strong force, as we shall discuss in Section 32–10.

So far we have discussed the particles that mediate the electromagnetic and strong nuclear forces. But there are four known types of force—or interaction—in nature. What about the other two: the weak nuclear force, and gravity? Theorists believe that these are also mediated by particles. The particles presumed to transmit the weak force are referred to as the W^+, W^-, and Z^0, and were only detected in 1983 (see Fig. 32–10). The quantum (or carrier) of the gravitational force, called the **graviton**, has not yet been found. A comparison of the four forces is given in Table 32–1, where they are listed according to their (approximate) relative strengths. Notice that although gravity may be the most obvious force in daily life (because of the huge mass of the Earth), on a nuclear scale, it is much the weakest of the four forces and its effect at the nuclear or atomic level can nearly always be ignored.

TABLE 32–1   The Four Forces in Nature

Type	Relative Strength (approx., for 2 protons in nucleus)	Field Particle
Strong nuclear	1	Gluons† (mesons)
Electromagnetic	10^{-2}	Photon
Weak nuclear	10^{-6}	$W^{\pm}$ and Z^0
Gravitational	10^{-38}	Graviton (?)

†Until the 1970s, thought to be mesons, now gluons (see Section 32–10).

32–4 Particles and Antiparticles

The positron, as we have seen, is basically a positive electron. That is, many of its properties are the same as for the electron, such as mass, but it has the opposite charge. The positron is said to be the **antiparticle** to the electron. After the positron was discovered in 1932, it was predicted that other particles also ought to have antiparticles. In 1955 the antiparticle to the proton was found, the **antiproton** ($\bar{\text{p}}$), which carries a negative charge; see Fig. 32–11. (The bar over the p is used to indicate antiparticle.) Soon after, the antineutron ($\bar{\text{n}}$) was found. Most other particles also have antiparticles. But the photon, the π^0, and a few other particles do not have distinct antiparticles—or we say that they are their own antiparticles.

Antiparticles are produced in nuclear reactions when there is sufficient energy available, and they do not live very long in the presence of matter. For example, a positron is stable when by itself, but if it encounters an electron, the two annihilate each other. The energy of their vanished mass, plus any kinetic energy they possessed, is converted to the energy of γ rays or of other particles. Annihilation also occurs for all other particle–antiparticle pairs.

32–5 Particle Interactions and Conservation Laws

One of the important uses of high-energy accelerators is to study the interactions of elementary particles with each other. As a means of ordering this subnuclear world, the conservation laws are indispensable. The laws of conservation of energy, of momentum, of angular momentum, and of electric charge are found to hold precisely in all particle interactions.

A study of particle interactions has revealed a number of new conservation laws which (just like the old ones) are ordering principles: they help to explain why some reactions occur and others do not. For example, the following reaction has never been found to occur:

$$\text{p} + \text{n} \nrightarrow \text{p} + \text{p} + \bar{\text{p}}$$

even though charge, energy, and so on, are conserved ($\bar{\text{p}}$ means an antiproton and $\nrightarrow$ means the reaction does not occur). To understand why such a reaction doesn't occur, physicists hypothesized a new conservation law, the conservation of **baryon number**. (Baryon number is a generalization of nucleon number, which we saw earlier is conserved in nuclear reactions and decays.) An important addition to this law is the proposal that whereas all nucleons have baryon number $B = +1$, all antinucleons (antiprotons, antineutrons) have $B = -1$. The reaction above does not conserve baryon number since on the left side we have $B = (+1) + (+1) = +2$, and on the right $B = (+1) + (+1) + (-1) = +1$. On the other hand, the following reaction does conserve B and *does* occur if the incoming proton has sufficient energy:

$$\text{p} + \text{n} \rightarrow \text{p} + \text{n} + \bar{\text{p}} + \text{p},$$

$$B = +1 + 1 = +1 + 1 - 1 + 1.$$

As indicated, $B = +2$ on both sides of this equation. From these and other reactions, the conservation of baryon number has been established as a basic law of physics.

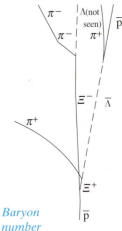

Baryon number

FIGURE 32–11 Liquid-hydrogen bubble-chamber photograph of an antiproton ($\bar{\text{p}}$) colliding with a proton, producing a Xi–anti-Xi pair ($\bar{\text{p}} + \text{p} \rightarrow \Xi^- + \Xi^+$) that subsequently decay into other particles. The drawing indicates the assignment of particles to each track, which is based on how or if that particle decays, and on mass values estimated from measurement of momentum (curvature of track in magnetic field) and energy (heaviness of track, for example). Neutral particle paths are shown by dashed lines since neutral particles produce no bubbles and hence no tracks.

Also useful are the conservation laws for the three **lepton numbers**, associated with weak interactions including decays. In ordinary β decay, an electron or positron is emitted along with a neutrino or antineutrino. In a similar type of decay, a muon is emitted instead of an electron. The neutrino (ν_e) that accompanies an emitted electron is found to be different from the neutrino (ν_μ) that accompanies an emitted muon. Each of these neutrinos has an antiparticle: $\bar{\nu}_e$ and $\bar{\nu}_\mu$. In ordinary β decay we have, for example,

$$n \rightarrow p + e^- + \bar{\nu}_e$$

but never $n \rightarrow p + e^- + \bar{\nu}_\mu$ or $n \rightarrow p + e^- + \bar{\nu}_e + \nu_e$. To explain why these do not occur, the concept of electron lepton number, L_e, was invented. If the electron (e^-) and the electron neutrino (ν_e) are assigned $L_e = +1$, and e^+ and $\bar{\nu}_e$ are assigned $L_e = -1$, whereas all other particles have $L_e = 0$, then all observed decays conserve L_e. For example, in $n \rightarrow p + e^- + \bar{\nu}_e$, $L_e = 0$ initially, and $L_e = 0 + (+1) + (-1) = 0$ after the decay. Decays that do not conserve L_e but would obey the other conservation laws are not observed to occur. Hence it is believed that L_e is conserved in all interactions.

In a decay involving muons, such as

$$\pi^+ \rightarrow \mu^+ + \nu_\mu,$$

a second quantum number, muon lepton number (L_μ), is conserved. The μ^- and ν_μ are assigned $L_\mu = +1$, and μ^+ and $\bar{\nu}_\mu$ have $L_\mu = -1$, whereas other particles have $L_\mu = 0$. It is believed that L_μ is also conserved in all interactions or decays. Similar assignments can be made for a third lepton number, L_τ, associated with the recently discovered τ lepton and its neutrino, ν_τ.

Keep in mind that antiparticles have not only opposite electric charge from their particles, but also opposite B, L_e, L_μ, and L_τ.

CONCEPTUAL EXAMPLE 32–3 | **Lepton number in muon decay.** Which of the following decay schemes is possible for muon decay: (a) $\mu^- \rightarrow e^- + \bar{\nu}_e$; (b) $\mu^- \rightarrow e^- + \bar{\nu}_e + \nu_\mu$; (c) $\mu^- \rightarrow e^- + \nu_e$? All of these particles have $L_\tau = 0$.

RESPONSE A μ^- has $L_\mu = +1$ and $L_e = 0$. This is the initial state, and the final state (after decay) must also have $L_\mu = +1$, $L_e = 0$. In (a), the final state has $L_\mu = 0 + 0 = 0$, and $L_e = +1 - 1 = 0$; L_μ would not be conserved and indeed this decay is not observed to occur. The final state of (b) has $L_\mu = 0 + 0 + 1 = +1$ and $L_e = +1 - 1 + 0 = 0$, so both L_μ and L_e are conserved. This is in fact the most common decay mode of the μ^-. Finally, (c) does not occur because L_e ($= +2$ in final state) is not conserved, nor is L_μ.

32–6 Particle Classification

In the decades following the discovery of the π meson in the late 1940s, a great many other subnuclear particles were discovered. Today they number in the hundreds. Much theoretical and experimental work has been done to try to understand this multitude of particles. An important aid to understanding is to arrange the particles in categories according to their

TABLE 32-2 Particles (stable under strong decay)[†]

Category	Particle Name	Symbol	Anti-particle	Spin	Rest Mass (MeV/c^2)	B	L_e	L_μ	L_τ	S	Lifetime (s)	Principal Decay Modes
Gauge bosons	Photon	γ	Self	1	0	0	0	0	0	0	Stable	
	W	W^+	W^-	1	80.33×10^3	0	0	0	0	0	3×10^{-25}	$e\nu_e, \mu\nu_\mu, \tau\nu_\tau$, hadrons
	Z	Z^0	Self	1	91.19×10^3	0	0	0	0	0	3×10^{-25}	$e^+e^-, \mu^+\mu^-, \tau^+\tau^-$, hadrons
Leptons	Electron	e^-	e^+	$\frac{1}{2}$	0.511	0	+1	0	0	0	Stable	
	Neutrino (e)	ν_e	$\bar{\nu}_e$	$\frac{1}{2}$	$0(<7.0 \times 10^{-6})^\ddagger$	0	+1	0	0	0	Stable	
	Muon	μ^-	μ^+	$\frac{1}{2}$	105.7	0	0	+1	0	0	2.20×10^{-6}	$e^-\bar{\nu}_e\nu_\mu$
	Neutrino (μ)	ν_μ	$\bar{\nu}_\mu$	$\frac{1}{2}$	$0(<0.17)^\ddagger$	0	0	+1	0	0	Stable	
	Tau	τ^-	τ^+	$\frac{1}{2}$	1777	0	0	0	+1	0	2.91×10^{-13}	$\mu^-\bar{\nu}_\mu\nu_\tau, e^-\bar{\nu}_e\nu_\tau$, hadrons $+ \nu_\tau$
	Neutrino (τ)	ν_τ	$\bar{\nu}_\tau$	$\frac{1}{2}$	$0(<24)^\ddagger$	0	0	0	+1	0	Stable	
Hadrons (selected)												
Mesons	Pion	π^+	π^-	0	139.6	0	0	0	0	0	2.60×10^{-8}	$\mu^+\nu_\mu$
		π^0	Self	0	135.0	0	0	0	0	0	0.84×10^{-16}	2γ
	Kaon	K^+	K^-	0	493.7	0	0	0	0	+1	1.24×10^{-8}	$\mu^+\nu_\mu, \pi^+\pi^0$
		K_S^0	$\bar{K}_S^0$	0	497.7	0	0	0	0	+1	0.89×10^{-10}	$\pi^+\pi^-, 2\pi^0$
		K_L^0	$\bar{K}_L^0$	0	497.7	0	0	0	0	+1	5.17×10^{-8}	$\pi^\pm e^{\mp}\overset{(-)}{\nu}_e, \pi^\pm\mu^{\mp}\overset{(-)}{\nu}_\mu, 3\pi$
	Eta	η^0	Self	0	547.5	0	0	0	0	0	5×10^{-19}	$2\gamma, 3\pi^0, \pi^+\pi^-\pi^0$
	and others											
Baryons	Proton	p	$\bar{p}$	$\frac{1}{2}$	938.3	+1	0	0	0	0	Stable	
	Neutron	n	$\bar{n}$	$\frac{1}{2}$	939.6	+1	0	0	0	0	887	$pe^-\bar{\nu}_e$
	Lambda	Λ^0	$\bar{\Lambda}^0$	$\frac{1}{2}$	1115.7	+1	0	0	0	-1	2.63×10^{-10}	$p\pi^-, n\pi^0$
	Sigma	Σ^+	$\bar{\Sigma}^-$	$\frac{1}{2}$	1189.4	+1	0	0	0	-1	0.80×10^{-10}	$p\pi^0, n\pi^+$
		Σ^0	$\bar{\Sigma}^0$	$\frac{1}{2}$	1192.6	+1	0	0	0	-1	7.4×10^{-20}	$\Lambda^0\gamma$
		Σ^-	$\bar{\Sigma}^+$	$\frac{1}{2}$	1197.4	+1	0	0	0	-1	1.48×10^{-10}	$n\pi^-$
	Xi	Ξ^0	$\bar{\Xi}^0$	$\frac{1}{2}$	1314.9	+1	0	0	0	-2	2.90×10^{-10}	$\Lambda^0\pi^0$
		Ξ^-	$\bar{\Xi}^+$	$\frac{1}{2}$	1321.3	+1	0	0	0	-2	1.64×10^{-10}	$\Lambda^0\pi^-$
	Omega	Ω^-	Ω^+	$\frac{3}{2}$	1672.5	+1	0	0	0	-3	0.82×10^{-10}	$\Xi^0\pi^-, \Lambda^0 K^-, \Xi^-\pi^0$
	and others											

[†]See also Table 32–4 for particles with charm and bottomness.

[‡]Experimental upper limits on neutrino masses are given in parentheses.

properties. One way of doing this is according to their interactions. Since not all particles interact by means of all four of the forces known in nature (though all interact via gravity), this is used as a classification scheme. Table 32–2 lists some of the more common particles classified in this way along with many of their properties. The particles listed are those that are stable, and many that are unstable. At the top of the table are the **gauge bosons** (so-named[†] after the theory that describes them, "gauge theory"), which include the *photon*, and the W and Z particles, that carry the electromagnetic and weak interactions, respectively.

Gauge bosons

Next in Table 32–2 are the **leptons**, which are particles that do not interact via the strong force but do interact via the weak nuclear force (as well as the much weaker gravitational force); those that carry electric charge also interact via the electromagnetic force. The leptons include the electron, the muon, and the tau (or τ, discovered in 1976 and more than 3000 times heavier than the electron), and three types of neutrino: the electron neutrino (ν_e), the muon neutrino (ν_μ), and the tau neutrino (ν_τ). They each have antiparticles, as indicated in Table 32–2.

Leptons

[†]Bosons are particles that are not governed by the Pauli exclusion principle (see Sections 28–7 and 32–10).

The third category of particle in Table 32–2 is the **hadron**. Hadrons are those particles that interact via the strong nuclear force. Hence they are said to be **strongly interacting particles**. They also interact via the other forces, but the strong force predominates at short distances. The hadrons include nucleons, pions, and a large number of other particles.

Baryons They are divided into two subgroups:[†] **baryons**, which are those particles that have baryon number $+1$ (or -1 in the case of their antiparticles); and

Mesons **mesons**, which have baryon number $= 0$.

Notice that the baryons Λ, Σ, Ξ, and Ω all decay to lighter-mass baryons, and eventually to a proton or neutron. All these processes conserve baryon number. Since there is no lighter particle than the proton with $B = +1$, if baryon number is strictly conserved, the proton itself cannot decay and is stable.

32–7 Particle Stability and Resonances

Lifetime depends on which force is acting Many of the particles listed in Table 32–2 are unstable. The lifetime of an unstable particle depends on which force is most active in causing the decay. When we say the strong nuclear force is stronger than the electromagnetic, we mean that two particles will interact more strongly and more quickly if this force is acting. When a stronger force influences a decay, that decay occurs more quickly. Decays caused by the weak force typically have lifetimes of 10^{-13} s or longer. Particles that decay via the electromagnetic force have much shorter lifetimes, typically about 10^{-16} to 10^{-19} s. (Exceptions to this scheme are the W and Z particles, which decay via the weak interaction but have very short lifetimes due to their special nature as exchange particles.) The unstable particles listed in Table 32–2 decay either via the weak or the electromagnetic interaction. Decays that involve a γ (photon) are electromagnetic (such as $\pi^0 \rightarrow 2\gamma$). The other decays shown take place via the weak interaction, often accompanied by a neutrino which interacts only via the weak interaction: examples are $\pi^+ \rightarrow \mu^+ \nu_\mu$ and $\Sigma^- \rightarrow n\pi^-$.

A great many particles have been found that decay via the strong interaction, and these are not listed in Table 32–2. Such particles decay into other strongly interacting particles (say, n, p, π, but not involving γ, e, ν, and so on) and their lifetimes are very short, typically about 10^{-23} s. In fact their

Very short-lived particles are inferred from their decay products lifetimes are so short that they do not travel far enough to be detected before decaying. Their decay products can be detected, however, and it is from them that the existence of such short-lived particles is inferred. To see how this is done, we consider the first such particle discovered (by Fermi). Fermi used a beam of π^+ directed through a hydrogen target (protons) with varying amounts of energy. A graph of the number of interactions (π^+ scattered) versus the pion's kinetic energy is shown in Fig. 32–12. The large

---

[†]Originally, particles were divided according to their mass into leptons (meaning light particles), baryons (meaning "heavy"), and those of intermediate mass, the mesons (meaning "middle"). The newer classification according to their interactions is not always consistent with this. For example, the J/ψ particles are very heavy but have $B = 0$ and so are classified as mesons (Section 32–9), and the τ is a lepton although it is heavier than some baryons.

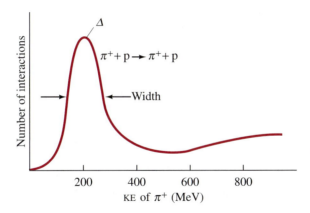

FIGURE 32-12 Number of π^+ particles scattered by a proton target as a function of the incident π^+ kinetic energy.

peak around 200 MeV was much higher than expected and certainly much higher than the number of interactions at neighboring energies. This led Fermi to conclude that the π^+ and proton combined momentarily to form a short-lived particle before coming apart again, or at least that they resonated together for a short time. Indeed, the large peak in Fig. 32–12 resembles a resonance curve (see Figs. 11–18 and 21–38), and this new "particle"—now called the Δ—is referred to as a **resonance**. Hundreds of other resonances have been found in a similar way. Many resonances are regarded as excited states of other particles such as of the nucleon (proton or neutron).

Resonance

The width of a resonance—in Fig. 32–12 the width of the Δ peak is on the order of 100 MeV—is an interesting application of the uncertainty principle. If a particle lives only 10^{-23} s, then its mass (i.e., its rest energy) will be uncertain by an amount $\Delta E \approx h/2\pi\Delta t \approx (6.6 \times 10^{-34}\,\text{J·s})/(6)(10^{-23}\,\text{s}) \approx 10^{-11}\,\text{J} \approx 100\,\text{MeV}$, which is what is observed. Actually, the lifetimes of $\approx 10^{-23}$ s for such resonances are inferred by the reverse process: from the measured width being $\approx 100\,\text{MeV}$.

32–8 Strange Particles

In the early 1950s, certain of the newly found particles, namely, the K, Λ, and Σ, were found to behave rather strangely in two ways. First, they were always produced in pairs. For example, the reaction

$$\pi^- + p \rightarrow K^0 + \Lambda^0$$

occurred with high probability, but the reaction $\pi^- + p \rightarrow K^0 + n$ was never observed to occur. This seemed strange because the unobserved reaction would not have violated any known conservation law, and plenty of energy was available. The second feature of these **strange particles** (as they came to be called) was that, although they were clearly produced via the strong interaction (that is, at a high rate), they did not decay at a rate characteristic of the strong interaction even though they decayed into strongly interacting particles (for example, $K \rightarrow 2\pi$, $\Sigma^+ \rightarrow p + \pi^0$). Instead of lifetimes of 10^{-23} s as expected for strongly interacting particles, strange particles have lifetimes of 10^{-10} to 10^{-8} s, which are characteristic of the weak interaction.

To explain these observations, a new quantum number, **strangeness**, and a new conservation law, conservation of strangeness, were introduced. By assigning the strangeness numbers (S) indicated in Table 32–2, the production of strange particles in pairs was readily explained. Antiparticles were assigned opposite strangeness from their particles: one of each pair was assigned $S = +1$ and the other $S = -1$ (see Table 32–2). For example, in the reaction $\pi^- + p \rightarrow K^0 + \Lambda^0$, the initial state has strangeness $S = 0 + 0 = 0$, and the final state has $S = +1 - 1 = 0$, so strangeness is conserved. But for $\pi^- + p \rightarrow K^0 + n$, the initial state has $S = 0$ and the final state has $S = +1 + 0 = +1$, so strangeness would not be conserved; and this reaction isn't observed.

To explain the decay of strange particles, it is assumed that strangeness is conserved in the strong interaction but is *not* conserved in the weak interaction. Thus, although strange particles were forbidden by strangeness conservation to decay to lower-mass nonstrange particles via the strong interaction, they could undergo such decay by means of the weak interaction. This would occur much more slowly, of course, which accounts for their longer lifetimes of 10^{-10} to 10^{-8} s.

The conservation of strangeness was the first example of a "partially conserved" quantity. In this case, the quantity strangeness is conserved by strong interactions but not by weak.

32–9 Quarks

We saw in our discussion of Table 32–2 that all particles, except the gauge bosons, fall into either of two categories: leptons and hadrons. The principal difference between these two groups is that the hadrons interact via the strong interaction, whereas the leptons do not. Another important difference that physicists had to deal with in the 1960s was that there were

TABLE 32–3 Properties of Quarks and Antiquarks

Quarks								
Name	Symbol	Spin	Charge	Baryon Number	Strangeness	Charm	Bottomness	Topness
Up	u	$\frac{1}{2}$	$+\frac{2}{3}e$	$\frac{1}{3}$	0	0	0	0
Down	d	$\frac{1}{2}$	$-\frac{1}{3}e$	$\frac{1}{3}$	0	0	0	0
Strange	s	$\frac{1}{2}$	$-\frac{1}{3}e$	$\frac{1}{3}$	-1	0	0	0
Charmed	c	$\frac{1}{2}$	$+\frac{2}{3}e$	$\frac{1}{3}$	0	$+1$	0	0
Bottom	b	$\frac{1}{2}$	$-\frac{1}{3}e$	$\frac{1}{3}$	0	0	-1	0
Top	t	$\frac{1}{2}$	$+\frac{2}{3}e$	$\frac{1}{3}$	0	0	0	$+1$

Antiquarks								
Name	Symbol	Spin	Charge	Baryon Number	Strangeness	Charm	Bottomness	Topness
Up	$\bar{u}$	$\frac{1}{2}$	$-\frac{2}{3}e$	$-\frac{1}{3}$	0	0	0	0
Down	$\bar{d}$	$\frac{1}{2}$	$+\frac{1}{3}e$	$-\frac{1}{3}$	0	0	0	0
Strange	$\bar{s}$	$\frac{1}{2}$	$+\frac{1}{3}e$	$-\frac{1}{3}$	$+1$	0	0	0
Charmed	$\bar{c}$	$\frac{1}{2}$	$-\frac{2}{3}e$	$-\frac{1}{3}$	0	-1	0	0
Bottom	$\bar{b}$	$\frac{1}{2}$	$+\frac{1}{3}e$	$-\frac{1}{3}$	0	0	$+1$	0
Top	$\bar{t}$	$\frac{1}{2}$	$-\frac{2}{3}e$	$-\frac{1}{3}$	0	0	0	-1

only four known leptons (e^-, μ^-, ν_e, ν_μ; the τ and ν_τ were not yet discovered), but there were well over a hundred hadrons.

The leptons are considered to be truly elementary particles since they do not seem to break down into smaller entities, do not show any internal structure, and have no measurable size. (Attempts to determine the size of leptons have put an upper limit of about 10^{-18} m.)

The hadrons, on the other hand, are more complex. Experiments indicate they do have an internal structure. And the fact that there are so many of them suggests that they can't all be elementary. To deal with this problem, M. Gell-Mann and G. Zweig in 1963 independently proposed that none of the hadrons so far observed, not even the proton and neutron, was elementary. Instead, they proposed that the hadrons are made up of combinations of three, more fundamental, pointlike entities called **quarks**.[†] Quarks, then, would be considered truly elementary particles, like leptons. The three quarks were labeled u, d, s, and given the names *up*, *down*, and *sideways* (or, more commonly now, *strange*). They were assumed to have fractional charge ($\frac{1}{3}$ or $\frac{2}{3}$ the charge on the electron—that is, less than the previously thought smallest charge). Other properties of quarks and antiquarks are indicated in Table 32–3. All hadrons known at the time could be constructed in theory from these three types of quark. Mesons would consist of a quark–antiquark pair. For example, a π^+ meson is considered a u$\bar{\text{d}}$ pair (note that for the u$\bar{\text{d}}$ pair, $Q = \frac{2}{3}e + \frac{1}{3}e = +1e$, $B = \frac{1}{3} - \frac{1}{3} = 0$, $S = 0 + 0 = 0$, as they must for a π^+); and a $K^+ = $ u$\bar{\text{s}}$, with $Q = +1$, $B = 0$, $S = +1$. Baryons, on the other hand, would consist of three quarks. For example, a neutron is n = ddu, whereas an antiproton is $\bar{\text{p}} = \bar{\text{u}}\bar{\text{u}}\bar{\text{d}}$. See Fig. 32–13.

Soon after the quark theory was proposed, physicists began looking for these fractionally charged particles, but direct detection has not been successful. Indeed, it may be that quarks are so tightly bound together that they may not ever exist singly in the free state.

In 1964 several physicists proposed that there ought to be a fourth quark. Their argument was based on the expectation that there exists a deep symmetry in nature, including a connection between quarks and leptons. If there are four leptons (as was thought in the 1960s), then symmetry in nature would suggest there should also be four quarks. The fourth quark was said to be *charmed*. Its charge would be $+\frac{2}{3}e$ and it would have another property to distinguish it from the other three quarks. This new property, or quantum number, was called **charm** (see Table 32–3). Charm was assumed to be like strangeness: it would be conserved in strong and electromagnetic interactions, but would not be conserved by the weak. The new charmed quark would have charm $C = +1$ and its antiquark $C = -1$. The first charmed particle, the J/ψ meson, was discovered in 1974.

Also in the 1970s strong evidence appeared for the tau (τ) lepton, with a mass of 1777 MeV/c^2. This lepton, like the electron and muon, presumably has a neutrino associated with it. Thus, the family of leptons is at present believed to have six members. This would upset the balance between leptons and quarks, the presumed basic building blocks of matter,

Hadrons can't all be elementary

Quarks

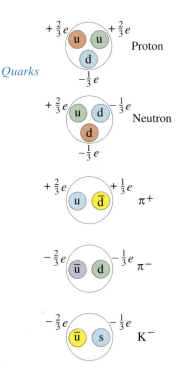

FIGURE 32–13 Quark compositions for several particles.

Charm

Charmed quark

Two more leptons

[†]Gell-Mann chose the word from a phrase in James Joyce's *Finnegans Wake*.

TABLE 32–4   Partial List of Hadrons Associated with Charm and Bottomness ($L_e = L_\mu = L_\tau = 0$)

Category	Particle	Anti-particle	Spin	Rest Mass (MeV/c^2)	Baryon Number	Strange-ness	Charm	Bottom-ness	Lifetime (s)	Principal Decay Modes
Mesons	D^+	D^-	0	1869.4	0	0	$+1$	0	10.6×10^{-13}	K + others, e + others
	D^0	$\overline{D}^0$	0	1864.6	0	0	$+1$	0	4.2×10^{-13}	K + others, μ or e + others
	D_S^+	D_S^-	0	1969	0	$+1$	$+1$	0	4.7×10^{-13}	K + others
	J/ψ (3097)	Self	1	3096.9	0	0	0	0	0.8×10^{-20}	Hadrons, e^+e^-, $\mu^+\mu^-$
	Υ (9460)	Self	1	9460.4	0	0	0	0	1.3×10^{-20}	Hadrons, $\mu^+\mu^-$, e^+e^-, $\tau^+\tau^-$
	B^-	B^+	0	5279	0	0	0	-1	1.5×10^{-12}	D^0 + others
	B^0	$\overline{B}^0$	0	5279	0	0	0	-1	1.5×10^{-12}	D^0 + others
Baryons	Λ_c^+	Λ_c^-	$\frac{1}{2}$	2285	$+1$	0	$+1$	0	2.0×10^{-13}	Hadrons (e.g., Λ + others)
	Σ_c^{++}	Σ_c^{--}	$\frac{1}{2}$	2453	$+1$	0	$+1$	0	?	$\Lambda_c^+ \pi^+$
	Σ_c^+	Σ_c^-	$\frac{1}{2}$	2454	$+1$	0	$+1$	0	?	$\Lambda_c^+ \pi^0$
	Σ_c^0	$\overline{\Sigma}_c^0$	$\frac{1}{2}$	2452	$+1$	0	$+1$	0	?	$\Lambda_c^+ \pi^-$
	Λ_b^0	Λ_b^0	$\frac{1}{2}$	5640	$+1$	0	0	-1	1.1×10^{-12}	$J/\psi\Lambda^0$, $pD^0\pi^-$, $\Lambda_c^+\pi^+\pi^-\pi^-$

t and b quarks

unless two additional quarks also exist. Indeed, theoretical physicists postulated the existence of a fifth and sixth quark, named **top** and **bottom**. (Some physicists prefer the names *truth* and *beauty* for these t and b quarks.) The names apply also to the new properties (quantum numbers) that distinguish the new quarks from the old quarks (see Table 32–3), and which (like strangeness) are conserved in strong, but not weak, interactions. New mesons involving b quarks were soon detected (Table 32–4). Strong evidence for the top quark was not forthcoming until 1995 (see photo at start of this chapter) after years of searching. Its extremely high mass of almost $200 \text{ GeV}/c^2$ contributed to its elusiveness because of the huge energy required to produce it.

Today, the "truly" elementary particles are considered to be the six quarks, the six leptons, and the gauge bosons that carry the fundamental forces. See Table 32–5, where the quarks and leptons are arranged in three groups (generations).

TABLE 32–5   The Elementary Particles[†] as Seen Today

	First generation	Second generation	Third generation
Quarks	u, d	s, c	b, t
Leptons	e, ν_e	μ, ν_μ	τ, ν_τ
Gauge bosons	γ(photon)	$W^\pm$, Z^0	gluons

[†]Note that the quarks and leptons are arranged into three generations each, and the gauge particles are arranged in groups for the forces they mediate.

EXAMPLE 32–4   Quark combinations. Find the baryon number, charge, and strangeness for the following quark combinations and identify the hadron particle that is made up of these quark combinations: (*a*) udd, (*b*) u$\overline{u}$, (*c*) uss, (*d*) sdd, and (*e*) b$\overline{u}$.

SOLUTION We use Table 32–3 to get the properties of the quarks, then Table 32–2 or 32–4 to find the particle that has these properties. (a) udd has $Q = +\frac{2}{3}e - \frac{1}{3}e - \frac{1}{3}e = 0$, $B = \frac{1}{3} + \frac{1}{3} + \frac{1}{3} = 1$, $S = 0 + 0 + 0 = 0$, $C = 0$, bottomness $= 0$, topness $= 0$; the only baryon ($B = +1$) that has $Q = 0$, $S = 0$, etc., is the neutron (Table 32–2).
(b) uū has $Q = \frac{2}{3}e - \frac{2}{3}e = 0$, $B = 0$, and all other quantum numbers $= 0$. Sounds like a π^0.
(c) uss has $Q = 0$, $B = +1$, $S = -2$, others $= 0$. This is a Ξ^0.
(d) sdd has $Q = -1$, $B = +1$, $S = -1$, so must be a Σ^-.
(e) bū has $Q = -1$, $B = 0$, $S = 0$, $C = 0$, bottomness $= -1$, topness $= 0$. This must be a B^- meson (Table 32–4).

32–10 The "Standard Model": Quantum Chromodynamics (QCD) and the Electroweak Theory

Not long after the quark theory was proposed, it was suggested that quarks have another property (or quality) called **color**. The distinction between the six quarks (u, d, s, c, b, t) was referred to as **flavor**. According to theory, each of the flavors of quark can have three colors, usually designated red, green, and blue. (These are the three primary colors which, when added together in equal amounts, as on a TV screen, produce white.) Note that the names "color" and "flavor" have nothing to do with our senses, but are purely whimsical—as are other names, such as charm, in this new field. (We did, however, "color" the quarks in Fig. 32–13.) The antiquarks are colored antired, antigreen, and antiblue. Baryons are made up of three quarks, one of each color. Mesons consist of a quark–antiquark pair of a particular color and its anticolor. Thus baryons can be said to be white, and mesons colorless.

Originally, the idea of quark color was proposed to preserve the Pauli exclusion principle (Section 28–7). Not all particles obey the exclusive principle. Those that do, such as electrons, protons, and neutrons, are called **fermions**. Those that don't are called **bosons**. These two categories are distinguished also in their spin (Section 28–6): bosons have integer spin (0, 1, 2, etc.) whereas fermions have half-integer spin, usually $\frac{1}{2}$, as for electrons and nucleons, but can also be $\frac{3}{2}, \frac{5}{2}$, etc. Matter is made up mainly of fermions, but the carriers of the forces (γ, W, Z, and gluons, as we'll see) are all bosons. Quarks are fermions (they have spin $\frac{1}{2}$) and therefore should obey the exclusion principle. Yet for three particular baryons (uuu, ddd, and sss), all three quarks would have the same quantum numbers, and at least two of them have their spin in the same direction (since there are only two choices, spin up [$m_s = +\frac{1}{2}$] or spin down [$m_s = -\frac{1}{2}$]). This would seem to violate the exclusion principle, but if quarks have an additional quantum number (color), which could be different for each quark, it would serve to distinguish them and the exclusion principle would hold. Although quark color, and the resulting threefold increase in the number of quarks, was thus originally an *ad hoc* idea, it also served to bring the theory into better agreement with experiment, such as predicting the correct lifetime of the π^0 meson. The idea of color soon became, in addition, a central feature of the theory as determining the force binding quarks together in a hadron. Each quark is assumed to carry a *color charge*, analogous to electric charge, and the strong force

FIGURE 32–14 (a) The force between two quarks holding them together as part of a proton, for example, is carried by a gluon, which in this case involves a change in color. (b) Strong interaction np → np with the exchange of a charged π meson (+ or −, depending on whether it is considered moving to the left or to the right). (c) Quark representation of the same interaction np → np. The wavy lines between quarks represent gluon exchanges holding the hadrons together.

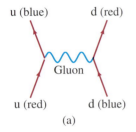

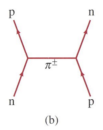

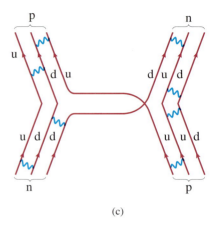

(a)

(b)

(c)

QCD

Gluons

between quarks is often referred to as the **color force**. This new theory of the strong force is called **quantum chromodynamics** (*chrome* = color in Greek), or **QCD**, to indicate that the force acts between color charges (and not between, say, electric charges). The strong force between two hadrons[†] is considered to be a force between the quarks that make them up, as suggested in Fig. 32–14. The particles that transmit the color force (analogous to photons for the EM force) are called **gluons** (a play on "glue"). They are included in Table 32–5. There are eight gluons, according to the theory, all massless, and six of them have color charge.[‡] Thus gluons have replaced mesons (Table 32–1) as the particles carrying the strong (color) force.

The color force has the interesting property that its strength increases with increasing distance (as does the force exerted by a coiled spring, Chapter 11); as two quarks approach each other very closely (equivalently, have high energy), the force between them becomes very small. This aspect is referred to as **asymptotic freedom**.

The weak force, as we have seen, is thought to be mediated by the W^+, W^-, and Z^0 particles. It acts between the "weak charges" that each particle has. Each elementary particle can thus have electric charge, weak charge, color charge, and gravitational mass, although one or more of these could be zero. For example, all leptons have color charge of zero, so they do not interact via the strong force.

FIGURE 32–15 Quark representation of the Feynman diagram for beta decay of a neutron into a proton.

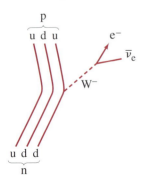

> **CONCEPTUAL EXAMPLE 32–5** **Beta decay.** Draw a Feynman diagram, showing what happens in beta decay using quarks.
>
> **RESPONSE** Beta decay is a result of the weak interaction, and the mediator is either a $W^\pm$ or Z^0 particle. What happens, in part, is that a neutron (udd quarks) decays into a proton (uud). Apparently a d quark (charge $-\frac{1}{3}e$) has turned into a u quark (charge $+\frac{2}{3}e$). Charge conservation means that a negatively charged particle, namely a W^-, was emitted by the d quark. Since an electron and an antineutrino appear in the final state, they must have come from the decay of the virtual W^-, as shown in Fig. 32–15.

[†]The strong force between hadrons appears feeble, however, in comparison to the force directly between quarks within hadrons.

[‡]Compare to the EM interaction, where the photon has no electric charge. Because gluons have color charge, they could attract each other and form composite particles (photons cannot). Such "glueballs" are being searched for.

To summarize, the latest theories propose that the truly elementary particles (Table 32–5) are the leptons, the quarks, and the gauge bosons (photon, W and Z, and the gluons). Some theories suggest there may be other bosons as well. The photon and the leptons are observed in experiments, and finally so too were the W^+, W^-, and Z^0. But so far only combinations of quarks (baryons and mesons) have been observed, and it seems likely that free quarks and gluons are unobservable.

One important aspect of new theoretical work is the attempt to find a unified basis for the different forces in nature. This was a long-held hope of Einstein, which he was never able to fulfill. A so-called **gauge theory** that unifies the weak and electromagnetic interactions was put forward in the 1960s by S. Weinberg, S. Glashow, and A. Salam. In this **electroweak theory**, the weak and electromagnetic forces are seen as two different manifestations of a single, more fundamental, *electroweak* interaction. The electroweak theory has had many successes, including the prediction of the $W^\pm$ particles as carriers of the weak force, with masses of $81 \pm 2\,\text{GeV}/c^2$ in excellent agreement with the measured values of $80.33 \pm 0.15\,\text{GeV}/c^2$ (and similar accuracy for the Z^0). The electroweak theory plus QCD for the strong interaction are often referred to today as the **standard model**. *Electroweak theory*

Standard model

Theoreticians have wondered why the W and Z have large masses rather than being massless like the photon. Electroweak theory suggests an explanation by means of a new **Higgs field** and its particle, the **Higgs boson**, which interact with the W and Z to "slow them down." In being forced to go slower than the speed of light, they must acquire mass.

32–11 Grand Unified Theories

With the success of the unified electroweak theory, attempts have recently been made to incorporate it and QCD for the strong (color) force into a so-called **grand unified theory** (GUT). One type of such a grand unified theory of the electromagnetic, weak, and strong forces has been worked out in which there is only one class of particle—leptons and quarks belong to the same family and are able to change freely from one type to the other—and the three forces are different aspects of a single underlying force. The unity is predicted to occur, however, only on a scale of less than about $10^{-30}\,\text{m}$. If two elementary particles (leptons or quarks) approach each other to within this **unification scale**, the apparently fundamental distinction between them would not exist at this level, and a quark could readily change to a lepton, or vice versa. Baryon and lepton numbers would not be conserved. The weak, electromagnetic, and strong (color) force would blend to a force of a single strength. *GUT*

Unification of forces

How could a lepton become a quark, or vice versa? The theory predicts the existence of particles, called X bosons, that can be exchanged between a quark and a lepton allowing one to change into the other, somewhat like the charged pion exchanged between the p and n in Fig. 32–14b allows the proton on the right to become a neutron. The mass of the proposed X boson, consistent with the uncertainty principle as applied earlier in this chapter (see Eq. 32–2 with $d \approx 10^{-30}\,\text{m}$), would be about $10^{14}\,\text{GeV}/c^2$, or 10^{14} times the proton mass. With such an incredibly large mass, there is little hope of seeing them in the laboratory. It is also this huge mass that would keep baryon and lepton numbers conserved in observed reactions, since the likelihood of producing such a massive particle, even as a virtual exchange particle, is extremely small at even the highest laboratory energies.

What happens between the unification distance of 10^{-30} m and more normal (larger) distances is referred to as **symmetry breaking**. As an analogy, consider an atom in a crystal. Deep within the atom, there is much symmetry—in the innermost regions the electron cloud is spherically symmetric (Chapter 28). Farther out, this symmetry breaks down—the electron clouds are distributed preferentially along the lines (bonds) joining the atoms in the crystal. In a similar way, at 10^{-30} m the force between elementary particles is theorized to be a single force—it is symmetrical and does not single out one type of "charge" over another. But at larger distances, that symmetry is broken and we see three distinct forces. (In the "standard model" of electroweak interactions, Section 32–10, the symmetry breaking between the electromagnetic and the weak interactions occurs at about 10^{-18} m.)

Symmetry breaking

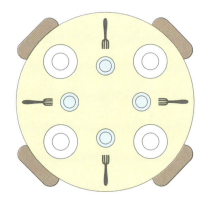

FIGURE 32–16 Symmetry around a table. Example 32–6.

CONCEPTUAL EXAMPLE 32–6 **Symmetry.** The table in Fig. 32–16 has four identical place settings. Four people sit down to eat. Describe the symmetry of this table and what happens to it when someone starts the meal.

RESPONSE The table has several kinds of symmetry. It is symmetric to rotations of 90°: that is, the table will look the same if everyone moved one chair to the left or to the right. It is also north–south symmetric and east–west symmetric, so that swaps across the table don't affect the way the table looks. It also doesn't matter whether any person picks up the fork to the left of the plate or the fork to the right. But once that first person picks up either fork, the choice is set for all the rest at the table as well. The symmetry has been *broken*. The underlying symmetry is still there—the water glasses could still be chosen either way—but some choice must get made and at that moment the symmetry of the diners is broken.

Connection with cosmology

An interesting prediction of unified theories relates to cosmology (see Chapter 33). It is thought that during the first 10^{-35} s after the theorized big bang that created the universe, the temperature was so extremely high that particles had energies corresponding to the unification scale. Baryon number would not have been conserved then, thus allowing an imbalance that might account for the observed predominance of matter ($B > 0$) over antimatter ($B < 0$) in the universe.

This last example is interesting, for it illustrates a deep connection between investigations at either end of the size scale: theories about the tiniest objects (elementary particles) have a strong bearing on the understanding of the universe as a whole. We will look at this more in the next chapter.

Even more ambitious than grand unified theories are attempts to also incorporate gravity, and thus unify all four forces in nature into a single theory. (Such theories are sometimes referred to misleadingly as **theories of everything**.) The only consistent theory so far that attempts to unify all four forces is called **string theory**, in which the elementary particles (Table 32–5) are imagined not as points but as one-dimensional strings about 10^{-35} m long.

String theory

Supersymmetry

A related idea is **supersymmetry**, which applied to strings is known as **superstring theory**. Supersymmetry predicts that interactions exist that would change fermions into bosons and vice versa, and that all known fermions have supersymmetric boson partners. Thus, for every quark there would be a *squark*, and for every lepton there would be a *slepton*. Likewise, for every known boson (photons and gluons, for example), there would be a supersymmetric fermion (*photinos* and *gluinos*). But why hasn't this "missing half" of the universe ever been detected? The best guess is that supersymmetric particles might be heavier than their conventional counterparts, perhaps too heavy to have been produced in today's accelerators. Until a supersymmetric particle is found, however, supersymmetry is just an elegant guess.

The world of elementary particles is opening new vistas. What happens in the near future is bound to be exciting.

◼ S U M M A R Y

Particle accelerators are used to accelerate charged particles, such as electrons and protons, to very high energy. High-energy particles have short wavelength and so can be used to probe the structure of matter at very small distances in great detail. High kinetic energy also allows the creation of new particles through collision (via $E = mc^2$). Van De Graaff and linear accelerators use high voltage to accelerate particles along a line. Cyclotrons and synchrotrons use a magnetic field to keep the particles in a circular path and accelerate them at intervals by high voltage.

An **antiparticle** has the same mass as a particle but opposite charge. Certain other properties may also be opposite: for example, the antiproton has **baryon number** (nucleon number) opposite to that for the proton. In all nuclear and particle reactions, the following conservation laws hold: momentum, angular momentum, mass–energy, electric charge, baryon number, and the three **lepton numbers**. Certain particles have a property, called **strangeness**, which is conserved by the strong force but not by the weak force. The more recently noted properties, **charm**, **bottomness**, and **topness**, also are believed to be conserved by the strong force but not by the weak.

Just as the electromagnetic force can be said to be due to an exchange of photons, the strong nuclear force is thought to be carried by *mesons* that have rest mass, or, according to more recent theory, by massless **gluons**. The W and Z particles carry the weak force. These fundamental force carriers (photon, W and Z, gluons) are called **gauge bosons**.

Other particles can be classified as either *leptons* or *hadrons*. **Leptons** participate in the weak and electromagnetic interactions. **Hadrons** participate in the strong interaction as well. The hadrons can be classified as **mesons**, with baryon number zero, and **baryons**, with nonzero baryon number.

All particles, except for the photon, electron, neutrinos, and proton, decay with measurable half-lives varying from 10^{-25} s to 10^3 s. The half-life depends on which force is predominant in the decay. Weak decays usually have half-lives greater than about 10^{-13} s. Electromagnetic decays have half-lives on the order of 10^{-16} to 10^{-19} s. The shortest lived particles, called **resonances**, decay via the strong interaction and live typically for only about 10^{-23} s.

The latest theories of elementary particle physics postulate the existence of **quarks** as the basic building blocks of the hadrons. Initially, three quarks were proposed (called **up**, **down**, and **strange**.) Recent evidence suggests that three additional quarks are needed, called the **charmed**, **bottom**, and **top** quarks. It is expected that there are the same number of quarks as leptons (six of each), and that quarks and leptons are the truly elementary particles along with the gauge bosons (γ, W, Z, gluons). Quarks are said to have **color**, and, according to **quantum chromodynamics** (QCD), the strong color force acts between their color charges and is transmitted by **gluons**. **Electroweak theory** views the weak and electromagnetic forces as two aspects of a single underlying interaction. QCD plus the electroweak theory are referred to as the **standard model**.

Grand unified theories of forces suggest that at very short distances (10^{-30} m) and very high energy, the weak, electromagnetic, and strong forces appear as a single force, and the fundamental difference between quarks and leptons disappears.

QUESTIONS

1. What limits the maximum energy attainable for protons in an ordinary cyclotron? How is this limitation overcome in a synchrotron?

2. Give a reaction between two nucleons, similar to Eq. 32–3, that could produce a π^-.

3. If a proton is moving at very high speed, so that its KE is much greater than its rest energy (m_0c^2), can it then decay via p $\rightarrow$ n + π^+?

4. What would an "antiatom," made up of the antiparticles to the constituents of normal atoms, consist of? What might happen if *antimatter*, made of such antiatoms, came in contact with our normal world of matter?

5. What particle in a decay signals the electromagnetic interaction?

6. Does the presence of a neutrino among the decay products of a particle necessarily mean that the decay occurs via the weak interaction? Do all decays via the weak interaction produce a neutrino?

7. Why is it that a neutron decays via the weak interaction even though the neutron and one of its decay products (proton) are strongly interacting?

8. Which of the four interactions (strong, electromagnetic, weak, gravitational) does an electron take part in? A neutrino? A proton?

9. Check that charge and baryon number are conserved in each of the decays in Table 32–2.

10. Which of the particle decays in Table 32–2 occur via the electromagnetic interaction?

11. Which of the particle decays in Table 32–2 occur by the weak interaction?

12. By what interaction, and why, does $\Sigma^\pm$ decay to Λ^0? What about Σ^0 decaying to Λ^0?

13. The Δ baryon has spin $\frac{3}{2}$, baryon number 1, and charge $Q = +2, +1, 0,$ or -1. Why is there no charge state $Q = -2$?

14. Which of the particle decays in Table 32–4 occur via the electromagnetic interaction?

15. Which of the particle decays in Table 32–4 occur by the weak interaction?

16. Quarks have spin $\frac{1}{2}$. How do you account for the fact that baryons have spin $\frac{1}{2}$ or $\frac{3}{2}$, and mesons have spin 0 or 1?

17. Suppose there were a kind of "neutrinolet" that was massless, had no color charge or electrical charge, and did not feel the weak force. Could you say that this particle even exists?

PROBLEMS

SECTIONS 32–1 AND 32–2

1. (I) What is the total energy of a proton whose kinetic energy is 8.50 GeV?

2. (I) Calculate the wavelength of 40-GeV electrons.

3. (I) What strength of magnetic field is used in a cyclotron in which protons make 2.4×10^7 revolutions per second?

4. (I) What is the time for one complete revolution for a very high-energy proton in the 1.0-km-radius Fermilab accelerator?

5. (I) If α particles are accelerated by the cyclotron of Example 32–2, what must be the frequency of the voltage applied to the dees?

6. (II) Is a 30-MeV alpha particle or 30-MeV proton better for picking out details of the nucleus? Compare each of their wavelengths with the size of a nucleon in a nucleus.

7. (II) (*a*) If the cyclotron of Example 32–2 accelerated α particles, what maximum energy could they attain? What would their speed be? (*b*) Repeat for deuterons ($^2_1$H). (*c*) In each case, what frequency of voltage is required?

8. (II) The voltage across the dees of a cyclotron is 45 kV. How many revolutions do protons make to reach a kinetic energy of 25 MeV?

9. (II) A cyclotron with a radius of 1.0 m is to accelerate deuterons ($^2_1$H) to an energy of 10 MeV. (*a*) What is the required magnetic field? (*b*) What frequency is needed for the voltage between the dees? (*c*) If the potential difference between the dees averages 25 kV, how many revolutions will the particle make before exiting? (*d*) How much time does it take for one deuteron to go from start to exit, and (*e*) how far does it travel during this time?

10. (II) What is the wavelength, and maximum resolving power attainable, using 900-GeV protons at Fermilab?

11. (II) Protons are injected into the 1.0-km-radius Fermilab Tevatron with an energy of 8.0 GeV. If they are accelerated by 2.5 MV each revolution, how far do they travel and approximately how long does it take for them to reach 900 GeV?

12. (II) The Fermilab Tevatron takes about 20 seconds to bring the energies of the stored protons from 150 GeV to 900 GeV. The acceleration is done once per turn. Estimate the energy given to the protons on each turn. (You can assume that the speed of the protons is essentially c the whole time.)

13. (III) What magnetic field intensity is needed at the 1.0-km-radius Fermilab synchrotron for 900-GeV protons? Use the relativistic mass.

14. (III) Show that the energy of a particle (charge e) in a synchrotron, in the relativistic limit ($v \approx c$), is given by E (in eV) $= Brc$, where B is magnetic field strength and r the radius of the orbit (SI units).

SECTIONS 32–3 TO 32–6

15. (I) Draw a Feynman diagram for $n + p \rightarrow n + p + \pi^0$.

16. (I) Draw a Feynman diagram for $p\bar{p}$ annihilation with the production of two pions.

17. (I) Draw a Feynman diagram for the photoelectric effect.

18. (I) Two protons are heading toward each other with equal speeds. What minimum kinetic energy must each have if a π^0 meson is to be created in the process? (See Table 32–2.)

19. (I) How much energy is released in the decay $\pi^+ \rightarrow \mu^+ + \nu_\mu$?

20. (I) About how much energy is released when a Λ^0 decays to $n\pi^0$? (See Table 32–2.)

21. (I) How much energy is required to produce a neutron–antineutron pair?

22. (I) Estimate the range of the strong force if the mediating particle were the kaon instead of the pion.

23. (II) Which of the following decays are possible? For those that are forbidden, explain which laws are violated.
    (a) $\Xi^0 \rightarrow \Sigma^+ + \pi^-$ (b) $\Omega^- \rightarrow \Sigma^0 + \pi^- + \nu$
    (c) $\Sigma^0 \rightarrow \Lambda^0 + \gamma + \gamma$.

24. (II) Estimate the range of the weak force using Eq. 32–2, given the masses of the W and Z particles as about 80 to 90 GeV/c^2.

25. (II) What are the wavelengths of the two photons produced when a proton and antiproton at rest annihilate?

26. (II) (a) Show, by conserving momentum and energy, that it is impossible for an isolated electron to radiate only a single photon. (b) With this result in mind, how can you defend the photon exchange diagram in Fig. 32–8?

27. (II) What would be the wavelengths of the two photons produced when an electron–positron pair, each with 300 keV of KE, annihilate?

28. (II) In the rare decay $\pi^+ \rightarrow e^+ + \nu_e$, what is the kinetic energy of the positron? Assume the π^+ decays from rest.

29. (II) What minimum kinetic energy must a neutron and proton each have if they are traveling at the same speed toward each other, collide, and produce a K^+K^- pair in addition to themselves? (See Table 32–2.)

30. (II) Calculate the KE of each of the two products in the decay $\Xi^- \rightarrow \Lambda^0\pi^-$. Assume the Ξ^- decays from rest.

31. (III) Could a π^+ meson be produced if a 100-MeV proton struck a proton at rest? What minimum KE must the incoming proton have?

32. (III) For the reaction $p + p \rightarrow 3p + \bar{p}$, where one of the initial protons is at rest, use relativistic formulas to show that the threshold energy is $6m_pc^2$, equal to three times the Q-value of the reaction, where m_p is the proton rest mass.

SECTIONS 32–7 TO 32–11

33. (I) The measured width of the J/ψ meson is 88 keV. Estimate its lifetime.

34. (I) The measured width of the ψ(3685) meson is 277 keV. Estimate its lifetime.

35. (I) What is the energy width (or uncertainty) of (a) η^0, and (b) Σ^0?

36. (I) Use Fig. 32–12 to estimate the energy width and then the lifetime of the Δ resonance.

37. (I) The B$^-$ meson is presumed to be a $b\bar{u}$ quark combination. (a) Show that this is consistent for all quantum numbers. (b) What are the quark combinations for B$^+$, B^0, $\overline{B}^0$?

38. (II) What are the quark combinations that can form (a) a neutron, (b) an antineutron, (c) a Λ^0, (d) a $\overline{\Sigma}^0$?

39. (II) What particles do the following quark combinations produce? (a) uud, (b) $\overline{uu}s$, (c) $\overline{u}s$, (d) $d\overline{u}$, (e) $\overline{c}s$.

40. (II) What is the quark combination needed to produce a D^0 meson ($Q = B = S = 0, C = +1$)?

41. (II) The D_S^+ meson has $S = C = +1, B = 0$. What quark combination would produce it?

42. (II) (a) Show that the so-called unification distance of 10^{-30} m in grand unified theory is equivalent to an energy of about 10^{14} GeV. Use either the uncertainty principle or de Broglie's wavelength formula, and explain how they apply. (b) Calculate what temperature this corresponds to.

43. (II) Draw possible Feynman diagrams using quarks (as in Fig. 32–14c) for the reactions (a) $\pi^-p \rightarrow \pi^0n$, (b) $\bar{p}p \rightarrow \pi^+\pi^-$.

44. (II) Draw a possible quark Feynman diagram (see Fig. 32–14c) for the reaction $K^-p \rightarrow K^-p$.

GENERAL PROBLEMS

45. There are typically 5×10^{13} protons at 900 GeV stored in the 1.0-km-radius ring of the Tevatron. (a) How much current (amperes) is carried by this beam? (b) How fast would a 1500 kg car have to move to carry the same kinetic energy as this beam?

46. The 4.25-km-radius LEP tunnel will be reused to house the magnets for the Large Hadron Collider (LHC), scheduled for completion around 2005. If the design calls for proton beams of energy 7.0 TeV, what magnetic field will be required?

47. (a) How much energy is released when an electron and a positron annihilate each other? (b) How much energy is released when a proton and an antiproton annihilate each other?

48. Which of the following reactions are possible, and by what interaction could they occur? For those forbidden, explain why.

    (a) $\pi^- p \rightarrow K^+ \Sigma^-$ (f) $\pi^- p \rightarrow K^0 p \pi^0$
    (b) $\pi^+ p \rightarrow K^+ \Sigma^+$ (g) $K^- p \rightarrow \Lambda^0 \pi^0$
    (c) $\pi^- p \rightarrow \Lambda^0 K^0 \pi^0$ (h) $K^+ n \rightarrow \Sigma^+ \pi^0 \gamma$
    (d) $\pi^+ p \rightarrow \Sigma^0 \pi^0$ (i) $K^+ \rightarrow \pi^0 \pi^0 \pi^+$
    (e) $\pi^- p \rightarrow p e^- \bar{\nu}_e$ (j) $\pi^+ \rightarrow e^+ \nu_e$

49. For the decay $\Lambda^0 \rightarrow p \pi^-$, calculate (a) the Q-value (energy released), and (b) the KE of the p and π^-, assuming the Λ^0 decays from rest. (Use relativistic formulas.)

50. Symmetry breaking occurs in the electroweak theory at about 10^{-18} m. Show that this corresponds to an energy that is on the order of the mass of the $W^\pm$.

51. The mass of a π^0 can be measured by observing the reaction $\pi^- + p \rightarrow \pi^0 + n$ at very low incident π^- kinetic energy (assume it is zero). The neutron is observed to be emitted with a KE of 0.60 MeV. Use conservation of energy and momentum to determine the π^0 mass.

52. Calculate the Q value for each of the reactions, Eq. 32–3, for producing a pion.

53. Calculate the maximum KE of the electron in the decay $\mu^- \rightarrow e^- + \bar{\nu}_e + \nu_\mu$. [Hint: In what direction do the two neutrinos move relative to the electron in order to give the latter the maximum KE? Both energy and momentum are conserved; use relativistic formulas.]

54. Calculate the Q-value for the reaction $\pi^- p \rightarrow \Lambda^0 K^0$, when negative pions strike stationary protons. Estimate the minimum pion KE needed to produce this reaction.

55. Draw the normal Feynman diagrams for (a) Compton scattering, (b) bremsstrahlung by an electron, and (c) electron–positron pair creation.

56. How many fundamental fermions are there in a water molecule?

Hubble Space Telescope photo of columns of gas and dust inside M16, the Eagle Nebula. The Eagle Nebula is thought to be a site of recent star formation.

ASTROPHYSICS AND COSMOLOGY

I n the previous chapter, we studied the tiniest objects in the universe—the elementary particles. Now we leap to the largest—stars and galaxies. These two extreme realms, elementary particles and the cosmos, are among the most intriguing and exciting subjects in science. And, surprising though it may seem, these distant realms are related in a fundamental way, as already hinted in Chapter 32.

Use of the techniques and ideas of physics to study the heavens is often referred to as **astrophysics**. At the base of our present theoretical understanding of the universe is Einstein's *general theory of relativity* and its theory of gravitation—for in the large-scale structure of the universe, gravity is the dominant force. General relativity serves also as the foundation for modern **cosmology**, which is the study of the universe as a whole. Cosmology deals especially with the search for a theoretical framework to understand the observed universe, its origin, and its future. The questions posed by cosmology are complex and difficult; the possible answers are often unimaginable. They are questions like "Has the universe always existed, or did it have a beginning in time?" Either alternative is difficult to imagine: time going back indefinitely into the past, or an actual moment when the universe began (but, then, what was there before?). And what about the size of the universe? Is it infinite in size? (It is hard to imagine infinity.) Or is it finite in size? (This is also hard to imagine, for if the universe is finite, it does not make sense to ask what is beyond it, because the universe is all there is.)

Our survey of astrophysics and cosmology will be necessarily brief and qualitative, but we will nonetheless touch on the major ideas. We begin with a look at what can be seen beyond the Earth.

33–1 Stars and Galaxies

According to the ancients, the stars, except for the few that seemed to move (the planets), were fixed on a sphere beyond the last planet. The universe was neatly self-contained, and we on Earth were at or near its center. But in the centuries following Galileo's first telescopic observations of the heavens in 1610, our view of the universe has changed dramatically. We no longer place ourselves at the center, and we view the universe as vastly larger. The distances involved are so great that we specify them in terms of the time it takes light to travel the given distance: for example, 1 light-second = $(3.0 \times 10^8 \text{ m/s})(1.0 \text{ s}) = 3.0 \times 10^8 \text{ m} = 3.0 \times 10^5 \text{ km}$; 1 light-minute = 18×10^6 km; and 1 **light-year** (ly) is

Light-year

$$1 \text{ ly} = (2.998 \times 10^8 \text{ m/s})(3.156 \times 10^7 \text{ s/yr}) = 9.46 \times 10^{15} \text{ m} \approx 10^{13} \text{ km}.$$

For specifying distances to the Sun and Moon, we usually use meters or kilometers, but we could specify them in terms of light. The Earth–Moon distance is 384,000 km, which is 1.28 light-seconds. The Earth–Sun distance is 1.50×10^{11} m, or 150,000,000 km; this is equal to 8.3 light-minutes. The most distant planet in the solar system, Pluto, is about 6×10^9 km from the Sun, or 6×10^{-4} ly. The nearest star to us, other than the Sun, is Proxima Centauri, about 4.3 ly away. (Note that the nearest star is 10,000 times farther from us than the farthest planet.)

On a clear moonless night, thousands of stars of varying degrees of brightness can be seen, as well as the elongated cloudy stripe known as the Milky Way (Fig. 33–1). It was Galileo who first observed, about 1610, that the Milky Way is comprised of countless individual stars. A century and a half later (about 1750), Thomas Wright suggested that the Milky Way was a flat disc of stars extending to great distances in a plane, which we call the **Galaxy** (Greek for "milky way").

Our Galaxy has a diameter of almost 100,000 light-years and a thickness of very roughly 2000 light-years. It has a bulging central nucleus and spiral arms (Fig. 33–2). Our Sun, which seems to be just another star, is located more than halfway from the center to the edge, about 28,000 ly from the center. Our Galaxy contains about 10^{11} stars. The Sun orbits the galactic center approximately once every 200 million years, so its speed is

FIGURE 33–1 A section of the Milky Way. The thin line is the trail of an artificial Earth satellite.

FIGURE 33–2 Our Galaxy, as it would appear from the outside: (a) "end view," in the plane of the disc; (b) "top view," looking down on the disc. (If only we could see it like this—from the outside!) (c) Infrared photograph of the inner reaches of the Milky Way, showing the central bulge of our Galaxy. This very wide angle photo extends over 180° of sky, and to be viewed properly it should be wrapped in a semicircle with your eyes at the center. The white dots are nearby stars.

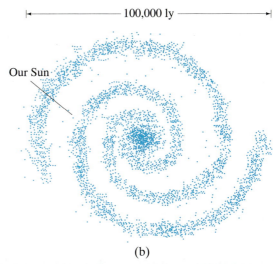

100,000 ly

Our Sun

(b)

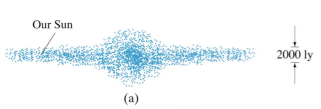

Our Sun

2000 ly

(a)

(c)

about 250 km/s relative to the center of the Galaxy. The total mass of all the stars in our Galaxy is about 3×10^{41} kg.

EXAMPLE 33–1 ESTIMATE **Our Galaxy's mass.** Estimate the total mass of our Galaxy using the orbital data of the Sun (including our solar system) about the center of the Galaxy. Assume that most of the mass of the Galaxy can be approximated as a uniform sphere of mass (the central bulge, Fig. 33–2a).

SOLUTION Our Sun and solar system orbit the center of the Galaxy, according to the best measurements as mentioned above, with a speed of about $v = 250$ km/s at a distance from the Galaxy center of about $r = 28{,}000$ ly. We use Newton's second law, $F = ma$, with a being the centripetal acceleration, $a = v^2/r$, and F being the universal law of gravitation (Chapter 5)

$$F = ma$$

$$G\frac{Mm}{r^2} = m\frac{v^2}{r}$$

where M is the mass of the Galaxy and m is the mass of our Sun and solar system. Solving this, we find

$$M = \frac{rv^2}{G} \approx \frac{(28{,}000 \text{ ly})(10^{16} \text{ m/ly})(2.5 \times 10^5 \text{ m/s})^2}{6.67 \times 10^{-11} \text{ N·m}^2/\text{kg}^2} \approx 3 \times 10^{41} \text{ kg}.$$

In terms of *numbers* of stars, if they are like our Sun ($m = 2.0 \times 10^{30}$ kg), there would be about $(3 \times 10^{41} \text{ kg})/(2 \times 10^{30} \text{ kg}) \approx 10^{11}$ or about 100 billion stars.

FIGURE 33–3 This globular star cluster is located in the constellation Hercules.

FIGURE 33–4 This gaseous nebula, found in the constellation Carina, is about 9000 light-years from us.

We can see by telescope, in addition to stars both within and outside the Milky Way, many faint cloudy patches in the sky which were all referred to once as "nebulae" (Latin for "clouds"). A few of these, such as those in the constellations Andromeda and Orion, can actually be discerned with the naked eye on a clear night. Some are **star clusters** (Fig. 33–3), groups of stars that are so numerous they appear to be a cloud. Others are glowing clouds of gas or dust (Fig. 33–4), and it is for these that we now mainly reserve the word **nebula**. Most fascinating are those that belong to a third category: they often have fairly regular elliptical shapes and seem to be a great distance beyond our Galaxy. Immanuel Kant (about 1755) seems to have been the first to suggest that these latter might be circular discs like our Galaxy: they appear elliptical because we see them at an angle, and are faint because they are so distant. At first it was not universally accepted that these objects were **extragalactic**—that is, outside our Galaxy. The very large telescopes constructed in this century revealed that individual stars could be resolved within these extragalactic objects and that many contained spiral arms. Edwin Hubble (1889–1953), who did much of this observational work in the 1920s using the 2.5-m (100-inch) telescope[†] on Mt. Wilson near Los Angeles, California, was also able to demonstrate that these objects were indeed extragalactic because of their great distances. The distance to the Andromeda nebula (a galaxy), for example, is over 2 million light-years, a distance 20 times greater than the diameter of our Galaxy. Thus it was determined that these nebulae are **galaxies** similar to ours. Today, the largest telescopes can see about 10^{11} galaxies. See Fig. 33–5. (Note that it is usual to capitalize the word galaxy only when it refers to our own.)

Galaxies tend to be grouped in **galaxy clusters**, with anywhere from a few to many thousands of galaxies in each cluster. Furthermore, clusters themselves seem to be organized into even larger aggregates: clusters of clusters of galaxies, or **superclusters**. The galaxies nearest us are about 2 million light-years away. The farthest detectable galaxies are thousands of times farther away, on the order of 10^{10} ly. (See Table 33–1.)

TABLE 33–1 Heavenly Distances

Object	Approx. Distance from Earth (ly)
Moon	4×10^{-8}
Sun	1.6×10^{-5}
Farthest planet in our solar system (Pluto)	6×10^{-4}
Nearest star (Proxima Centauri)	4.3
Center of our Galaxy	3×10^{4}
Nearest galaxy	2×10^{6}
Farthest galaxies	10^{10}

[†]2.5 m (= 100 inches) refers to the diameter of the curved objective mirror. The bigger the mirror, the more light it collects and the less diffraction there is, so more and fainter stars can be seen. See Chapter 25.

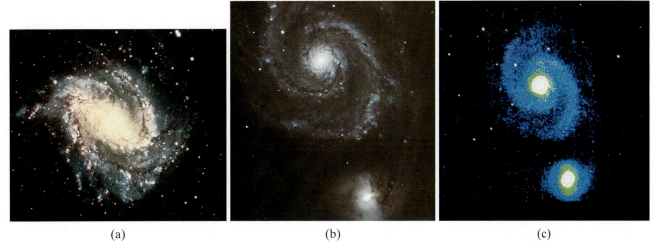

(a) (b) (c)

FIGURE 33–5 Photographs of galaxies. (a) Spiral galaxy in the constellation Hydra. (b) Two galaxies: the larger and more dramatic one is known as the Whirlpool galaxy. (c) The same galaxies as in (b), but this is a false-color infrared image which shows the arms of the spiral as being more regular than in the visible light photo (b); the different colors correspond to different light intensities.

From this perspective of a vast universe, we living creatures on Earth may feel very small and anonymous—a far cry from the view held only a few centuries ago that placed us at the very center of the universe. If this modern view of the universe seems too stark, however, we can take a philosophical approach: all of these observations and interpretations have been made here on Earth, and the ideas were created here on Earth. Who is to say that we are in any way unimportant in the universe, even though we may not hold a geometrically central position?

Besides the usual stars, clusters of stars, galaxies, and clusters and super-clusters of galaxies, the universe contains a number of other interesting objects. Among these are stars known as *red giants*, *white dwarfs*, *neutron stars*, *black holes* (at least theoretically), and exploding stars called *novae* and *supernovae*. In addition there are *quasars* ("quasistellar radio sources"), which, if we judge their distance correctly, are galaxies thousands of times brighter than ordinary galaxies. Furthermore, there is radiation that reaches the Earth but does not emanate from the bright pointlike objects we call stars: it is a background radiation that seems to arrive uniformly from all directions in the universe. We discuss all these phenomena in due course.

We have talked about the vast distances of objects in the universe. But how do we measure these distances? One basic technique employs simple geometry to measure the **parallax** of a star. By parallax we mean the apparent motion of a star, against the background of more distant stars,

How astronomical distances are measured

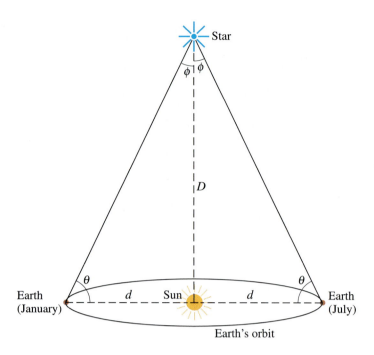

Star

ϕ | ϕ

D

θ

Earth
(January)

d Sun d

θ

Earth
(July)

Earth's orbit

FIGURE 33–6 Distance to a star determined by parallax. (Not to scale.)

due to the Earth's motion about the Sun. As shown in Fig. 33–6, the sighting angle of a star relative to the plane of Earth's orbit (angle θ) is measured at different times of the year. Since we know the distance d from Earth to Sun, we can reconstruct the right triangles shown in the figure and can determine[†] the distance D to the star.

EXAMPLE 33–3 ESTIMATE Distance to a star using parallax. Estimate the distance D to a star if the angle θ in Fig. 33–6 is measured to be 89.99994°.

SOLUTION The angle ϕ in Figure 33–6 is 0.00006°, or about 1.0×10^{-6} radians. Since $\tan \phi = d/D$, then the distance D to the star is

$$D = \frac{d}{\tan \phi} = \frac{d}{\phi} = \frac{1.5 \times 10^8 \text{ km}}{1.0 \times 10^{-6} \text{ rad}} = 1.5 \times 10^{14} \text{ km},$$

or about 15 ly. (We can use $\tan \phi \approx \phi$ since ϕ is very small.)

The distances to stars are often specified in terms of parallax angle given in seconds of arc: 1 second (1″) is $\frac{1}{60}$ of a minute (′) of arc, which is $\frac{1}{60}$ of a degree, so $1″ = \frac{1}{3600}$ of a degree. The distance, then, is specified in parsecs (meaning *par*allax angle in *sec*onds of arc), where the **parsec** (pc) is defined as $1/\phi$ and ϕ is given in seconds. In the Example we just did, $\phi = (6 \times 10^{-5})° (3600) = 0.22″$ of arc, so we would say the star is at a distance of $1/0.22″ = 4.5$ pc. It is easy to show that the parsec is given by

Parsec   1 pc = 3.26 ly.

Parallax can be used to determine the distance to stars as far away as 100 light-years (≈ 30 parsecs). Beyond that distance, parallax angles are too small to measure. For greater distances, more subtle techniques must

[†]This is essentially the way the heights of mountains are determined, by "triangulation."

be employed. We can, for example, compare the apparent brightnesses of galaxies, and—using the inverse square law (intensity drops off as the square of the distance)—estimate their relative distances. This method cannot be very precise since we cannot expect all galaxies to have the same intrinsic brightness. A better estimate compares the brightest stars in galaxies or the brightest galaxies in galaxy clusters. It is reasonable to assume, for example, that the brightest stars in all galaxies are similar and have about the same intrinsic luminosity. Consequently, their *apparent* brightness would be a measure of how far away they are.

Another important technique for estimating distance is via the "redshift" in the line spectra of elements, which is related to the expansion of the universe, as will be discussed in Section 33–4.

As we look farther and farther away, the measurement techniques are less and less reliable, so there is more and more uncertainty in the measurements of large distances.

33–2 Stellar Evolution: The Birth and Death of Stars

The stars appear unchanging. Night after night the heavens reveal no significant variations. Indeed, on a human time scale, the vast majority of stars (except novae and supernovae) change very little. Although stars *seem* fixed in relation to each other, many move sufficiently for the motion to be detected. Indeed, the speeds of stars relative to neighboring stars can be hundreds of km/s, but at their great distance from us, this motion is detectable only by careful measurement. Furthermore, there is a great range of brightness among stars. The differences in brightness are due both to differences in the amount of light stars emit and to their distances from us.

A useful parameter for a star or galaxy is its **absolute luminosity**, L, by which we mean the total power radiated in watts. Also important is the **apparent brightness**, l, defined as the power crossing unit area perpendicular to the path of the light at the Earth. Given that energy is conserved, and ignoring any absorption in space, the total emitted power L at a distance d from the star will be spread over a sphere of surface area $4\pi d^2$. If we let d be the distance from the star to the Earth, then L must be equal to $4\pi d^2$ times l (power per unit area at Earth). That is,

Luminosity and brightness of stars

$$l = \frac{L}{4\pi d^2}. \qquad (33\text{–}1)$$

EXAMPLE 33–4 **Apparent brightness.** Suppose a particular star has absolute luminosity equal to that of the Sun but is 10 pc away from Earth. By what factor will it appear dimmer than the Sun?

SOLUTION The luminosity L is the same for both stars, so the apparent brightness depends only on their relative distances. We use the inverse square law as stated above: the star appears dimmer by a factor

$$\frac{l_1}{l_2} = \frac{d_2^2}{d_1^2} = \frac{(1.5 \times 10^8 \text{ km})^2}{(10 \text{ pc})^2 (3.26 \text{ ly/pc})^2 (10^{13} \text{ km/ly})^2} \approx 2 \times 10^{-13},$$

where d_1 and d_2 are the distances from Earth to the star and to the Sun, respectively.

Careful study of nearby stars has shown that for most stars, the absolute luminosity depends on the mass: *the more massive the star, the greater its luminosity*. Another important parameter of a star is its surface temperature, which can be determined from the spectrum of electromagnetic frequencies it emits, just as for a blackbody (Section 27–2). As we saw in Chapter 27, the spectrum of hotter and hotter bodies shifts from predominantly lower frequencies (and longer wavelengths, such as red) to higher frequencies (and shorter wavelengths such as blue). Quantitatively, the relation is given by Wien's law (Eq. 27–2): the peak wavelength λ_P in the spectrum of light emitted by a blackbody (and stars are fairly good approximations to blackbodies) is inversely proportional to its kelvin temperature T; that is, $\lambda_P T = 2.90 \times 10^{-3}$ m·K. The surface temperatures of stars typically range from about 3500 K (reddish) to perhaps 50,000 K (bluish).

Luminosity increases with star's mass

An important astronomical discovery, made near the turn of the century, was that for most stars, the color is related to the absolute luminosity and therefore to the mass. A useful way to present this relationship is by the so-called Hertzsprung–Russell (H–R) diagram. On the H–R diagram, the horizontal axis shows the temperature T whereas the vertical axis is the luminosity L, and each star is represented by a point on the diagram, Fig. 33–7. Most stars fall along the diagonal band termed the **main sequence**. Starting at the lower right we find the coolest stars, reddish in color; they are the least luminous and therefore of low mass. Farther up toward the left we find hotter and more luminous stars that are yellowish-white, like our Sun. Still farther up we find still more massive and more luminous stars, bluish in color. Stars that fall on this diagonal band are called *main-sequence stars*. There are also stars that fall outside the main sequence. Above and to the right we find extremely large stars, with high luminosities but with low (reddish) color temperature: these are called *red giants*. At the lower left, there are a few stars of low luminosity but with high temperature: these are the *white dwarfs*.

H–R diagram

Main-sequence stars

Red giants

White dwarfs

Before considering the different types of stars, let us first look at a couple of Examples that show how physical principles can be used to gain information about stars.

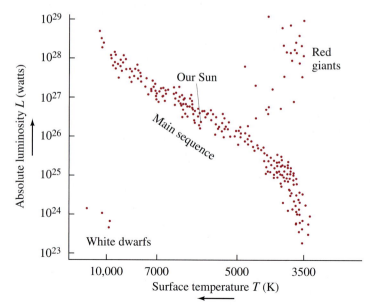

FIGURE 33–7
Hertzsprung–Russell (H–R) diagram.

EXAMPLE 33–5 **Determining star temperatures and size.** Suppose that the distances to two nearby stars can be reasonably estimated and this data, together with their measured apparent brightnesses, suggests that the two stars have about the same absolute luminosity, L. The spectrum of one of the stars peaks at about 700 nm (so it is reddish). The spectrum of the other peaks at about 350 nm (bluish). Use Wien's law (Eq. 27–2) and the Stefan–Boltzmann law (Section 14–9) to determine (a) the surface temperature of each star, and (b) how much larger one star is than the other.

SOLUTION (a) Wien's law (Eq. 27–2) states that $\lambda_p T = 2.90 \times 10^{-3}$ m·K. So the temperature of the reddish star is

$$T_r = \frac{2.90 \times 10^{-3} \text{ m·K}}{700 \times 10^{-9} \text{ m}} = 4140 \text{ K}.$$

The temperature of the bluish star will be double this since its peak wavelength is half (350 nm vs 700 nm); just to check:

$$T_b = \frac{2.90 \times 10^{-3} \text{ m·K}}{350 \times 10^{-9} \text{ m}} = 8280 \text{ K}.$$

(b) The Stefan-Boltzmann law, which we discussed in Chapter 14 (see Eq. 14–4), states that the power radiated *per unit area* of surface from a body is proportional to the fourth power of the kelvin temperature, T^4. Now the temperature of the bluish star is double that of the reddish star, so the bluish one must radiate $(2)^4 = 16$ times as much energy per unit area. But we are given that they have the same luminosity (the same total power output), so the surface area of the blue star must be $\frac{1}{16}$ that of the red one. Since the surface area of a sphere is $4\pi r^2$, we conclude that the radius of the reddish star is $\sqrt{16} = 4$ times larger than the radius of the bluish star (or $4^3 = 64$ times the volume).

EXAMPLE 33–6 **ESTIMATE** **Distance to a star using H–R and color.** Suppose that detailed study of a certain star suggests that it most likely fits on the main sequence of an H–R diagram. Its measured apparent brightness is $l = 1.0 \times 10^{-12}$ W/m^2, and the peak wavelength of its spectrum is $\lambda_p \approx 600$ nm. Estimate its distance from us.

SOLUTION The star's temperature, from Wien's law (Eq. 27–2), is

$$T \approx \frac{2.90 \times 10^{-3} \text{ m·K}}{600 \times 10^{-9} \text{ m}} \approx 4800 \text{ K}.$$

A star on the main sequence of an H–R diagram at this temperature has absolute luminosity of about $L \approx 1 \times 10^{26}$ W, read off of Fig. 33–7. Then, from Eq. 33–1,

$$d = \sqrt{\frac{L}{4\pi l}} \approx \sqrt{\frac{1 \times 10^{26} \text{ W}}{4(3.14)(1.0 \times 10^{-12} \text{ W/m}^2)}} \approx 3 \times 10^{18} \text{ m}.$$

Its distance from us in light-years and parsecs is

$$d = \frac{3 \times 10^{18} \text{ m}}{10^{16} \text{ m/ly}} \approx 300 \text{ ly} \approx \frac{300 \text{ ly}}{3.26 \text{ ly/pc}} \approx 90 \text{ pc}.$$

Why are there different types of stars, such as red giants and white dwarfs, as well as main-sequence stars? Were they all born this way, in the beginning? Or might each different type represent a different age in the life cycle of a star? Astronomers and astrophysicists today believe the latter is most likely the case. Note, however, that we cannot actually follow any but the tiniest part of the life cycle of any given star since they live for ages vastly greater than ours, on the order of millions or billions of years. Nonetheless, let us follow the process of **stellar evolution** from the birth to the death of a star, as astrophysicists have theoretically reconstructed it today.

Birth of a star

Stars are born, it is believed, when gaseous clouds (mostly hydrogen) contract due to the pull of gravity. A huge gas cloud might fragment into numerous contracting masses, each mass centered in an area where the density was only slightly greater than that at nearby points. Once such *"globules"* formed, gravity would cause each to contract in toward its center of mass. As the particles of such a *protostar* accelerate inward, their kinetic energy increases. When the kinetic energy is sufficiently high, the Coulomb repulsion that keeps the hydrogen nuclei apart can be overcome and nuclear fusion can take place. In a star like our Sun, the "burning" of hydrogen[†] occurs via the proton–proton cycle (Section 31–3, Eqs. 31–3), in which four protons fuse to form a $_2^4$He nucleus with the release of γ rays and neutrinos. These reactions require a temperature of about 10^7 K, corresponding to an average KE (kT) of about 1 keV. In more massive stars, the carbon cycle produces the same net effect: Four $_1^1$H produce a $_2^4$He— see Section 31–3. The fusion reactions take place primarily in the core of a star, where T is sufficiently high. (The surface temperature is, of course, much lower—on the order of a few thousand kelvins.) The tremendous release of energy in these fusion reactions produces a pressure sufficient to halt the gravitational contraction, and our protostar, now really a young *star*, stabilizes on the main sequence. Exactly where the star falls along the main sequence depends on its mass. The more massive the star, the farther up (and to the left) it falls on the diagram. To reach the main sequence requires perhaps 30 million years, if it is a star like our Sun, and (assuming our theory is right) it will remain there[‡] about 10 billion years (10^{10} yr). Although most stars are billions of years old, there is evidence that stars are actually being born at this moment.

Fusion begins when T (and $\overline{\text{KE}}$) is large enough

Reaching the main sequence

As hydrogen "burns"—that is, fuses to form helium—the helium that is formed is denser and tends to accumulate in the central core where it was formed. As the core of helium grows, hydrogen continues to "burn" in a shell around it, Fig. 33–8. When much of the hydrogen within the core has been consumed, the production of energy decreases and is no longer sufficient to prevent the huge gravitational forces from once again causing the core to contract and heat up. The hydrogen in the shell around the core then "burns" even more fiercely because of this rise in temperature,

FIGURE 33–8 A shell of "burning" hydrogen (fusing to become helium) surrounds the core where the newly formed helium gravitates.

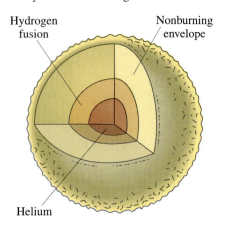

Hydrogen fusion

Nonburning envelope

Helium

[†]The word "burn" is put in quotation marks because these high-temperature fusion reactions occur via a *nuclear* process, and must not be confused with ordinary burning (of, say, paper, wood, or coal) in air, which is a chemical reaction at the *atomic* level (and at a much lower temperature).

[‡]More massive stars, since they are hotter and the Coulomb repulsion is more easily overcome, "burn" much more quickly, and so use up their fuel faster, resulting in shorter lives. A star 10 times more massive than our Sun, for example, will remain on the main sequence only for about 10^7 years. Stars less massive than our Sun live much longer than our Sun's 10^{10} yr.

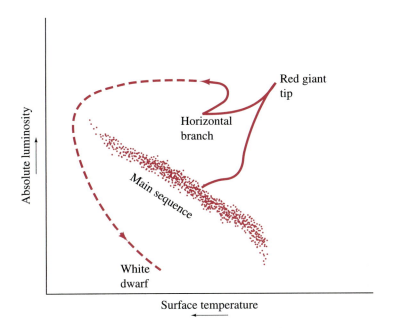

FIGURE 33–9 Evolution of a star like our Sun represented on an H–R diagram.

causing the outer envelope of the star to expand and to cool. The surface temperature, thus reduced, produces a spectrum of light that peaks at longer wavelength. By this time the star has left the main sequence. It has become redder, and as it has grown in size, it has become more luminous. So it will have moved to the right and upward on the H–R diagram, as shown in Fig. 33–9. As it moves upward, it enters the **red giant** stage. Thus, *Red giants* theory explains the origin of red giants as a natural step in a star's evolution. Our Sun, for example, has been on the main sequence for about $4\frac{1}{2}$ billion years. It will probably remain there another 4 or 5 billion years. (We can take comfort in that!) When our Sun leaves the main sequence, it is expected to grow in size (as a red giant) until it occupies all the volume out to approximately the present orbit of Earth.

As a star's envelope expands, the core is shrinking and heating up. When the temperature reaches about 10^8 K, helium nuclei, with their greater charge and hence greater electrical repulsion, can then undergo fusion. The reactions are

$$^4_2\text{He} + {}^4_2\text{He} \rightarrow {}^8_4\text{Be} + \gamma,$$
$$^4_2\text{He} + {}^8_4\text{Be} \rightarrow {}^{12}_6\text{C} + \gamma. \qquad (33\text{--}2) \qquad \textit{Nucleosynthesis}$$

These two reactions occur in quick succession, and the net effect is

$$3\,{}^4_2\text{He} \rightarrow {}^{12}_6\text{C}.$$

The burning of helium causes a rapid and major change in the star, and the star moves rapidly to the "horizontal branch" on the H–R diagram (see Fig. 33–9). Further reactions are possible, creating elements of higher and higher Z, up to $Z = 10$ or 12. If the mass of the star is great enough, higher Z elements can be formed. The star can get even hotter and in the *Production of* range $T = 2.5\text{--}5 \times 10^9$ K, nuclei as heavy as ${}^{56}_{26}\text{Fe}$ and ${}^{56}_{28}\text{Ni}$ can be made. *heavy elements* But here the process of **nucleosynthesis**, the formation of heavy nuclei from lighter ones by fusion, ends. As we saw in Fig. 30–1, the average binding energy per nucleon begins to decrease for A greater than about

60. Further fusions would *require* energy, rather than release it. (Elements heavier than Ni are probably formed mainly by neutron capture, in supernova explosions, as we shall discuss shortly.)

What happens next depends on the mass of the star. If the star has a residual mass less than about 1.4 solar masses[†] (known as the *Chandrasekhar limit*), no further fusion energy can be obtained and the star collapses under the action of gravity. As it shrinks, the star cools and typically follows the route shown in Fig. 33–9, according to theory, descending from the upper left downward, becoming a **white dwarf**. A white dwarf with a mass equal to that of the Sun would be about the size of the Earth. What determines its size is that it collapses to the point at which the electron clouds of the atoms start to overlap: at this point it collapses no further because, as the Pauli exclusion principle claims, no two electrons can be in the same quantum state. A white dwarf continues to lose internal energy, decreasing in temperature and becoming dimmer and dimmer until its light goes out. It has then become a *black dwarf*, a dark cold chunk of ash.

White dwarf

More massive stars (residual mass greater than 1.4 solar masses) are thought to follow a quite different scenario. A star with this great a mass can contract under gravity and heat up even further, reaching extremely high temperatures. The KE of the nuclei is then so high that fusion of elements heavier than iron is possible even though the reactions are energy-requiring. Furthermore the high-energy collisions can also cause the breaking apart of iron and nickel nuclei into He nuclei, and eventually into protons and neutrons:

$$^{56}_{26}\text{Fe} \rightarrow 13 \ ^{4}_{2}\text{He} + 4\text{n}.$$

$$^{4}_{2}\text{He} \rightarrow 2\text{p} + 2\text{n}.$$

These are energy-requiring (endoergic) reactions, but at such extremely high temperature and pressure there is plenty of energy available, enough even to force electrons and protons together to form neutrons in inverse β decay:

$$\text{e}^- + \text{p} \rightarrow \text{n} + \nu.$$

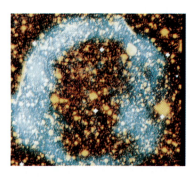

FIGURE 33–10 Computer-enhanced false-color image of the shell of material expanding outward from "Tycho's supernova" (seen by Tycho Brahe in 1572).

Neutron stars

As the core contracts under the huge gravitational forces, the tremendous mass becomes essentially an enormous nucleus made up almost exclusively of neutrons. The size of the star is no longer limited by the exclusion principle applied to electrons, but rather applied to neutrons, and the star begins to contract rapidly toward forming an enormously dense **neutron star**. The contraction of the core would mean a great reduction in gravitational potential energy. Somehow this energy would have to be released. Indeed, it was suggested in the 1930s that the final core collapse to a neutron star may be accompanied by a catastrophic explosion whose tremendous energy could form virtually all elements of the periodic table and blow away the entire outer envelope of the star (Fig. 33–10), spreading its contents into interstellar space. Such explosions are believed to produce one type of observed **supernova** (there may be other mechanisms as well). Indeed, the presence of heavy elements on Earth and in our solar system suggests that our solar system may have formed from the debris from at least one supernova.

Supernovae

[†]This cutoff refers to the mass of the star remaining *after* mass loss from the red giant stage. A star whose residual mass is 1.4 solar masses had an original mass of perhaps 6 to 8 solar masses.

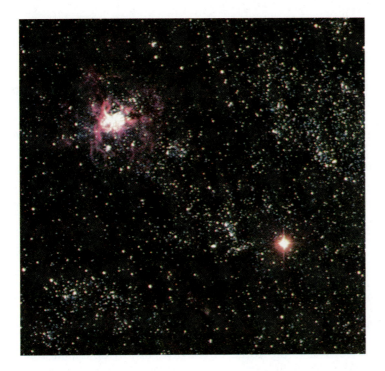

In a supernova explosion, a star's brightness is observed to suddenly increase billions of times in a period of just a few days and then fade away over the next few months. In February 1987 such a supernova occurred in an appendage to our Galaxy known as the Large Magellanic Cloud, only 170,000 ly away, and was visible to the naked eye in the southern hemisphere (Fig. 33–11). Supernovae are seen relatively often in distant galaxies, but this recent one, called SN1987a, is the first to have occurred close enough to be visible to the naked eye since the famous one of 1604 observed by Kepler and Galileo (and thus SN1987a is the first since the invention of the telescope). A supernova visible to the naked eye was observed by Chinese astronomers in A.D. 1054, and its remains are still visible in the sky (in the Crab Nebula), in the midst of which there is now a **pulsar**. Pulsars are astronomical objects that emit sharp pulses of radiation at regular intervals, on the order of a second. They are now believed to be neutron stars which, because of conservation of angular momentum, increase greatly in rotational speed as their moment of inertia decreases during their contraction. The intense magnetic field of such a rapidly rotating star (perhaps as high as 10^6 to 10^{10} T) could trap and accelerate charged particles, which then give off radiation in a beam that rotates with the star. The discovery of pulsars in 1967 has lent support to the theory that neutron stars are the result of supernova explosions.

Pulsars

The core of a neutron star contracts to the point at which all neutrons are as close together as they are in a nucleus. That is, the density of a neutron star is on the order of 10^{14} times that of normal solids and liquids on Earth. A neutron star that has a mass 1.5 times that of our Sun would have a diameter of only about 10 km.

If the mass of a neutron star is less than about two or three solar masses, its subsequent evolution is thought to be similar to that of a white dwarf. If the mass is greater than this, the gravitational force may become so strong that the star contracts to an even smaller diameter and an even greater density, overcoming, in a sense, even the neutron exclusion principle. Gravity would then be so strong that even light emitted from it could not escape—it would be pulled back in by the force of gravity. In other words, the escape velocity from such a star is greater than c, the speed of light. Since no radiation could escape from such a star, we could not see it—it would be black. A body may pass by it and be deflected by its gravitational field, but if it came too close, it would be swallowed up, never to escape. This is a **black hole**. Black holes are predicted by theory to exist. Experimentally, evidence for their existence is strong but not everyone agrees it is sufficient to fully confirm the existence of black holes. One possibility is there may be a giant black hole at the center of our Galaxy—its mass is estimated at several million times that of the Sun—and it seems possible that many or all galaxies have black holes at their centers.

Black holes

33–3 General Relativity: Gravity and the Curvature of Space

We have seen that the force of gravity plays a dominant role in the processes that occur in stars and, in general, in the evolution of the universe as a whole. The reasons gravity, and not one of the other of the four forces in nature, plays the dominant role in the universe are (1) it is long-range and (2) it is always attractive. The strong and weak nuclear forces act over very short distances only, on the order of the size of a nucleus; hence they do not act over astronomical distances (although of course they act between nuclei and nucleons in stars to produce nuclear reactions). The electromagnetic force, like gravity, acts over great distances; but it can be either attractive or repulsive. And since the universe does not seem to contain large areas of net electric charge, a large net force does not occur. But gravity acts as an attractive force between *all* masses, and there are large accumulations in the universe of only the one "sign" of mass (not + and − as with electric charge).

However, the force of gravity as Newton described it in his law of universal gravitation shows discrepancies on a cosmological scale. Einstein, in his general theory of relativity, developed a theory of gravity that resolves these problems and forms the basis of cosmological dynamics.

In the *special theory of relativity* (Chapter 26), Einstein concluded that there is no way for an observer to determine whether a given frame of reference is at rest or is moving at constant velocity in a straight line. Thus, Einstein said, the laws of physics must be the same in different inertial reference frames. (As we saw, the laws can be the same even though the specific *paths* of objects are different in different reference frames.) To consider only uniformly moving reference frames is somewhat restricting. What about the general case of motion, where reference frames can be *accelerating*?

It is in the **general theory of relativity** that Einstein tackled the problem of accelerating reference frames and developed a theory of gravity. The mathematics of general relativity is very complex, so our discussion will be mainly qualitative.

We begin with Einstein's famous **principle of equivalence**, which states that

> **No observer can determine by experiment whether he or she is accelerating or is rather in a gravitational field.**

If some observers sensed that they were accelerating (as in a vehicle speeding around a sharp curve), they could not prove by any experiment that in fact they weren't simply experiencing the pull of a gravitational field. Conversely, we might think we are being pulled by gravity when in fact we are undergoing an "inertial" acceleration having nothing to do with gravity. For example, pilots making a steeply banked turn in fog often have this experience, and cannot tell in which direction the Earth lies.

As a thought experiment, consider a person in a freely falling elevator near the Earth's surface. If our observer held out a book and let go of it, what would happen? Gravity would pull it downward toward the Earth, but at the same rate ($g = 9.8 \, \text{m/s}^2$) at which the person and elevator were falling. So the book would hover right next to the person's hand (Fig. 33–12). The effect is exactly the same as if this reference frame were at rest and *no* forces were acting. On the other hand, suppose the elevator were far out in space where there is no gravitational field. If the person released the book, it would float, just as it does in Fig. 33–12. If, instead, the elevator (out in space) were accelerating upward at an acceleration of $9.8 \, \text{m/s}^2$, the book as seen by our observer would fall to the floor with an acceleration of $9.8 \, \text{m/s}^2$, just as if it were falling because of gravity. According to the principle of equivalence, the observer could do no experiment to determine whether the book fell because the elevator was accelerating upward at $a = 9.8 \, \text{m/s}^2$ in the absence of gravity, or because a gravitational field with $g = 9.8 \, \text{m/s}^2$ was acting downward and he was at rest (say on the Earth). The two descriptions are equivalent.

The principle of equivalence is related to the concept of mass and to the idea that there are two types of mass. For any force, Newton's second law says that $F = ma$, where m is the mass—or more precisely, the **inertial mass**. The more inertial mass a body has, the less it is affected by a given force (the less acceleration it undergoes). You might say that inertial mass represents resistance to any type of force whatever. The second type of mass is **gravitational mass**. When one body attracts another by the gravitational force (Newton's law of universal gravitation, $F \propto mm'/d^2$, Chapter 5), the strength of the force is proportional to the product of the gravitational masses of the two bodies. This is much like the electric force between two bodies that is proportional to the product of their electric charges. The electric charge of a body is clearly not related to its inertial mass; so why should we expect that a body's gravitational mass (call it gravitational charge if you like) be related to its inertial mass? We have, up to now, assumed they were the same. Why? Because no experiment—not even high-precision experiments—has been able to discern any measurable difference between inertial and gravitational mass. This, then, is another way to state the equivalence principle: *gravitational mass is equivalent to inertial mass.*

The principle of equivalence can be used to show that light ought to be deflected due to the gravitational force of a massive body. Let us consider a thought experiment to get the idea. Consider an elevator in free space where no gravity acts. If there is a hole in the side of the elevator

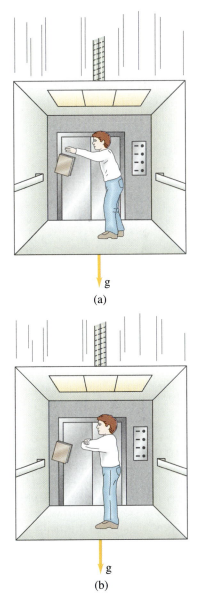

FIGURE 33–12 A freely falling elevator. The released book hovers next to the owner's hand.

(a)

(b)

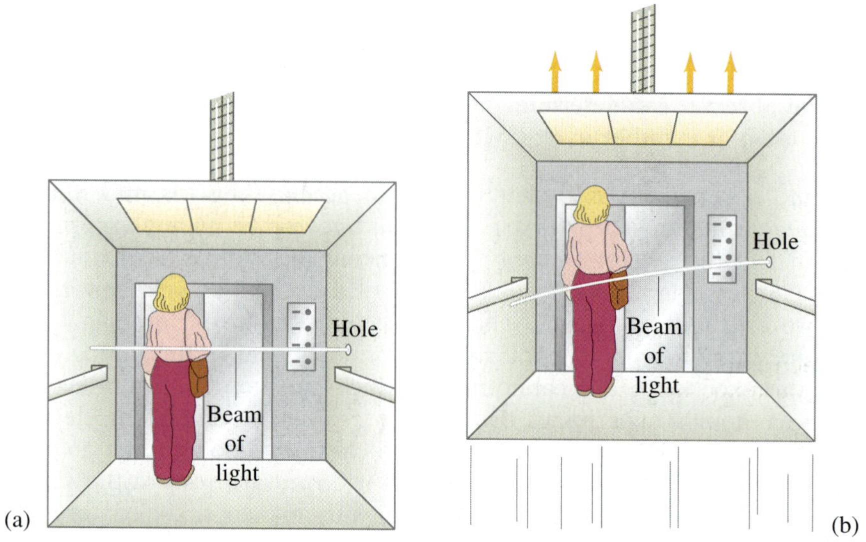

FIGURE 33–13 (a) Light beam goes straight across an elevator not accelerating. (b) The light beam bends (exaggerated) in an elevator accelerating in an upward direction.

(a)

Hole

Beam of light

(b)

Hole

Beam of light

and a beam of light enters from outside, the beam travels straight across the elevator and makes a spot on the opposite side if the elevator is at rest (Fig. 33–13a). If the elevator is accelerating upward as in Fig. 33–13b, the light beam still travels straight across in a reference frame at rest. In the upwardly accelerating elevator, however, the beam is observed to curve downward. Why? Because during the time the light travels from one side of the elevator to the other, the elevator is moving upward at ever-increasing speed. Now, according to the equivalence principle, an upwardly accelerating reference frame is equivalent to a downward gravitational field. Hence, we can picture the curved light path in Fig. 33–13b as being the effect of a gravitational field. Thus we expect gravity to exert a force on a beam of light and to bend it out of a straight-line path!

That light is affected by gravity is an important prediction of Einstein's general theory of relativity. And it can be tested. The amount a light beam would be deflected from a straight-line path must be small even when passing a massive body. (For example, light near the Earth's surface after traveling 1 km is predicted to drop only about 10^{-10} m, which is equal to the diameter of a small atom and not detectable.) The most massive body near us is the Sun, and it was calculated that light from a distant star would be deflected by 1.75″ of arc (tiny but detectable) as it passed near the Sun (Fig. 33–14). However, such a measurement could be made only

FIGURE 33–14 (a) Three stars in the sky. (b) If the light from one of these stars passes very near the Sun, whose gravity bends the rays, the star will appear higher than it actually is.

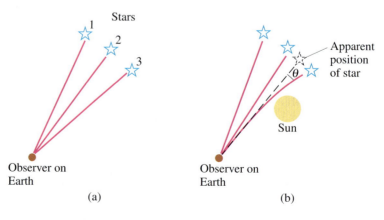

during a total eclipse of the Sun, so the Sun's tremendous brightness would not overwhelm the starlight passing near its edge. An opportune eclipse occurred in 1919 and scientists journeyed to the South Atlantic to observe it. Their photos of stars around the Sun revealed shifts in accordance with Einstein's prediction. One newspaper flaunted the headline "Light Caught Bending." Newton's universal law of gravitation, coupled with the equivalence principle, predicts a gravitational deflection of light even when light is considered as photons of mass $m = E/c^2$, but the predicted deflection is half that predicted by Einstein's general relativity; and it is the latter that is observed. (See Fig. 33–15.)

Now a light beam must travel by the shortest, most direct, path between two points. If it didn't, some other object could travel between the two points in a shorter time and thus have a greater speed than the speed of light—a clear contradiction of the special theory of relativity. If a light beam can follow a curved path (as discussed above), then this curved path must be the shortest distance between the two points—which suggests that *space itself is curved* and that it is the gravitational field that causes the curvature. Indeed, the curvature of space—or rather, of four-dimensional space-time—is a basic aspect of general relativity.

What is meant by *curved space*? To understand, let us recall that our normal method of viewing the world is via Euclidean plane geometry. In Euclidean geometry, there are many axioms and theorems we take for granted, such as that the sum of the angles of any triangle is 180°. Other geometries, non-Euclidean which involve curved space, have also been imagined by mathematicians. Now it is hard enough to imagine three-dimensional curved space, much less curved four-dimensional space–time. So let us explain the idea of curved space by using two-dimensional surfaces.

Consider, for example, the two-dimensional surface of a sphere. It is clearly curved, Fig. 33–16, at least to us who view it from the outside—from our three-dimensional world. But how would hypothetical two-dimensional creatures determine whether their two-dimensional space were flat (a plane) or curved? One way would be to measure the sum of the angles of a triangle. If the surface is a plane, the sum of the angles is 180°, as we learn in plane geometry. But if the space is curved, and a sufficiently large triangle is constructed, the sum of the angles will *not* be 180°. To construct a triangle on a curved surface, say a sphere, we must use the equivalent of a straight line: that is, the shortest distance between two points, which is called a **geodesic**. On a sphere, a geodesic is an arc of a great circle (an arc contained in a plane passing through the center of the sphere) such as the Earth's equator and the Earth's longitude lines. Consider, for example, the large triangle of Fig. 33–16 whose sides are two "longitude" lines passing from the "north pole" to the equator, a part of which forms the third side. The two longitude lines make 90° angles with the equator (look at a world globe to see this more clearly). If they make, say, a 90° angle with each other at the north pole, the sum of these angles is $90° + 90° + 90° = 270°$. This is clearly *not* a Euclidean space. Note, however, that if the triangle is small in comparison to the radius of the sphere, the angles will add up to nearly 180°, and the triangle (and space) will seem flat.

Another way to test the curvature of space is to measure the radius r and circumference C of a large circle. On a plane surface, $C = 2\pi r$. But on a two-dimensional spherical surface, C is *less* than $2\pi r$, as can be seen in

Light bending

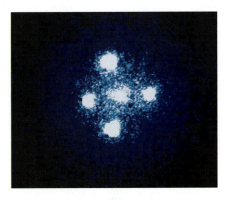

FIGURE 33–15 Photograph, taken by the Hubble Space Telescope, of the so-called "Einstein cross" (or Huchra's cross after its discoverer, John Huchra). It is thought to represent "gravitational lensing": the central spot is a relatively nearby galaxy, whereas the four other spots are thought to be images of a single quasar *behind* the galaxy, and the galaxy bends the light coming from the quasar to produce the four images. See also Fig. 33–14. [If the shape of the nearby galaxy were perfectly symmetric, we would expect the "image" of the distant quasar to be a circular ring or halo, instead of the four spots seen here.]

FIGURE 33–16 On a two-dimensional curved surface, the sum of the angles of a triangle may not be 180°.

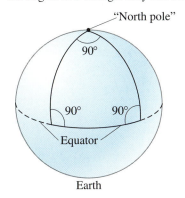

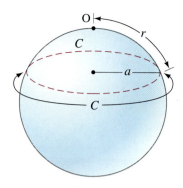

FIGURE 33–17 On a spherical surface, a circle of circumference C is drawn about point O as the center. The radius of the circle (not the sphere) is the distance r along the surface. (Note that in our three-dimensional view, we can tell that $2\pi a = C$; since $r > a$, then $2\pi r > C$.)

The universe: open or closed?

FIGURE 33–18 Example of a two-dimensional surface with negative curvature.

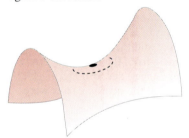

Fig. 33–17. The proportionality constant between C and r is *less* than 2π. Such a surface is said to have *positive curvature*. On the saddlelike surface of Fig. 33–18, the circumference of a circle is greater than $2\pi r$, and the sum of the angles of a triangle is less than 180°. Such a surface is said to have a *negative curvature*.

Now, what about our universe? On a large scale (not just near a large mass), what is the overall curvature of the universe? Does it have positive curvature, negative curvature, or is it flat (zero curvature)? In the nineteenth century, Carl Friedrich Gauss (1777–1855) tried to determine whether our natural three-dimensional space deviated from Euclidean space by measuring the angles of a triangle formed by three mountain peaks using light rays as sides of the triangle. He was unable to detect any deviation from 180°, nor have experiments today detected any deviation.

Nonetheless, the question of the curvature of space in the real world is an important one in cosmology. And the answer is still not known. If the universe has a positive curvature, then the universe is *finite*, or *closed*. This does not mean that in such a universe the stars and galaxies would extend out to a certain boundary, beyond which there is empty space. Rather, galaxies would be spread throughout the space, and the space would fold back and "close on itself." There is no boundary or edge in such a universe. If a particle were to move in a straight line in a particular direction, it would eventually return to the starting point—albeit eons of time later. To ask "What is beyond such a closed universe?" is futile, since the space of the universe is all there is. If the curvature of space is zero or negative, on the other hand, the universe would be *open*. It would just go on and on and never fold back on itself. An open universe would be *infinite*. Whether the universe is open or closed depends, in part, on how much total mass there is in the universe, as we will discuss in Section 33–7. If the mass is great enough, it bends space into a positively curved, closed, and finite space.

According to Einstein's theory, space–time is curved, especially near massive bodies. To visualize this, we might think of space as being like a thin rubber sheet; if a heavy weight is hung from it, it curves as shown in Fig. 33–19. The weight corresponds to a huge mass that causes space (space itself!) to curve. Thus, in Einstein's theory[†] we do not speak of the

FIGURE 33–19 Rubber-sheet analogy for space–time curved by matter.

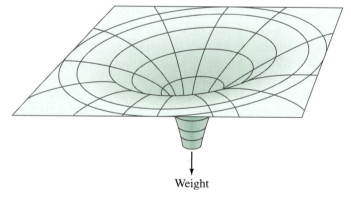

Weight

[†]Alexander Pope (1688–1744) wrote an epitaph for Newton:
"Nature, and Nature's laws lay hid in night:
God said, *Let Newton be!* and all was light."
Sir John Squire (1884–1958), perhaps uncomfortable with Einstein's profound thoughts, added:
"It did not last: the Devil howling '*Ho!*
Let Einstein be!' restored the status quo."

FIGURE 33-20 Artist's conception of how a black hole, which is one star of a binary pair, might pull matter from the other (more normal) star.

"force" of gravity acting on bodies. Instead we say that bodies and light rays move along geodesics (the equivalent of straight lines in plane geometry) in curved space–time. Thus, a body at rest or moving slowly near the great mass of Fig. 33–19 would follow a geodesic toward that body.

The extreme curvature of space–time shown in Fig. 33–19 could be produced by a **black hole**. A black hole, as we saw in the previous Section, is so dense that even light cannot escape from it. To become a black hole, a body of mass M must undergo **gravitational collapse**, contracting by gravitational self-attraction to within a radius called the **Schwarzschild radius**:

Black holes

$$R = \frac{2GM}{c^2},$$

where G is the gravitational constant and c is the speed of light.

How might we observe black holes? We cannot see them because no light can escape from them. They would be black objects against a black sky. But they do exert a gravitational force on nearby bodies. The suspected black hole at the center of our Galaxy was inferred by examining the motion of matter in its vicinity. Another technique is to examine stars which appear to be rotating as if they were members of a *binary system* (two stars rotating about their common center of mass), although the companion is invisible. If the unseen star is a black hole, it might be expected to pull off gaseous material from its visible companion (Fig. 33–20). As this matter approached the black hole, it would be highly accelerated and should emit X-rays of a characteristic type. One of the candidates for a black hole is in the binary star Cygnus X-1.

33-4 The Expanding Universe

We discussed in Section 33–2 how individual stars evolve from their birth to their death as white dwarfs, neutron stars, and black holes. But what about the universe as a whole: is it static or does it evolve? The evolution of stars suggests that the universe as a whole evolves.

One of the most important scientific ideas of this century proposes that distant galaxies are racing away from us, and that the farther they are from us, the faster they are moving away. How astronomers arrived at this astonishing idea and what it means for the past history of the universe as well as its future, will occupy us for the remainder of the book.

Hubble and the redshift

The idea that the universe is expanding was first put forth by Hubble in 1929. It was based on observations of the Doppler shift of light emitted by stars. In Chapter 12 we discussed how the frequency and wavelength of sound are altered if the source is moving toward or away from an observer. If the source moves toward us, the frequency is higher and the wavelength is shorter. If the source moves away from us, the frequency is lower and the wavelength is longer. The **Doppler effect** occurs also for light, but the shifted wavelength or frequency is given by a formula slightly different from that for sound because for light (according to special relativity), we can make no distinction between motion of the source and motion of the observer. (Recall that sound travels in a medium, such as air, but light does not; according to relativity, there is no ether.) According to special relativity, the formula is (see Problem 31)

$$\lambda' = \lambda \sqrt{\frac{1 + v/c}{1 - v/c}}, \qquad (33\text{--}3)$$

where λ is the emitted wavelength as seen in the source's reference frame, and λ' is the wavelength measured in a frame moving with velocity v away from the source along the line of sight. Note that this relation depends only on the relative velocity v. (For relative motion toward each other, $v < 0$ in

Redshift

this formula.) Thus, when a source emits light of a particular wavelength and the source is moving away from us, the wavelength appears longer to us: the color of the light (if it is visible) is shifted toward the red end of the visible spectrum, an effect known as a **redshift**. If the source moves toward us, the color shifts toward the blue (shorter wavelength) end of the spectrum. The amount of shift depends on the velocity of the source (Eq. 33–3). For speeds not too close to the speed of light, it is easy to show (Problem 30) that the fractional change in wavelength is proportional to the speed of the source to or away from us (as was the case for sound):

$$\frac{\Delta\lambda}{\lambda} = \frac{\lambda' - \lambda}{\lambda} \approx \frac{v}{c}. \qquad [v \ll c] \quad (33\text{--}4)$$

In the spectra of stars and galaxies, lines are observed that correspond to lines in the known spectra of particular atoms (see Section 27–9). What Hubble found was that the lines seen in the spectra of galaxies were generally *redshifted*, and that the amount of shift seemed to be approximately proportional to the distance of the galaxy from us. That is, the velocity, v, of a galaxy moving away from us is proportional to its distance, d, from us:

Hubble's law

$$v = Hd. \qquad (33\text{--}5)$$

This is **Hubble's law**, one of the most important astronomical ideas of this century. The constant H is called the **Hubble parameter**. Hubble's law does not work well for nearby galaxies—in fact some are actually moving toward us ("blueshifted"); but this is believed to merely represent random motion of the galaxies. For more distant galaxies, the velocity of recession (Hubble's law) is much greater than that of random motion, and so is dominant.

The value of H is not known very precisely. It is generally taken to be about

$$H \approx 80 \text{ km/s/Mpc}$$

Hubble's constant

(that is, 80 km/s per megaparsec of distance). If we use light-years for distance, then $H \approx 25$ km/s per million light-years of distance. However, H could be as low as 50 km/s/Mpc or as high as about 100 km/s/Mpc (15 to 30 km/s/Mly). This rather large uncertainty arises mainly from the uncertainty in distance measurement for remote galaxies.

What does it mean that distant galaxies are all moving away from us, and with ever greater speed the farther they are from us? It is as if there had been a great explosion at some distant time in the past. And at first sight we seem to be in the middle of it all. But we aren't, necessarily. The expansion appears the same from any other point in the universe. To understand why, see Fig. 33–21. In Fig. 33–21a we have the view from Earth (or from our Galaxy). The velocities of surrounding galaxies are indicated by arrows, pointing away from us, and greater for galaxies more distant from us. Now what if we were on the galaxy labeled A in Fig. 33–21a. From Earth, galaxy A appears to be moving to the right at a velocity, call it $\mathbf{v}_A$, represented by the arrow pointing to the right. If we were *on* galaxy A, Earth would appear to be moving to the left at velocity $-\mathbf{v}_A$. To determine the velocities of other galaxies relative to A, we vectorially add the velocity vector, $-\mathbf{v}_A$, to all the velocity arrows shown in Fig. 33–21a. This yields Fig. 33–21b, where we see clearly that the universe is expanding away from galaxy A as well; and the velocities of galaxies receding from A are proportional to their distance from A.

Thus the expansion of the universe can be stated as follows: all galaxies are racing away from *each other* at an average rate of about 80 km/s per megaparsec of distance between them (or 25 km/s per million light-years of separation). The ramifications of this idea are profound, and we discuss them in a moment.

There is, however, a class of objects called **quasars**, or "quasistellar radio sources," that do not seem to conform to Hubble's law. Quasars are as bright as nearby stars but display very large redshifts. If quasars are normal participants in the general expansion of the universe according to Hubble's law, their large redshifts suggest they are very distant. If they are so far away, they must be incredibly bright—sometimes thousands of times brighter than normal galaxies. On the other hand, it has been suggested that some quasars, at least, are not of abnormal brightness because

Quasars

FIGURE 33–21 Expansion of the universe looks the same from any point in the universe.

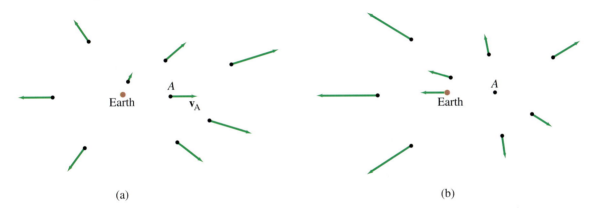

(a)

(b)

they are nearer than their redshifts suggest. In the first case we would have an unresolved brightness problem, in the second case an unresolved redshift problem. An interesting piece of evidence is that the population density of quasars seems to increase with distance from us. If they are at great distances from us—as their redshifts suggest—this would merely mean they were more common in the early universe than they are now. But if instead they are much closer to us—as their brightness suggests—it would seem that *we* are in a special place in the universe, the place where quasars are least populous. Most astronomers are unwilling to accept this, for it would violate a basic assumption of uniformity, the so-called *cosmological principle*, which we discuss in the next paragraph. A recent suggestion that could resolve the quasar question is that these mysterious objects may be cores of galaxies that emit enormous amounts of energy when they draw in nearby material, for example when galaxies "collide." If the center of a galaxy is a black hole, a large amount of matter accelerated toward it would give off a prodigious amount of energy. It seems possible, then, that quasars are powered by black holes.

Cosmological principle   A basic assumption in cosmology has been that on a large scale, the universe looks the same to observers at different places at the same time. In other words, the universe is both isotropic (looks the same in all directions) and homogeneous (would look the same if we were located elsewhere, say in another galaxy). This assumption is called the **cosmological principle**. On a local scale, say in our solar system or within our Galaxy, it clearly does not apply (the sky looks different in different directions). But it has long been thought to be valid if we look on a large enough scale, so that the average population density of galaxies and clusters of galaxies ought to be the same in different areas of the sky. The expansion of the universe (Fig. 33–21) is consistent with the cosmological principle, and the uniformity of the cosmic microwave background radiation (discussed in the next Section) supports it. But matter (i.e., galaxies), even on the largest scale, seems not to be homogeneous but tends to clump, as already discussed. Although astronomers are reluctant to give up the cosmological principle, there is now some doubt about its validity. One possible resolution might be that over 90 percent of the universe may be nonluminous dark matter, which might be uniformly distributed. In any case, the cosmological principle has aided us in treating the universe as a single evolving entity and not as a random collection of material bodies. Another way of stating the cosmological principle is this: There is nothing special about the Earth on a cosmological scale; our large-scale observations are no different from those that might be made elsewhere in the universe. What a change has occurred since that time, only a few hundred years ago, when we saw ourselves at the center of the universe!

The expansion of the universe, as described by Hubble's law, strongly suggests that galaxies must have been closer together in the past than they are now. Further, Hubble's law is consistent with all galaxies having been quite close together at some time in the past. This is, in fact, the basis of the *Big Bang* theory of the origin of the universe (discussed in the next section) which pictures the beginning of the universe as a great *Age of the universe*   explosion. Let us see what can then be said about the age of the universe.

We can estimate the age of the universe using the Hubble parameter, and let us choose the midpoint between the accepted extremes of $H = 15$ to $30\,\text{km/s/Mly}$, namely $H \approx 22\,\text{km/s}$ per million light-years. Then the time required for the galaxies to arrive at their present separations would be approximately (starting with $v = d/t$ and using Eq. 33–5):

$$t = \frac{d}{v} = \frac{d}{Hd} = \frac{1}{H} \approx \frac{(10^6 \text{ ly})(10^{13} \text{ km/ly})}{(22 \text{ km/s})(3 \times 10^7 \text{ s/y})} \approx 15 \times 10^9 \text{ yr,}$$

or roughly 15 billion years. The age of the universe calculated in this way is called the *characteristic expansion time* or "Hubble age." It is not very precise since we don't know the value of H precisely, and may be an overestimate since the galaxies have not been moving with fixed velocities but have been slowing down under the action of their mutual gravitational attraction.

There are two other independent checks on the age of the universe. The first is determination of the age of the Earth (and solar system) from radioactivity, primarily using uranium, which places the age of the solar system at about $4\frac{1}{2}$ billion years. Second, by using the theory of stellar evolution, the ages of stars have been estimated to be about 10 to 15 billion years. These independent and unrelated determinations are consistent with a Big Bang occurring, according to our best estimates today, 10 to 15 billion years ago. The lower value determined from radioactivity is consistent since we would expect the origin of the Earth (and the solidification of rocks) to have occurred somewhat after the origin of the universe as a whole.

Before discussing the Big Bang theory in more detail, we briefly discuss one of the alternatives to the Big Bang—the **steady-state model**— *Steady-state model* which assumes that the universe is infinitely old and on the average looks the same now as it always has.[†] Thus, according to the steady-state model, no large-scale changes have taken place in the universe as a whole, particularly no Big Bang. To maintain this view in the face of the recession of galaxies away from each other, mass–energy conservation must be violated. That is, matter is assumed to be created continuously, keeping the density of the universe constant. The rate of mass creation required is very small— about one nucleon per cubic meter every 10^9 years. This is much too small to be detected, and thus cannot be tested.

The steady-state model provided the Big Bang model with healthy competition in the 1940s and 50s. However, the discovery of the cosmic microwave background radiation (next Section) pushed the Big Bang model to the forefront.

33–5 The Big Bang and the Cosmic Microwave Background

The expansion of the universe seems to suggest, as we have seen, that the matter of the universe was once much closer together than it is now. This is the basis for the idea that the universe began 10 to 15 billion years ago with a huge explosion known affectionately as the **Big Bang**. *The Big Bang*

If there was a Big Bang, it must have occurred simultaneously at all points in the universe. If the universe is *finite*, the explosion would have taken place in a tiny volume approaching a point. However, this point of extremely dense matter is not to be thought of as a concentrated mass in the midst of a much larger space around it. Rather, the initial dense point

[†]That the universe should be uniform in time as well as in space is essentially an extension of the cosmological principle and is sometimes referred to as the *perfect cosmological principle.*

was the universe—the entire universe. There wouldn't have been anything else. If, on the other hand, the universe is *infinite*, then the explosion would have occurred at *all* points in the universe since an infinite universe, even if smaller at an earlier time, would still have been infinite. In either case, when we say the universe was once smaller than it is now, we mean that the average separation between galaxies was less. Thus, it is the *size of the universe itself* that has increased since the Big Bang.

What is the evidence supporting the Big Bang? First, the age of the universe as calculated from the Hubble expansion, from stellar evolution, and from radioactivity, all point to a consistent time of origin for the universe, as we saw in the last Section. Another, and crucial, piece of evidence was the discovery in the 1960s of the **cosmic microwave background** radiation (or CMB), which came about as follows.

The 2.7 K cosmic microwave background radiation

In 1964, Arno Penzias and Robert Wilson were experiencing difficulty with what they assumed to be background noise, or "static," in their radio telescope (a large antenna device for detecting radio waves from the heavens, Fig. 33–22). Eventually, they became convinced that it was real and that it was coming from outside our Galaxy. They made precise measurements at a wavelength $\lambda = 7.35$ cm, which is in the microwave region of the electromagnetic spectrum. (See Fig. 22–10; this radiation is called "microwave" because the wavelength, though much greater than that of visible light, is somewhat smaller than wavelengths for ordinary radio waves, which are typically meters or hundred of meters.) The intensity of this radiation was found initially not to vary by day or night or time of year, nor to depend on direction. It came from all directions in the universe with equal intensity (within less than one part per thousand). It could only be concluded that this radiation came from beyond our Galaxy, from the universe as a whole. The remarkable uniformity of the cosmic microwave background radiation was in accordance with the cosmological principle. But theorists also felt that there needed to be some small inhomogeneities in the CMB that would have provided "seeds" around which galaxy formation could have started. Such small inhomogeneities, on the order of parts per million, were finally detected in 1992.

FIGURE 33–22 Robert Wilson (left) and Arno Penzias, and behind them their "horn antenna."

The intensity of this radiation as measured at $\lambda = 7.35$ cm corresponds to blackbody radiation (see Section 27–2) at a temperature of about 3 K, now measured more precisely to be 2.7 ± 0.1 K. When radiation at other wavelengths was measured, the intensities were found to fall on a blackbody curve (see Fig. 27–4), with the peak of the curve a little above 0.1 cm corresponding to a temperature of 2.7 K.

Importance of CMB: the Big Bang

The discovery of the CMB at a temperature of 2.7 K ranks as one of the two most significant cosmological discoveries of this century. (The other was Hubble's expanding universe.) It is highly significant because it provides strong evidence in support of the Big Bang, and it gives us some idea of conditions in the very early universe. In fact, in the late 1940s, George Gamow and his collaborators calculated that a Big Bang origin of the universe should have generated just such a microwave background radiation.

To understand this, let us look at what a Big Bang might have been like. There must have been a tremendous release of concentrated energy. The temperature must have been extremely high, so high that there could not have been any atoms in the very early stages of the universe. Instead, the universe must have consisted solely of radiation (photons) and elementary particles. The universe would have been opaque—the photons in a sense "trapped," since as soon as they were emitted they would have been scattered or absorbed, primarily by electrons. Indeed, the microwave background radiation is strong evidence that matter and radiation were once in equilibrium at a very high temperature. As the universe expanded, the energy would have spread out over an increasingly larger volume and the temperature would have dropped. Only when the temperature had reached about 3000 K some 300,000 years later, could nuclei and electrons have stayed together as atoms. With the disappearance of free electrons, as they combined with nuclei to form atoms, the radiation would have been freed—"decoupled" from matter, we could say—to spread throughout the universe. As the universe expanded, so too the wavelengths of the radiation expanded, redshifting to longer wavelengths that correspond to lower temperature (recall Wien's law, $\lambda_{max} T = $ constant, Section 27–2), until they would have reached the 2.7 K background radiation we observe today.

Although the total energy associated with the cosmic microwave background is much larger than that from other radiation sources (such as the light emitted by stars), it is small compared to the energy, and mass (remember $E = mc^2$), associated with matter. In fact today, radiation is believed to make up less than $\frac{1}{1000}$ of the energy of the universe: today the universe is *matter-dominated*. But it was not always so. The cosmic microwave background radiation strongly suggests that early in its history the universe was *radiation-dominated*.[†] But, as we shall see in the next Section, that period lasted less than $\frac{1}{10,000}$ of the history of the universe (thus far).

33–6 The Standard Cosmological Model: The Early History of the Universe

It is now almost generally agreed that the evolution of the universe must have been determined in the first few moments of the Big Bang. In the last decade or two, a convincing theory of the origin and evolution of the universe has developed, now known as the **standard model**. Although a few cosmologists hold other views, most favor the standard model. Much of this theory is based on recent theoretical and experimental advances in elementary particle physics. Indeed, in the last few years cosmology and elementary particle physics have cross-fertilized to a surprising extent.

Let us go back now to the earliest of times—as close as possible to the Big Bang—and follow a standard model scenario of events as the universe expanded and cooled after the Big Bang. Initially, we will be talking of extremely small time intervals, as well as extremely high temperatures, far

---

[†]If there had not been such intense radiation in the first few minutes of the universe, nuclear reactions might have produced a much larger percentage of heavy nuclei than we see. Instead, nearly 75 percent of visible matter is hydrogen, presumably because the intense radiation immediately blasted apart any heavy nuclei that formed into their constituent protons and neutrons.

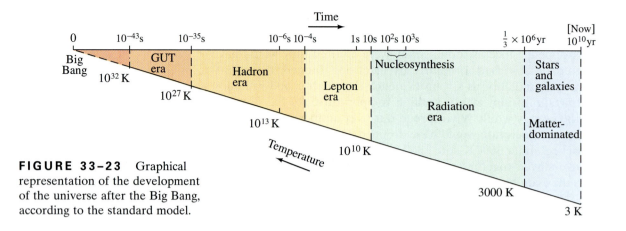

FIGURE 33–23 Graphical representation of the development of the universe after the Big Bang, according to the standard model.

beyond anything in the universe today. Figure 33–23 is a graphical representation of the events, and it may be helpful to consult it as we go along.

The standard-model "scenario" of the history of the universe after the Big Bang

We begin at a time only a minuscule fraction of a second after the Big Bang, 10^{-43} s. Although this is an unimaginably short time, predictions as early as this can be made based on present theory, albeit somewhat speculatively. Earlier than this instant, we can say nothing since we do not yet have a theory of quantum effects on gravity which would be needed for the incredibly high densities and temperatures then. It is imagined, however,

All four forces unified

that prior to 10^{-43} s, the four forces in nature were unified—there was only one force. The temperature would have been about 10^{32} K, corresponding to particles moving about every which way with an average kinetic energy of 10^{19} GeV (see Eq. 13–8):

$$\text{KE} \approx kT \approx \frac{(1.4 \times 10^{-23}\,\text{J/K})(10^{32}\,\text{K})}{1.6 \times 10^{-19}\,\text{J/eV}} \approx 10^{28}\,\text{eV} = 10^{19}\,\text{GeV}.$$

(Note that the factor $\frac{3}{2}$ in Eq. 13–8 is usually ignored in such calculations.) At $t = 10^{-43}$ s, a kind of "phase transition" is believed to have occurred dur-

Symmetry broken (gravity condensed out)

ing which the gravitational force, in effect, "condensed out" as a separate force. This, and subsequent phase transitions, are somewhat analogous to the phase transitions water undergoes as it cools from a gas, condenses into a liquid, and with further cooling freezes into ice. The symmetry of the four forces was broken, but the strong, weak, and electromagnetic forces were still unified. At this point, the universe entered the so-called **grand unified era** (after grand unified theory—see Chapter 32). There was no distinction between quarks and leptons; baryon and lepton numbers were not conserved. Very shortly thereafter, as the universe expanded considerably and the temperature had dropped to about 10^{27} K, there was another phase transition during which the strong force condensed out. This probably occurred about 10^{-35} s after the Big Bang. Now the universe was filled with a soup of leptons and quarks. The leptons included electrons, muons, taus, neutrinos, and all their antiparticles. The quarks were initially free (something we have not seen in our present universe), but soon they began to "condense" into more normal particles: nucleons and the other hadrons and their antiparti-

Quark confinement (hadron era)

cles. With this **confinement of quarks**, the universe entered the **hadron era**.

We can think of this "soup" as a grand mixture of particles and antiparticles, as well as photons—all in roughly equal numbers—colliding with one another frequently and exchanging energy.

By the time the universe was only about a microsecond (10^{-6} s) old, it had cooled to about 10^{13} K, corresponding to an average KE of 1 GeV, and the vast majority of hadrons disappeared. To see why, let us focus on the most familiar hadrons: nucleons and their antiparticles. When the average kinetic energy of particles was somewhat higher than 1 GeV, protons, neutrons, and their antiparticles were continually being created out of the energies of collisions involving photons and other particles, such as

Most hadrons disappear

$$\text{photons} \rightarrow \text{p} + \bar{\text{p}}, \qquad \text{n} + \bar{\text{n}}.$$

But just as quickly, particles and antiparticles would annihilate: for example

$$\text{p} + \bar{\text{p}} \rightarrow \text{photons or leptons.}$$

So the processes of creation and annihilation of nucleons were in equilibrium. The numbers of nucleons and antinucleons were high—roughly as many as there were electrons, positrons, or photons. But as the universe cooled and the average kinetic energy of particles dropped below about 1 GeV, which is the minimum energy needed to create nucleons and anti-nucleons (about 940 MeV each) in a typical collision, the process of nucleon creation could not continue. The process of annihilation could continue, however, with antinucleons annihilating nucleons, until there were almost no nucleons left. But not quite zero. To explain our present world, which consists mainly of matter (nucleons and electrons) with very little antimatter in sight, we must suppose that earlier in the universe, perhaps around 10^{-35} s after the Big Bang, a slight excess of quarks over antiquarks was formed.[†] This would have resulted in a slight excess of nucleons over antinucleons. And it is these "leftover" nucleons that we are made of today. The excess of nucleons over antinucleons was about one part in 10^9. Earlier, during the hadron era, there should have been about as many nucleons as photons. After it ended, the "leftover" nucleons thus numbered only about one nucleon per 10^9 photons, and this ratio has persisted to this day. Protons, neutrons, and all other heavier particles were thus tremendously reduced in number by about 10^{-6} s after the Big Bang. The lightest hadrons, the pions, disappeared as the nucleons had; because they are the lightest mass hadrons (140 MeV), they were the last hadrons to go, around 10^{-4} s after the Big Bang. Lighter particles, including electrons, positrons, neutrinos, photons—in roughly equal numbers—dominated, and the universe entered the **lepton era**.

Why is there matter now?

Lepton era

By the time the first full second had passed (clearly the most eventful second in history!), the universe had cooled to about 10 billion degrees, 10^{10} K. The average KE was about 1 MeV. This was still sufficient energy to create electrons and positrons and balance their annihilation reactions, since their masses correspond to about 0.5 MeV. So there were about as many e^+ and e^- as there were photons. But within a few more seconds, the temperature had dropped sufficiently so that e^+ and e^- could no longer be formed. Annihilation ($e^+ + e^- \rightarrow$ photons) continued. And, like nucleons before them, electrons and positrons all but disappeared from the universe—except for a slight excess of electrons over positrons (later to join with nuclei to form atoms). Thus, about $t = 10$ s after the Big Bang, the universe entered the **radiation era**. Its major constituents were now pho-

[†]An alternative possibility is that there was perfect symmetry between quarks and antiquarks, matter and antimatter, but that the universe somehow separated into domains, some containing only matter, others only antimatter. If this were true, we would expect antiparticles from such distant domains to reach us, at least occasionally, in cosmic rays; but none has ever been detected.

tons and neutrinos. But the neutrinos, partaking only in the weak force, rarely interacted. So the universe, until then experiencing an energy balance between matter and radiation, became **radiation-dominated**: much more energy was contained in radiation than in matter, a situation that would last hundreds of thousands of years (Fig. 33–23).

Radiation-dominated universe

Meanwhile, during the next few minutes, crucial events were taking place. Beginning about 2 or 3 minutes after the Big Bang, nuclear fusion began to occur. The temperature had dropped to about 10^9 K, corresponding to $\overline{\text{KE}} \approx 100$ keV, where nucleons could strike each other and be able to fuse (Section 31–3), but now cool enough that newly formed nuclei would not be immediately broken apart by subsequent collisions. Deuterium, helium, and very tiny amounts of lithium nuclei were probably made. But the universe was cooling too quickly, and larger nuclei were not made. After only a few minutes, probably not even a quarter of an hour after the Big Bang, the temperature dropped far enough that nucleosynthesis stopped, not to start again for millions of years (in stars). Thus, after the first hour or so of the universe, matter consisted mainly of bare nuclei of hydrogen (about 75 percent) and helium (about 25 percent)[†] and electrons. But radiation (photons) continued to dominate.

Making He

Our story is almost complete. The next important event is presumed to have occurred about 300,000 years later. The universe had expanded to about $\frac{1}{1000}$ of its present size, and the temperature had cooled to about 3000 K. The average KE of nuclei, electrons, and photons was less than an electron volt. Since ionization energies of atoms are on the order of eV, then as the temperature dropped below this point, electrons could orbit the bare nuclei and remain there (without being ejected by collisions), thus forming atoms. With the birth of atoms, the photons which were continually scattering from the free electrons became much freer to spread throughout the universe. The total energy contained in radiation had been decreasing (redshifting as the universe expanded), until at this point it was about equal to the total energy contained in matter. As the universe continued to expand, the radiation cooled further (to 2.7 K today, forming the cosmic microwave background radiation we detect from everywhere in the universe), and lost energy. But the mass of material particles did not decrease, so beginning at about this point the energy of the universe became increasingly concentrated in matter rather than in radiation: the universe became **matter-dominated**, as it remains today.[‡]

Birth of stable atoms

Matter-dominated universe

Shortly after the birth of atoms, stars and galaxies formed—probably by self-gravitation around mass concentrations (inhomogeneities). This transpired about a million years after the Big Bang. The universe continued to evolve (see Section 33–2) until today, 10 to 15 billion years later.

*   *   *

[†]This standard model prediction of a 25 percent primordial production of helium is fully in accord with what we observe today—the universe *does* contain about 25 percent He—and it is strong evidence in support of the standard Big Bang model. Furthermore, the theory says that 25 percent He abundance is fully consistent with there being three neutrino types, which is the number we observe so far. And it sets an upper limit of four to the maximum number of possible neutrino types. Actually, the fourth could be another type of low-mass particle, a *photino* or a *gravitino*, for example. Here we have a situation where cosmology actually makes a specific prediction about fundamental physics.

[‡]Although today matter contains more of the energy of the universe than does radiation, there are many more photons (perhaps 10^9 times more) than atoms, nuclei, and electrons. But each photon (at $T \approx 2.7$ K) has very little energy.

This scenario is by no means "proven" in any sense. Nor does it answer all questions. But it does provide a tentative picture, for the first time, of how the universe may have begun and evolved. It does have problems, however, some of which have been resolved by a modification first proposed in the early 1980s known as the **inflationary scenario**. According to the inflationary scenario, at the earliest stages of cosmological evolution, around 10^{-35} s after the Big Bang, the universe underwent a very rapid exponential expansion associated with the phase transition (symmetry breaking) that separated the strong force from the electroweak (as discussed earlier in this Section). After the brief inflationary period, the scenario settles back to the standard expansion already discussed. The appeal of the inflationary scenario is that it provides natural explanations for a number of problems not resolved by the standard model, such as why the universe is as close to being as flat as it seems to be, and why the cosmic microwave background radiation is so uniform.

Inflation

There are, however, some questions we haven't yet treated such as: was there a stage before the Big Bang, or did time just begin with the Big Bang? And what of the future of the universe? We'll look at these questions next, in the final Section of this book.

33–7 The Future of the Universe?

According to the standard Big Bang model, the universe is evolving and changing. Individual stars are evolving and dying as white dwarfs, neutron stars, black holes. At the same time, the universe as a whole is expanding. One important question is whether the universe will continue to expand forever. This question is connected to the curvature of space–time (Section 33–3) and to whether the universe is open (and infinite) or closed (and finite). There are three possibilities as shown in Fig. 33–24. If the curvature of the universe is *negative*, the expansion of the universe will never stop, although it should decrease due to the gravitational attraction of its parts. Such a universe would be *open* and infinite. If the universe is *flat* (no curvature), it will still be open and infinite but its expansion will slowly approach a zero rate. Finally, if the universe has *positive* curvature, it will be *closed* and finite. The effect of gravity in this case would be strong enough so that the expansion would eventually stop and the universe would then begin to contract. All matter eventually would collapse back onto itself in a **big crunch**. If, in this last case, the maximum expansion of the universe corresponded to, say, an intergalactic separa-

An open or closed universe?

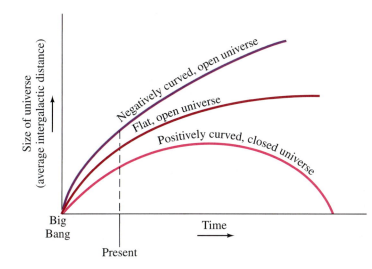

FIGURE 33–24 Three future possibilities for the universe.

tion twice what it is now, the maximum expansion would occur about 30 or 40 billion years from now. Then, as the universe began to contract, the big crunch would occur about 100 billion years after the Big Bang.

Whether we live in an open and continually expanding universe, or a closed one that eventually will contract, is a basic question in cosmology. But we don't know the answer. How might we find out? One way is to determine the average mass density in the universe. If the average mass density is above a critical value known as the **critical density**, estimated to be about

Critical density of the universe

$$\rho_c \approx 10^{-26} \, \text{kg/m}^3$$

(i.e., a few nucleons/m^3 on average throughout the universe), then gravity will prevent expansion from continuing forever, and will eventually pull the universe back into a big crunch. To say it another way, if $\rho > \rho_c$, there would be sufficient mass that gravity would give space–time a positive curvature. If instead the actual density is equal to the critical density, $\rho = \rho_c$, the universe will be flat and open. If the actual density is less than the critical density, $\rho < \rho_c$, the universe will have negative curvature and be open, expanding forever.

Great efforts have gone into measuring the actual density of the universe. Estimates of the amount of visible matter in the universe put the actual density at between one and two orders of magnitude less than the critical density, thereby suggesting an open universe. However, there is evidence for a significant amount of nonluminous matter in the universe, often referred to as the **missing mass** or **dark matter**, enough to bring the density to almost exactly ρ_c. For example, observations of the rotation of galaxies suggest that they rotate as if they had considerably more mass than we can see. And observations of the motion of galaxies within clusters also suggest that they have considerably more mass than can be seen. Furthermore, the highly regarded inflationary theory offers a strong argument in favor of ρ being precisely equal to ρ_c, and thus that space on a grand scale is Euclidean. If there is nonluminous matter in the universe, what might it be? One suggestion is that the missing mass consists of previously unknown weakly interacting massive particles (WIMPS), or perhaps small primordial black holes made in the early stages of the universe. Another possibility for the missing mass (or part of it) are MACHOS (massive compact halo objects), which would be chunks of matter in the form of large planets (like Jupiter) or stars that are too small to sustain fusion; they might glow faintly due to energy released from gravitational contraction (thus they are sometimes referred to as *brown dwarfs*), but too faintly to be seen. Evidence for MACHOS was found in 1993, inferred from their (assumed) gravitational effect on light passing by them from a distant star.

"Missing" mass or dark matter

Another proposal for the missing mass is that neutrinos, once believed to be massless, may actually have nonzero rest mass. Since the universe probably contains roughly as many neutrinos of each type as it does photons (that is, about 10^9 times the number of nucleons, although this neutrino background has yet to be detected), neutrino masses of only a few eV could help to bring the actual density of the universe up to the critical density. The supernova of 1987 offered a fine opportunity to measure the electron neutrino mass. If neutrinos have nonzero rest mass, then their velocities are less than the speed of light ($v < c$). High-energy neutrinos emitted from the supernova should in this case have arrived at Earth earlier than lower-energy (and therefore slower) neutrinos emitted at the same instant. Since they traveled a distance of about 170,000 ly from SN1987a, the time difference ought to be measurable. To get an idea of the size of the effect, let us consider the following Example.

EXAMPLE 33–7 **ESTIMATE** **Neutrino mass estimate from supernova.**
[A challenging Example.] Suppose two neutrinos from SN1987a were detected on Earth (via the reaction $\bar{\nu}_e + p \rightarrow n + e^+$) 10 seconds apart, with measured kinetic energies of about 20 MeV and 10 MeV. (a) Estimate the rest mass of the neutrino, m_0, using this data, assuming both neutrinos were emitted at the same time. (b) Theoretical models of supernova explosions suggest that the neutrinos are emitted in a burst that lasts from a second or two up to perhaps 10 s. If we assume the neutrinos are not emitted simultaneously but rather at any time over a 10-s interval, what then can we say about the neutrino mass based on the data given above?

SOLUTION (a) We expect the neutrino mass to be less than 100 eV (from laboratory measurements). Since our neutrinos have KE of 20 MeV and 10 MeV, we can make the approximation $m_0 c^2 \ll E$, where E (the total energy) is essentially equal to the kinetic energy. From Eqs. 26–4 and 26–6, we have

$$E = mc^2 = \frac{m_0 c^2}{\sqrt{1 - v^2/c^2}}.$$

We solve this for v, the velocity of a neutrino with energy E:

$$v = c\left(1 - \frac{m_0^2 c^4}{E^2}\right)^{\frac{1}{2}} = c\left(1 - \frac{m_0^2 c^4}{2E^2} + \cdots\right),$$

where we have used the binomial expansion $(1 + x)^{\frac{1}{2}} = 1 + \frac{1}{2}x + \cdots$ (see Appendix A), and we ignore higher-order terms since $m_0^2 c^4 \ll E^2$. The time t for a neutrino to travel a distance $d (= 170,000 \text{ ly})$ is

$$t = \frac{d}{v} = \frac{d}{c\left(1 - \frac{m_0^2 c^4}{2E^2}\right)} \approx \frac{d}{c}\left(1 + \frac{m_0^2 c^4}{2E^2}\right),$$

where again we used the binomial expansion $[(1 + x)^{-1} = 1 - x + \cdots]$. The difference in arrival times for our two neutrinos of energies $E_1 = 20$ MeV and $E_2 = 10$ MeV is

$$t_2 - t_1 = \frac{d}{c}\frac{m_0^2 c^4}{2}\left(\frac{1}{E_2^2} - \frac{1}{E_1^2}\right).$$

We solve this for $m_0 c^2$ and set $t_2 - t_1 = 10$ s:

$$m_0 c^2 = \left[\frac{2c(t_2 - t_1)}{d}\frac{E_1^2 E_2^2}{E_1^2 - E_2^2}\right]^{\frac{1}{2}}$$

$$= \left[\frac{2(3 \times 10^8 \text{ m/s})(10 \text{ s})}{(1.7 \times 10^5 \text{ ly})(10^{16} \text{ m/ly})}\frac{(400 \text{ MeV}^2)(100 \text{ MeV}^2)}{(400 \text{ MeV}^2 - 100 \text{ MeV}^2)}\right]^{\frac{1}{2}}$$

$$= 22 \times 10^{-6} \text{ MeV} = 22 \text{ eV}.$$

We thus estimate the mass of the neutrino to be 22 eV, but there would of course be experimental uncertainties.
(b) If the two neutrinos of energy $E_1 = 20$ MeV and $E_2 = 10$ MeV were emitted at unknown times over a 10-s interval, then the 10-s difference in their arrival times could be due to a 10-s difference in their emission time. In this case our data are consistent with zero rest mass and it puts an approximate *upper limit* on the neutrino mass of 22 eV.

In the actual experiments, the most sensitive detector consisted of several thousand tons of water in an underground chamber. It detected 11 events in 12 seconds, probably via the reaction $\bar{\nu}_e + p \rightarrow n + e^+$. There was not a clear correlation between energy and time of arrival. Nonetheless, a careful analysis of this experiment and others, has set a rough upper limit on the electron anti-neutrino mass of about 7 eV. The upper limits for the masses of the other neutrinos are much higher (see Table 32–2), so we cannot yet rule out neutrinos as being able to "close" the universe.

Deceleration parameter

Another factor that could provide the answer to the question of an open or closed universe, if we could measure it accurately enough, is the so-called **deceleration parameter**. It is a measure of the rate at which the expansion of the universe is slowing. But to measure this rate requires looking far back in time, to the galaxies farthest away, whose light we now receive was emitted at a time closer to the beginning of the universe. At that time, the rate of expansion was much faster than today. Unfortunately, we would have to know the distance to these galaxies more precisely than is now possible. So this method does not at present yield an answer to whether the universe is open or closed. A recent related analysis of the number of galaxies as a function of redshift (i.e., distance) gives a result consistent with $\rho = \rho_c$, although the experimental uncertainty allows fairly wide limits on ρ. Nonetheless, the suggestion of the simplest universe of all, with no overall curvature, is appealing.

If the universe is open, how would it evolve in the future? According to the latest theories, which rely to a large extent on elementary particle theory, after about 10^{18} years, galaxies would have much of their matter knocked away and scattered throughout the universe by collisions with other stars. The remaining matter would eventually condense into massive "galactic black holes." Clusters of these would then coalesce into extremely massive "supergalactic black holes." Finally, the black holes themselves would "evaporate"—the matter within them, through the slow quantum-mechanical process of tunneling (see Section 30–12), would "leak out." This process is so slow that it would take on the order of 10^{100} years. The universe would then be mainly a thin gas of electrons, positrons, neutrinos, and photons.

On the other hand, if the universe is closed, it might turn around and begin to contract even before all the stars have burnt out. As the universe contracts, the background radiation would increase in energy and temperature. The universe might simply retrace its steps, if it weren't for black holes. As density increases and the universe rushes toward its inevitable end in the "big crunch," black holes might gobble up more and more matter until the entire universe coalesced into a single supermassive black hole—which would then be the universe.

If the universe is closed, what happens after the big crunch? We don't know, of course. What is possible, though, is a "bounce." That is, the dense fiery nucleus of the big crunch might explode again, resulting once more in an expanding universe. Thus the universe might be cyclic as shown in Fig. 33–25. Such a **cyclic** or **pulsating** universe proposes a possible answer to one of our favorite "unanswerable" questions: what happened before the Big Bang? In this model there was simply a previous cycle. But left unanswered would be a number of other questions such as "When did it all begin?"

The questions raised by cosmology can seem absurd at times, they are so removed from everyday "reality." We can always say: the Sun is shining, it's going to burn on for an unimaginably long time, all is well. Nonethe-

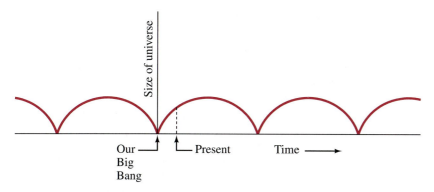

FIGURE 33–25 Cyclic model of the universe. Although the cycles shown here are the same, they could have different periods and different expansion rates.

less, the questions of cosmology are deep ones that fascinate the human intellect. One aspect that is especially intriguing is this: calculations on the formation and evolution of the universe have been performed that deliberately varied the values—just slightly—of certain fundamental physical constants. The result? A universe in which life as we know it could not exist. [For example, if the difference in mass between proton and neutron were zero, or small (less than $0.5 \, \text{MeV}/c^2$), there would be no atoms: electrons would be captured by protons never to be freed again.] Such results have given rise to the so-called **Anthropic principle**, which says that if the universe were even a little different than it is, we couldn't be here. It's as if the universe were exquisitely tuned, almost as if to accommodate us.

SUMMARY

The night sky contains myriads of stars including those in the Milky Way, which is a "side view" of our **Galaxy** looking along the plane of the disc. Our Galaxy includes about 10^{11} stars. Beyond our Galaxy are billions of other galaxies. Astronomical distances are measured in **light-years** ($1 \, \text{ly} \approx 10^{13} \, \text{km}$). The nearest star is about 4 ly away and the nearest other galaxy is 2 million ly away. Our galactic disc has a diameter of about 100,000 ly. Distances are often specified in **parsecs**, where 1 parsec = 3.26 ly.

Stars are believed to begin life as collapsing masses of hydrogen gas (protostars). As they contract, they heat up (PE is transformed to KE). When the temperature reaches 10 million degrees, nuclear fusion begins and forms heavier elements (**nucleosynthesis**), mainly helium at first. The energy released during these reactions balances the gravitational force, and the young star stabilizes as a **main-sequence** star. The tremendous luminosity of stars comes from the energy released during these thermonuclear reactions. After billions of years, as helium is collected in the core and hydrogen is used up, the core contracts and heats further. The envelope expands and cools, and the star becomes a **red giant** (larger diameter, redder color). The next stage of stellar evolution depends on the mass of the star. Stars of residual mass less than about 1.4 solar masses cool further and become **white dwarfs**, eventually fading and going out altogether. Heavier stars contract further due to their greater gravity: the density approaches nuclear density, the huge pressure forces electrons to combine with protons to form neutrons, and the star becomes essentially a huge nucleus of neutrons. This is a **neutron star**, and the energy released from its final core contraction is believed to produce **supernovae**. If the star's residual mass is greater than two or three solar masses, it may contract even further and form a **black hole**, which is so dense that no matter or light can escape from it.

In the **general theory of relativity**, the **equivalence principle** states that an observer cannot distinguish acceleration from a gravitational field. Said another way, gravitational and inertial mass are the same. The theory predicts gravitational bending of light rays to a degree consistent with experiment. Gravity is treated as a curvature in space and time, the curvature being greater near massive bodies. The universe as a whole may be curved. If there is sufficient mass, the curvature of the universe is positive, and the universe is **closed** and **finite**; otherwise, it is **open** and **infinite**.

Distant galaxies display a **redshift** in their spectral lines, interpreted as a Doppler shift. The universe seems to be expanding, its galaxies racing away from each other at speeds (v) proportional to the distance (d) between them:

$$v = Hd,$$

which is known as **Hubble's law** (H is the **Hubble parameter**). This expansion of the universe suggests an explosive origin, the **Big Bang**, which probably occurred 10 to 15 billion years ago. **Quasars** are objects with a large redshift (suggesting great distance) and high luminosity (suggesting closeness or, more likely, extraordinary energy output). The **cosmological principle** assumes that the universe, on a large scale, is homogeneous and isotropic.

Important evidence for the Big Bang model of the universe was the discovery of the **cosmic microwave background** radiation (CMB), which conforms to a blackbody radiation curve at a temperature of 2.7 K. The **standard model** of the Big Bang provides a possible scenario as to how the universe developed as it expanded and cooled after the Big Bang. Starting at 10^{-43} seconds after the Big Bang, according to this model, there was a series of **phase transitions** during which previously unified forces of nature "condensed out" one by one. The **inflationary scenario** assumes that during one of these phase transitions, the universe underwent a brief but rapid exponential expansion. Until about 10^{-35} s, there was no distinction between quarks and leptons. Shortly thereafter, quarks were **confined** into hadrons (the **hadron era**). About 10^{-6} s after the Big Bang, the majority of hadrons disappeared, introducing the **lepton era**. By the time the universe was about 10 s old, the electrons too had mostly disappeared and the universe became **radiation-dominated**. A couple of minutes later, nucleosynthesis began, but lasted only a few minutes. It was then several hundred thousand years before the universe was cool enough for electrons to combine with nuclei to form atoms. Also about this time, the background radiation had expanded and cooled so much that its total energy equaled the energy in matter. As the radiation cooled further, losing energy, the universe became **matter-dominated** (not in numbers, but in energy). Then stars and galaxies formed, producing a universe not much different than it is today—10 or 15 billion years later.

If the universe is **open**, it will continue to expand indefinitely. If it is **closed**, gravity is sufficiently strong to halt expansion and the universe will eventually begin to collapse back on itself, ending in a "big crunch." Whether the universe is open or closed depends on whether its average mass density is above or below a critical density. If the universe is closed, it may rebound from the big crunch and perhaps reexpand in a **cyclic** manner.

QUESTIONS

1. The Milky Way was once thought to be "cloudy," but no longer is considered so. Explain.

2. Give an explanation for why some galaxies have arms.

3. If you were measuring star parallaxes from the Moon instead of Earth, what corrections would you have to make? What changes would occur if you were measuring parallaxes from Mars?

4. A star is in equilibrium when it radiates at its surface all the energy generated at its core. What happens when it begins to generate more energy than it radiates? Less energy? Explain.

5. Describe a red giant star. List some of its properties.

6. Select a point on the H–R diagram. Mark several directions away from this point. Now describe the changes that would take place in a star moving in each of these directions.

7. Does the H–R diagram reveal anything about the core of a star?

8. Why do some stars end up as white dwarfs, and others as neutron stars or black holes?

9. What is a geodesic? What is its role in general relativity?

10. If it were discovered that the redshift of spectral lines of galaxies was due to something other than expansion, how might our view of the universe change? Would there be conflicting evidence? Discuss.

11. All galaxies appear to be moving away from us. Are we therefore at the center of the universe? Explain.

12. If you were located in a galaxy near the boundary of our observable universe, would galaxies in the direction of the Milky Way appear to be approaching you or receding from you? Explain.

13. What is the difference between the Hubble age of the universe and the actual age? Which is greater?

14. Compare an explosion on Earth to the Big Bang. Consider such questions as: would the debris spread at a higher speed for more distant particles, as in the Big Bang? Would the debris come to rest? What type of universe would this correspond to, open or closed?

15. When the primordial nucleus exploded, thus creating the universe, into what did it expand? Discuss.

16. If nothing, not even light, escapes from a black hole, then how can we tell if one is there?

17. Explain what the 2.7 K cosmic microwave background radiation is. Where does it come from? Why is its temperature now so low?

18. The birth of atoms—that is, the combination of electrons with nuclei about which they orbit—occurred when the universe had cooled to about 3000 K and is generally called *recombination*. Why is this term misleading?

19. Why were atoms unable to exist until hundreds of thousands of years after the Big Bang?

20. Explain why today the universe is said to be matter-dominated, yet there are probably 10^9 times as many photons as there are massive particles.

21. If the universe is open, what will eventually happen to the cosmic microwave background radiation? If it is closed, what will happen to it?

22. Under what circumstances would the universe eventually collapse in on itself?

▮ PROBLEMS

SECTIONS 33–1 AND 33–2

1. (I) The parallax angle of a star is 0.00017°. How far away is the star?

2. (I) A star exhibits a parallax of 0.28 seconds of arc. How far away is it?

3. (I) Using the definitions of the parsec and the light-year, show that $1\,pc = 3.26\,ly$.

4. (I) A star is 24 parsecs away. What is its parallax angle? State (*a*) in seconds of arc, and (*b*) in degrees.

5. (I) What is the parallax angle for a star that is 42 light-years away? How many parsecs is this?

6. (I) A star is 25 parsecs away. How long does it take for its light to reach us?

7. (I) If one star is twice as far away from us as a second star, will the parallax angle of the first star be greater or less than that of the second star? By what factor?

8. (II) We saw earlier (Chapter 14) that the rate energy reaches the Earth from the Sun (the "solar constant") is about $1.3 \times 10^3\,W/m^2$. What is (*a*) the apparent brightness l of the Sun, and (*b*) the absolute luminosity L of the Sun?

9. (II) What is the apparent brightness of the Sun as seen on Jupiter? (Jupiter is 5.2 times farther from the Sun than the Earth.)

10. (II) Estimate the angular width that our Galaxy would subtend if observed from the nearest galaxy to us. Compare to the angular width of the Moon from Earth.

11. (II) When our Sun becomes a red giant, what will be its average density if it expands out to the orbit of Earth ($1.5 \times 10^{11}\,m$ from the Sun)?

12. (II) When our Sun becomes a white dwarf, it is expected to be about the size of the Moon. What angular width will it subtend from Earth?

13. (II) Calculate the density of a white dwarf whose mass is equal to the Sun's and whose radius is equal to the Earth's. How many times larger than Earth's density is this?

14. (II) A neutron star whose mass is 1.5 solar masses has a radius of about 11 km. Calculate its average density and compare to that for a white dwarf (Problem 13) and to that of nuclear matter.

15. (II) Calculate the Q-values for the He burning reactions of Eqs. 33–2. (The mass of $^8_4$Be is 8.005305 u.)

16. (II) In the later stages of stellar evolution, a star (if massive enough) will begin fusing carbon nuclei to form, for example, magnesium:

$$^{12}_6C + {}^{12}_6C \rightarrow {}^{24}_{12}Mg + \gamma.$$

(*a*) How much energy is released in this reaction (see Appendix F). (*b*) How much kinetic energy must each carbon nucleus have (assume equal) in a head-on collision if they are just to touch (use Eq. 30–1) so that the strong force can come into play? (*c*) What temperature does this KE correspond to?

17. (II) Repeat Problem 16 for the reaction

$$^{16}_8O + {}^{16}_8O \rightarrow {}^{28}_{14}Si + {}^4_2He.$$

18. (II) Suppose two stars of the same apparent brightness l are also believed to be the same size. The spectrum of one star peaks at 800 nm whereas that of the other peaks at 400 nm. Use Wien's law (Section 27–2) and the Stefan-Boltzmann law (Eq. 14–4) to estimate their relative distances from us. [*Hint*: See Examples 33–5 and 33–6.]

19. (III) Stars located in a certain cluster are assumed to be about the same distance from us. Two such stars have spectra that peak at $\lambda_1 = 500$ nm and $\lambda_2 = 700$ nm, and the ratio of their apparent brightness is $l_1/l_2 = 0.091$. Estimate their relative sizes (give ratio of their diameters). [*Hint*: Use the Stefan-Boltzmann law, Eq. 14–4.]

SECTION 33–3

20. (I) Describe a triangle, drawn on the surface of a sphere, for which the sum of the angles is (*a*) 360°, and (*b*) 180°.

21. (I) Show that the Schwarzschild radius for a star with mass equal to that (*a*) of our Sun is 2.95 km, and (*b*) of Earth is 8.9 mm.

22. (II) What is the Schwarzschild radius for a typical galaxy?

23. (II) What is the maximum sum-of-the-angles for a triangle on a sphere?

SECTION 33–4

24. (I) If a galaxy is traveling away from us at 1.0 percent of the speed of light, roughly how far away is it?

25. (I) The redshift of a galaxy indicates a velocity of 2500 km/s. How far away is it?

26. (I) Estimate the speed of a galaxy (relative to us) that is near the "edge" of the universe, say 10 billion light-years away.

27. (I) Make an approximate calculation for the age of the universe using Hubble's constant, assuming (*a*) $H = 50$ km/s per Mpc and (*b*) 100 km/s per Mpc.

28. (II) Estimate the observed wavelength for the 656-nm line in the Balmer series of hydrogen in the spectrum of a galaxy whose distance from us is (*a*) 1.0×10^6 ly, (*b*) 1.0×10^8 ly, (*c*) 1.0×10^{10} ly.

29. (II) Estimate the speed of a galaxy, and its distance from us, if the wavelength for the hydrogen line at 434 nm is measured on Earth as being 610 nm.

30. (II) Starting from Eq. 33–3, show that the Doppler shift in wavelength is $\Delta\lambda/\lambda \approx v/c$ (Eq. 33–4) for $v \ll c$. [*Hint*: Use the binomial expansion—see Appendix A.]

31. (III) Derive the Doppler shift formula for light (Eq. 33–3) using the special theory of relativity. [*Hint*: Proceed as in Section 12–8 for sound, but invoke special relativity including time dilation.]

SECTIONS 33–5 TO 33–7

32. (I) Calculate the wavelength at the peak of the blackbody radiation distribution at 2.7 K using the Wien law.

33. (II) The critical density for closure of the universe is $\rho_c \approx 10^{-26}$ kg/m³. State ρ_c in terms of the average number of nucleons per cubic meter.

34. (II) The size of the universe (the average distance between galaxies) at any one moment is believed to have been inversely proportional to the absolute temperature. Estimate the size of the universe, compared to today, at (*a*) $t = 10^6$ yr, (*b*) $t = 1$ s, (*c*) $t = 10^{-6}$ s, and (*d*) $t = 10^{-35}$ s.

35. (II) At approximately what time had the universe cooled below the threshold temperature for producing (*a*) kaons ($M \approx 500$ MeV/c^2), (*b*) Y ($M \approx 9500$ MeV/c^2), and (*c*) muons ($M \approx 100$ MeV/c^2)?

GENERAL PROBLEMS

36. Suppose that three main-sequence stars could undergo the three changes represented by the three arrows, *A*, *B*, and *C*, in the H–R diagram of Fig. 33–26. For each case, describe the changes in temperature, luminosity, and size.

37. Assume that the nearest stars to us have an intrinsic luminosity about the same as the Sun's. Their apparent brightness, however, is about 10^{11} times fainter than the Sun. From this, estimate the distance to the nearest stars. (Newton did this calculation, although he made a numerical error of a factor of 100.)

38. Use conservation of angular momentum to estimate the angular velocity of a neutron star which has collapsed to a diameter of 10 km, from a star whose radius was equal to that of our Sun (7×10^8 m), of mass 1.5 times that of the Sun, and which rotated (like our Sun) about once a month.

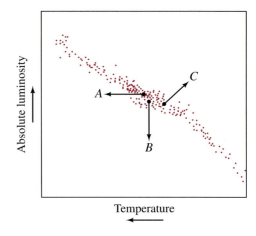

FIGURE 33–26 Problem 36.

39. By what factor does the rotational KE change when the star in Problem 38 collapses to a neutron star?

40. A certain pulsar, believed to be a neutron star of mass 1.5 times that of the Sun, with diameter 10 km, is observed to have a rotation speed of 1.0 rev/s. If it loses rotational KE at the rate of 1 part in 10^9 per day, which is all transformed into radiation, what is the power output of the star?

41. Estimate the rate of which hydrogen atoms would have to be created, according to the steady-state model, to maintain the present density of the universe of about 10^{-27} kg/m^3, assuming the universe is expanding with Hubble constant $H = 80$ km/s/Mpc.

42. Estimate what neutrino rest mass (in eV) would provide the critical density to close the universe. Assume the neutrino density is, like photons, about 10^9 times that of nucleons, and that nucleons make up only (*a*) 2 percent of the mass needed, or (*b*) 5 percent of the mass needed.

43. Two stars, whose spectra peak at 600 nm and 400 nm, respectively, both lie on the main sequence. Use Wien's law, the Stefan-Boltzmann law, and the H–R diagram (Fig. 33–7) to estimate the ratio of their diameters. [*Hint*: See Examples 33–5 and 33–6.]

44. The farthest we can measure with parallax is about 30 parsecs. What is our minimum angular resolution (in degrees), based on this information?

45. Through some coincidence, the Balmer lines from singly ionized helium in a distant star happen to overlap with the Balmer lines from hydrogen in the Sun. How fast is the star receding from us?

46. What is the temperature that corresponds to the 1.8 TeV collisions at the Fermilab collider? To what era in cosmological history does this correspond?

47. Astronomers have recently measured the rotation of gas around what might be a supermassive black hole of about 2 billion solar masses at the center of a galaxy. If the radius from the galactic center to the gas clouds is 60 light-years, what Doppler shift $\Delta\lambda/\lambda$ do you estimate they saw?

48. Astronomers use an *apparent magnitude* (*m*) scale to describe apparent brightness. It uses a logarithmic scale, where a higher number corresponds to a less bright star. (For example, the Sun has magnitude -27, whereas most stars have positive magnitudes.) On this scale, a change in apparent magnitude by $+5$ corresponds to a decrease in apparent brightness by a factor of 100. If Venus has an apparent magnitude of -4.4, whereas Sirius has an apparent magnitude of -1.4, which is brighter? What is the ratio of the apparent brightness of these two objects?

A MATHEMATICAL REVIEW

A–1 Relationships, Proportionality, and Equations

One of the important aspects of physics is the search for relationships between different quantities—that is, determining how one quantity affects another. For example, how does temperature affect the air pressure in a tire? Or how does the net force on an object affect its acceleration? Sometimes a given quantity is affected by two or more quantities; for instance, the acceleration of an object is related to both its mass and the applied force. If it is suspected that a relationship exists between two or more quantities, one can try to determine the precise nature of this relationship. This is done by varying one of the quantities and measuring how the other varies as a result. If it is likely that a particular quantity will be affected by more than one factor or quantity, only one of the latter is varied at a time, while the others are held constant.[†]

Direct proportion

As a simple example, the ancients found that if one circle has twice the diameter of a second circle, the first also has twice the circumference. If the diameter is three times as large, the circumference is also three times as large. In other words, an increase in the diameter results in a proportional increase in the circumference. We say that the circumference is *directly proportional to* the diameter. This can be written in symbols as $C \propto D$, where "$\propto$" means "is proportional to," and C and D refer to the circumference and diameter of a circle, respectively. The next step is to change this proportionality to an equation, which will make it possible to link the two quantities numerically. This merely entails inserting a proportionality constant, which in many cases is determined by measurement. (In some cases it can be chosen arbitrarily, if it involves only the definition of a new unit.) The ancients found that the ratio of the circumference to the diameter of any circle was 3.1416 (to keep only the first few decimal places). This number is designated by the Greek letter π. It is the constant of proportionality for the relationship $C \propto D$, and to obtain an equation we insert it into the proportion and change the $\propto$ to $=$. Thus, $C = \pi D$.

Other kinds of proportionality occur as well. For example, the area of a circle is proportional to the *square* of its radius. That is, if the radius is doubled, the area becomes four times as large; and so on. In this case we can write $A \propto r^2$, where A stands for the area and r for the radius of the circle.

Inverse proportion

Sometimes two quantities are related in such a way that an increase in one leads to a proportional *decrease* in the other. This is called *inverse proportion*. For example, the time required to travel a given distance is in-

[†]When one quantity affects another, we often use the expression "is a function of" to indicate this dependence; for example, we say that the pressure in a tire is a function of the temperature.

versely proportional to the speed of travel. The greater the speed, the less time it takes. We can write this inverse proportion as: time $\propto$ 1/speed. The larger the denominator of a fraction, the smaller the fraction is as a whole. For example, $\frac{1}{4}$ is smaller than $\frac{1}{2}$. Thus, if the speed is doubled, the time is halved, which is what we want to express by this inverse proportionality relationship.

Whatever kind of proportion is found to hold, it can be changed to an equality by insertion of the proper proportionality constant. Quantitative statements or predictions about the physical world can then be made with the equation.

A–2 Exponents

When we write 10^4 we mean that you multiply 10 times itself four times: $10^4 = 10 \times 10 \times 10 \times 10 = 10{,}000$. The superscript 4 is called an *exponent*, and 10 is said to be raised to the fourth power. Any number or symbol can be raised to a power; special names are used when the exponent is 2 (a^2 is "a squared") or 3 (a^3 is "a cubed"). For any other power, we say a^n is "a to the nth power." If the exponent is 1, it is usually dropped: $a^1 = a$, since no multiplication is involved.

The rules for multiplying numbers expressed as powers are as follows:

$$(a^n)(a^m) = a^{n+m}. \tag{A–1}$$

That is, the exponents are added. To see why, consider the result of the multiplication of 3^3 by 3^4:

$$(3^3)(3^4) = (3)(3)(3) \times (3)(3)(3)(3) = (3)^7.$$

Here the sum of the exponents is $3 + 4 = 7$, so rule A–1 works. Notice that this rule works only if the base numbers (a in Eq. A–1) are the same. Thus we *cannot* use the rule of summing exponents for $(6^3)(5^2)$; these numbers would have to be written out. However, if the base numbers are different but the exponents are the same, we can write a second rule:

$$(a^n)(b^n) = (ab)^n. \tag{A–2}$$

For example, $(5^3)(6^3) = (30)^3$ since

$$(5)(5)(5)(6)(6)(6) = (30)(30)(30).$$

The third rule involves a power raised to another power: $(a^3)^2$ means $(a^3)(a^3)$, which is equal to $a^{3+3} = a^6$. The general rule is then

$$(a^n)^m = a^{nm}. \tag{A–3}$$

In this case, the exponents are multiplied.

Negative exponents are used for reciprocals. Thus,

$$\frac{1}{a} = a^{-1}, \qquad \frac{1}{a^3} = a^{-3},$$

and so on. The reason for using negative exponents is that it allows us to use the multiplication rules given above. For example, $(a^5)(a^{-3})$ means

$$\frac{(a)(a)(a)(a)(a)}{(a)(a)(a)} = a^2.$$

Rule A–1 gives us the same result:

$$(a^5)(a^{-3}) = a^{5-3} = a^2.$$

What does an exponent of zero mean? That is, what is a^0? Any number raised to the zeroth power is defined as being equal to 1:

$$a^0 = 1.$$

This definition is used because it follows from the rules for adding exponents. For example,

$$a^3 a^{-3} = a^{3-3} = a^0 = 1.$$

But *does* $a^3 a^{-3}$ actually equal 1? Yes, because

$$a^3 a^{-3} = \frac{a^3}{a^3} = 1.$$

Fractional exponents are used to represent *roots*. For example, $a^{\frac{1}{2}}$ means the square root of a; that is $a^{\frac{1}{2}} = \sqrt{a}$. Similarly, $a^{\frac{1}{3}}$ means the cube root of a, and so on. The fourth root of a means that if you multiply the fourth root of a by itself four times, you again get a:

$$(a^{\frac{1}{4}})^4 = a.$$

This is consistent with rule A–3 since $(a^{\frac{1}{4}}) = a^{\frac{4}{4}} = a^1 = a$.

A–3 Powers of 10, or Exponential Notation

Writing out very large and very small numbers such as the distance of Neptune from the Sun, 4,500,000,000 km, or the diameter of a typical atom, 0.00000001 cm, is inconvenient and prone to error. It also leaves in question (see Section 1–4) the number of significant figures. (How many of the zeros are significant in the number 4,500,000,000 km?) We therefore make use of the "powers of 10," or exponential notation. The distance from Neptune to the Sun is then expressed as 4.50×10^9 km (assuming that the value is significant to three digits) and the diameter of an atom 1.0×10^{-8} cm. This way of writing numbers is based on the use of exponents, where a^n signifies a multiplied by itself n times. For example, $10^4 = 10 \times 10 \times 10 \times 10 = 10,000$. Thus, $4.50 \times 10^9 = 4.50 \times 1,000,000,000 = 4,500,000,000$. Notice that the exponent (9 in this case) is just the number of places the decimal point is moved to the right to obtain the fully written out number (4.500,000,000.)

When two numbers are multiplied (or divided), you first multiply (divide) the simple parts and then the powers of 10. Thus, 2.0×10^3 multiplied by 5.5×10^4 equals $(2.0 \times 5.5) \times (10^3 \times 10^4) = 11 \times 10^7$, where we have used the rule for adding exponents (Appendix A–2). Similarly, 8.2×10^5 divided by 2.0×10^2 equals

$$\frac{8.2 \times 10^5}{2.0 \times 10^2} = \frac{8.2}{2.0} \times \frac{10^5}{10^2} = 4.1 \times 10^3.$$

For numbers less than 1, say 0.01, the exponent power of 10 is written with a negative sign: $0.01 = 1/100 = 1/10^2 = 1 \times 10^{-2}$. Similarly, $0.002 = 2 \times 10^{-3}$. The decimal point has again been moved the number of places expressed in the exponent. Thus, $0.020 \times 3600 = 72$; in exponential notation $(2.0 \times 10^{-2}) \times (3.6 \times 10^3) = 7.2 \times 10^1 = 72$.

Notice also that $10^1 \times 10^{-1} = 10 \times 0.1 = 1$, and by the law of exponents, $10^1 \times 10^{-1} = 10^0$. Therefore, $10^0 = 1$.

When writing a number in exponential notation, it is usual to make the simple number be between 1 and 10. Thus it is conventional to write 4.5×10^9 rather than 45×10^8, although they are the same number.[†] This notation also allows the number of *significant figures* to be clearly expressed. We write 4.50×10^9 if this value is accurate to three significant figures, but 4.5×10^9 if it is accurate to only two.

A–4 Algebra

Physical relationships between quantities can be represented as equations involving symbols (usually letters of the alphabet) that represent the quantities. The manipulation of such equations is the field of algebra, and is used a great deal in physics. An equation involves an equals sign, which tells us that the quantities on either side of the equals sign have the same value. Examples of equations are

$$3 + 8 = 11$$

$$2x + 7 = 15$$

$$a^2b + c = 6.$$

The first equation involves only numbers, so is called an arithmetic equation. The other two equations are algebraic since they involve symbols. In the third equation, the quantity a^2b means the product of a times a times b: $a^2b = a \times a \times b$.

Solving for an Unknown

Often we wish to solve for one (or more) symbols, and we treat it as an *unknown*. For example, in the equation $2x + 7 = 15$, x is the unknown; this equation is true, however, only when $x = 4$. Determining what value (or values) the unknown(s) can have to satisfy the equation(s) is called *solving the equation*. To solve an equation, the following rule can be used:

An equation will remain true if any operation performed on one side is also performed on the other side: for example, (*a*) addition or subtraction of a number or symbol; (*b*) multiplication or division by a number or symbol; (*c*) raising each side of the equation to the same power, or taking the same root (such as square root).

[†]Another convention used, particularly with computers, is that the simple number be between 0.1 and 1. Thus we could write 4,500,000,000 as 0.450×10^{10}.

EXAMPLE A–1 Solve for x in the equation

$$2x + 7 = 15.$$

SOLUTION We first subtract 7 from both sides:

$$2x + 7 - 7 = 15 - 7$$

$$2x = 8.$$

Then we divide both sides by 2 to get

$$\frac{2x}{2} = \frac{8}{2}$$

$$x = 4,$$

and this solves the equation.

EXAMPLE A–2 (*a*) Solve the equation

$$a^2 b + c = 24$$

for the unknown a in terms of b and c. (*b*) Solve for a assuming that $b = 2$ and $c = 6$.

SOLUTION (*a*) We are trying to solve for a, so we first subtract c from both sides:

$$a^2 b = 24 - c,$$

then divide by b:

$$a^2 = \frac{24 - c}{b},$$

and finally take square roots:

$$a = \sqrt{\frac{24 - c}{b}}.$$

(*b*) If we are given that $b = 2$ and $c = 6$, then

$$a = \sqrt{\frac{24 - 6}{2}} = 3.$$

To check a solution, we put it back into the original equation (this is really a check that we did all the manipulations correctly). In the equation

$$a^2 b + c = 24,$$

we put in $a = 3, b = 2, c = 6$ and find

$$(3)^2(2) + (6) \stackrel{?}{=} 24$$

$$24 = 24,$$

which checks.

Two or More Unknowns

If we have two or more unknowns, one equation is not sufficient to find them. In general, if there are n unknowns, n independent equations are needed. For example, if there are two unknowns, we need two equations. If the unknowns are called x and y, a typical procedure is to solve one equation for x in terms of y, and substitute this into the second equation.

EXAMPLE A–3 Solve the following pair of equations for x and y.

$$3x - 2y = 19$$

$$x + 4y = -3.$$

SOLUTION We solve the second equation for x in terms of y by subtracting $4y$ from both sides:

$$x = -3 - 4y.$$

We substitute this expression for x into the first equation, and simplify:

$$3(-3 - 4y) - 2y = 19$$

$$-9 - 12y - 2y = 19 \quad \text{(carried out the multiplication by 3)}$$

$$-14y = 28 \quad \text{(added 9 to both sides)}$$

$$y = -2 \quad \text{(divided both sides by } -14\text{).}$$

Now that we know $y = -2$, we substitute this into the expression for x:

$$x = -3 - 4y$$

$$= -3 - 4(-2) = -3 + 8 = 5.$$

Our solution is $x = 5$, $y = -2$. We check this solution by putting these values back into the original equations:

$$3x - 2y \stackrel{?}{=} 19$$

$$3(5) - 2(-2) \stackrel{?}{=} 19$$

$$15 + 4 \stackrel{?}{=} 19$$

$$19 = 19 \quad \text{(it checks)}$$

and

$$x + 4y \stackrel{?}{=} -3$$

$$5 + 4(-2) \stackrel{?}{=} -3$$

$$-3 = -3 \quad \text{(it checks).}$$

Other methods for solving two or more equations, such as the method of determinants, can be found in an algebra textbook.

The Quadratic Formula

We sometimes encounter equations that involve an unknown, say x, that appears not only to the first power, but squared as well. Such a *quadratic equation* can be written in the form

$$ax^2 + bx + c = 0.$$

The quantities a, b, and c are typically numbers or constants that are given.[†] The general solutions to such an equation are given by the *quadratic formula*:

Quadratic formula

$$x = \frac{-b \pm \sqrt{b^2 - 4ac}}{2a}.$$

(A–4)

The $\pm$ sign indicates that there are two solutions for x: one where the plus sign is used, the other where the minus sign is used.[‡]

EXAMPLE A–4 Find the solutions for x in the equation

$$3x^2 - 5x = 2.$$

SOLUTION First we write this equation in the standard form

$$ax^2 + bx + c = 0$$

by subtracting 2 from both sides:

$$3x^2 - 5x - 2 = 0.$$

In this case, a, b, and c in the standard formula take the values $a = 3$, $b = -5$, and $c = -2$. The two solutions for x are

$$x = \frac{+5 + \sqrt{25 - (4)(3)(-2)}}{(2)(3)} = \frac{5 + 7}{6} = 2$$

and

$$x = \frac{+5 - \sqrt{25 - (4)(3)(-2)}}{(2)(3)} = \frac{5 - 7}{6} = -\frac{1}{3}.$$

In this Example, the two solutions are $x = 2$ and $x = -\frac{1}{3}$. In physics problems, it sometimes happens that only one of the solutions corresponds to a real-life situation; in this case, the other solution is discarded. In other cases, both solutions may correspond to physical reality.

Notice, incidentally, that b^2 must be greater than $4ac$, so that $\sqrt{b^2 - 4ac}$ yields a real number. If $(b^2 - 4ac)$ is less than zero (negative), there is no real solution. The square root of a negative number is called *imaginary*.

[†]Or one or more of them could be variables, in which case additional equations are needed.

[‡]A second-order equation—one in which the highest power of x is 2—has two solutions: a third-order equation—involving x^3—has three solutions, and so on.

A–5 The Binomial Expansion

Sometimes we end up with a quantity of the form $(1 + x)^n$. That is, the quantity $(1 + x)$ is raised to the nth power. This can be written as an infinite sum of terms, known as a *series expansion*, as follows:

$$(1 + x)^n = 1 + nx + \frac{n(n - 1)}{2!}x^2 + \cdots. \qquad \textbf{(A–5)}$$

This formula is useful for us mainly when x is very small compared to one ($x \ll 1$). In this case, each successive term is much smaller than the preceding term. For example, if $x = 0.01$, and $n = 2$, say, then whereas the first term equals 1, the second term is $nx = (2)(0.01) = 0.02$, and the third term is $[(2)(1)/2](0.01)^2 = 0.0001$, and so on. Thus, when x is small, we can ignore all but the first two (or three) terms and can write

$$(1 + x)^n \approx 1 + nx. \qquad \textbf{(A–6)}$$

This approximation often allows us to easily solve an equation that otherwise might be very difficult. Some examples are

$$(1 + x)^2 \approx 1 + 2x$$

$$\frac{1}{1 + x} = (1 + x)^{-1} \approx 1 - x$$

$$\sqrt{1 + x} = (1 + x)^{\frac{1}{2}} \approx 1 + \tfrac{1}{2}x$$

$$\frac{1}{\sqrt{1 + x}} = (1 + x)^{-\frac{1}{2}} \approx 1 - \tfrac{1}{2}x$$

where $x \ll 1$.

As a numerical example, let us evaluate $\sqrt{1.02}$ using the binomial theorem since $x = 0.02$ is much smaller than 1:

$$\sqrt{1.02} = (1.02)^{\frac{1}{2}} = (1 + 0.02)^{\frac{1}{2}} \approx 1 + \tfrac{1}{2}(0.02) = 1.01.$$

You can check with a calculator (and maybe not even more quickly) that $\sqrt{1.02} \approx 1.01$.

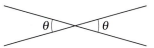

FIGURE A–1

A–6 Plane Geometry

We review here a number of theorems involving angles and triangles that are useful in physics.

1. *Equal angles.* Two angles are equal if any of the following conditions are true:
   (a) They are vertical angles (Fig. A–1); *or*
   (b) the left side of one is parallel to the left side of the other, and the right side of one is parallel to the right side of the other (the left and right sides are as seen from the vertex, where the two sides meet; Fig. A–2); *or*
   (c) the left side of one is perpendicular to the left side of the other, and the right sides are likewise perpendicular (Fig. A–3).

2. *The sum of the angles* in any plane triangle is 180°.

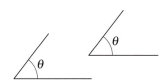

FIGURE A–2

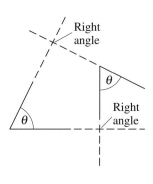

FIGURE A–3

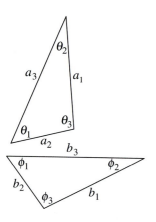

FIGURE A–4

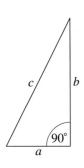

FIGURE A–5

3. *Similar triangles.* Two triangles are said to be similar if all three of their angles are equal (in Fig. A–4, $\theta_1 = \phi_1$, $\theta_2 = \phi_2$, $\theta_3 = \phi_3$). Similar triangles thus have the same basic shape but may be different sizes and have different orientations. Two useful theorems about similar triangles are:

(*a*) Two triangles are similar if any two of their angles are equal. (This follows because the third angles must also be equal since the sum of the angles of a triangle is 180°.)

(*b*) The ratio of corresponding sides of two similar triangles are equal. That is (Fig. A–4),

$$\frac{a_1}{b_1} = \frac{a_2}{b_2} = \frac{a_3}{b_3}.$$

4. *Congruent triangles.* Two triangles are congruent if one can be placed precisely on top of the other. That is, they are similar triangles and they have the same size. Two triangles are congruent if any of the following holds:

(*a*) The three corresponding sides are equal.

(*b*) Two sides and the enclosed angle are equal ("side-angle-side").

(*c*) Two angles and the enclosed side are equal ("angle-side-angle").

5. *Right triangles.* A right triangle has one angle that is 90° (a *right angle*); that is, the two sides that meet at the right angle are perpendicular. The two other (acute) angles in the right triangle add up to 90°.

6. *Pythagorean theorem.* In any right triangle, the square of the length of the hypotenuse (the side opposite the right angle) is equal to the sum of the squares of the lengths of the other two sides. In Fig. A–5,

$$c^2 = a^2 + b^2.$$

A–7 Areas and Volumes

Object	Surface Area	Volume
Circle, radius r	πr^2	—
Sphere, radius r	$4\pi r^2$	$\frac{4}{3}\pi r^3$
Right circular cylinder, radius r, height h	$2\pi r^2 + 2\pi rh$	$\pi r^2 h$
Right circular cone, radius r, height h	$\pi r^2 + \pi r\sqrt{r^2 + h^2}$	$\dfrac{\pi r^2 h}{3}$

A–8 Trigonometric Functions and Identities

(See Fig. A–6.)

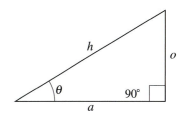

$$\sin \theta = \frac{o}{h} \qquad\qquad \csc \theta = \frac{1}{\sin \theta} = \frac{h}{o}$$

$$\cos \theta = \frac{a}{h} \qquad\qquad \sec \theta = \frac{1}{\cos \theta} = \frac{h}{a}$$

$$\tan \theta = \frac{o}{a} = \frac{\sin \theta}{\cos \theta} \qquad\qquad \cot \theta = \frac{1}{\tan \theta} = \frac{a}{o}$$

$$a^2 + o^2 = h^2 \qquad\qquad \text{(Pythagorean theorem).}$$

FIGURE A–6

Figure A–7 shows the signs (+ or −) that cosine, sine, and tangent take on for angles θ in the four quadrants (0° to 360°). Note that angles are measured counterclockwise from the x axis as shown; negative angles are measured from *below* the x axis, clockwise: for example, $-30° = +330°$, and so on. The following are some useful identities among the trigonometric functions:

$$\sin^2 \theta + \cos^2 \theta = 1, \quad \sec^2 \theta - \tan^2 \theta = 1, \quad \csc^2 \theta - \cot^2 \theta = 1$$

$$\sin 2\theta = 2 \sin \theta \cos \theta$$

$$\cos 2\theta = \cos^2\theta - \sin^2 \theta = 2 \cos^2 \theta - 1 = 1 - 2 \sin^2 \theta$$

$$\tan 2\theta = \frac{2 \tan \theta}{1 - \tan^2 \theta}$$

$$\sin (A \pm B) = \sin A \cos B \pm \cos A \sin B$$

$$\cos (A \pm B) = \cos A \cos B \mp \sin A \sin B$$

$$\tan (A \pm B) = \frac{\tan A \pm \tan B}{1 \mp \tan A \tan B}$$

$$\sin (180° - \theta) = \sin \theta$$

$$\cos (180° - \theta) = -\cos \theta$$

$$\sin (90° - \theta) = \cos \theta$$

$$\cos (90° - \theta) = \sin \theta$$

$$\sin \tfrac{1}{2}\theta = \sqrt{\frac{1 - \cos \theta}{2}}, \quad \cos \tfrac{1}{2}\theta = \sqrt{\frac{1 + \cos \theta}{2}}, \quad \tan \tfrac{1}{2}\theta = \sqrt{\frac{1 - \cos \theta}{1 + \cos \theta}}$$

$$\sin A \pm \sin B = 2 \sin \left(\frac{A \pm B}{2}\right) \cos \left(\frac{A \mp B}{2}\right).$$

For any triangle (see Fig. A–8):

$$\frac{\sin \alpha}{a} = \frac{\sin \beta}{b} = \frac{\sin \gamma}{c} \qquad\qquad \text{(Law of sines)}$$

$$c^2 = a^2 + b^2 - 2ab \cos \gamma. \qquad\qquad \text{(Law of cosines)}$$

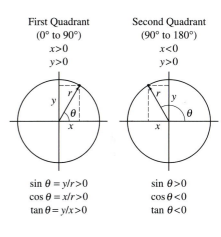

First Quadrant
(0° to 90°)
$x > 0$
$y > 0$

$\sin \theta = y/r > 0$
$\cos \theta = x/r > 0$
$\tan \theta = y/x > 0$

Second Quadrant
(90° to 180°)
$x < 0$
$y > 0$

$\sin \theta > 0$
$\cos \theta < 0$
$\tan \theta < 0$

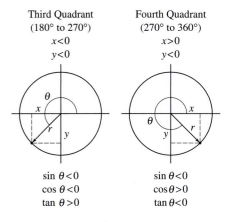

Third Quadrant
(180° to 270°)
$x < 0$
$y < 0$

$\sin \theta < 0$
$\cos \theta < 0$
$\tan \theta > 0$

Fourth Quadrant
(270° to 360°)
$x > 0$
$y < 0$

$\sin \theta < 0$
$\cos \theta > 0$
$\tan \theta < 0$

FIGURE A–7

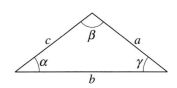

FIGURE A–8

A–9 | Logarithms

Logarithms are defined in the following way:

$$\text{if } y = A^x, \qquad \text{then } x = \log_A y.$$

That is, the logarithm of a number y to the base A is that number which, as the exponent of A, gives back the number y. For *common logarithms*, the base is 10, so

$$\text{if } y = 10^x, \qquad \text{then } x = \log y.$$

The subscript 10 on $\log_{10}$ is usually omitted when dealing with common logs. Another base sometimes used is the natural number $e = 2.718 \cdots$.[†] Such logarithms are called *natural logarithms* and are written ln. Thus,

$$\text{if } y = e^x, \qquad \text{then } x = \ln y.$$

For any number y, the two types of logarithm are related by

$$\ln y = 2.3026 \log y.$$

Some simple rules for logarithms are as follows:

$$\log (ab) = \log a + \log b. \tag{A–7}$$

This is true because if $a = 10^n$ and $b = 10^m$, then $ab = 10^{n+m}$. From the definition of logarithm, $\log a = n$, $\log b = m$, and $\log (ab) = n + m$; hence, $\log (ab) = n + m = \log a + \log b$. In a similar way, we can show that

$$\log \left(\frac{a}{b}\right) = \log a - \log b \tag{A–8}$$

and

$$\log a^n = n \log a. \tag{A–9}$$

These three rules apply not only to common logs but to natural or any other kind of logarithm.

Logs were once used as a technique for simplifying certain types of calculation. Because of the advent of electronic calculators and computers, they are not often used any more for this purpose. However, logs do appear in certain physical equations, so it is helpful to know how to deal with them. If you do not have a calculator that calculates logs, you can easily use a *log table*, such as the small one shown here (Table A–1). The number N is given to two digits (some tables give N to three or more digits); the first digit is in the vertical column to the left, the second digit is in the horizontal row across the top. For example, the table tells us that $\log 1.0 = 0.000$, $\log 1.1 = 0.041$, and $\log 4.1 = 0.613$. Note that the table does not include the decimal point—it is understood. The table gives logs for numbers between 1.0 and 9.9; for larger or smaller numbers we use rule A–7:

$$\log (ab) = \log a + \log b.$$

For example,

$$\log (380) = \log (3.8 \times 10^2) = \log (3.8) + \log (10^2).$$

From the table, $\log 3.8 = 0.580$; and from rule A–9, $\log (10^2) = 2 \log (10) = 2$, since $\log (10) = 1$. [This follows from the definition of the

---

[†]The natural number e can be written as an infinite series:

$$e = 1 + \frac{1}{1} + \frac{1}{1 \cdot 2} + \frac{1}{1 \cdot 2 \cdot 3} + \frac{1}{1 \cdot 2 \cdot 3 \cdot 4} + \cdots.$$

TABLE A-1 Short Table of Common Logarithms

N	0.0	0.1	0.2	0.3	0.4	0.5	0.6	0.7	0.8	0.9
1	000	041	079	114	146	176	204	230	255	279
2	301	322	342	362	380	398	415	431	447	462
3	477	491	505	519	531	544	556	568	580	591
4	602	613	623	633	643	653	663	672	681	690
5	699	708	716	724	732	740	748	756	763	771
6	778	785	792	799	806	813	820	826	833	839
7	845	851	857	863	869	875	881	886	892	898
8	903	908	914	919	924	929	935	940	944	949
9	954	959	964	968	973	978	982	987	991	996

logarithm: if $10 = 10^1$, then $1 = \log(10)$.] Thus,

$$\log(380) = \log(3.8) + \log(10^2)$$
$$= 0.580 + 2$$
$$= 2.580.$$

Similarly,

$$\log(0.081) = \log(8.1) + \log(10^{-2})$$
$$= 0.908 - 2 = -1.092.$$

Sometimes we need to do the reverse process: find the number N whose log is, say, 2.670. This is called "taking the antilogarithm." To do so, we separate our number 2.670 into two parts, making the separation at the decimal point:

Antilog

$$\log N = 2.670 = 2 + 0.670$$
$$= \log 10^2 + 0.670.$$

We now look at Table A–1 to see what number has its log equal to 0.670; none does, so we must *interpolate*: we see that $\log 4.6 = 0.663$ and $\log 4.7 = 0.672$. So the number we want is between 4.6 and 4.7, and closer to the latter by 7/9. Approximately we can say that $\log 4.68 = 0.670$. Thus

$$\log N = 2 + 0.670$$
$$= \log(10^2) + \log(4.68) = \log(4.68 \times 10^2),$$

so $N = 4.68 \times 10^2 = 468$. If the given logarithm is negative, say, -2.180, we proceed as follows:

$$\log N = -2.180 = -3 + 0.820$$
$$= \log 10^{-3} + \log 6.6 = \log 6.6 \times 10^{-3},$$

so $N = 6.6 \times 10^{-3}$. Notice that what we did was to add to our given logarithm the next largest integer (3 in this case) so that we have an integer, plus a decimal number between 0 and 1.0 whose antilogarithm can be looked up in the table.

APPENDIX

B DIMENSIONAL ANALYSIS

When we speak of the *dimensions* of a quantity, we are referring to the type of units that must be used. The dimensions of area, for example, are always length squared (abbreviated $[L^2]$, using square brackets) and the units can be square meters, square feet, and so on. Velocity, on the other hand, can be measured in units of km/h, m/s, and mi/h, but the dimensions are always a length $[L]$ divided by a time $[T]$, that is, $[L/T]$. The formula for a quantity may be different in different cases, but the dimensions remain the same. For example, the area of a triangle of base b and height h is $A = \frac{1}{2}bh$, whereas the area of a circle of radius r is $A = \pi r^2$. The formulas are different in the two cases, but the dimensions in both cases are the same: $[L^2]$.

When we specify the dimensions of a quantity, we usually do so in terms of basic quantities such as length $[L]$, time $[T]$, mass $[M]$, and electric current $[I]$. Thus, force, which by Newton's second law has the same units as mass $[M]$ times acceleration $[L/T^2]$, has dimensions of $[ML/T^2]$.

Dimensions can be used as a help in working out relationships, and such a procedure is referred to as *dimensional analysis*. One useful technique is the use of dimensions to check a relationship for correctness. Two simple rules apply here. First, we can add or subtract quantities only if they have the same dimensions (we do not add centimeters and pounds); second, the quantities on each side of an equals sign must have the same dimensions.

For example, suppose that you derived the equation $v = v_0 + \frac{1}{2}at^2$, where v is the speed of an object after a time t, when it starts with an initial speed v_0 and undergoes an acceleration a. Let us do a dimensional check to see if this equation can be correct. We write a dimensional equation as follows, remembering that the dimensions of speed are $[L/T]$ and of acceleration are $[L/T^2]$:

$$\left[\frac{L}{T}\right] \overset{?}{=} \left[\frac{L}{T}\right] + \left[\frac{L}{T^2}\right][T^2]$$

$$\overset{?}{=} \left[\frac{L}{T}\right] + [L].$$

The dimensions are incorrect: on the right side, we have the sum of two quantities whose dimensions are not the same. Thus, we conclude that an error was made in the derivation of the original equation.

If such a dimensional check does come out correct, it does not prove that the equation is correct; for example, a dimensionless numerical factor (such as $\frac{1}{2}$ or 2π) could be wrong. Thus a dimensional check can only tell you when a relationship is wrong; it cannot tell you if it is completely right.

Another use of dimensional analysis is for a quick check on an equation you are not sure about. For example, suppose that you cannot remember whether the equation for the period T of an oscillating mass m on the end of a spring with constant k is $T = 2\pi\sqrt{k/m}$ or is $T = 2\pi\sqrt{m/k}$. A dimensional check can tell you. The dimensions of k, since from Hooke's law $k = \text{force}/\text{distance}$, are $[ML/T^2/L] = [M/T^2]$. Thus the formula $T = 2\pi\sqrt{m/k}$ is correct:

$$[T] = \sqrt{\frac{[M]}{[M/T^2]}},$$

whereas the formula $T = 2\pi\sqrt{k/m}$ is not:

$$[T] \neq \sqrt{\frac{[M/T^2]}{[M]}}.$$

Finally, an important use of dimensional analysis, but one with which much care must be taken, is to obtain the *form* of an equation. That is, we may want to determine how one quantity depends on others. To take a concrete example, let us try to find an expression for the period T of a simple pendulum. First, we try to figure out what T could depend on, and make a list of these variables. It might depend on its length l, on the mass m of the bob, on the angle of swing θ, and on the acceleration due to gravity, g. It might also depend on air resistance (we would use the viscosity of air), the gravitational pull of the Moon, and so on; but everyday experience suggests that the Earth's gravity is the major force involved, so we ignore the other possible forces. So let us assume that T is a function of l, m, θ, and g, and that each of these factors is present to some power:

$$T = Cl^w m^x \theta^y g^z.$$

C is a dimensionless constant, and w, x, y, and z are exponents we want to solve for. We now write down the dimensional equation for this relationship:

$$[T] = [L]^w [M]^x [L/T^2]^z;$$

because θ has no dimensions (a radian is a length over a length—see Eq. 8–1), it does not appear. We simplify and obtain

$$[T] = [L]^{w+z} [M]^x [T]^{-2z}.$$

To have dimensional consistency, we must have

$$1 = -2z$$

$$0 = w + z$$

$$0 = x.$$

We solve these equations and find that $z = -\frac{1}{2}$, $w = \frac{1}{2}$, and $x = 0$. Thus our desired equation must be

$$T = C\sqrt{l/g}\, f(\theta), \tag{B–1}$$

where $f(\theta)$ is some function of θ that we cannot determine using this technique. Nor can we determine in this way the dimensionless constant C. (Of course, to obtain C and f, we would have to do an analysis such as that in Chapter 11 using Newton's laws, which reveals that $C = 2\pi$ and $f \approx 1$ for small θ). But look what we *have* found, using only dimensional consistency. We obtained the form of the expression that relates the period of a simple pendulum to the major variables of the situation, l and g (see Eq. 11–11a) and saw that it indeed does not depend on the mass m.

How did we do it? And how useful is this technique? Basically, we had to use our intuition as to which variables were important and which were not. This is not always easy, and often requires a lot of insight. As to usefulness, the final result in our example could have been obtained from Newton's laws, as in Chapter 11. But in many physical situations, such a derivation from other laws cannot be done. In those situations, dimensional analysis can be a powerful tool.

In the end, any expression derived by the use of dimensional analysis (or by any other means, for that matter) must be checked against experiment. For example, in our derivation of Eq. B–1, we can compare the periods of two pendula of different lengths, l_1 and l_2, whose amplitudes (θ) are the same. For, using Eq. B–1, we would have

$$\frac{T_1}{T_2} = \frac{C\sqrt{l_1/g}\, f(\theta)}{C\sqrt{l_2/g}\, f(\theta)} = \sqrt{\frac{l_1}{l_2}}.$$

Because C and $f(\theta)$ are the same for both pendula, they cancel out, so we can experimentally determine if the ratio of the periods varies as the ratio of the square roots of the lengths. This comparison to experiment checks our derivation, at least in part. C and $f(\theta)$ could be determined by further experiments.

ROTATING FRAMES OF REFERENCE; INERTIAL FORCES; CORIOLIS EFFECT

Inertial and Noninertial Reference Frames

In Chapters 5 and 8 we examined the motion of bodies, including circular and rotational motion, from the outside, as observers fixed on the ground. Sometimes it is convenient to place ourselves (in theory, if not physically) into a rotating system. Let us, for example, examine the motion of objects from the point of view, or frame of reference, of persons seated on a rotating platform such as a merry-go-round. It looks to them as if the rest of the world is going around *them*. But let us focus attention on what they observe when they place a tennis ball on the floor of the rotating platform, which we assume is frictionless. If they put the ball down gently, without giving it any push, they will observe that it accelerates from rest and moves outward as shown in Fig. C–1a. According to Newton's first law, an object initially at rest should stay at rest if no force acts on it. But, according to the observers on the rotating platform, the ball starts moving even though there is no force applied to it. To observers on the ground, this is all very clear: the ball has an initial velocity when it is released (because the platform is moving), and it simply continues moving in a straight-line path as shown in Fig. C–1b, in accordance with Newton's first law.

But what shall we do about the frame of reference of the observers on the rotating platform? Clearly, Newton's first law, the law of inertia, does not hold in this rotating frame of reference. For this reason, such a frame is called a **noninertial reference frame**. An **inertial reference frame** is one in which the law of inertia—Newton's first law—does hold, and so do Newton's second and third laws. In a noninertial reference frame, such as our rotating platform, Newton's second law also does not hold. For instance in the situation described above, there is no net force on the ball; yet, with respect to the rotating platform, the ball accelerates.

Fictitious (Inertial) Forces

Because Newton's laws do not hold when observations are made with respect to a rotating frame of reference, calculation of motion can be complicated. However, we can still apply Newton's laws in such a reference frame if we make use of a trick. The ball on the rotating platform of Fig. C–1a flies outward when released (as if a force were acting on it—though as we saw above, no force actually does act on it); so the trick we use is to write down the equation $\Sigma F = ma$ as if a force equal to mv^2/r (or $m\omega^2 r$) were acting radially outward on the object in addition to any other forces that may be acting. This extra force, which might be designated as "centrifugal force" since it *seems* to act outward, is called a **fictitious force** or **pseudoforce**. It is a pseudoforce ("pseudo" means "false") because there is no object that exerts

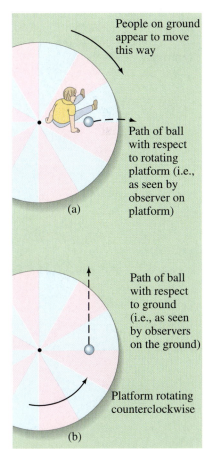

People on ground appear to move this way

Path of ball with respect to rotating platform (i.e., as seen by observer on platform)

(a)

Path of ball with respect to ground (i.e., as seen by observers on the ground)

Platform rotating counterclockwise

(b)

FIGURE C–1 Path of a ball released on a rotating merry-go-round (a) in the reference frame of the merry-go-round, and (b) in a reference frame on the ground.

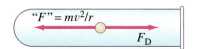

FIGURE C-2 The forces on a particle in a rotating centrifuge, in the reference frame of the centrifuge.

this force. Furthermore, when viewed from an inertial reference frame, the effect doesn't exist at all. We have made up this pseudoforce so that we can make calculations in a noninertial frame using Newton's second law, $\Sigma F = ma$. Thus the observer in Fig. C–1a can determine the motion of the ball by assuming that a force equal to mv^2/r acts on it. Such pseudoforces are also called **inertial forces** since they arise only because the reference frame is not an inertial one.

We can examine the motion of a particle in a centrifuge (Section 5–5) from the frame of reference of the rotating rotor. In this frame of reference, the particles move in a more-or-less straight path down the tube. (From the reference frame of the Earth, the particles go round and round.) The acceleration of a particle with respect to the rotor can then be calculated using $F = ma$ if we include a pseudoforce, "F," equal to $m\omega^2 r = m(v^2/r)$ acting down the tube, in addition to the drag force F_D exerted by the fluid on the particle (Fig. C–2) up the tube.

In Section 5–3 we discussed the forces on a person in a car going around a curve (Fig. 5–11) from the point of view of an inertial frame. The car, on the other hand, is not an inertial frame. Passengers in such a car could interpret this being pressed outward as the effect of a "centrifugal" force. But they need to recognize that it is a pseudoforce because there is no identifiable object exerting it. It is an effect of being in a noninertial frame of reference.

The Earth itself is rotating on its axis. Thus, strictly speaking, Newton's laws are not valid on the Earth. However, the effect of the Earth's rotation is usually so small that it can be ignored, although it does influence the movement of large air masses and ocean currents. Because of the Earth's rotation, the material of the Earth is concentrated slightly at the equator. The Earth is thus not a perfect sphere but is slightly fatter at the equator than it is at the poles.

Coriolis Effect

In a reference frame that rotates at a constant angular speed ω (relative to an inertial system), there exists another pseudoforce known as the *Coriolis force*. It appears to act on a body in a rotating reference frame only if the body is moving relative to that reference frame, and it acts to deflect the body sideways. It, too, is an effect of the reference frame being noninertial and hence is referred to as an *inertial force*. To see how the Coriolis force arises, consider two people, A and B, at rest on a platform rotating with angular speed ω, as shown in Fig. C–3a. They are situated at distances r_A and r_B from the axis of rotation (at O). The woman at A throws a ball with a horizontal velocity $\mathbf{v}$ (in her reference frame) radially outward toward the man at B on the outer edge of the platform. In Fig. C–3a, we view the situation from an inertial reference frame. The ball initially has not only the velocity $\mathbf{v}$ radially outward, but also a tangential velocity $\mathbf{v_A}$ due to the rotation of the platform. Now Eq. 8–4 tells us that $v_A = r_A\omega$, where r_A is the woman's radial distance from the axis of rotation at O. If the man at B had this same velocity v_A, the ball would reach him perfectly. But his speed is greater than v_A (Fig. C–3a) since he is farther from the axis of rotation. His speed is $v_B = r_B\omega$, which is greater than v_A because $r_B > r_A$. Thus, when the ball reaches the outer edge of the platform, it passes a point that the man at B has already passed because his speed in that direction is greater than the ball's. So the ball passes behind him.

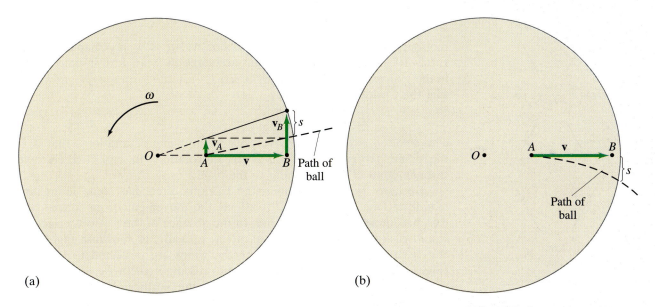

(a)

(b)

Figure C–3b shows the situation as seen from the rotating platform as frame of reference. Both A and B are at rest, and the ball is thrown with velocity $\mathbf{v}$ toward B, but the ball deflects to the right as shown and passes the behind B as previously described. This is not a centrifugal-force effect, for the latter acts radially outward. Instead, this effect acts sideways, perpendicular to $\mathbf{v}$, and is called a **Coriolis acceleration**; it is said to be due to the Coriolis force, which is a fictitious, inertial force. Its explanation as seen from an inertial system was given above: it is an effect of being in a rotating system, wherein points that are farther from the rotation axis have higher linear speeds. On the other hand, when viewed from the rotating system, we can describe the motion using Newton's second law, $\Sigma\mathbf{F} = m\mathbf{a}$, if we add a "pseudoforce" term corresponding to this Coriolis effect.

Let us determine the magnitude of the Coriolis acceleration for the simple case described above. (We assume v is large and distances short, so we can ignore gravity.) We do the calculation from the inertial reference frame (Fig. C–3a). The ball moves radially outward a distance $r_B - r_A$ at speed v in a time t given by

$$r_B - r_A = vt.$$

During this time, the ball moves to the side a distance s_A given by

$$s_A = v_A t.$$

The man at B, in this time t, moves a distance

$$s_B = v_B t.$$

The ball therefore passes behind him a distance s (Fig. C–3a) given by

$$s = s_B - s_A = (v_B - v_A)t.$$

We saw earlier that $v_A = r_A\omega$ and $v_B = r_B\omega$, so

$$s = (r_B - r_A)\omega t.$$

We substitute $r_B - r_A = vt$ (see above) and get

$$s = \omega v t^2. \tag{C–1}$$

This equals the sideways displacement as seen from the noninertial rotating system (Fig. C–3b). We see immediately that Eq. C–1 corresponds to

FIGURE C–3 The origin of the Coriolis effect. Looking down on a rotating platform, (a) as seen from a nonrotating inertial system, and (b) as seen from the rotating platform as frame of reference.

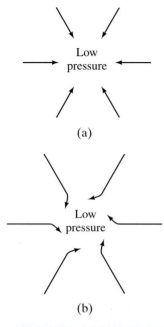

(a)

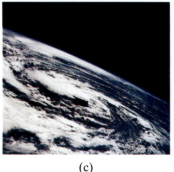

(b)

(c)

FIGURE C–4 (a) Winds (moving air masses) would flow directly toward a low-pressure area if the Earth did not rotate; (b) and (c): because of the Earth's rotation, the winds are deflected to the right in the Northern Hemisphere (as in Fig. C–3) as if a fictitious (Coriolis) force were acting.

motion at a constant acceleration. For as we saw in Chapter 2 (see Eq. 2–10b), $y = \frac{1}{2}at^2$ for a constant acceleration (with zero initial velocity in the y direction). Thus, if we write Eq. C–1 in the form $s = \frac{1}{2}a_{\text{cor}}t^2$, we see that the Coriolis acceleration a_{cor} is

$$a_{\text{cor}} = 2\omega v. \tag{C–2}$$

This relation is valid for any velocity in the plane of rotation—that is, in the plane perpendicular to the axis of rotation (in Fig. C–3, the axis through point O perpendicular to the page).

Because the Earth rotates, the Coriolis effect has some interesting manifestations on the Earth. It affects the movement of air masses and thus has an influence on weather. In the absence of the Coriolis effect, air would rush directly into a region of low pressure, as shown in Fig. C–4a. But because of the Coriolis effect, the winds are deflected to the right (Fig. C–4b), so that there tends to be a counterclockwise wind pattern around a low-pressure area. This is true for the Northern Hemisphere—for which the situation of Fig. C–3b applies since the Earth rotates from west to east. The reverse is true in the Southern Hemisphere. Thus cyclones rotate counterclockwise in the Northern Hemisphere and clockwise in the Southern Hemisphere. The same effect explains the easterly trade winds near the equator: any winds heading south toward the equator will be deflected toward the west (that is, as if coming from the east).

The Coriolis effect also acts on a falling body. A body released from the top of a high tower will not hit the ground directly below the release point, but will be deflected slightly to the east. Viewed from an inertial frame, this is because the top of the tower revolves with a slightly higher speed than does the bottom of the tower.

GAUSS'S LAW D

A n important relation in electricity is Gauss's law, developed by the great mathematician Karl Friedrich Gauss (1777–1855). It relates electric charge and electric field, and is a more general and elegant version of Coulomb's law.

Gauss's law involves the concept of *electric flux*, which refers to the electric field passing through a given area. For a uniform electric field **E** passing through an area A, as shown in Fig. D–1a, the electric flux Φ_E is defined as

$$\Phi_E = EA \cos \theta, \qquad \text{(D–1)}$$

where θ is the angle between the electric-field direction and a line drawn perpendicular to the area, as shown. The flux can be written equivalently as

$$\Phi_E = E_\perp A = EA_\perp,$$

where $E = E \cos \theta$ is the component of **E** perpendicular to the area (Fig. D–1b), and, similarly, $A = A \cos \theta$ is the projection of the area A perpendicular to the field **E** (Fig. D–1c).

Electric flux has a simple intuitive interpretation in terms of field lines. We saw in Section 16–8 that field lines can always be drawn so that the number (N) passing through unit area perpendicular to the field (A) is proportional to the magnitude of the field (E): that is, $E \propto N/A$. Hence,

$$N \propto EA_\perp = \Phi_E, \qquad \text{(D–2)}$$

so the flux through an area is proportional to the number of lines passing through that area.

Gauss's law involves the *total* flux through a closed surface. For any such surface, such as that shown in Fig. D–2, we divide the surface up into many tiny areas, ΔA_1, ΔA_2, ΔA_3, $\cdots$, and so on. We make the division so that each ΔA is small enough that it can be considered flat and so that the electric field can be considered constant within each ΔA. Then the *total* flux through the entire surface is the sum over all the individual fluxes through each of the tiny areas:

$$\Phi_E = E_1 \Delta A_1 \cos \theta_1 + E_2 \Delta A_2 \cos \theta_2 + \cdots$$
$$= \sum E \, \Delta A \cos \theta = \sum E_\perp \Delta A,$$

where the symbol Σ means "sum of." We saw in Section 16–8 that the number of field lines starting on a positive charge or ending on a negative charge is proportional to the magnitude of the charge. Hence, the *net* number of lines N pointing out of any closed surface (number of lines pointing out minus the number pointing in) must be proportional to the *net* enclosed

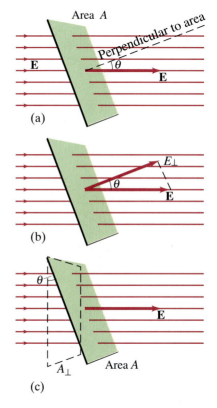

FIGURE D–1 (a) A uniform electric field **E** passing through a flat area A. (b) $E_\perp = E \cos \theta$ is the component of **E** perpendicular to the plane of the area A. (c) $A_\perp = A \cos \theta$ is the projection (shown dashed) of the area A perpendicular to the field **E**.

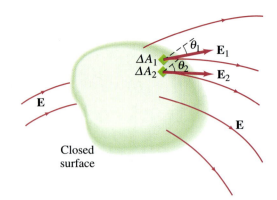

FIGURE D–2 Electric field lines passing through a closed surface. The surface is divided up into many tiny areas, ΔA_1, $\Delta A_2, \cdots$, and so on, of which only two are shown.

charge, Q. But from Eq. D–2, we have that the net number of lines N is proportional to the total flux Φ_E. Therefore,

$$\Phi_E = \sum_{\substack{\text{closed} \\ \text{surface}}} E_\perp \, \Delta A \propto Q.$$

The constant of proportionality is $1/\epsilon_0$, as we shall see, so we have

Gauss's law

$$\sum_{\substack{\text{closed} \\ \text{surface}}} E_\perp \Delta A = \frac{Q}{\epsilon_0}, \tag{D–3}$$

where the sum (Σ) is over any closed surface, and Q is the net charge enclosed within that surface. This is **Gauss's law**.

This argument to justify Gauss's law may seem rather abstract. So let us show, for a particular example—that of a single point charge—that Gauss's law is equivalent to Coulomb's law, and at the same time show that the proportionality constant is indeed $1/\epsilon_0$. In Fig. D–3, we have a single charge Q, and for our "gaussian surface," we choose an imaginary sphere of radius r centered on the charge. Because Gauss's law is supposed to be valid for any surface, we have chosen one that will make our calculation easy, namely a sphere centered on the charge. Since the charge is at the center of the sphere, the magnitude of the electric field E at all points on the sphere is the same, namely

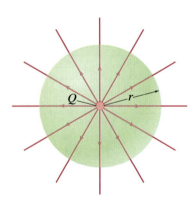

FIGURE D–3 A single point charge Q at the center of an imaginary sphere of radius r (our "gaussian surface"—that is, the closed surface we choose to use for applying Gauss's law in this case).

$$E = \frac{1}{4\pi\epsilon_0} \frac{Q}{r^2},$$

from Coulomb's law (Eq. 16–4b). The direction of $\mathbf{E}$ at all points is perpendicular to the sphere, so $E_\perp = E$. Hence, the total flux is

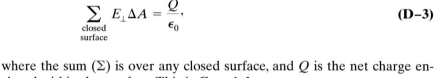

$$\sum_{\substack{\text{closed} \\ \text{surface}}} E_\perp \, \Delta A = E \sum \Delta A = \frac{1}{4\pi\epsilon_0} \frac{Q}{r^2} \sum \Delta A.$$

Because E is the same through each of the areas ΔA, we were able to factor it out of the sum. Now the sum over the entire area of the spherical surface, $\Sigma \Delta A$, is simply the surface area of a sphere, $4\pi r^2$. Hence,

$$\sum E_\perp \, \Delta A = \frac{1}{4\pi\epsilon_0} \frac{Q}{r^2} (4\pi r^2) = \frac{Q}{\epsilon_0},$$

which is just Gauss's law, as we set out to show.

We can argue that Gauss's law will hold for any closed surface other than a sphere by noting that the number of lines N passing through it would be the same as through our sphere (Fig. D–4). Since the flux through the surface is proportional to N, then $\Sigma E_\perp \Delta A = Q/\epsilon_0$ is valid for any surface surrounding a point charge.

If we have many charges, $Q_1, Q_2, \cdots$, then for each charge individually,

$$\sum E_{i\perp} \Delta A = \frac{Q_i}{\epsilon_0}$$

where the subscript i refers to any one of the charges. The sum of all the charges is the total charge $Q \,(= \Sigma Q_i)$. And, by the superposition principle, the total electric field E is equal to the sum of the fields due to each separate charge: $E = \Sigma E_i$. Hence,

$$\sum E_\perp \Delta A = \frac{\sum Q_i}{\epsilon_0} = \frac{Q}{\epsilon_0}.$$

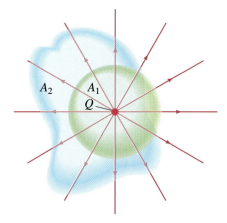

FIGURE D–4 A single point charge Q, surrounded by a spherical surface A_1 and an irregular surface A_2.

So Gauss's law is consistent with Coulomb's law for any number of charges enclosed within a closed surface of any shape.

Coulomb's law and Gauss's law are equivalent for electric fields produced by static charges. However, electric fields can also be produced by (changing) magnetic fields, as was discussed in Chapter 21. Coulomb's law does not apply for such electric fields, but Gauss's law does apply. Hence, Gauss's law is considered a more general law than Coulomb's law.

Coulomb's law and Gauss's law can be used to determine the electric field due to a given (static) charge distribution. Gauss's law is useful when the charge distribution is simple and symmetrical. However, we must choose the "gaussian" surface very carefully so we can determine **E**. We normally choose a surface that has just the symmetry needed so E will be constant on all or on parts of its surface.

EXAMPLE D–1 **Spherical shell.** A thin spherical shell of radius r_0 possesses a total net charge Q that is uniformly distributed on it, Fig. D–5. Determine the electric field at points (a) outside the shell, and (b) inside the shell.

SOLUTION (a) Because the charge is distributed symmetrically, the electric field must also be symmetric. Thus the field must be directed radially outward (inward if $Q < 0$) and must depend only on r. The electric field will thus have the same magnitude at all points on an imaginary gaussian surface, which we draw as a sphere of radius r, concentric with the shell, and shown in Fig. D–5 as a dashed circle labeled A_1. Because **E** is perpendicular to this surface, Gauss's law gives

$$\sum E_\perp \Delta A = E \sum \Delta A = E(4\pi r^2) = \frac{Q}{\epsilon_0},$$

or

$$E = \frac{1}{4\pi\epsilon_0} \frac{Q}{r^2}. \qquad\qquad [r > r_0]$$

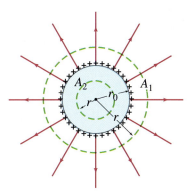

FIGURE D–5 Uniform distribution of charge on a thin spherical shell of radius r_0. Example D–1.

Thus the field outside a uniformly charged spherical shell is the same as if all the charge were concentrated at the center as a point charge.

(b) Inside the shell, the field must also be symmetric. So E must again have the same value at all points on a spherical gaussian surface (A_2 in Fig. D–5) concentric with the shell. Thus, E can be factored out of the sum and we have

$$\sum E_\perp \, \Delta A = E \sum \Delta A = E(4\pi r^2) = \frac{Q}{\epsilon_0} = 0$$

since the charge Q inside the shell is zero. Hence

$$E = 0 \qquad\qquad\qquad [r < r_0]$$

inside a uniform spherical shell of charge.

The useful results of Example D–1 also apply to a uniform *solid* spherical conductor that is charged, since all the charge would lie in a thin layer at the surface (Section 16–9).

EXAMPLE D–2 **Long uniform line of charge.** A very long straight wire possesses a uniform charge per unit length, λ. Calculate the electric field at points near (but outside) the wire, far from the ends.

SOLUTION Because of the symmetry, we expect the field to be directed radially outward (if $Q > 0$) and to depend only on the perpendicular distance, r, from the wire. Because of this cylindrical symmetry, the field will be the same at all points on a gaussian surface that is a cylinder with the wire along its axis, Fig. D–6. **E** is perpendicular to this surface at all points. But for Gauss's law, we need a closed surface, so we must include the flat ends of the cylinder. Since **E** is parallel to the ends, there is no flux through the ends (the $\cos\theta$ in Gauss's law is $\cos 90° = 0$). So Gauss's law tells us

$$\sum E_\perp \, \Delta A = E \sum \Delta A = E(2\pi r l) = \frac{Q}{\epsilon_0} = \frac{\lambda l}{\epsilon_0},$$

where l is the length of our chosen gaussian surface ($l \ll$ length of wire), and $2\pi r$ is its circumference. Hence

$$E = \frac{1}{2\pi\epsilon_0} \frac{\lambda}{r}.$$

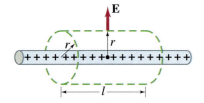

FIGURE D–6 Calculation of **E** due to a very long line of charge. Example D–2.

FIGURE D–7 Electric field near the surface of a conductor. Two small cylindrical boxes are shown dashed. Either one can serve as our gaussian surface. See Example D–3.

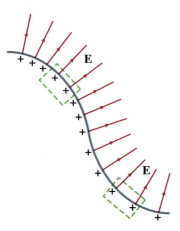

EXAMPLE D–3 *E* **at surface of conductor.** Show that the electric field just outside the surface of any good conductor of arbitrary shape is given by

$$E = \frac{\sigma}{\epsilon_0},$$

where σ is the surface charge density (Q/A) on the conductor at that point.

SOLUTION We choose as our gaussian surface a small cylindrical box, very small in height so that one of its circular ends is just above the conductor (Fig. D–7); the other end is just below the conductor's surface, and the sides are perpendicular to it. The electric field is zero inside a conductor and is perpendicular to the surface just outside it (Section 16–9), so electric flux passes only through the outside end of our cylindrical box. We choose the area A (of the flat cylinder end) small enough so that E is essentially uniform over it. Then Gauss's law gives

$$\sum E_\perp \, \Delta A = EA = \frac{Q}{\epsilon_0} = \frac{\sigma A}{\epsilon_0},$$

so that

$$E = \frac{\sigma}{\epsilon_0}. \qquad \text{[at surface of conductor]}$$

This useful result applies for any shape conductor, including a large uniformly charged flat sheet: the electric field will be constant and equal to σ/ϵ_0.

This last Example also gives us the field between the two parallel plates we discussed in Fig. 16–29d, and also relative to Eq. 17–6. If the plates are large compared to their separation, then the field lines are perpendicular to the plates and, except near the edges, they are parallel to each other. Therefore the electric field (see Fig. D–8, which shows the same gaussian surface as Fig. D–7) is also

$$E = \frac{\sigma}{\epsilon_0} = \frac{Q/A}{\epsilon_0} \qquad \text{(D–4)}$$

where $Q = \sigma A$ is the charge on one of the plates.

We saw in Section 16–9 that in the static situation, the electric field inside any conductor must be zero. (Otherwise, the free charges in the conductor would move—until the net force on each, and hence $\mathbf{E}$, was zero.) We also saw there that any net electric charge on a conductor must all reside on its outer surface. This is readily seen using Gauss's law. Consider any charged conductor of any shape, such as that shown in Fig. D–9, which carries a net charge Q. We choose the gaussian surface, shown dashed in the diagram, so that it lies below the surface of the conductor. Our gaussian surface can be arbitrarily close to the surface, but still *inside* the conductor. The electric field is zero at all points on this gaussian surface since it is inside the conductor. Hence, from Gauss's law, Eq. D–3, the net charge within the surface must be zero. Hence, there can be no net charge within the conductor. Any net charge must lie on the surface of the conductor.

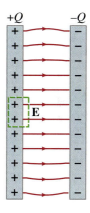

FIGURE D–8 The electric field between two parallel plates is uniform and equal to $E = \sigma/\epsilon_0$.

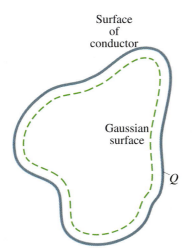

FIGURE D–9 An insulated charged conductor of arbitrary shape, showing (dashed) the gaussian surface just below the surface of the conductor.

E GALILEAN AND LORENTZ TRANSFORMATIONS

We now examine in detail the mathematics of relating quantities in one inertial reference frame to the equivalent quantities in another. In particular, we will see how positions and velocities *transform* (that is, change) as we go from one frame of reference to another.

We begin with the classical or Galilean viewpoint. Consider two reference frames S and S' which are each characterized by a set of coordinate axes, Fig. E–1. The axes x and y (z is not shown) refer to S, and x' and y' refer to S'. The x' and x axes overlap one another, and we assume that frame S' moves to the right (in the x direction) at speed v with respect to S. And for simplicity let us assume the origins O and O' of the two reference frames are superimposed at time $t = 0$.

Now consider an event that occurs at some point P (Fig. E–1) represented by the coordinates x', y', z' in reference frame S' at the time t'. What will be the coordinates of P in S? Since S and S' overlap precisely initially, after a time t, S' will have moved a distance vt'. Therefore, at time t', $x = x' + vt'$. The y and z coordinates, on the other hand, are not altered by motion along the x axis; thus $y = y'$ and $z = z'$. Finally, since time is assumed to be absolute in Galilean–Newtonian physics, clocks in the two frames will agree with each other; so $t = t'$. We summarize these in the following **Galilean transformation equations**:

Galilean

transformations

$$x = x' + vt'$$
$$y = y'$$
$$z = z'$$
$$t = t'.$$

(E–1)

These equations give the coordinates of an event in the S frame when those in the S' frame are known. If those in the S system are known, then

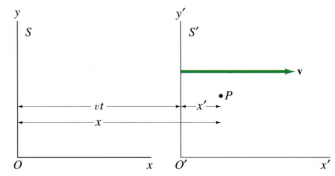

FIGURE E–1 Inertial reference frame S' moves to the right at speed v with respect to frame S.

the S' coordinates are obtained from

$$x' = x - vt, \qquad y' = y, \qquad z' = z, \qquad t' = t.$$

These four equations are the "inverse" transformation and are very easily obtained from Eqs. E–1. Notice that the effect is merely to exchange primed and unprimed quantities and replace v by $-v$. This makes sense because from the S' frame, S moves to the left (negative x direction) with speed v.

Now suppose that the point P in Fig. E–1 represents an object that is moving. Let the components of its velocity vector in S' be u'_x, u'_y, and u'_z (we use u to distinguish it from the relative velocity of the two frames, v). Now $u'_x = \Delta x'/\Delta t'$, $u'_y = \Delta y'/\Delta t'$, and $u'_z = \Delta z'/\Delta t'$, where all quantities are as measured in the S' frame. For example, if at time t'_1 the particle is at x'_1 and a short time later, t'_2, it is at x'_2, then

$$u'_x = \frac{x'_2 - x'_1}{t'_2 - t'_1} = \frac{\Delta x'}{\Delta t'}.$$

Now the velocity of P as seen from S will have components u_x, u_y, and u_z. We can show how these are related to the velocity components in S' by using Eqs. E–1. For example,

$$
\begin{aligned}
u_x &= \frac{\Delta x}{\Delta t} = \frac{x_2 - x_1}{t_2 - t_1} = \frac{(x'_2 + vt'_2) - (x'_1 + vt'_1)}{t'_2 - t'_1} \\
&= \frac{(x'_2 - x'_1) + v(t'_2 - t'_1)}{t'_2 - t'_1} \\
&= \frac{\Delta x'}{\Delta t'} + v = u'_x + v.
\end{aligned}
$$

For the other components, $u'_y = u_y$ and $u'_z = u_z$, so we have

$$u_x = u'_x + v,$$

$$u_y = u'_y, \qquad\qquad\qquad\qquad \textbf{(E–2)}$$

$$u_z = u'_z.$$

Galilean

velocity

transformations

These are known as the **Galilean velocity transformation** equations. We see that the y and z components of velocity are unchanged, but the x components differ by v. This is just what we have used before when dealing with relative velocity. For example, if S' is a train and S the Earth, and the train moves with speed v with respect to Earth, a person walking toward the front of the train with speed u'_x will have a speed with respect to the Earth of $u_x = u'_x + v$.

The Galilean transformations, Eqs. E–1 and E–2, are valid only when the velocities involved are not relativistic (Chapter 26)—that is, much less than the speed of light, c. We can see, for example, that the first of Eqs. E–2 will not work for the speed of light; for light traveling in S' with speed $u'_x = c$ will have speed $c + v$ in S, whereas the theory of relativity insists it must be c in S. Clearly, then, a new set of transformation equations is needed to deal with relativistic velocities.

Relativity theory

We will derive the required equations in a simple way, again looking at Fig. E–1. We assume the transformation is linear and of the form

$$x = \gamma(x' + vt'), \qquad y = y', \quad z = z'.$$

That is, we modify the first of Eqs. E–1 by multiplying by a factor γ which is yet to be determined. But we assume the y and z equations are unchanged because we expect no length contraction in these directions. We won't assume a form for t, but will derive it. The inverse equations must have the same form with v replaced by $-v$. (The principle of relativity demands it, since S' moving to the right with respect to S is equivalent to S moving to the left with respect to S'.) Therefore

$$x' = \gamma(x - vt).$$

Now if a light pulse leaves the common origin of S and S' at time $t = t' = 0$, after a time t it will have traveled along the x axis a distance $x = ct$ (in S), or $x' = ct'$ (in S'). Therefore, from the equations for x and x' above,

$$ct = \gamma(ct' + vt') = \gamma(c + v)t',$$

$$ct' = \gamma(ct - vt) = \gamma(c - v)t.$$

We substitute t' from the second equation into the first and find $ct = \gamma(c + v)\gamma(c - v)(t/c) = \gamma^2(c^2 - v^2)t/c$. We cancel out the t on each side and solve for γ to find

$$\gamma = \frac{1}{\sqrt{1 - v^2/c^2}}.$$

Now that we have found γ, we need only find the relation between t and t'. To do so, we combine $x' = \gamma(x - vt)$ with $x = \gamma(x' + vt')$:

$$x' = \gamma(x - vt) = \gamma[\gamma(x' + vt') - vt].$$

We solve for t and find $t = \gamma(t' + vx'/c^2)$. In summary,

Lorentz

transformations

$$x = \frac{1}{\sqrt{1 - v^2/c^2}}(x' + vt')$$

$$y = y'$$

$$z = z'$$

$$t = \frac{1}{\sqrt{1 - v^2/c^2}}\left(t' + \frac{vx'}{c^2}\right).$$

(E–3)

These are called the **Lorentz transformation** equations. They were first proposed, in a slightly different form, by Lorentz in 1904 to explain the null result of the Michelson–Morley experiment and to make Maxwell's equations take the same form in all inertial systems. A year later, Einstein derived them independently based on his theory of relativity. Notice that not only is the x equation modified as compared to the Galilean transformation, but so is the t equation. Indeed, we see directly in this last equation, as well as in the first, how the space and time coordinates mix.

The relativistically correct velocity equations are readily obtained. For example, using Eqs. E–3 (we let $\gamma = 1/\sqrt{1 - v^2/c^2}$),

$$u_x = \frac{\Delta x}{\Delta t} = \frac{\gamma(\Delta x' + v\,\Delta t')}{\gamma(\Delta t' + v\,\Delta x'/c^2)} = \frac{(\Delta x'/\Delta t') + v}{1 + (v/c^2)(\Delta x'/\Delta t')}$$

$$= \frac{u'_x + v}{1 + vu'_x/c^2}.$$

The others are obtained in the same way and we collect them here:

$$u_x = \frac{u_x' + v}{1 + vu_x'/c^2}$$

$$u_y = \frac{u_y'\sqrt{1 - v^2/c^2}}{1 + vu_x'/c^2}$$ **(E–4)** *Relativistic velocity*

$$u_z = \frac{u_z'\sqrt{1 - v^2/c^2}}{1 + vu_x'/c^2}.$$ *transformations*

The first of these equations is Eq. 26–9, which we used in Section 26–11 where we discussed how velocities do not add in our commonsense (Galilean) way, because of the denominator $(1 + vu_x'/c^2)$. We can now also see that the y and z components of velocity are also altered and that they depend on the x' component of velocity.

EXAMPLE E–1 **Length contraction.** Derive the length-contraction formula, Eq. 26–2, from the Lorentz transformation equations.

SOLUTION Let an object of length L_0 be at rest on the x axis in S. The coordinates of its two end points are x_1 and x_2, so that $x_2 - x_1 = L_0$. At any instant in S', the end points will be at x_1' and x_2' as given by the Lorentz transformation equations. The length measured in S' is $L = x_2' - x_1'$. An observer in S' measures this length by measuring x_2' and x_1' at the same time (in the S' frame), so $t_2' = t_1'$. Then, from the first of Eqs. E–3,

$$L_0 = x_2 - x_1 = \frac{1}{\sqrt{1 - v^2/c^2}}(x_2' + vt_2' - x_1' - vt_1').$$

Since $t_2' = t_1'$, we have

$$L_0 = \frac{1}{\sqrt{1 - v^2/c^2}}(x_2' - x_1') = \frac{L}{\sqrt{1 - v^2/c^2}},$$

or

$$L = L_0\sqrt{1 - v^2/c^2},$$

which is Eq. 26–2.

EXAMPLE E–2 **Time dilation.** Derive the time-dilation formula, Eq. 26–1, from the Lorentz transformation equations.

SOLUTION The time Δt_0 between two events that occur at the same place $(x_2' = x_1')$ in S' is measured to be $\Delta t_0 = t_2' - t_1'$. Since $x_2' = x_1'$, then from the last of Eqs. E–3, the time Δt between the events as measured in S is

$$\Delta t = t_2 - t_1 = \frac{1}{\sqrt{1 - v^2/c^2}}\left(t_2' + \frac{vx_2'}{c^2} - t_1' - \frac{vx_1'}{c^2}\right)$$

$$= \frac{1}{\sqrt{1 - v^2/c^2}}(t_2' - t_1') = \frac{\Delta t_0}{\sqrt{1 - v^2/c^2}},$$

which is Eq. 26–1. Notice that we chose S' to be the frame in which the two events occur at the same place, so that $x_1' = x_2'$, and the terms containing x_1' and x_2' cancel out.

F SELECTED ISOTOPES

(1) Atomic Number Z	(2) Element	(3) Symbol	(4) Mass Number A	(5) Atomic Mass[†]	(6) % Abundance (or Radioactive Decay Mode)	(7) Half-life (if radioactive)
0	(Neutron)	n	1	1.008665	β^-	10.4 min
1	Hydrogen	H	1	1.007825	99.985%	
	Deuterium	D	2	2.014102	0.015%	
	Tritium	T	3	3.016049	β^-	12.33 yr
2	Helium	He	3	3.016029	0.000137%	
			4	4.002602	99.999863%	
3	Lithium	Li	6	6.015121	7.5%	
			7	7.016003	92.5%	
4	Beryllium	Be	7	7.016928	EC, γ	53.29 days
			9	9.012182	100%	
5	Boron	B	10	10.012936	19.9%	
			11	11.009305	80.1%	
6	Carbon	C	11	11.011433	β^+, EC	20.385 min
			12	12.000000	98.90%	
			13	13.003355	1.10%	
			14	14.003242	β^-	5730 yr
7	Nitrogen	N	13	13.005738	β^+	9.965 min
			14	14.003074	99.63%	
			15	15.000108	0.37%	
8	Oxygen	O	15	15.003065	β^+, EC	122.24 s
			16	15.994915	99.76%	
			18	17.999160	0.20%	
9	Fluorine	F	19	18.998404	100%	
10	Neon	Ne	20	19.992435	90.48%	
			22	21.991383	9.25%	
11	Sodium	Na	22	21.994434	β^+, EC, γ	2.6088 yr
			23	22.989767	100%	
			24	23.990961	β^-, γ	14.9590 h
12	Magnesium	Mg	24	23.985042	78.99%	
13	Aluminum	Al	27	26.981538	100%	
14	Silicon	Si	28	27.976927	92.23%	
			31	30.975362	β^-, γ	157.3 min

[†]The masses given in column (5) are those for the neutral atom, including the Z electrons.

(1) Atomic Number Z	(2) Element	(3) Symbol	(4) Mass Number A	(5) Atomic Mass	(6) % Abundance (or Radioactive Decay Mode)	(7) Half-life (if radioactive)
15	Phosphorus	P	31	30.973762	100%	
			32	31.973908	β^-	14.262 days
16	Sulfur	S	32	31.972071	95.02%	
			35	34.969033	β^-	87.51 days
17	Chlorine	Cl	35	34.968853	75.77%	
			37	36.965903	24.23%	
18	Argon	Ar	40	39.962384	99.600%	
19	Potassium	K	39	38.963708	93.2581%	
			40	39.964000	0.01117%	
					β^-, EC, γ, β^+	1.28×10^9 yr
20	Calcium	Ca	40	39.962591	96.941%	
21	Scandium	Sc	45	44.955911	100%	
22	Titanium	Ti	48	47.947947	73.8%	
23	Vanadium	V	51	50.943962	99.750%	
24	Chromium	Cr	52	51.940511	83.79%	
25	Manganese	Mn	55	54.938048	100%	
26	Iron	Fe	56	55.934940	91.72%	
27	Cobalt	Co	59	58.933198	100%	
			60	59.933820	β^-, γ	5.2714 yr
28	Nickel	Ni	58	57.935346	68.077%	
			60	59.930789	26.233%	
29	Copper	Cu	63	62.929599	69.17%	
			65	64.927791	30.83%	
30	Zinc	Zn	64	63.929144	48.6%	
			66	65.926035	27.9%	
31	Gallium	Ga	69	68.925580	60.108%	
32	Germanium	Ge	72	71.922079	27.66%	
			74	73.921177	35.94%	
33	Arsenic	As	75	74.921594	100%	
34	Selenium	Se	80	79.916519	49.61%	
35	Bromine	Br	79	78.918336	50.69%	
36	Krypton	Kr	84	83.911508	57.0%	
37	Rubidium	Rb	85	84.911793	72.17%	
38	Strontium	Sr	86	85.909266	9.86%	
			88	87.905618	82.58%	
			90	89.907737	β^-	29.1 yr
39	Yttrium	Y	89	88.905847	100%	
40	Zirconium	Zr	90	89.904702	51.45%	
41	Niobium	Nb	93	92.906376	100%	
42	Molybdenum	Mo	98	97.905407	24.13%	
43	Technetium	Tc	98	97.907215	β^-, γ	4.2×10^6 yr
44	Ruthenium	Ru	102	101.904348	31.6%	
45	Rhodium	Rh	103	102.905502	100%	
46	Palladium	Pd	106	105.903481	27.33%	

(1) Atomic Number Z	(2) Element	(3) Symbol	(4) Mass Number A	(5) Atomic Mass	(6) % Abundance (or Radioactive Decay Mode)	(7) Half-life (if radioactive)
47	Silver	Ag	107	106.905091	51.839%	
			109	108.904754	48.161%	
48	Cadmium	Cd	114	113.903359	28.73%	
49	Indium	In	115	114.903876	95.7%; β^-, γ	4.41×10^{14} yr
50	Tin	Sn	120	119.902197	32.59%	
51	Antimony	Sb	121	120.903820	57.36%	
52	Tellurium	Te	130	129.906228	33.87%	2.5×10^{21} yr
53	Iodine	I	127	126.904474	100%	
			131	130.906111	β^-, γ	8.04 days
54	Xenon	Xe	132	131.904141	26.9%	
			136	135.90721	8.9%	
55	Cesium	Cs	133	132.905436	100%	
56	Barium	Ba	137	136.905816	11.23%	
			138	137.905236	71.70%	
57	Lanthanum	La	139	138.906346	99.9098%	
58	Cerium	Ce	140	139.905434	88.43%	
59	Praseodymium	Pr	141	140.907647	100%	
60	Neodymium	Nd	142	141.907718	27.13%	
61	Promethium	Pm	145	144.912745	EC, γ, α	17.7 yr
62	Samarium	Sm	152	151.919728	26.7%	
63	Europium	Eu	153	152.921226	52.2%	
64	Gadolinium	Gd	158	157.924099	24.84%	
65	Terbium	Tb	159	158.925344	100%	
66	Dysprosium	Dy	164	163.929172	28.2%	
67	Holmium	Ho	165	164.930320	100%	
68	Erbium	Er	166	165.930292	33.6%	
69	Thulium	Tm	169	168.934213	100%	
70	Ytterbium	Yb	174	173.938861	31.8%	
71	Lutecium	Lu	175	174.940772	97.4%	
72	Hafnium	Hf	180	179.946547	35.100%	
73	Tantalum	Ta	181	180.947993	99.988%	
74	Tungsten (wolfram)	W	184	183.950929	30.7%	
75	Rhenium	Re	187	186.955746	62.60%; β^-	4.35×10^{10} yr
76	Osmium	Os	191	190.960922	β^-, γ	15.4 days
			192	191.961468	41.0%	
77	Iridium	Ir	191	190.960585	37.3%	
			193	192.962916	62.7%	
78	Platinum	Pt	195	194.964765	33.8%	
79	Gold	Au	197	196.966543	100%	
80	Mercury	Hg	199	198.968253	16.87%	
			202	201.970617	29.86%	
81	Thallium	Tl	205	204.974401	70.476%	

(1) Atomic Number Z	(2) Element	(3) Symbol	(4) Mass Number A	(5) Atomic Mass	(6) % Abundance (or Radioactive Decay Mode)	(7) Half-life (if radioactive)
82	Lead	Pb	206	205.974440	24.1%	
			207	206.975871	22.1%	
			208	207.976627	52.4%	
			210	209.984163	β^-, γ, α	22.3 yr
			211	210.988734	β^-, γ	36.1 min
			212	211.991872	β^-, γ	10.64 h
			214	213.999798	β^-, γ	26.8 min
83	Bismuth	Bi	209	208.980374	100%	
			211	210.987254	α, γ, β^-	2.14 min
84	Polonium	Po	210	209.982848	α, γ	138.376 days
			214	213.995177	α, γ	164.3 μs
85	Astatine	At	218	218.00868	α, β^-	1.6 s
86	Radon	Rn	222	222.017571	α, γ	3.8235 days
87	Francium	Fr	223	223.019733	β^-, γ, α	21.8 min
88	Radium	Ra	226	226.025402	α, γ	1600 yr
89	Actinium	Ac	227	227.027749	β^-, γ, α	21.773 yr
90	Thorium	Th	228	228.028716	α, γ	1.9131 yr
			232	232.038051	100%; α, γ	1.405×10^{10} yr
91	Protactinium	Pa	231	231.035880	α, γ	3.276×10^4 yr
92	Uranium	U	232	232.037131	α, γ	68.9 yr
			233	233.039630	α, γ	1.592×10^5 yr
			235	235.043924	0.720%; α, γ	7.038×10^8 yr
			236	236.045562	α, γ	2.3415×10^7 yr
			238	238.050784	99.2745%; α, γ	4.468×10^9 yr
			239	239.054289	β^-, γ	23.50 min
93	Neptunium	Np	239	239.052932	β^-, γ	2.355 days
94	Plutonium	Pu	239	239.052157	α, γ	24,119 yr
95	Americium	Am	243	243.061373	α, γ	7380 yr
96	Curium	Cm	245	245.065484	α, γ	8500 yr
97	Berkelium	Bk	247	247.07030	α, γ	1380 yr
98	Californium	Cf	249	249.074844	α, γ	351 yr
99	Einsteinium	Es	254	254.08802	α, γ, β^-	275.7 days
100	Fermium	Fm	253	253.085174	EC, α, γ	3.00 days
101	Mendelevium	Md	255	255.09107	EC, α, γ	27 min
102	Nobelium	No	255	255.09324	α, γ, EC	3.1 min
103	Lawrencium	Lr	257	257.0995	α, EC	0.646 s
104	Rutherfordium[†]	Rf	261	261.10869	α	65 s
105	Dubnium[†]	Db	262	262.11376	α, EC	34 s
106	Seaborgium[†]	Sg	263	263.1182	α, fission	0.9 s
107	Bohrium[†]	Bh	262	262.1231	α	0.10 s
108	Hassium[†]	Hs	264	264.1285	α, fission	0.08 ms
109	Meitnerium[†]	Mt	266	266.1378	α	3.4 ms
110			269	269	α	0.27 ms
111			272	272	α	1.5 ms
112			277	277	α	0.24 ms

[†]Provisional names for elements 104–109.

ANSWERS TO ODD-NUMBERED PROBLEMS

CHAPTER 1

1. (a) 1×10^{10} yr; (b) 3×10^{17} s.

3. (a) 1.156×10^6; (b) 2.18×10^2;
(c) 6.8×10^{-3};
(d) 2.7635×10^1;
(e) 2.1×10^{-1}; (f) 2.2×10^1.

5. 11%.

7. (a) 1×10^1%; (b) 1%; (c) 0.2%.

9. 1.8×10^5 s.

11. 6%.

13. (a) 0.086 6 m; (b) 0.000 035 V;
(c) 0.860 g;
(d) 0.000 000 000 600 s;
(e) 0.000 000 000 000 012 5 m;
(f) 250,000,000,000 volts.

15. 1.8 m.

17. (a) 3.9×10^{-9} in;
(b) 1.0×10^8 atoms.

19. (a) 0.621 mi/h; (b) 3.28 ft/s;
(c) 0.278 m/s.

21. (a) 9.46×10^{15} m;
(b) 6.31×10^4 AU;
(c) 7.20 AU/h.

23. (a) 10^4; (b) 10^3; (c) 10^{-3};
(d) 10^{10}.

25. ≈ 20%.

27. $\approx 1 \times 10^5$ cm³.

29. (a) 500 dentists.

31. $\approx 3 \times 10^8$ kg/yr.

33. (a) 3.16×10^7 s;
(b) 3.16×10^{16} ns;
(c) 3.17×10^{-8} yr.

35. 50,000 chips.

37. 1.4×10^3 gumballs.

39. ≈ 3 ft ≈ 1 m.

41. $\approx 4 \times 10^3$ km.

43. ≈ 4 yr.

CHAPTER 2

1. 70.8 km/h.

3. 61 m.

5. (a) 203 mi; (b) 61 mi/h.

7. (a) 4.29 m/s; (b) 0.

9. 2.7 min.

11. 55 km/h, 0.

13. 4.3 m/s².

15. (a) 7.41 m/s²;
(b) 9.60×10^4 km/h².

17.

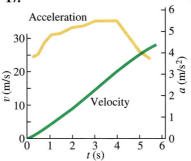

t(s)	v(m/s)	t(s)	a(m/s²)
0.0	0.0		
0.125	0.44		
0.375	1.4	0.25	3.8
0.625	2.4	0.50	4.0
0.875	3.5	0.75	4.5
1.25	5.36	1.06	4.9
1.75	7.85	1.50	5.0
2.25	10.5	2.00	5.2
2.75	13.1	2.50	5.3
3.25	15.9	3.00	5.5
3.75	18.7	3.50	5.6
4.25	21.4	4.00	5.5
4.75	23.9	4.50	4.8
5.25	25.9	5.00	4.1
5.75	27.8	5.50	3.8

19. $a = 2.2$ m/s², 1.1×10^2 m.

21. 1.5×10^2 m.

23. 62.5 m.

25. (a) 1.6×10^2 m; (b) 25 s;
(c) 12 m, 10 m.

27. (a) 103 m; (b) 64 m.

29. 24.0 s, 240 km/h.

31. 3.1 s.

33. 2.6 s.

35. (a) 8.81 s; (b) 86.3 m/s.

37. 1.5 s.

39.

(a)

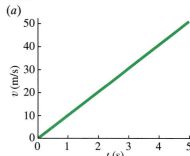

(b)

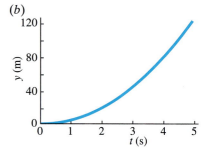

41. 5.22 s.

45. 0.038 s.

47. 52 m.

49. (a) 5.33 s; (b) 40.2 m/s;
(c) 89.7 m.

51. (a) 0.28 m/s; (b) 1.2 m/s;
(c) 0.28 m/s; (d) 1.6 m/s;
(e) -1.0 m/s.

53. (a) 50 s; (b) 90 s to 107 s; (c) 0 to 20 s, and 90 s to 107 s; (d) 75 s.

55. (a) 4.7 m/s^2; (b) 2.2 m/s^2; (c) 0.3 m/s^2; (d) 1.6 m/s^2.

57.

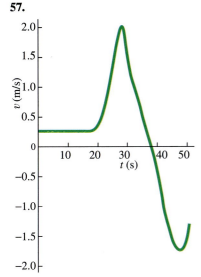

59. (a) negative; (b) speeding up; (c) negative; (d) positive; (e) speeding up; (f) positive; (g) not moving, zero.

61. $6h_{Earth}$.

63. 209.5 km/h.

65. (a) 3.88 s; (b) 73.9 m; (c) 38.0 m/s, 48.4 m/s;

67.

(a)

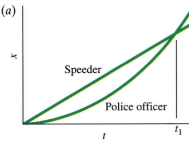

(b) 22.9 s; (c) 2.67 m/s^2; (d) 61.1 m/s = 220 km/h.

69. 3.3 m.

71. The passing attempt should not be made.

73. 2 poles.

CHAPTER 3

1. 177 km, 15° S of W.

5. 31 m, 44° N of E.

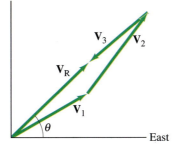

7. (a) 2.76 in the $+x$-direction; (b) 13.86 in the $+x$-direction; (c) 13.86 in the $-x$-direction.

9. (a) 614 km/h, 489 km/h; (b) 1.84×10^3 km, 1.47×10^3 km.

11. (a) 35.9, 17.3; (b) 39.9, 25.8° above $+x$-axis.

13. (a) 80.7, 1.56° above $-x$-axis; (b) 80.7, 1.56° below $+x$-axis.

15. (a) 117, 72.1° below $-x$-axis; (b) 226, 35.5° below $+x$-axis.

17. 5032 m.

19. 5.6 m.

21. 14° and 76°.

23. Unsuccessful, 34.7 m.

27. 12.9 m.

29. Six times as far.

31. 5.71 s.

35. (a) 92.6 m; (b) 8.69 s; (c) 539 m; (d) 68.0 m/s, 24.2° above the horizontal.

39. $\theta = \tan^{-1}(gt/v_0)$ below the horizontal.

41. 2.9 m/s, 20° from the river bank.

43. (a) 2.59 m/s, 62.4° from the shore; (b) 7.77 m at 62.4° to the shore.

45. (a) 435 km/h, 9.36° east of south; (b) 12 km.

47. 1.41 m/s.

49. (a) 1.66 m/s; (b) 3.19 m/s.

51. (a) 120 m; (b) 150 s (2.5 min).

53. 42.9° N of E.

55. 58 km/h, 31°, 58 km/h opposite to $\mathbf{v}_{12}$.

57. 109 km/h.

59. (a) Parallel; (b) perpendicular; (c) $\mathbf{V}_2 = 0$.

61. -2.4 m/s^2 (opposite to the truck's motion); -1.4 m/s^2 (down).

63. $v_r = v_T/\tan\theta$.

65. 1.8×10^2 s, 4.8 km, 21 s, 0.56 km.

67. 1.6 m/s^2.

69. 2.7 s, 1.9 m/s.

71. 61.8° below the horizontal.

73. 59 m/s, 78° above the horizontal.

1. 69.0 N.

3. 8.8×10^2 N.

5. (a) 6.5×10^2 N; (b) 1.1×10^2 N; (c) 2.4×10^2 N; (d) 0.

7. 3.4×10^3 N, opposite to the velocity.

9. $m > 1.5$ kg.

11. 2.1×10^2 N.

13. 3.5 m/s^2 (down).

15. a (downward) ≥ 2.2 m/s^2.

17. 0.557 m/s^2.

19. (a) 2.2 m/s^2; (b) 18 m/s; (c) 93 s.

21. (a) 9.4 m/s; (b) 3.3×10^3 N up.

23. (a) 40 N; (b) 10 N; (c) 0.

25.

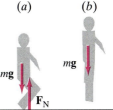

27. (a) 19.0 N, 0.702 m/s^2, both 57.5° below $-x$-axis; (b) 14.0 N, 0.519 m/s^2, both 51.0° above $+x$-axis.

29. (a) 11 m/s^2; (b) 4.3 m/s.

31. (a) 8.01×10^4 N; (b) 1.25×10^4 N.

35. (a)

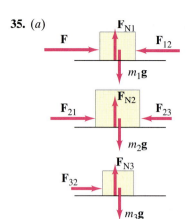

(b) $a = F/(m_1 + m_2 + m_3)$;
(c) $F_{net1} = m_1 F/(m_1 + m_2 + m_3)$,
$F_{net2} = m_2 F/(m_1 + m_2 + m_3)$,
$F_{net3} = m_3 F/(m_1 + m_2 + m_3)$;
(d) $F_{12} = F_{21} = F(m_1 + m_2)/$
$(m_1 + m_2 + m_3)$,
$F_{23} = F_{32} = Fm_3/(m_1 + m_2 +$
$m_3)$; (e) 2.67 m/s², F_{net} = 32 N,
$F_{21} = F_{12}$ = 64 N,
$F_{32} = F_{23}$ = 32 N.

37. 1.74 m/s², F_{T1} = 22.6 N,
F_{T2} = 20.9 N.

39. (a) μ_s = 0.82; (b) μ_k = 0.74.

41. μ_s = 0.41.

43. 6.7 kg.

45. (a) 2.5 m/s²; (b) 4.4 × 10² N.

47. −7.4 m/s².

49. μ_k = 0.40.

51. 1.8 s, no change.

53. (a) 1.2 m up the plane; (b) 1.6 s.

55. 23 m/s (85 km/h).

57. Too steep.

59. θ_{max} = 31°.

61. (a) $a = (m_2 - m_1 \sin\theta)g/$
$(m_1 + m_2)$; (b) a down requires
$m_2 < m_1 \sin\theta$, a up requires
$m_2 > m_1 \sin\theta$.

63. μ_k = 0.64.

65. 2.0 × 10⁻² N.

67. (b) Will slide.

69. 35 m/s (130 km/h).

71. (a) m_{max} = 4.6 kg;
(b) m_{max} = 3.8 kg.

73. (a) 4.1 m/s², 3.2 m/s²;
(b) a_1 = 4.1 m/s²,
a_2 = 3.2 m/s²; (c) 3.5 m/s².

75. (a) $\mu_k = (v_0^2/2gd$
$\cos\theta) - \tan\theta$; (b) $\mu_s \geq \tan\theta$.

77. 8.8°.

79. 82 m/s (300 km/h).

81. (a) 11.6 kg; (b) 0.879 m/s².

CHAPTER 5

1. 4.25g up.

3. 5.93 × 10⁻³ m/s² toward the
Sun, 3.55 × 10²² N toward the
Sun, Sun.

7. 23 m/s, independent of the
mass.

9. $\mu_s \geq$ 0.84.

11. 0.16.

13. (a) 5.8 × 10³ N;
(b) 4.1 × 10² N; (c) 31 m/s.

17. 8.0 × 10³ N down the slope.

19. 5.91°, 14.3 N.

21. 1.1 km.

23. a_{tan} = 6.29 m/s²,
a_R = 19.7 m/s², $\mu_s \geq$ 2.11.

25. 1.52 × 10³ N.

27. 1.6 m/s².

29. (a) 2.10 kg on both;
(b) w_{Earth} = 20.6 N,
w_{planet} = 25.2 N.

31. 6.7 × 10¹² m/s².

33. 4.4 × 10⁷ m/s².

35. 2.0 × 10⁻⁸ N toward center of
the square.

37. 6.4 × 10²³ kg.

39. 6.3 × 10³ m/s.

41. 11 s.

43. 2.0 h.

45. 5 h 35 min, 19 h 50 min.

47. (a) 20 N (toward the Moon);
(b) 1.8 × 10² N (away from the
Moon).

49. 5.5 × 10²⁹ kg.

51. 0.0587 days (1.41 h).

53. 1.6 × 10² yr.

55. 5.4 × 10¹² m, in the Solar
System, Pluto.

57. (a) 1.90 × 10²⁷ kg;
(b) 1.90 × 10²⁷ kg,
1.89 × 10²⁷ kg, 1.89 × 10²⁷ kg.

59. (a) 5.2 yr; (b) No.

61. 9.0 Earth-days.

63. 6.1 m/s.

65. 2.1 × 10² N.

67. 3.46 × 10⁸ m from Earth's
center.

69. 3.3 m/s² upward.

71. $M_P = g_P r^2/G$.

73. 74 km/h < v < 140 km/h.

75. (a) Attractive gravitational
force; (b) 9.6 × 10²⁶ kg.

77. 2.3g_{Earth}.

CHAPTER 6

1. 7.35 × 10³ J.

3. 1.3 × 10⁴ J.

5. 36.1 m.

7. (a) 8.8 × 10⁵ J; (b) 1.6 × 10⁶ J.

9. 23 J.

11. (a) 1.10Mg; (b) 1.10Mgh.

13. (a) 2.8 × 10³ J; (b) 2.1 × 10³ J.

15. 1.9 × 10² J.

17. 484 m/s.

19. −1.64 × 10⁻¹⁸ J.

21. 23%.

23. 3.4 × 10² N in the direction of
the motion of the ball.

25. 27 m/s (97 km/h or 60 mi/h),
mass cancels.

27. 7.1 m/s, 3.5 m/s.

29. 0.34 m.

31. 2.5 × 10³ J.

33. (a) 8.1 × 10⁵ J; (b) 8.1 × 10⁵ J;
(c) yes.

35. h = 1.6 m, no.

37. 5.14 m/s.

39. (a) 9.2 m/s; (b) 0.35 m.

41. 199 m/s.

43. (a) 8.03 m/s; (b) 3.44 m.

45. $E = \frac{1}{2}kx_0^2 = \frac{1}{2}mv^2 + mgx$.

47. 3.7 × 10³ N/m.

49. 4.5 × 10⁶ J.

51. (a) 22 m/s; (b) 2.9 × 10² m.

53. 20 m/s.

55. 0.48.

57. (a) 1.07 × 10³ km/h;
(b) 2.1 × 10³ N.

59. 5.4 × 10² N.

61. (b) 360 kWh; (c) 1.3 × 10⁹ J;
(d) $43.20, independent of rate.

63. 8.1 × 10⁶ J.

65. 4.57 W.

67. 2.6 m/s.

69. 1.8 × 10⁶ W.

71. (a) -3.6×10^4 J;
(b) -3.3×10^3 N;
(c) -2.5×10^5 J.

73. 1.5×10^3 J.

75. (a) 19 m/s; (b) -1.4 N.

77. (a) 1.1×10^6 J;
(b) 60 W $= 0.081$ hp;
(c) 4.0×10^2 W $= 0.54$ hp.

79. (a) $h = 2.5r$; (b) $11mg$; (c) $5mg$;
(d) mg.

81. (a) 40 m/s; (b) 2.6×10^5 W.

85. (a) 29°; (b) 6.4×10^2 N;
(c) 9.2×10^2 N.

CHAPTER 7

1. 0.18 kg·m/s.

3. 5.2×10^7 N, up.

5. 12.3 m/s.

7. 0.491 m.

9. 6.7×10^3 m/s.

11. 0.39 m/s.

13. 140 kg.

15. 1.32×10^4 N.

17. 2.1 N·s.

19. (a) ≈ 5.0 N·s; (b) 83 m/s.

21. $v_1' = 1.23$ m/s, $v_2' = 4.93$ m/s.

23. $v_1' = -3.00$ m/s (rebound),
$v_2' = 2.00$ m/s.

25. (a) $v_2' = 1.8$ m/s;
(b) $m_2 = 1.1$ kg.

27. (a) $m_2 = 0.840$ kg; (b) 0.750.

29. (b) $v_1' \approx -v_1$, $v_2' \approx 0$;
(c) $v_1' \approx v_1$, $v_2' \approx 2v_1$;
(d) $v_1' \approx 0$, $v_2' \approx v_1$.

31. (a) $+M/(m + M)$; (b) 0.964.

33. 15 m/s.

35. 9.

37. 1.08×10^{-22} kg·m/s, 30.1° from
the direction opposite to the
electron's.

39. (a) $m_A v_A + 0 = m_A v_A' \cos \theta_A' +$
$m_B v_B' \cos \theta_B'$, $0 + 0 =$
$m_A v_A' \sin \theta_A' - m_B v_B' \sin \theta_B'$;
(b) 0.808 m/s, 33.0°.

41. 141°.

43. $\theta_1' = 76°$, $v_n' = 5.1 \times 10^5$ m/s,
$v_{He}' = 1.8 \times 10^5$ m/s.

47. 2.74 m from the front of the car.

49. 1.10 m (east), -1.10 m (south).

51. Along the line joining the
centers $0.07R$ outside the hole.

53. 19% of the height.

55. 9.4% of the body height, yes.

57. (a) 6.2 m; (b) 3.4 m; (c) 3.8 m.

59. (a) $7D/3$, or $2D/3$ closer to the
launch site; (b) $5D$, or $2D$
farther from the launch site.

61. 8.3×10^2 N, about the same.

63. 1.4×10^4 N, 43.3°.

65. A "scratch shot".

67. 4.00 m.

69. 59 mi/h.

71. (a) 2.5×10^{-13} m/s;
(b) 1.7×10^{-17}; (c) 0.19 J.

73. (b) 0.93 N·s; (c) 4.2 g.

75. $m \le M/3$.

77. 29.6 km/s.

CHAPTER 8

1. (a) $\pi/6$ rad $= 0.524$ rad;
(b) $19\pi/60 = 0.995$ rad;
(c) $\pi/2 = 1.571$ rad;
(d) $2\pi = 6.283$ rad;
(e) $7\pi/3 = 7.330$ rad.

3. $\theta_{Sun} = 9.30 \times 10^{-3}$ rad (0.53°),
$\theta_{Moon} = 9.06 \times 10^{-3}$ rad (0.52°).

5. $D_{spot} = 6.8$ km.

7. 33 m/s, 6.2×10^3 m/s^2.

9. (a) 0.105 rad/s;
(b) 1.75×10^{-3} rad/s;
(c) 1.45×10^{-4} rad/s; (d) zero.

11. 9.5 cm.

13. $D_{Moon} \approx 3.8 \times 10^3$ km.

15. (a) 464 m/s;
(b) 185 m/s; (c) 328 m/s.

17. (a) 3.1 rad/s^2;
(b) $a_R = 1.9 \times 10^2$ m/s^2,
$a_{tan} = 1.1$ m/s^2.

19. (a) 1.8×10^{-4} rad/s^2;
(b) $a_R = 1.2 \times 10^{-2}$ m/s^2,
$a_{tan} = 7.7 \times 10^{-4}$ m/s^2.

21. 2.75×10^4 rev.

23. (a) $\alpha = 0.070$ rad/s^2;
(b) 40 rpm.

27. (a) -4.4 rad/s^2; (b) 7.9 s.

29. 92 m·N.

31. 1.1 m·N (clockwise).

33. 2.7×10^2 N, 1.8×10^3 N.

35. 0.139 kg·m^2, distance of its mass
from the axis is so small.

37. 1.2×10^{-10} m.

39. 10 N.

41. 62 m·N.

43. 8.80×10^2 rev, 10.6 s.

45. (a) 95.2 rad/s^2; (b) 9.4×10^2 N.

47. (a) 2.9 m; (b) 3.6 m.

49. (a) 10.8 rad/s^2; (b) 13.0 m/s^2;
(c) 653 m/s^2; (d) 4.77×10^3 N;
(e) 1.14°.

51. 1.11×10^4 J.

53. 1.70×10^4 J.

55. 3.22 m/s.

57. 2.64 kg·m^2/s.

59. 0.38 rev/s.

61. (a) 2×10^{17} J;
(b) 5×10^{20} kg·m^2/s.

63. (a) 7.1×10^{33} kg·m^2/s;
(b) 2.7×10^{40} kg·m^2/s.

65. 4.2 rev/s.

67. 3×10^{-18}.

69. 4.8×10^{-2} rad/s,
$KE_2 = 2.0 \times 10^4$ KE_1.

71. (a) $\omega_P = -(I_W/I_P)\omega_W$ (down);
(b) $\omega_P = -(I_W/2I_P)\omega_W$ (down);
(c) $\omega_P = (I_W/I_P)\omega_W$ (up);
(d) $\omega_P = 0$.

73. 8.21×10^{-6}.

75. (a) 0.84 m/s; (b) 0.964.

77. 2.0×10^9 rev/day.

79. (a) 7.3 m; (b) 6.8 s.

81. $F_{min} = Mg[h(2R - h)]^{1/2}/(R - h)$.

83. $h_{min} = 2.7R_0$.

85. $v_{CM} = (3gL/4)^{1/2}$.

CHAPTER 9

1. 370 N, 116°.

3. 1.8×10^3 m·N.

5. 9.2×10^2 N.

7. 2.94×10^3 N down,
1.52×10^3 N down.

9. 2.3 N.

11. 3.3 m from the adult.

13. $F_{T1} = 1.8 \times 10^2$ N,
$F_{T2} = 2.4 \times 10^2$ N.

15. $F_1 = -2.94 \times 10^3$ N (down),
$F_2 = 1.47 \times 10^4$ N.

17. Top hinge: $F_{Ax} = 55.2$ N,
$F_{Ay} = 63.7$ N; bottom hinge:
$F_{Ax} = -55.2$ N, $F_{Ay} = 63.7$ N.

19. $F_1 = -1.8 \times 10^3$ N (down),
$F_2 = 2.4 \times 10^3$ N (up).

21. 4.38 kg.

23. $F_T = 1.9 \times 10^2$ N,
$\mathbf{F_W} = 1.9 \times 10^2$ N, 50° above
the horizontal.

25. 89.5 cm from the feet.

27. $F_T = 2.5 \times 10^2$ N,
$F_{AH} = 2.5 \times 10^2$ N,
$F_{AV} = 2.0 \times 10^2$ N.

29. 6.5×10^2 N.

31. 0.50.

33. 1.0×10^2 N.

35. 9.9×10^2 N.

37. 2.7×10^3 N.

39. 1.4×10^3 N (up), 2.1×10^3 N
(down).

41. 2.7×10^3 N.

43. (b) beyond the table;
(c) $D = L \sum_{i=1}^{n} \frac{1}{2i}$; (d) 32 bricks.

45. (a) 1.2×10^5 N/m^2;
(b) 2.4×10^{-6}.

47. (a) 1.3×10^5 N/m^2;
(b) 6.6×10^{-7}; (c) 0.0062 mm.

49. 997 cm^3.

51. 9.0×10^7 N/m^2, 9.0×10^2 atm.

53. 0.017 J.

55. 5.1×10^4 N.

57. (a) Not break; (b) 1.3 mm.

59. $A_{min2} = 7.7 \times 10^{-3}$ m^2,
$A_{min1} = 1.4 \times 10^{-3}$ m^2.

61. 2.5 cm.

63. 2.5×10^6 N.

65. $+2.3 \times 10^9$ m·N, will not
topple.

67. (a) $Mg[h/(2R - h)]^{1/2}$;
(b) $Mg[h(2R - h)]^{1/2}/(R - h)$.

69. (a) 3.5×10^8 N/m^2; (b) will
break; (c) 8.2×10^6 N/m^2, will
not break.

71. $F_{T1} = 4.54mg$, $F_{T2} = 4.96mg$,
$h = 158$ m.

73. 36 kg.

75. 3.8.

77. 6.3 m.

79. 1.3 m.

81. (a) 2.6×10^2 N; (b) 88 N.

CHAPTER 10

1. 2.7×10^{11} kg.

3. 4.3×10^2 kg.

5. 0.8477.

7. 1.65×10^4 N/m^2.

9. (a) 3.1×10^5 N (down);
(b) 3.1×10^5 N.

11. 1.96×10^3 kg.

13. 13 m.

15. 8.0 km.

17. 9.0×10^5 N/m^2 (gauge); 92 m.

19. (a) 0.34 kg; (b) 1.5×10^4 N
(up).

21. 983 kg/m^3.

23. 0.199.

25. 12 kg.

27. (a) 1.03×10^3 kg/m^3;
(b) $\rho_{liquid} =$
$[(m - m_{apparent})/m]\rho_{object}$.

29. 0.85.

31. 7.9×10^2 kg.

35. 3.8 m/s.

37. 1.2×10^5 N/m^2 = 1.2 atm.

39. 1.4×10^5 N.

41. 9.7×10^4 N/m^2.

43. 2.22 m/s, 1.8 atm (gauge).

45. 0.072 Pa·s.

47. 9.9×10^2 N/m^2.

49. 0.12 m.

51. 0.96 N/m^2/cm.

53. 1×10^{-20}.

55. 0.036 N/m.

57. Would not remain on the
surface.

59. 3.

61. (a) 0.88 m; (b) 0.55 m;
(c) 0.24 m.

63. 1.5×10^2 N $\le F \le 2.2 \times 10^2$ N.

65. 4×10^2 atm.

67. 0.63 N.

69. 1.4×10^2 N.

71. 1.1 m.

73. 2.44×10^7 kg.

75. 1.1 W.

77. 1.13×10^4 N/m^2.

79. (a) 2.8×10^2 m/s;
(b) 1.03×10^5 N/m^2 (1.01 atm).

CHAPTER 11

1. 0.74 Hz.

3. 1.00 m.

5. Cosine wave.

7. (a) 12 N/m; (b) 1.1 Hz.

9. (a) 2.8 m/s; (b) 2.1 m/s;
(c) 2.0 J; (d) (0.15 m)
cos $[2\pi(3.0$ Hz$)t]$.

11. $(2k/m)^{1/2}/2\pi$.

13. 10.3 m/s.

15. (a) 4.2×10^2 N/m; (b) 3.3 kg.

17. (a) 0.45 m; (b) 1.34 Hz; (c) 3.6 J;
(d) 2.0 J, 1.6 J.

19. (a) $x = (0.280$ m$)\sin[(29.0 s^{-1})t]$;
(b) maximum at 0.0542 s,
0.271 s, 0.488 s, $\cdots$, minimum at
0.163 s, 0.379 s, 0.596 s, $\cdots$.

21. 557 m/s.

23. (a) $0.707A$; (b) $0.866A$.

29. (a) 0.993 m; (b) slow.

31. (a) 0.61 Hz; (b) 0.53 m/s;
(c) 0.044 J.

33. (a) $-15°$, (b) $+4.6°$, (c) $+15°$.

35. 1.26 m.

37. (a) 1.4×10^3 m/s;
(b) 4.1×10^3 m/s;
(c) 5.1×10^3 m/s.

39. 0.33 s.

41. 2.1 km.

43. 1.0 m.

45. (a) 5.0×10^9 J/m^2·s;
(b) 5.0×10^{10} W.

47. 1.41.

49. 0.56.

51. 440 Hz, 880 Hz, 1320 Hz,
1760 Hz.

53. 70 Hz, 140 Hz, 210 Hz.

55. 0.097 m.

57. 300 Hz, 600 Hz, 900 Hz.

61. 4.

63. 5.4 km/s.

65. 41°.

67. 0.16 m/s.

69. 0.38 s.

71. 4.0 Hz.

73. (a) 1.4 Hz; (b) 15 J.

75. (a) 784 Hz, 1176 Hz; 880 Hz,
1320 Hz; (b) 1.26; (c) 1.12;
(d) 0.794.

77. (a) $v_2/v_1 = (\mu_1/\mu_2)^{1/2}$;
(b) $\lambda_2/\lambda_1 = (\mu_1/\mu_2)^{1/2}$;
(c) lighter cord.

79. $k = \rho_{water}gA$.

81. Slow, -12.3 s.

CHAPTER 12

1. 2.6×10^2 m.

3. (a) 1.7 cm $\leq \lambda \leq 17$ m;
(b) 3.4×10^{-5} m.

5. 5.4×10^2 m.

7. 1200 m, 300 m.

9. 1.0 W/m^2, 1.0×10^{-10} W/m^2.

11. 1.3.

13. 6.3×10^5.

15. (a) 5.0×10^{-12} W;
(b) 6.3×10^3 yr.

17. (a) 7.9×10^2 W;
(b) 2.5×10^2 m.

19. (a) 136 dB; (b) 99 dB; (c) 57 dB.

21. 150 dB.

23. 25 dB.

25. (a) 2×10^8; (b) 2×10^{11}.

27. 0.17 m.

29. (a) 570 Hz; (b) 860 Hz.

31. (a) 76.6 Hz, 230 Hz, 383 Hz,
536 Hz; (b) 153 Hz, 306 Hz,
459 Hz, 612 Hz.

33. 8.0 Hz, Yes.

35. (a) 0.289 m; (b) 869 Hz.

37. 198 m/s.

39. (a) 283 overtones;
(b) 284 overtones.

41. $0.64, 0.20, -2$ dB, -7 dB.

43. 15 Hz, not audible, 3.8 Hz,
audible.

45. 346 Hz.

47. 2.2 Hz.

49. (a) 5.6 Hz; (b) 61 m.

51. (a) 1950 Hz; (b) 1640 Hz.

53. 2 Hz.

55. 9 Hz.

57. 13 Hz.

59. 90 beats/min.

61. (a) 1.1×10^2 m/s;
(b) 2.6×10^2 m/s.

63. (a) 26°; (b) 23 s.

65. (a) 2.46°; (b) 11.1°.

67. 10.

69. 30 dB.

71. 333 Hz.

73. (a) 2.8×10^2 m/s, 44 N;
(b) 19.5 cm; (c) 880 Hz,
1320 Hz.

75. 635 Hz.

77. (a) 0; (b) 12 Hz; (c) 0.

79. 6.4×10^2 N.

81. $\Delta d = (\lambda/4d)[(\ell/2)^2 + d^2]^{1/2}$.

83. 780 Hz.

85. 11.5 m.

CHAPTER 13

1. 0.548.

3. (a) 20°C; (b) 3272°F.

5. 58°C, -89°C.

7. 65°.

9. 0.50 cm.

11. 5/9.

15. 5.1 mL.

17. 0.27 cm^3.

23. -2.8×10^{-3} (0.28%).

25. 3.5×10^7 N/m^2.

27. (a) 27°C; (b) 5.2×10^3 N.

29. -459.7°F.

31. 0.854 m^3.

33. 2.32 atm.

35. (a) 0.439 m^3; (b) -63°C.

37. -0.074 (7.4%).

39. 0.588 kg/m^3, water vapor is not
an ideal gas.

41. 5.50 cm^3.

43. 55.6 mol, 3.34×10^{25}
molecules.

45. 6.4×10^3 N.

47. 6.1×10^3 m/s.

49. 5.2.

51. 31.7°C.

57. (a) Solid or vapor;
(b) 5.11 atm $< P <$ 73 atm,
-56.6°C $< T <$ 31°C.

59. (a) Vapor; (b) solid.

61. 0.69 atm.

63. 11°C.

65. 3.1 kg.

67. 120°C.

69. 29%.

71. 0.28 s, 5.4×10^{-5} m/s,
3.1×10^2 m/s.

73. (b) Nitrogen, 6.9%.

75. -3.8 mL (1.3%).

77. $P = \rho RT/M$.

79. 3×10^2 molecules/cm^3.

81. 11 L, not advisable.

83. 1.4×10^5 K.

85. (a) 9.2×10^2 kg;
(b) 1.1×10^2 kg.

87. 1.1×10^{44} molecules.

89. (a) 6.09 cm; (b) $L =$
$4V_0(\beta_{mercury} - \beta_{glass})\Delta T/\pi d^2$.

91. 3.0×10^{-10} J.

93. 7.6×10^2 m.

CHAPTER 14

1. 6.7×10^6 J.

3. 10.6°C.

7. 2.1×10^2 kg/h.

9. 2.6×10^3 J/kg·C°.

11. 10.7 : 4.65 : 1.

13. 0.124 kg.

15. 61 C°.

17. 87°C.

19. 2.3×10^3 J/kg·C°.

21. 5.8 min.

23. 0.334 kg (0.334 L).

25. 9.6 g.

27. 4.7×10^3 kcal.

29. 1.12×10^4 J/kg.

31. 360 m/s.

33. (a) 95 W; (b) 33 W.

35. 4.2×10^{30} W, 1.2×10^4 P_{Sun} .

37. 22 bulbs.

39. 164°C.

41. 10 C°.

43. 36 W.

45. 14 h.

47. 9.8×10^3 kg.

49. (a) 3.2×10^{26} W;
(b) 1.1×10^3 W/m^2.

51. 0.80 C°.

53. 1.96 cm, rod vaporizes.

55. 4.3 kg.

59. (a) 44 C°; (b) no melting.

61. 1.1×10^{-6} kg/s (4.1 g/h).

CHAPTER 15

1. -1.8×10^3 J.

3. P (atm)

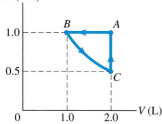

5.

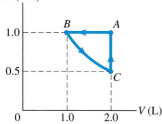

7. (a) 0; (b) $+1350$ J; (c) rise.

9. (a) 3.5×10^2 J; (b) $\Delta U = 0$;
(c) $+3.5 \times 10^2$ J (into the gas).

11. (a) -76 J; (b) $+24$ J; (c) $+52$ J;
(d) $+28$ J; (e) $+23$ J.

13. (a) $+57$ J; (b) $+57$ J; (c) 0;
(d) 48%.

15. 1.6×10^2 W.

17. 28%.

19. 30.5%.

21. 1.7×10^{13} J/h.

23. (a) 28%; (b) 7.46×10^4 W,
1.52×10^9 J/h
$(3.64 \times 10^5$ kcal/h).

25. 238°C.

27. 4.3×10^8 kg/h.

29. 7.0.

31. (a) 2.1×10^2 J; (b) 3.5×10^2 J.

33. 3.2 min.

35. $+0.31$ kcal/K.

37. $+11$ kcal/K.

39. $+2.2$ cal/K.

41. (a) 75.8% of ideal;
(b) $+0.270$ J/K; (c) 0.

43. (b), (a), (c), (d).

45. (a) 1/379; (b) $1/1.59 \times 10^{11}$.

47. 15 MW.

49. 3.3×10^{13} J.

51. -401 C°.

53. (a) 2.2×10^5 J; (b) -1.7×10^5 J;
(c) P (atm)

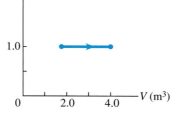

55. 69%.

57. 2.9×10^3 J/K.

59. (a) 17; (b) 6.1×10^7 J/h.

61. 0.15 J/K.

63. 17 km³/day, 83 km².

CHAPTER 16

1. 1.88×10^{14} electrons.

3. 11.5 cm.

5. 2.7×10^{-3} N.

7. 2.5×10^3 N.

9. 6.8×10^3 C.

11. -1.4×10^2 N (left),
$+5.3 \times 10^2$ N (right),
-3.9×10^2 N (left).

13. 6.20×10^5 N away from the
center of the square.

15. $F_{electric} = 8.2 \times 10^{-8}$ N,
$F_{gravitational} = 3.6 \times 10^{-47}$ N,
2.3×10^{39}.

17. (a) $Q_1 = \frac{1}{2}Q_T$;
(b) Q_1 (or Q_2) = 0.

19. 0.91 m beyond the negative
charge.

21. 1.05×10^{14} m/s², opposite to
the direction of the electric
field, independent of the
velocity.

23. $+2.0 \times 10^5$ N/C (south).

25. 3.30×10^6 N/C (up).

27. 7.12×10^{-10} N/C (south).

29. $2kQa/(a^2 + x^2)^{3/2}$ parallel to
the line of the charges.

31. 3.80×10^6 N/C away from the
positive charge.

33. (a) $1.73kQ/L^2$ 60° below the
$-x$-axis; (b) kQ/L^2 30° below
the $+x$-axis.

35. 0.10 N/C.

37. $Q_2 = 4Q_1$.

39. (a) In the direction of the
velocity, to the right;
(b) 5.1×10^2 N/C.

41. 5.08 m.

43. 1.02×10^{-7} N/C (up).

45. 1.0×10^7 electrons.

47. 2.4×10^{-10} N (attraction).

49. Positive, 2.2×10^{-7} C.

51. (a) 0.83 mm; (b) 2.2 ns.

53. $5.4 \ \mu$C.

55. $+0.19$ N (right).

CHAPTER 17

1. -6.5×10^{-4} J (done by the
field).

3. 3.4×10^{-15} J, 21 keV.

5. 4.2×10^4 V/m.

7. -32.5 kV.

9. $V_a - V_b = 269$ V.

11. 7.32×10^7 m/s.

13. 2.40×10^5 V.

15. $+1.08$ J.

17. 4.2 MV.

19. (a) $+6.1 \times 10^3$ V;
(b) 8.8×10^4 N/C, 53° N of E.

21. 3.0×10^3 m/s.

23. $2kq(2b - d)/b(d - b)$.

25. (a) 0.036 V; (b) 0.025 V;
(c) -0.025 V.

27. 5.2×10^{-20} C.

29. Zero, resultant force.

31. 0.79 pF.

33. 108 μC.

35. 1.5×10^{-10} F.

37. 26.3 nC.

39. 4.5×10^4 V/m.

41. 56 μF.

43. 1.09×10^{-3} J.

45. (a) 3.6 pF; (b) 32 pC;
(c) 90 V/m; (d) 1.5×10^{-10} J;
(e) capacitance, charge, and
work done by the battery.

47. (a) $4\times$; (b) $4\times$; (c) $\frac{1}{2}\times$.

49. (a) 2.73×10^{-3} J;
(b) 1.10×10^{-3} J;
(c) -1.63×10^{-3} J; (d) the
stored potential energy is not
conserved.

51. 5.4×10^5 V/m.

53. 9.96×10^{-8} J/m^3.

55. (a) 0.039 eV; (b) 0.039 eV; (c) 0.3 keV; (d) 0.029 eV.

57. 24 mC.

59. 0.14 μC.

61. 10.9 V.

63. (a) 4.9 cm from the negative charge, and 7.4 cm from the positive charge; (b) 0.56 cm from the negative charge toward the positive charge, 2.1 cm from the negative charge away from the positive charge.

65. $Q_1 = Q_0 C_1/(C_1 + C_2)$, $Q_2 = Q_0 C_2/(C_1 + C_2)$, $V = Q_0/(C_1 + C_2)$.

67. 1.03×10^6 m/s.

69. 0.10 m/s.

71. (a) 0.649 μF; (b) 23 C; (c) 4.0×10^8 J.

CHAPTER 18

1. 1.4×10^5 C.

3. 2.5×10^{-11} A.

5. 35 Ω.

7. 2.1×10^{21} electron/min.

9. 24 Ω, 3.6×10^2 J.

11. 0.029 Ω.

13. $R_{Al} = 1.6 R_{Cu}$.

15. 1/8 the length, 8.75 Ω, 1.25 Ω.

17. 58.3°C.

19. 1.8×10^3 °C.

21. (a) 0.28 Ω; (b) 2.9×10^2 A; (c) 31 V, 49 V.

23. 17 Ω.

25. 26 V.

27. (a) 240 Ω, 0.50 A; (b) 32.7 Ω, 3.67 A.

29. 0.092 kWh, 18¢/month.

31. (a) 8.6 Ω, 1.1 W; (b) 4×.

33. (a) 0.23 W; (b) 32 mA.

35. 71%.

37. (a) 26.2 Ω; (b) 19 s; (c) 0.12¢.

39. 26 A, 0.46 Ω.

41. 0.545 A, 0.386 A.

43. 5.3×10^2 V.

45. 6.0 hp, 13 A.

47. 7.8×10^{-10} m/s.

49. 6.25 mol/m^3.

51. 7.0×10^6 V/m.

53. 5×10^{-11} J, 5×10^{-5} W.

55. 3.60×10^3 C.

57. 12 h.

59. 0.23 S.

61. $R_2 = \frac{1}{4} R_1$.

63. 4.0 mm.

65. 8.4%.

67. 9.00 Ω.

69. 1.5 kW.

71. 0.258 mm, 38.8 m.

73. 2.1×10^{-3} F.

75. 5.5×10^{-7} m.

CHAPTER 19

1. 560 Ω, 35 Ω.

3. 30 Ω, 80 Ω, 19 Ω.

5. Connect three 1.0-Ω resistors in series.

7. 4.6 kΩ.

9. 450 Ω, 2.6%.

11. 960 Ω in parallel.

13. 105 Ω.

15. (a) V_1 and V_2 increase; V_3 and V_4 decrease; (b) I_1 and I_2 increase; I_3 and I_4 decrease; (c) increases; (d) 0, 0.300 A, 0.150 A, 0.338 A, 0.113 A.

17. 50 V.

19. 0.23 Ω.

21. 0.060 Ω.

23. 0.16 A.

25. 17.4 V, 13.3 V.

27. 43 V, 77 V.

29. $I_1 = 0.60$ A, $I_2 = -0.33$ A, $I_3 = 0.93$ A.

31. $I_1 = 0.15$ A right, $I_2 = 0.33$ A left, $I_3 = 0.18$ A up.

33. 0.46 A up.

35. 1.30 A.

37. 22 μF, 0.62 μF.

39. 10,560 pF, yes.

41. 300 pF in parallel.

43. 0.053 J.

47. $2C_0 K_1 K_2/(K_1 + K_2)$.

49. (a) 6.1 μF; (b) 7.2 V, 16.8 V, 24 V.

51. (a) 41 μC; (b) 0.63 A, 0.21 A, 0.42 A.

53. 93 ms.

55. 7.50 MΩ.

57. (a) 50×10^{-6} Ω in parallel; (b) 20 MΩ in series.

59. 1.0 kΩ in series, 100 Ω/V.

61. 5.54 mA.

63. (a) 2.00 Ω, 3.00 Ω, circuit 1 is better; (b) 99 Ω, 101 Ω, both circuits give about the same inaccuracy; (c) 3.3 kΩ, 5.0 kΩ, circuit 2 is better.

65. $R_V \geq 240$ kΩ.

67. $R_1 = 9.4$ kΩ, $R_2 = 6.8$ kΩ.

69. 288 Ω, 144 Ω.

71. 7.3 mA.

73. 0.11 MΩ.

75. (a) 0.12 A; (b) 0.12 A; (c) 64 mA.

77. 26.4 m.

79. (a) Parallel; (b) 6.2 pF $\leq C \leq$ 37 pF.

81. 10 0.22-kΩ resistors in series, 10 22-kΩ resistors in parallel.

83. 13 Ω in series, 30 kΩ in parallel.

85. (a) 25 kΩ; (b) between b and c.

CHAPTER 20

1. (a) 7.8 N/m; (b) 5.5 N/m.

3. 2.68 A.

5. 7.45×10^{-13} N south.

7. (a) left; (b) left; (c) up; (d) in; (e) zero; (f) down.

9. 1.3 T.

11.

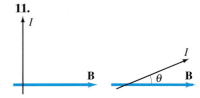

$\theta = 27°$.

13. 4.47 keV.

17. 2.0 A, down.

19. 2.0×10^{-5} T, 36%.

21. 0.18 N attraction.

23. 3 A.

25. $\theta = 34°$ W of N.

27. 1.6×10^{-17} T.

29. 1.0×10^{-6} T up, 2% of the Earth's field.

31. 5.8×10^{-5} up, 3.4×10^{-5} N/m $60°$ below the line toward C, 3.4×10^{-5} N/m $60°$ below the line toward B.

33. 4.1×10^{-5} T.

35. 5.18×10^{-5} T $10.9°$ below the plane parallel to the two wires.

37. Between long, thin and short, fat.

41. $73.3\ \mu$A.

43. decrease by 28%.

47. (a) 2.2×10^{-4} V/m;
(b) 2.7×10^{-4} m/s;
(c) 4.6×10^{28} electrons/m^3.

49. (a) determine the polarity of the emf; (b) 0.43 m/s.

51. 70 u, 72 u, 73 u, 74 u.

53. (a) 2.08×10^{-3} T; (b) out of the page; (c) 5.8×10^{7} Hz.

57. 3.0 T up.

59. 42.8.

61. 0.93 N.

63. (a) 4.1 cm; (b) 5.4×10^{-7} s.

65. 1.50×10^{4} A.

67. (a) 6.6 mm; (b) 1.7 mm.

69. (c) 48 MeV; (d) resonance

71. (a) 0.512 mm; (b) 0.103 Ω.

1. 0.050 V.

3. Counterclockwise.

5. -3.2×10^{2} V.

7. (a) Counterclockwise;
(b) clockwise.

9. Counterclockwise.

11. (a) 14.4 mV;
(b) 0.120 V/m down.

13. (a) 0.84 m/s;
(b) 0.758 V/m down.

15. 0.367 N.

17. $B\ell vA/\rho(2vt + \ell)$.

19. 31.0 V.

23. 0.71 kV, 120 rev/s.

25. 100 V.

27. 13 A.

29. (a) 220 V; (b) 156 V.

31. Step-down, $V_S/V_P = 0.285$, $I_S/I_P = 3.50$.

33. 55, 4.0 V.

35. (a) Step-up; (b) 2.4.

37. 451 V, 56.4 A.

41. 2.35 cm.

43. 0.14 H.

45. 3.4×10^{3} turns.

47. 8.29 V.

49. (a) Consistent; (b) 0.18 mH.

51. $M = \mu_0 N_1 N_2 A/\ell$.

53. 24.3 Ω, 0.463 H.

55. 7.5×10^{-5} J, 1.08 s.

57. (a) 4.47 ms. (b) 1.12 H.

59. (a) 2.3; (b) 4.6; (c) 6.9.

61. 1.5 kHz.

63.

65. 10.1 kΩ, 23.9 mA.

67. (a) 7.6 kΩ; (b) 0.37 A.

69. (a) 18 mV; (b) 50 mV.

71. (a) 1.6 kΩ; (b) 1.5 kΩ.

73. 1.6 kHz.

75. (a) 65.5 mA; (b) $+10.4°$;
(c) 7.73 W; (d) $V_R = 118$ V, $V_L = 22$ V.

77. 18 Ω.

79. 228 Hz.

81. 3.8×10^{5} Hz.

83. (a) 0.41 μF; (b) 11A.

85. 4 Ω.

87. (a) Yes; (b) immediately;
(c) when the current in the first loop reaches the steady state;
(d) counterclockwise; (e) yes;
(f) away from the other loop.

89. 7.33×10^{-3} J.

93. (a) 0.66 A; (b) 9.3.

95. 82 V.

97. 0.10 H.

99. 13.8 Ω, 43.5 mH.

101. 14 Ω, 75 mH.

103. 2.2×10^{3} Hz, 1.1×10^{4} Hz.

CHAPTER 22

1. 1.4×10^{-8} A.

3. 1.2×10^{15} V/m·s.

7. 5.25 V/m.

9. 80.0 kHz, 2.33 V/m, horizontal north-south line.

11. 3.0×10^{18} Hz.

13. 3.33×10^{14} Hz, infrared.

15. 8.33 min.

17. 3.4×10^{3} rad/s.

19. 9.47×10^{15} m.

21. 8.5×10^{3} wavelengths, 3.5 fs.

23. 4.24 months.

25. 0.796 W/m^2, 17.3 V/m.

27. 1.24×10^{3} V/m, 4.13×10^{-6} T.

29. (a) 2.5×10^{2} J;
(b) 2.5×10^{9} V/m.

31. 1.5 pF.

33. (a) 1.22 μH; (b) 106 pF.

35. 441 m.

37. AM station, 100×.

39. 8.33 min.

41. 1.43 s.

43. 38.7 V/m, 1.29×10^{-7} T.

45. Downward, 614 V/m, 2.05×10^{-6} T.

47. (a) 0.40 W; (b) 12 V/m; (c) 12 V.

49. 1.5×10^{11} W.

51. (a) Parallel;
(b) 8.9 pF $\leq C \leq$ 80 pF;
(c) 1.05 mH $\leq L \leq$ 1.22 mH.

CHAPTER 23

1. 3.0 m.

5. 5°.

9. 34.0 cm.

13. 3.9 m.

15. Convex, virtual, 0.13 m.

17. (a) Center of curvature;
(b) real; (c) inverted; (d) -1.

23. (a) $f = \infty$; (b) $d_i = -d_o$; (c) $+1$;
(d) yes.

25. (a) Concave mirror; (b) erect and virtual; (c) 173 cm.

27. 1.31.

29. $(2.997026 \pm 0.000030) \times 10^{8}$ m/s.

31. 64.0°.

33. 61.5°.

35. 4.6 m.

37. (a) 26.9°; (b) 31.2°; (c) 31.2°.

41. 1.42.

43. 3.9×10^{-4} m.

45. $n \geq 1.414$.

47. (a) 1.414; (b) no; (c) 1.88.

49. (a) Converging lens; (b) 5.41 D.

51. (a) 3.39 D, converging;
(b) -16.0 cm, diverging.

53. (a) -0.26 mm; (b) -0.47 mm;
(c) -1.9 mm; (d) 28 mm:
wide-angle lens, 200 mm:
telephoto lens.

55. -7.6 cm (virtual image behind
the lens), 0.54 mm (upright).

57. (a) 75.0 mm; (b) 25.0 mm.

59. (a) -11.1 cm (diverging),
virtual; (b) $+184$ cm
(converging).

61. Midway between the object and
screen.

63. (b) 110.8 cm; (c) 110.8 cm.

65. Real, 7.4 cm beyond second lens.

69. 1.54.

71. 13.1 cm.

73. -43.1 cm (concave).

75. $+1.3$ D.

77. 5.2 m.

79. 0.101 m, -2.7 m.

81. $+12$ cm.

83. $0 < -d_o < -f$.

85. (b) Real, 20 cm beyond second
lens, upright, same size as
object.

87. (c) $\Delta d = (d_T^2 - 4d_T f)^{1/2}$,
$\{[d_T + (d_T^2 - 4d_T f)^{1/2}]/$
$[d_T - (d_T^2 - 4d_T f)^{1/2}]\}^2$.

89. -3.25 m.

91. $1/f' = [(n/n')-1][(1/R_1)+(1/R_2)]$;
$(1/d_o) + (1/d_{i2}) = 1/f'$,
where $1/f' = [(n/n')-1]/f(n-1)$;
$m = h_i/h_o = -d_i/d_o$.

CHAPTER 24

3. 7.5 μm.

5. 3.4 cm.

7. 0.085 mm.

9. Reverse of the usual double-slit
pattern.

11. 2.40 mm.

13. 3.0%.

15. 0.22°.

17. 0.61 cm.

19. 952 nm.

21. 5.61 μm.

23. 0.0230 mm.

25. (a) $D_{\max} = \lambda$; (b) 400 nm.

27. 357 nm.

29. 518 nm, 593 nm, 737 nm.

31. One full order.

33. 6.67×10^3 lines/cm.

35. 10.0°.

37. 704 nm, 1030 nm.

39. 1.08 cm.

41. 228 nm.

43. 9.05 μm.

45. 471 nm.

49. 699 nm.

51. 0.250 mm.

53. 0.289 mm.

55. 56.7°.

57. 52°, 38°.

59. (a) 30°; (b) 18°; (c) 5.7°.

61. 78.5%.

63. 24 polarizers, with each at an
angle of 3.75° with the next.

65. 31° on either side of the
normal.

67. 12,500 lines/cm.

69. (a) Constructive;
(b) destructive.

71. (a) 355 nm.

73. (a) 0; (b) $0.094I_0$; (c) no light
gets transmitted.

75. 810 nm.

77. 658 nm, 938 lines/cm.

79. $\frac{1}{2}(m - \frac{1}{2})\lambda/n_{\text{film}}, m = 1,2,3,\cdots$;
$\frac{1}{2}m\lambda/n_{\text{film}}, m = 0,1,2,3,\cdots$;
$\frac{1}{2}m\lambda/n_{\text{film}}, m = 0,1,2,3,\cdots$;
$\frac{1}{2}(m - \frac{1}{2})\lambda/n_{\text{film}}, m = 1,2,3,\cdots$.

81. (a) $0.21I_0$; (b) second or third
polarizer; (c) second and third
polarizers.

83. 0.47 m.

CHAPTER 25

1. 2.5 mm $\leq D \leq$ 39 mm.

2. (1/60) s.

7. 17 mm.

9. 54 mm.

11. $+2.3$ D.

13. Glasses would be better.

15. (a) -1.43 D; (b) 28 cm.

17. -26.8 cm.

19. 2.1×.

21. (a) 17 cm; (b) 10 cm.

23. (a) 3.78×; (b) 11.7 mm;
(c) 6.62 cm from the lens.

25. (a) -81 cm; (b) 9.5×; (c) 2.9×.

27. 17 cm, 100 cm.

29. 3.2 cm, 83 cm.

31. -34×.

33. 12×.

35. 4.2 m, 8.4 m.

37. 6.4×.

39. 1.5 cm.

41. (a) 0.85 cm; (b) 230×.

43. (a) 14.4 cm; (b) 137×.

45. (a) 15.9 cm; (b) 14.8 cm;
(c) 1.1 cm; (d) 0.19 cm.

47. 51.6°, 216 nm.

49. 3.8×10^{11} m.

51. 0.46.

53. (a) 16 km; (b) 0.42′.

55. 14.0°.

57. (a) 52.6°; (b) 0.19 nm.

59. (a) 1; (b) 2.7, 1.

61. $I_2 = 16I_1$.

63. 100 mm, 200 mm.

65. 2.9×, 4.1×.

67. (a) -2.25×; (b) 4.5 D.

69. $+3.6$ D.

71. 14 cm, 5.1×10^{-6} rad.

73. -16×.

75. 42.7 m.

CHAPTER 26

1. (a) 1.000; (b) 0.99995; (c) 0.995;
(d) 0.436; (e) 0.141; (f) 0.0447.

3. 2.07×10^{-6} s.

5. 5.57 m, 6.68 m.

7. 0.96c.

9. (a) 2.3 yr; (b) 11 yr.

11. (a) 7.12 m, 1.20 m; (b) 16.3 s;
(c) 0.580c; (d) 16.3 s.

13. 3.8×10^{-27} kg.

15. $0.417c$.

17. (a) 1.5 m/s less than c; (b) 30 cm.

19. 5.36×10^{-13} kg.

21. 8.20×10^{-14} J, 0.511 MeV.

23. 9×10^2 kg.

27. (a) 6.3×10^7 kg;
(b) 1.4×10^{17} kg.

29. 7.49×10^{-19} kg·m/s.

31. $0.941c$.

33. $2m_0/[1 - (v^2/c^2)]^{1/2}$, no loss,
$2m_0c^2(\{1/[1 - (v^2/c^2)]^{1/2}\} - 1)$.

35. $0.866c$, 4.73×10^{-22} kg·m/s.

37. 55 MeV (8.7×10^{-12} J),
1.7×10^{-19} kg·m/s, -8%, -6%.

39. $0.855c$.

41. 3.0 T.

45. $0.80c$.

47. $0.65c$.

49. Not relativistic.

51. (a) $0.73c$; (b) 5.9 yr.

53. 9.3×10^{-8} s.

55. 1.02 MeV (1.64×10^{13} J).

57. 2.2 mm.

59. (a) 4.4×10^9 kg/s;
(b) 4.3×10^7 yr;
(c) 1.4×10^{13} yr.

61. 28.3 MeV.

63. (a) 1.32×10^{-18} kg·m/s; (b) 0;
(c) 7.45×10^{-18} kg·m/s.

65. Not possible.

CHAPTER 27

1. 5.8×10^4 C/kg.

3. 5.

5. 7.07×10^3 K.

7. (a) 10.6 μm, near infrared;
(b) 967 nm, infrared;
(c) 0.73 mm, far infrared.

9. (a) 6.0×10^{-34} J;
(b) 1.5×10^{35}; (c) 6.8×10^{-36},
not measurable.

11. 9.4 μm.

13. 4.23×10^{-7} eV.

15. 2.4×10^{13} Hz, 1.2×10^{-5} m.

17. 400 nm.

19. Copper and iron.

21. 0.62 eV.

23. (a) 1.08 eV; (b) no ejected
electrons.

25. 3.75 eV.

27. 1.9×10^{-24} kg·m/s,
6.3×10^{-33} kg.

29. 212 MeV, 5.85×10^{-15} m.

31. (a) 2.43×10^{-12} m;
(b) 1.32×10^{-15} m.

33. (a) 5.90×10^{-3}, 1.98×10^{-2},
3.89×10^{-2}; (b) 60.8 eV, 204 eV,
401 eV.

35. (a) 9.21×10^{-17} J (576 eV);
(b) 0.105 nm.

37. 7.2×10^{-12} m.

39. 8.6×10^{-12} m, diffraction
effects are negligible.

43. 2.6×10^{-11} m.

47. 3.2×10^{-3}.

49. 3.4 eV.

51. 122 eV.

53. 52.5 nm.

55. $n = 1$ to $n = 3$.

57.

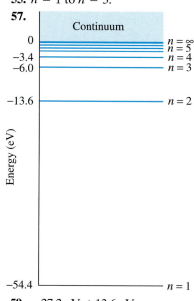

59. -27.2 eV, $+13.6$ eV.

61. Justified.

65. (a)

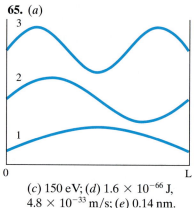

(c) 150 eV; (d) 1.6×10^{-66} J,
4.8×10^{-33} m/s; (e) 0.14 nm.

67. 3.28×10^{15} Hz.

71. 2.78×10^{21} photons/s·m^2.

73. 8.3×10^6 photons/s.

75. 4.7×10^{-14} m.

77. 10.2 eV.

79. 4.4×10^{-40}, yes.

81. 14.2 MeV.

85. 5×10^{-12} m.

CHAPTER 28

1. 3.6×10^{-7} m.

3. $\pm 1.3 \times 10^{-11}$ m.

5. 6.6×10^{-8} eV.

7. 3.0×10^{-10} eV/c^2.

9. 1.4×10^{-3} m, 9.1×10^{-33} m.

13. (b) Infinite uncertainty.

15. $m_\ell = -4, -3, -2, -1, 0, 1, 2, 3,$
$4; m_s = -\frac{1}{2}, +\frac{1}{2}$.

17. $(1, 0, 0, -\frac{1}{2}), (1, 0, 0, +\frac{1}{2}), (2, 0, 0,$
$-\frac{1}{2}), (2, 0, 0, +\frac{1}{2}), (2, 1, -1, -\frac{1}{2}),$
$(2, 1, -1, +\frac{1}{2}), (2, 1, 0, -\frac{1}{2})$.

19. 32 states.

21. $\ell \geq 3, n \geq \ell + 1$ (minimum 4),
$m_s = -\frac{1}{2}, +\frac{1}{2}$.

25. (a) $1s^2 2s^2 2p^6 3s^2 3p^6 3d^{10} 4s^2 4p^4$;
(b) $1s^2 2s^2 2p^6 3s^2 3p^6 3d^{10} 4s^2 4p^6$
$4d^{10} 4f^{14} 5s^2 5p^6 5d^{10} 6s^1$;
(c) $1s^2 2s^2 2p^6 3s^2 3p^6 3d^{10} 4s^2 4p^6 4d^{10}$
$4f^{14} 5s^2 5p^6 5d^{10} 6s^2 6p^6 5f^3 6d^1 7s^2$.

27. (a) 6; (b) -0.378 eV;
(c) 5.78×10^{-34} kg·m^2/s;
(d) $m_\ell = -5, -4, -3, -2, -1,$
$0, 1, 2, 3, 4, 5$.

29. (a) 1.56; (b) 1.4×10^{-10} m.

33. 41 kV.

35. 0.18 nm.

37. 0.061 nm, partial shielding
provided by the $n = 2$ shell.

39. 0.018 J, 5.8×10^{16} photons.

41. r_1, Bohr radius.

43. 0, 4.72×10^{-34} kg·m^2/s.

45. (a) 3.1×10^{-34} m;
(b) 1.8×10^{-32} kg·m/s;
(c) 1.6×10^{-30} m.

47. 3.8×10^3 C°/min.

49. (a) 45°, 90°, 135°; (b) 35.3°,
65.9°, 90°, 114.1°, 144.7°;
(c) 30.0°, 54.7°, 73.2°, 90°,
106.8°, 125.3°, 150.0°; (d) 5.71°,
0.0573°.

CHAPTER 29

1. 5.13 eV.

3. 4.6 eV.

5. 1.4 eV.

9. (a) 1.79×10^{-4} eV;
(b) 7.16×10^{-4} eV, 1.73 mm.

11. 2.35×10^{-10} m.

15. (a) 680 N/m; (b) 2.1 μm.

17. 0.283 nm.

21. 1.1 μm.

23. Could be used, 1.09 μm.

25. 9.2×10^5.

27. 0.89 μm.

29. 8.6 V.

33. (a) 38 mA; (b) 76 mA.

35. 0.53 V.

37. (a) 2.5×10^2; (b) 2.3×10^4.

39. 13 eV.

41. (a) -5.3 eV; (b) 5.0 eV.

43. (a) 0.094 eV; (b) 0.63 nm.

45. (a) 146 V $\leq V \leq$ 326 V;
(b) 3.34 k$\Omega \leq R_{load} < \infty$.

47. 2.0×10^{-9}.

CHAPTER 30

1. 3727 MeV/c^2.

3. 1.9×10^{-15} m.

5. (a) 0.99945; (b) 1.2×10^{-14};
(c) 2.3×10^{17} kg/m^3, $10^{14}\times$.

7. 28 MeV.

9. $^{31}_{15}$P.

11. (a) 1.8×10^3 MeV; (b) 730 MeV.

13. 7.48 MeV.

15. 18.7 MeV.

17. (a) 12.1 MeV; (b) 15.7 MeV,
repulsive electric force.

19. 59.93525 u.

21. 0.783 MeV.

23. β^+, $^{22}_{11}$Na $\rightarrow$ $^{22}_{10}$Ne + β^+ + ν,
1.82 MeV.

25. $^{228}_{90}$Th, 228.02883 u.

27. (a) $^{32}_{16}$S; (b) 31.97207 u.

29. 0.861 MeV.

31. 0.0718 MeV, 4.27 MeV.

33. 0.960 MeV, 0.960 MeV, 0.

35. 3.0 h.

37. 2.1 min.

39. 0.0625.

41. (a) 0.0625; (b) 0.0442.

43. (a) 2.44×10^{12} decays/s;
(b) 2.43×10^{12} decays/s;
(c) 4.48×10^5 decays/s.

45. 5.0×10^{10} decays/s.

47. (a) 1.38×10^{-13} s^{-1};
(b) 5.38×10^8 decays/min.

49. 34 decays/s.

51. 3.4 mg.

53. 8.6×10^{-7}.

55. $^{211}_{82}$Pb.

57. $N_0(1 - e^{-\lambda t})$.

59. (a) 2.29×10^{17} kg/m^3;
(b) 180 m; (c) 2.58×10^{-10} m.

61. 28.6 eV.

63. 41 yr.

65. 6×10^3 yr.

67. $6.65 T_{1/2}$.

71. (a) $^{191}_{77}$Ir;
(b)

$^{191}_{76}$Os

β^-(0.14 MeV)

γ(0.042 MeV) $^{191}_{77}$Ir*
               $^{191}_{77}$Ir*

γ(0.129 MeV)

             $^{191}_{77}$Ir

73. (a) 550 MeV; (b) 2.5×10^{12} J.

75. (a) 2.5×10^5 yr; (b) no change.

CHAPTER 31

1. $^{28}_{13}$Al, β^-, $^{28}_{14}$Si.

3. Yes.

5. 18.000953 u.

7. (a) Yes; (b) 0.53 MeV.

9. (a) Yes; (b) 6.49 MeV.

11. 41.6 MeV.

13. (a) Incident ^{3}He picks up a
neutron; (b) $^{11}_6$C; (c) 1.857 MeV,
exothermic.

17. 126.5 MeV.

19. 6.3×10^{18} fissions/s.

21. 482 kg.

23. 1.49.

29. 0.120 g.

31. (a) 13.93 MeV; (b) 4.3 MeV;
(c) 3.3×10^{10} K.

33. 3.23×10^9 J, 65$\times$.

35. (a) 4.8×10^{31} particles/m^3;
(b) 6×10^{-12} s.

37. 1000 rad.

39. 35 J.

41. 4 days.

43. 8.9×10^{-5} rad/day.

45. (a) $^{131}_{53}$I $\rightarrow$ $^{131}_{54}$Xe + $^{0}_{-1}$e + $\bar{\nu}$;
(b) 27 days; (c) 8.0×10^{-9} g.

47. (a) $^{218}_{84}$Po; (b) radioactive,
α-emission, β^- emission,
3.1 min; (c) chemically reacting;
(d) 150 μCi, 0.63 μCi.

49. (a) $^{12}_6$C; (b) 5.70 MeV.

51. 1.0043.

53. 55 mrem/yr.

55. 4.5 m.

57. (a) 1.54×10^3 kg/yr;
(b) 4.8×10^6 Ci.

59. (a) 4.0×10^{26} J/s;
(b) 3.8×10^{38} protons/s;
(c) 10^{11} yr.

61. (a) 3.7×10^3 decays/s;
(b) 5.2×10^{-2} rem, ≈ 0.15
background.

CHAPTER 32

1. 9.44 GeV.

3. 1.6 T.

5. 13 MHz.

7. (a) 8.7 MeV, 2.0×10^7 m/s;
(b) 4.3 MeV, 2.0×10^7 m/s;
(c) 13 MHz (both).

9. (a) 0.65 T; (b) 4.9 MHz;
(c) 200 rev; (d) 41 μs;
(e) 0.63 km.

11. 2.2×10^6 km, 7.5 s.

13. 3.0 T.

15.

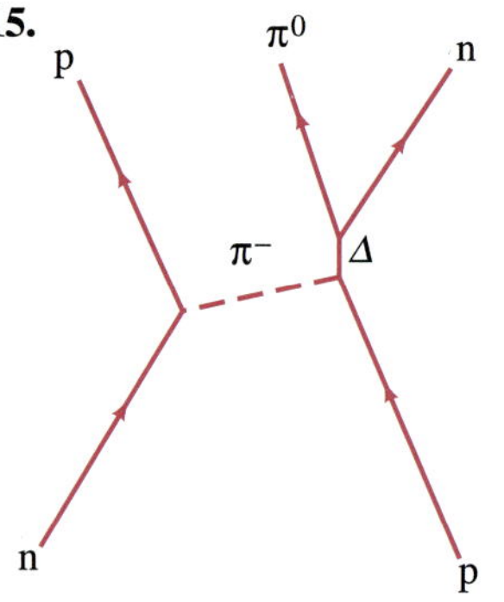

19. 33.9 MeV.

21. 1.879 GeV.

23. (*a*) Forbidden; (*b*) forbidden; (*c*) possible.

25. 1.32×10^{-15} m.

27. 1.53×10^{-3} nm.

29. $KE_n = 494.0$ MeV, $KE_p = 493.4$ MeV.

31. No, 292.4 MeV.

33. 7.5×10^{-21} s.

35. (*a*) 1.3 keV; (*b*) 8.9 keV.

37. (*b*) $B^+ = \bar{b}u$, $B^0 = b\bar{d}$, $\bar{B}^0 = \bar{b}d$.

39. (*a*) p; (*b*) $\bar{\Sigma}^+$; (*c*) K^-; (*d*) π^-; (*e*) D_s^-.

41. $c\bar{s}$.

43.
(*a*)

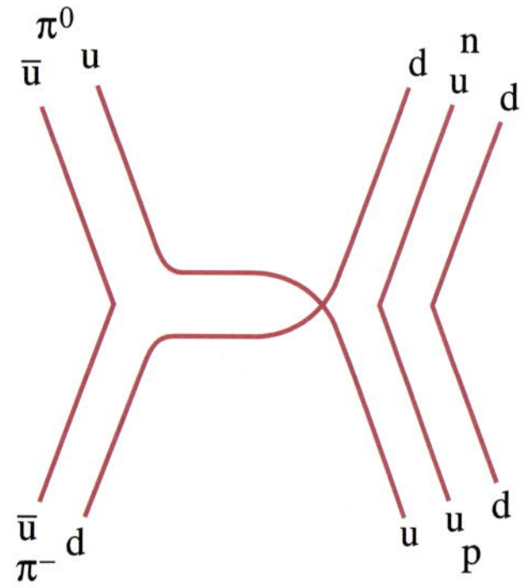

(*b*)

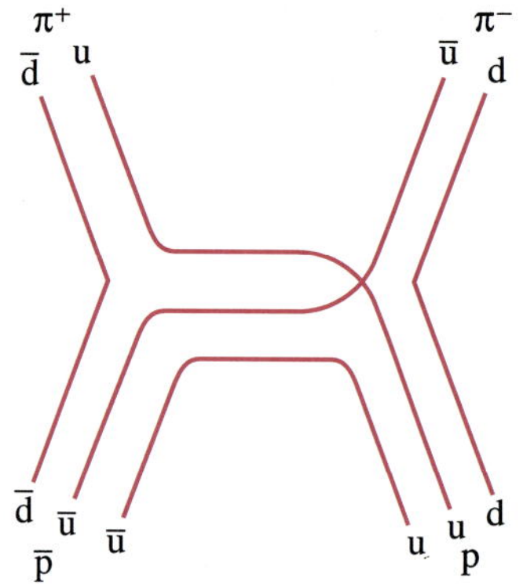

45. (*a*) 0.38 A; (*b*) 98 m/s (≈350 km/h).

47. (*a*) 1.002 MeV; (*b*) 1876.6 MeV.

49. (*a*) 37.8 MeV; (*b*) $KE_p = 5.37$ MeV, $KE_\pi = 32.4$ MeV.

51. 133.5 MeV/c^2.

53. 52.4 MeV.

55. (*a*)

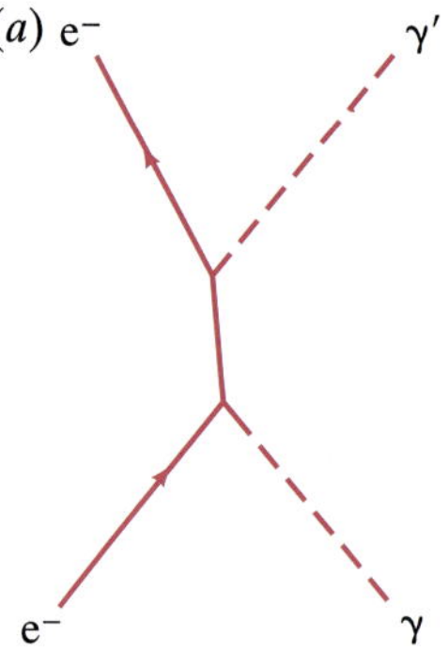

CHAPTER 33

1. 5.3 ly.

5. 0.078″, 13 pc.

7. $\phi_1 = \frac{1}{2}\phi_2$.

9. 48 W/m^2.

11. 1.4×10^{-4} kg/m^3.

13. 1.8×10^9 kg/m^3, $3 \times 10^5\times$.

15. −0.0941 MeV, 7.365 MeV.

17. (*a*) 9.595 MeV; (*b*) 7.6 MeV; (*c*) 6×10^{10} K.

19. $d_2 = 6.5 d_1$.

23. 540°.

25. 1.0×10^8 ly.

27. (*a*) 2×10^{10} yr; (*b*) 1×10^{10} yr.

29. 0.328*c*, 4.0×10^9 ly.

33. 6 nucleons/m^3.

35. (*a*) 10^{-5} s; (*b*) 10^{-7} s; (*c*) 10^{-4} s.

37. 5.0 ly.

39. 2×10^{10}.

41. 1 H atom/km^3 every 7 yr.

43. 2.

45. 0.88*c*.

47. $\pm 2.3 \times 10^{-3}$.

INDEX

PHOTO CREDITS

17–7, D. Giancoli; **17–12**, Paul Silverman/Fundamental Photographs; **17–20**, Jon Feingersh/Stock Market; **CO–18**, Mahaux Photography/Image Bank; **18–1**, Franca Principe/Science Museum, Florence; **18–2**, J. L. Charmet/Science Photo Library/Photo Researchers; **18–3(a), (b)**, Burndy Library; **18–12(a)**, T. J. Florian/Rainbow; **18–14**, Takeshi Takahara/Photo Researchers; **18–33**, Gamma-Liaison; **CO–19**, McGlynn/Image Works; **19–24, 19–28**, Paul Silverman/Fundamental Photographs; **CO–20**, Manfred Kage/Peter Arnold; **20–4**, Richard Megna/Fundamental Photographs; **20–7**, Mary Teresa Giancoli; **20–19**, Jack Finch/Science Photo Library/Photo Researchers; **20–40**, Richard Megna/Fundamental Photographs; **CO–21, 21–10**, Werner H. Muller/Peter Arnold; **21–15**, Tomas D. W. Friedmann/Photo Researchers; **21–20**, Jon Feingersh/Comstock; **21–27(b)**, USGS National Earthquake Information Center **CO–20**, Mulvehill/The Photo Works; **22–1**, Emilio Segrè/Visual Archives/American Institute of Physics; **22–11**, Image Works; **CO–23, 23–5**, D. Giancoli; **23–10(a)**, Mary Teresa Giancoli & Suzanne Saylor, **23–10(b)**, **23–20**, Mary Teresa Giancoli; **23–25**, Hank Morgan/Photo Researchers; **23–27(c), (d), 23–28(a)**, D. Giancoli; **23–28(b)**, S. Elleringmann/Bilderberg/Aurora & Quanta Productions; **23–30**, D. Giancoli & Howard Shugat; **23–36(a), (b), 23–42**, D. Giancoli; **23–47**, Mary Teresa Giancoli; **CO–24**, Richard Megna/Fundamental Photographs; **24–4(a)**, John M. Dunay IV/Fundamental Photographs; **24–9(a)**, Bausch & Lomb; **24–13**, Science Photo Library/David Parker/Photo Researchers; **24–16(b)**, D. Giancoli; **24–18(a)**, P. M. Rinard, from *American Journal of Physics*, 1976, p70; **24–18(b), (c)**, Ken Kay/Fundamental Photographs; **24–27(a), (c), (d)** Wabash Instrument Corp/Fundamental Photographs; **24–27(b), (c)**, Bausch & Lomb; **24–28(a), (c)**, Paul Silverman/Fundamental Photographs; **24–28(b)**, Richard Megna/Fundamental Photographs; **24–30(b)**, Richard Megna/Fundamental Photographs; **24–33.** Kristen Brochmann/Fundamental Photographs; **24–38(a), (b)**, D. Giancoli; **24–42**, Diane Schiumo/Fundamental Photographs; **CO–25, 25–2, 25–3(a), (b)**, Mary Teresa Giancoli; **25–5**, Leonard Lessin/Peter Arnold; **25–13**, Mary Teresa Giancoli; **25–15(a), (b)**, Franca Principe/Instituto e Museo di Storia della Scienza, Florence, Italy; **25–19**, Yerkes Observatory; **25–20(c)**, Palomar/Caltech; **25–20(d)**, Roger Ressmeyer/Starlight; **25–22**, Olympus America Inc, Precision Instrument Division, **25–24(a), (b)**, Springer-Verlag; **25–30**, Space Telescope Science Institute; **25–32**, National Astronomy & Ionosphere Center, Cornell University, Arecibo, Puerto Rico; **25–33**, Burndy Library; **25–36(b)**, Bausch & Lomb; **25–41**, Rosalind Franklin/Photo Researchers; **25–47(a)**, Martin M. Rotker/Taurus Photos; **25–47(b)**, Simon Fraser/Science Photo Library/Photo Researchers; **CO–26, 26–1**, Einstein Archives; **CO–27**, Prof. P. Motta/Department of Anatomy, University "La Sapienza," Rome/Science Photo Library/Photo Researchers; **27–2(a)**, Granger Collection; **27–6**, Ullstein Bilderdienst; **27–13, 27–14**, AIP Niels Bohr Library; **27–15**, Education Development Center, Inc; **27–18(a)**, CNRI/Science Photo Library/Photo Researchers; **27–18(b)**, Secchi-Lecaque/Roussel-UCLAF/CNRI/Science Photo Library/Photo Researchers; **27–20**, Driscoll, Youngquist, & Baldeschwieler, Caltech/Science Photo Library/Photo Researchers; **27–22**, Emilio Segrè/AIP Niels Bohr Library; **27–25**, Richard Megna/Fundamental Photographs; **27–26(a), (b), (c)**, Wabash Instrument Corp/Richard Megna/Fundamental Photographs; **CO–28**, Patricia Peticolas/Fundamental Photographs; **28–1**, Niels Bohr Library; **28–2**, Emilio Segrè/AIP Niels Bohr Library; **28–4**, Advanced Research Lab, Hitachi, Ltd; **28–15**, Paul Silverman/Fundamental Photographs; **28–21**, Yoav Levy/Phototake NYC; **CO–29**, Michael W. Davidson/Photo Researchers; **CO–30**, Michael Collier/Stock Boston; **30–3**, Center for the History of Chemistry; **30–6**, University of Chicago/AIP Niels Bohr Library; **30–15(a)**, Fermilab; **30–15(b)**, Brookhaven National Laboratory; **CO–31**, Will & Deni McIntyre/Photo Researchers; **31–4**, Gary Sheahan/Chicago Historical Society; **31–7**, Novosti/Gamma-Liaison; **31–8**, Los Alamos National Laboratory; **31–9**, Bettmann; **31–13(a)**, LLNL/Science Source/Photo Researchers; **31–13(b)**, Gary Stone/LLNL; **31–14**, R. Turgeon; **31–16**, J. Van't Hof.; **31–21(b)**, Siu/Peter Arnold; **31–23**, Mehau Kulyk/Science Photo Library/Photo Researchers; **CO–32**, Philippe Plailly/Science Photo Library/Photo Researchers; **32–1(b)**, Brookhaven National Laboratory; **32–2**, Science Service, Watson Davis/AIP Niels Bohr Library; **32–4(a), (b)**, Fermilab; **32–5**, Stanford Linear Accelerator Center/US Department of Energy; **32–6**, Scientific American, 6/90; **32–10(a)**, David Parker/Science Photo Library/Photo Researchers; **32–10(b)**, Science Photo Library/Photo Researchers; **32–11**, Brookhaven National Laboratory; **CO–33, 33–1**, NASA; **33–2(c)**, Photo Researchers; **33–3**, Hansen Planetarium; **33–4**, NOAO; **33–5(a)**, ©1979 R. J. Dufour, Rice University, Hansen Planetarium; **33–5(b)**, US Naval Observatory; **33–5(c), 33–10, 33–11**, NOAO; **33–15**, Space Telescope Science Institute/NASA/Mark Marten/Photo Researchers; **Appendix C–4**, NASA

Periodic Table of the Elements§

Transition Elements

Legend:
Symbol — Cl 17 — Atomic Number
Atomic Mass§ — 35.4527
Electron Configuration (outer shells only) — 3p⁵

$$\text{Cl} \quad 17 \quad 35.4527 \quad 3p^5$$

Group I	Group II					Transition Elements											Group III	Group IV	Group V	Group VI	Group VII	Group VIII
H 1 1.00794 $1s^1$																						He 2 4.002602 $1s^2$
Li 3 6.941 $2s^1$	Be 4 9.012182 $2s^2$																B 5 10.811 $2p^1$	C 6 12.011 $2p^2$	N 7 14.00674 $2p^3$	O 8 15.9994 $2p^4$	F 9 18.9984032 $2p^5$	Ne 10 20.1797 $2p^6$
Na 11 22.989768 $3s^1$	Mg 12 24.3050 $3s^2$																Al 13 26.98154 $3p^1$	Si 14 28.0855 $3p^2$	P 15 30.97376 $3p^3$	S 16 32.066 $3p^4$	Cl 17 35.4527 $3p^5$	Ar 18 39.948 $3p^6$
K 19 39.0983 $4s^1$	Ca 20 40.078 $4s^2$	Sc 21 44.955910 $3d^14s^2$	Ti 22 47.88 $3d^24s^2$	V 23 50.9415 $3d^34s^2$	Cr 24 51.9961 $3d^54s^1$	Mn 25 54.93805 $3d^54s^2$	Fe 26 55.847 $3d^64s^2$	Co 27 58.93320 $3d^74s^2$	Ni 28 58.6934 $3d^84s^2$	Cu 29 63.546 $3d^{10}4s^1$	Zn 30 65.39 $3d^{10}4s^2$						Ga 31 69.723 $4p^1$	Ge 32 72.61 $4p^2$	As 33 74.92159 $4p^3$	Se 34 78.96 $4p^4$	Br 35 79.904 $4p^5$	Kr 36 83.80 $4p^6$
Rb 37 85.4678 $5s^1$	Sr 38 87.62 $5s^2$	Y 39 88.90585 $4d^15s^2$	Zr 40 91.224 $4d^25s^2$	Nb 41 92.90638 $4d^45s^1$	Mo 42 95.94 $4d^55s^1$	Tc 43 (97.9072) $4d^55s^2$	Ru 44 101.07 $4d^75s^1$	Rh 45 102.90550 $4d^85s^1$	Pd 46 106.42 $4d^{10}5s^0$	Ag 47 107.8682 $4d^{10}5s^1$	Cd 48 112.411 $4d^{10}5s^2$						In 49 114.818 $5p^1$	Sn 50 118.710 $5p^2$	Sb 51 121.757 $5p^3$	Te 52 127.60 $5p^4$	I 53 126.90447 $5p^5$	Xe 54 131.29 $5p^6$
Cs 55 132.90543 $6s^1$	Ba 56 137.327 $6s^2$	57–71† $57-71$	Hf 72 178.49 $5d^26s^2$	Ta 73 180.9479 $5d^36s^2$	W 74 183.84 $5d^46s^2$	Re 75 186.207 $5d^56s^2$	Os 76 190.23 $5d^66s^2$	Ir 77 192.22 $5d^76s^2$	Pt 78 195.08 $5d^96s^1$	Au 79 196.96654 $5d^{10}6s^1$	Hg 80 200.59 $5d^{10}6s^2$						Tl 81 204.3833 $6p^1$	Pb 82 207.2 $6p^2$	Bi 83 208.98037 $6p^3$	Po 84 (208.9824) $6p^4$	At 85 (209.9871) $6p^5$	Rn 86 (222.0176) $6p^6$
Fr 87 (223.0197) $7s^1$	Ra 88 (226.0254) $7s^2$	89–103‡	Rf* 104 (261.10869) $6d^27s^2$	Db* 105 (262.11376) $6d^37s^2$	Sg* 106 (263.1182) $6d^47s^2$	Bh* 107 (262.1231) $6d^57s^2$	Hs* 108 (264.1285) $6d^67s^2$	Mt* 109 (266.1378) $6d^77s^2$	110 (269)	111 (272)	112 (277)											

†**Lanthanide Series**

La 57 138.9055 $5d^16s^2$	Ce 58 140.115 $4f^15d^16s^2$	Pr 59 140.90765 $4f^35d^06s^2$	Nd 60 144.24 $4f^45d^06s^2$	Pm 61 (144.9127) $4f^55d^06s^2$	Sm 62 150.36 $4f^65d^06s^2$	Eu 63 151.965 $4f^75d^06s^2$	Gd 64 157.25 $4f^75d^16s^2$	Tb 65 158.92534 $4f^95d^06s^2$	Dy 66 162.50 $4f^{10}5d^06s^2$	Ho 67 164.93032 $4f^{11}5d^06s^2$	Er 68 167.26 $4f^{12}5d^06s^2$	Tm 69 168.93421 $4f^{13}5d^06s^2$	Yb 70 173.04 $4f^{14}5d^06s^2$	Lu 71 174.967 $4f^{14}5d^16s^2$

‡**Actinide Series**

Ac 89 (227.0278) $6d^17s^2$	Th 90 232.0381 $6d^27s^2$	Pa 91 231.03588 $5f^26d^17s^2$	U 92 238.0289 $5f^36d^17s^2$	Np 93 (237.0482) $5f^46d^17s^2$	Pu 94 (244.0642) $5f^66d^07s^2$	Am 95 (243.0614) $5f^76d^07s^2$	Cm 96 (247.07035) $5f^76d^17s^2$	Bk 97 (247.07030) $5f^96d^07s^2$	Cf 98 (251.0796) $5f^{10}6d^07s^2$	Es 99 (252.0984) $5f^{11}6d^07s^2$	Fm 100 (257.0951) $5f^{12}6d^07s^2$	Md 101 (258.0984) $5f^{13}6d^07s^2$	No 102 (259.10011) $5f^{14}6d^07s^2$	Lr 103 (262.1098) $5f^{14}6d^17s^2$

§ Atomic mass values averaged over isotopes in percentages they occur on Earth's surface. For many unstable elements, mass of the most stable known isotope is given in parentheses. 1993 revisions.

* Recommended symbols and names (see Appendix F) as of 1997.

Trigonometric Table

Angle in Degrees	Angle in Radians	Sine	Cosine	Tangent	Angle in Degrees	Angle in Radians	Sine	Cosine	Tangent
0°	0.000	0.000	1.000	0.000					
1°	0.017	0.017	1.000	0.017	46°	0.803	0.719	0.695	1.036
2°	0.035	0.035	0.999	0.035	47°	0.820	0.731	0.682	1.072
3°	0.052	0.052	0.999	0.052	48°	0.838	0.743	0.669	1.111
4°	0.070	0.070	0.998	0.070	49°	0.855	0.755	0.656	1.150
5°	0.087	0.087	0.996	0.087	50°	0.873	0.766	0.643	1.192
6°	0.105	0.105	0.995	0.105	51°	0.890	0.777	0.629	1.235
7°	0.122	0.122	0.993	0.123	52°	0.908	0.788	0.616	1.280
8°	0.140	0.139	0.990	0.141	53°	0.925	0.799	0.602	1.327
9°	0.157	0.156	0.988	0.158	54°	0.942	0.809	0.588	1.376
10°	0.175	0.174	0.985	0.176	55°	0.960	0.819	0.574	1.428
11°	0.192	0.191	0.982	0.194	56°	0.977	0.829	0.559	1.483
12°	0.209	0.208	0.978	0.213	57°	0.995	0.839	0.545	1.540
13°	0.227	0.225	0.974	0.231	58°	1.012	0.848	0.530	1.600
14°	0.244	0.242	0.970	0.249	59°	1.030	0.857	0.515	1.664
15°	0.262	0.259	0.966	0.268	60°	1.047	0.866	0.500	1.732
16°	0.279	0.276	0.961	0.287	61°	1.065	0.875	0.485	1.804
17°	0.297	0.292	0.956	0.306	62°	1.082	0.883	0.469	1.881
18°	0.314	0.309	0.951	0.325	63°	1.100	0.891	0.454	1.963
19°	0.332	0.326	0.946	0.344	64°	1.117	0.899	0.438	2.050
20°	0.349	0.342	0.940	0.364	65°	1.134	0.906	0.423	2.145
21°	0.367	0.358	0.934	0.384	66°	1.152	0.914	0.407	2.246
22°	0.384	0.375	0.927	0.404	67°	1.169	0.921	0.391	2.356
23°	0.401	0.391	0.921	0.424	68°	1.187	0.927	0.375	2.475
24°	0.419	0.407	0.914	0.445	69°	1.204	0.934	0.358	2.605
25°	0.436	0.423	0.906	0.466	70°	1.222	0.940	0.342	2.747
26°	0.454	0.438	0.899	0.488	71°	1.239	0.946	0.326	2.904
27°	0.471	0.454	0.891	0.510	72°	1.257	0.951	0.309	3.078
28°	0.489	0.469	0.883	0.532	73°	1.274	0.956	0.292	3.271
29°	0.506	0.485	0.875	0.554	74°	1.292	0.961	0.276	3.487
30°	0.524	0.500	0.866	0.577	75°	1.309	0.966	0.259	3.732
31°	0.541	0.515	0.857	0.601	76°	1.326	0.970	0.242	4.011
32°	0.559	0.530	0.848	0.625	77°	1.344	0.974	0.225	4.331
33°	0.576	0.545	0.839	0.649	78°	1.361	0.978	0.208	4.705
34°	0.593	0.559	0.829	0.675	79°	1.379	0.982	0.191	5.145
35°	0.611	0.574	0.819	0.700	80°	1.396	0.985	0.174	5.671
36°	0.628	0.588	0.809	0.727	81°	1.414	0.988	0.156	6.314
37°	0.646	0.602	0.799	0.754	82°	1.431	0.990	0.139	7.115
38°	0.663	0.616	0.788	0.781	83°	1.449	0.993	0.122	8.144
39°	0.681	0.629	0.777	0.810	84°	1.466	0.995	0.105	9.514
40°	0.698	0.643	0.766	0.839	85°	1.484	0.996	0.087	11.43
41°	0.716	0.656	0.755	0.869	86°	1.501	0.998	0.070	14.301
42°	0.733	0.669	0.743	0.900	87°	1.518	0.999	0.052	19.081
43°	0.750	0.682	0.731	0.933	88°	1.536	0.999	0.035	28.636
44°	0.768	0.695	0.719	0.966	89°	1.553	1.000	0.017	57.290
45°	0.785	0.707	0.707	1.000	90°	1.571	1.000	0.000	∞